Manufacturing Engineering

Manufacturing Engineering

K. Hiraniya Singh
Formerly Joint Director
Foremen Training Institute, DGET
Bengaluru

I.K. International Pvt. Ltd.

NEW DELHI

Published by

I.K. International Pvt. Ltd.
4435-36/7, Ansari Road, Daryaganj
New Delhi–110 002(India)
E-mail: info@ikinternational.com
Website: www.ikbooks.com

ISBN 978-93-82332-48-0

Reprint 2023

Published by Krishan Makhijani for I.K. International Pvt. Ltd., 4435-36/7, Ansari Road, Daryaganj, New Delhi–110 016 and Printed by Rekha Printers Pvt. Ltd., Okhla Industrial Area, Phase II, New Delhi–110 020

My parents, teachers who taught me, and to my students who inspired me to bring out this venture

Preface

This book is designed to provide basic technical knowledge about the tools, gadgets, instruments and other devices used during the manufacture of workpieces by machining, machining operations and their physical principles. It deals with the manufacture of workpiece on the machine, with the measuring and testing of these parts with necessary calculations involved.

The book is intended for those who are directly occupied with machine tools, but particularly for those connected with metal trade, who are, however, not supposed to know all details about the machining operations, but they should have working knowledge about machine tools, cutting tools, their design and performance.

Moreover, the students entering polytechnics in mechanical engineering branch, whose field of activity predominantly comprises machine tools and appliances, must be well acquainted with the ways of manufacture. The students can, with the help of this book, make initial and close-to-practice acquaintance with the various tools and cutting methods.

This book mainly covers the theory of manufacture and not the description of the skill itself, as the skill can only be acquired in the workshop and practice.

More than 1000 line diagrams is the most outstanding feature of this book.

Hiraniya Singh

Contents

Part One

Dimensional Metrology and Measurement

Chapter 1

From birth to death we measure so many things, like weight, height, time of birth, height and weight of the would-be bride, the colour, the looks, etc.

As such, a primary concern of the metalwork industry is dimensional metrology, i.e., measurement of length and geometry. Geometry measurements include measurement of angles, roundness, surface texture, concentricity, straightness, taper, pitch, and so on. With the advent of high speed electronic counters, digital computers, lasers, electron microscopes, radars, gyroscopes, interferometers, and other highly sophisticated measuring tools, metrology has assumed great significance in manufacturing engineering.

1.1 FUNDAMENTAL UNITS AND STANDARDS

Measurement is the means whereby data or information is generated about an item or an event.

The first basic requirement of measurement is that the units, in terms of which things are to be measured, are consistent. In other words, each unit should have one and only one value, and that value should be universally acceptable.

Two major systems of units were in use, i.e., the English system and the metric system. English units are now defined in terms of the metric units (e.g., one inch is defined as 25.4 mm), making the latter the primary system of units. In this way, consistency of units is assured. It should be noted that the metric system, by virtue of its greater simplicity, has supplanted the English system in many countries, even in England, where the system originated. At present, over 90 per cent of the people in the world use the metric system, and is expected that within the next 50 years the metric system will prevail universally.

The Government of India realized the importance of easy·measurement and entrusted the job of framing new system of Indian Standards Institute (ISI); as early as 1967 and the ISI approved the **centimetre, gram and second (CGS)** system.

For linear measurement, the unit is foot, for weight it is pound, and for time, it is second; hence the name FPS. This system was evolved by the British Standard Institute and is known as the BSI system.

Under the British System, The Units Are:

Units of Linear Measurement of Length

8 gauges = 1 inch

12 inch = 1 foot

3 feet = 1 yard

220 yards = 2 furlong

8 furlong (or 1760 yards) = 1 mile

Units of Weight Measurement

16 drams = 1 ounce

16 ounces = 1 pound

28 pounds = 1 quarter

4 quarters = 1 hundred weight

20 hundred weight = 1 tonne

Measurement of Capacity

4 gills = 1 pint

2 pints = 1 quarter

4 quarters = 1 gallon.

Other Units

8 *ratti* = 1 *masha*

12 *masha*s = 1 *tola*

5 *tola*s = *chhatank*

16 *chhatank*s = 1 *seer*

40 *seer*s = 1 *maund*

Measurement of Time

60 seconds = 1 minute

60 minutes = 1 hour

24 hours = 1 day

1.2 CGS (CENTIMETRE, GRAM AND SECOND) SYSTEM

This system is also known as French or Metric System. This is an international system for measurement and weighing physical things. The Indian Standards Institute has approved this system. Under this system, for measurement of length the unit is centimetre, for weight, gram and for time, second. Under CGS System the following units are used:

Measurement of Length

10 millimetres = 1 centimetre

10 centimetres = 1 decimetre

10 decimetres = 1 metre

10 metres = 1 dekametre

10 dekametres =1 hectometre

10 hectometres = kilometre (1000 metres)

Measurement of Weight

10 milligrams = 1 centigram

10 centigrams = 1 decigram

10 decigrams = 1 gram

10 grams = 1 dekagram

10 hectograms = 1 kilogram

10 dekagram = 1 hectagram

100 kilograms = 1 quintal

10 quintals = 1 metric tonne

Measurement of Capacity

10 millilitres = 1 centilitre

10 centilitres = 1 decilitre

10 decilitres = 1 litre

10 litere = 1 hectolitre

10 hectolitres = 1 kilolitre

Measurement of Time

60 seconds = 1 minute

60 minutes = 1 hour

24 hours = 1 day

7 days = 1 week

30 days = 1 month

12 months (or 365 days) = 1 year

Apart from FPS and CGS systems, MKS system is also prevalent in India. This system is almost similar to CGS system. For measuring length, metre, for weighing, kilogram, and for time, second is used.

Area Measurement (British System)

144 sq. inch = 1 sq. foot

9 sq. feet = 1 sq. yard

4840 sq. yards = 1 acre

640 acres = 1 sq. mile

Area Measurement (Metric System)

100 sq. millimetres = 1 sq. centimetre

100 sq. centimetres = 1 sq. decimetre

100 sq. metres = 1 acre

100 acres = 1 hectare (or 10 000 sq.m)

100 hectares = 1sq. kilometre (or 1, 00, 000 sq. acres)

Measurement of Volume (British System)

1728 cubic inches = 1 cubic foot

27 cubic feet = 1 cubic yard

Measurement of Volume (Metric system)

1000 cubic mm = 1 cubic centimetre

1000 cubic cm =1 cubic decimetre

1000 cubic decimetres = 1 cubic metre

1000 cubic metres = 1 dekametre

1000 cubic dekametres = 1 cubic hectometre

1000 cubic hectometres = 1 cubic kilometre.

Units of Temperature

Temperature indicates the heat level of an article. Organs of our body cannot measure the heat of things. If we put our finger in hot water and then put it into cool water, we observe that our finger remains hot for some time even after putting the finger into the cold water. Therefore, we depend on our body organs for knowing the warmth of an article only. For measuring this warmth thermometers containing mercury are used. Normally celsius (centigrade) and Fahrenheit thermometers are in use. In centigrade thermometers the melting point of water at atmospheric pressure is 0°C and the boiling point 100°C. These degrees are divided into equal parts. In the same way, at atmospheric pressure in Fahrenheit thermometer the melting point of ice is 32°F and boiling point of water is 212°F. For meeting the technical and other needs these units are converted into one another. For this purpose, the following formula is applied.

Temperature Conversion Formula

From Centigrade to Fahrenheit — F° to C° Deduct 32, then multiply by 5, then divide by 9.

$$C^\circ = 5/9(F^\circ - 32)$$

From Fahrenheit to Centigrade, — C° to F° Multiply by 9, then divide by 5, then add 32

$$F^\circ = 5/9(C^\circ - 32)$$

Example (i) Conversion of 32° centigrade into Fahrenheit:

$$32^\circ C = 5/9(F^\circ - 32)$$

$$F - 32\ 32 \times 9/5$$

$$= 57' + 32 = 89^\circ\ 6 \leq F.$$

Example (ii) Conversion 110° Fahrenheit into centigrade

$$C/5 = (110 - 32/9) = 78/9$$

$$9C = \frac{78 \times S}{9} = \frac{390}{9} = C = 43.33^\circ\ C$$

1.3 INTERNATIONAL SYSTEM OF UNITS (SI)

It is the system established in 1960 by the General Conference of Weights and Measures (CGPM) and abbreviated as SI (System Internationale d'units) in all languages. In India, we switched over to metric system of weights and measures conforming to SI units by an Act of Parliament in 1956. This system like traditional metric system is based on decimal arithmetic. For each physical quantity, units of different sizes are formed by multiplying or dividing a single base value by powers of 10. Obviously this offers great advantage because the changes can be made by adding zeros or shifting decimal point in the metric system we have been following so far. This simplicity of a series of units linked by powers of 10 is limited for plain quantities like length, and this simplicity is lost when more complex units like energy, etc., are encountered. For example, energy is now represented by several units like kgm, H.P. kW etc. In contrast the SI provides only one basic unit for each physical quantity, and universality is thus achieved. This system is superior to the systems and also more convenient as it is coherent, rational and comprehensive.

The SI is a coherent system, in the sense that the product or quotient of any two unit quantities in the system is the unit of the resultant quantity, e.g., If the unit of length is metre, then the unit of area will be square metre and not acres or *bighas* etc. It is a rational system since it has absorbed in itself the rationalized MKSA system. It is also comprehensive because its seven-base units cover all disciplines.

1.4 DECIMAL RELATIONSHIP

The metric unit for length is the metre, which approximates a man's pace (or a yard in our customary system). While this unit is suitable for measuring distances of football field, it may not lend itself well to expressing greater distance, e.g., the distance from New York to Los Angeles. Similarly, metre would be of little help in measuring the size of molecules, or even the granulation of flour. But rather than using entirely different units for measuring different lengths, as we presently do in India, we use powers of 10 of the same basic units, just as powers of 10 are used in our system of numbers.

From the definition of algebra, powers of 10 always have 10 as their base. Only integers (whole numbers) are used as exponents in connection with SI units. Thus, 10^4 is shorthand for "ten multiplied by ten four times equals ten thousand," or $10^4 = 10 \times 10 \times 10 \times 10 = 10\,000$.

The seven-base SI units established by the General Conference of Weights and Measures are:

Quantity	Unit	Symbol	Base of Definition
Length	Metre	m	Wavelength of red line in krypton 86
Time	Second	s	Cycles of radiation of cesium
Mass	Kilogram	kg	Platinum–cylinder prototype
Temperature	Kelvin	k	Absolute zero and water
Electric current	Ampere	A	Force between two conducting wires
Luminous intensity	Candela	cd	Intensity of an area of platinum
Quantity of substance	Mole	mol	Amount of atoms in carbon 12

All derived units are constructed from basic SI units. Simple derived units are developed from one type of basic unit. Both the measurement of area and of volume are derived from the measurement of length. Square metre and cubic metres are derived units. But basic units with prefixes, such as "cm" are not derived units; they are multiples and submultiples of base units.

Compound derived units utilize several SI units to arrive at other units. When measuring density, pressure, speed, acceleration, and many other characteristics, it is necessary to employ more than one basic SI unit. Hence, we measure:

Speed in m/s

Acceleration in m/s^2

Electrical field strength in V/m,

Many more derived units are in use. Additional ones can be created as required for particular tasks.

Some derived units are stated in the basic units from which they were derived example acceleration (linear) m/s^2. Others have names of their own, for example, force is not called m kg/s2, but N (newton). Usually these proper names for derived units are simpler and easier to use than the more complex arrays of basic units from which they were developed. Major disadvantage of the adopted unit names is the fact that they have to be remembered, and that the derivation of the unit is no longer obvious.

The coherency of SI system makes it possible, on occasion, that a derived unit can be defined in various ways. The power unit W (watt) can be derived from mechanical unit J/s. The same unit can also be obtained by multiplying the electrical units, i.e., V (volt) and A (ampere).

Supplementary Units

These are the units that can be expressed by base units but have nevertheless been made available as a convenience to users. The only authorized supplementary units are measures for planes and spherical angels. Angles can be stated by using the relationships of the sides and the laws of trigonometry. Therefore, linear measures could be used to specify plane angels, and area measures for solid angles. But the procedure being cumbersome use of supplementary angle units to simplify computations involving angles is preferred. It is worth noting that the familiar, universally known angular degrees did not become SI supplementary units. Radians and steradians are the SI units, but CGPM allows the use of degree measures in connection with other SI units.

Coherency of SI Units

A coherent system is defined as a system of units in which the product or quotient of any two unit quantities is a unit of the resulting quantity. The coherency in SI has the following other SI units.

1. The same system and unit of measurement are used regardless of discipline, industry, or trade involved.
2. A minimum of conversion factors other than powers of 10 are needed.

Lengths are measured in metres, or power of 10 of metres, regardless of whether we are measuring a wavelength of radiation or the dimensions of a continent. Force is measured in newtons, regardless of how this force is developed – by electricity, magnetism, hydraulic, mechanics, or a human pulling on a rope.

Power is measured in watts. With the electrical input and the mechanical output of an electric motor stated in the same units, we can immediately recognize the efficiency of the motor. This is not possible with incompatible units. At present the input of an electric motor is stated in watts but the output is rated in horsepower (hp). The input of thermal engines is stated in such units as pounds of steam, gallons of liquid fuel, and pounds of solid fuel. The output of such engines is rated in horsepower. Complex calculations are needed to compute efficiency. Compatible units facilitate such computations.

While no conversion factors other than powers of 10 are usually required in connection with SI, there are a few exceptions. Most of these exceptions are associated with natural phenomena which lack decimal relationships and which cannot be decimally incorporated into the measuring system.

It is conceivable that time measurements could be decimalized by replacing arbitrary measures such as seconds, minutes, and hours. This was actually done in France for a short period after the revolution of 1792. But arbitrary decisions cannot alter natural events. For example, the day is based on the earth's period of rotation on its axis. The year is based on the revolution of the earth around the sun. Furthermore, these periods are related not by a power of 10 but by a inconvenient factor, a mixed number which starts with 365.2.

Also, other astronomical bodies have constants of their own. This must be considered in space exploration. The presence and innavigability of such conversions based on nature plus the traditional worldwide familiarity with smaller non-decimal time units may be reasons for permitting "day" and "hour" in conjunction with SI units.

Among other natural constants entering computations are the speed of light and of electricity of 299.8 Mm/s. The speed of sound in dry air under atmospheric pressure of 100 kPa is 331 M/s the circular conversion factor *pi* shows up in computations involving circles regardless of the system of units that is used for other parameters. The SI supplementary units "rad" and 'sr" for measuring angles is also related to *pi*.

The fact that 1 dm^3 of water at standard conditions has a mass of 1 kg eliminates computations. Customary units of weight and volume, however, are not always related. 1 ft^3 of water weighs 62.4 lb.

Another convenient fact is that normal atmospheric pressure at the surface of the earth, i.e., 100 kPa corresponds to a water column of 10 m, approximately.

Examples for Compatibility of Unit

1. A 50-mm ID pipe connects a pump and a processing machine. If water can be pumped efficiently with the available equipment at a speed of 25 cm/s, what is the capacity of this pipe in cm^3/s?
   (a) Pipe diameter in units of required answer: 50 mm = 5 cm
   (b) area of pipe:
   $$\frac{rd^2}{4} = \frac{r.5^2}{4} = 19.6 \text{ cm}^2$$
   (c) Delivery: Area times velocity – 19.6 cm^2 × 25 cm/s = 490 cm^3/s

   Consider the computations necessary had the questions been asked with conventional units; pipe diameter in velocity in ft/min, delivery in gal/min.
2. A workpiece on a lathe rotates at 80 rev/min and the machined circumference of the work has a diameter of 95 mm. What is the cutting speed?
   (a) Rotational speed in s
   $$\frac{80 \text{ rev. min}}{60 \text{ min. s}} = 1.33 \text{ rev/s}$$
   (b) Diameter in m: 95 mm = 0.095 m
   (c) Circumference: 0.095 π = 0.298 m, 0.298 m × 1.33 m/s = 0.397 m/s

Basic Units

Quantity	Unit	SI Symbol
Length	Metre	m
Mass	Kilogram	kg
Time	Second	s
Temperature	Kelvin	K
Electric current	Ampere	A
Luminous intensity	candela	cd
Quantity of substance	Mole	Mol

Supplementary Units

Plane angle	radian	rad
Solid angle	steradian	sr

Derived Units

Acceleration	Metre per second squared	m/s^2
Activity of a radioactive source	Disintegration per second	Disintegration/sec
Angular acceleration	Radian per second squared	rad/s^2
Angular velocity	Radian per second	rad/s^2

Quantities	Unit	SI Symbol	Formula
Area	Square metre	…	m^2
Density	Kilogram per cubic metre	…	kg/m^3
Electric capacitance	farad	F	A. s/v
Electric conductance	siemens	S	A/v
Electric field strength	Volt per metre	…	V/m
Electric inductance	henry	H	V.s/A
Electric potencial difference	volt	v	W/A
Electric resistance	ohm	Ω	V/A
Electromotive force	volt	V	W/A
Energy	Joule	J	N.m
Entropy	Joule per Kelvin	…	J/k
Force	Newton	N	…
Frequency	hertz	Hz	Cycle/s
Illuminance	Lux	lx	lm/m^2
Luminance	Candela per sq. metre	…	Cd/m^2
Luminous flux	lumen	lm	cd.sr
Megnetic field strength	Ampere per metre	…	A/m
Megnetic flux	Weber	Wb	v.s
Megnetic flux density	Tesla	T	Wb/m sq.
Megnetomotive force	Ampere	A	…
Power	Watt	W	I/s
Pressure	Pascal	Pa	N/m^2
Quantity of electricity	Coulomb	C	A.s
Quantity of heat	Joule	J	N.m
Radiant intensity	Watt per steradian	…	w/st
Specific heat	Joule per kilogram kelvin	…	J/kg.k
Stress	Pascal	Pa	N/m^2
Thermal conductivity	Watt per metre-kelvin	…	W/mk
Velocity	Metre per second	…	m/s
Velocity dynamic	Pascal second	…	Pa.s
Velocity kinematics	Square metre per second		m/s
Voltage	Volt	V	W/A
Volume	Cubic metre	…	m^3
Wave number	Reciprocal metre	…	Wave/m
Work	joule	J	N.m

Multiplication factors	Prefix	SI symbols
$1\,000\,000\,000\,000 = 10^{12}$	Tera	T
$1\,000\,000\,000 = 10^9$	Giga	G
$1\,000\,000 = 10^6$	Mega	M
$1000 = 10^3$	Kilo	k

$100 = 10^2$	Hecto	h
$10 = 10^1$	Deka	da
$0.1 = 10^{-1}$	Deci	d
$0.01 = 10^{-2}$	Centi	c
$0.001 = 10^{-3}$	Milli	m
$0.000\ 001 = 10^{-6}$	Micro	μ
$0.000\ 000\ 001 == 10^{-9}$	Nano	n
$0.000000\ 000\ 001 = 10^{-12}$	Pica	p
$0.000000000000001 = 10^{-15}$	Femto	f
$0.000\ 000\ 000000\ 000\ 001 = 10^{-18}$	Alto	a

Rules Followed

The two major advantages of coherency of SI units are:

(i) Same system and unit of measurement are used regardless of industry, trade, or discipline; and

(ii) Minimum of conversion factors are needed (other than powers of 10).

In SI units, some rules of style, abbreviation, writing, and drafting, practices are adopted, they are:

(a) No dots, commas, etc., are used after SI symbols. For example 32 metres will be written as 32 m.

(b) Plurals are never used, for example, 32 metres will be written as 32 m and not as 32 ms which would mean 32 milliseconds.

(c) Decimal fractions always start with 0, for example, half a metre will be written as 0.5 m.

(d) Multiplication or times signs are: X. This is used between the number to be multiplied and between unit symbols in derived units where two unit symbols adjoin for the purpose of clarity, e.g. Unit of torque may be written as m.N (meter.newton), if written as mN can be misunderstood as millinewton.

(e) All symbols and prefixes are lowercase letters, except symbols derived from proper names, like W for watt, M.G. and T for the largest three powers of 10 prefixes. All symbols should be used to avoid any confusion.

(f) All units may be written either in full or using the agreed symbols. There is only one acceptable symbol in all languages for a metric unit. For example, kilogram be written as kg and not as keg, etc.

(g) The product of two or more units is preferably indicated by a dot which can be dispensed with when there is no risk of confusion with another unit symbol, e.g., 63 Nm is correct but not 63 mN.

(h) There is a mixture of capital and lower case letters in the symbols for the prefixes, but the full names of the prefixes commence with lower case letter only, e.g., 200 MW is written in 200 megawatts.

(i) No degree mark (°) is used with Kelvin, the unit of temperature. A temperature interval can also be expressed in degree Celsius.

(j) Double prefixes should not be used, e.g., one kilo megawatt cannot be written as kMW, but as GW.

(k) Algebraic symbols representing quantities are written in italics while symbols for units are written in upright characters.

(l) The expression "per" in symbols of derived units is always indicated by a fraction line as m/s of m but word "per" or letter "p" should not be used for this purpose.

(m) Always leave one space between numerical values and the symbol. For example, 92 N is commitment 92 N is not.

(n) Only the numerator should be multiplied by powers of 10 in compound derived units and the denominator should always remain the base unit.

For example 0.032 m/s may be put as 32 mm/s and not as 0.000032 m/Mts. (meter per millisecond).

(o) Numbers may be grouped in clusters of three in both directions from the decimal mark, and gap may be given for clarity and comma should not be used there, e.g. 153297.3 m may be written as 153 297.3 m and not 153,297.3 m. A sequence of four figures is generally not broken in groups.

(p) Units with names of scientists should be capitalised when written in full.

(q) According to SI recommendations, litre is a special name given to cubic decimetre. Hence, the world litre should not be used for expressing results of high precision volume measurements.

(r) Some units which are strictly incompatible with SI units, have been allowed initially, like km/hr, rev/min.

(s) The appropriate integers, multiples and sub-multiples to which a unit is expressed is between 0.1 and 1000.

(t) Figures are written in groups of three to the left and to the right of the decimal sign.

(u) Thousands are separated by the space and NOT by a comma e.g., 60 00 000 kl is correct but 6,000,000 kl is not correct.

(v) In some countries, a point (dot) is used as a decimal marker, and in some countries a comma is used: both are acceptable.

EXERCISE

1. Which is the approved measurement and weight system in India?
2. What do you understand by FPS and CGS ?
3. What is the formula for converting inches into millimetres ?
4. Which formula is used for converting centigrade into Fahrenheit ?
5. Convert 30° centigrade into Fahrenheit.
6. For what reason are the rough dimensions of the price list or order-card transferred to the rough part without oversize in length?
7. What is the measuring accuracy that can be obtained by: a) with the steel straight edge and b) with the sliding calliper?

Measuring Tools

Chapter 2

Procure a metre stick or a measuring tape. The metre stick or the tape has a scale and in some cases even two scales, i.e., the English and the metric scale. The metric scale, which has the smaller divisions and goes up to 100, (or for tapes even to 150) is the one which you have to use for your measurements in science. The other rulers available in the market are made of wood or plastic and are used by the school children.

According to the decimal system, a metre is subdivided into 10 equal parts, called decimetres and written as dm.

Each decimetre is again subdivided into 10 equal parts, called centimetres, and written as cm.

Finally, the centimetre is again divided into 10 equal parts, called millimetres, and written as mm.

Therefore,

$$\begin{aligned} 1 \text{ m} &= 10 \text{ decimetres} \\ &= 100 \text{ centimetres} \\ &= 1000 \text{ millimetres} \\ 1 \text{ m} &= 10 \text{ dm} \\ &= 100 \text{ cm} \\ &= 1000 \text{ mm} \\ 1 \text{ mm} &= 0/001 \text{ m} \\ 1 \text{ cm} &= 0.01 \text{ m} \\ 1 \text{ dm} &= 0.1 \text{ m} \end{aligned}$$

The other side has markings in inches; an inch may be divided into eight, sixteen, thirty -two or sixty -four divisions.

1. Try to find out the thickness of one page of your book. You are supposed to use your ruler only and a bit of brain. **(Fig. 2.1)**
2. With any of the rulers it is not possible to measure the thickness of the sheet of a paper. It is so thin that the markings may not give the correct measurement.
3. 100 sheets of the book may be taken. The thickness of these sheets may be found out with the help of any of the simple rulers. Measure the thickness of 100 sheets and then calculate the thickness of one sheet. How accurate will be the measurement?

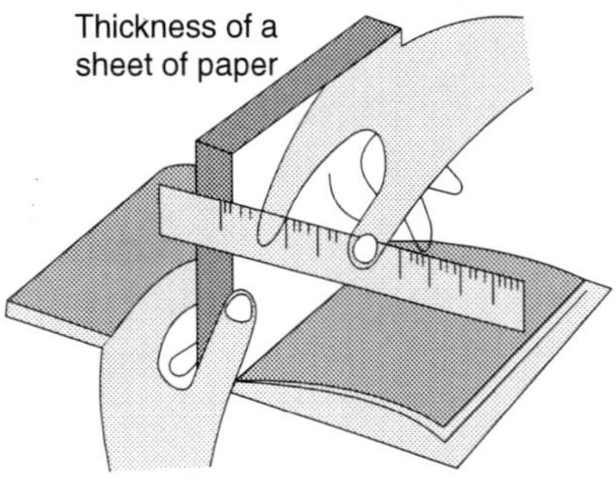

Fig. 2.1

1. Special devices and precision tools are necessary for this purpose.

Measuring tools can be further divided into the following categories:

(i) Direct measurement tools.

(ii) Indirect measurement tools.

The above two categories of measuring tools can further be divided as under:

(i) Simple measuring tools.

(ii) Precision measuring tools.

All the above categories can be included into the following tools:

1. Foot rule.
2. Outside calliper.
3. Inside calliper.
4. Outside micrometer.
5. Inside micrometer.
6. Depth calliper.
7. Vernier calliper.
8. Vernier height gauge.
9. Vernier depth gauge.
10. Combination set
11. Vernier bevel protractor.
12. Dial test indicator.

2.1 DIRECT MEASURING TOOLS

2.1.1 Foot Rule

Foot rule is used for measuring any article physically. These are made of wood, cardboard, plastic, aluminium, spring steel or stainless steel. In engineering, generally foot rules made of spring steel or stainless steel are used. Foot rules, whatever material these are made of, are of the size of 6″, 12″ 36″ and 48″. Foot rules of

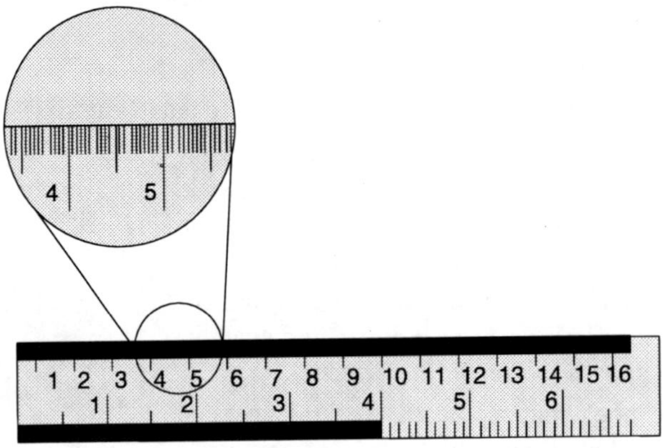

Fig. 2.2 *Steel rule*

12″ size are most widely used. Foot rules indicate inches or millimetres or both, and division of inches and millimetres. Generally, foot rules indicate division markings at a distance of an inch. One inch is divided into 8 equal parts. In this way, each marking represents 1/8th of an inch, known as *soot.* This 1/8th is divided into two parts making it 1/16th of an inch. Further, this1/16th is divided making it 1/32th and 1/64th part of an inch.

This foot rule can be used to measure minimum up to 1/64th of an inch. The graduation can be indicated on one edge or both the edges of the foot rule. Some foot rules have signs on one edge only and on the other edge there is marking of metric system. These foot rules are more useful as we can measure in British as well as metric system. In the edge where millimetre graduation is indicated, markings at every centimetre are indicated. Each centimetre is divided into ten equal parts to indicate a millimetre. Each sign of millimetre is also divided into two equal parts. Thus, we can measure up to ½ millimetre from this edge.

Types of foot rules

1. Flexible steel rule
2. Hook rule
3. Narrow rule
4. Shrink rule
5. Short rule
6. Key-set rule
7. Calliper rule
8. Steel tape rule
9. Folding rule
10. Canvas tape rule

2.1.2 Flexible Rule

This rule is used for measuring spherical or hemispherical objects. These are 6″ to 12″ in size. These are made from spring steel sheet of 1/64″ thickness so that it attains elasticity.

This rule contains measurement both in inches and millimetres.

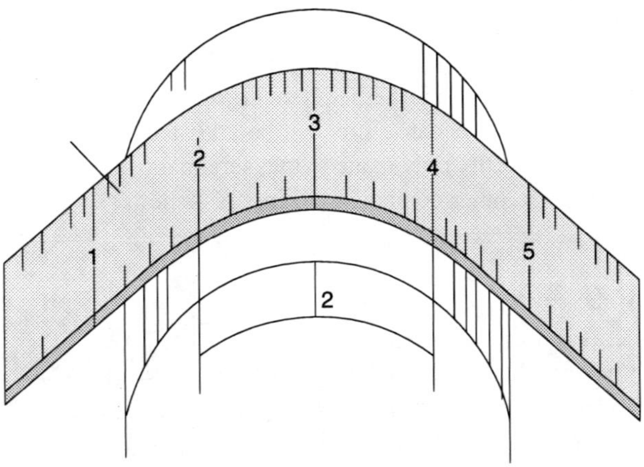

Fig. 2.3 *Flexible rule*

First two inches are divided into 32 equal parts and the third inch is divided into 64 equal parts. On rest of the rule each inch is divided into 8 equal parts.

In the same way, other side of the rule markings indicate centimetres. Each centimetre is divided into ten equal parts ,i.e., one millimetre. This part is further divided into two equal parts.

2.1.3 Hook Rule (Fig. 2.8)

This rule is also made of spring steel with a hook attached to its one end. It is used to measure special type of spaces. Suppose there is a groove inside a hole and we have to measure the remaining part (leaving aside the groove) or to measure the remaining part of a round job leaving aside its step.

Hook rule can be of the size of 6″ 12″ or more.

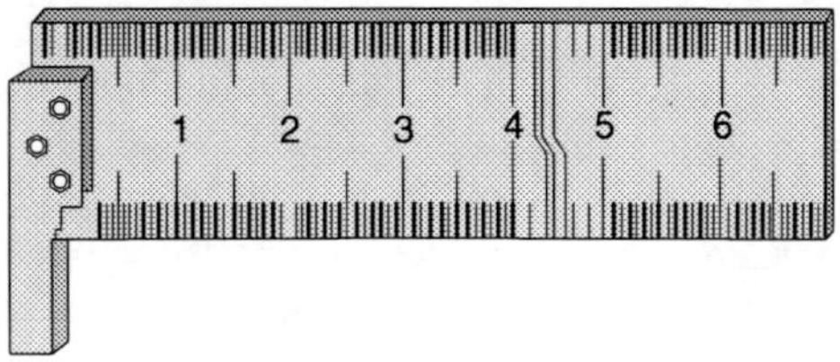

Fig. 2.4

2.1.4 Narrow Rule

Narrow rules are like a flexible rule, it is smaller and its breadth is narrow, about 6 mm. A clip is attached to it with which it is kept in the pocket. Both inches and millimetres are marked on it. Generally, it is used to measure slot, slit, or thin pipe.

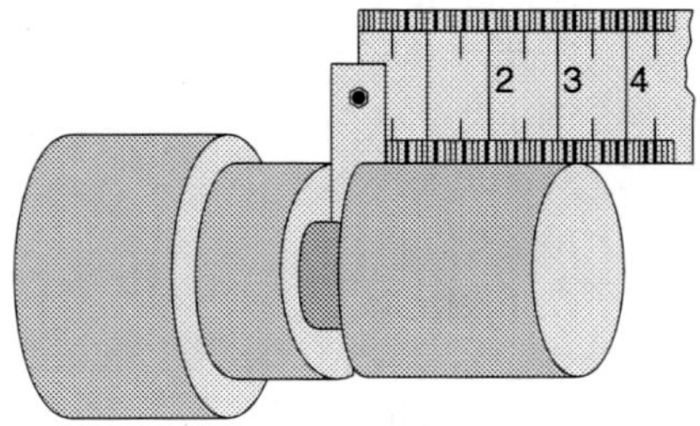

Fig. 2.5 Use of Hook scale

2.1.5 Shrink Rule

This rule is generally used by pattern makers while making patterns in order to allow for the shrinkage of the metal in which the pattern is to be cast. The marking is slightly more than actual, in proportion; because of its enlarged size patterns get shrink allowance.

2.1.6 Short Rule

This rule is used by the machinists and die-makers for measuring small gaps, slots, etc. It is a set of small rules, made of steel leaves.

2.1.7 Key Seat Rule (Fig. 2.6)

Its shape is somewhat different from an ordinary rule. It is almost like angle iron, having both beams at 90° Spherical and hemispherical jobs which cannot be measured with ordinary rules because this can be adjusted on those round things easily. These are particularly made to measure marking of key way rules one is plain and the other contains marking of inches or millilitres.

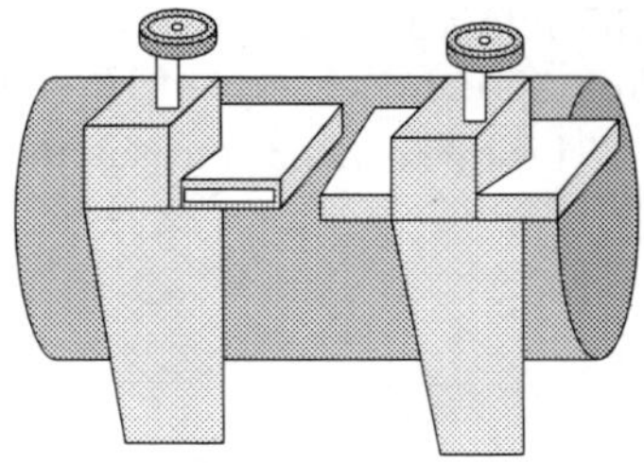

Fig. 2.6

2.1.8 Calliper Rule (Fig. 2.7)

This is a special type of rule. It is used to take inside and outside measurements of the round articles. A narrow rule and a rule of normal breadth are used in it. In the broader rule, the narrow rule can be slided. Both the rules have a jaw on one end. The tip is so small that it can go into a hole.

A screw is fixed on the broader rule with which the narrow rule can be adjusted and can be locked in position.

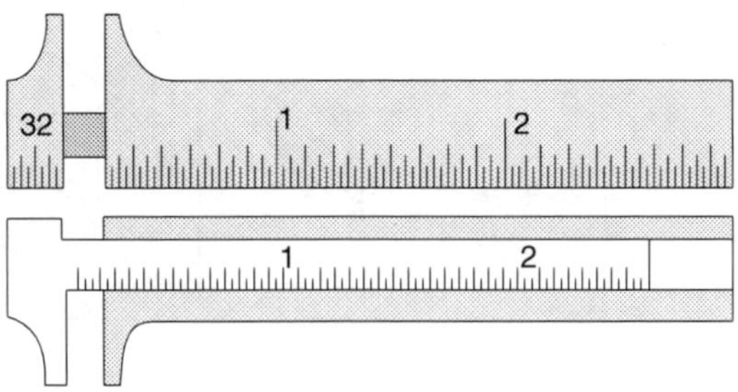

Fig. 2.7 Calliper rule

These rules contain both the Metric system and the British systems of graduation.

2.1.9 Steel Tape Rule (Fig. 2.8)

It is often difficult or rather impractical to carry long foot rules. For the purpose of convenience steel tape rules are available.

This is made of spring steel and is elastic. It can be rolled and packed in a round tiny box (case). Its breadth is 1″ and length is 6″. It contains inch as well as millimetre graduation.

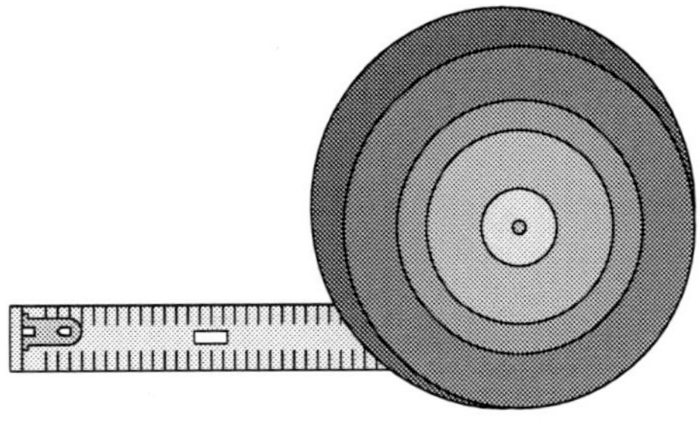

Fig. 2.8 Steel tape rule

2.1.10 Canvas Tape (Fig. 2.9)

This rule is used for measuring greater lengths. This is made of a long canvas tape which can be rolled and packed in a small leather or steel box. A ring is attached to the one end. By pulling this ring we can pull out the tape. For folding it a small handle is provided. These tape rules are generally of 50 to 100 feet in length.

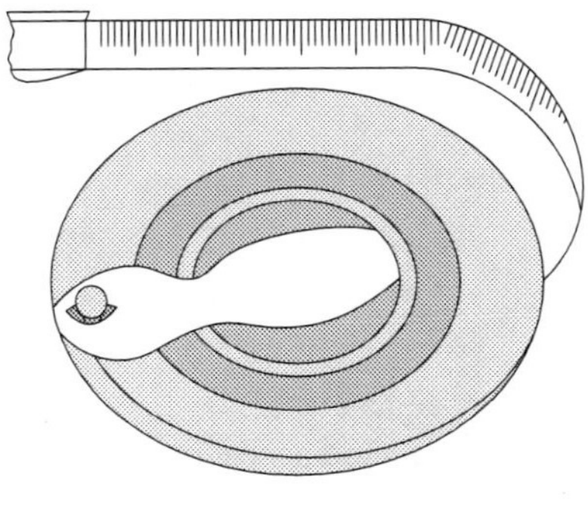

Fig. 2.9 Canvas tape

2.1.11 Folding Rule (Fig. 2.10)

This rule is made by joining four pieces of 6 inches. These pieces are either wooden or of steel. It is generally used by the carpenters and masons, by unfolding this rule we can take measurement up to 2 feet. When we fold it, it is reduced to 6″. Graduation in inches and millimetres is indicated on it.

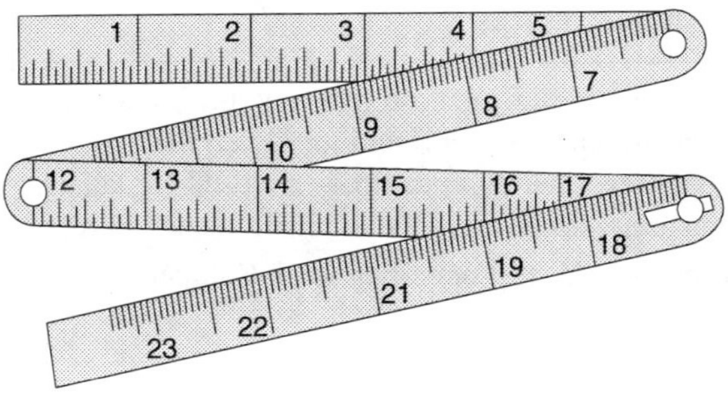

Fig. 2.10 Folding rule

2.2 MEASURING WITH THESE TOOLS

The method of using a trisquare and a steel rule for accurate measurement is illustrated. For accurate readings, it is necessary to read vertically and avoid errors arising out of parallax.

Some important points to be remembered while using the steel rule:

1. Measurement should be taken by keeping the steel rule erect on the job and not by laying it down on the job.
2. As far as possible measurement should be taken from the corner of the steel rule.
3. Prevent the steel rule from getting rusted.
4. Steel rules should be kept separate from other tools so that it does not get any scratches.

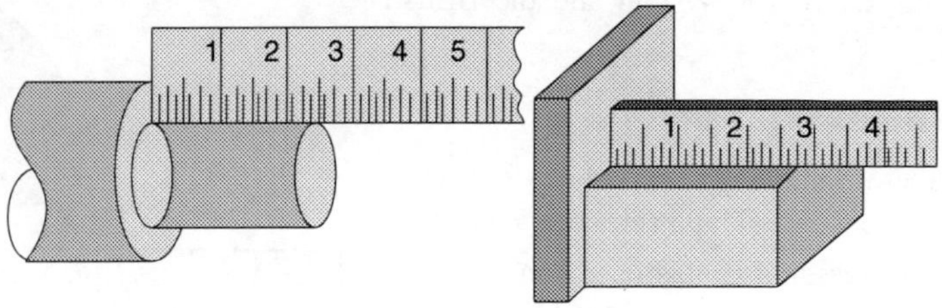

Fig. 2.11

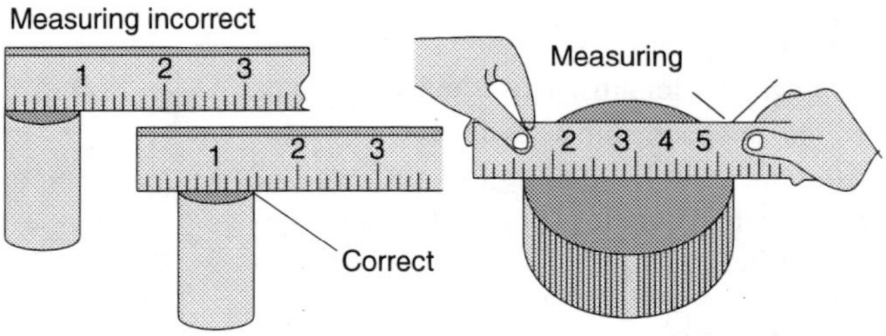

Fig. 11a

5. Do not use a steel rule as a screw driver otherwise its corners would be spoiled.
6. Do not hit the steel rule at a hard thing.

After use. The steel rule should be cleaned properly and kept at its proper place.

2.3 INDIRECT MEASURING TOOLS

Measurements can easily be taken of a rectangular job with any rule but it requires a great deal of experience to measure outside and inside measurements of a round job. Even then there is doubt reading the accuracy of measurement taken in this manner. Therefore, callipers are used for taking measurement of round jobs. Calliper is one of the measuring tools, which takes measurement in an indirect way.

The calliper is generally made of mild steel or carbon steel. Callipers are of two kinds:

1. Outside calliper
2. Inside calliper

Callipers have two legs, which are joined with the help of an ordinary rivet or a spring leaf. Their names are based on shape of their legs. Their leg points are hard so that they do not easily wear out.

Callipers joined with rivets are called firm joint callipers. Their legs can be adjusted manually. The legs or spring joint callipers are joined with a pin at the centre, which has a spring in the from of a leaf. Its legs are regulated by a nut mounted on its screw, its special future is that the measurement remains static unless its nut is moved. Callipers are available in 6″ and 8″ size.

2.3.1 Outside Callipers

It is used for gauging outside measurements or size of round jobs. Gauging the breadth and thickness is also possible. Both its legs are opened in a round shape. Between these legs the measurement of a job is taken.

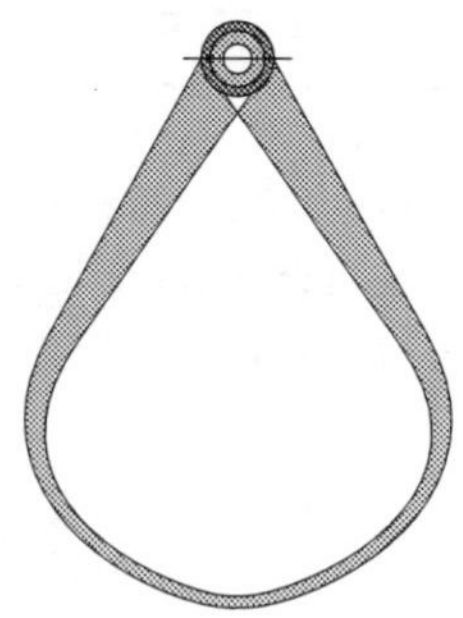

Fig. 2.12 Outside calliper

Measuring a cubical and cylindrical job (Fig. 2.13)

The points at the end of the legs are opened to cover the round shape. Between these legs the measurement of the job is taken, for gauging inside diameters. It can also gauge the breadth and depth of a slot or slit groove etc.

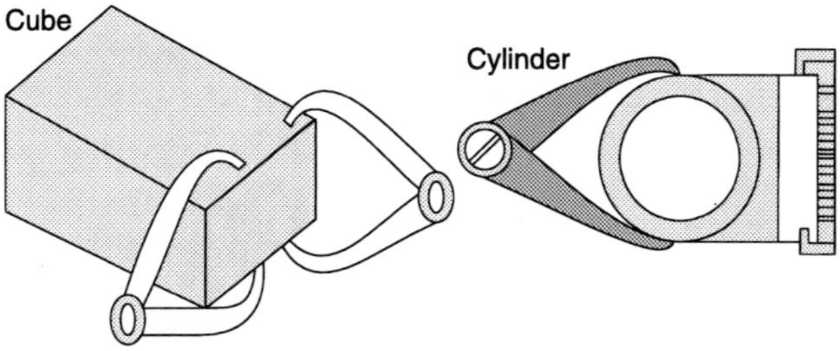

Fig. 2.13

2.3.2 Inside Callipers

It has the points at the end of the legs turned outward instead of inward, so that it may be used of gauging inside diameters, but we can also gauge the breadth and depth of a slot or slit groove etc.

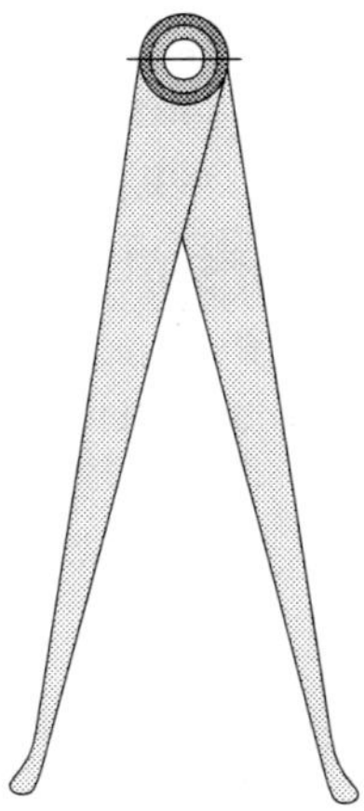

Fig. 2.14 Inside calliper

Use of Callipers and Precautions

1. Select the callipers according to the job.
2. Open the calliper according to the job.
3. Set the calliper in such a manner that its legs have only a mild touch with the job.
4. While opening and closing the calliper with firm joint it should be ensured that we do not strike it on any other metal except wood. Otherwise its points would be damaged and would not be able to measure correctly.
5. Spring callipers should be handled (picked up) in such a way that we are able to move its nut with our thumb and finger.
6. After taking correct measurement we should read from the steel rule.
7. Measurement with callipers should be taken to get actual measurement.
8. Do not measure a fixed part on a moving machine, the calliper may fall down and get damaged.
9. At times, for certain purpose, measurement has to be transferred from one calliper to another. In such a case, ensure that the tips of the legs of both the callipers meet at the same point.
10. Do not tighten or loosen the joints of callipers.

2.3.3 Spring Callipers (Fig. 2.15)

The legs of spring callipers are made from suitable alloy steel, measuring faces properly heat treated and hardened to a hardness of 65+50 HV. All the component parts are smoothly finished. The right part which is made from carbon spring steel is properly hardened and tempered to a hardness of 0 to 520 HV and all the components are manufactured and assembled in such a way as to achieve and maintain the smooth adjustment and efficient operation.

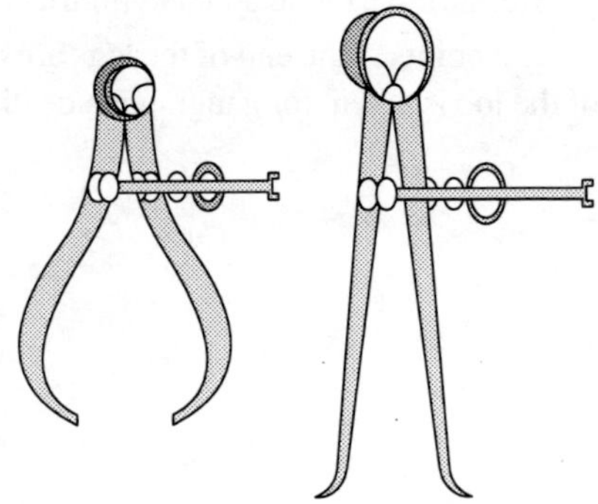

Fig. 2.15 Spring calliper

The working ends of the legs of each spring calliper should be identical in shape and have contact points of equal distance from the fulcrum. The measuring ends of each pair are so shaped that all sizes within its capacity may be readily transferred from measuring scales or gauges. The cross section to the legs may be either of rectangular or circular shape.

The callipers are adjusted to a set dimension by means of either a knurled solid nut or a knurled quick action released nut operating on a finely threaded adjusting screw, the latter one being preferred. The top portions of the legs are located in flanged fulcrum roller and held in position by a spring in order to maintain the alignment of the working ends. The spring provides sufficient tension to hold the legs rigid at all points of the adjustment. The friction between the adjusting nut and the leg is minimized by using a separate washer. One end of the adjusting screw is securely hinged to a leg of the calliper and a steel ball positively fixed to the free end of the screw for the purpose of retaining the adjusting nut. The inside and outside spring callipers are available in sizes of 75, 100, 150, 200, 250 and 300 mm. A good calliper should be able to retain its size when adjusted to three pieces near the lower and upper extremes of its range and in the midle.

Uses of Spring Callipers

Callipers are used along with steel rules, and the accuracy is limited to 0.5 mm. Parallelism of jobs, etc. can be checked with higher accuracy by using callipers.

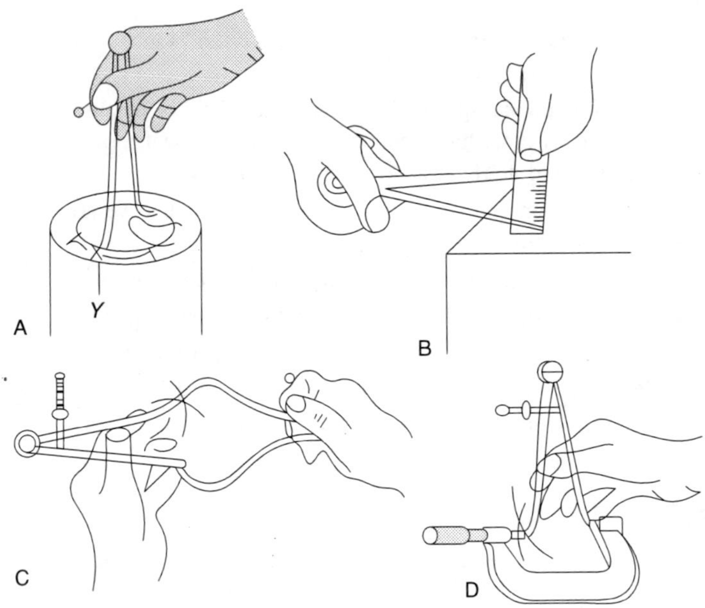

Fig. 2.16

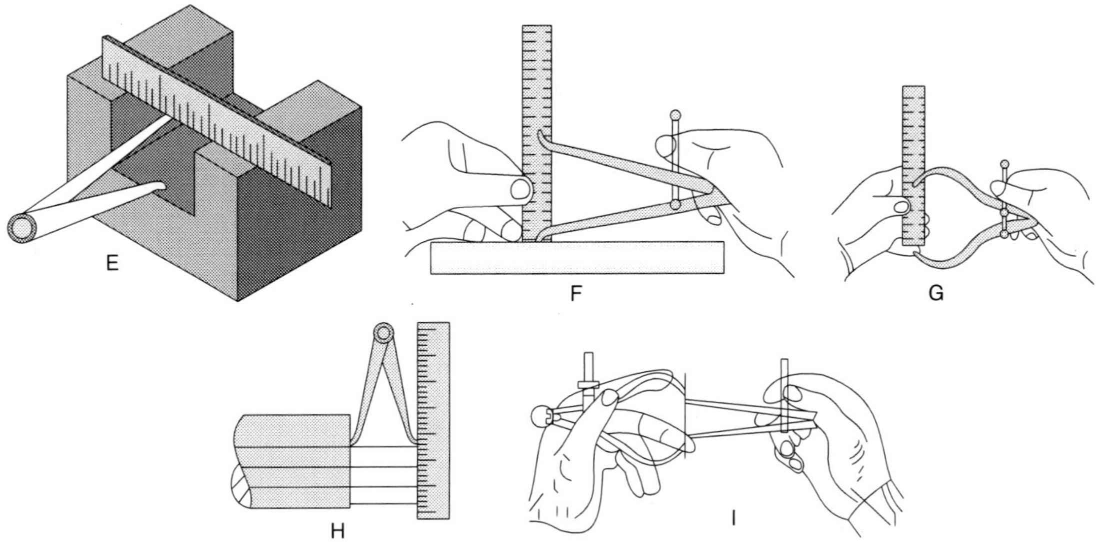

Fig. 2.16a

EXERCISE

1. In how many categories the tools used in the fitter trade have been divided? List them.
2. What are the different types of foot rule? To what minimum extent graduation is indicated on the steel foot rule ?
3. Name the different types of callipers?
4. What is the method of taking measurement with outside and inside callipers?
5. What function is performed by Jenny Calliper? How does it function? What is its other name?
6. For what reason are the rough dimensions of the piece list or order-card transferred to the rough part without oversize in length?
7. What is the measuring accuracy that can be obtained: (a) with the steel straight edge and (b) the sliding caliper?
8. What is the difference between the accuracy of measurement and accuracy of reading?
9. Describe the sliding calliper that allows a measurmg accuracy up to 1/10th of a millimetre.
10. How can measuring mistakes be avoided?
11. How are measuring instruments protected against fall, shock, heat or rust?

Semi Precision Measuring Tools: Vernier Calliper

Chapter 3

3.1 VERNIER CALLIPER

It is an instrument for measuring precision jobs. We can take measurement in Metric as well as in British system. It can take measurement of internal and external length and breadth etc., up to 0.001″ in British system and up to 0.02 mm in Metric system. This instrument was invented in 1630 by a great mathematician Parry Vernier in France. The instrument has been named after him. These are made in the range of 6″ to 48″ i.e., 150 mm to 1200 mm such as 6″ 8″, 10″ etc. i.e., with difference of 2″ each and so on and 150, 200, 250 mm etc., i.e., with difference 50 mm each.

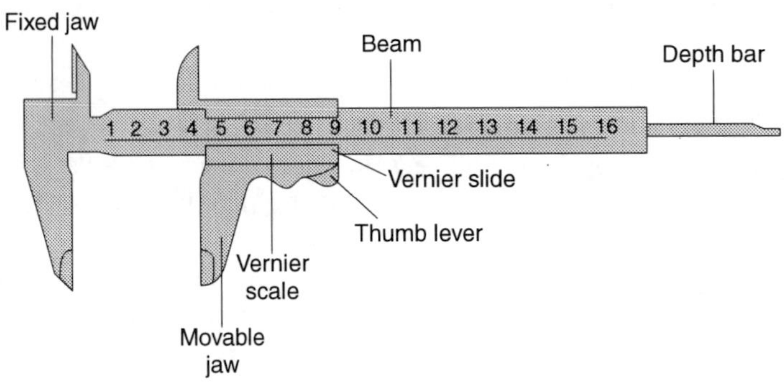

Fig. 3.1

3.1.1 Parts of Vernier Calliper

Fixed Jaws Fixed jaws are part of the beam scale. One jaw is used for taking external measurements, and the other for internal measurements.

Movable Jaws Movable jaws are part of the vernier slide. One jaw is used for external measurement and the other for internal measurement **(Figs. 3.2 & 3.3).**

Vernier Slide A vernier slide moves over the beam and can be set in any position by means of a spring-loaded thumb lever.

Beam The slide, and the depth bar attached to it, slide over the beam. The graduations on the beam are called the main scale divisions.

Depth Bar (Fig. 3.4) The depth bar is attached to the vernier slide and is used for depth measurements.

Thumb Lever The thumb lever is a spring-loaded lever which helps to set the vernier slab in any position on the beam scale.

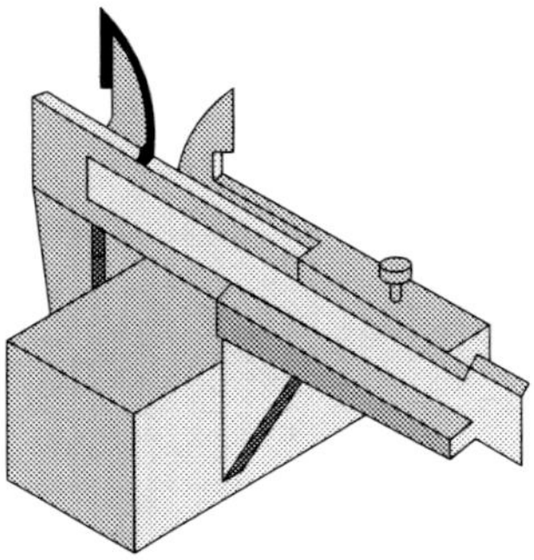

Fig. 3.2

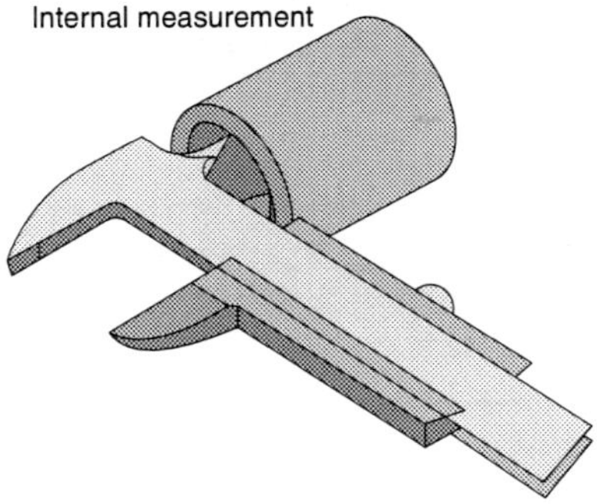

Fig. 3.3

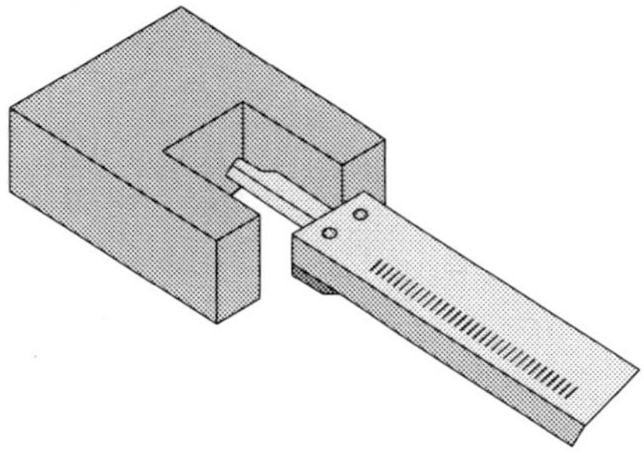

Fig. 3.4

Vernier Scale The vernier scale is graduation marked on the vernier slide. The divisions of this scale are called vernier divisions.

Main Scale Main scale graduations or divisions are marked on the beam.

3.1.2 Vernier Callipers: Sizes

Vernier callipers are available in sizes of 150 mm, 225 mm, 900 mm and 1200 mm. The selection of the size depends on the measurements to be taken. Vernier calipers are precision instruments, and extreme care should be taken while handling them.

These are available with different accuracies. The selection of the vernier callipers depends on the accuracy needed and the size of the job to be measured.

This accuracy/least count is determined by the graduations on the main scale and the vernier scale divisions.

3.1.3 Determining Least Count of Vernier Callipers

The main scale division (9 mm) shown in the Fig. 3.5 is divided into 10 equal parts in the vernier scale, i.e., one main scale division (MSD) = 1 mm.

One vernier scale division (VSD) = 9/10

Least count is 1mm --9/10 = 1/10 mm

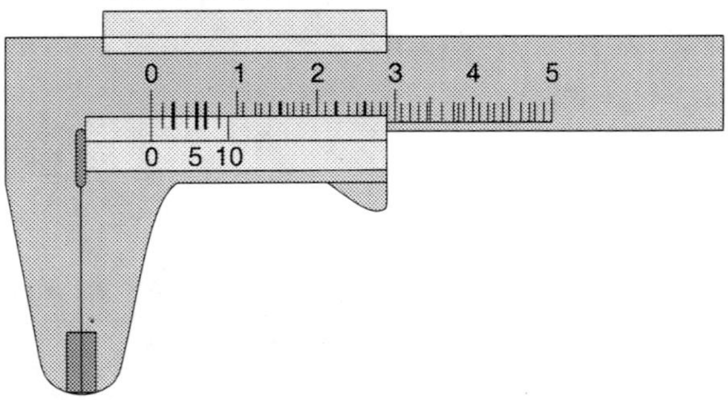

Fig. 3.5

3.1.4 Reading Vernier Measurements

Vernier callipers are available with different graduations and least counts. For reading measurements with a vernier callipers, the least count should be determined first. (The least count of callipers is sometimes marked on the vernier slide.)

Figure 3.5 shows the graduations of a common type of vernier calliper with a least count of 0.02 mm. In this 50 divisions of the vernier scale occupy 49 divisions (49 mm) on the main scale.

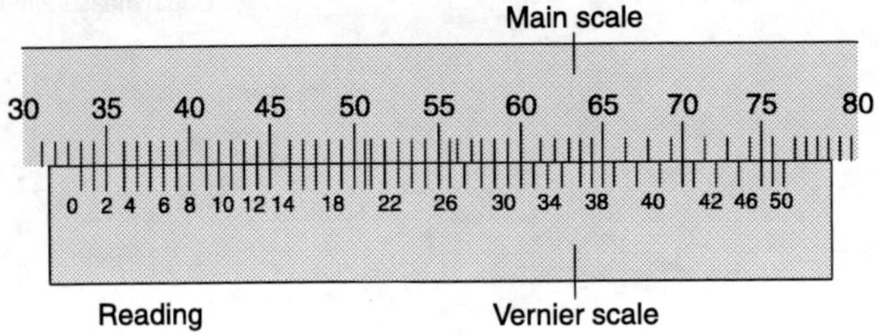

Fig. 3.6 Reading vernier measurement

3.2 DIAL VERNIER CALLIPERS

A dial calliper is a direct reading instrument which resembles the vernier callipers; it is easier to read a dial calliper than the traditional vernier callipers. The beam scale is graduated into 5 mm increments on 0.05 accuracy calliper.

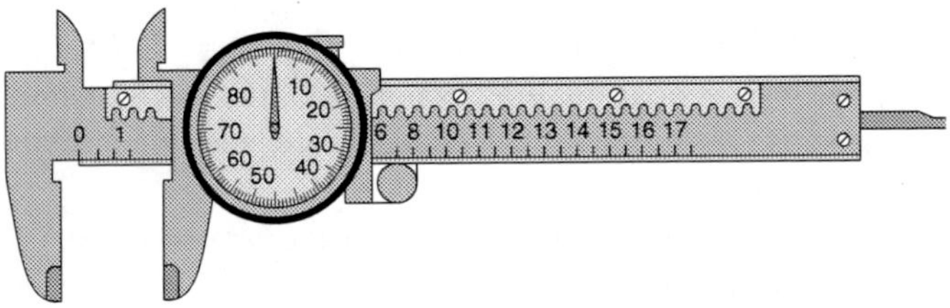

Fig. 3.7

The calliper dial on the movable jaw is graduated into 100 equal divisions. The hand of the dial makes one complete revolution for each 5 mm, therefore, each dial graduation represents 1/100th of 5 mm or 0.05 mm.

For reading a measurement (Fig. 3.8)

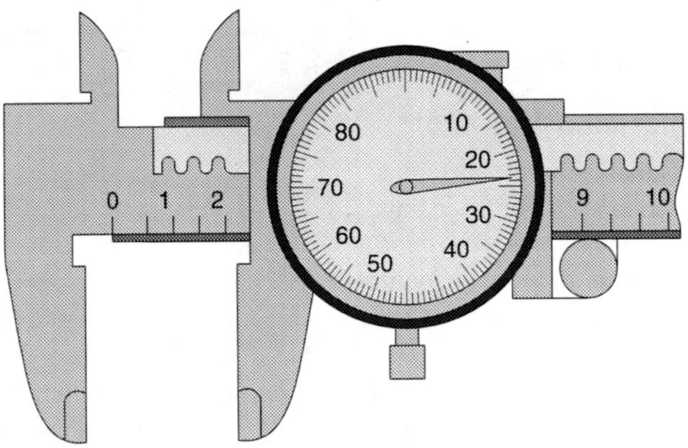

Fig. 3.8

Read the beam scale reading (25 mm) and add the reading shown by the hand of the dial. 24 × 0.05 = 1.2 mm.

Reading = 25 + 1.2 mm = 26.2 mm.

All parts are made of nickel-chromium steel, heat treated and ground. They are machined to a high accuracy. They are stabilized to avoid distortion due to temperature variations.

3.2.1 Constructional Features

The beam is the main part and the main scale gradations are marked on it. The markings are in millimetres and every tenth line is drawn a little longer and brighter than the other graduations and numbered 1, 2, 3.

To the left of the beam the fixed jaws for external and internal measurements are fixed as integral part. The vernier unit slides over the beam.

At the bottom face of the beam a groove is machined for its full length permitting the blade to slide in the groove.

At the bottom right hand end, a unit is fixed serving as a support for the blade when it slides in the groove.

The vernier unit has got the vernier graduations marked on it. The movable jaws for both external and internal measurements are integral with this.

The fixed and movable jaws are knife-edged to have better accuracy during measurement. When the fixed and movable jaws are made to contact each other, the zero of the vernier scale coincides with the zero of the main scale. At this position the blade will align with the right end edge of the beam.

When the vernier scale unit slides over the beam, the movable jaws of both the measurements as well as the blade advance to make the reading.

To slide the vernier unit, the thumb lever is pressed and pulled or pushed according to the direction of movement of the vernier unit.

Least Count

In the vernier scale illustrated here, 19 mm are divided into 10 equal parts. The value of 1 vernier scale division will then be 19/10 = 1.9 mm.

The difference of the two main scale divisions and 1 vernier scale division gives the least count and it is equal to 2 × 1 – 1.9 mm = 0.1 mm.

For more accuracy, a 49 mm space is divided into 50 equal parts on the vernier scale so that one vernier scale division will be: 49/50 = 0.98 mm.

Here the least count will be:

1 main division-1 vernier scale division

= 1 mm-0.98 mm

= 0.02 mm.

Advantage No need to have separate precision instruments for taking external, internal and depth measurements.

Disadvantages Accuracy of the reading depends on the skill of the operator, it loses its accuracy by constant usage as slackness in the sliding develops.

It cannot be used to measure components having dimensions less than 0.02 mm.

Possibility of parallax error during noting down the conciding line.

To read a measurement Note the number of graduations on the main scale passed by the zero of the vernier. This gives the full mm.

Note down, which of the vernier scale division coincides with any one line on the main scale.

Multiply this number with the least count.

Add the multiplied value to the main scale reading.

3.3 DIGITAL CALLIPER

Advantages Easy reading of digital display with large 4, 7 mm high characters with decimal point, minus sign.

Zero settings at any position: Within measuring range by conveniently placed button.

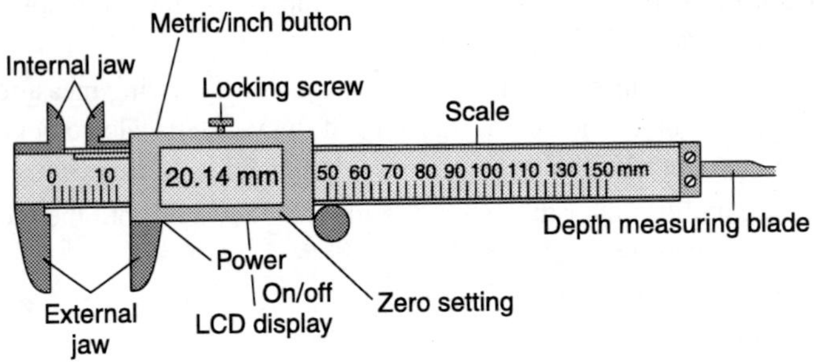

Fig. 3.9

3.4 ELECTRONIC DIGITAL CALLIPER

Electronic digital calliper is the most advanced electronic digital measuring instrument for fast, accurate and reliable measurements and can even be used by semiskilled and non-technical persons on the shop floor. It measures outside and inside diameters, depth and steps etc. Capacitive measurements system and with the revolutionary new chip permitting interface with computer print-out facilities are provided in it. The powerful large-scale integration (LSI) microchip gives full scale computer functions to measure, monitor, and memory. The clear easily read LCD reads up to 0.01 mm.

Salient Features

- Error free, accurate and fast reading.
- Correct and accurate inch/metric conversion.
- Floating zero allows comparative measurements.
- Display memory facilities.
- HOLD function freeze display reading.
- No vernier scale – rack or pinion, and there is no backlash.

- Unaffected by oil or dust and ideally suited for shop floor use.
- Low power consumption, (one button cell of 1.5 volt) for one year continuous use

The frame jaws and depth rod are of hardened stainless steel precision ground.

3.5 UNIVERSAL VERNIER CALLIPERS (FIG. 3.10)

- Beam
- Fixed jaw for external measurement
- Movable jaw for external measurement
- Movable jaw for internal measurement
- Blade for depth measurement
- Main scale
- Vernier scale
- Fine adjustment screw
- Set of locking screws.

All parts are made of nickel-chromium steel, heat-treated and ground. They are machined to a high accuracy. They are stabilized to avoid distortion due to temperature variations.

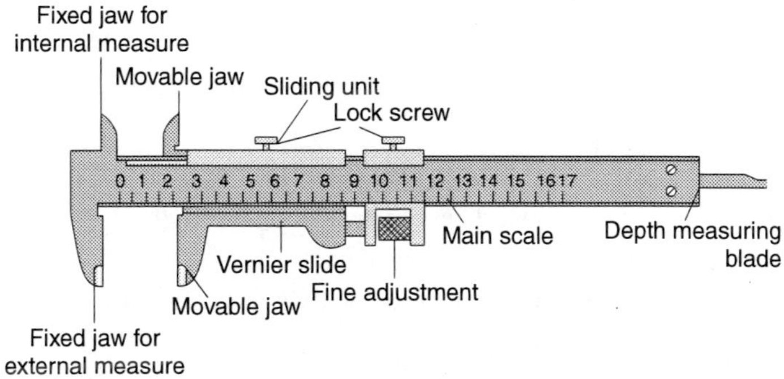

Fig. 3.10

3.5.1 Constructional Features

The beam is the main part and the main scale graduations are marked on it. The markings are in millimetres and every tenth line is drawn a little longer and brighter than the other graduations and numbered 1, 2, 3,..

To the left of the beam the fixed jaws for external and internal measurements are fixed as integral parts. The vernier unit slides over the beam.

At the bottom face of the beam a key way – like groove is machined for its full length, permitting the blade to slide in the groove.

At the bottom right hand end, a unit is fixed to serve as a support for the blade when it slides in the groove.

The vernier unit has got the vernier graduations marked on it. The movable jaws for both external and internal measurements are integral with this.

The fixed and movable jaws are knife-edged to have better accuracy during the measurement. When the fixed and movable jaws are made to contact each other, the zero of the vernier scale coincides with the zero of the main scale.

At this position the blade will be in line with the right hand edge of the beam.

When the vernier scale unit slides over the beam, the movable jaws of both the measurements as well as the blade advance to make the reading.

To slide the vernier unit, the thumb lever is pressed and pulled or pushed according to the direction of movement of the vernier unit.

Least Count

In the vernier scale illustrated here, 19 mm are divided into 10 equal parts on the vernier scale. The value of 1 vernier scale division will then be:

$$19/10 = 1.9 \text{ mm}$$

The difference of the two main scale divisions and 1 vernier scale division gives the least count and it is equal to 2×1 mm – 1.9 mm = 0.1 mm

For better accuracy, a 49 mm space is divided into 50 equal parts on the vernier scale so that one vernier scale division value will be 49/50 = 0.98 mm.

Here the least count will be 1 main scale division – vernier scale division = 1 mm – 0.98 = 0.02.

The application of the universal vernier calliper is taking external, internal and depth measurements.

Advantage No need to have separate precision instruments for taking external, internal and depth measurements.

Disadvantages Accuracy of reading depends on the skill of the operator.

Loses its accuracy by constant usage as slackness in the sliding develops.

It cannot be used to measure components having deviations less than 0.02 mm.

Possibility of parallax error during noting down the coinciding line may cause the reading of the measuring to be wrong.

To read a measurement Note the number of graduations on the main scale passed by the zero of the vernier. This gives full mm.

Note, which of the vernier scale division coincides with any one line on the main scale.

Multiply this number with the least count.

Add the multiplied value to the main scales reading.

Precautions while measuring with vernier calliper Clean the contacting faces and sliding parts with a piece of cloth. Do not use cotton waste.

Close the jaws and examine the gap between the jaws through light.

Check zero error.

Confirm the function of the locking screws.

Reading the vernier Read the correct vernier scale division which coincides with the main scale division. Avoid error of parallax.

Use a magnifying glass, if required.

Place the jaws at right angle to the axis of the workpiece.

Maintain contact between the jaws and the workpiece. (Close the jaw over the workpiece in such a way that the jaws just slip from the point of contact.)

Lock the vernier calliper before taking it out from the workpiece.

Measuring round rods Measure the round rods with the vernier calliper jaws touching at right angles.

Hold the workpiece in the middle of the jaws.

While measuring large diameters, the relationship of the diameter of job (*D*) with the length of jaws (*a*) should be a =1/2 *D*.

If <1/2 *d*, measure by letting the beam contact the ends of the object being measured.

Measuring the inside diameter Measure by matching the nibs of the vernier calliper, to the axial centre.

Ensure correct contact of the nibs. Insert nibs in the hole as deep as possible. If insertion is not sufficiently deep, the measuring surface of the vernier calliper will tilt, resulting in incorrect measurements.

Measuring the depth of a hole The depth reference surface of the depth bar of the calliper must be in contact with the surface of the object.

The depth bar should not tilt with respect to the object being measured.

The tip of the depth bar should not touch the corner of the job.

When no direct measurement is possible, bridge with an object which can be used as a reference for measurement.

3.6 VERNIER CALLIPER – BRITISH SYSTEM

The principle of the British vernier calliper for reading the measurements to an accuracy of one thousandth of an inch (1/1000) is as follows.

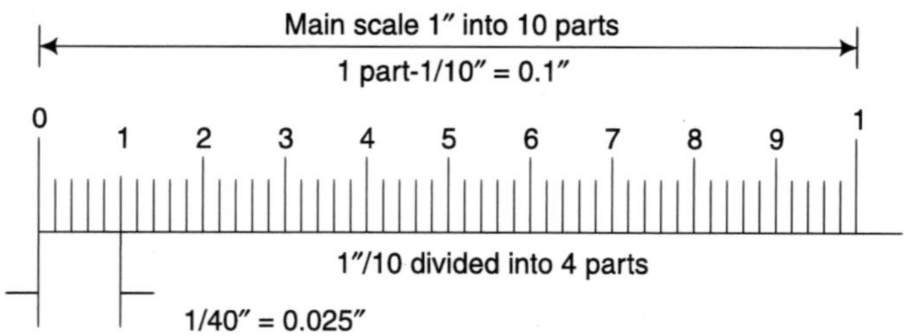

Fig. 3.11

One inch of the main scale is divided into 10 equal parts, and graduated from 0 to 9. Each one-tenth of an inch of the main scale is further subdivided into 4 equal parts, thus making the value of each subdivision of the main scale equal to one-fortieth of an inch (1/40″) = 0.025″.

The 14th line on vernier add 0.014″

$$1.0 + 0.1 + 0.025 + 0.014 + 1.139''$$

The vernier scale is made from 24 such fortieth divisions (1/40 × 24), and divided into 25 equal parts. So the value of each vernier scale division works out to be 1/25 of the total length.

$$1/40 \times 24/25 = 24/40 \times 25 = 24/100\ c = 0.024 \text{ inch.}$$

Therefore the least count = difference between one main scale division and one vernier scale division = 0.025″ (1/40″) – 0.024″ (24/1000″) = 0.001″ (1/1000″).

When the ‘0’ graduation of the main scale and the ‘0’ graduation of vernier scale, coincide, the 24th division on the main scale coincides with the 25th division of the vernier scale. The first main scale division will be ahead of the first vernier scale division by 0.001″, the 2nd main scale division will be ahead of the 2nd vernier scale division by 0.002″ and so on till the 24th division of the main scale which will be ahead by 0.024″ from the 24th division of the vernier scale. So the 25th division of the vernier scale will exactly coincide with the 24th division of the main scale.

Determining the reading of an object with the help of a British vernier calliper

Read the main scale to the left of the zero gradution of the vernier scale. Note the tenth (1/10″ = 0.100″) and (1/40″) one-fortieth divisions of the main scale. Look at the coinciding division of the vernier scale with the main scale division. The coinciding number of the vernier scale gives the number of thousandths of the inch to add to the main scale reading. The method of reading the British vernier calliper is shown in Fig. 3.12 &12 A). 14th line on the vernier add 0.014 inch; 1.0 + 0.1 + 0.025 + 0.014 = 1.139 inch.

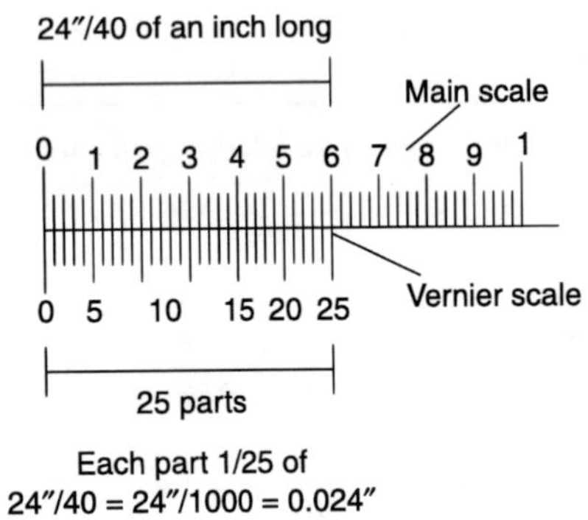

Fig. 3.12 British vernier

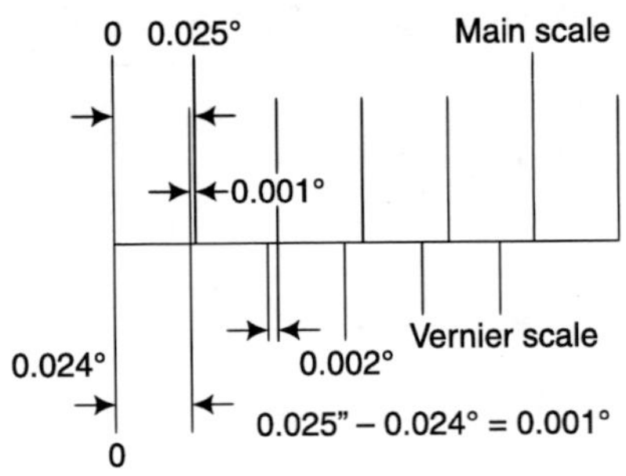

Fig. 3.12A Reading British vernier

3.7 VERNIER HEIGHT GAUGE

It is used for taking accurate measurement of the height of a job or for marking. It is almost similar to vernier calliper but it is used by attaching some additional attachments. Beam remains fitted on a base in length form. Offset scriber is fitted on the beam itself with which height of a job is measured or marking is done.

(i) Solid base (**Fig. 3.13**).

(ii) Moveable base (**Fig. 3.14**).

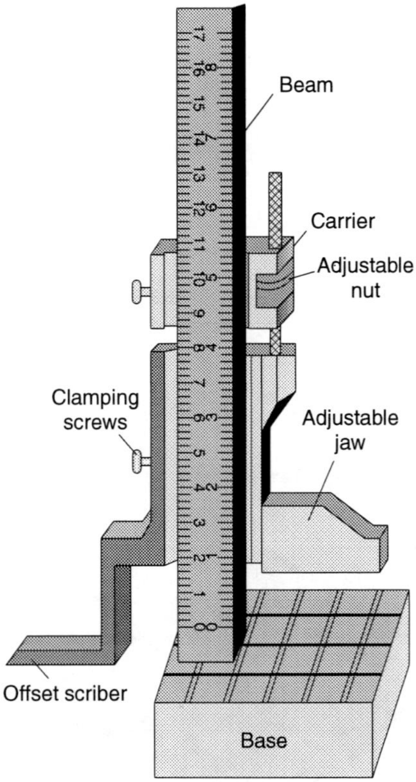

Fig. 3.13

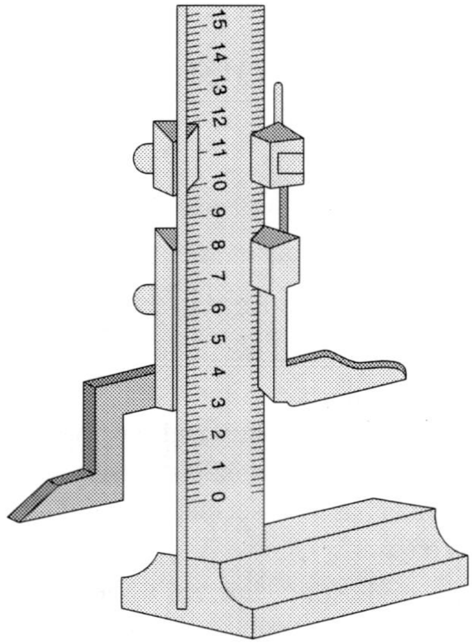

Fig. 3.14

In vernier height gauge, slide base remains joined with the beam permanently (**Fig. 3.13**).

In these types of vernier height gauges there is no facility to set the beam or base according to the requirement of the job. In movable base vernier this facility exists. This type of vernier height gauge is in the form of a set, which has base, vernier calliper, offset scriber, fixing screw, etc.

3.7.1 Constructional Features

The construction of a vernier height gauge is similar to that of the vernier calliper except that it is vertical with a rigid base. It is graduated on the same vernier principle which is applied to the vernier calliper.

It consists of an upright steel bar fastened to a steel base. The beam is graduated with the main scale in mm as well as in inches. The main slide carries a jaw to which various attachments may be clamped. The jaw is an integral part of the main slide.

The vernier scale is attached to the main slide which is graduated, to read metric dimensions as well as the inch dimensions. The main slide is attached with a fine adjusting slide. The movable jaw is most widely used with the chisel pointed scriber blade for accurate marking as well as for checking the height, steps, etc. Care should be taken to allow for the thickness of the jaw depending on whether the attachment is clamped on the top or under the jaw for this purpose. The thickness of the jaw is marked on the instrument. As in a vernier calliper, the least count of this instrument is also 0.02 mm. An offset scriber is also used on the movable jaw. It is required to take measurement from the lower planes.

The complete sliding attachment along with the jaw can be arrested on the beam to the desired height with the help of the lock screws. The vernier height gauges are available in the range from zero to 1000 mm.

The movable jaw is mostly used with the chisel pointed scriber blade for accurate marking as well as for checking the height, steps, etc. Care should be taken to allow for the thickness of the jaw depending on whether the attachment is clamped on the top or under the jaw for this purpose. The thickness of the jaw is marked on the instrument.

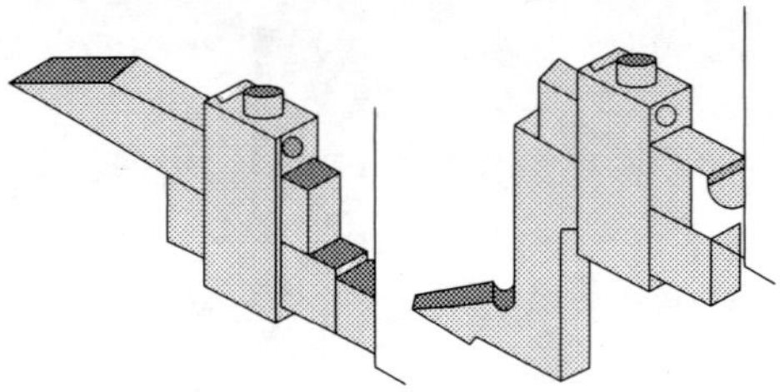

Fig. 3.15

Like the vernier calliper, the least count of the instrument is also 0.02 mm.

An offset scriber is also used on the movable jaw when it is required to take measurement from the lower planes (**Fig. 3.15**).

3.7.2 Functional Features

Vernier height gauges are used in conjunction with the surface plate. In order to move the main slide, both the locking screws of the slide and the finer adjusting slide have to be loosened. The main slide along with the chisel pointed scriber has to be set manually, for any approximate height as required.

The finer adjusting slide has to be locked in position, for an approximate height as required. To get an exact markable height, the finer adjustments have to be carried out on the slide with the help of the adjusting nut. After obtaining the exact markable dimension the main slide is locked in position.

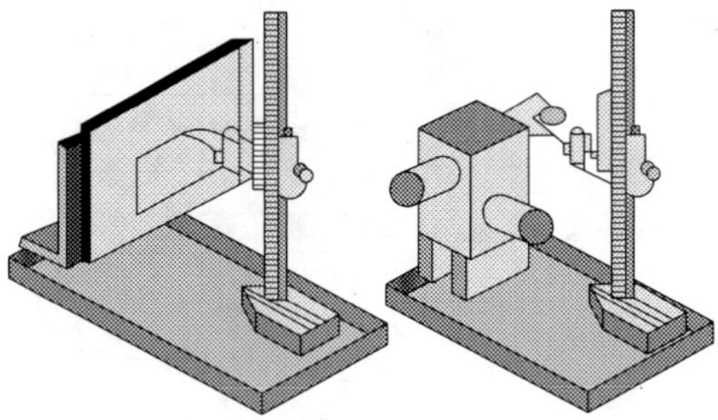

Fig. 3.16

Modern vernier height gauges are designed on the screw rod principle. In these height gauges, the screw rod may be operated with the help of the thumb screw at the base. In order to have quick settings of the main slide, it is designed with a quick releasing manual mechanism. With the help of this, it is possible to bring the slide to a desired approximate height without wastage of time. For all other purposes, these height gauges work as ordinary height gauges. In order to set the zero graduation of the main scale for the initial reading. This minimizes the possible errors in reading the various sizes in the same settings.

Another kind of modern vernier height gauge has a rack and pinion set up for operating the sliding unit.

Applications

- It is used for measuring the width of the slot and external dimension.
- The vernier height gauge is used with the dial indicator to check the hole location, pitch dimensions, concentricity and eccentricity.
- It is also used for measuring depth, with a depth attachment.
- It is used to measure size from the lower plane with the help of an offset scriber.

- Accurate measurements are important in layout and inspection work.
- Vernier height gauges are particularly suitable for marking accurate distance, and centre locations.

3.8 VERNIER DEPTH GAUGE

A vernier depth gauge is a precision instrument for measuring holes, recesses, slots and steps.

Its construction and method of reading are similar to those of a vernier calliper.

The base (anvil) is the fixed unit and serves as a datum for measurement. It also carries the vernier scale and the fine adjustment mechanism.

The beam with the main scale graduations is the sliding member or part.

Fine adjustments for measurement are made after tightening the clamping screw and the fine adjustment mechanism.

While taking measurements the base should be firmly held against the reference surface.

The relief given at the end of the beam of some vernier height gauges is to avoid the seating in the corners of slots and to ensure correct reading.

Remove burrs, if any before taking measurement, excessive pressure on the beam, while taking measurements will lift the base from the reference surface and will show wrong measurement.

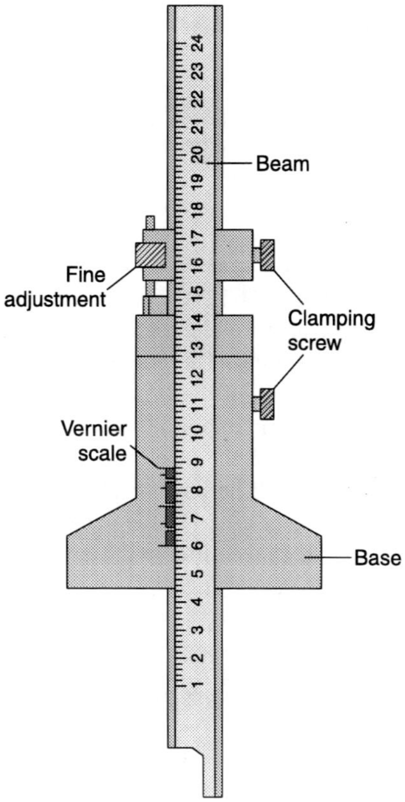

Fig. 3.17

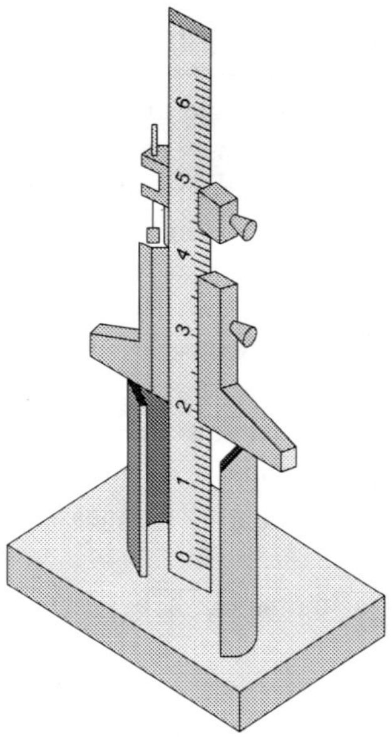

Fig. 3.17a

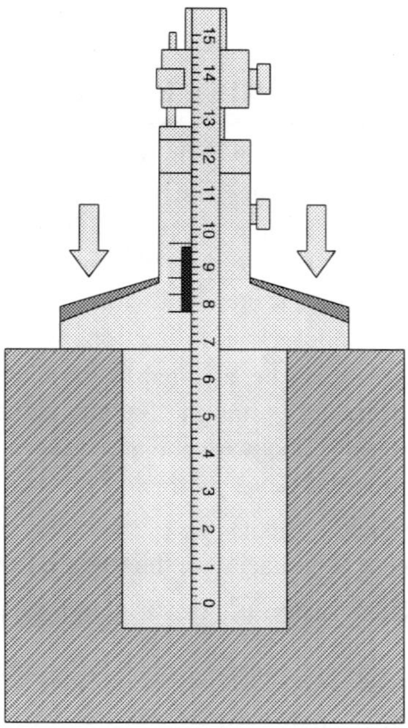

Fig. 3.17b

3.8.1. Measurement of Angles

The unit of an angle

For angular measurement a complete circle is divided into 360 equal parts. Each division is called a degree. (A half circle will have 180°.)

Subdivision of an angle

For more precise angular measurements, one degree is further divided into 60 equal parts, this division is one minute. (')

The minute is used to represent a fractional part of a degree and is written as 30° 15°. One minute is further divided into smaller units known as second ("). There are 60 seconds in a minute. An angular measurement written in degrees, minutes and seconds would read as 30° 15' 20".

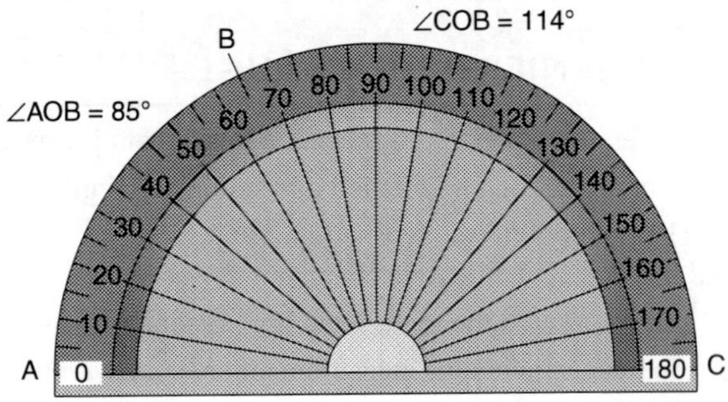

Fig. 3.18

One minute is further divided into smaller units known as seconds.

("). There are 60 seconds in a minute.

An angular measurement written in degrees, minutes and seconds would read as 30° 15′ 20″.

Example for angular divisions:

1 complete circle = 360°

1/2 circle = 180°

1/4 of a circle (right angle) = 90°. Subdivisions 1 degree or 1° = 60 mts or 60 '. 1 min or 1' min = 60 secs or 60 ".

3.9 VERNIER BEVEL PROTRACTOR (FIG. 3.19)

The vernier bevel protractor is able to read the angle to an accuracy of 5 minutes (1/12°) with the help of a vernier scale. The vernier bevel protractor consists of an outer disc, an inner disc, blade and base. The base is the integral part of the outer disc and marked with the main scale in degrees. The graduations are '0' to 90° to 0° then 90° and back to '0'. Thus, the main scale is capable of reading from 0° to 360°. The inner disc is attached with the vernier scale which can be slid on the circular main scale in both clockwise and anticlockwise directions with the help of a knob. The centre knob is fitted with the blade called bevel which can be adjusted to the required position with the help of a locking screw. The base of the vernier protractor is straight, flat and parallel with reference to the graduations of the outer disc. For the purpose of reading clockwise and anticlockwise directions, two vernier scales are fitted to the inner disc. In the case of a universal bevel protractor, the acute angle attachment is equipped to measure the acute angles.

Stock This is one of the contacting surfaces during the measurement of an angle. Preferably it should be kept in contact with the surface from which the inclination is measured.

Disc The disk is an integrated part of the stock. It is circular in shape, and the edge is graduated in degrees.

Dial It is pivoted to the disc and can be rotated through 360°. The vernier scale of the instrument is attached to the dial. The dial is locked to the disc while reading the measurement.

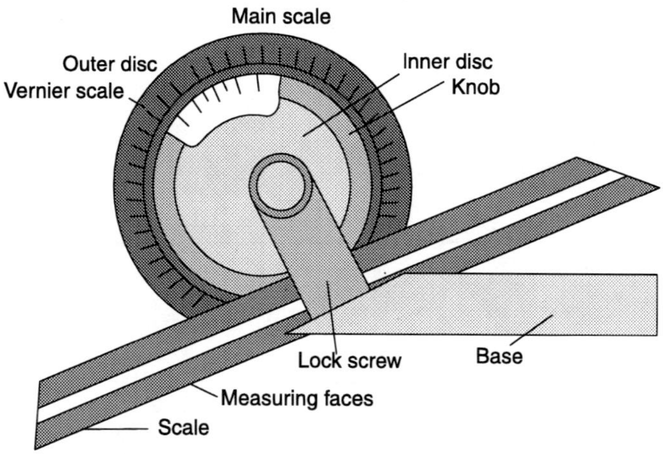

Fig. 3.19

Blade This is the other contacting surface of the instrument that contacts the workpiece during measurement, preferably the inclined surfaces. It is fixed to the dial with the help of a clamping lever. A parallel groove is provided in the centre of the blade to enable it to be longitudinally positioned whenever nesessary.

Locking Screws Two knurled locking screws are provided, one to lock the dial to the disc, and the other to lock the blade to the dial.

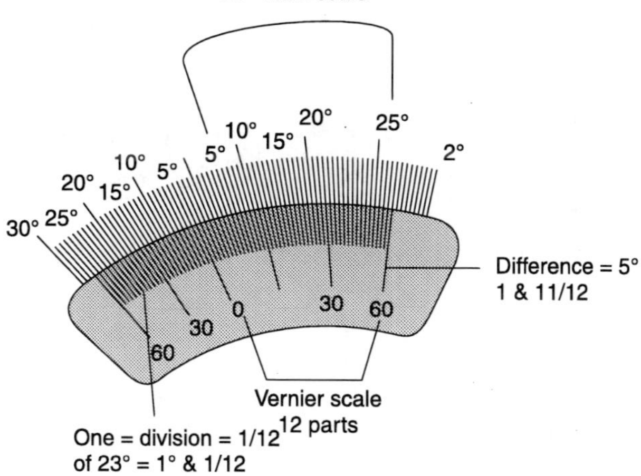

Fig. 3.20

All parts are made of good quality alloy steel, properly heat-treated and highly finished. A magnifying glass is sometimes fitted for clear reading of the graduations.

Applications

Figure (3/21 A) shows the method of checking the interior angle of a hexagon. B) Shows the method of checking the taper angle of a lathe collet C) and D) show the method of checking the angle of a dovetail machined part.

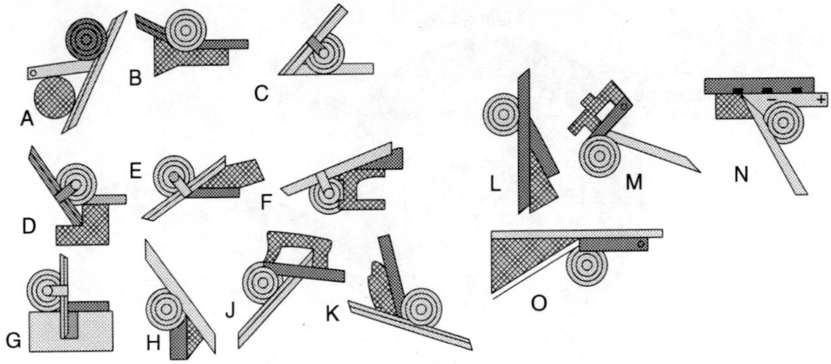

Fig. 3.21

E) Checking the cone angle of a turned part, F) Checking the acute angle of a machined part with the help of an acute angle attachment, G) the method of checking the straightness with reference to the other surface and H), J), K), L), M), N) and O). show some more examples of the use of vernier bevel protractor.

3.10 COMBINATION SET

Combination sets can be used for different types of work, like layout work, measurement and checking of angles.

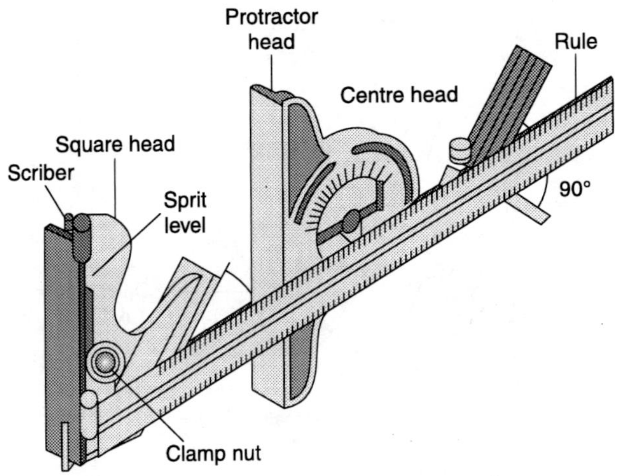

Fig. 3.22

The combination set has a (1) protractor head, (2) square head, (3) centre head, and (4) rule.

Protractor Head The protractor head can be rotated and set to any required angle. The protractor head is used for marking and measuring angles with an accuracy of 1° . The spirit level attached to this is useful for setting jobs in a horizontal plane.

It is made of steel. It is fitted with the blade with the help of a knurled nut which is at the centre of the protractor dial. It contains two dials. One contains graduation up to 180°. At every tenth part lock scale is indicated. This dial revolves in the other outer dial. Two screws are fitted on it, through which the inner dial is set. On the outer dial a line is marked, which acts as a basis for setting the angles. Through this we can measure an angle of the job or mark it. On the inner dial, there is a spirit label through which surface cut in angle can be measured without blade.

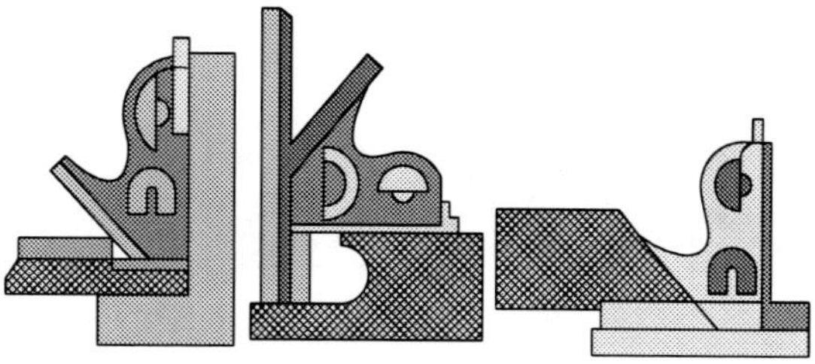

Fig. 3.23 Square head

Square Head The square head has one measuring face at 90° and another at 45°. It is also used to set the workpieces on the machines and measure the depth of slots (**Fig. 3.23**).

This attachment can be used for various functions. One of its arms is built at 45° the other at 90°. Through it we can verify 45° and 90° angle of a job. It can also be used as depth gauge and height gauge. A scribbler is also attached to it, which is fitted in the hole of the head. Therefore the head can be set at the proper place and marking can be done. It also contains a spirit. With its help we can verify the accuracy of any surface.

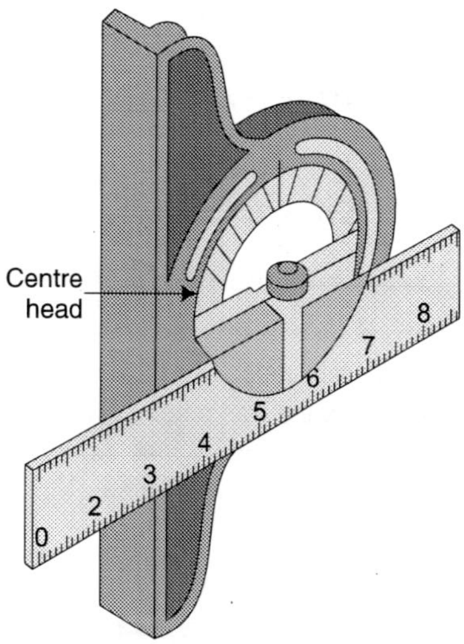

Fig. 3.24

Centre Head This along with the rule is used for locating the centre of cylindrical jobs (**Fig. 3.24**).

Its shape is like English letter 'V'. The angle between its two legs is of 90°. It can be set anywhere on the blade through lock screw. With it, we can find the central point of round jobs. Radial lines can be drawn on the job.

For ensuring accurate results, the combination set should be cleaned well after use and should not be mixed with cutting tools, either while using or storing.

Blade or Rule It is made of steel and is tempered. Its length ranges from 6″ to 24″. Graduation in metric and British system is there on it as on a steel foot rule. A groove to the extent of entire length is cut on one side of this blade. Because of this any head can be fixed on it through turret thumb screw, its breadth is I″ and thickness 1/16″.

EXERCISE

1. For what reasons are the rough dimensions of the piece list or order-card transferred to the rough part without oversize in length?
2. What is the measuring accuracy that can be obtained: a) with the steel straight edge and b) with the sliding calliper?
3. What is the difference between accuracy of measurement and accuracy of reading?
4. Describe the sliding calliper that allows a measuring accuracy up to $1/10^{th}$ of a millimetre.
5. How can measuring mistakes be avoided?
6. How are measuring instruments protected against fall, shock, heat and rust?

Precision Measuring Tools

Chapter 4

4.1 MICROMETER

A micrometer is a precision instrument used to measure a job, generally with an accuracy of 0.01 mm. Micrometers which take the outside measurements are known as outside micrometers.

4.1.1 History of Micrometer

Micrometer was invented in 1848 by a scientist named Jean Palmer in France who had settled in America. Production of micrometer was taken up in 1870 in England by Brown and Sharp Company and named it after the name of its inventor "Palmer Micrometer". In 1877 it was available in improved form and a model or sample of micrometer was made on this basis.

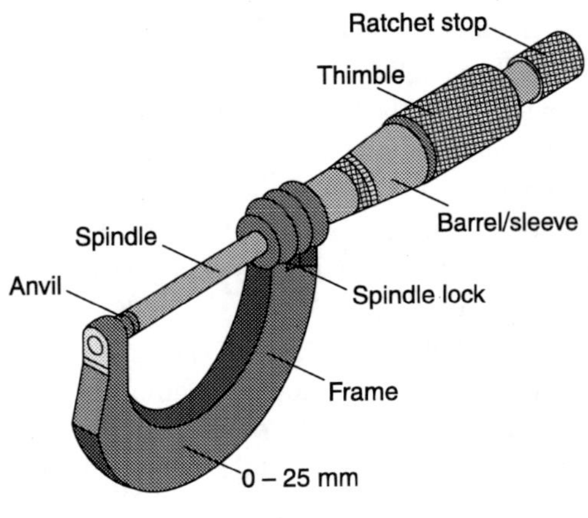

Fig. 4.1

4.1.2 Principle of Micrometer

Micrometer is made on the basis of nut and screw bolt. You must be familiar with the screw bolt. It contains threads cut at an equal distance; similarly, threads are also there at equal distance in the nut. When we fit the nut on the bolt and revolve it, it starts advancing forward and backward. The distance from a point on a screw thread to a corresponding point on the next thread is called the pitch of the screw.

In a single thread the pitch is the amount of advance move forward or backward in one revolution.

The longitudinal movement of the spindle during one rotation is equal to the pitch of the screw. The movement of the spindle to the distance of the pitch or its fractions can be accurately measured on the barrel and thimble.

In other words, the "pitch" of a "single" type thread is the distance of the centre of one thread from the centre of an adjoining one. The bolt having a single start and its lead and pitch are equal. The distance a screw advances when given a single complete turn is called lead. In actual micrometer this principle has been adopted.

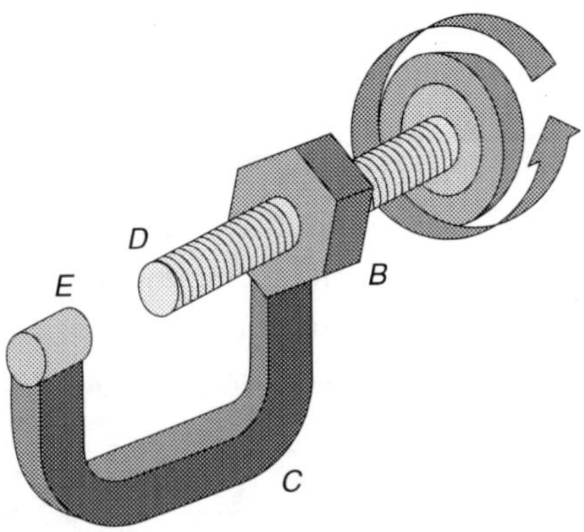

Fig. 4.2

In **Fig. 4.2**, bolt "A" is fitted in nut "B". With nut 'B', frame 'C' is joined. When the head of bolt 'A' completes one turn, the end 'D' of the bolt moves towards or opposite the anvil 'E' in linear motion. Thus, space for measuring of job is created between the bolt edge D and anvil 'E'.

4.1.3 Graduations

In metric micrometers the pitch of the spindle thread is 0.5 mm.

Thereby, in one rotation of the thimble, the spindle advances by 0.5 mm.

On the barrel a 25 mm long datum line is marked. This line is further graduated to millimetres and half millimetres (i.e., 1 mm and 0.5 mm). The graduations are numbered as 0, 5, 10, 15, 20 and 25 mm.

The circumference of the bevel edge of the thimble is graduated into 50 divisions and marked 0-5-10-15, 45-50 in a clockwise direction.

The distance moved by the spindle during one rotation of the thimble is 0.5 mm.

Movement of one division of the thimble = 0.5 × 1/50 = 0.01 mm.

Accuracy or least count of a metric outside micrometer is 0.01 mm.

4.1.4 Construction of Micrometer

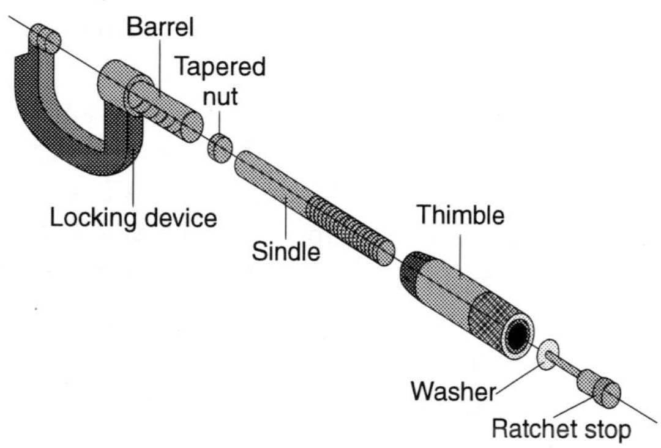

Fig. 4.3

The micrometer (outside), has five parts.

1. Frame.
2. Anvil.
3. Spindle.
4. Sleeve or barrel.
5. Thimble.
6. Lock nut.
7. Ratchet stop.

1. **Frame** The frame is in "U" shape or in a shape according to the needs. This is made of cast steel, aluminium, and alloy or cast-iron.

   It is galvanized or it has some other anti-corrosive treatment to save it from rusting. At one end on the frame there is an anvil and the other end has a hole, from where the spindle comes out. The range of measurement of the frame is indicated on the frame itself. At times, it also contains a conversion table of inches into decimals for convenience.
2. **Anvil** It is a small round part made of hard steel. It is finished with high accuracy. The job is measured by placing it between the anvil and the spindle. Both have tips made of chromium to avoid wear out. It protrudes from the frame for a distance of at least 3 mm.
3. **Spindle** The spindle is according to the diameter of the anvil but is more than anvil length. Nearly half of the spindle is outside the frame and the rest has threads cut of 0.5 mm pitch. On the part having thread, a thimble with the ratchet is fixed.
4. **Sleeve or barrel** It is a hollow part which meets at frame towards the spindle. It has threads at the scale of 0.5 mm pitch in metric measure and 40 threads per inch (TPI) in the British system. On these threads the spindle moves forward or backwards.

   Readings: The graduation on the barrel is in two parts, namely one above the reference line and the other below. The graduation above the reference line is graduated in 1 mm intervals. The first and every fifth are long and numbered 0, 5, 10, 15, 20, and 25. The lower graduations are in 1 mm intervals but each graduation shall be placed at the middle of the two successive upper graduations to be read 0.5 mm.

5. **Thimble** It is a tubular cover fastened with the spindle and it moves with the spindle. The bevelled edge of the thimble is divided into 50 equal parts, every fifth being numbered. On the other end knurling is provided which enables the thimble to rotate. As the thimble rotates the spindle also revolves with it and it gains linear motion, i.e., it moves forward or backward.
6. **Lock nut** Lock nut is provided to keep the micrometer fixed so that reading can be taken easily and without any difficulty. It is fixed on the edge, on the spindle side of the frame in the form of asterisk round ring. It is knurled towards the outside. By using a lock nut, we do not need to set micrometer again and again for the same measurement.
7. **Ratchet stop** At the last end of the micrometer the ratchet is fixed, duly knurled. When the correct measurement of the job is indicated on the micrometer (after the movement of the thimble) the ratchet functions thus avoiding further advancement of the instrument. In the absence of the ratchet stop the reading may be by exerting more pressure on the thimble or the pressure may be less. In this case, there is a possibility of difference in measurement. Ratchet stop ensures same measurement irrespective of the person who does the measurement.

The micrometer screw has a pitch of 0.5 mm, while the thimble has a scale of 50 divisions round its circumference. Thus, on going through one complete turn, the thimble moves forward or backward by one thread pitch of 0.5 mm, and one division of its scale is, therefore, equivalent to a longitudinal movement of 0.5 × 1/50 mm = 0.01 mm. It is the value of one division on the thimble which is the least that can be correctly read with the help of a micrometer and is called the least count.

The job is measured between the end of the spindle and the anvil which is fitted to the frame. When the micrometer is closed, the line marked "0" on the thimble coincides with the line marked "0" on the barrel. If the "0" graduation does not coincide then the micrometer requires adjustment.

To take the reading from the micrometer note:

1. The number of main divisions in mm above the reference line,
2. The number of subdivisions below the reference line exceed only the upper graduation, and
3. The number of divisions in the thimble.

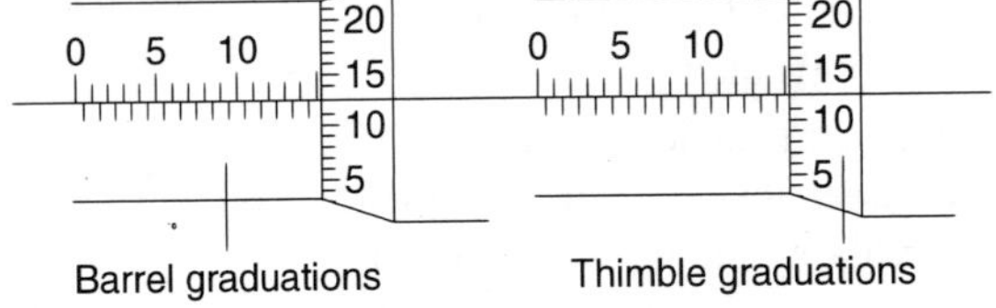

Fig. 4.4

For example:

1. First note the minimum range of the outside micrometer. While measuring with a 50 to 75 mm micrometer, note it as 50 mm.
2. Then read the barrel graduations. Read the value of the visible lines on the left of the thimble edge, i.e.,

   13.00 mm + 0.50 mm = 13.50 mm

Next read the thimble graduations.

Read the thimble graduation in line with the barrel datum line, 13th division

Multiply this value with 0.01 mm (least count)

$$13 \times 0.01 \text{ mm} = 0.13 \text{ mm}.$$

Add minimum range	50.00 mm
Barrel reading	13.50 mm
Thimble reading	00.13 mm
Total	63.63 mm

The micrometer reading is 63.63 mm.

The external micrometers have a range of 25 mm and are available with the following ranges of measurement: 0 to 25, 25 to 50, 50 to 100, 100 to 125, and so on, with 575 to 600 mm as the largest one. The measuring range of micrometers greater than 300 mm in size can be extended by means of interchangeable and adjustable anvils. In using micrometers over 0 to 25 mm, the initial gap between the measuring faces must be added with the reading obtained from the micrometer.

4.1.5 English Micrometer

The minimum measurement	= 0.0
Main division 2 points	= 0.2000
Subdivision 2 points	= 0.0500
0 division of thimble	= 0.0000
Division of vernier	= 0.0007
Reading	= 0.2507″

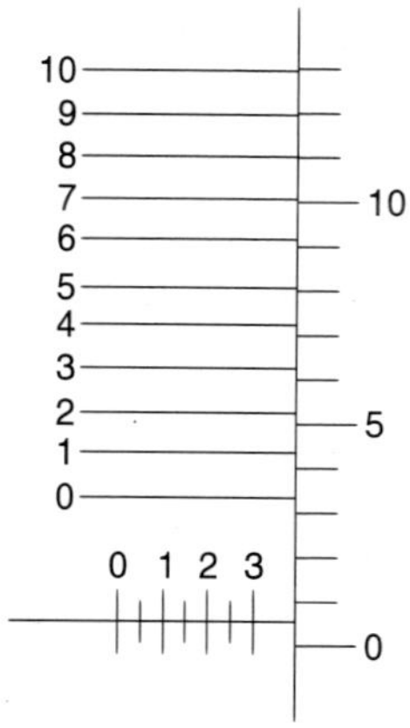

Fig. 4.5

Outside micrometers have a range of 1 in, from 0 to 1 in, from 1 to 2 in, and so on, with 5 to 6 in, as the largest one. Therefore, when we use a 2 to 3 inch micrometer we must not forget that the size is over 2 in, and that 2 inch has to be added to the reading obtained from the barrel and the thimble.

4.1.6 Vernier Micrometer

Ordinary metric micrometers can measure only upto an accuracy of ±.0.001 mm.

Some micrometers measure 1/10000 inch by means of additional graduations on the barrel. This is called vernier micrometer. The vernier scale is used in reading a micrometer whenever the longitudinal line on the spindle does not coincide, that is, it is not in alignment with a line on the thimble.

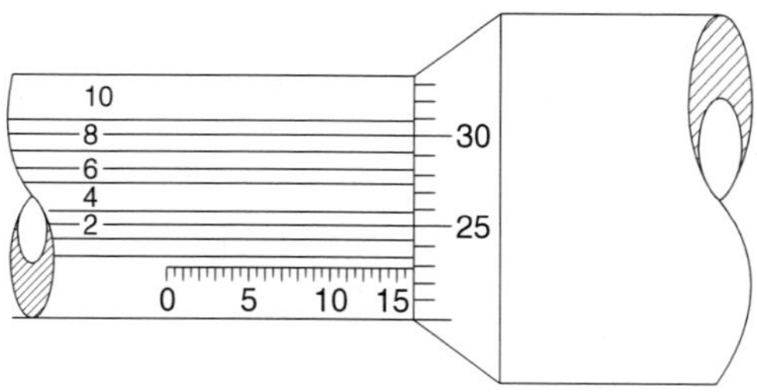

Fig. 4.6

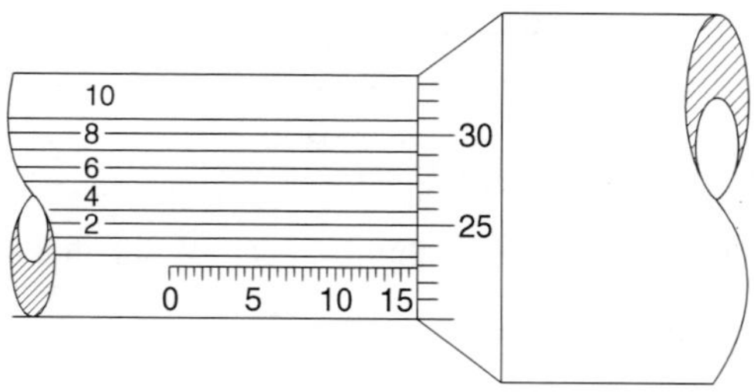

Fig. 4.6a

The vernier scale consisting of 10 spaces running parallel to the longitudinal lines on the barrel (as shown in Fig. (4.6), covers exactly 9 of the spaces of the thimble. Since the 10 spaces of the vernier equal the overall spaces of the 9 divisions of the thimble, then 1 division of the vernier equals.

1/10 of 9/1000 = 9/10000 inch.

Graduations on the thimble equal to 1/1000 or 10/10000 inch. The difference between the division on the thimble and one division on the barrel is, therefore, 10/10000 – 9/10000 = 1/10000 = 0.0001 inch.

Construction and Graduation Vernier micrometers are similar to ordinary micrometers in construction. The difference is in the graduation. These micrometers have additional, equally spaced graduations (vernier graduations) given above the datum line. There are ten such vernier graduation lines marked parallel above the datum line (**Fig. 4.6).** The space between these 10 lines is equal to 9 divisions in the thimble.

The value of 10 vernier divisions is .01 mm × 9 = 09 mm

The value of a vernier division = .09/10 mm = 0.009 mm

Reading a Vernier Micrometer 1/10 of 9/1000 = 9/10000 inch.

Graduations on the thimble equal to 1/1000 or 10/10000 inch. The difference between the division on the thimble and one division on the barrel is, therefore, 10/10000 – 9/10000 = 1/10000 = 0.0001 inch.

Reading a Vernier Micrometer

Full division in mm.	15 mm
Note the half divisions, if any visible on the barrel.	No half divisions.
Read the thimble divisions below the datum line	23 divisions
Note the vernier division coinciding with the thimble 3rd divisions	
Add up all the readings together:	
The range of micrometer is 0 to 25 mm	
Full mm division visible before the thimble edge	15 = 15.000 mm
Half mm division visible after the full mm division on thimble nil = 0.000 mm	
Thimble division below the index line	23 = 0.230 mm
Vernier division coinciding with thimble division	3 = 0.003 mm
Reading	15 = 15.233 mm

Micrometer Correction When zero on the thimble is below the index line, then the zero error must be deducted from the reading.

Example

When the micrometer is closed let the reading on the thimble be 2

(Zero error = + 0.002 inch)

Let the micrometer reading with an object between the anvil and the spindle = 0.502 inch.

Deduct the zero error = 0.002 inch

Actual thickness = 0.500 inch.

When the zero of the thimble is above the index line, then the zero error must be added to the reading.

Example

When the micrometer is closed let the reading on the thimble be 2

(Zero error = + 0.002 inch.)

Let the micrometer reading with an object between the anvil and the spindle = 0.502 inch.

Deduct the zero error = 0.002 inch

Actual thickness = 0.500 inch.

When the zero of the thimble is above the index line, then the zero error must be added to the reading.

Example

When the micrometer is closed let the reading on the thimble be 23

(Zero error – 25-23-0.002 inch)

Let the micrometer reading = 0.498 inch

Add zero error = 0.002 inch

Therefore, the actual reading = 0.500 inch

4.1.7 Screw Thread Micrometer (Figs. 4.7 & 4.8)

The screw thread micrometer is an instrument used for measuring the effective diameter of the screw threads. This is very similar to the ordinary micrometer in construction but has facilities to change the anvils. The anvils are replaceable and are changed according to the profile and pitch of the different systems of threads.

This consists of the same parts as that of outside micrometer with the only difference that the movable spindle is pointed and the end of the anvil is of the same form as the screw thread to be measured. In measuring screw threads, the angle of the spindle point and sides of the anvil 'vee' contact the surfaces of the thread so that the reading of the instrument indicates the pitch diameter of the thread. The zero on the thimble represents a line drawn through AB.

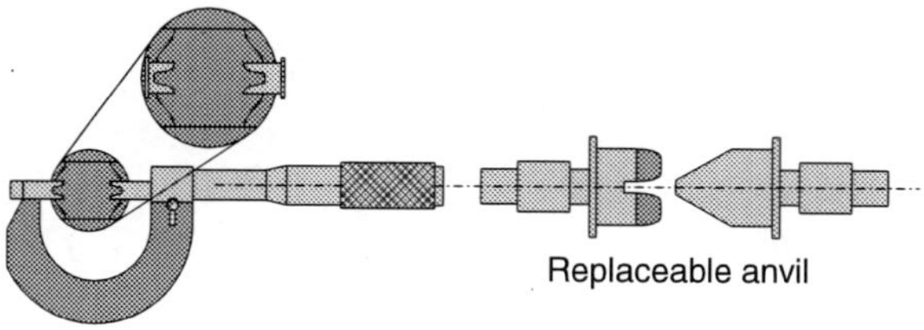

Fig. 4.7

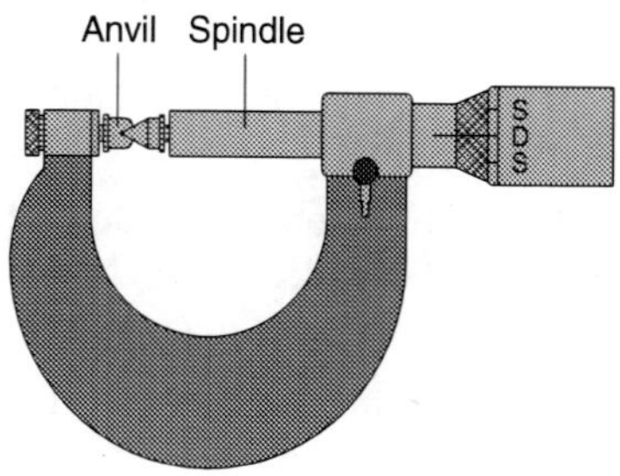

Fig. 4.8

The graduations on the sleeve and the thimble are similar to that of an external micrometer.

The anvils are replaceable and are changed according to the profile and pitch of the different systems of threads.

Three-wire Method This method uses three wires of the same diameter for checking the effective diameter and the flank form. The wires are finished with a high degree of accuracy.

The size of the wires depends on the pitch of the thread to be assured.

For measuring the effective diameter three wires suitable for the thread pitch are placed between the threads (**Fig. 4.9**).

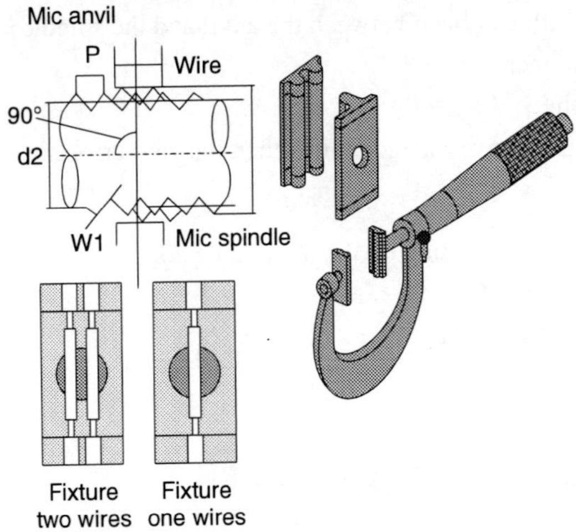

Fig. 4.9

The measuring wires are fitted in wire-holders which are supplied in pairs. One holder has provision to fix one wire and the other for two wires.

While measuring the screw thread, the holder with one wire is placed on the spindle of the micrometer and the other holder with two wires is fixed on the fixed anvil.

Selection of 'best wire' The best wire is one which, when placed in the thread groove, makes contact at the nearest to the effective diameter. The selection of the wire is based on the type of thread and pitch to be measured (**Fig. 4.10**).

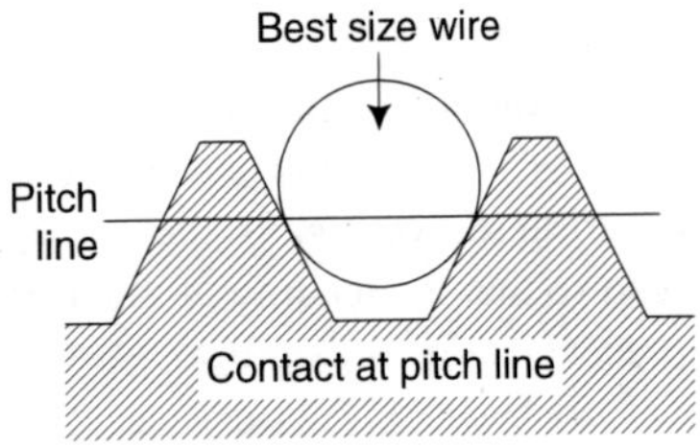

Fig. 4.10

The selection of wire can be calculated and determined but ready-made charts are available from which the selection can be made.

Measurement with measuring wires

Metric threads with coarse pitch (M)

Thread designation	Pitch Pmm	Basic Measure d_1 mm	Measuring wire dia W_1 mm	Dimension Over wire M_1 mm
M1	0.25	0.838	0.15	1.072
M1.2	0.25	1.038	0.15	1.272
M1.4	0.3	1.205	0.17	1.456
M1.6	0.35	1.373	0.2	1.671
M1.8	0.35	1.573	0.22	1.870
M2	0.4	1.740	0.22	2.055
M2.2	0.45	1.908	0.25	2.270
M2.5	0.45	2.908	0.25	2.569
M3	0.5	2.675	0.3	2.569
M3.5	0.6	3.110	0.35	3.143
M4	0.7	3.545	0.4	3.642

M4.5	0.75	4.013	0.45	4.140
M5	0.8	4.480	0.45	4.751
M6	1.2	5.350	0.6	5.139
M8	1.25	7.188	0.7	6.285.
M10	1.752	9.026	0.85	10.279
M12	2	10.863	1.0	12.350
M14	2.5	12.701	1.15	14.421
M16	2.5	14.701	1.15	16.420
M18	2.5	16.376	1.45	18.564
M20	2.5	18.376	1.45	20.563
M22	2.5	20.376	1.45	22.563
M24	3	22.051	1.75	24.706
M27	3.5	22.051	1.75	27.705
M30	3.5	27.727	2.05	30.848

4.1.8 Inside Micrometer

The inside micrometer is intended for internal measurement to an accuracy of 0.01 mm. In principle, it is similar to an external micrometer and is used for measuring holes with a diameter over 50 mm.

The inside micrometer is similar to an ordinary outside micrometer but without the 'U' frame. The measurement is taken over the contact points. As the thimble opens or closes, the contact points get opened or closed.

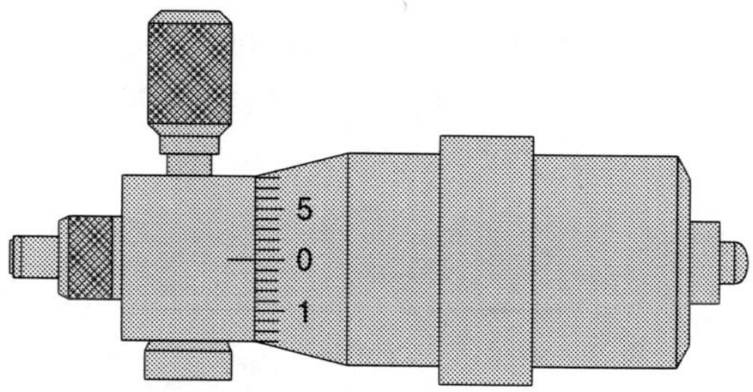

Fig. 4.11

The inside micrometer consists of a sleeve, thimble, anvils, a spacing collar and extension rods. It is also equipped with a handle to measure deep bores. The least count of the instrument is also 0, 01 mm. The inside micrometer is equipped with a 12 mm spacing collar and 4 extension rods for measuring holes of ranges 50-75 mm, 75-100 mm, 100-125 mm and 125-150 mm. The sleeve is marked with the main scale and the thimble with the thimble scale.

The barrel has a limited adjustment of 13 mm. When the inside micrometer is closed (when zero of thimble coincides with the zero of the barrel), it is capable of reading the minimum dimension of 25 mm. In addition to this, it is possible to read up to 38 m with the thimble opening to the extreme right. In order to read further higher ranges, a standard spacing collar of 12 mm width is to be added. This facilitates the micrometer to read a maximum range of 50 mm.

Similarly, each extension rod has to be used without the collar for measuring, a minimum range up to 13 mm variation and with the collar for a maximum range of measurements. A clamping screw is also provided to clamp the extension rod firmly.

Determining the Distance Between Internal Parallel Surfaces

While checking parallelism between the two surfaces of a deep bore, a handle must be used along with the inside micrometer. In order to ascertain the parallelism, a minimum of two readings have to be taken one at the top surface of the deep bore and the other at the bottom surface of the bore. If there is no difference in the

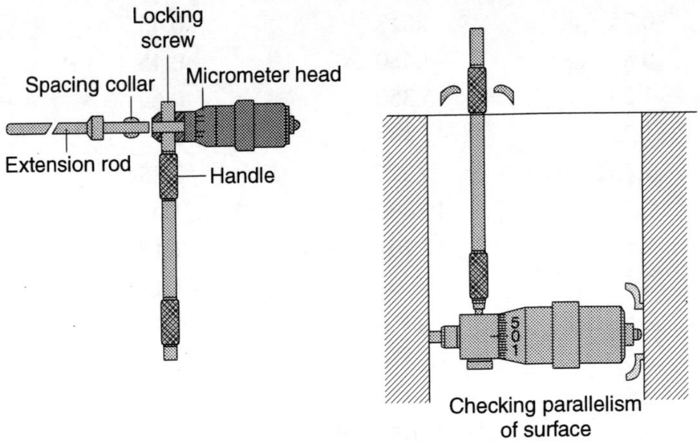

Fig. 4.12

two readings, we may take it for granted that the surfaces are perfectly parallel. Any variations in the reading show the bore has an error between the two surfaces.

4.1.9 Three-point Internal Micrometer

The three-point internal micrometers are useful for:

1. Measuring the diameters of through and blind holes
2. Checking cylindricity and roundness of bores.

The commonly used three-point internal micrometers have a least count of 0.005 mm.

Parts

- Measuring head consisting of three measuring anvils
- Ratchet stop
- Thimble
- Barrel

A three-point internal micrometer is used for direct measurement of an internal diameter accurately and efficiently. It is also used to measure the diameter of a deep hole, the end of a blind hole, internal recess, etc.

The instrument is checked for its zero error with a master ring gauge.

This micrometer has a cone spindle which advances when the thimble is rotated clockwise. The movement of the cone spindle makes the measuring anvils to move forward and backward uniformly. The three measuring anvils facilitate self-alignment of the instrument within the bore.

Three-point internal micrometers are available in different sizes permitting measurement within a range.

The ratchet stop permits uniform pressure between the anvils and the work-surface being measured.

These micrometers are provided with one or more zero setting rings.

Before taking measurement, the zero setting has to be checked using the setting ring.

The position of the anvils can be reset by loosening the barrel using a screwdriver.

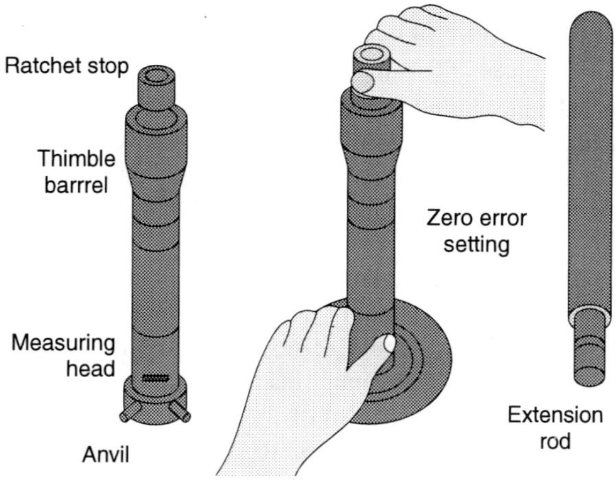

Fig. 4.13

Depending on the depth of the bore the length of the micrometer can be varied using the extension rod. A spanner is provided for changing the extension rods.

These instruments are available in various sizes for different uses.

They are also available in analog or digital read-outs

For the measurement of larger internal dimensions, inside micrometer is employed. It consists of four parts:

(i) Measuring head (micrometer unit)

(ii) Extension rods

(iii) Spacing collars

(iv) Handle

Using one range of extension rods, distance between the contacting surfaces is varied by rotating the thimble up to the extent of the screw length. Greater distances are obtainable by the use of different extension rods and spacing collars. The measuring tip on the thimble is adjustable one in order to compensate for the wear. Collars are used when those dimensions are needed which cannot be obtained by the use of extension rod and micrometer unit. For using this instrument first the diameter of the bore is measured approximately by a scale. Extension rod is then chosen to the nearest one centimetre and inserted in the micrometer head. Then zero error is checked by taking the dimension on standard sizes specimen. The micrometer is then adjusted at a dimension slightly smaller than the diameter of bore. One micrometer head is then held firmly against the bore and the other contact surface is adjusted by moving the thimble till correct feel is sensed. One contact surface must always be moved sideways up and down in order to ensure that the diameter is being measured and not the chord. The micrometer is then removed and reading taken. Dimension from the micrometer head is noted down and to this, lengths of extension rod and collar are added up.

By suitable base arrangements, inside micrometer can be converted into height gauge.

It is used for measuring the inside diameters of cylinders, rings, etc., and also for measuring parallel surface and for setting callipers, etc.

The desired range is obtained by assembling measuring rods and lapped gauges or spacing collars to the micrometer held.

The description of various units of internal micrometer is given below the micrometer unit.

The body of the measuring head is made of suitable steel and its spindle and terminal measuring faces are made of high grade tool steel. The measuring face of the anvil is so designed as to have a hardness of about 800 HV (63.5 HRC). It is desirable that the measuring face be tipped with tungsten carbide or other suitable hard material or be faced with a deposit of hard chromium. The radius of curvature of the measuring face is made slightly less than one-half of the smallest measuring range of the micrometer. The micrometer screw has a pitch of 0.5 mm and the threads of the screw and the thread of the screw and nut are truncated so as to continue contact to the flanks of the thread. The screw should travel smoothly throughout the length of its travel. An adjusting nut is provided to arrange for the thimble to be sufficiently tight for the micrometer to retain its reading after being set. There should be no perceptible backlash between the spindle screw and the nut. When the micrometer is at its maximum reading, there should be full engagement of the nut and micrometer screw. It is desirable that the travel of the micrometer be continued for at least one turn of the screw beyond each end of the graduated scale on the barrel. Generally, means are also provided for compensation of wear between the screw and nut. The diameter of the graduated edge of the thimble is generally not less than 10 mm. For ease of reading, the surfaces of the thimble and barrel should have a dull finish and the graduation blinds blackened.

Generally, a micrometer is provided with means for adjusting the zero setting. The means of adjustment should be such that after resetting, the parts are secured, and the original accuracy of the instrument is not impaired. Suitable spanners and keys are provided for this adjustment.

Measuring range (mm)	Maximum permissible error in length
25-150	±0.005
150-300	±0.010
300-450	±0.015
450-600	±0.020

The measuring head with its associated extension rods and collars should conform to the following limits of error when tested at standard reference temperature of 20°C .

The error in travels of the micrometer screw should not exceed 0.003 mm, whether errors are progressive or periodic in character.

Extension Rods These are provided in order to enable any measurement to be made throughout the range specified for the set, spacing collars being used whenever necessary. These are made of suitable material hardened up to 800 HV and the measuring races are tipped with tungsten carbide or other hard material. The measuring faces and the contact face are finished by lapping operation. The extension rods may either incorporate their own measuring anvils or these may take the form of different pieces to which a separate common anvil piece may be fitted. When the extension rods have their own measuring faces, a suitable adjustment for zero setting is incorporated in it (e.g., the attachment of the anvil by means of a friction-tight fine pitch thread) and there is no need of providing any such adjustment on the micrometer head. If the extension rods have separate anvil pieces for measuring faces then the adjustment for zero setting may be incorporated either in the anvil pieces or in the measuring head. The range of measurement for which an extension rod is applicable is always marked on it.

Spacing Collars These are intended for smaller adjustments in the range of measurement. The ends of the spacing collars are well finished by lapping and are flat and mutually parallel and square to the axis. No sharp edges are there and at least one face of the spacing collar is provided with dirt clearance grooves.

Handle The micrometer set for the measuring range 25 to 50 mm is generally supplied with a suitable detachable handle so that the micrometer head can be easily lowered into deep holes.

The micrometer assembly with any combination of extension rods and spacing collars should be rigid. When the assembled unit is supported at it ends, then the deflection of mid-point under an applied load of 0.5 kg should not exceed 0.05 mm for an assembly 150 mm long with an increase of 0.05 mm for each increase in the length of 25 mm. The maximum deflection allowed for an assembly 1 m long is 1.75 mm.

4.1.10 Self-Centring Inside Micrometer

Measuring through bores, blind holes, registers.

It consists of micrometer screw and screw in interchangeable measuring head.

Components

Measuring spindle precision ground, hardened throughout. Ratchet stop. Heat insulator. Moving parts and scales satin chrome finished, threaded for convenient mounting of measuring heads.

Measurements Head

Three laterally protruding fingers, positioned at a 120° angle to each other, provide self-centring action and firm positioning, positively no tilting or tipping. From 10 mm and up the fingers extend over the face of the head to allow measurements through the entire bore length. Individual measuring heads.

4.1.11 Stick Micrometers

Stick micrometers are designed for the measurement of longer internal lengths. These comprise the following parts:

(i) A 150 mm or 300 mm micrometer unit fitted with a micrometer of 25 mm range and having rounded terminal faces.

(ii) A series of extension rods which together with the micrometer unit permit of a continuous range of measurements up to the maximum length required.

How the stick micrometer will look like is shown **Fig. (4.14 and 4.14A).** Its accuracy is of the order of ±0.005 mm throughout the range. It is essential that the radius of curvature of the terminal faces of the micrometer unit be slightly less than one-half the smallest measuring range of the micrometer unit, and abutment faces should have minimum radial width of 2 mm and properly lapped; these should be parallel to each other and normal to the axis of the micrometer unit to within 0.005 mm across their diameters.

It may be noted that screwed joints are used for joining the end piece, extension rod and the measuring unit. The screw units generally have threads of 0.05 mm pitch. The extension rod is generally hollow and has a minimum external diameter of 14 mm. In the design of extension rod, shown in Fig. (4.14), it will be noted that the contact faces come together by screwing and the pressure between the contact faces depends upon the heavy or light tightening of the screw. In order to overcome this difficulty, the extension rods are generally employed. This type consists of length standards which are mounted in protective bushing and when these extension rods are combined, the contact faces of the length standards come together under constant pressure and thus the measurement is not affected by heavy tightening of the screw. The accuracy of this instrument is of the order of ± 0.005 mm, i.e., the micrometer unit with terminal piece in position should record correctly throughout its range to within ± 0.005 mm. In this type of micrometer, there should be sufficient play between the external and internal threads of the joint to permit the abutment faces of the various parts of the micrometer to butt together solidly.

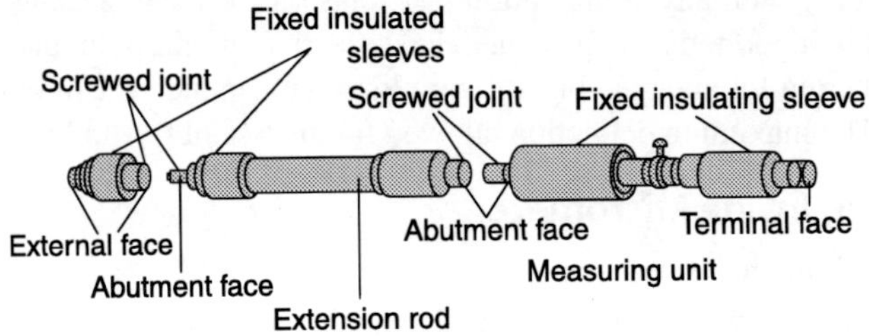

Fig. 4.14 End piece.

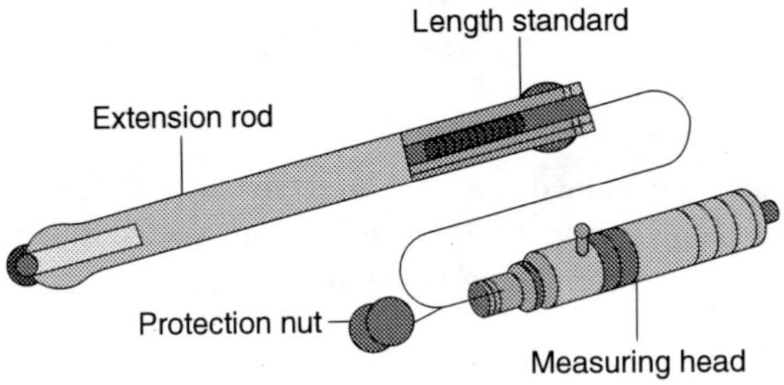

Fig. 4.14a End piece.

4.1.12 Method of Testing Internal Micrometers

The accuracy of the internal micrometer reading is checked in two parts, namely, checking the accuracy of the traverse of the measuring head and the checking of the accuracy or the overall lengths when the measuring head, set to zero, is associated with the various extension rods in turn.

The accuracy of the traverse of the measuring head is determined by clamping it in the Vee-block with its axis in line with a sensitive indicator and the feeler (with a flat contact face) of the indicator touching the rounded contact face of measuring head. The feeler is set initially such that it reads zero when the micrometer head is also reading zero and a slip gauge of convenient dimension has been introduced between the two faces. The micrometer head is then set at the same reading and correspondingly the slip gauge size is reduced. Any error in the reading is revealed by a corresponding departure of the indicator pointer from its initial position. In this way the accuracy is tested at various readings of the micrometer head. In order to check both progressive and periodic errors, and in order that both types of errors may be revealed, the reading in micrometer head of 0-25 mm range may be taken at the following values 2.4, 5.1, 7.7, 10.3, 12.5, 14.9, 17.6, 20.2, 22.8, 25.0 mm. The accuracy of the overall length is determined by using a vertical comparator. The indicator of the comparator is set at a height corresponding to that of the overall length to be measured, from the flat base. After this setting the internal micrometer is placed under it in the maximum position and the error noted down.

4.1.13 Inside Micrometer Callipers

This micrometer calliper has no U-shape frame and spindle. The measuring tips are constituted by the jaws with contact surfaces which are hardened and ground to a radius. One of the jaws is held stationary at the end and the second one moves by the movement of the thimble. A lock nut is provided to check the movement of the movable jaw. This facilitates the inspection of small internal dimensions.

This is not widely used as other types have various advantages over it. Its range is from 5 to 50 mm.

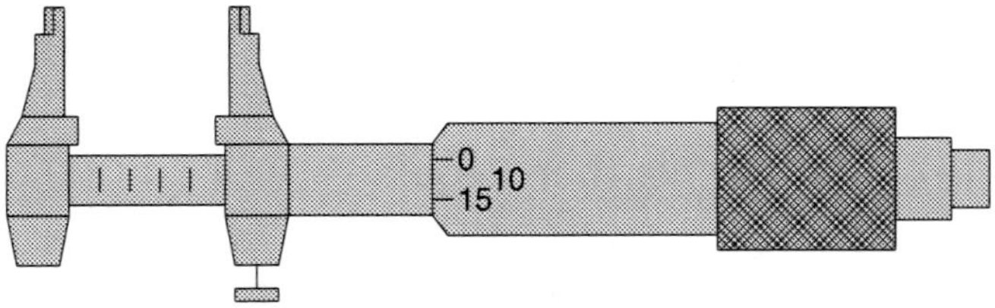

Fig. 4.15

4.1.14 Micrometer Depth Gauge

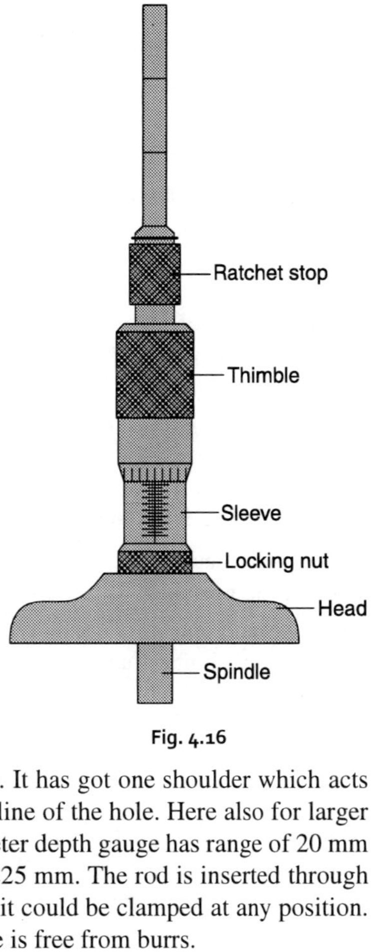

Fig. 4.16

Depth micrometer is designed to measure accurately the depth of grooves, bores, counter bores, recesses and holes. The graduations are read in the same manner as in the case of regular micrometers. Larger ranges depth can be measured by inserting an extension rod through the top of the micrometer. The graduations are in the reversed direction to those of an outside micrometer.

It is an instrument used for measuring depth of holes to an accuracy of 0.01 mm and 0.001 in. Depth micrometer can only be used in places where there is a satisfactory seating for the instrument head, and the bottom of the hold being measured is parallel with the seating.

In consists of the following parts **(Fig. 4.16)**

1. Micrometer head
2. Screwed spindle. This spindle does the actual measuring.
3. Graduated sleeve. This has the datum line and fixed graduations which are numbered 0, 9, 8,7 etc., reading from the left to right.
4. A tubular core. This is known as the thimble fastened with the spindle at its outer end and moves with the spindle.
5. Ratchet stop. It is a small extension to the thimble.
6. Locking ring. This is to lock the instrument at any desired setting.

It is used for measuring the depths of holes, slots and recessed areas. It has got one shoulder which acts as reference surfaces and is held firmly and perpendicular to the centre line of the hole. Here also for larger ranges of measurements, extension rods are used. The screw of micrometer depth gauge has range of 20 mm or 25 mm. The length of the micrometer depth gauge varies from 0 to 225 mm. The rod is inserted through the top of the micrometer. The rod is marked after every 10 mm so that it could be clamped at any position. In using this instrument, first it must be ensured that the edge of the hole is free from burrs.

The scale here is calibrated in the reverse direction. The accuracy again depends upon the sense of touch.

In use, the spindle is screwed back to a point where it will clear the bottom of the hold being gauged. The instrument head is then placed across the top of the hole and held there whilst the spindle is screwed down until it just contacts the work.

The principle of measurement is similar to that of an external micrometer. Each depth micrometer is supplied with three interchangeable spindles and thus has the measuring ranges 0 to 25 mm 25 to 50 mm, 50 to 75 mm and 75 to 100 mm.

Depth micrometers are available with measuring ranges of 50 to 75, 75 to 175, 175 to 600, 650 to 1,250, 800 to 2,500, 1250 to 4,000, 2,500 to 6,000, 4,000 to 10,000 mm.

4.1.15 Types of Micrometers Other than Regular

In addition to regular micrometers, there are several other types of micrometers, with the same fundamental principle, but specifically designed to meet the various particular applications. such as external, internal, depth measurement, etc.

1. Screw thread micrometer
2. Tube micrometer
3. Digital micrometer
4. Depth micrometer
5. Flange micrometer
6. Ball micrometer
7. Stick micrometer
8. External micrometer with interchangeable anvils
9. Key way depth micrometer.

4.1.16 Screw Thread Micrometer

A screw thread micrometer is similar to an outside micrometer except that the spindle is pointed to fit between 60° V threads, and the anvil is shaped to fit over 60° V thread. It is used to measure the pitch diameter of the thread. Screw thread micrometers are available in many sizes depending on the pitch of the thread to be measured.

4.1.17 Tube Micrometer

A tube micrometer is specially designed to measure the thickness of the material of piping, tubing and other parts of similar shapes.

A tube micrometer is specially designed to measure the thickness of the material of piping, tubing and other parts of similar shapes.

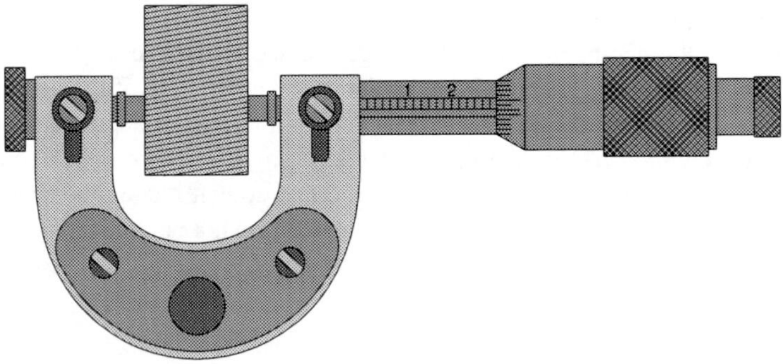

Fig. 4.17

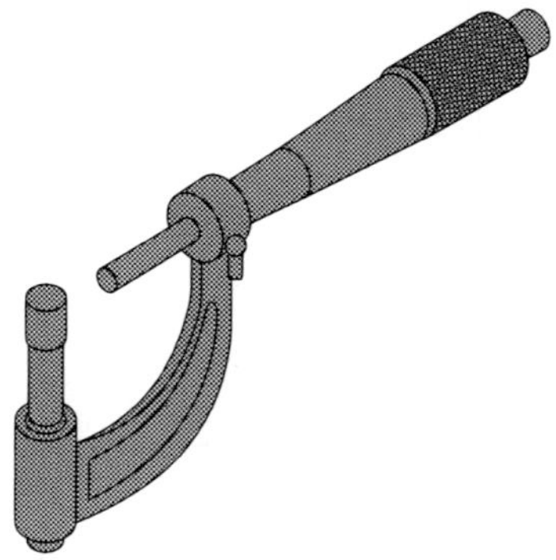

Fig. 4.18

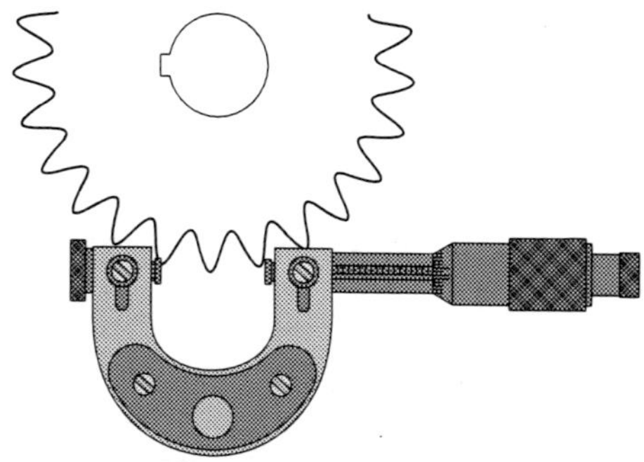

Fig. 4.19

4.1.18 Digital Micrometer (Fig. 4.20)

This type of micrometer has got a dial on the frame of the micrometer and an illuminated screen below it. The dial pointer has an internal connection with the micrometer screw for measuring. The graduations on the sleeve and thimble are the same as on a regular micrometer. This micrometer is used to measure the dimensions similar to those measured by the outside micrometers, and the reading can be noted.

The advantage of this micrometer is that the readings are seen on the screen or the dial directly, without any difficulty. We need not look on the sleeve or the thimble scale. This avoids errors in reading and saves time. A layman can also read the measurement directly.

Advantages

Easy readings of digital display with large 4, 7 mm high characters with decimal point minus sign.

Zero setting at any position Within measuring range by conveniently placed button.

- **One hand operation:** Only two function buttons.
- **No misreading,** due to extremely high measuring speed. Max speed 1.5 m/s

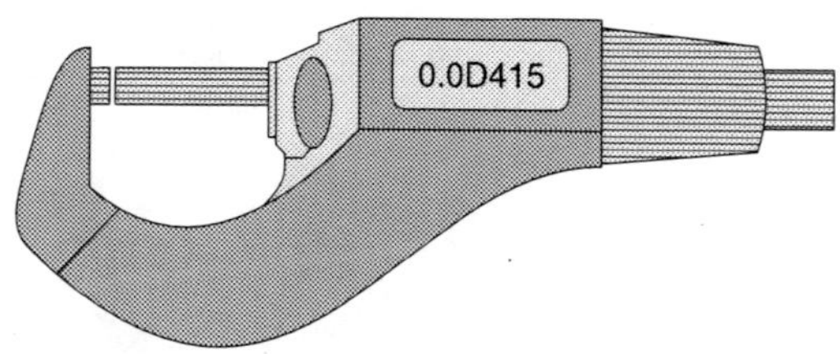

Fig. 4.20

- **Low power consumption:** Battery has life of 1 year continuous operation.
- **Data exit:** For connection of printers and computers for evaluation and processing of measuring results. Data acquisition is initiated by button on slide or by foot switch.

Positive locking screw for slide This is of advantage for many applications.

4.1.19 Flange Micrometer

A flange micrometer is similar to regular micrometer and is equipped with two flanges in the place of the anvil and spindle. This is used to measure chordal thickness of the gear teeth and the thickness of the fins of an engine and the collar thickness of the job.

4.1.20 Ball Micrometer

In this form of micrometer, hemispherical balls are fitted at the anvil and spindle. Measurement is similar to that in a regular micrometer.

It is used to measure a sphere where the point of contract comes in between.

4.1.21 Stick Micrometer (Fig. 4.14)

Stick micrometers are designed for the measurement of longer internal lengths.

It comprises 150 mm or 300 mm micrometer unit, fitted with a micrometer of 25 mm range and having rounded terminal faces

Series of extension rods, which together with the micrometer unit, permits a continuous range of measurement up to the maximum length required.

Secured joints are used for joining the end piece, extension rod and the measuring unit. The screw unit generally has threads of 0.5 mm pitch. The extension rod is generally hollow and has a minimum external diameter of 14 mm. In this type of micrometer, there should be sufficient play between the external and internal threads of the joint to permit the abutment forces of the various parts of the micrometer to butt together solidly.

4.1.22 External Micrometer with Interchangeable Anvils

It is nothing but an external micrometer. The advantage in this micrometer is the range of the micrometer can be increased by merely changing the different anvils.

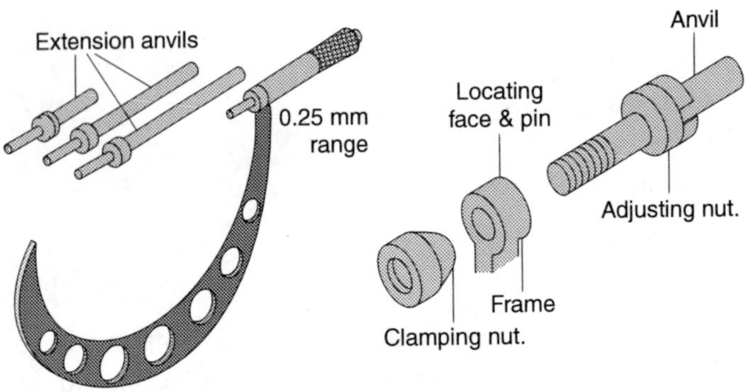

Fig. 4.20a

A set of replaceable anvils is supplied in a box and the size of the anvil is marked on each anvil. Depending upon the size of the job the anvil size can be changed, and reading can be taken. Thus, it is an economy micrometer, i.e., one micrometer itself. Long ranges can be accommodated. To fix the anvils to the frame, a guide is provided and locked by a nut.

4.1.23 Key Way Depth Micrometer

It is similar to a depth micrometer except that the frame has 120° inclined butting surfaces to rest on the circumference on a cylindrical job. It is used for measuring depth of key ways on a cylindrical shaft. While measuring the depth of the key way, first take the measurement on a cylindrical job opposite to the key

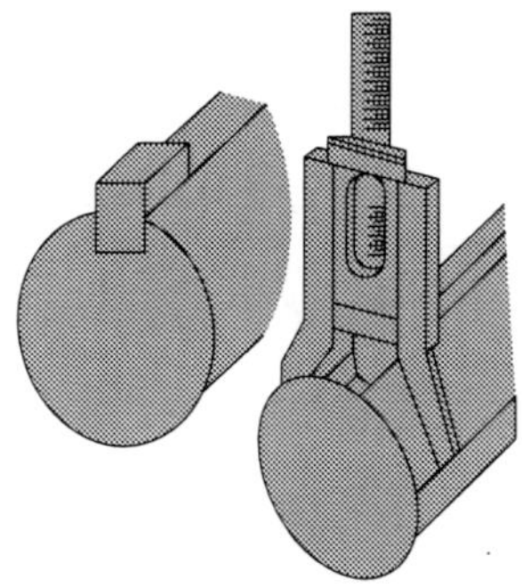

Fig. 4.21

way; then take the measurement of the key way depth, subtract the initial measurement from the final measurement to know the exact depth of the key way.

For the adjustment of the depth of the key way, the index ring on the spindle for the height adjustment of the table is used.

Testing of Key Way

The width of the key way can be tested with slip gauges, For measuring the key way depth, the depth gauge for key ways is suited. The centre position of the key way is tested with slip gauges and dial indicator.

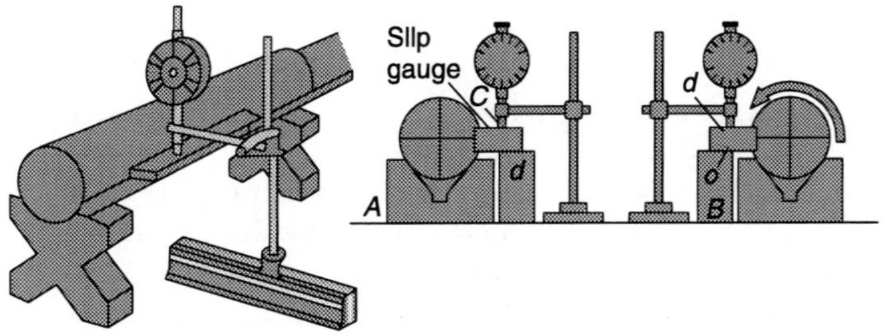

Fig. 4.21a

EXERCISE

1. What is the working principle of a micrometer?
2. Draw a sketch of a micrometer and show its parts.
3. What do you understand by zero error?
4. Explain micrometer least count.
5. Draw a sketch showing reading of 0.792″ in an outside micrometer.
6. What is the least count of vernier micrometer. Explain.
7. What is the functlon of depth gauge micrometer? How is it different from an outside micrometer?
8. Explain with the help of a sketch, the make-up' of inside micrometer and how to use it.
9. Explain the principle of vernier calliper with the help of a sketch.
10. What are the various kinds of vernier callipers? Name them along with their uses.
11. How to find the least count of a vernier calliper. Explain.
12. What is the difference between a vernier calliper and a micrometer?
13. Comment on telescopic gauge.
14. Explain in details of the combination set.
15. Draw a sketch of vernier bevel protractor showing its parts.
16. How reading is taken from a vernier bevel protractor? Explain with examples.
17. Why vernier bevel protractor is used?
18. How can we find the least count of vernier bevel protractor?

Measurement of Surface Finish

Chapter 5

5.1 SURFACE ROUGHNESS, SPECIFICATION AND EVALUATION

The proper functioning of a machine part is in many instances largely dependent on the quality of its surface. A thorough understanding of surface quality is of importance to the designer so that he will be able to specify correctly those surfaces which will provide optimum properties for service, life, appearance, performance, and other mechanical functions. It is no less important that manufacturing and inspection personnel should be equally well acquainted with the subject, but with special emphasis on the capabilities of the various manufacturing methods and means of inspection at their disposal.

Surface characteristics with which this section deals are roughness, waviness, lay, and integrity. Definitions, recommended symbols, and values presented here are in accordance with ANSI B46. 1-1962 (RI971), "Surface Texture."

Because of the widely varying conditions under which different surfaces must operate, no single set of criteria can be used to determine which surface or combination of surfaces may be best for a given purpose. It is believed that experience with surfaces serving similar functions to the one in question or experimentation should provide the basis for specification in new design.

5.2 NATURE OF STRUCTURE ROUGHNESS

The surface of hard materials is composed of a pattern of irregularities, which may be the surface produced by casting, extruding, or flame cutting, or by a mechanical process such as shaping, sandblasting, turning, or grinding. The pattern of surface irregularities is usually very complex, as each of the various factors in the finishing machine adds its own contribution to the finished surface.

Following are the salient features of surfaces, their ratings, and related symbols. The definitions are taken from ANSI B46. 1-1962.

5.2.1 Surface

The surface of an object is the boundary which separates that object from another object or substance. This standard relates to surfaces produced by such means as machining, abrading, extruding, casting, moulding, forging, rolling, coating, plating, blasting, or burnishing.

5.2.2 Nominal Surface

The theoretically accurate design, the shape and extent of which are usually shown and dimensioned on a drawing or descriptive specification.

5.2.3. Surface Irregularities

Deviations from the nominal surface including roughness and waviness are irregularities.

5.2.4 Profile

The contour of a section perpendicular to a surface, unless some other angle is specified is called the profile.

5.3 ROUGHNESS

Roughness is relatively finely spaced surface irregularities, the height, width, and direction of which establish the predominant surface pattern. Irregularities are produced by cutting.

Edges and machine tool feed may be considered roughness. The height is rated in micro inches arithmetical average deviation from the mean line, the symbol AA being used as the descriptive abbreviation. The width is rated in inches as the maximum permissible spacing between repetitive units of the surface pattern. The roughness width should be used to establish the roughness width cutoff value which, unless otherwise specified, should always be greater than the roughness width.

Roughness should be specified by referring to the arithmetical average (AA), which is also known as the centre line average (CLA) in British Standard 1134:1950. The arithmetical average is the average deviation of the surface, expressed in micro inches, from a mean line or centre line. This line is parallel to the general direction of the profile throughout the roughness width cutoff length such that the sums of the areas contained between it and those parts of the profile which lie on either side of it are equal. The arithmetical average deviation from the mean line is:

$$Y_{AA} = \frac{1}{l} + \int_{x=0}^{x=l} |y|\, dx$$

where Y_{AA} = arithmetical average deviation from the mean line.

y = ordinate of curve of profile

l = length over which average is taken

An approximation of the average roughness may be obtained by adding the *y* increments shown in (Fig. 5.2) without regard to sign and dividing the sum by the number of increments taken:

$$Y_{AA} = \frac{Ya + Yb + Yc + Yd + Ye + \ldots + Yn}{n}$$

Roughness has also been specified by the term root mean square average. Many earlier instruments now in use employ it. The root mean square average is the square root of the average square of a series of measurements of surface deviation from the mean line referred to above, expressed in micro inches. It is defined mathematically as follows:

$$Y_{ms} = \left(\frac{1}{l}\int_{x=0}^{x=l} y^2\, dx\right)^{1/2}$$

This type of average is common for electrical measurements. Roughness measuring instruments calibrated for rms average will usually be found to read 11 per cent higher on a given surface than those calibrated for arithmetical average. Instruments originally calibrated to read rms average can, in many cases, be adjusted to read arithmetical average. The difference is in such a direction that continued use of drawings which originally specified rms measurement and the inspection of surface to those values with instruments calibrated to arithmetical average as called for in the standard would theoretically allow the acceptance of surfaces up to 11 per cent greater roughness than originally intended. This difference, however, is of lesser magnitude than the variations from point to point on a machined surface and is considerably less than the variation expected from one piece to another. For these reasons, and because the absolute limit of roughness for satisfactory functioning is usually even more indefinite, many manufacturers consider it practical to adopt arithmetical average ratings without changing the roughness values on previous part drawings.

An approximation of the root mean square average roughness may be adding to the squares of the *y* increments shown in (**Fig. 5.1**), dividing the sum by the number of increments taken, and extracting the square root:

$$Y_{rms} = \frac{\sqrt{ya^2 + yb^2 + yc^2 + yd^2 + ye^2\ 1 \ldots yn^2}}{n}$$

In such a process greater accuracy is attained through increasing the number of *y* measurements taken. Modern electronic instruments take into account an infinite number of such deviations in determining the roughness values.

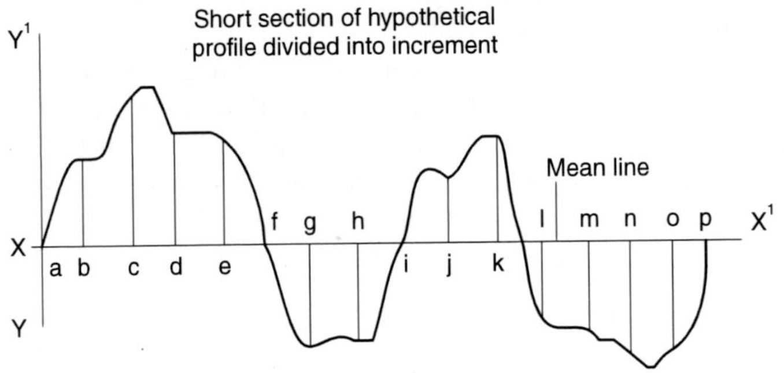

Fig. 5.1

Waviness refers to those irregularities of the nominal surface which are of greater spacing than roughness. These irregularities may result from such factors as machine or work deflections, vibration, heat-treatment, or warping strains. The height is rated in inches as the peak-to-valley distance. The width is rated in inches as the spacing of adjacent waves.

In addition to the measurement of maximum peak-to-valley height in inches (total indicator reading) for waviness, certain applications require waviness to be measured in micro inches average departure from the nominal surface. Instrumentation for this type of average height measurement is currently in wide use in the manufacture of bearing races and other precision surfaces of rotation. Designation of average waviness height measurement on a drawing should consist of a specific note directed to the surface.

Roughness width cutoff is the maximum width, in inches, of a surface irregularity to be included in the measurement of roughness height. Roughness may be considered superposed on a wavy surface. (The American "roughness-width cutoff" is equivalent to the British "sampling length." The American "sampling length" is equivalent to the British "traversing length.").

The effect of variation in roughness width cutoff can be understood better by referring to **Fig. (5.2)**. The profile at the top is a trace of the true movement of a stylus over a surface, and the profiles below are interpretations of the same surface with roughness width cutoff value settings of 0.030, 0.010, and 0.003, respectively. It can be seen that the trace based on 0.030 roughness-width cutoff includes most of the coarse irregularities and all the fine irregularities of the surface; that the trace based on 0.010 excludes the coarser irregularities but includes the fine and medium fine; and that the trace based on 0.003 roughness width cutoff

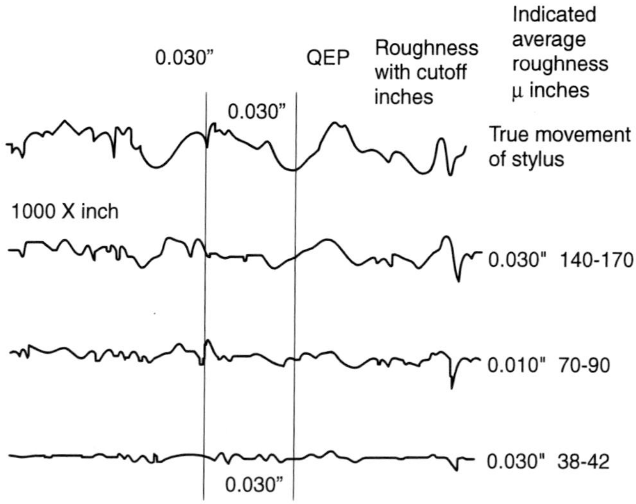

Fig. 5.2

includes only the very fine irregularities. In this example the effect of reducing the roughness width cutoff has been to reduce the roughness height indication. However, had the surface been made up of irregularities no greater than those of the bottom trace, the roughness height indications would have been the same for all three roughness width cutoff settings.

In other words, all irregularities having a spacing less than the value of the roughness width cutoff used are included in a measurement. Obviously, if the roughness width cutoff value is too small to include coarser irregularities of a surface, the measurements will not agree with those taken with a wider roughness width cutoff. For this reason care must be taken to choose a roughness width cutoff value which will include all the surface irregularities it is desired to assess.

Standard roughness width cutoff values in inches are as follows:

0.003 0.010 0.030 0.100 0.300 1.000

The value of 0.030 is preferred for most surfaces and is to be used unless otherwise specified.

In general, surfaces contain irregularities with a large range of roughness widths. Instruments are designed to respond only to irregularity spacing less than the roughness width cutoff. In some cases, such as surfaces in which actual contact area with a mating surface is important, the largest convenient roughness width cutoff will be used. In other cases, such as surfaces subject to fatigue failure, only the irregularities of small width will be important, and more significant values will be obtained when a short roughness width cutoff is used. In still other cases, such as for identifying chatter marks on machined surfaces, information is needed on only the widely spaced irregularities. For such measurements an instrument with a long roughness width cutoff and a large radius stylus should be used, so that there will be no response to the more closely spaced irregularities.

If the precautions outlined above are observed, and an instrument conforming to the specifications in the standard is employed, reliable and reproducible measurements within the overall accuracy of the instrument will be obtained, independent of the exact type of instrument used.

Lay is the direction of the predominant surface pattern, produced by tool marks or grains of the surface. Lay is ordinarily established by the production method used.

Symbols used for indicating direction of lay are:

= Parallel to the boundary line of the nominal surface indicated by symbol

⊥ - Perpendicular to the boundary line of the surface indicated by symbol

× X - Angular in both direction to the boundary line of the surface indicated by symbol

M – Multidirectional (random, as in case of lapped surface)

C - Approximately circular relative to the centre of the surface indicated by the symbol

R - Approximately radial relative to the centre of the surface indicated by the symbol

5.4 FLAWS

Irregularities which occur at one place, or at relatively infrequent intervals in the surface, e.g., a scratch, ridge, hole, peak, crack, or check.

Micro Inch Micro inch is one-millionth of an inch (0.000001 in). For written specifications or reference to surface finish requirements, micro inches may be abbreviated or symbolized as μ in, or mu inch.

5.5 DESIGNATION OF SURFACE ROUGHNESS

Where no surface control is specified, it is to be assumed that the surface produced by the operation will be satisfactory. If the surface is critical, the quality of surface desired should be indicated by means of a surface designating symbol. The standard symbol used to designate surface irregularities is the check mark with the horizontal extension, which may be applied. The point of the symbol should be on the line indicating the surface, on the witness line, or on a leader pointing to the surface. The long leg and extension should be to the right as the drawing is read. Where roughness height only is indicated, it is permitted to omit the horizontal extension.

Where the symbol is used with a dimension, it affects all surfaces defined by the dimension.

Areas of transition, such as chamfers and fillets, should conform to the roughest adjacent finished area unless otherwise indicated.

Where surface roughness control of several operations is required within a given area, or on a given surface, qualities may be designated.

Roughness and waviness ratings unless otherwise specified apply in a direction which gives the maximum reading, normally across the lay.

Surface roughness symbols, unless otherwise specified, apply to the completed surface.

Drawings or specifications for plated or coated parts should definitely indicate whether the surface roughness symbols apply before plating, after plating, or both before and after plating.

To ensure proper interpretation of surface roughness designations, drawing title blocks should include a general note. For parts requiring extensive and uniform surface roughness control, a general note may be added to the face of the drawing, which specifies the maximum allowable roughness for the given part.

Even with such general note it is permissible to specify closer or wider surface finish tolerance by using additional specific symbols on critical surfaces.

The standard surface roughness symbol is a multipurpose one within which roughness, waviness, and lay rating may be inserted that will adequately specify the desired surface. Only those ratings required to control the surface quality should be indicated in the symbol; to do otherwise will add additional cost needlessly and may confuse rather than clarify requirements for those not fully informed. The full value of the symbol may be utilized when required through the proper placement of the numerical designations as follows:

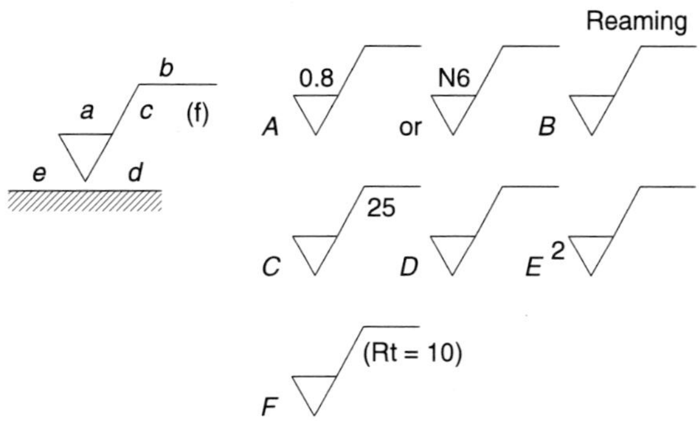

Fig. 5.3

1. Roughness height rating is placed at the left of the long leg. The specification of only one rating will indicate the maximum average and any lesser average will be acceptable.
2. The specification of maximum average and minimum average roughness height ratings indicates permissible range of average rating.
3. Maximum waviness (peak-to-valley) height rating is placed above the horizontal extension. Any lesser rating will be acceptable.
4. Maximum waviness width rating is placed above the horizontal extension and to the right of the waviness height rating. Any lesser rating will be acceptable.
5. To specify the contact area, when waviness must be controlled, the percentage value should be placed above the extension line.
6. Lay designation is indicated by the appropriate lay symbol placed to the right of the long leg.
7. Roughness width cutoff rating is placed below the horizontal extension.
8. Where required, maximum roughness width rating should be placed at the right of the lay symbol. Any lesser rating will be acceptable.

The surface finish (roughness or smoothness) of components is expressed by anyone of the following.

- Roughness number
- Grade number
- Roughness symbol.

The most accepted method of representing roughness is by Grade Number or Roughness Number. The use of roughness symbol is not a common practice though some manufacturing concerns still use them.

5.6 ROUGHNESS VALUE – GRADE NUMBER

The roughness value represents the average departure of the surface from perfection over a prescribed length, and is expressed in micrometers. One micrometer will be equal to 0.001 mm. (μ.m) and 1 μm. = 0.0001 mm.

It is possible to achieve surfaces of any given roughness value. But for standardization purposes only a few values are recommended in the Indian Standard. The roughness values and corresponding grade numbers are given in **Table 5.1**.

Table 5.1

Roughness values Ra values in microns (Micromtere)	Roughness grade number
50	N12
25	N11
12.5	N10
6.3	N9
3.2	N8
1.6	N7
0.8	N6
0.4	N5
0.2	N4
0.1	N3
0.05	N2
0.025	N1

5.7 ROUGHNESS SYMBOLS

These symbols indicate the practice followed in the industry, and are given here for guidance only. I.S.3073 - 1967 (Table 5.2) states this as Assessment of Surface Roughness

Table 5.2

Roughness Grade	Roughness Value Ra μ m	Roughness Symbol
N12	50	
N11	25	
N10	12.5	
N9	6.3	
N8	3.2	
N7	1.6	
N6	0.8	
N5	0.4	
N4	0.2	
N3	0.1	
N2	0.05	
N1	0.0025	

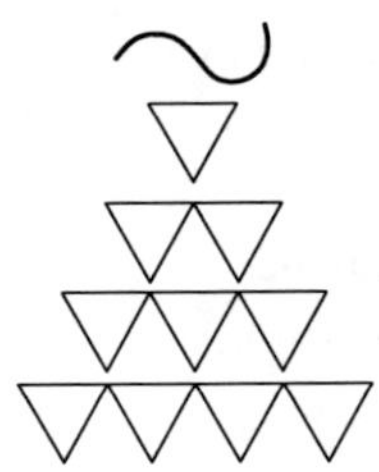

In order to obtain the required surface quality, it is necessary to choose the appropriate manufacturing process.

The Table 5.3 gives an idea about the different processes and range of surface quality attainable.

5.8 POSITION OF THE SPECIFICATION OF SURFACE TEXTURES IN THE DRAWING

The specification of surface roughness and other surface texture features should be placed relative to the symbol as shown in **Fig. 5.4**.

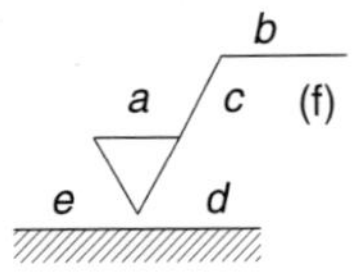

Fig. 5.4

5.9 SURFACE FINISH STANDARD

One method of determining the surface roughness is by using a surface finish standard.

A = Roughness value Ra or Roughness grade No.N1 to N12

B = Production method, treatment or coating

C = Sampling length

D = Direction of lay ¢

E = Machining allowance

F = Other roughness values in brackets.

This box consists usually of 20 blocks, each of a specific surface finish attained by a specific machining operation. The type of machining operation is marked on each block together with the surface roughness number for height and width. Using a surface finish standard, we can make comparisons between the machined surface and the standard surface using our sense of touch.

However, this method is sometimes not accurate enough and the individual must be very sensitive to the different surface roughness.

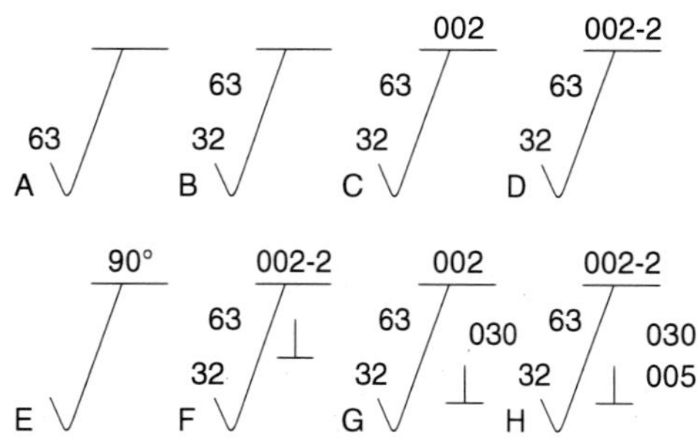

Fig. 5.5

Instrument is inevitable.

Note: For more detailed information on surface texture, symbols and their representations refer to IS : 10719.

5.10 METHODS OF MEASURING SURFACE FINISH

There are two methods used for measuring the finish of machined part:

1. Surface inspection of comparison methods.
2. Direct instrument measurements.

In comparative methods, the surface texture is assessed by observation of the surface.

But these methods are not reliable as they can be misleading if comparison is not made with surfaces produced by the same techniques. The various methods available under comparison method are:

1. Touch inspection,
2. Visual inspection,
3. Scratch inspection,
4. Microscopic inspection,
5. Surface photographs,
6. Micro-interferometer,
7. Wallace surface dynamometer
8. Reflected light intensity.

5.10.1 Touch Inspection

The main limitation of this method is that the degree of surface roughness can't be assessed. Also the minute flaws can't be detected. This method can simply tell which surface is rougher. In this method, the fingertip is moved along the surface at a speed of about 25 mm per second and the irregularities as small as 0.01 mm can be easily detected. A modification of it is possible by using a table tennis ball, which is rubbed over the surface and vibrations from the ball transmitted to hand and surface roughness judged thereby.

5.10.2 Visual Inspection

Visual inspection by naked eye is always likely to be misleading particularly when surfaces having high degree of finish are inspected. The method is, therefore, limited to rougher surfaces and results vary from person to person. More accurate inspection can be done by using illuminated magnifiers.

5.10.3 Scratch Inspection

In this method, a softer material like lead babbitt or plastic is rubbed over the surface to be inspected. By doing so it carries the impression of the scratches on the surfaces which can be easily visualized.

5.10.4 Microscopic Inspection

This is probably the best method for examining the surface finish but suffers due to limitation that only a small portion of the surface can be inspected at a time. Thus, several readings are required to get an average value. In this method, a master finished surface is placed under the microscope and compared with the surface under inspection. In another method a straight edge is placed on the surface to be inspected and a beam of light projected at about 60° to the work. Thus, the shadows cast into the surface scratches are magnified and the surface irregularities can be studied.

5.10.5 Surface Photographs

In this method magnified photographs of the surface are taken with different types of illumination. In case we use vertical illumination, then defects like irregularities and scratches appear as dark spots and flat portion of the surface appears as bright area. In case of oblique illumination, reverse is the case. Photographs with different illuminations are compared and the results assessed.

5.10.6 Micro Interferometer

In this method, an optical flat is placed on the surface to be inspected and illuminated by a monochromatic source of light. Interference bands are studied through a microscope. Defects, i.e., scratches in the surface appear as interference lines extending from the dark bands into the bright bands. The depth of the defect is measured in terms of the fraction of the interference band.

5.10.7 Wallace Surface Dynamometer

This is a sort of friction meter and consists of a pendulum in which the testing shoes are clamped to a bearing surface and a predetermined spring pressure can be applied. In this method the pendulum is lifted to its initial starting position and allowed to saving over the surface to the tested. If the surface is smooth, then there will be less friction and pendulum swings for a longer period. Thus, time of swing is a direct measure of surface finish.

5.10.8 Reflected Light Intensity

In this method beams of light of known quantity is projected upon the surface. This light is reflected in several directions as beams of lesser intensity and the change in light intensity in different directions is measured by a photocell. The measured intensity changes are already calibrated by means of reading taken from the surface of known roughness by some other suitable method.

5.11 OTHER METHODS OF MEASURING SURFACE ROUGHNESS

5.11.1 Taper Sectioning

In this method, a section is cut through the surface to be examined at a shallow angle θ thus magnifying height variations by a factor of cot θ, and the section is examined by optical microscope. It is very accurate method but is a destructive method.

5.11.2 Gloss Measurement

There is a variation in the optical properties of different rough surfaces as judged subjectively by the human eye and various instruments have been developed to try and quantify this variation more objectively. In one instrument the reflection of a series of vertical bars in the surface is reflected at the specific angle. This angle is gradually decreased until the bars can no longer be distinguished. The roughness is inversely proportional to the cosine of the angle of extinction.

5.11.3 Diffraction Techniques

Speckle is produced when a rough surface is illuminated with spatially coherent light and observed with an optical system of finite aperture. Light is received at each point in the image from several different points on the object because of the resolution limitations of the system. The path length of the light from each point on the surface depends on the height of the surface contour at that point. If this height varies significantly across the width of the point spread function, interference effects will appear as speckle. Surface roughness can be deduced from measurements of speckle pattern.

Other methods are tactile tests, pneumatic gauging, optical sections, thermal comparator, capacitance gauges, etc.

5.12 DIRECT INSTRUMENT MEASUREMENTS

These methods enable to determine a numerical value of the surface finish of any surface. Nearly all instruments used are stylus probe type of instruments. These operate on electrical principles. Further, these electrical instruments can be of two kinds.

In the first type, they operate on the carrier-modulating principle. The movements of the stylus exploring the surface are caused to modulate a high frequency carrier current.

The second type includes those operating on voltage-generating principle. In these the movements of the stylus are caused to generate a voltage signal. In both these types the output has to be amplified and the amplified output is used to operate a recording or indicating instrument.

The carrier modulated frequency types of instruments have the advantage that the signal fed to the recorder depends only upon the position of the stylus. While in the voltage generating type when the oscillatory movement of the stylus stops, the output falls to zero, no matter where the stylus is.

Some pneumatic instruments are also used for measuring surface finish.

5.13 STYLUS PROBE INSTRUMENTS

This type of instrument generally consists of the following units:

(i) A skid or shoe which is drawn slowly over the surface either by hand or by motor drive. The skid when moved over the surface follows its general contours and provides a datum for the measurements.

In case a skid is not used and only a probe is used then the probe will trace the actual profile, but upward and downward movement of the probe will be dependent upon the setting of the work under probe. But since the roughness of the surface does not depend upon the position of the work, it will be necessary to choose a datum from which the measurement is to be taken. A line touching the crest of the profile, that is, the envelope line which defines the macro-geometrical form is generally chosen as datum line and this is obtained by using a skid of such a size which can span a large number of surface undulations. A lot of work has been done on the shape of skid, and different types of skids are available for different purposes by which the true macro-geometrical form may be obtained.

(ii) A stylus or probe which moves over the surface with the skid. The stylus for R measurement on new instrument can have a radius of 10 microns ± 30%. when in use, tip radius is allowed to vary ± 50% (5 to 10 /Am). However, for calibration purposes, stylus should have 10 μm nominal size radius ± 20%. The stylus should be cone shaped with a spherical tip. This records the micro-geometrical form of the surface. It moves vertically up and down relative to skid movement due to roughness of the surface. Generally, it is desired that if the skid is moving up then the stylus must also be moving up. But when the pitch between the skid and stylus is comparable to that of half the wavelength of surface, then record will not be desirable one, e.g., in the skid and stylus position skid is moving up, while stylus is giving downward indication.

(iii) An amplifying device for magnifying the stylus movement and an indicator.

(iv) A recording device to produce a trace or record of the surface profile. Usually the vertical movement is magnified more in comparison to the horizontal movement, thus the record will not give the actual picture of surface roughness but a distorted trace obtained.

(v) A means for analysing the trace is obtained. The analysis can be done separately or some automatic device may be incorporated in the instrument for analysis.

5.14 PROFILOMETER

This instrument is most commonly used in the USA for direct measurement of surface quality. This is a dynamic instrument similar in principle to a gramophone pick-up. A finely pointed stylus mounted in the pick-up unit is traversed across the surface either by hand or by motor drive. The instrument records the rectified output from the pick-up which is amplified further and operates an indicating device. Thus, this records the average height of the surface roughness. In this instrument, roughness together with waviness and flaws comprises the irregularities found on the surface. An indication is obtained only when the pick-up is moving. This instrument is best suited for measuring surface finish of deep bores.

5.15 THE TOMLINSON SURFACE METER

This instrument was designed by Dr. Tomlinson. This instrument uses mechanical-cum-optical means for magnification.

The diamond stylus on the surface finish recorder is held by spring pressure against the surface of a lapped steel cylinder. The stylus is also attached to the body of the instrument by a leaf spring and its height is adjustable to enable the diamond to be positioned conveniently.

The lapped cylinder is supported on one side by the stylus and on the other side by two fixed rollers as shown in **Fig. 5.5a**. The stylus is restrained from all motions except the vertical one by the tensions in coil and leaf spring. The tensile forces in these two springs also keep the lapped steel cylinder in position between the

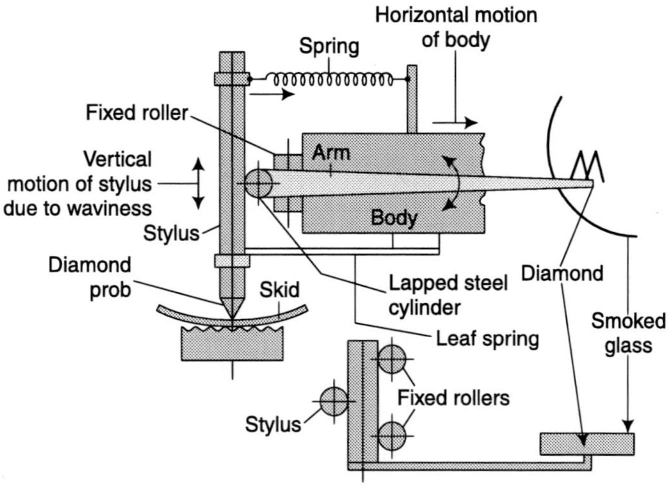

Fig. 5.5a Tomlinson surface meter

stylus and a pair of fixed rollers. A light spring steel arm is attached to the horizontal lapped steel cylinder and it carries at its tip a diamond scriber which bears against a smoked glass.

When measuring surface finish, the body is traversed across the surface by a screw rotated by a synchronous motor. Any vertical movement of the stylus caused by the surface irregularities causes the horizontal lapped steel cylinder to roll. By its rolling, the light arm attached to its end provides a magnified movement on a smoked glass plate. This vertical movement coupled with the horizontal movement produces a trace on the glass magnified in vertical direction and there being no magnification in the horizontal direction. The smoked glass trace is then, further projected at × 50 or × 100 magnification for examination. This instrument is comparatively cheaper and gives reliable results.

5.16 THE TAYLOR-HOBSON TALYSURF

The Talysurf is an electronic instrument working on carrier modulating principle. This instrument also gives the same information as the previous instrument, but much more rapidly and accurately. This instrument as also the previous one records the static displacement of the stylus and is dynamic instrument like Profilometer

The measuring head of this instrument consists of a diamond stylus of about 0.002 mm tip radius and skid or shoe which is drawn across the surface by means of a motorized driving unit (gearbox) which provides three motorized speeds giving respectively × 20 and × 100 horizontal magnification and a speed suitable for average reading. A neutral position in which the pick-up can be traversed manually is also provided. In this case the arm carrying the stylus forms an armature which pivots about the centre piece of E shaped stamping as shown in **Fig. 5.6**. On two legs of (outer pole pieces) the E-shaped stamping there are coils carrying a.c. current. These two coils with other two resistances form an oscillator. As the armature is pivoted about the central leg, any movement of the stylus causes the air gap to vary and thus the amplitude of the original a.c. current flowing in the coils is modulated. The output of the bridge thus consists of modulation only as shown in **Fig. 5.6**. This is further demodulated so that the current is directly proportional to the vertical displacement of the stylus only.

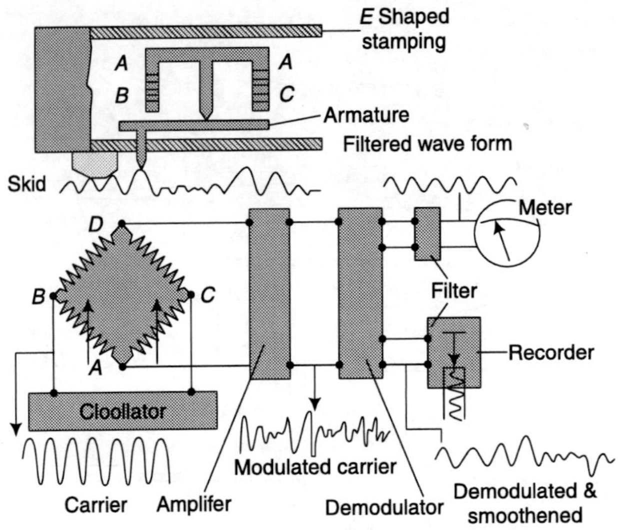

Fig. 5.6 Talysurf

The demodulated output is caused to operate a pen recorder to produce a permanent record and a meter to give a numerical assessment directly. In recorder of this statement the marking medium is an electric discharge through a specially treated paper which blackens at the point of the stylus, so this has no distortion due to drag and the record strictly rectilinear one.

Nowadays microprocessors have made available complete statistical multi-trace systems measuring several places over a given area and can provide standard deviations and average over area-type readings and define complete surface characterization. These systems lend themselves to research applications where specialized programming can achieve auto correlation, power spectrum analysis and peak curvature.

5.17 STYLUS

Phonograph needles, though used in some cases are found to be too large and too heavily loaded. It also causes damage. Diamond styli are used universally. Some of them are cones of 90° include dangle and tip radius 4-12 μm. A popular stylus with truncated pyramid is shown in **Fig. 5.7**. The angle between the faces is 90°. The short edge is parallel to the direction of motion. Thus, this stylus cannot resolve a wavelength shorter than 6 μm, and integrates over a narrow strip of surface 8 μm wide.

It may be noted that this pick-up has finite dimensions, and it is constrained to move in a nearly vertical plane, relative to the moving pick-up. Thus, the stylus cannot record re-entrant features an unimportant drawback for engineering investigations as re-entrant structures are absent on most machined surfaces. This stylus will fail to follow peaks and valley faithfully and produce a distorted record of the surface.

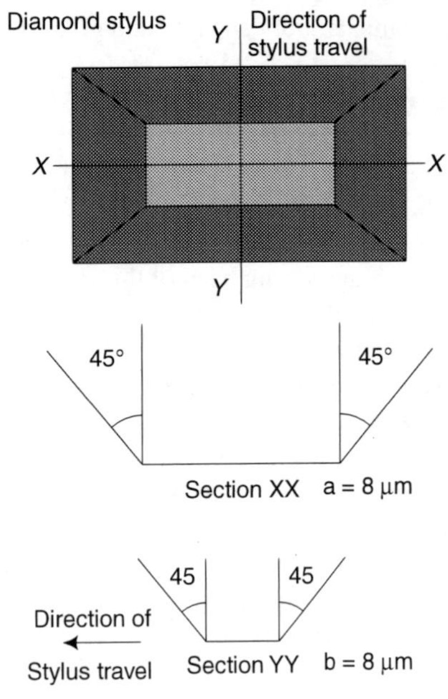

Fig. 5.7

Since the dimensions of the stylus are finite, so also is the load on it. The load is of the order of 70 mg force. But as the area of contact is too small, the local pressure may be sufficiently high to cause significant local elastic downward deformation of the surface under examination.

5.18 REPLICA METHOD

When the surface is not readily accessible to the probe, then a replica of surface is made by dampening a piece of cellulose acetate film in acetone to dampen it all pressing it against the surface till it hardens. By this method, 80% fidelity can be obtained. Modern techniques using epoxy and other resins approach 100% fidelity.

5.19 SAMPLE LENGTH OR CUTOFF LENGTH

Difficulties are generally experienced, when measuring the surface of irregular surfaces, in choosing the sample length. The nature of the surface many a time is such that the good indication of the surface can't be obtained when the selected length is different, e.g., in **Fig. 5.8** different values would be obtained for various measures of surface irregularities if determined for lengths, L1 and L2 etc. This is because the macrogeometrical irregularities affect each to a varying extent. Then the next problem is as to which type of irregularity should be allowed to influence the assessment of other. Sampling length should bear some relation to the type of profile. It is found that the required length can be related to the manufacturing process. In a surface there are different kinds of irregularities having different frequencies and it is never clear which has got more importance.

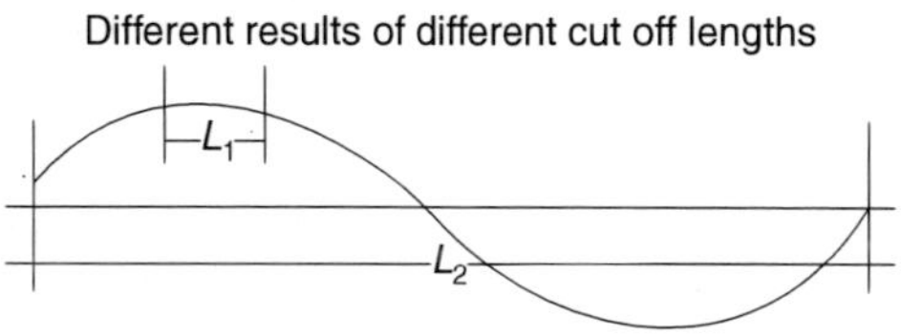

Fig. 5.8

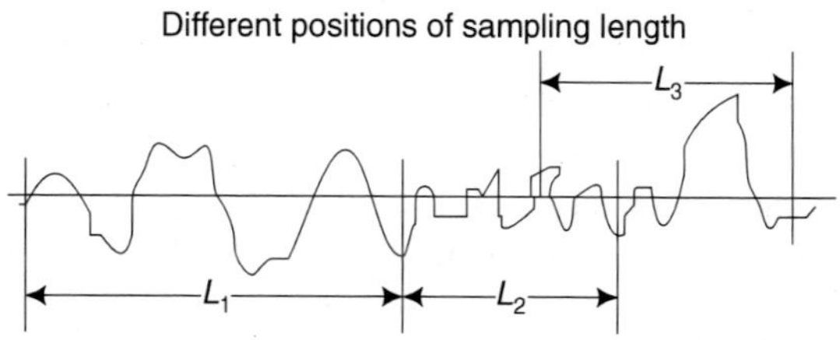

Fig. 5.9

Thus, the best thing would be to standardize some sampling length and that can be always associated with the surface roughness value. To standardize the sampling length is also not an easy task as the wavelengths vary considerably to the way in which the surface is produced. Further referring to **Fig. 5.9.** Another problem is as to where the sample length must be chosen, i.e., position of sample length on surface will also give different results.

This effect is taken into account by taking the readings at several places and taking the mean value. Sampling length is standardized for various operations considering all the above facts as follows:

Process	Symbol	Sample	Length in mm	Spread	
Milling	M	—	0.75	2.5	7.5
Turning	T	0.25	0.75	2.5	—
Grinding	G	0.25	0.75	2.5	—
Planning	.P	—	—	2.5	—
Lapping	L_p	0.25	0.75	—	—
Polishing	P_g	—	—	2.5	—

5.20 ANALYSIS OF SURFACE TRACES

A numerical assessment is assigned to indicate the degree of smoothness (roughness) in a number of ways. In practice three roughness measures have shown themselves to be particularly useful.

5.21 MAXIMUM PEAK TO VALLEY HEIGHT OF ROUGHNESS

This is obviously the most common measure of roughness but is not by any means a complete definition of roughness, e.g., in the two cases in **Fig. 5.10** peak to valley height is same, but frequencies of irregularities are different and the second surface is more rough in comparison to the first one. But, since this is a relatively simple method of analysis, this will be a satisfactory measure where it is desired to control the cost of finishing for checking the rough machining. This method is also very advantageous in cases where the condition of the surface is likely to exert an important influence on such properties as fatigue resistance and it is intended to clear the surface of the irregularities left by a previous operation.

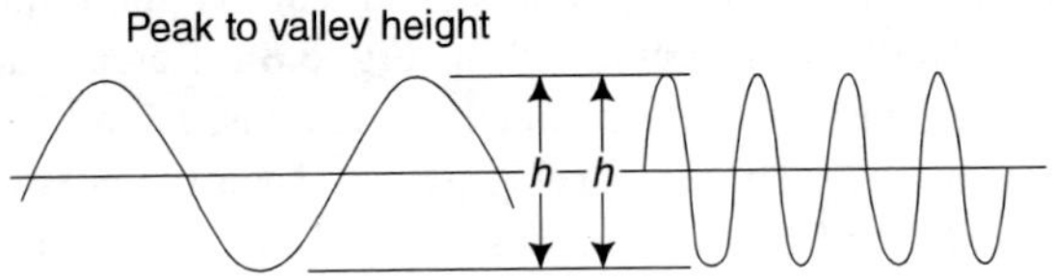

Fig. 5.10

5.22 ROOT MEAN SQUARE VALUE (R.M.S. VALUE)

This measure was in use earlier but now superseded by Centre Line Average measure, as the latter has the properties of being easily measured. R.M.S. value is defined as the square root of the mean of the squares of the ordinates of the surface measured from a mean line.

Referring to **Fig. 5.11**, the selected length L is divided into n equal parts. Ordinates are erected at the points 1, 2, 3,4,, *n*, whose heights are (say) h_1. h_2, h_3, h_4, h_n. Then

$$h_{rms} = “\ \mathbf{h_1 + h_2 + h_3 + \ldots + h_n/n}$$

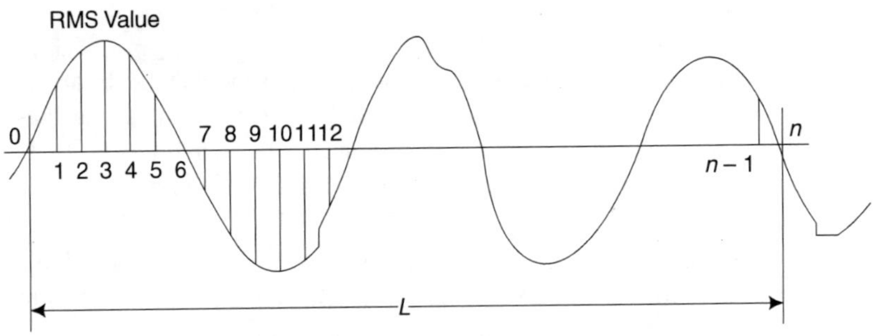

Fig. 5.11

5.23 CENTRE LINE AVERAGE METHOD (C.L.A. VALUE)

This is defined as the average height from a mean line of all ordinates of the surface regardless of the sign. Thus referring to **Fig. 5.12**, C.L.A. = $- h_1 + h_2 + h_3 + h_4 + h_n$

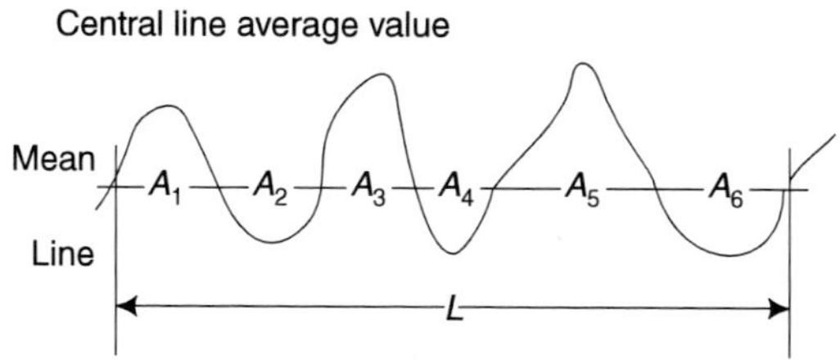

Fig. 5.12

But to find C.L.A. value like this will be a laborious job. Also by this method, spacing chosen may be such that important ordinates are likely to be missed.

Things can be much simplified by using a planimeter which can find out the area of any curve. Referring to **Fig. 5.13**, let us say that somehow or other the mean line is exactly known. Then

$$\text{C. L. A.} = \frac{A_1 + A_2 + A_3}{L} = \frac{\Sigma A}{L}$$

One has to take care of units carefully to find the C.L.A. value in micron.

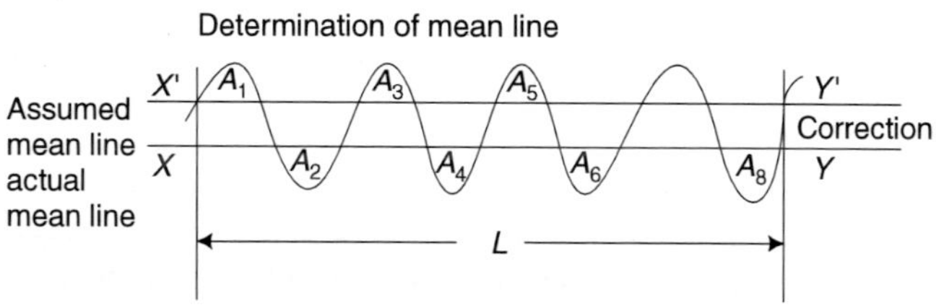

Fig. 5.13

5.24 HOW TO DETERMINE MEAN LINE

For it first the mean line is estimated by eye judgment. Then the total area above and below the assumed mean line are measured and a correction is applied to the assumed mean line to get the correct mean line. Error [ó A (above) - ó A (below)]/L is applied to the assumed mean line to get the correct mean line. Thus in **Fig. 5.13**, X' Y' is the assumed mean line

$$\text{Then correction} = \frac{(A_1 + A_3 + A_5 + A_7) - (A_2 + A_4 + A_6 + A_8)}{L}$$

And this correction added to the assumed line algebraically gives the actual mean line. The C. L. A. value can be found out as described earlier.

Talysurf has got built-in arrangement for integrating the areas and the average value is directly given.

C.L.A. value does not give any idea regarding the greatest extent and the nature of the surface irregularities. It is likely to give identical values for surfaces of vastly different characteristics. So this is the main disadvantage of CLA. value. But when the characteristics of a surface have been found out to be satisfactory and such conditions are produced which do not allow the surface to change radically in manufacturing, then CLA. value provides a workable control.

5.25 ASSESSMENT OF SURFACE ROUGHNESS AS PER INDIAN STANDARDS

There are various methods of evaluating the surface roughness but the most prominent and commonly used methods are the M [Centre Line Average method (CLA) or the mean line] system and the E (Envelope system) system. The M system expresses the arithmetical average departure of the actual surface both above and below a mean line within a specified sampling length and in a plane substantially normal to the direction of surface. A similar system, also based on the mean line, expresses the departure of the actual surface as a root mean squared value. CLA system is more popular than RMS value method. To eliminate the effects of secondary texture the profile is split into a number of sampling lengths and a separate mean line computed for each length. In this way, the mean lines of adjacent sampling length may not coincide even. The E system expresses the arithmetical departure of a surface both above and below a 'mean' curve. This mean curve is developed from a contacting envelop by displacing it to a position, where the areas enclosed by the profile above and below the mean curve are equal. The contacting envelop referred to above is obtained by rolling across the surface a sphere of radius 25 mm and displacing the locus of the centre for this circle towards the surface by an amount equivalent to the radius.

The merits and demerits of the two systems (i.e., M and E) are given below. The M system is very widely used and all the instruments for measuring surface roughness available are designed to measure in this system. It is more useful and a satisfactory means of controlling, at the point of production, the consistency of results from a process when the production parameters have been established.

This system is unable to control the functional qualification of a surface when associated with a machine process.

The E system, though at present in its early stage of development is going to be widely accepted in future because of its better definition of peak measure and the fact that it is more easily applied to the surface finish instruments based on the interference principle.

5.26 M SYSTEM

The surface roughness in this system is expressed by Ra value which is useful for comparing various surfaces obtained by similar operation; of course, it does not indicate the limits of irregularity R_{max} value (measure of peak value) which is perceived by looking at the profile and by touching the surface is probably the most direct method of expressing the surface roughness value. This value is also useful in identifying the surfaces for comparison and record purposes but there is no direct relationship between R_{max} and Ra value and as such R_{max} value alone can't be specified in drawings.

5.27 INFORMATION TO BE GIVEN IN THE STATEMENTS OF SURFACE ROUGHNESS

The various roughness grade numbers N 1 to N 12 in 5 groups are specified as under by BIS. The relationship between the roughness grade numbers and the commonly used and generally accepted system of indicating surface roughness by symbol is given below.

The values of the surface roughness expected from various manufacturing processes are indicated below.

		Ra value in μ mm
	Manufacturing process	5 to 50
	Sand casting	0. to 6.3
	Permanent mould casting	0.8 to 3.2
Casting	Die-casting	0.32 to 2
	High pressure casting	2.5 to 50
	Hot rolling	1.6 to 25

	Forging	0.16 to 5
Hot working	Extrusion	6.3 to 1.00
		1 to 6.3
	Flame cutting, sawing & chipping	3.6 to 25
	Radial cut off	1.6 to 25
	Hand grinding	0.25 to 25
	Filing	1.6 to 50
	Planning	1.6 to 25
Machining operation	Shaping	1.6 to 20
	Drilling	0.032 to 25
	Turning & milling	0.4 to 6.3
	Boring	0.4 to 6.3
	Reaming	0.4 to 3.2
	Broaching	0.4 to 3.2
Grinding and Super finishing operation	Hobbing	0.063 to 5
	Surface grinding	0.063 to 5
	Honing	0.025 to 0.4
	Lapping	0.012 to 0.16
	Polishing	0.04 to 0.16
	Burnishing	0.04 to 0.8
	Super finishing	0.16 to 0.32

5.28 LAY

This is the direction of the predominant surface pattern and is determined by the production method used. The surface roughness is generally measured across the direction of the lay.

The various directions of lays depending upon the production process is shown in **Fig. 5.14**.

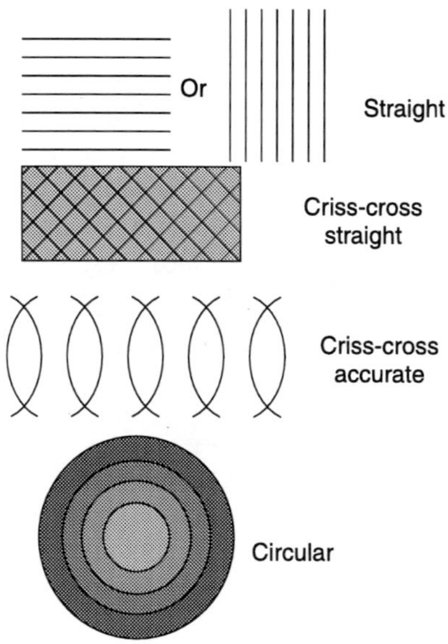

Fig. 5.14

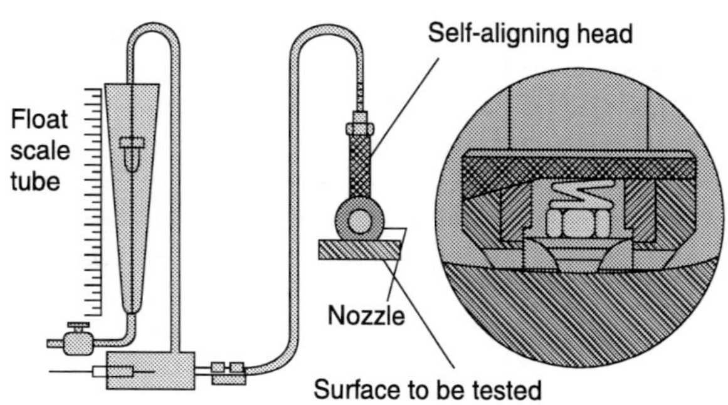

Fig. 5.15

5.29 PREFERRED VALUES OF R_a AND R_z

Preferred values for arithmetical mean deviation R_a in μm are selected from:

0.025, 0.05, 0.1, 0.2, 0.4, 0.8, 1.6, 3.2, 6.3, 12.5 and 25 and the preferred values for ten point height of irregularities R_z in μm are selected from:

0.05, 0.1, 0.2, 0.4, 0.8, 1.6, 3.2, 6.3, 12.5, 25, 50 and 100.

5.30 SELECTION OF SAMPLING LENGTH

For measuring surface roughness, the value of the sampling length l is selected from the following series depending upon the process of manufacturing:

0.08, 0.25, 0.8, 2.5, 8, 10 and 25 mm.

It is usual practice to choose shorter value of the finer and larger value for the coarser grade of a given process when more than one value is given.

5.31 MEASUREMENT

It may be emphasized here again that the surface texture represents the combined effects of several causes, of which the roughness, waviness and the error of form are more predominant. If only the total height (i.e., the difference between the highest and lowest points on the plot) is considered then idea of above three characteristics can be had by considering different sample lengths. If we are interested only in the measurement of roughness of the surface then the sampling length should be neither too big as to include the waviness, nor too small as to ignore the occasional deep scratches.

The best thing is, therefore, to limit the measurement to a sufficiently short length of the surface and in order to take into consideration the considerable variation of roughness from place to place, several readings at various places are taken. When the surface roughness has a directional quality, the readings should be taken in a direction perpendicular to the direction of lay as this way the effective spacing of the crests will be at short distances. The effective spacing of crests goes on decreasing as the direction of measurement is reduced from 90° to direction of lay and in the extreme case when measurement is made parallel to the lay, a minimum result of roughness is obtained.

5.32 DETERMINATION OF R_a VALUE

The R_a value can be determined either by graphical methods or by direct reading instruments. Whenever the surface texture has a directional quality, the direction in which the measurement is made is approximately at right angles to the 'lay'. It is always preferable to compute the R_a value by taking mean results from the measurement of several sampling lengths taken consecutively along the profile so that it gives better indication and is the correct representation of the whole of surface roughness.

5.33 GRAPHICAL METHODS

In this method, it is presumed that the surface is nominally flat and that the record is produced in rectilinear coordinates in which a truly flat surface is represented as a straight line. First of all, mean line of successive sampling lengths is determined. In case the profile is not nominally straight, then mean line of appropriate curvature could be assumed; for the record could be developed such that base line is straightened. For such cases, the profile records are made on plastic replicas by taking from curved surfaces and flattening out: or such instruments may be used which operating on a curved surface under appropriate conditions produced records in which the nominally curved base line is automatically straightened.

The mean line is determined by first drawing any straight line parallel to the general course of the record over the sampling length. Then the sum of areas above this line and below it are computed and their difference divided by the sampling length gives the distance by which the above drawn line is to be shifted in order to obtain the central line. Then the arithmetical sum of all the areas above and below the central line divided by the sampling length and the vertical magnification of the record gives the Ra value in mm, if areas are computed in mm^2 and sampling length in mm. This value multiplied by 1000 gives Ra value in microns.

5.34 SOME NOTES ON STYLUS AND SKID METHOD

It is obvious that the stylus which rests on the surface and traverses across it responds to the irregularities of the surface relative to a selected datum provided by a skid having a large radius of curvature in the direction of the traverse. The skid simply follows the general contour, riding over the crests of smaller irregularities without responding to them individually. This method is subjected to some limitations in giving exact response corresponding to the shape of the irregularities. It is firstly due to the impossibility of having a mathematically sharp point for the stylus, which at the most can be finished with a small but finite radius. The effect of this is that the full depth of the groove is reduced, the trace is smoothed out over the sharp corner and crests, and the trace offers no evidence of fine deep scratches in the actual surface. However, it may be appreciated that for an average reading the loss in the valley is fully offset by the gain resulting from the apparent rounding of the crests. Further detailed studies indicated that the surface irregularities being relatively shallow compared with their spacing, the errors due to finite radius of the stylus are not very serious and that a nominal radius of 0.003 mm for the tip of the stylus gives quite satisfactory results. Secondly, the error is also introduced due to the displaced position of the stylus in relation to the skid, because it is possible that the stylus touches a valley when the skid is at its lowest point and vice versa. In order to reduce the chances of meeting extreme phase conditions, it is common practice to have two skids having a small radius and both touching the surface with the stylus acting midway between them.

Keeping the above in view, the stylus type instruments employed for determining the surface roughness should comply with the following requirements.

5.35 STYLUS

For graphical recording, nominal radius of the tip of stylus should be 0.003 mm. For determination of Ra value by direct reading instruments the nominal radius of the tip of stylus should be 0.01 mm. The variation in the tip radius of the stylus should be within 30% of its nominal dimension or 0.003 mm whichever is greater. The shank may be either conical or pyramidal and in either case the included angle of the cone should not exceed 90°.

The maximum force between a stylus and the surface under test should not be more than 0.016 × (tip radius in micron)^2 grams.

5.36 SKID

If a single skid is employed its radius in the direction of traverse should not be less than 50 times the sampling length. In case of two skids, the radii should not be less than 8 times the sampling length.

5.37 ROUGHNESS COMPARISON SPECIMENS

The roughness comparison specimens are used for qualitative assessment of the surface roughness. In this qualitative assessment of the roughness of the finished product is evaluated by visual or tactile comparison (i.e., lightly moving a finger nail across the lay of the roughness) with roughness comparison specimen of known roughness value finished by similar machining process. The roughness comparison specimens are produced in a variety of materials (steel having a hardness of about 400 HB) by direct application of the production process or by copying techniques and are available in various grades of roughness. The copying techniques consist of (a) electro-forming positive replicas of master surfaces and (b) making positive replicas of master surfaces in plastic materials by coating or otherwise and having the feel and appearance of the natural machined metal surface. It is general practice to produce the specimen surfaces as replicas of master surfaces in a hard material like nickel alloy or metallised plastic.

5.38 OTHER METHODS FOR EVALUATING SURFACE ROUGHNESS

The roughness comparison specimens are available either in flat form or segmented cylindrical form. The flat specimens have the size: length-25 mm, width-16 mm, thickness 1.6 to 10 mm. The specimen is rigidly supported on a base to prevent it from being warped or distorted during use. The segmented cylindrical specimens will have either convex surface or concave surface having a surface radius of about 125 mm.

These specimens have also proved to be very good devices for training the personnel to judge the R value of the feel and appearance of the surfaces. While these are being used for direct comparison purposes, it is important to note that

(i) Two surfaces being compared should resemble each other in material, shape and method of finish;

(ii) Two surfaces must be closely alike in general character in the sense that the cutting technique and machine feeds used for making the specimen should resemble those used for the products;

(iii) Surfaces being compared should be clean and dry.

5.38.1 Pneumatic Methods of Evaluating Surface Finish

This method has been successfully used for evaluating the micro irregularities. This method is popularly known as air-leakage method. The instrument works on the following principle. The compressed air is made to pass through a self-aligning nozzle which moves on the surfaces to be tested. The gap between the nozzle tip of the instrument and the tested surface depends upon the height of the micro irregularities. The size of the gap, in turn, affects the air discharge through the nozzle. The air discharge is measured by a special variable area air flow meter, (a rota meter) and determines the height of the micro irregularities.

It will be observed that it may be difficult to calibrate this instrument directly in terms of surface micro irregularities. It is; therefore used to test the surface with instrument by the comparative method. Certified samples are used as reference for comparison. Under these circumstances, the instrument is best suited for inspection of parts in mass production.

The diagram of the instrument and the design of the nozzle are shown in **Fig. 5.16**. Air is first passed through a tube in which there is a float which can freely move up and down in the tube depending upon the flow of air through tube. The air from this tube is led to the measuring nozzle of the self-aligning head through a flexible hose. The air leakage between the nozzles and the test surface will be same as in the tube. To allow the air to pass, the float is lifted by the air stream to a definite height depending upon the flow of air through the tube or the amount of air passing through the nozzle. The float height is read on the scale placed by the side of the tube and it indicates the quality of surface finish.

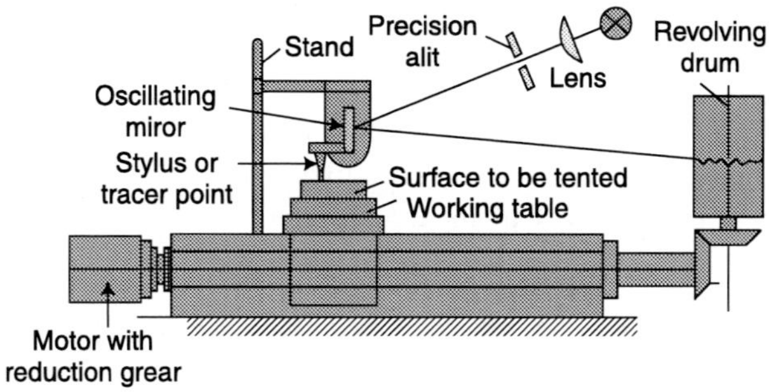

Fig. 5.16 Profilogram-tracer type

5.38.2 Tracer Type Profilogram

The principle of the tracer-type profilogram is shown in **Fig. 5.17**. The surface to be tested is placed on a working table. The table can move to and fro because there is a nut at the bottom of table, which fits into a lead screw driven by a motor. The stylus or tracer point which is made to rest on the surface to be tested is pivoted with mirror as shown in the figure. The pivot point is fixed in space with the stand and its position can be changed initially. When the surface to be tested moves, the oscillations of the tracer are transmitted to the mirror. The beam of light (received from light source and through the lens and precision slit) strikes on the oscillating mirror and is reflected on to the revolving drum, upon which sensitized film is arranged. The drum is also rotated from the same lead-screw that traverses the table through bevel gears.

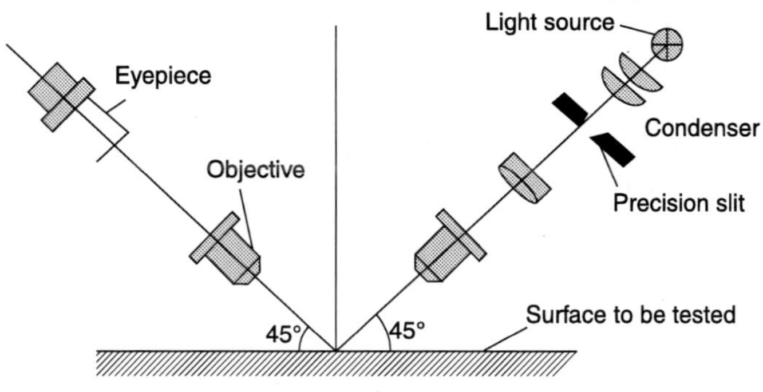

Fig. 5.17 Light cross-section method

The trace obtained on the drum is very much magnified one and can be further assessed by various methods.

5.38.3 Light Cross-Section Method

The principle of this method is that when a thin film of light strikes the surface to be tested at an angle of 45°, a band of reflected light will appear on the surface. The edge of this band will reproduce the profile of the surface irregularities or so to say it will show an optical cross section. This profile is magnified and observed by an optical device, arranged at an angle of 45°, a band of reflected light will appear on the surface. The edge of this band will reproduce the profile of the surface irregularities or so to say it will show an optical cross section. This profile is magnified and observed by an optical device, arranged at an angle of 45° (perpendicular to the light source).

A double microscope working on the above principle is shown in **Fig. 5.18**, and is used to evaluate the surface finish by this method. The general view and optical system of the microscope is shown in **Fig. 5.18**. From a light source, a beam of light passes through the condenser and precision slit and is directed at an angle of 45° to the surface to be tested. The observing microscope having objective and eyepiece is also inclined at an angle of 45° with the tested surface. The field of view of the eyepiece contains a reading scale called an eyepiece micrometer, which is used to measure the height of surface irregularities.

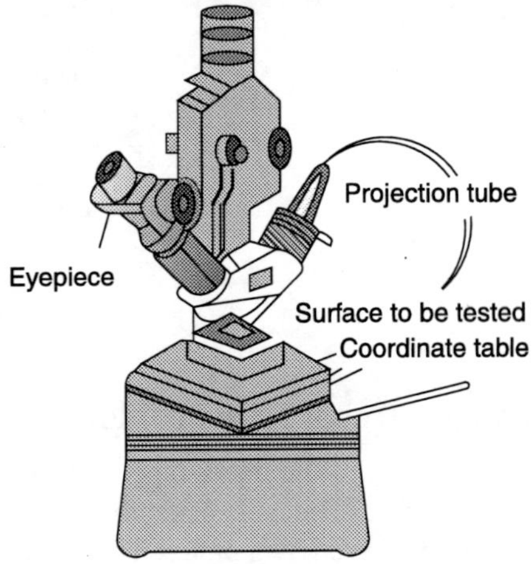

Fig. 5.18 Double microscope

5.38.4 Mechanical Roughness Indicator (MECRIN)

It is a recent development from the tactile method. It is suitable for non-regular surfaces where the frictional properties measured compare with those provided by electronic stylus instrument, such as ground, honed, lapped, hand filed and emery clothed surfaces, etc., to approximately 4.5 μm Ra and on the finer grades of turning and milling to 0.8 μ m Ra. On regular surfaces of rougher values, it is used as a comparator against surface roughness blocks to decide which is rougher. This instrument assesses the surface through its frictional properties and the average slope of the irregularities. It is a common experience that a thin metallic blade when pushed against a surface at a certain angle will either slide or buckle depending on (a) the surface roughness, and (b) the angle of attack, i.e., the angle of the blade to the surface. At angles below the critical angle the blade tip will simply slide along the surface but if the angle is successively increased, at a certain angle the blade just begins to buckle. This angle is the critical angle of attack. The 'smoother the surface the greater the critical angle and thus this angle is a function of the degree of roughness encountered.

Figure 5.19 shows such an indication in which the angle of attack is measured with a gravity dial indicator which incorporates a pendulum and gear mechanism and the buckling blade is observed through the transparent cover. The dial is calibrated in terms of the N grades and the corresponding central Ra values. Calibration of the instrument is against hardened carbon steel specimens having nominal values of 0.1 and 0.4 μm Ra which are accurate to within ± 12%. It can be used to provide μm Ra values for relatively smooth irregular surfaces; as a comparator for two surfaces as a limit gauge for roughness; to establish the direction of the lay of a surface to determine the direction of maximum roughness if the surface has a definite lay; to measure short lengths down to 1 mm which may be inaccessible to stylus instruments.

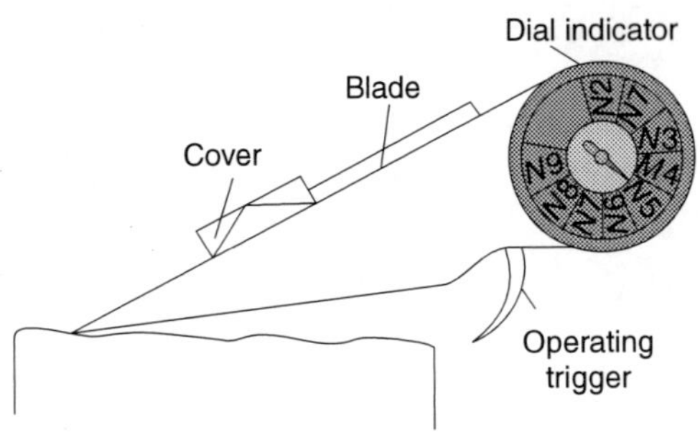

Fig. 5.19

EXERCISE

1. Calculate the CLA (Ra) value of a surface for which the sampling length is 0.8 mm.
2. In the instrument of surface roughness, height of 20 successive peak and troughs were measured from a datum and were 35, 25, 40, 22, 65, 18, 42, 25, 65, 22, 36, 18, 42, 22, 32, 21, 37, 18, 35, 20 microns.

3. If these measurements were obtained over a length of 20 mm, determine the CLA (Ra) and RMS value of the rough surface.
4. What is surface metrology?
5. Why surface finish is important in engineering applications?
6. Define surface roughness.
7. How is surface roughness assessed? Define M system.
8. Discuss what do you understand by the following terms in connection with surface finish measurement:

   (i) Waviness (ii) Lay Envelope method (iii) Crest line method.
9. Why is the assessment of surface texture important? Describe two methods to obtain a numerical value of the texture from a given graphical record.
10. What is the difference between primary texture and secondary texture? Describe an instrument that may be used to obtain a graphical record of primary texture.
11. (a) Describe an instrument for measuring surface finish.
12. (b) Discuss the relative merits and demerits of the M (mean line) and the E (envelop) system of measurement of surface finish.
13. What is meant by roughness and waviness of machined surfaces?
14. Name the methods used for measuring surface roughness.
15. How the Tomlinson surface recorder and Talysur machine work? What are their relative merits?
16. How CLA Index Number is determined? Explain why CLA Index number alone is not sufficient to specify the surface texture required and to make the information complete, what else is to be specified? How the required texture is specified on drawing?
17. Describe various methods of measuring surface texture giving their relative advantages.
18. State the possible causes of each of the various types of irregularities found in surface texture. Show how surfaces having the same numerical assessment may have different properties and textures.
19. What are roughness comparison specimens and how they assess surface roughness? What are the limitations?
20. Describe briefly some recent advances in the field of surface texture measurement.
21. Which of the methods is recommended by IS: 3073-1967 for specifying the surface texture on machined parts?
22. Calculate the *Ra* value of a surface for which the sample length was 8.1 mm, the graph was drawn to a vertical magnification of 1000 and the areas above and below the datum line were:

Above:	180	90	155	55 mm^2
Below:	70	90	170	150 mm^2

23. State the possible causes of each of the various types of irregularities found in surface texture. Show how surfaces having the same numerical assessment may have different properties and textures ?
24. What do you mean by 'surface quality'? State the characteristics of the forms of surfaces.
25. Explain the difference between roughness and waviness.
26. State different methods of surface evaluation.

Marking Tools

Chapter 6

Castings, forging, weldment, and rolled sections are laid out in order to establish reference lines or planes which are used to set up the part on the machine, to outline the surfaces that are to be machined, and to show the approximate amount of metal to be removed from each surface. Castings and forging in particular, have excess material on all surfaces which is to be machined, including cored holes. One purpose of the layout is to determine if enough metal exists on all of these surfaces to "clean up" or "machine out." Sometimes, a core will shift in making a casting, resulting in insufficient metal to "machine out." When the core shift is only a small amount this can be corrected in the layout; at other times, the casting must be scrapped. However, it is best to know that the casting will be scraped before any expense is encountered in machining.

Two other reasons for laying out workpieces that are to be machined are to locate and outline the position of holes that are to be drilled and to outline key ways that are to be cut. Although castings, forging, and weldment are usually laid out before machining any surface, additional layout work is often done after one or more machining operations have been finished. In some cases, usually involving rolled material, the initial layout operation is performed after several machining operations are completed. Not all workpieces are laid out. Parts made from bar stock and those parts having a relatively simple shape are often machined directly, without making a layout.

The layout is usually made from a reference surface which is a plane.

Surface plates, layout plates, layout tables, floor plates, and the top of machine tool tables are used in making layouts. They must be kept clean and free from scratches, nicks, and burrs that would impair the accuracy of the reference surface and the layout work done upon it.

For making any new part or product (job), correct marking is necessary, according to the drawing. If the marking is not exact, the job will not be according to the measurements in the drawing. We can imagine the difficulty in using a machine part made with wrong measurements. A lot of time is wasted in fitting a wrong part. For exact marking, the following points should be kept in mind:

1. Before marking, drawing of the job should be carefully read.
2. While reading the drawings its measurement should be carefully noted.
3. According to the measurement, the required tools should be kept ready in advance.
4. Only correct and genuine tools should be used for marking.
5. Before making the marking permanent through dot-punching, the marking lines should be checked once again.
6. The place, at which drilling is to be done, should be marked with centre punch. For different kinds of jobs, different types of marking are needed. The following kinds of tools and instruments are used for different markings:

(a) Marking table
(b) Surface plate
(c) Angle plate
(d) Odd leg or Jenny calliper_
(e) Divider
(f) Scriber
(g) Surface gauge
(h) Universal surface gauge
(i) Trammel
(j) 'V' Block
(k) Punch.

Apply marking media on the job before marking.

The different marking media are whitewash, Prussian blue, copper sulphate and cellulose lacquer.

Purpose: To get a clear and thin line, fine lime powder is made into a thin paste by adding water.

Copper sulphate solution: Add copper sulphate crystals into water (e.g., 200 ml) till the solution is saturated. Add thereafter 10-20 drops of sulphuric or nitric acid. Apply the solution with a brush. Copper sulphate coating should be dried well before commencing marking as otherwise the solution may stick on the instruments used for marking.

Caution: Copper sulphate solution and acids are poisonous.

.**Whitewash:** Whitewash is prepared in many ways. Chalk powder mixed with water. Chalk mixed with methylated spirit. White lead powder mixed with turpentine. Whitewash is applied to rough forging and castings with oxidized surfaces. Whitewash is not recommended for workpieces of high accuracy.

Prussian blue: This is used on filed or machine-finished surfaces. This gives very clear lines but takes more time for drying than the other marking media.

Cellulose lacquer: This is a commercially available marking medium. It is made in different colours, and dries very quickly.

The selection of marking medium for a particular job depends on the surface finish and the accuracy of the workpiece.

Marking off: Marking off or layout is carried out to indicate the locations of operation to be done, and provide guidance during rough machining or filing.

Witness marks: The line marked on metal surfaces is likely to be erased due to handling. To avoid this, permanent marks are made by placing punch marks at convenient intervals along the marked line. Punch marks act as a witness against inaccuracies in machining and hence, they are known as witness marks.

6.1 MARKING TABLE

A marking table (marking-off table) is used as a reference surface for marking on workpieces.

Marking tables are of a rigid construction with accurately finished top surfaces. The edges are also finished at right angles to the top surface.

Marking tables are made of cast iron or granite, and are available in various sizes. These tables are also used for setting measuring instruments, and for checking sizes, parallelism and angles.

Its breadth is from 2 to 5 feet and length is 6 to 10 feet. The upper part of this table is generally made plain. This table is used for simple and routine marking jobs. Where general tools such as steel foot-rule, scriber and odd-leg calliper is used. In addition, it is also used for punching and chipping job.

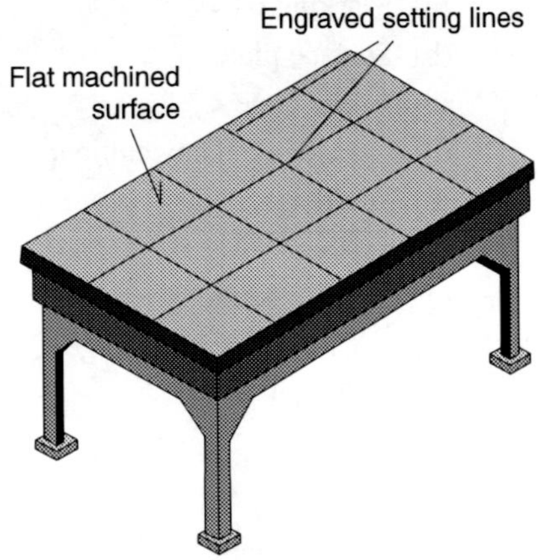

Fig. 6.1

6.2 SURFACE PLATE

For jobs which require exact markings, surface plate is used. Like the marking table, this table is also either rectangular or circular. These plates are made of close grain cast iron and these are kept out at least for one year for the purpose of seasoning.

Their upper surface has fine finish which may contain 10 per cent to 20 per cent high spots. For convenience there are two handles on opposite sides. There are ribs on the inner side of the plates, which makes them stronger. The main plate extends beyond its base frame **(Fig. 6.2).**

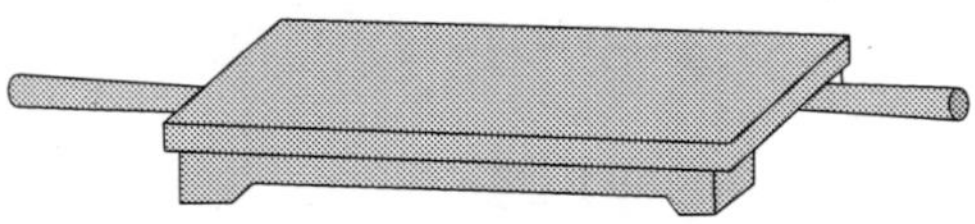
Fig. 6.2

Considerable care is required for making a surface plate. After its moulding and seasoning, it is cut or scratched according to the size on the planer machine. Its surface is almost made plain and through scrapping, it gives its true form. Master plate is used for this purpose. Lead oxide is applied on the surface plate and then rubbed on the master plate.

The spots, which are higher than the basic principle/ original surface, there lead oxide is wiped and their high spots are levelled by scrapping. Similarly, lead oxide is applied to two new surface plates and their high spots are planed after detecting them.

For the purpose of steadiness and convenience in levelling, a three-point suspension is given.

Smaller surface plates are placed on benches while the larger surface plates are placed on stands.

6.3 OTHER MATERIALS USED

Granite is also used for manufacturing surface plates. Granite is a dense and stable material. Surface plates made of granite retain their accuracy, even if the surface is scratched. Burrs are not formed on these surfaces.

Classification and uses: Surface plates used for machine shop work are available in three grades: 1, 2 and 3. The grade 1 surface plate is more acceptable than the other two grades.

Specifications: Cast iron surface plates are designated by their length, breadth, grade and the Indian Standard number.

Example Cast iron surface plate 2000 × 1000 Gr 1. I.S. 2285.

6.4 USE OF SURFACE PLATE

Surface plate is generally used for jobs which are required to be made accurate. Straight or parallel lines are drawn on any job with the help of vernier, height gauge and surface gauge. A pair of "V" block is kept on it for marking of centre of a round job is also done. Besides these, surface plate is used for the following jobs:

1. To measure the smoothness of the job with the main surface of the plate.
2. To measure the right angle of a job.
3. To measure or mark angle of a job with the help of sine bar.
4. To check the jobs produced, in less time.
5. To measure a job with a dial test indicator.

Precautions

Before and after using the surface plate, the following precautions should be taken:

1. Before marking or measuring a job, its surface should be properly cleaned with a soft paper.
2. While using it, it should not be hit as there are chances of damage to its surface.
3. In order to make the marking of the job permanent it should not be used as a table.
4. Cutting tool should never be used on it.
5. It should be kept separately in the furnace.
6. After using it, oil or grease should be applied and it should be covered with a lid and kept safely.

A marking table is very precise equipment, and should be protected from damage and rust.

After use, the marking table should be cleaned with a soft cloth.

The surface of the marking table, made of cast iron, should be protected by applying a thin layer of oil.

6.5 ANGLE PLATE

This plate is made of good cast iron. It is built at 90°. It has rectangular grooves. Both of its outer surfaces are accurately planed by the scraper. If we place it on the surface plate, it makes an angle of 90°. It is used to provide support to the job from behind while measuring. In heavy jobs, nut bolts are fitted in its grooves and marking is done or they are tied on the angle plate for other machining activity is done.

Angle plate is of two types in big and small size as shown in **Fig. 6.3.**

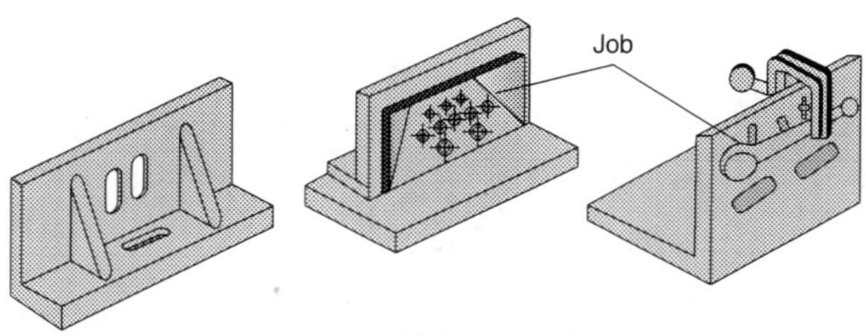

Fig. 6.3

1. Fixed Angle Plate

   It is used to give support to the job at the right angle. Job is fixed on it for carrying other machining activities.

   In some types long slots are made to fix the job with bolts and nuts.

2. Adjustable Angle Plate (**Fig. 6.4**)

   Adjustable angle plates are available to support jobs in different angles.

   It is used for doing angular marking or measurement.

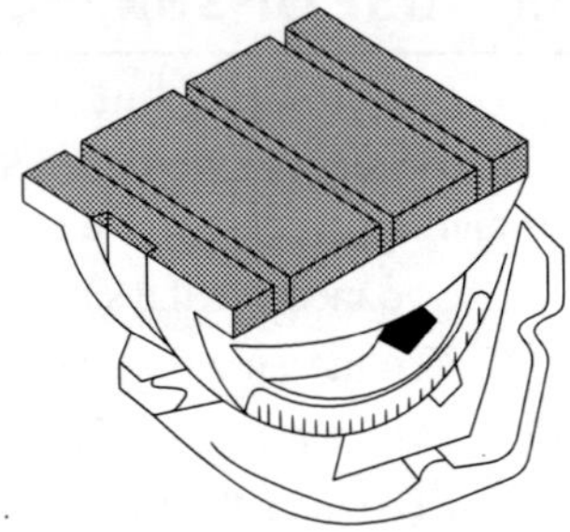

Fig. 6.4

6.6 SWIVEL TYPE ANGLE PLATE

This is adjustable so that the two surfaces can be kept at an angle. The two machined surfaces are on two separate pieces which are assembled. Graduations are marked on one to indicate the angle of tilt with respect to the other. When both the zeros coincide, the two plane surfaces are at 90° to each other. Bolts and nuts are provided for locking in position.

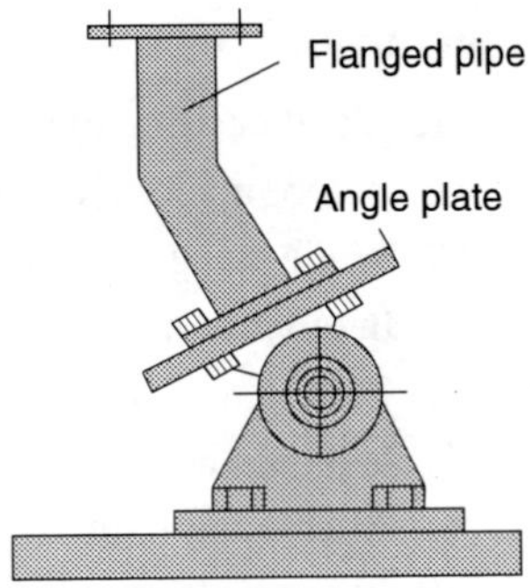

Fig. 6.5

6.7 BOX ANGLE PLATES

The applications are similar to those of other angle plates. After setting, the work can be turned over with the box enabling further marking out or machining. This is a significant advantage. This has all the faces machined square to each other.

Angle plates are available in two grades: Grade 1 and Grade 2. The Grade 1 angle plates are more accurate and are used for very accurate tool room type of work. The Grade 2 angle plates are used for general machine shop work. In addition to the above two grades of angle plates, precision angle plates are also available for inspection work.

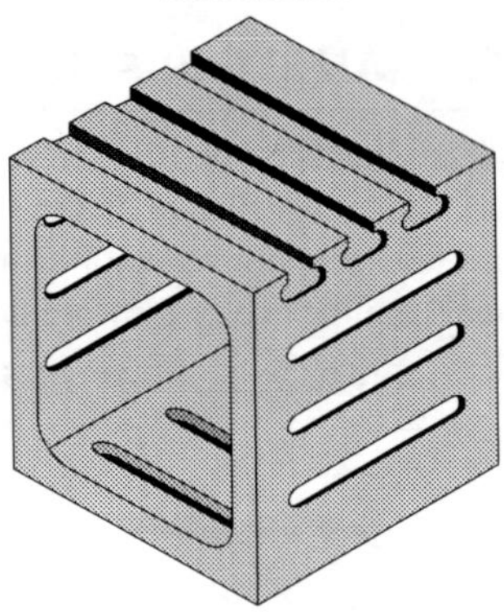

Fig. 6.6

6.8 SIZES

Angle plates are available in different sizes. The sizes are indicated by numbers. Table 1 gives the number of the sizes and the corresponding size proportions of the angle plates.

Specification of angle plates

(a) Size 6 Grade 1

Box plate will be designated as box angle plate 6 Gr 1 IS 623.

(b) Size 2 Grade 2

Angle plate will be designated as angle plate 2 Gr 2 I.S 623.

Size No.	L	B	H
1.	125	75	100
2.	175	100	125
3.	250	150	175

4.	350	200	250
5.	450	300	450
6.	600	400	450
7.	700	420	700
8.	600	500	1000
9.	1500	900	1500
10.	2800	900	2200

Angle plates should be carefully handled and maintained. Any nicks or scratches can spoil the accuracy of the angle plates.

6.9 ODD LEG CALLIPER

It is also known as Jenny calliper or Hermaphrodite calliper. One of its legs is straight and the other is bent about 1/8 to ½ inches towards inside as shown in **Fig. 6.7.**

Jenny callipers are used for marking lines parallel to inside and outside edges **(Fig. 6.8).**

Locating the centre of a round job using a Jenny calliper is illustrated in **Fig. 6.9**.

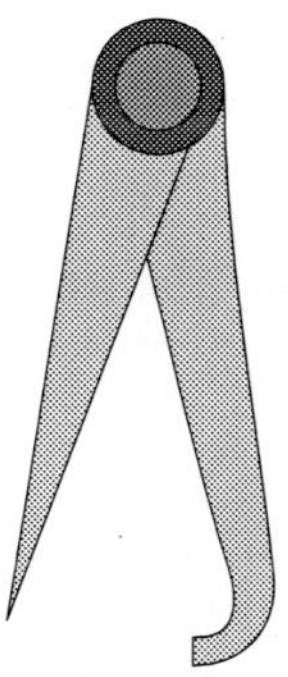

Fig. 6.7

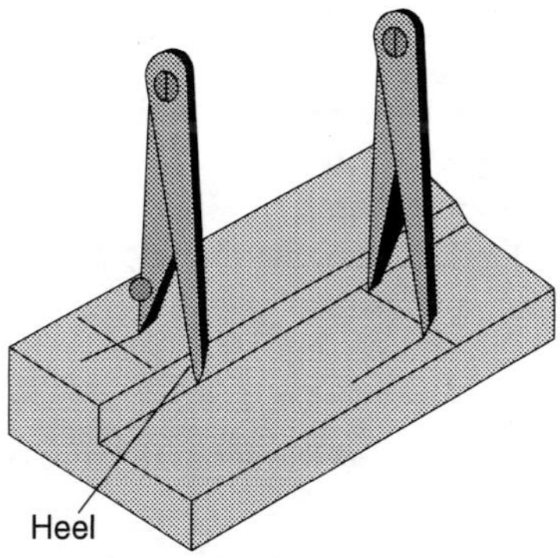

Fig. 6.8

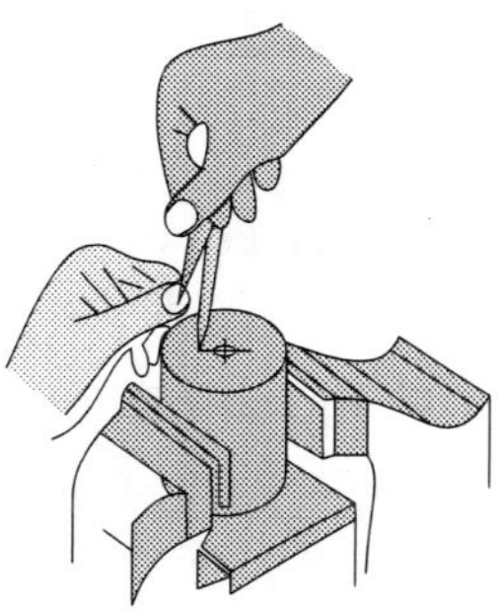

Fig. 6.9

Callipers are available with the usual bent leg or with a heel. Callipers, with ordinary bent legs, are used for drawing lines parallel along an inside edge, and the heel type is used for drawing parallel lines along outer edges.

Jenny callipers can also be used for scribing lines along curved edges. While setting dimensions and scribing lines, both legs should be of equal length **(Fig. 6.10).**

The Jenny callipers should be slightly inclined while scribing lines (**Fig. 6.11**)

While setting dimensions for accurate setting the Jenny calliper point should 'click' into the graduation **(Fig 6.12).**

6.10 DIVIDERS

A divider is similar in construction to a calliper except that both the legs are straight with sharp hardened points at the ends (**Fig. 6.13**). The tool is used for transferring dimensions, scribing circles, and doing general

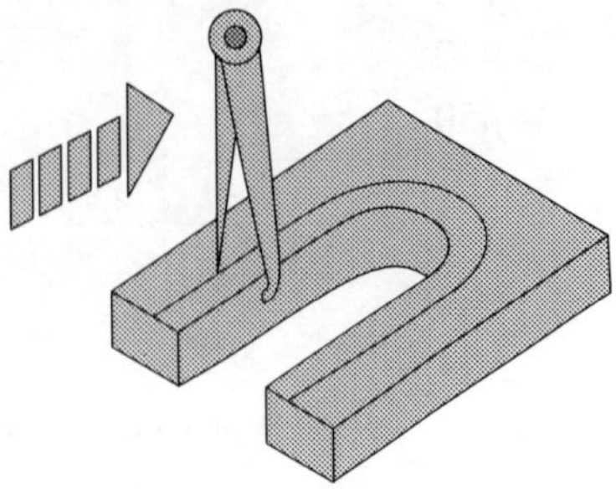

Fig. 6.10

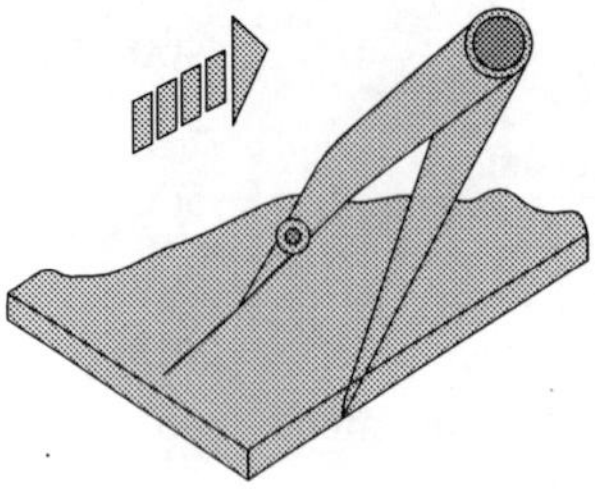

Fig. 6.11

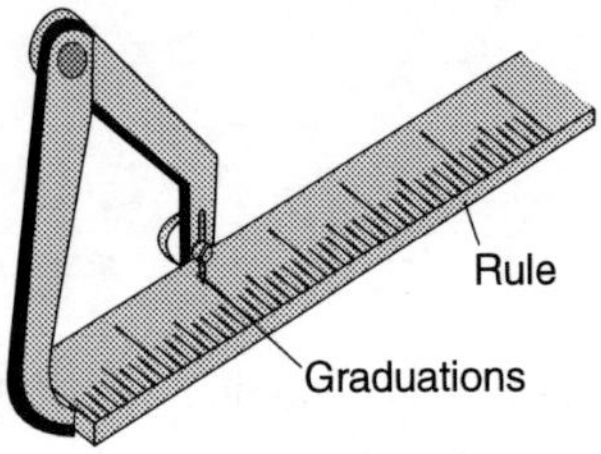

Fig. 6.12

layout work. In practice, one point is placed in the centre pop making the exact centre, and the circle or any other arc may be scribed on the job with the other point. The size is measured by the greatest distance it can be opened between the legs. Thus, a 100 mm divider open 100 mm between points. A steel scale must also be used with this divider.

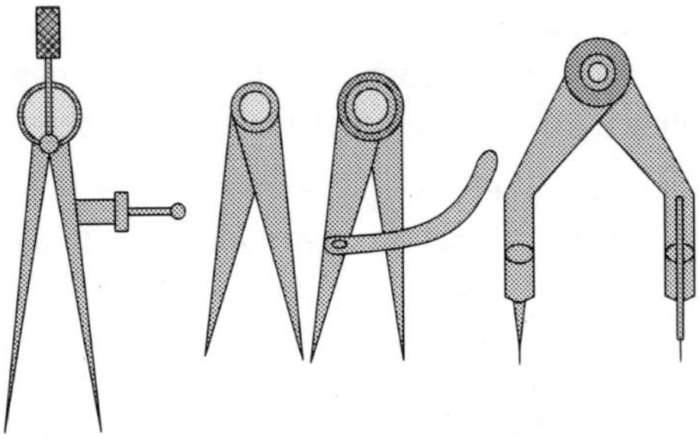

Fig. 6.13

6.11 DIFFERENT TYPES OF DIVIDERS

Dividers are used to draw circles on a job, chop or divide a line into equal parts and to mark them. These are generally made of mild steel. Its pointed ends are case hardened so that they do not wear out easily. On the basis of combination of their legs, they are classified into two parts.

6.11. 1 Firm Joint Dividers

In this type of dividers their legs are joined by a rivet and washer at the top. But the rivet is not fixed tightly so that its legs can be opened easily.

The size of dividers ranges between 50 mm to 200 mm. The distance from the point to the centre of the fulcrum roller (pivot) is the size of the divider.

For the correct location and seating of the divider legs, prick punch marks of 30° are used.

Both the legs of the divider should always be of equal length. Dividers are specified by the type of their joints and length.

The divider point should be kept sharp in order to produce fine lines. Frequent sharpening with an oil stone is better than sharpening by grinding. Sharpening by grinding will make the points soft.

Do not sharpen the divider points on grinding wheels.

6.11.2 Spring Joint Dividers

In this type of dividers, both the legs are tightened on a pin with a spring. Their legs are also joined with each other by an extra screw and split nut. Their legs can be easily opened or closed as required, by moving its nut. A knurled stud is often fixed in the centre of the spring. We hold it at the time of marking. After once setting it we can check the job several times because its size does not change itself easily. It saves time and makes the job accurate.

6.12 SCRIBER

For drawing straight lines on jobs made of metal or wood scriber is used. It is made of a steel or carbon steel wire of 3 mm to 5 mm, whose one end is straight and the other is bent at about 90° angle. It is used along with engineer square, steel foot rule, and straight edge. Its length varies from 150 mm to 300 mm.

Both its ends are pointed. For special use in various jobs, other types of improved scribers are also used. These are of the following five types as shown in (**Fig. 6.14**).

1. Ordinary scriber
2. Improved scriber
3. Adjustable sleeve scriber
4. Pocket scriber
5. Knock scriber

6.12.1 Uses and Precautions

Before drawing lines on a job coating of Prussian blue, copper sulphate or chalk should be done on its surface and it should be allowed to get dried. We should cover the point, which is not to be used with a piece of cloth, cork, etc., so that its point does not hurt our fingers. We should check the point of the legs; if it needs to be sharpened we must sharpen it with a file before using it. We should not exert excess pressure at the time of drawing lines. Scriber should be slightly bent in the direction in which the line is being extended.

Scriber was fixed by fitting a snug bolt in this pillar. For making use of it, it was adjusted with hand and drawn near the steel foot rule. It used to take more time and there was no provision of placing it on a round job. Moreover its pillar (rod) used to remain erect. As such it could not be set in an angular position. Thus, several difficulties were experienced in its use.

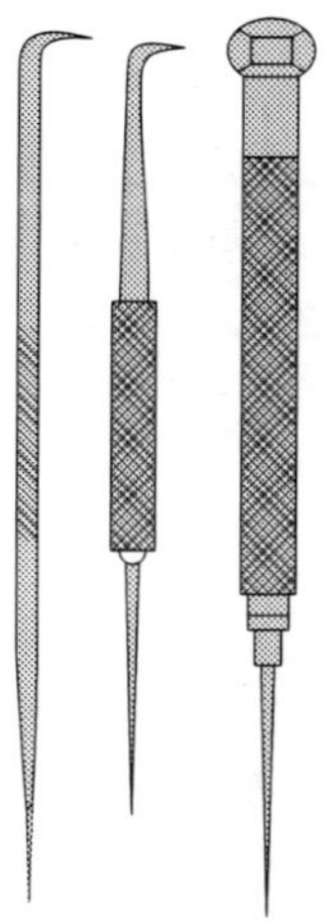

Fig. 6.14

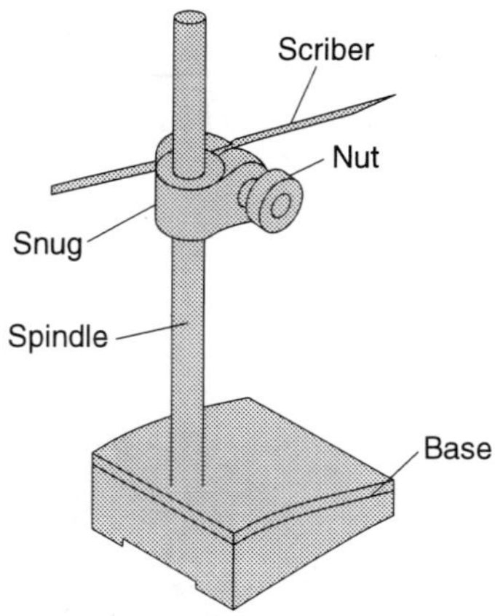

Fig. 6.15

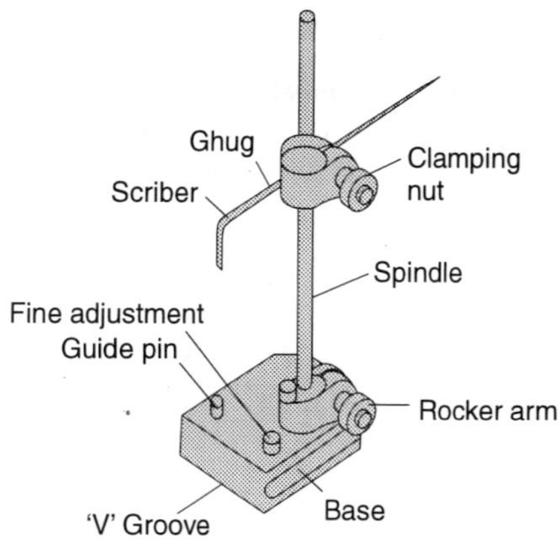

Fig. 6.16

6.13 UNIVERSAL SURFACE GAUGE

For accurate marking in less time and without difficulty universal surface gauges were made and came into use. With this gauge accurate marking can be done on almost all types of jobs (**Fig. 6.16**).

The base of universal surface gauge is made of good cast iron in rectangular form. It is slotted in "V" shape throughout its length. On the upper surface of its base, two guide pins are fixed. These pins are used for marking like an odd leg calliper for sticking it a machined surface. Because of "V" shaped slots this does not stick to the surface plate thus it can be easily pushed ahead or pulled backwards. Besides, this dirt and filth of the surface plate does not have any effect on the marking because that dirt can be transferred to the slots. It can be easily placed on the round rod. On one side of the base rocker arm is tightened with a screw. On one side of this rocker arm, pillar is fixed with a snug bolt. With the help of bolt this pillar can be bent at any angle. Scriber's point can be easily matched with any part of the steel foot rule, through the adjusting screw. Scriber is fitted in the pillar with the help of the snug bolt, which does the actual marking. Swivel post snug can be set on the pillar at any height. For this purpose a clamping nut is fixed.

Universal surface gauge is not only used for drawing lines on vertical surface. Mainly it is used for the following purposes:

1. To draw lines parallel to its base.
2. To find out the centre of cylindrical and square rods.
3. To draw lines parallel to plain arm of a job.
4. To properly focus the job on the lathe.
5. To draw parallel line like height gauge on normal job.

Precautions

1. We should keep the swivel post snug of surface gauge tight.
2. Scriber should be bent (inclined) at less than 90°.
3. While drawing similar parallel line, we should check them from time to time.
4. Scriber should not be kept excessively outside the snug.

6.14 TRAMMEL

Like a divider, a trammel is also used for drawing a circle chop or two parallel lines. It has a round or square rod, which is called a trammel bar or beam.

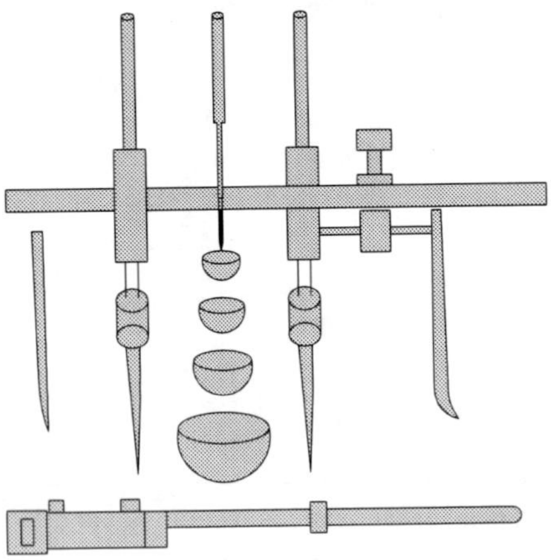

Fig. 6.17

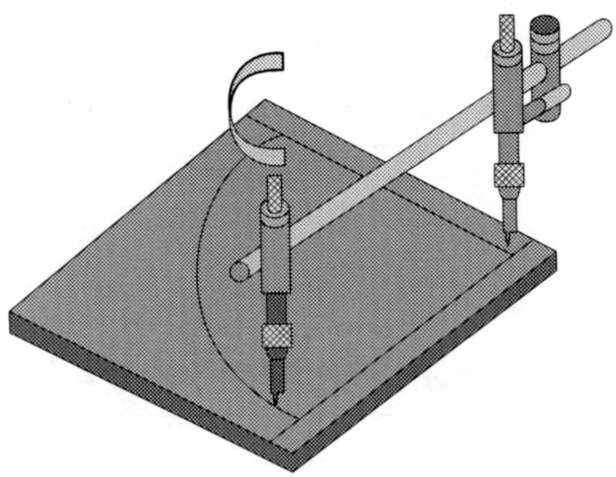

Fig. 6.17a

There are two movable trams on this beam. It can also be used as an inside or outside calliper by fixing points of different ·types. It has the following main parts.

1. Beam
2. Tram
3. Clamping nut
4. Chuck
5. Point
6. Carrier
7. Adjusting screw
8. Swivel handle

Beam is generally 15 mm to 50 mm in length. In good trammels graduation is indicated in mm. Ball point is fitted in its tram; from a bored centre a circle or chop is drawn. For extraordinary big jobs additional beams are also attached.

6.15 'V' BLOCK

For marking on a round or rectangular job, 'V' block is used to clamp it. With its help, drilling, punching and other machining operations are easily carried out on such jobs. It is made of cast iron or tool steel. Its surfaces are well machined and ground. A 'V' shaped slot of 90° angle is made in the centre on both sides. On the remaining two sides rectangular grooves are made in which clamp is fixed. These are generally of the following three types and are available in pairs.

6.15.1 Simple 'V' Block

This is rectangular block which has a 'V' shaped groove on its upper surface. Heavy round jobs are kept over it for marking or machining. There is no provision to clamp the job because of heavy weight of the job, it remains fixed.

The included angle of the 'V' is 90° in all cases. 'V' blocks are finished to a high accuracy in respect of dimension, flatness and squareness.

Different types of 'V' blocks are available. As per B18, there are four types, as listed below.

***Single level single groove 'V' Block* (Fig. 6.18a)**

This type has only one 'V' groove, and has single groove (slots) on either side. These grooves are for accommodating the holding clamps.

***Single level double groove 'V' Block* (Fig. 6.18b)**

This type will have one 'V' groove, and two grooves (slots) on either side for clamping in two positions.

***Double level single groove 'V' Block* (Fig. 6.19)**

In this case, the 'V' block will have two 'V' grooves on the top and bottom, and a single groove for clamping on either side.

***Matched pair 'V' Block* (Figs. 6.20 and 6.21)**

These blocks are available in pairs which have the same size and the same grade of accuracy. They are identified by the number or letter given by the manufacturer. These sets of blocks are used for supporting long shafts, parallel on machine tables or marking off tables.

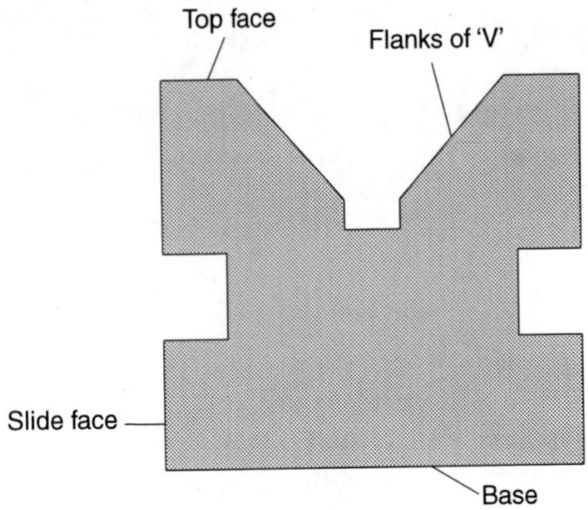

Fig. 6.18a

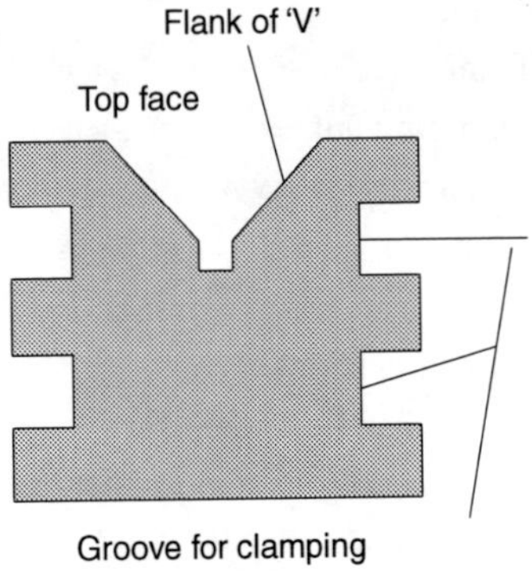

Fig. 6.18b

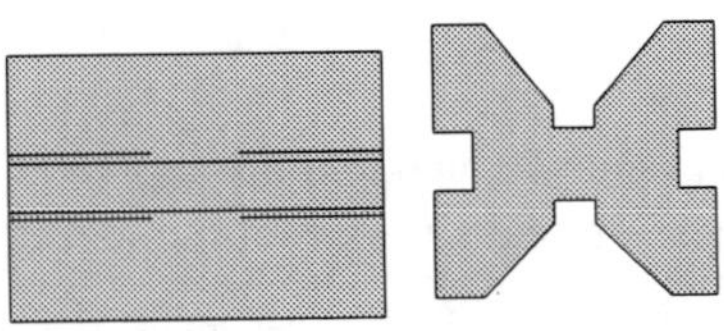

Fig. 6.19

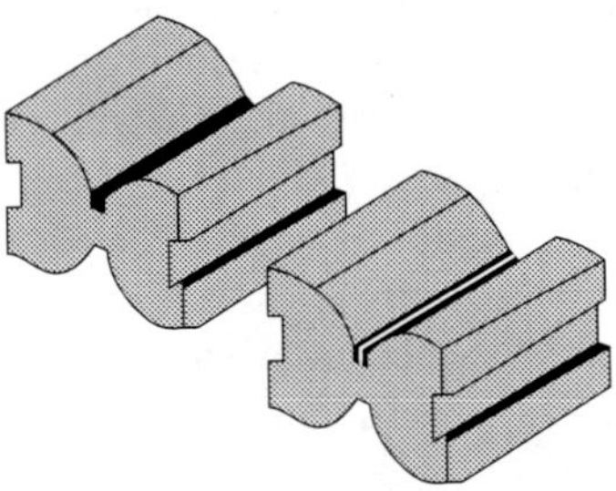

Fig. 6.20

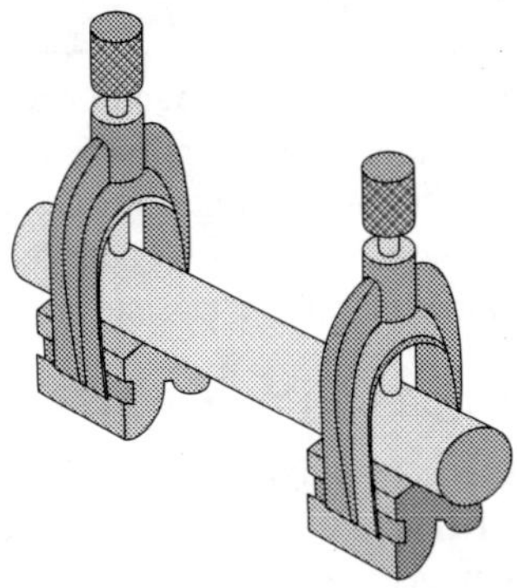

Fig. 6.21

6.16 GRADES AND MATERIALS

'V' blocks are available in Grade A and Grade B.

6.16.1 Grade A 'V' Blocks

These are more accurate, and are available only up to 100 mm length. They are made of high quality steel.

6.16.2 Grade B 'V' Blocks

These blocks are not as accurate as the ones in Grade A. These blocks are used for general machine shop work.

These blocks are available up to 300 mm length. These 'V' blocks are made of closely grained cast iron.

6.17 CLAMPING DEVICES FOR 'V' BLOCKS

'U' clamps are provided for holding cylindrical jobs firmly on 'V' blocks **(Fig. 6.22).**

Designation

'V' blocks are designated by the nominal size (length) and the minimum and maximum diameter of the workpiece capable of being clamped, and the grade and the number of the corresponding B.1.8. Standard.

In the case of matched pairs, it should be indicated by the letter M.

For 'V' blocks with clamps it should be indicated as, 'with clamps'.

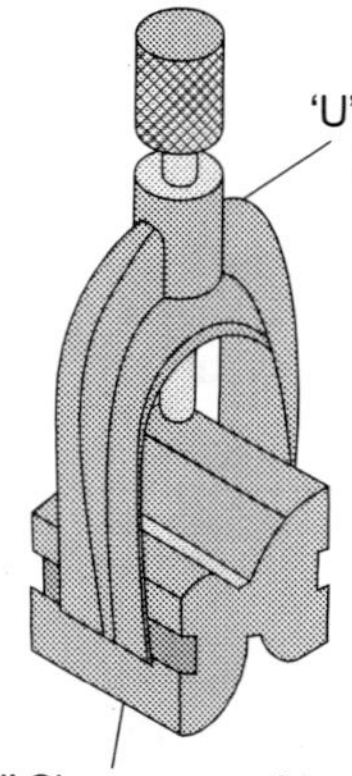

Fig. 6.22

Example

A 50 mm long (nominal size) 'V' block capable of clamping workpieces between 5 to 40 mm in diameter and of Grade A will be designated as:

'V' Block 50/5-40 A - 1.8.2949.

In the case of a matched pair, it will be designated as 'V' Block M 50/5-40 A 1.8.2949.

For 'V' Block supplied with clamps, the designation will be 'V' Block with clamp 50/5 40 A 1.8. 2949.

6.18 MAGNETIC 'V' BLOCK

In this 'V' block which looks like an ordinary 'V' block, magnetic power is provided. Because of magnetic effect the job remains stuck to the block.

A switch is provided to start or put off the magnetic effect. When we put the switch in "on" condition the job gets stuck to the block and in "off" condition the magnetic effect ends.

6.19 MARKING ON SHEET METAL

Sheet metal work is generally regarded as the working of metal from 16 gauges down to 30 gauges with hand tools and simple machines to shape and join. A large number of hand tools are used in sheet metal work.

6.19.1 Steel Square

A steel square is an 'L' shaped piece of hardened steel with graduation marks on the edges for measuring and marking. It is used for marking in the perpendicular direction to any base line **(Fig. 6.23)**. The short arm of the square or blade is called the tongue and the long arm is called the body or blade and the corner is called the heel.

Though 'L' square is normally 90°, squares are also available in 60°, 45° and 30° in different sizes.

The size of an 'L' square is specified by the length of the body and the angle.

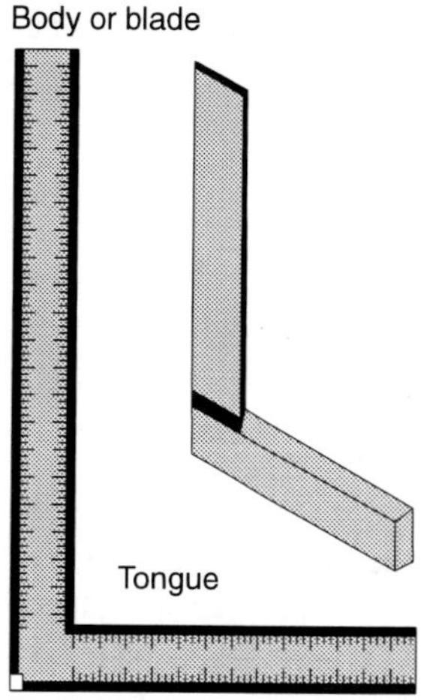

Fig. 6.23

6.19.2 Scratch Awl and Scriber

These are used for marking or layout purposes. A scratch awl is used where the marking lines are creeper, and the pressure is exerted over the wooden handle **(Fig. 6.24)**.

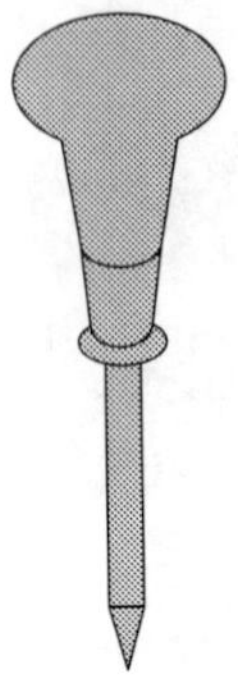

Fig. 6.24

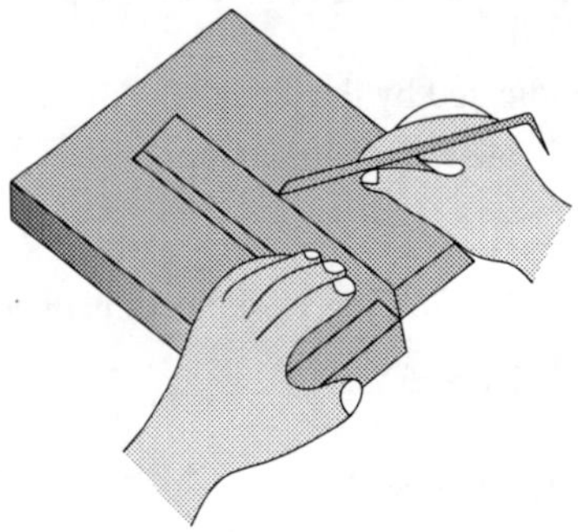

Fig. 6.25

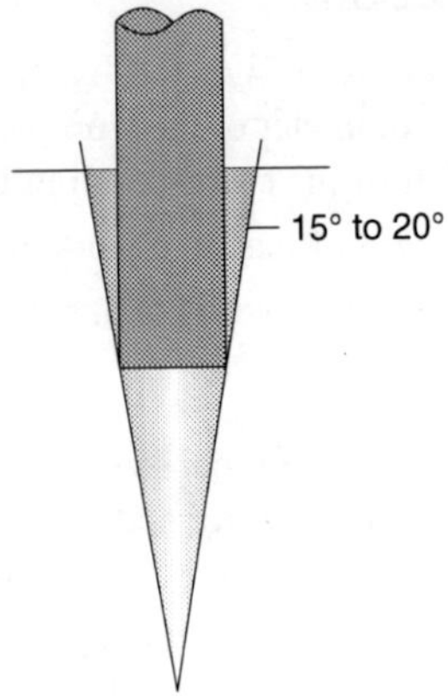

Fig. 6.26

The scriber is used to make sharp clean lines on sheet metal surfaces. It is sometimes called the metal worker's pencil. The body of the scriber is knurled for a proper grip (**Fig. 6.25**).

The tips of the scriber and scratch awl are hardened and tempered. They are tapered to 15° to 20° to get a sharp point (**Fig. 6.26**).

6.19.3 Compasses

Compasses of different types are used in sheet metal work. They are used for marking dimensions accurately. Compasses are used for transferring the dimensions from the rule to the work (**Fig. 6.27**).

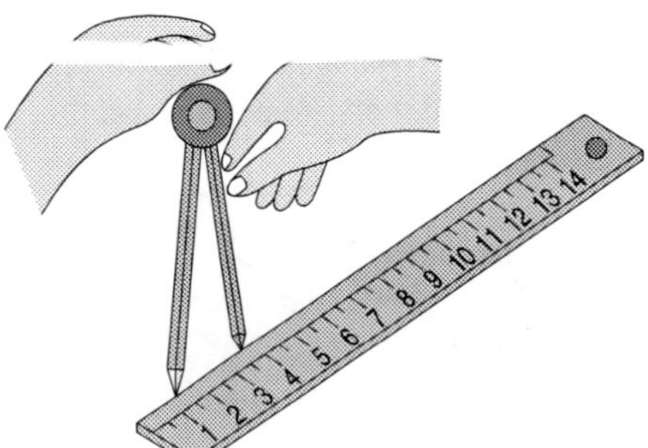

Fig. 6.27

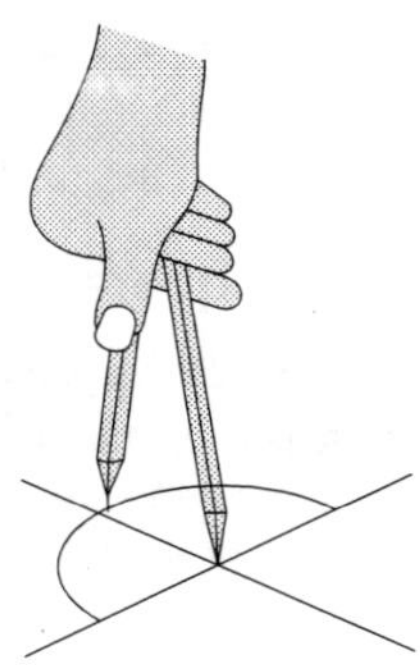

Fig. 6.28

6.19.4 Punches

After drawing marking lines on a job by scriber or surface gauge, different kinds of punches are used to make these markings permanent, because there is possibility of the lines marked with a scriber getting effaced. These punches are generally made of carbon steel or tool steel. Mainly the following punches are used:

1. Centre punch
2. Dot punch
3. Prick punch
4. Solid punch

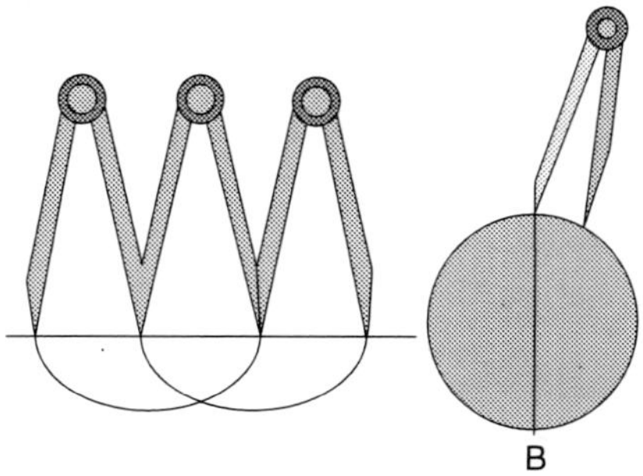

Fig. 6.29

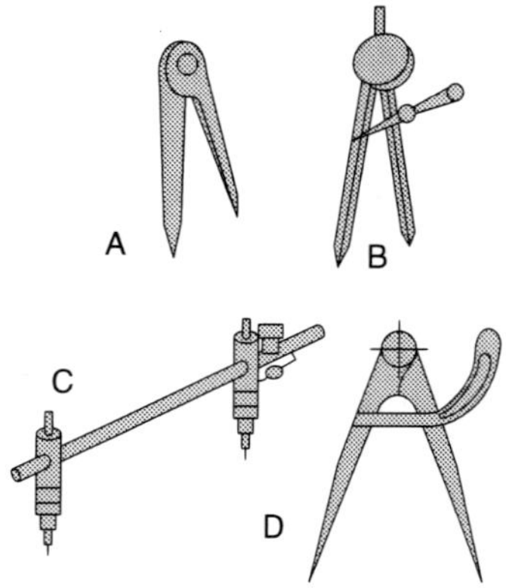

Fig. 6.30

5. Hollow punch
6. Pin punch
7. Bell punch
8. Automatic punch

Centre Punch (Fig. 6.31): These are made of high carbon steel. These are made hard and given temper treatment. Their point is made at 90° angle and the handling part is knurled. Its length is from 35 mm to 125 mm and its width is from 4 to 12 mm. Their use is generally made to make a centre at the drilling point so that drill bits point can rest there properly. If this centre is not made by centre punch, there are chances of slipping of the point and drilling at a wrong point.

Dot Punch (Fig. 6.32): This is by and large similar to the centre punch. The main difference is that its point is at 60° angle. It is used to make permanent the marking done by scriber so that these lines do 'not obliterate

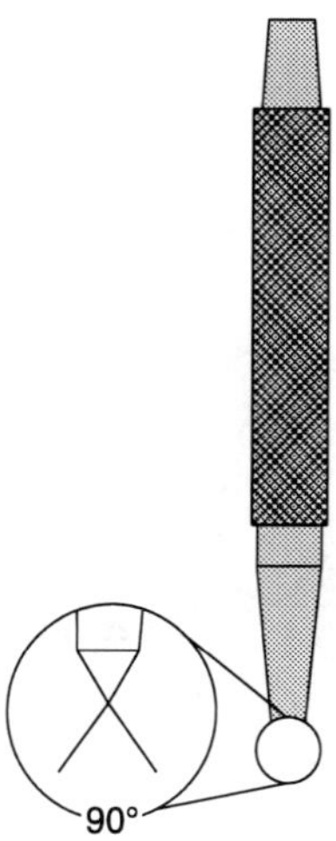

Fig. 6.31

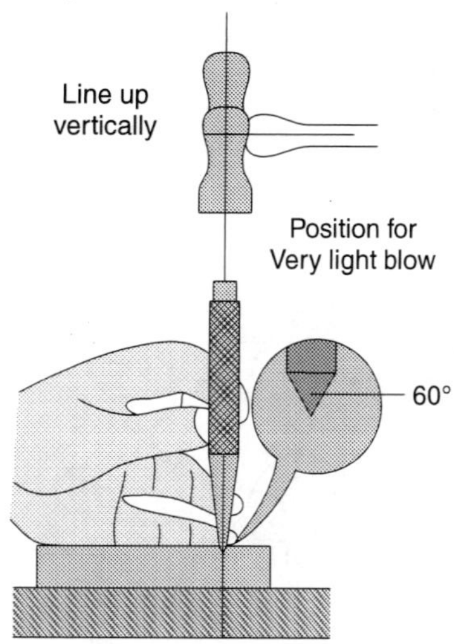

Fig. 6.32

at the time of work. Its length is from 80 to 100 mm and its thickness from 3 to 5 mm. They are made of tool steel or high carbon steel, hardened, tempered and ground. In the dot punch, the depth of penetration is less. This punch is used for drawing radius, arc, etc.

A dot punch resembles a centre punch, but differs in the included angle. The rest of the parts are similar to those of a centre punch. The included angle of the dot punch is 60°.

Hollow Punch (Fig. 6.33): It is a special type of a hardened steel punch with hollowness. It is used for cutting holes in metal, cardboard, fabric, non-metallic materials. These are available in different sets for holes of different sizes. As they are hollow, the metal which is cut with them falls out through the grooves built on its side. Thus, we can make a number of holes in little time and the holes are quite clear. The body is knurled for good grip. In the front it is hollow. Provision is made to remove the waste pieces (i.e., below the body).

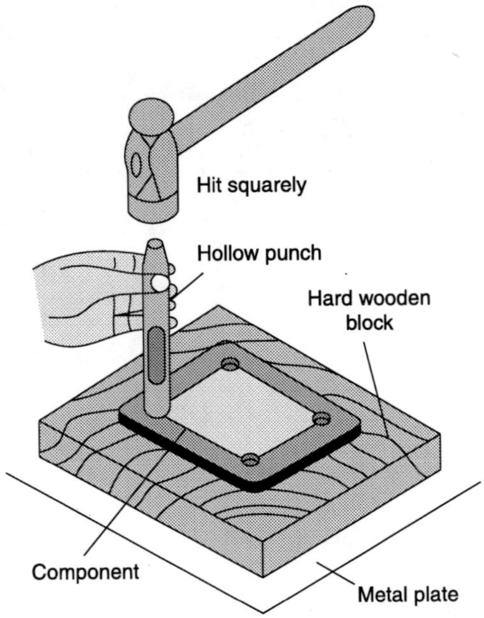

Fig. 6.33

Hollow punches are generally used on cork sheets, leather sheets, rubber sheets, etc., holes are made by using a hammer to punch to pierce the material to be pinched.

They are available in sizes ranging from 1.5 mm to 13 mm in steps of approximately 1 mm.

Prick Punch: This is also made of high carbon steel. Prick punch is also made hard and tempered.

Its point is made at 30° angle. It is used for marking on jobs of thin metal sheets or to make holes in them.

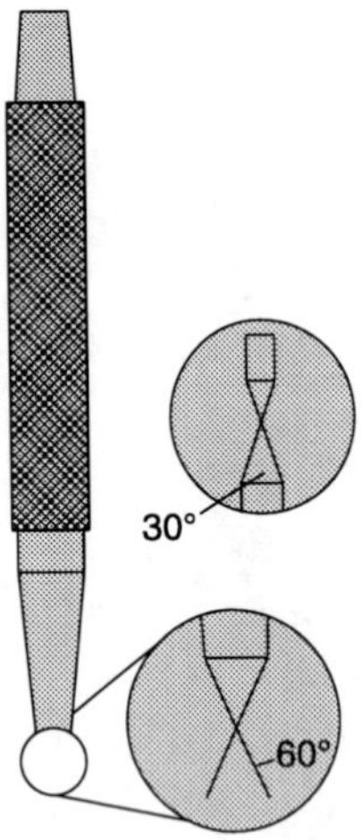

Fig. 6.34

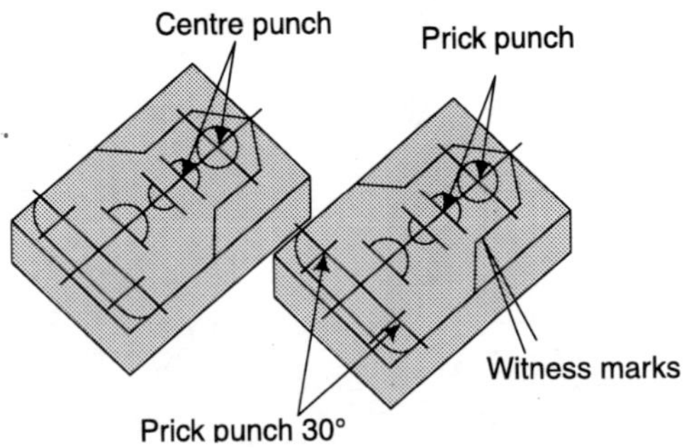

Fig. 6.35

The angle of the prick punch is 30° or 60°. The 30° point punch is used for marking light punch marks needed to position dividers. The divider leg will get a proper seating in the punch mark. The 60° prick punch is used for marking witness marks.

The witness marks should not be too close to one another.

Solid Punch: Solid punch is used for boring holes in blacksmithy and sheet metal. These are used in different measurements according to the jobs. These are made of high carbon steel, hardened and tempered. While using these on hot jobs it is essential to cool them in cold water. These are used in rough works because holes made with solid punch are not exactly according to measurements and also not clean.

Pin Punch (Fig. 6.36): This punch is somewhat different from other punches. It is comparatively larger in length. Instead of being pointed in shape these are parallel. It is used for tight fitting pins.

Pin punches are made out of high carbon steel, hardened, tempered and ground to the correct diameter. The pin punch has a knurled body and shank portion which helps to drive in or drive out dowel pins or locating pins.

These punches are available in standard sizes, ranging from 3 to 12 mm. The length ranges from 100 to 200 mm.

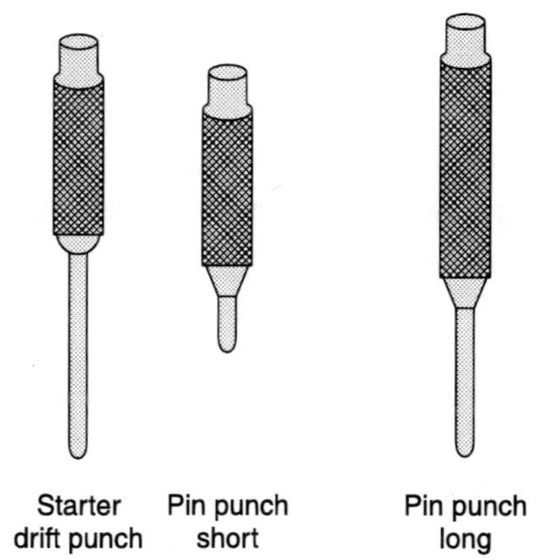

Fig. 6.36

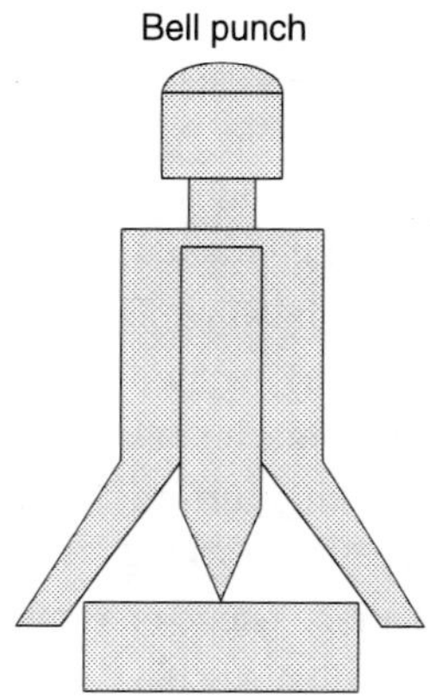

Fig. 6.37

Bell Punch (Fig. 6.37): As per its name it is like a small bell, in shape. Its size is considered according to the circumference of the bell. It is used to find out the centre of round jobs and for light marking. In this circumference a punch is fixed with the help of a spring, which is like a centre punch. For using it its head is hit with a hammer and because of punch spring it raises alone.

It is provided with a plunger compression spring, body, guide, etc. Different diameters can be accommodated and the centre can be located very easily. Finding centres is easiest and quickest on cylindrical shafts.

Automatic Punch: This is also a special kind of punch. It is made of high carbon steel which is hardened or tapered. There is no need for a hammer, etc., for using this punch.

A main punch is fitted with its body with a spring; spring remains settled through a lever head. When marking is to be done, its lever head is pressed. Through this pressure in punch would be pressed and do the marking: The blow of the punch can be limited or controlled with the lever head. Its main punch is built at 60° or 90° angle.

Precautions

In order to ensure correct and accurate marking with any type of punch, the following precautions should be taken:

1. Selection of punch should be made according to the make-up of the job or metal.
2. Light hammer should be used for a light punch.
3. Do not use an ordinary punch or a hard metal.
4. Use a punch after seeing its angle.
5. If point of the punch is damaged, it should be repaired and then the punch should be used.
6. Hollow punch should not be used on thick sheets.

6.20 METHOD OF MARKING

The efficiency of a job depends to a considerable extent, on the marking done on a job, before it is machined. If marking is not accurate, the job would be of wrong size, it may become slightly more or less than the size required. In a way, marking job is a matter of experience and different methods have been evolved for marking on different types of jobs. There are different methods for marking a round and a square job. Although no specific rules can be framed for marking yet for general marking the following method is adopted:

1. Keep the drawing of job, which we are supposed to make and understand it properly.
2. Before marking, two adjoining sides and a surface of the job should be prepared after filling, scraping and other machining.
3. The marking surface should be coated with a thin paste of Prussian blue, copper sulphate or chalk and dried before marking.
4. If a surface gauge or vernier gauge to be used for marking the job be properly placed on the surface plate, with angle plate and 'V' block.
5. Marking scriber or punch should be pointed.
6. Marking tools should be selected according to the job.
7. In case a chop or circle is to be drawn on the job, a deep mark should be made at the centre of the job with the help of a centre punch otherwise there are chances of slipping of divider.
8. If drill marking is necessary, we should draw a circle of the same size outside the centre and from the same centre another bigger circle should also be drawn so that we can know that drill is being done at the proper place.
9. If a line has to come up to a cut or slot then we should fill this gap with lead, etc.
10. For marking of a circle, four marks should be made opposite each other with a dot punch at its circumference.
11. After marking is done according to the drawing, it should be checked again.
12. When marking is done accurately, it should be made permanent with a dot punch or any other punch, as the case may be. But before making the marking permanent other helping lines should be erased.

6.21 DATUM LINE AND CUMULATIVE ERROR

When the marking is done progressively **(Fig. 6.38),** the maximum distance from the datum line AB to the 5th hole is 75.5 mm. The minimum distance from the datum line AS is 74.5 mm.

There is a tolerance of 1 mm in size. Assembly of matching parts and the interchange ability may not be possible in such cases.

To avoid this error, each feature of the component should be marked out individually from the edge AB, as shown in **Fig. 6.39**. In this case no hole will be more than the specified deviation of ± 6.1 mm and there is no possibility of build up or cumulative error.

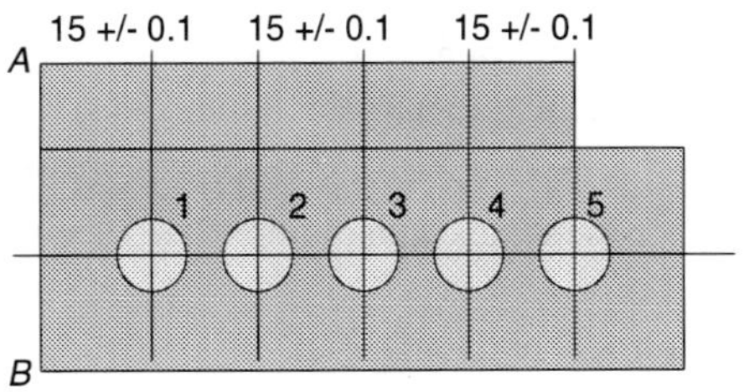

Fig. 6.38

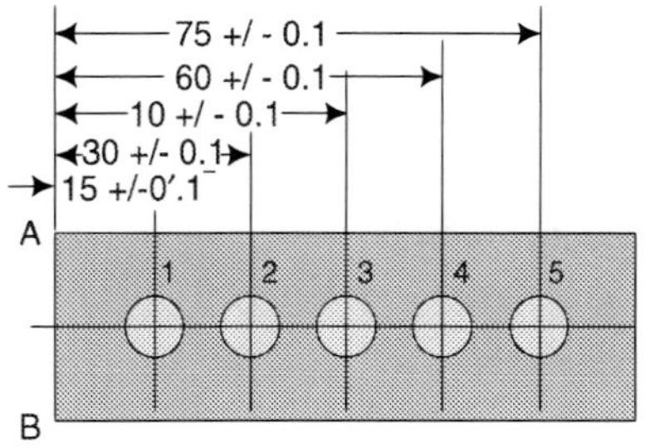

Fig. 6.39

Datum line and datum surface

Datum line is a reference line from which all other lines are marked.

This can also be a reference surface or plane for marking or measurement.

For most components two datum lines are required, one horizontal, and the other vertical; the intersecting point provides the common position **(Fig. 6.40).**

The surface plate or marking table top surface provides the datum surface. All measurements are made from this surface. All lines scribed by the scribing block or vernier height gauge will be parallel to this surface **(Fig. 6.40).**

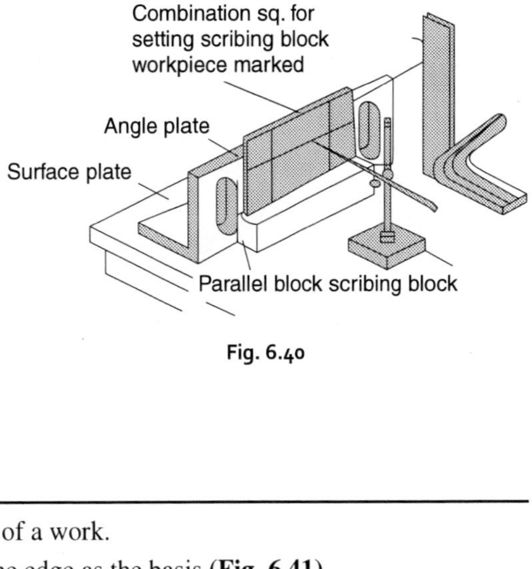

Fig. 6.40

6.22 TYPES OF REFERENCE

Edge datum: The datum surface in this case is the edge of a work.

All markings and measurements are carried out using the edge as the basis **(Fig. 6.41).**

Point datum: In this case, the reference is a point for marking and measurement **(Fig. 6.42).**

Line datum: Use of a centre line as a datum is an example of this **(Fig. 6.43).**

Letter and number punch: These hardened and tempered steel punches are used to stamp identifying symbols, letters or numbers as required on the work.

They are obtainable with symbols ranging in size from 0.8 to 13 mm.

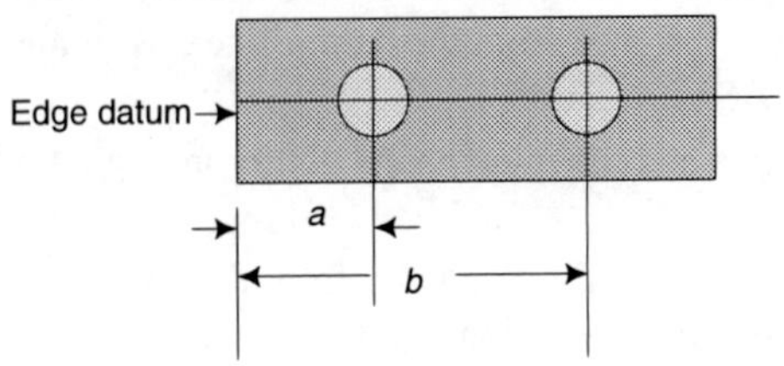

Fig. 6.41

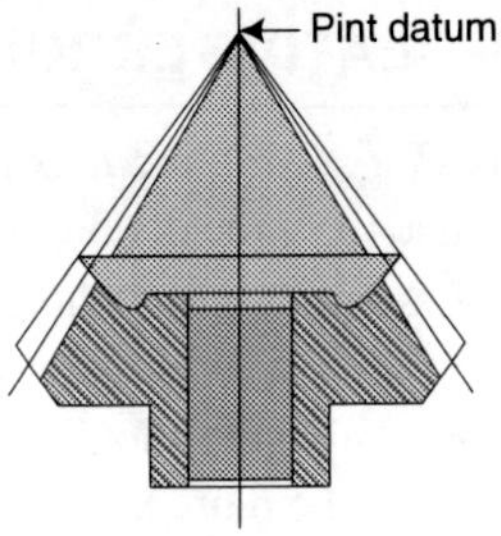

Fig. 6.42

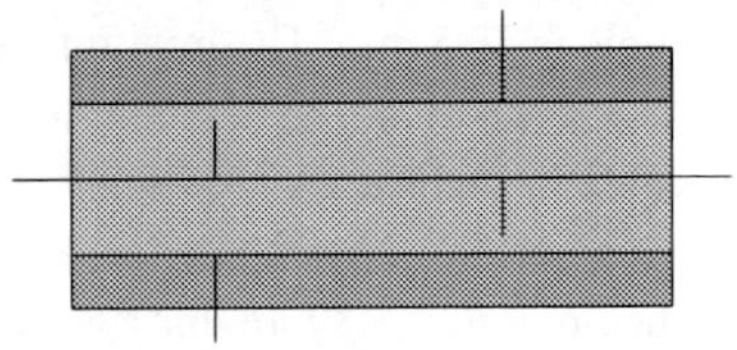

Fig. 6.43

They are kept in boxed sets.

Use a file on the work to be stamped to check the work is softer than the punch. Any attempt to stamp the hard material would damage the punch. Use an electric pencil or acid etching to mark hard materials **(Fig. 6.44).**

Each symbol must be made with a single blow. A second blow gives a distorted second impression.

Letters such as M and W may require firmer blows to produce the same depth of impression such as letters I and T can make.

The depth of impression for a given blow varies with the softness of the material.

Position the punch so that the symbol will be in line, square, correctly spaced and the correct way up **(Fig. 6.45).**

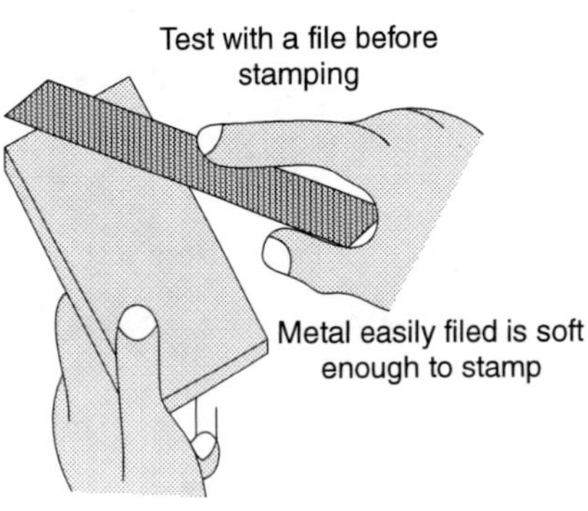

Fig. 6.44

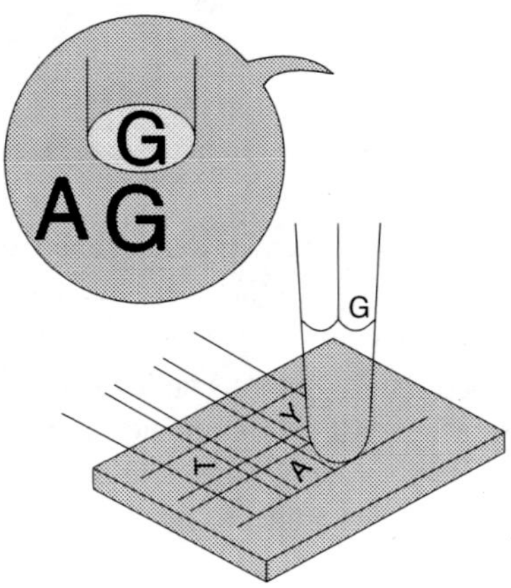

Fig. 6.45

Hold the punch in a vertical position **(Fig. 6.46).**

Hold the hammer vertically above the punch (**Fig. 6.46).**

Watch the point of the punch.

Strike the punch squarely with one firm blow.

Wad punch (Fig. 6.47): Wad punches are used for large diameter holes. They are similar to the hollow punches in their application.

The following steps will help in using these punches in a correct way.

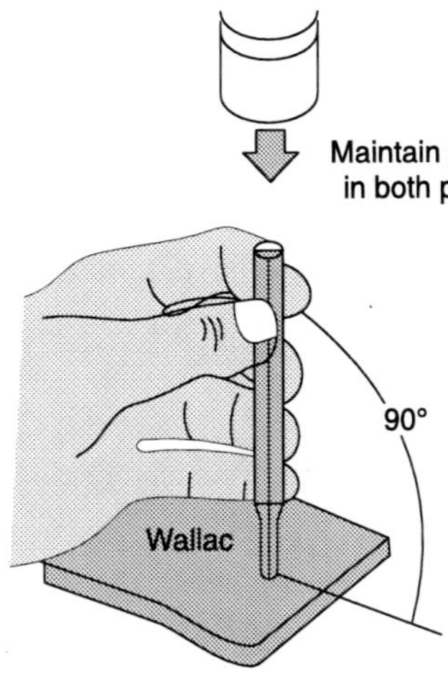

Fig. 6.46

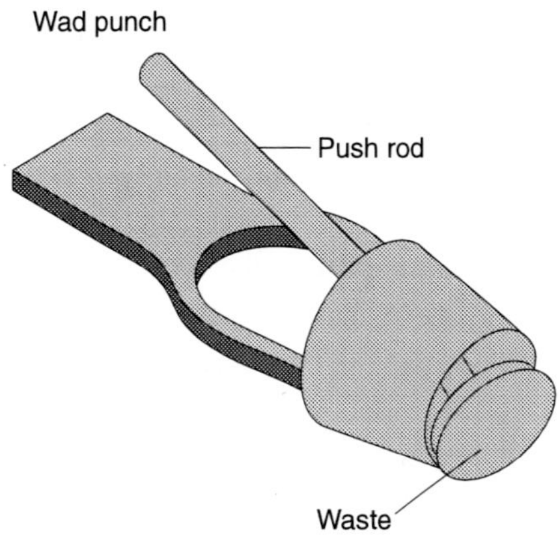

Fig. 6.47

Place the material on a rigidly supported block of wood (with the end grain up), or lead.

Position the punch accurately.

Hit the punch until the disc is cut from the material.

After cutting several holes, use a metal rod to push out the waste discs (**Fig. 6.47**).

EXERCISE

1. What is the importance of marking tools in making the job?
2. List the marking tools used by a fitter.
3. Draw a sketch of surface plate and explain it.
4. Explain the uses of an odd leg calliper.
5. Draw a sketch of trammel and explain its functions.
6. How do we find out the centre of a stick by using a divider?
7. In what ways a divider is different from a trammel?
8. Explain briefly the different types of punches.
9. What are the functions and advantages of a universal surface gauge?
10. Why the use of "V" block is essential?
11. What are the important points to be kept in mind while marking?
12. What do you understand by 'line standards' and 'end standards'?
13. How are 'end standards' derived from 'line standards'? Give examples of these two types of standards.
14. What are measuring, precision measuring, and gauging?
15. Describe the following: (1) spring calliper; (2) hermaphrodite calliper; (3) transfer calliper and (4) telescopic gauge. Give their uses.

16. Sketch and describe the working principles and uses of the following: (1) outside micrometer, (2) inside micrometer, and (3) depth micrometer. What is the limitation of a micrometer?
17. Explain the theory of a vernier scale. What is the principal advantage of a vernier calliper?
18. What is meant by measurement by comparison? Outline briefly the principles of working of: (1) a mechanical, (2) an optical and (3) a pneumatic comparator. Also discuss briefly their relative merits.
19. Describe the use of: (1) slip gauges, and (2) angle gauges, for measuring and comparing lengths and angles.
20. Write notes on: (1) combination set, (2) dividing head, (3) bevel gauge, (4) bevel protractor and (5) sine bar.
21. Explain how a spirit level can be used for angular measurement.
22. Write brief explanatory notes on 'autocollimator'.
23. Sketch and explain the optical path of a 'profile projector'. Discuss the uses of such projectors.
24. Describe the principle and application of interferometry.
25. How would you measure the surface roughness of fine finish surfaces? State and explain the working of an optical flat.
26. Discuss how would you proceed to check; (1) flatness, (2) squares, (3) roundness, and (4) concentricity.
27. What do you understand by limit gauging? Distinguish between workshop and inspection gauges.
28. Describe the use of limit gauges with suitable sketches.
29. Describe the ways air gauges operate. What are their advantages?

Cutting Tools

Chapter 7

7.1 CHISELLING

7.1.1 Fundamental of Chiselling

Plates, pieces of bar steel, profile, and tool steel and metal parts can be separated by chiselling. Rivet heads give way to the pressure of a chisel, and a few strokes of a hammer cleave non-detachable nuts of rust-in screws.

Rough workpieces can be rough-treated by chip removal by means of a chisel. Rivets can be hollowed and castings as well as welding seams can be cleaned and burred with a chisel.

7.1.2 Operation of a Chisel

Separating and chip removal by means of chisel is based on the wedge effect.

Under the pressure of the hammer strokes, the cutting edge or the chisel vertically penetrates into the material. The material withstands the side faces of the penetrating chisel edge and this resistance increases with the degree of tensile strain of the material. The particles of the laterallly displaced material upsets like a bulb (**Fig. 7.1**).

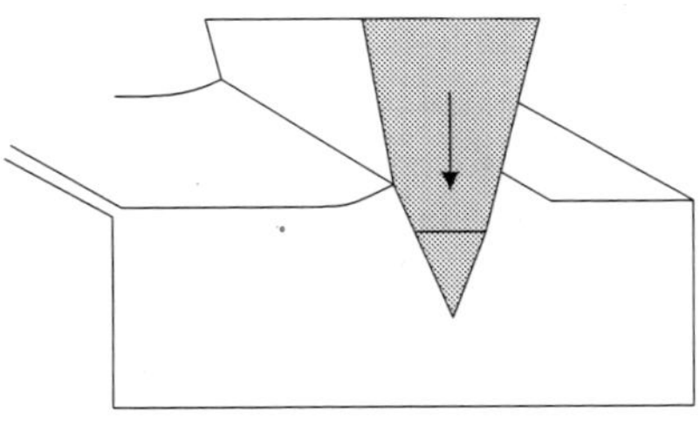

Fig. 7.1

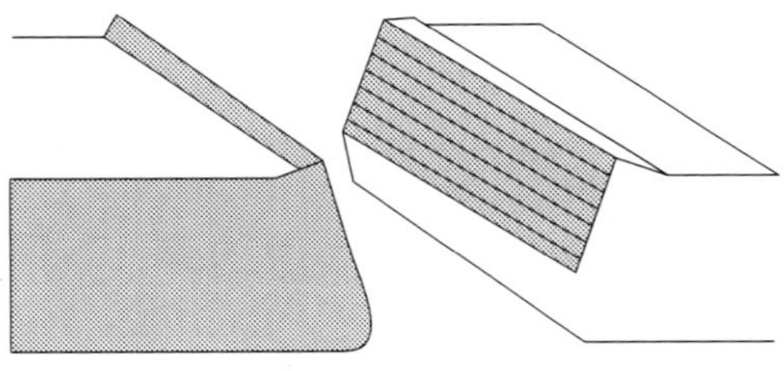

Fig. 7.2

With deeper penetration, the separating power of the cutting edge of the chisel is greater than the coherence power of the material, so that the workpiece tears at the separating line (**Fig. 7.2**).

The slim chisel to be driven into a material requires less power for displacing a small quantity of material than a blunt wedge does. A slim wedge, however, can easily break (**Figs. 7.3 and 7.4**).

In order to avoid breaking of the conically shaped cutting edge of the chisel, the lip angle (**Fig. 7.5**) at the chisel cutter should be adapted to the hardness of the material.

Penetration of a broad chisel cutter requires great efforts, particularly when cutting hard materials.

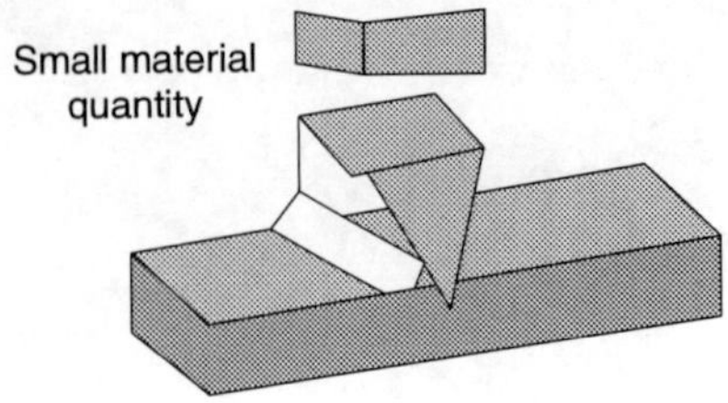

Fig. 7.3

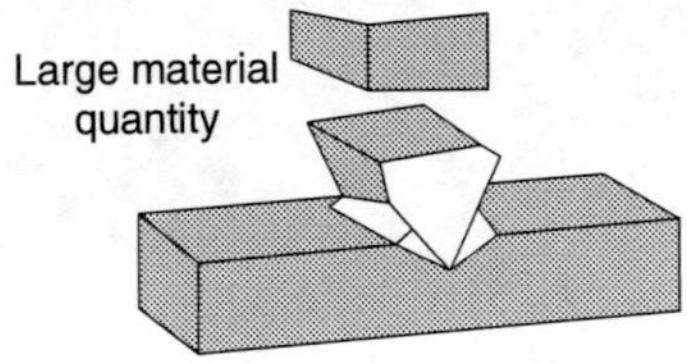

Fig. 7.4

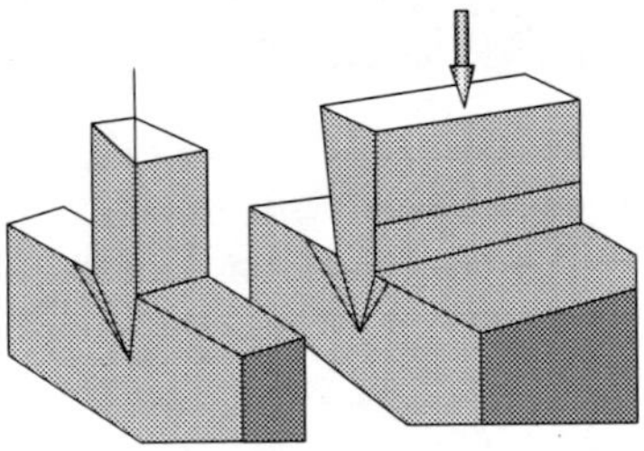

Fig. 7.5

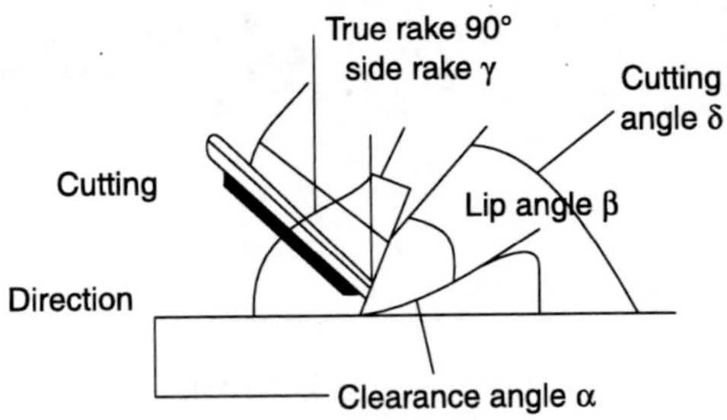

Fig. 7.6

Inclined penetration of the conically cutting edge of the chisel into the material and simultaneous movement parallel to the cutting surface produces a chip (**Fig. 7.6**). Thereby the material in front of the cutter is upset and pulled down. The rake of the cutting edge, that is, the cutting face (true rake) removes the chip and separates it. In this operation the cutting edge is kept at a determined angle to the workpiece.

7.1.3 The Angles at the Cutting Edge During the Cutting Operation

A cutting effect comes about if the cutting angle δ (clearance angle α plus lip angle β is smaller than 90°. The clearance angle α which should be about 8° to reduce the friction between the rake and the cut surface reduces heating up and hence, a longer service life of the cutting edge.

If the chisel is set at a too inclined position (large starting angle or clearance angle α it penetrates too deeply into the material (**Fig. 7.6**).

With a too low position of the chisel (too small clearance angle α it will slide out from the workpiece (**Fig. 7.7**). The result will be a surface that is not neat because of the frequent attacks of the chisel.

The thickness of the chip depends on the depth of penetration of the chisel cutter. The degree of the side rake (γ) between the cut surface and the imaginary face perpendicular to the cutting surface is a determining

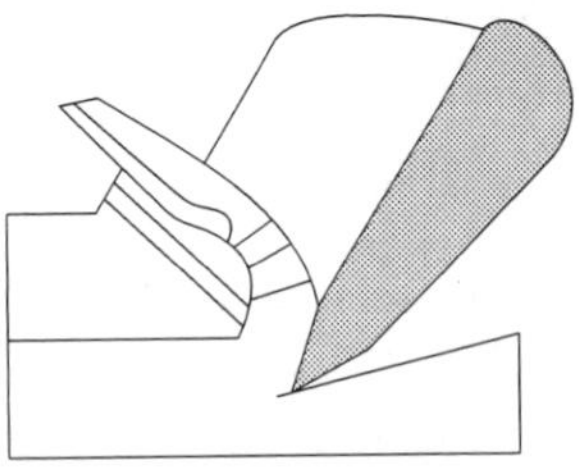

Chisel with too sharp inclination

Fig. 7.7

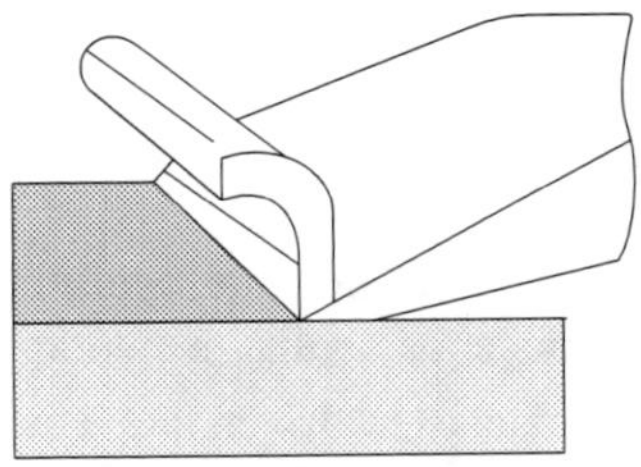

Chisel position too low

Fig. 7.8

factor for the surface quantity of the treated face. A small side rake ($\gamma = 0°$ to $8°$) produces a marked upsetting of the material in front of the cutting edge. The small chip, loosens from the material in the form of "tearing chips". The surface is rough.

Chipping at a large side rake from 20° to 30° results in a coherent "flowing chip". The treated surface is smooth.

For making any new job, after marking, it is prepared in the desired form with the help of cutting tools. Besides this, cutting tools are also used in repairing jobs for cutting, scraping or chipping. Different kinds of cutting tools are used according to the jobs. Exactness of the job largely depends on the use of proper and right kind of cutting tools. Therefore, it is essential to have adequate knowledge about the different kinds of cutting tools. For cutting, scraping, chipping or drilling a fitter, a turner or machinist uses the following tools:

1. Chisel
2. File
3. Scraper
4. Hacksaw
5. Drill
6. Tap
7. Die
8. Reamer
9. Stud extractor

7.1.4 Chisel

Chisel is used to cut flat round or angle iron and l inch thick metal sheets. It is also used to remove the unwanted metal from the surface of a job by cutting it in bits and small pieces. This act is known as chipping. Cutting done by a chisel is rough. Therefore, there is a need for finishing after chipping done by chisel. It saves time. Chisels are used both on hot and cold metals; as such chisels have been divided into two categories.

Hot Chisel (Fig. 7.10): Hot chisel is used for cutting hot metals in blacksmith's workshop. There is a hole in a chisel in which a wooden handle is fixed. Sometimes an iron handle is also used.

The cutting edge is made at an angle of 30°. While using this chisel, it is frequently dipped into water for making it cold so that its edge remains intact.

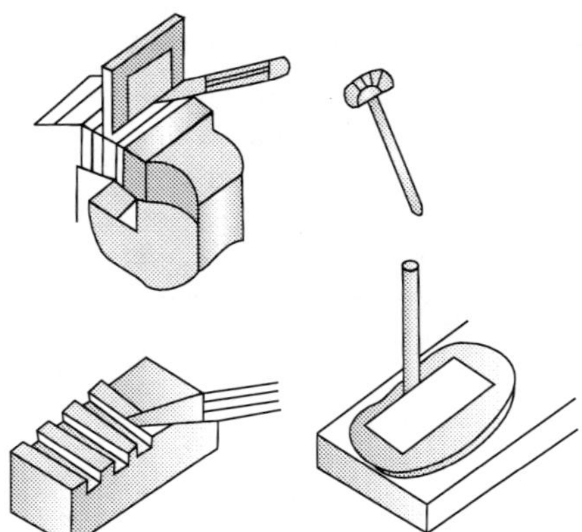

Fig. 7.9

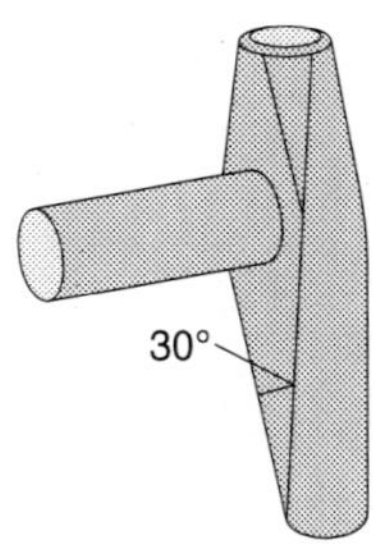

Fig. 7.10

Cold Chisel (Fig. 7.10): Cold chisels are generally used in sheet metal and fitter. With cold chisel cutting or chipping of metals can be done when these metals are in cold condition. This is available with 4 sides, 6 sides or 8 sides. These are made of high carbon steel, which contains 75% to 100% of carbon or chrome vanadium steel. The cutting edge is hardened and tempered. Its length for various jobs varies from 50 mm to 200 mm. Generally, chisel of 150 mm length is used. Practically, chisel's measurement is known from the width of chisel's cutting edge, **(Fig. 7.11).**

7.1.5 Parts of a Chisel

1. Head
2. Body
3. Forging angle
4. Cutting edge
5. Cutting angle

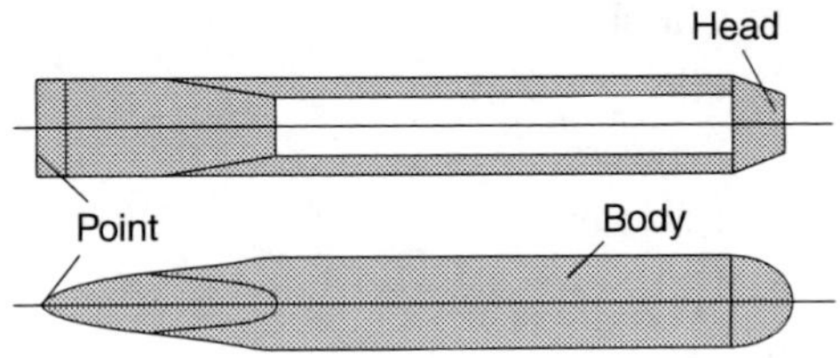

Fig. 7.11

Head: The topmost part of a chisel is called "head". It is a bit sunk in the middle so that the blow of the hammer strikes at the centre of the chisel.

Body and head of the chisel are not tempered. Thus, because of repeated blows, its head becomes like a mushroom.

In this case, chisel's head should be set right by grinding. While making the head, the upper part of the body should be at 10° to 20°.

Body: The part of a chisel where we hold it is called the body or the stock. It is made of square, hexagon or octagon bars so that it is easy to hold. Some chisels are round but it is difficult to hold it then. Body is made somewhat flat from the bottom.

Forging Angle: Some part of the flat position below a chisel body is tapered. This angle of taper is called forged angle. It is also known as draw angle. These angles are different, according to the use of the chisel.

Cutting Edge: The part of chisel that performs the function of cutting is called the cutting edge. Its width is taken as chisel measurement. Different types of chisels have different cutting edges.

Cutting Angle: Angle at which a cutting edge of chisel is ground is known as its cutting angle. The cutting is highly an important job. For cutting different types of metals, different suitable angles are recommended. For cutting hard metals, cutting angle is more and for cutting soft metals the cutting angle is less. The correct point/cutting angle of the chisel depends on the material to be chipped. Sharp angles are given for soft materials, and wide angles for hard materials.

The correct point angle and angle of inclination generate the correct rake and clearance angles. While grinding chisels for cutting various kinds of metals their angles are recommended as:

Material to be cut	*Point angle*	*Inclination*
High carbon steel	65°	39.5°
Cast iron	60°	37°
Mild steel	55°	34.5°
Brass	50°	32°
Copper	45°	29.5°
Aluminium	30°	22°

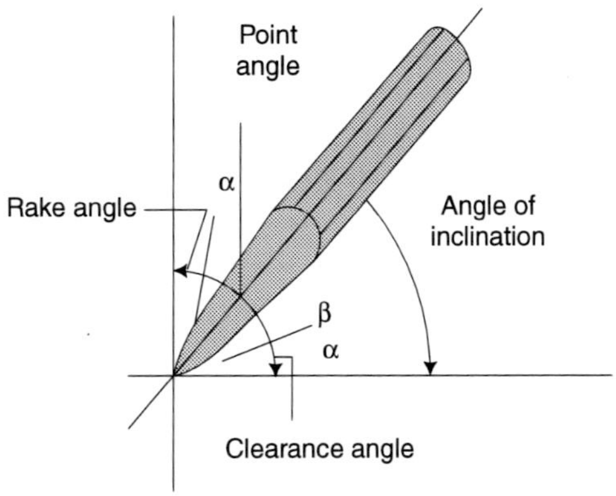

Fig. 7.12

7.1.6 Rake Angle

Rake angle (α) is the angle between the top face of the cutting point, and the normal to the work surface at the cutting edge.

If the clearance angle is too low or zero, the rake angle increases. The cutting edge cannot penetrate into the work. The chisel will slip (**Fig. 7.13**).

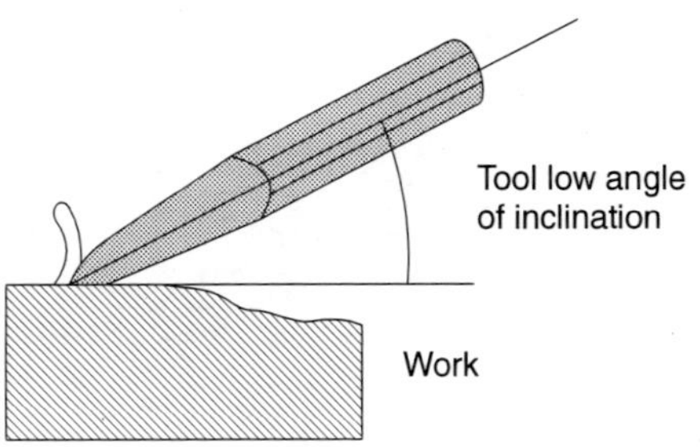

Fig. 7.13

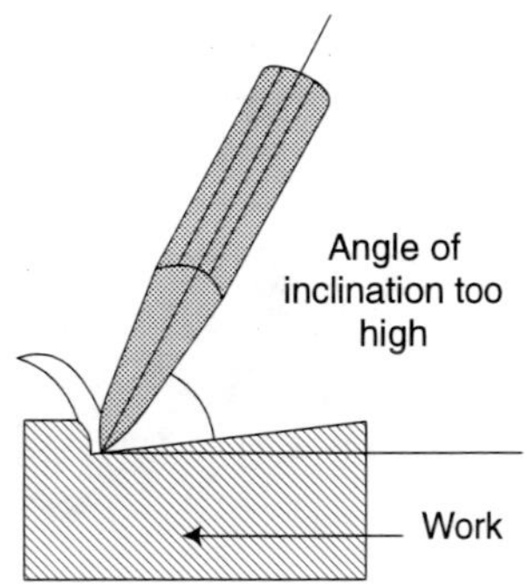

Fig. 7.14

If the clearance angle is too great, the rake angle reduces. The cutting edge digs in, and the cut progressively increases (**Fig. 7.14**).

7.1.7 Grinding of Cutting Edge

Unless the cutting edge is properly ground it will take longer time to cut anything with it and it would be damaged as well. While grinding the cutting edge, it should be ensured that the edge should be ground a bit in convex shape. When we hit the chisel with a hammer its protruding part touches the surface first of all. There is no likelihood of its slipping from the surface as well as its edge-ends do not break. While grinding an old edge, first its old edge should be levelled. While doing so it should be ensured that there should be enough pressure on the chisel so that it remains in contact with the grinding wheel because it should not be allowed to jump otherwise some scars are left on the chisel. While making a new edge, the tool rest of the grinding wheel should be brought quite close to the wheel. The chisel should be properly placed at the tool rest and grinding should be done at appropriate angle. For having a proper edge it is essential that at the time of grinding, the chisel should first be brought into contact with the centre of the grinding wheel. Then we should move the chisel on both the ends of the wheel turn by turn while placing it on the tool rest. The edge would take a round shape. Again, we should repeat this process after changing the side of the chisel. Chisel edge should be made cold intermittently by water.

7.1.8 Hardening and Tempering of Chisel

In order to make the cutting edge last longer even the tapered part of a chisel is made hard and the edge is tempered. The chisel should be heated at appropriate temperature up to half of its length and this edge part of the chisel should be cooled in clean water. This would make that particular part of the chisel hard. Then its edge part should be cleaned with sandpaper. Again it should be heated at appropriate temperature. Now the colour of the part would start changing and when it turns into appropriate colour, it should be cooled down in lime water. This would temper the chisel's cutting edge and it would work for longer period. This requires considerable experience.

7.1.9 Flat Chisel

It is an all-metal chisel used for obtaining a flat surface on metal by chipping. This chisel is used to cut wires, thin sheets and round rods, etc. This chisel's edge is in 'V' shape and is flat, which is raised from the centre. Because of this projected part, it does not get stuck into the cut surface, while chipping. Old rivets and jammed nut and bolts are cut with this chisel.

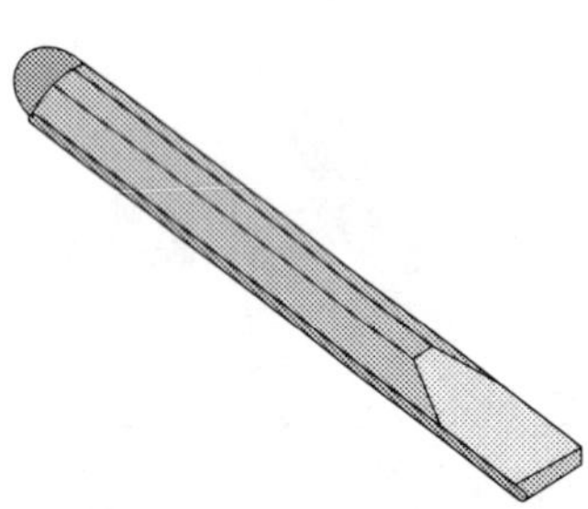

Fig. 7.15

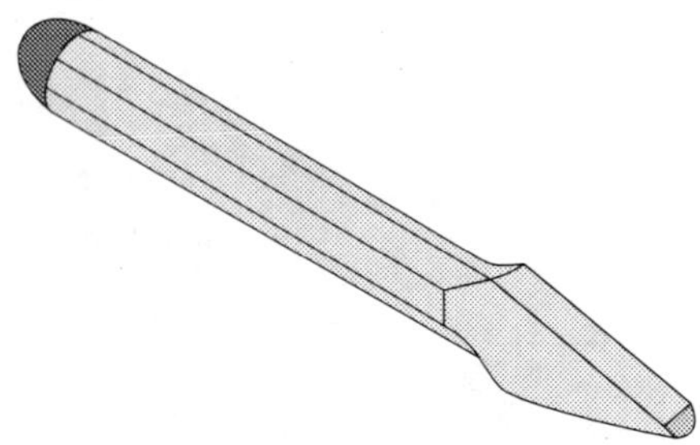

Fig. 7.16

7.1.10 Cross Cut Chisel

It is mainly used to cut square slits and channels. Key way on a shaft is made with this chisel. Its cutting edge is from 3 mm to 12 mm in width. In order to ensure that the chisel does not get stuck into the grooves and slits, the part behind its edge is made consequently thin.

7.1.11 Side Cut Chisel

Where other types of chisels cannot be used easily, side cut chisels are used. The key-way and cotter-way is cleaned with this chisel. It is almost similar to the flat chisel but the part which performs the function of cutting is turned to one side. It is turned at 90° angle and then forged. Its cutting angle is generally at 60° and forging angle at 25°.

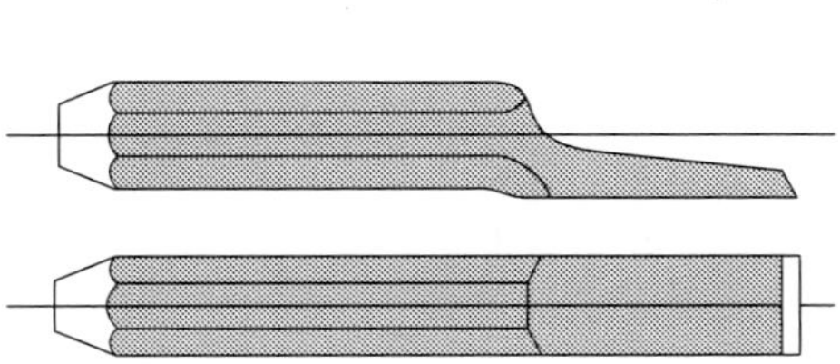

Fig. 7.17

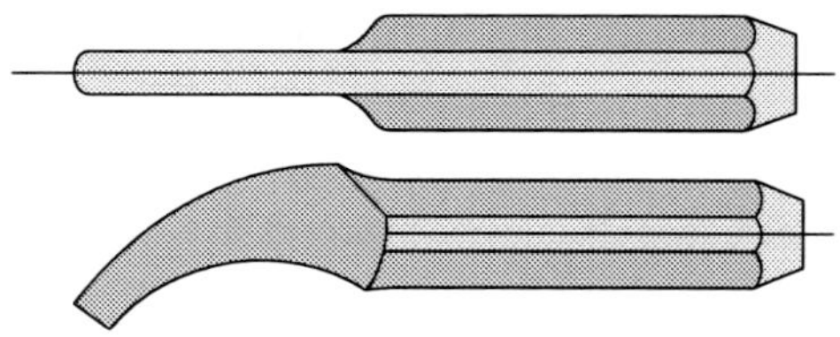

Fig. 7.18

7.1.12 Round Chisel

Its edge is round. It is used for making an oil groove in a bearing and for cutting slits at corners. Its forging angle is at 35° and cutting angle is of 35° to 40°. In such jobs where drill is dislocated at the time of drilling and some substance is left, the leftover metal is set right with this chisel.

7.1.13 Diamond Point

This is made of good quality steel. Its shape is square and it is tapered. Its edge is pointed and of the shape of a diamond. On the basis of its square ends, its measurement is determined. Square corners and triangular corners are cleaned with it. We can make groove in plain surface with this chisel.

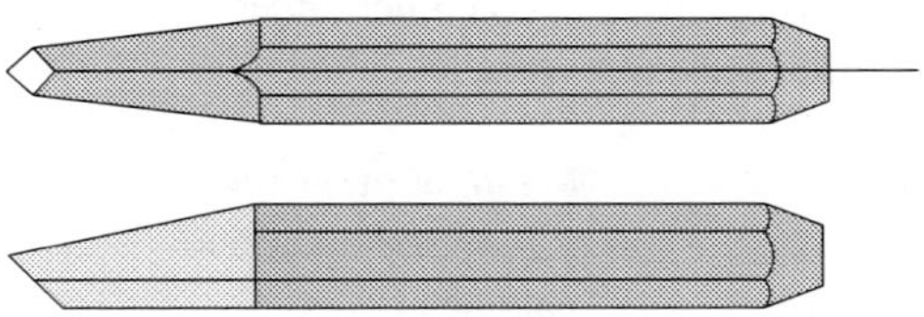

Fig. 7.19

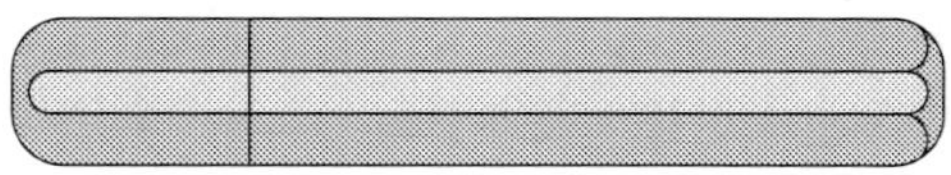

Fig. 7.20

7.1.14 Cow Mouth Chisel

Its shape is like that of a broad mouth of a cow. That is why it is known as cow mouth chisel. Edges of the protruding parts of plain metal and rivet heads and big round holes are made with it. Its forging angle is 25°. (Fig. 7.20).

7.1.15 Method of Using Chisel

Chisels are used for cutting and chipping a thin metal sheet, plate, round rod and such other materials. (**Fig. 7.21**). The following points should be kept in mind while using a chisel:

1. Accurate marking should be done on the metal to be cut.
2. Always hold the chisel properly with your left hand and the hammer in your right hand.
3. Set the chisel properly on the marked line then strike the hammer on its head.
4. Before striking the second blow on the chisel, it should be set again.
5. While cutting with a chisel, concentrate on its cutting edge.
6. Strike the hammer blow at the centre of the head of chisel.

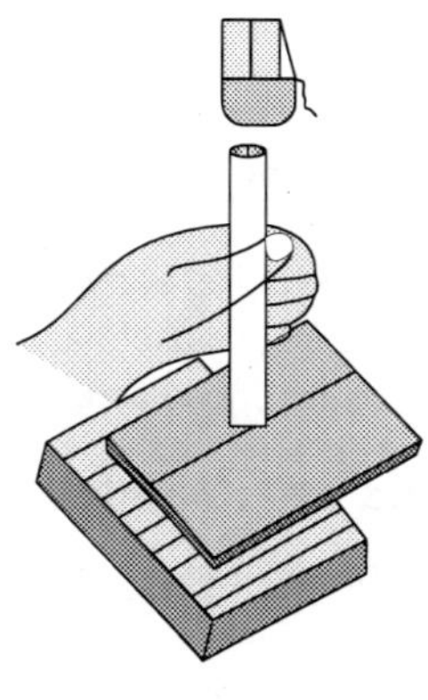

Fig. 7.21

7.1.16 Method of Chipping

Chipping is done to chip a metal about 6 mm deep. For this purpose cold chipping is done. The following points should be borne in mind while carrying out chipping:

1. Select the chisel with appropriate cutting angle according to the metal.
2. If the job is small in size, and it has to be held in a vice use a wooden block under the job.
3. While chipping, stand to the left of the vice and near the bench.
4. Hold the handle of the hammer from its end, raise it up to the height of your neck and then strike a blow.
5. Chisel should be inclined towards the user at an angle of 40°, while chipping.
6. At a time do not chip any metal more than 2 or 3 mm. Otherwise the edge of chisel may be damaged.
7. If the surface to be chipped is broad, make a groove on the job by making a cut.
8. Only a flat chisel should be used for chipping.
9. While chipping after hitting the chisel with a hammer three or four times, withdraw for a while, and then chip.
10. While chipping a job held in the vice, strike the hammer blow in the direction of its stationary jaw.

According to the required shape, the workpiece is worked with chisel and hammer either by separating or by chipping to achieve maximum precision as can be obtained by using a chisel.

Rule	Danger of accidents	Means of prevention
Remove the burr from the head of the chisel.	Splinters will cause wounds.	By removal of the burr regrinding chisel head
Do not look at the chisel head.	Unsteady strokes cause injuries.	By looking at the cutting edge of the chisel.
Hammer must be firmly seated on the handle.	Serious injuries by coming-off hammer.	Fastening the hammer with a wedge in inclined position to press on the wood on all sides.
Strokes must be vertically onto the chisel.	Inclined strokes will fling the chisel out of the hand.	Guide the hammer in such a way that the centre of the flat side meets vertically with the head of the chisel.
Towards the completion of the chisel work reduce the force of strokes.	Coming-off chips cause injuries.	By decreasing the force of the hammer strokes and setting the shield.

The chisel is mostly guided by hand and driven into the material by the force of the hand hammer or the pneumatic chisel hammer. The chisel is firmly held by the left hand while the left hand gripping the handle at its lever end actuates the force of the hammer strokes.

7.1.17 Maintenance of Tools

Permanent use, but also faulty use (chisel work on the anvil plate or on the vice jaws) blunt chisel cutters. Blunt chisel cutters press upward from the material when chipping work is done require much strength and cause overstress on the chisel and cutting tools.

The chisel cutter will become sharp again by sharpening on the grinding wheel.

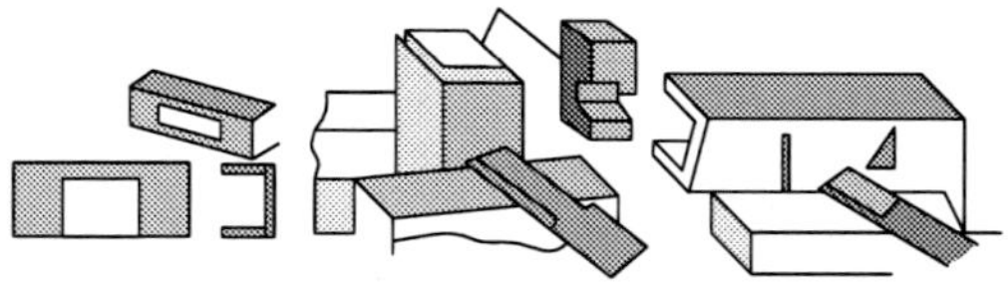

Fig. 7.22 Fixture to hold job

The eye is exposed to a number of injuries by penetration of foreign matters, e.g., steel particles. Great care should therefore be taken when doing chiselling and grinding work.

While small particles of dust and small insects can be removed from the eye by slight rubbing with a clean cloth, one should avoid practising this method for removing steel particles into the eye during the work with a chisel or the like.

These minute steel particles are pointed like a needle, sharp like a knife and pronged. They strike the eye cover at a high speed and tear it. If they stay in the cornea, it is possible to remove them with a magnet. If they are deeper into the eye, they may be retained by the iris. If the foreign matter also cuts the crystalline lens, the injury is serious.

Consult an ophthalmist in case of an injury. Wear goggles when working with chisel and grinding tools.

7.1.18 Special Chisel Works

A rectangular opening of 34 × 60 mm is to be worked in a sharp edge "U" steel 40 × 40 × 3.

Openings are pre-drilled and worked out.

The rectangular opening is first rough drilled on two sides. The transverse webs remaining after completion of the work, which are the thinner the smaller the distance from one hole to the other is cut out with the backing-out punch.

After reclamping, the remaining two edges are separated with the scraper. This separating operation is assisted by the edge of the workpiece as well as by the rear vice jaw. Preference should be given to the scraper because the chisel often hooks into small thicknesses of the workpiece.

7.1.19 Chiselling Small Surfaces

- By using a flat chisel it is possible to remove a chip over the entire width of a workpiece.
- To avoid breaking of the workpiece edge, when punching the work piece is reclamped and the remaining chip is chiselled from the opposite side.

Broad surfaces of more than 20 mm cannot be chiselled with one chip.

7.1.20 Chiselling Broad Surfaces

At first a groove is punched out by means of a cross chisel. With broader workpieces, several grooves are punched.

The remaining ridges not removed by the cross chisel can now be easily chiselled by means of a flat chisel.

7.1.21 Chiselling Round Surfaces

- The notches are made with a hewing chisel in a way that the marking line remains unchanged.
- By short strokes with the hammer, the waste part is separated from the plate placed on the edge of the anvil.

7.1.22 Judging Chiselling Work

Mistakes	Causes and relief
The chiselled surface is heavily undulating.	Position of the chisel was alternately too sharp inclined or too even. Clearance angle $\alpha = 8°$ was not observed
The chiselled face is not in compliance with the mark.	The workpiece was not secured against gliding downward during the chiselling operation.
The workpiece edge breaks off during the chiselling operation.	Chiselled down (before removing the last small chips re-clamp the workpiece and work from the opposite end.
The chisel penetrates into the material when using special force.	Cutting edge of the chisel is blunt because of working on the anvil plate without using a base resharpening the cutter with the grinding wheel.

7.1.23 Precautions

1. At the time of chipping we must always use safety goggles.
2. Keep your face towards the wall, while chipping.
3. If someone is working in your front, fix the chipping guard at the rear of vice so that chips do not hit the other person.
4. Mushroom head chisel should never be used for chipping.
5. There should not be any greasy substance on the chisel head and face of the hammer being used for chipping.
6. An extraordinarily long chisel should not be used for chipping.
7. While chipping keep on applying grease on the edge of the chisel.

7.2 SAWING

7.2.1 Fundamentals of Sawing

Sawing is a chip removing working operation. It is used for separating workpieces to rough lengths, for preparing grooves to be chiselled off and for making notches and incisions.

7.2.2 Operation of the Saw

The cutting tool is a thin and short saw blade made of tool steel or high-speed cutting steel; or it is a steel band of some length provided with cutting teeth on one or both sides of the narrow edge. Every tooth is like the cutting edge of a chisel. Chip removal by means of a saw comes about by a number of chisel-shaped teeth. They are milled or punched into the blade one adjacent to the other. There is a throat between every two teeth. The teeth are the cutting tools and the throats have to take the chips and remove them.

7.2.3 Hacksaw

In order to cut metal sheet, rods, pipes, plates or sheets of varied thickness a hacksaw is used. It can be operated by hand or by power. Less metal is wasted by cutting it with a hacksaw and the cut edges are cleaner than those cut with a chisel. Hacksaw is built by fitting two parts:

1. Hacksaw frame
2. Hacksaw blade.

Hacksaw Frame: In shape it looks like the English letter **"e"** with both its arms bent at the right angle. On the one end of these arms a handle is fixed and on the other end a pin is fixed. A pin is fixed on handle and sliding screws itself in which hacksaw blade is fitted. These pins are also known as "prongs". Hacksaw blade is fitted on these pins and then it is tightened according to the need with fly nut or wing nut. Mainly hacksaw frames are of two types:

Fixed Hacksaw Frame: For making this type of hacksaw frame, one piece of flat iron strip or pipe is bent at right angle. In these frames only hacksaw blades of a specific size can be fixed, not of short size or long size. Two types of handles are used: 1) Straight handle and 2) pistol type **(Fig. 7.23).**

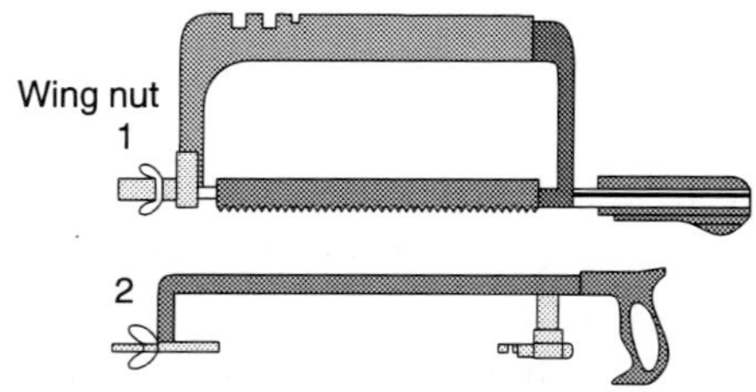

Fig. 7.23

Adjustable Hacksaw: Its structure is a bit different from a fixed frame. The frame is made in two parts. These parts can be adjusted at different distances. Therefore in these adjustable hacksaws, big and small hacksaw blades of different standards can be used. There is adequate provision of pin and grooves to pull, adjust and set them. These types of handles are used in this hacksaw:

1. Straight handle
2. Pistol type handle
3. Tubular type handle (**Fig. 7.24**)

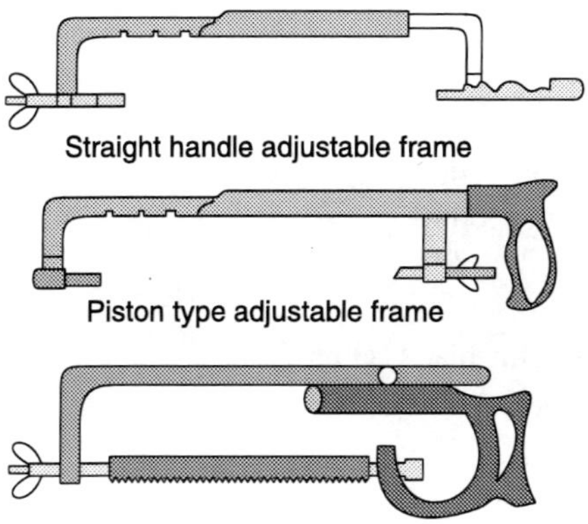

Fig. 7.24

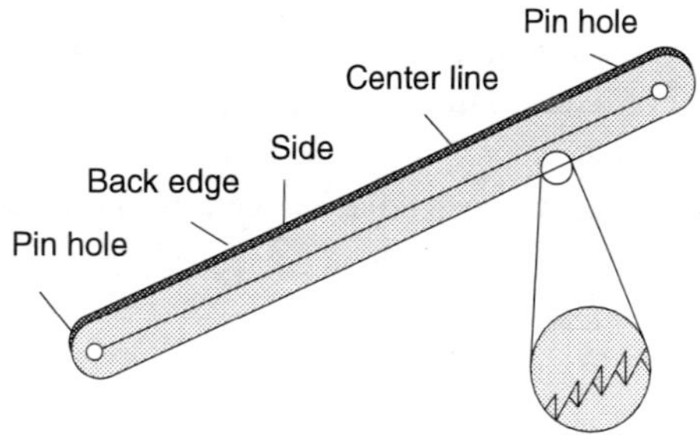

Fig. 7.25

Tubular type adjustable frame: These frames are generally made of flat strip or thin pipes and their handles are made of wood or aluminium.

Hacksaw Blade Design (Fig. 7.25): A hacksaw blade must be held at each end with tension proportionate to feed pressure and guided down through the work. The blade must be of a quality that will permit high speed and heavy feed pressure without excessive deflection or breakage. Four different types of steel are used in the manufacture of hacksaw blades: high-speed tungsten steel, all hard; high-speed molybdenum, all hard; flexible tough-alloy steel back with high-speed steel teeth and hard cutting edge; and semi- high-speed steel. The composite or welded-edge blades, with high-speed steel teeth supported by or welded to tough-alloy-steel backings, are comparable to carbide-tipped tools backed by tough-alloy steel. They are recommended for best results.

On the basis of metal of the job, different types of hacksaw blades are used in a hacksaw. These are made of tungsten steel or high speed steel and are hardened and tampered. Their classification depends on three points: 1) size of the blade, 2) number of dents cut on the blade per inch, and 3) nature of the blade.

Length of hacksaw blade is generally 8", 10" or 12". Dents of four different grades are cut on them.

Course Grade: Hacksaw blade of this grade is used for cutting thick or mild steel, copper, aluminium and brass, etc. It contains 14 to 18 teeth per inch.

Medium Grade: Hacksaw blade of this type is used for cutting all kinds of metals such as cast iron, tool steel, aluminium, brass, high carbon steel, etc. From 20 to 24 teeth per inch are cut.

Fine Grade: This type of hacksaw blade is mainly used to cut thin pipes, sheets, tubes, etc. It has 24 to 30 dents per inch.

Superfine Grade: For cutting extraordinary solid metals and thin metal sheets, this type of hacksaw blade is used. There are 30 to 32 dents per inch.

In all the hacksaw blades of various grades mentioned above, teeth are cut on one or both the edges. These blades are hardened in two different processes and on these depends their nature.

All Hard Blades: Blades of this nature are hardened and tampered on sides except the ends having holes. These are used for cutting articles of cast iron or moulded iron, etc.

Flexible Blade: In blades of this nature only the cutting teeth and nearby part is hardened and tempered. By this process they become elastic and there is less risk of them being broken in the event of a shock. These blades are used for cutting thin sheets, pipes, curves, etc.

7.2.4 Types of Teeth

There are two types of hacksaw teeth. The more common being the straight face (**Fig. 7.26**) which has a straight 56° angle back, the other is the 10° undercut face tooth at **Fig. 7.26(B)** with a 15° cutting-edge clearance and a rounder back closely resembling a milling-cutter tooth. This type of blade is made only in the coarser pitches, such as 2 or 2 ½ teeth per inch.

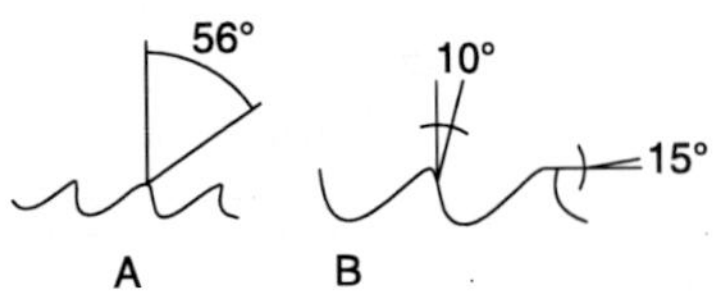

Fig. 7.26

The distance between adjacent teeth is known as the pitch of the blades (**Fig. 7.27**).

Hacksaw blades are designated according to their length, pitch and type.

Classification	Pitch
Coarse	1.8 mm
Medium	1.4 mm & 1.0 mm
Fine	0.8 mm

7.2.5 Setting of Teeth

To prevent the saw blade bending when penetrating into the material, and to allow free movement of the blade, the cut is to be broader than the thickness of the saw blade. This is achieved by the setting of the saw teeth.

Teeth of hacksaw blade are also bent to left or right as in the case of ordinary saw used for cutting wood. Bending of teeth is known as setting of teeth. With this setting of teeth the hacksaw blade cuts slot in the metal of slightly thicker than that of the blade. Therefore, because of this the blade does not get stuck up while in operation. If this setting is not done, there is great risk that the blade would get stuck in the metal and get broken. Setting of teeth in a blade is done in three ways.

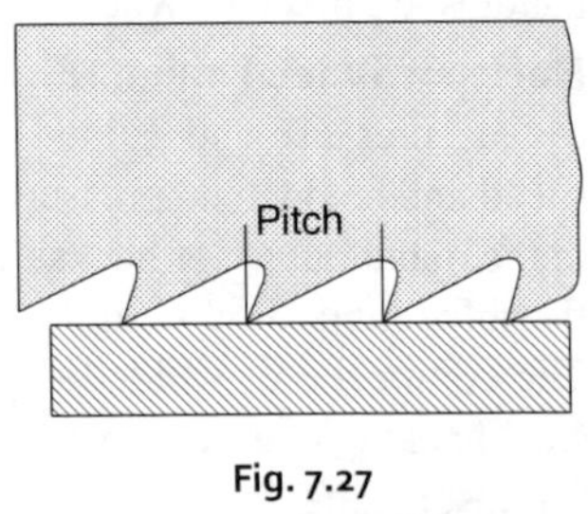

Fig. 7.27

There are three different types of tooth settings (**Fig. 7.28**).

1. Straight set in which the teeth are flared alternately, i.e., right, left, right, left, etc. (**Fig. 7.28 A**).
2. Raker set (**Fig. 7.28 B**), in which every third tooth is straight i.e., right, left, straight, right, left, straight, etc.
3. Wavy set (**Fig. 7.28 C**) used only in fine-pitch blades of 24 or 32 teeth per inch, in which groups of teeth are set alternately to provide clearance.

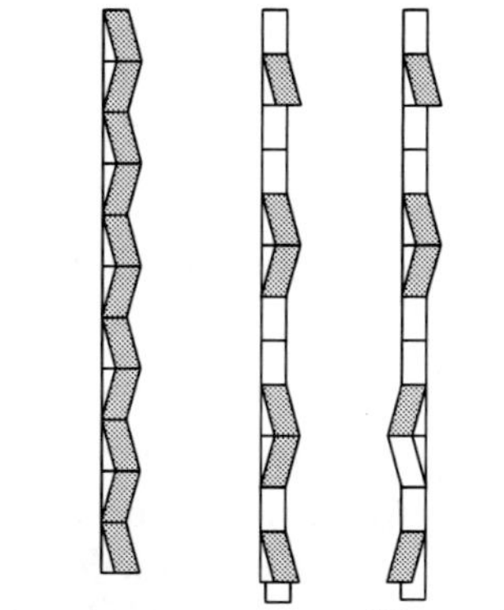

Fig. 7.28

In blade specification the following four points should be considered: number of teeth to the inch, gague or thickness of blade, length and width of blade, and type of blade.

In general, as coarse a tooth as possible should be used. Two or three teeth should always be in contact with the stock being cut. For average sawing conditions, blades with six teeth per inch are recommended for bars or sections 1 through 5 inch in diameter, four teeth per inch for bars or sections 5 through 8 inch in diameter, and three teeth per inch for bars or sections larger than 8 inch in diameter.

The blade should be as short as the size of the work permits, and wide and thick enough to withstand the necessary feed pressure applied and provide the necessary rigidity to control the flexing and quivering actions of the blade. Shorter, heavier blades assure straight cuts, give more cuts per blade and per hour, and reduce blade breakage.

The adjustment of speeds and feeds to the point where a proper balance is found between productive cutting and satisfactory blade life is of vital importance. The harder the stock, the finer the teeth and the softer the stock, the coarser the teeth should be. A deeper gullet is required to carry off the chips on soft material.

7.2.6 Accuracy and Finish

The hacksaw blade should be regarded as a roughing tool. Even when more rigid machines are used, a few thousandths of an inch should be allowed for finishing operations. The kerf, or width of the cut, varies from 0.055 to 0.130 in with the regular blades, and up to 0.135 in with the extra heavy blades.

Regular or Single Setting: One tooth is bent to the left and the next tooth to the right and every fifth tooth is kept straight. This tooth helps in cleaning the metal particles from the slot.

Double Alternate Setting: Two teeth are turned to the .right and two to the left and one dent between them is left straight, which performs the function of clearing the metal particles from the slot.

Zigzag or Wave Setting: A few teeth are bent to the right and a few to the left. Thus it looks like a wave. This type of setting is done on blades of fine grade and super fine grades.

Selection of blade: The selection of the blade depends on the shape and hardness of the material to be cut.

Pitch selection (Fig. 7.29): For soft materials such as bronze, brass, soft steel, cast iron, heavy angles, etc., use a 1.8 mm pitch blade (**A**).

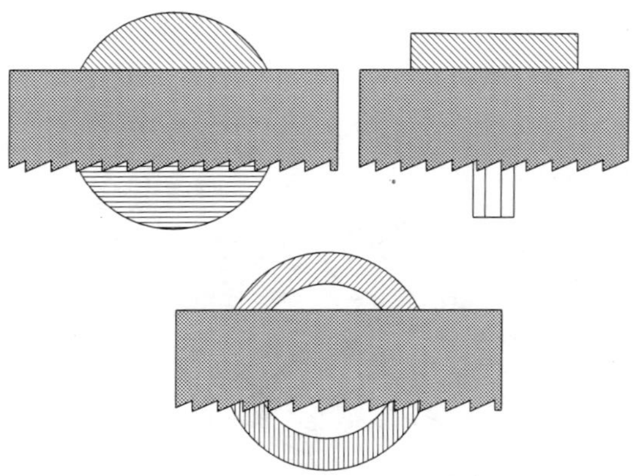

Fig. 7.29

For tool steel, high carbon, high speed steel, etc., use a 1.4 mm pitch. For angle iron, brass tubing, copper, iron pipe, etc., use a 1 mm pitch blade **(B).**

For conduit and other thin tubing, sheet metal work, etc., use a 0.8 mm pitch. **(C).**

7.2.7 Method of Using Hacksaw

Pipes, plates, sheets, etc., made of metal can be easily and clearly cut with a hacksaw. But for efficient cutting the following points should be kept in mind:

1. Selection of hacksaw blade should be done according to the metal for which it is to be used.
2. While fixing the blade on the frame it should be ensured that its teeth should cut the metal when they are driven forward.
3. Job should be held in the vice in such a way that its cutting lines are clearly visible.
4. Before starting a cut with hacksaw blade, the blade should be kept on the marked line; the left hand thumb should be placed with its support (**Fig. 7.30**). It would ensure that the blade would move only on the line, there would be no risk of its slip.
5. After marking with the support of thumb we should hold the handle with the right hand palm and fingers and with the left hand palm and fingers we should hold the other end of the hacksaw (**Fig. 7.31**).
6. Stand on the left hand of the vice and keep your right foot backward and the left foot a little ahead of the right.
7. Contact between the blade and the job should be in such a way that at least two teeth should remain in contact with the surface of the job.
8. The speed of operating hacksaw should be 40 to 50 strokes per minute.
9. Pressure should be exerted on forward stroke and it should be withdrawn on backward stroke.
10. Before finishing with a hacksaw the cutting speed should be slowed down.
11. Hacksaw blade should be fitted on the frame carefully. It should neither be very tight nor loose.

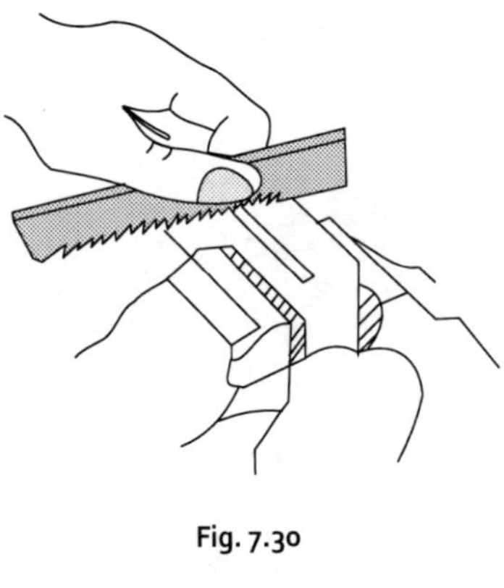

Fig. 7.30

Fig. 7.31

12. The job which is to be cut should not be held in the vice much high otherwise there would be vibrations in it.
13. While cutting thin metal sheet, packing of wood, plastic or any other appropriate material should be used in its front and back.
14. Hacksaw blade should not be allowed to become slanting while using a hacksaw, there are chances that it may break.
15. If the blade starts cutting in a slanting manner we should start a fresh cut. Operating a hacksaw in an old cut may break it.
16. If a blade gets broken while working, then we should replace it with a blade which has been used almost for the same period rather than using a brand new blade. The new blade can break if it is operated in that old slot.
17. If necessary, water should be used as a coolant.

After using a hacksaw either its blade should be taken out or it must be made loose.

7.2.8 Shape and Angle of the Saw Tooth

In order to obtain a good chip formation, the shape of the tooth must be adapted to the mechanical properties of the material to be cut. The ideal shape of tooth has a lip angle (β) of 65° a side rake (γ) of 5° and a clearance angle: of 20° (**Fig. 7.32**). But it is not possible to cut all the materials easily. Hard materials, e.g., steel and grey cast iron require an even larger cutting angle (γ), as otherwise, the teeth would clasp and could break because of the high resistance of the material to be sawn.

The teeth of the usual handsaw blades for steel working have a lip angle (β) of 50° and a clearance angle (α) 40° in this case the side rake is fully abandoned (**Fig. 7.33**). When working with light metals, preference is given to the tooth shape shown in (**Fig. 7.34**).

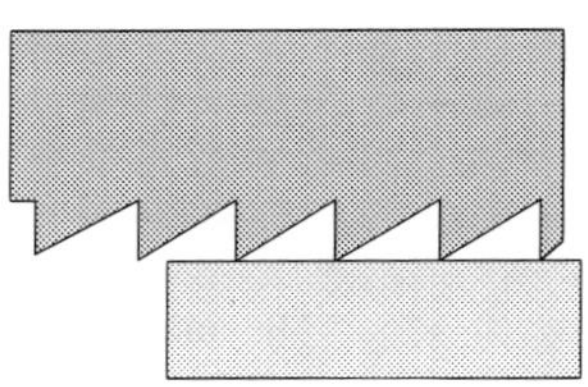
Fig. 7.32

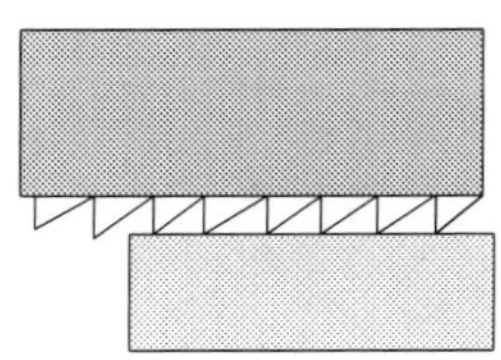
Fig. 7.33

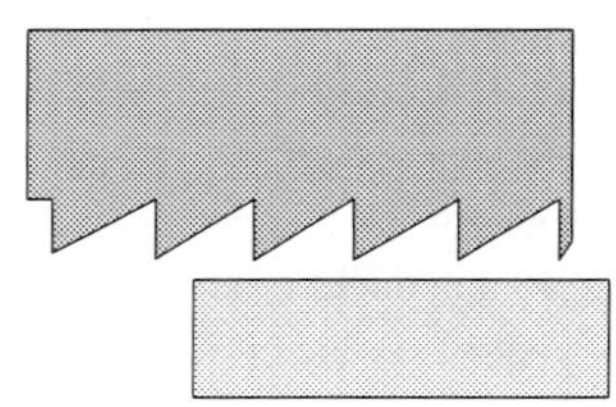
Fig. 7.34

7.2.9 Free Cutting Action of the Saw

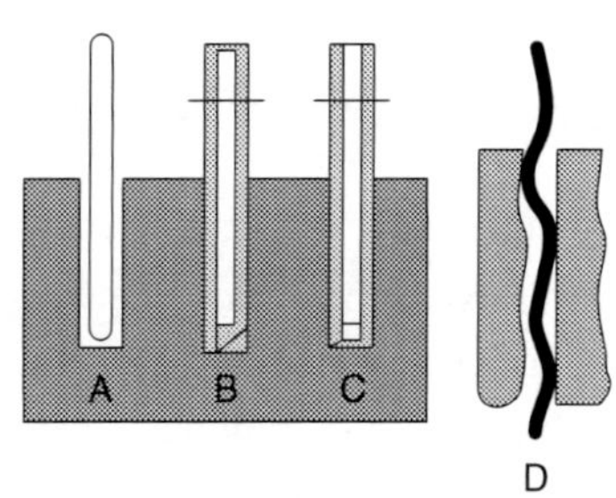

Fig. 7.35

Similar to the cross-chisel where the cutter is broader than the body of the chisel in order to avoid the chisel getting clamped in the material, the teeth of the saw are also arranged or shaped in a way that the saw kerfs is broader than the thickness of the blade. This ensures free cutting; and avoids locking of the blade. When the working material which can easily be cut, e.g., light metal or wood, free cutting of the saw blade is obtained by setting the teeth, which means that the teeth will be alternately bent in an uniform manner to the left and to the right.

Iron cutting saws for cutting metals harder than light metal have very small teeth, which means that setting each tooth would require too much time. In order to obtain free cutting of such saw blades the teeth sides receive wavy setting. With this method, five or six teeth are alternately waved to the left and to the right thus becoming flat bow (bent outwards) **(Fig. 7.35).** The width of cut of such a blade is determined by the most projecting teeth. The blade does not choke.

7.2.10 Kinds of Saws, Spacing of Teeth with Regard to Material

The distance of the teeth of a saw blade varies according to the hardness and thickness of the material thus, spacing of teeth is coarse medium or fine. Spacing is indicated by the number of teeth per 25 mm saw blade **(Fig. 7.36).**

For soft materials, blades with coarse teething are preferred whereas fine toothed blades are used for hard tough materials.

The teeth of the saw penetrate only a little into hard materials and the amount of removed chips is small. It is, however, small when soft materials are worked. The size of the throat must therefore be large enough to receive the larger chips and to pull them out of the saw kerfs. The chips quantity is particularly large in case of long notches, and the chips can be pressed in between the teeth if the throat is too small and the blade will choke and the teeth may break. Coarse spaced saw blades are therefore used for long notches, whereas fine teething is preferred for sawing thin worked pieces, e.g., tubes, plates and profiles, so that a maximum number of teeth will be engaged. The starting side of most of the usual blades for handsaws has fine teething which gets coarser and coarser in the remaining part of the blade. Such blades provide for a good sawing start.

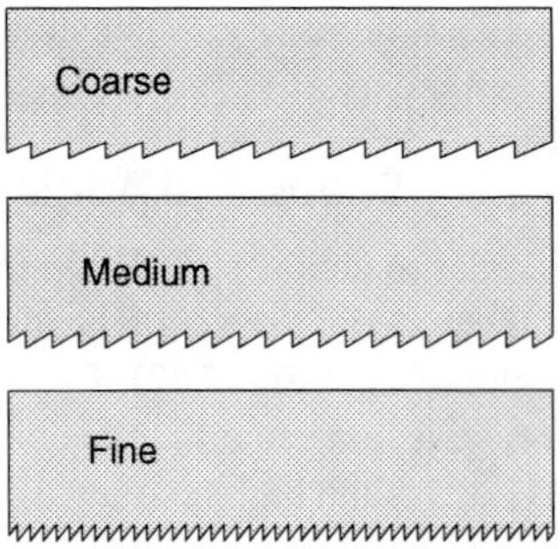

Fig. 7.36

Steel Marking Tool

Application	No. of teeth on 25 mm length	
For soft materials such as aluminium, tin, copper, plastics; for long saw notches, e.g., round and square bars of more than 40 mm thickness made of structural steel.	16	Coarse
For general sawing in steel and metal of medium hardness in particular for tubes. Profile tool steel as well as for bars of less than 20 mm thickness.	22	Medium
For hard materials, e.g., tool steel as well as for cables, plates, thin-walled tubes, wire and sheet metal profiles.	32	Fine

Spacing of teeth of hand saw blades, for iron with the generally used blade length of 300 mm:

Example

How large is the distance between two teeth of a saw blade with:

(a) Coarse teething: 16 teeth on 25 mm distance for one tooth = 25/16 = 1.56 mm.

(b) Medium teething: 22 teeth on 25 mm distance for one tooth = 25/22 = 1.14 mm.

(c) Fine teething : 32 teeth on 25 mm distance for one tooth = 25/32 = 0.78 mm

7.2.11 Working Method for Sawing

Marking before sawing: After the cutting line, for sawing has been determined by measuring with a steel straight edge, folding ruler, pull-push rule or the marking gauge, the marking tool is led along the square angle. Mitre square, protractor, ruler or along the template depending on the marked line.

Principally it is possible to trace all marking lines on metals and on light metals for notches and incisions as well as for cutting off by sawing A brass making tool is used only for provisional lines needed to determine the marking line cutting line and which are not removed by the saw notch or by subsequent working. A pencil can also be used for marking light metals. This will avoid the formation of dents which may occur when steel marking tools are used and which may lead to fractures of highly stressed work pieces.

When cutting rough lengths, it is necessary to observe not only the rough measure stated in the piece list but also the shape of the workpiece.

Example

The piece list stated a rough length of 250 mm for:

1. U-rail 25 × 14 × 25 of aluminium. For an economic working of the material and in order to reduce the working time, the U-rail should not be cut at a right angle at the beginning but direct in an angle of 45°.

1. What rough length is needed for cutting 40 pieces U-rails if cutting is done;
    - According to the rough measure of the piece list.
    - According to drawing.

    Machining allowance per soar cut when cutting according to drawing is 1.5 mm.
2. What percentage of material can be saved when cutting according to drawing.

Solution:

Material required when cutting in accordance with rough measures of piece list: rough length multiplied by number of pieces = 0.250 mm × 40 = 10.000 m

1 b) Material required when cutting according to drawing:

$0.223\text{ m} + 19 \times 0.446\text{ m} + 0.248. + 41 \times -0.0015\text{ m}$

$0.223 + 8.474 + 0.248 + 0.0062 = 9.007 \sim 9\text{ m}.$

2. Consequently, 1 m of material can be saved when cutting according to drawing. = 10 % material has been saved.

Clamping of Workpieces

For sawing operations, the marked workpieces are commonly clamped in the parallel vice or in the vice with lever stopper. Short and firm clamping avoids springiness and getting loose when particularly thin work pieces are sawn. Already treated workpieces are protected against the impression of the teeth of the vice jaws by intermediate protecting jaws of aluminium, lead, soft steel or wood.

7.2.12 The Sawing Operation

Sawing is done with bow saws (**Fig. 7.37**) which consist of the bow, the handle, the clamp dog and the thumb screw for blade tensioning, and whose blade is generally 300 mm long. Sawing is done by hand under slight pressure which is only exerted when advancing the saw whereas the return movement is done without pressure, lifting the saw slightly. The saw blade consisting of tool steel or of high speed cutting steel is tightly stretched in a straight line whereby the teeth of the blade are directed to the front to the tensioning screw. The clamp dogs of the bow saw are generally provided with cross slits for perpendicular and horizontal setting.

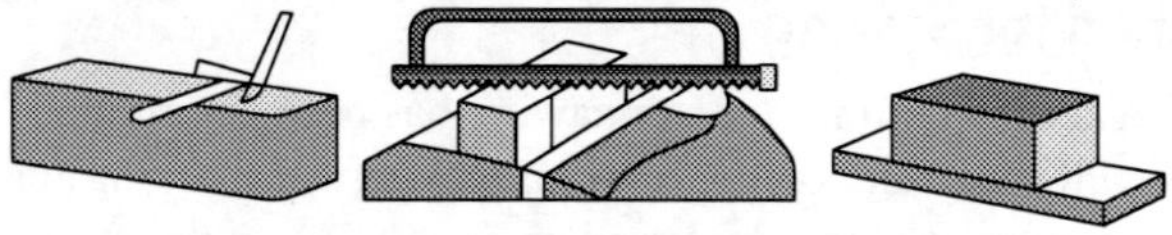

Fig. 7.37

7.2.13 Spot Sawing and Sawing Out of Thick Workpieces

A. **Marking:** The block is marked by means of the square angle and the marking tool.

B. **Spot sawing:** The block is clamped as short as possible. The saw is placed on the front edge about 0.5 mm beside the marking line. The saw is guided with the thumb or a score made on the marking line by means of a three square file. The saw and the surface of the workpiece form a small angle.

C. **Sawing out:** In order to maintain the marking line on both sides of the workpiece, when cutting thick workpieces, with long saw notches, it is advisable to work in accordance with the cross-hatched surface. The marking line must still be visible on completion of the sawing process; this will facilitate exact size when filing.

7.2.14 Maintenance of Tools

The saw blades will get blunt by permanent use but also by faulty manipulation. Working with blunted ·saw blades requires longer working time and gives inexact sawing results.

Blunted saw blades with rough teething can be resharpened. Saw blades with some teeth broken out, can again be used after the corresponding spots have been ground in an arch-like bent.

When sawing long notches the friction of the saw blade can be diminished by oiling.

After completion of the work, the saw should be released to avoid a stretching loss of the saw bow caused by the fatigue of the material.

7.2.15 Judgment on Sawing Work

Mistakes	Causes
The saw claps when spot sawing.	Spot sawing point was not spot filed; teeth spacing of the blade is too coarse, or the workpiece face too small for the teeth when edge-wise sawing of the flat workpieces
Teeth break out or wear out too much.	Too hasty sawing when cutting through in case of hard material; too fine toothing for long saw notch; throats are filed with chips.
Unequal wearing or the saw blade.	The saw was not guided over the workpiece with a full stroke.

An angled profile is cut in deeply: For producing deep incisions the saw is set in perpendicular position to the saw bow to prevent its inner edge from pushing against the workpiece. The sawing surface is increased by an inclined position of the saw.

Spot sawing a tube: At first the cutting point is marked in such a way that the marking tool is guided around the tube and along the tape-rule. Then a small kerf is filed into the clamped tube and close to the marking line and by means of a three-square file so that at least three teeth can attack simultaneously. Then the tube will be sawn from the upper side till reaching the hollow space.

Sawing up of a tube

The incised tube will be turned and sawing continued. The last sawing motions should be done slowly and without exercising any pressure as otherwise the teeth would clasp and break.

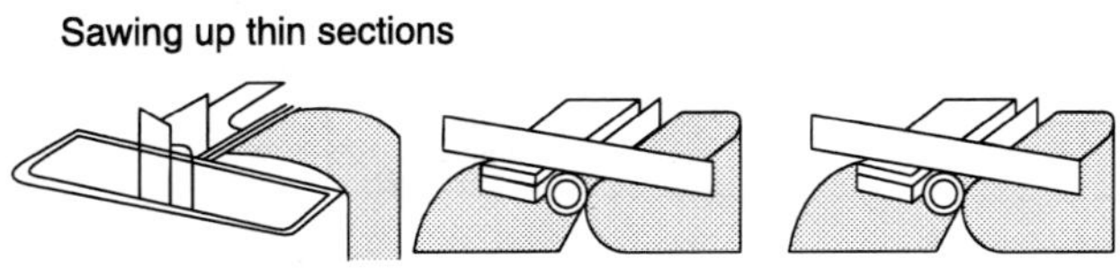

Fig. 7.38

7.2.16 Metal-Cutting Saws

Metal-cutting saws of different types are used. The most commonly used are:

Power saw

Horizontal band-saw

Circular saw

Contour band-saw.

Power saw: This is the most commonly used metal-cutting saw.

Horizontal band-saw (Fig. 7.39): This has a saw frame on which a motor is fitted. There are two pulley wheels on which an endless band saw passes. Speed variation is obtained through the stepped pulleys on the motor.

The roller-guide brackets provide the rigidity for the blade in the cutting area and also prevent wandering of the blade while cutting. The blade tension is maintained by using the adjusting handle, provided for this purpose. A vice is provided for holding the metal stocks. The vice is adjustable for angular cutting.

This machine has the advantage of continuous cutting ability, and is much faster than a power saw. It may be noted that a power saw cuts only in every alternate stroke.

Circular saw (Fig. 7.41): This type of cutting machine is used when cutting materials have a large cross-section. The circular saw has a continuous cutting action and is economical in production work where heavy section metals are used.

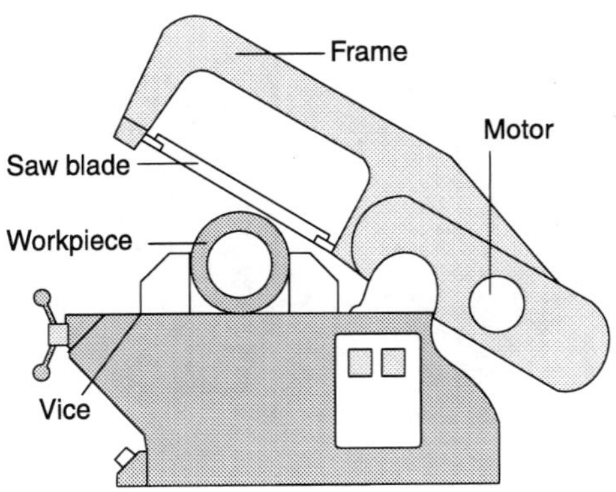

Fig. 7.39

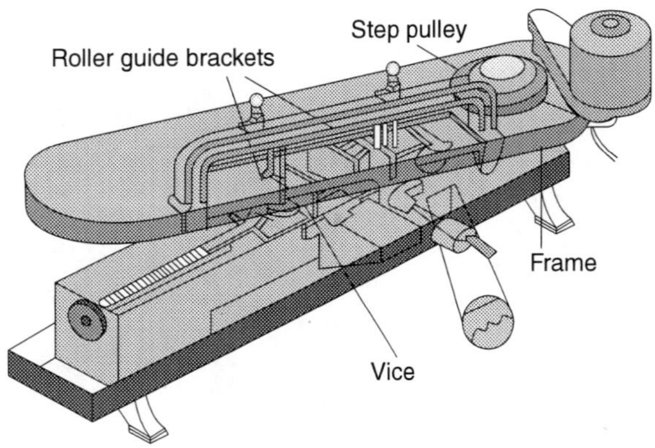

Fig. 7.40

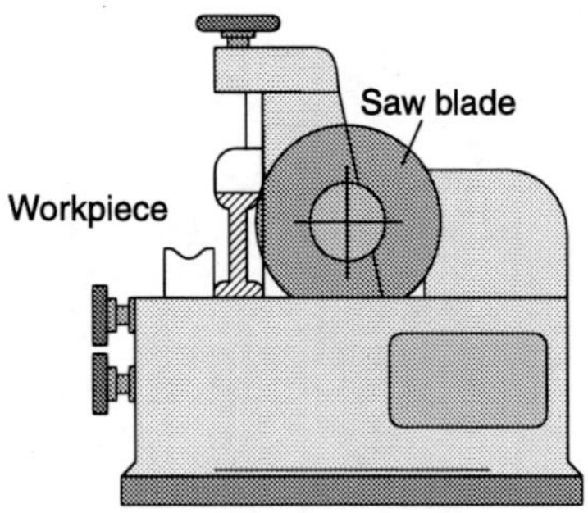

Fig. 7.41

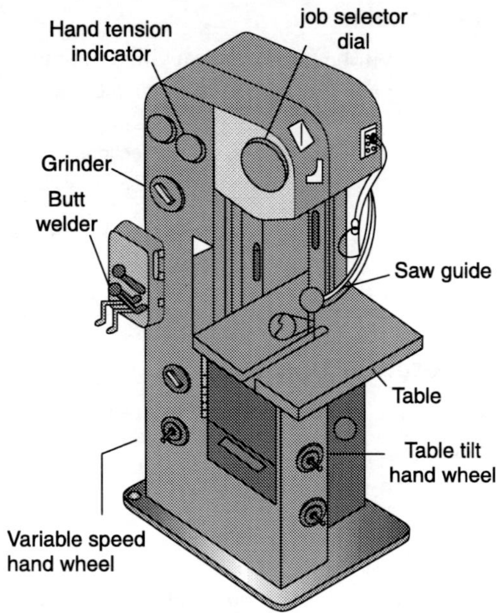

Fig. 7.42

Contour saw (Fig. 7.42): In this saw, a metal band saw blade is used, and the contour saw has a continuous cutting motion.

These machines are also used for cutting metals different profiles (**Fig. 7.46**).

Different speeds can be obtained while cutting, with the help of variable speed pulleys.

For repairing broken contour saw blades, these machines are fitted with a shear for trimming the blade ends, a butt-welding machine for joining the ends and the small grinder to finish the welded joint.

The table can be tilted to any angle for angular cutting.

The blade passes through a guide which prevents the blades from wandering and keeps it rigidly.

These machines are widely used for tool-room work, and not as a machine for cutting raw material stock.

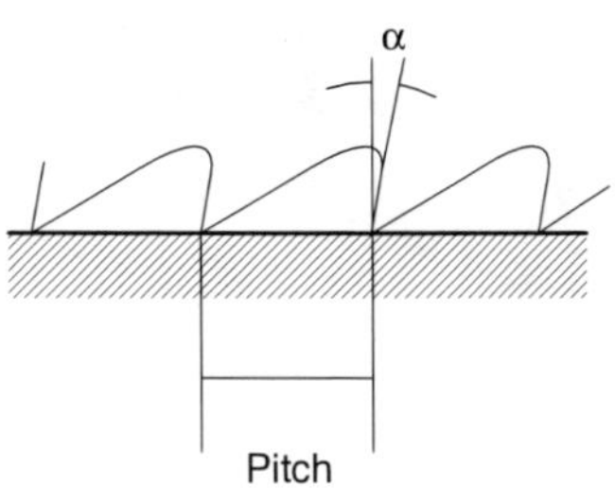

Fig. 7.43

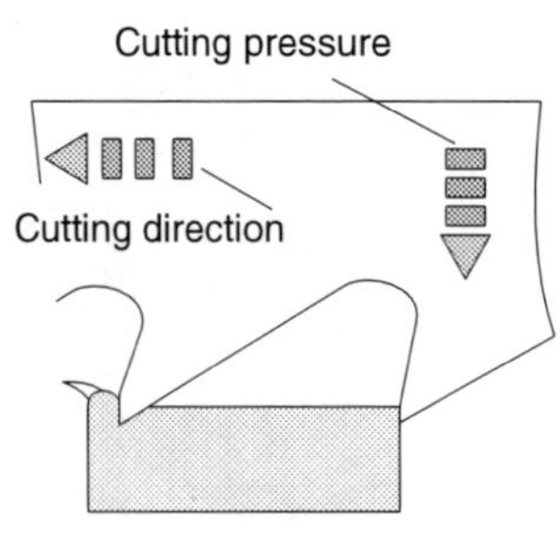

Fig. 7.44

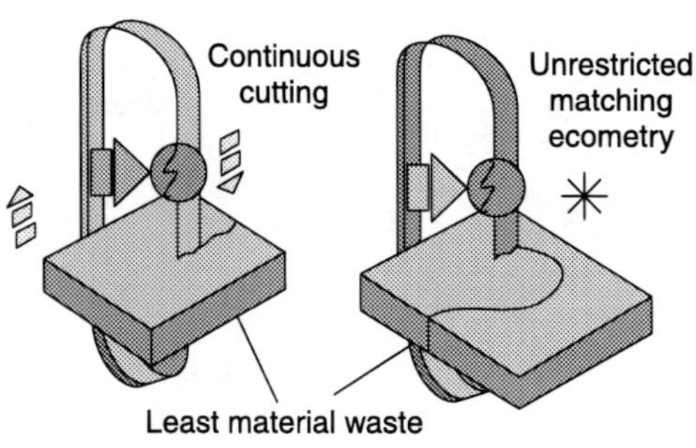

Fig. 7.45

Power Hacksaw (Fig. 7.44): Cut-off saws are used to cut metal stock roughly to the required length. The commonly used cut-off saw in small scale industries is a power saw.

The power saw works like a hand hacksaw, and has an arrangement for cutting during the movement in one direction and releasing pressure on the non-cutting stroke. The rotary motion of the motor is converted into linear motion by a crank mechanism.

The required cutting pressure is obtained hydraulically or by an adjustable weight. During the non-cutting motion the blade will be lifted away from the work. A clamping device/vice holds the work firmly.

Power hacksaw blades: The saw blades are selected, depending on the machine and the type of work on hand. The blades are made of high-speed steel and are fully hardened.

For different materials, blades of different pitches are used (number of teeth per 25 mm length). **(Fig. 7.43)**

Teeth with large pitch can accommodate large chip **(Fig. 7.44)**

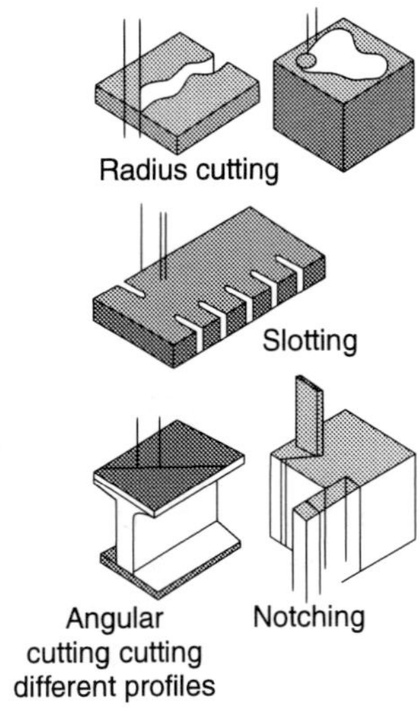

Fig. 7.46

Specification of Power Hach Saw Blade (Fig. 7.47)

While specifying power hacksaw blades, it is necessary to state:

- the length (the distance between centres of holes,
- width
- thickness and
- teeth pitch.

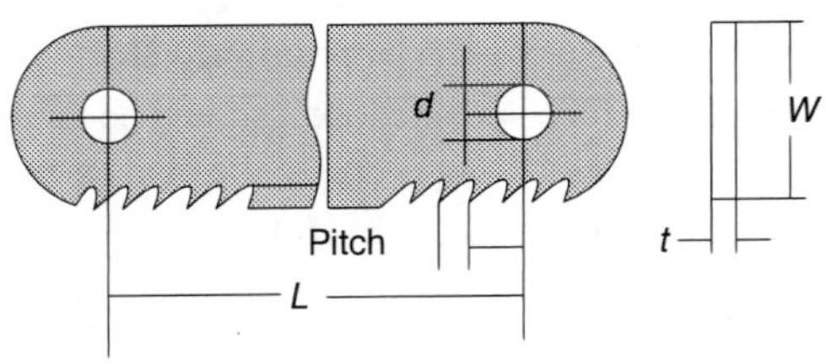

Fig. 7.47

7.2.17 Power Hack Sawing

The oldest and most economical sawing method is power hack sawing. Hack sawing machines typically have a four-sided, closed saw frame, saw-frame guide, and arch **(Fig. 7.48)**. The saw frame is usually mounted horizontally, and it reciprocates on fully enclosed ball or roller bearings which are adjusted and pre-loaded to assure the frictionless rigidity of the frame. The saw frame is reciprocated by a shaper-type crank lever, which provides a uniform cutting stroke and a quick return stroke. A blade backup plate controls the bowing of the blade for extra blade rigidity and a safety switch is provided to stop the machine if a blade should break. Blade feed pressure is applied progressively and is automatically adjusted to suit the workpiece being cut. In some machines, feed pressure is indicated with a gague. Dual feed cylinders control the vertical movement and cutting pressure on hydraulic machines, while mechanical machines employ a mechanical feed unit mounted at the top of the arch.

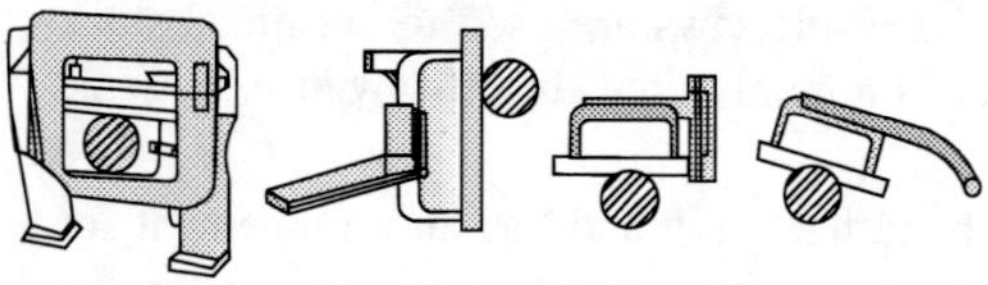

Fig. 7.48

The number of strokes per minute at which any hacksaw can be operated is limited by the peak blade speed. Therefore, a slower cutting stroke and a faster return stroke allow the blade to be operated at the greatest number of strokes per minute without exceeding the blade speed (in fpm) at which the blade will break down. Hacksaw machines are commonly provided with variable speed transmissions so that cutting speed can be changed to suit the material being cut.

The completely automatic power hacksaw machine performs the following steps:

1. Feeding the stock through the open vice
2. Gauging length to the required dimension
3. Closing and locking the clamping vice
4. Feeding the blade through the work
5. Raising the blade at the end of the cut
6. Opening the clamping vice

This cycle continues until the final cut, after which a limit switch stops the machine.

For some applications, a vertical power hacksaw machine is advantageous.

In it the blade reciprocates vertically and is fed into and away from the work hydraulically or mechanically. Designed chiefly for cutting tangs in large die blocks, this type of machine has a large, open work table, convenient for crane loading, upon which the work can be positioned at any required angle with the blade.

In selecting the proper hacksaw machine for a specific job, it should be remembered that the feed pressure on the blade must be applied progressively and in proportion to the blade size.

To prevent blade injury, the feed pressure must accurately follow the amount of blade penetration. Narrow kerfs or slots (as little as 0.055 inches in cutting through 6-inch-diameter bar stock) can be held accurately if the blade is properly guided.

A modern hacksaw employs a dual power feed. With hydraulic feed its capacity is sections 18 inches in diameter and larger; it is used with mechanical feed on sections 10 inches in diameter and smaller. The mechanical feed simultaneously controls both pressure and depth of feed directly in proportion to the number of blade teeth in contact with the work. In this way, the blade is automatically forced to cut as deeply as possible and practical on every cutting stroke without operator attention, resulting in cutoff in the least number of strokes regardless of the changing area or cut configuration of the workpiece. This mechanical screw-feed mechanism is simple, rugged, and easily understood and it is so nearly automatic that high cutting speeds and long blade life are obtained, even though an inexperienced operator may seldom change feed settings, regardless of the varying sizes and types of steels or nonferrous metals which may be cut.

7.2.18 Cutting Speeds

In power sawing, recommended speeds range from 50 to 160 strokes per minute. A speed up to the maximum of 160 may be employed in the cutting of soft metals. When the materials to be cut are hard and tough, the use of a lower speed will be found more efficient from the standpoint of production and is most economical because the blade life will be greatly lengthened. The Table below presents cutting speeds, pitch of blade, and cutting fluid recommended.

7.2.19 Power Hacksaw Recommendations

Work materials	Saw pitch	Speed-sfpm	Recommended cutting fluid
Aluminium and alloys	4-6	150	M or K
Angle iron (light)	6-10	125-140	W
Angle iron (heavy)	6	125-140	W
Armour plate	4-6	120-140	M or W
Babbitt	4-64	140	K
Brass	4-6	140	D or W
Bronze castings	4-6	120	W
Bronze, manganese	4-6	70-100	D or M
Cast iron	4	70-100	D
Channel (light)	6-10	120-140	W
Channel (heavy)	46	120-140	W
Copper bars	4-6	100	K or M
Copper tubing	46-10	100-120	W or M
Die blocks	4-6	50-100	S or W
Drill rod	46-10	90-120	S
Forging stock alloy	4-6	90-100	S
Forging stock milled	4-6	100-140	S or W
Magnesium alloys	2-4	130-150	S or W
Malleable iron	4-6	90-120	W or M
Monel metal	4-6	80-100	S or W
Nickel alloys	4-6	80-125	S or W
Pipe iron	6-10	100-130	W
Rails	6-10	60-125	W or S
Steel (carbon tool)	6-10	90-130	S or W
Steel (high speed)	4-6	85-100	S or W
Steel (machinery)	4-6	60-100	W or S
Steel (stainless)	46-10	60-100	S or W
Steel (structural)	6	100-140	W

D, dry: W, soluble (emulsifying) oil, 1 part oil to 5 water; S. synthetic water-soluble cutting fluid with wetting agent and anti-weldment properties mixed with 5 parts water: K. kerosene: 1. mineral lard oil sulfurized and chlorate for anti-weldment properties.

Note: Where two or more teeth per inch are recommended for any type of material, the correct number of teeth and strokes per minute should be determined by the size and type of machine and size of material to be cut.

The size of section, the rolling temperature, and the atmospheric temperature at the time of rolling all influence the final physical properties of any steel and therefore affect the efficiency of the cutting time and blade life.

7.2.20 Large-Capacity Hacksaw

The modern power hacksaw has proved the most efficient method of sawing large diameters on cross sections of ferrous and some nonferrous metals in sizes from 18 × 18 inches through 25 × 25 inches. Unlike the small saws with capacities through 10 × 10 inches, these large, heavy duty machines are of unique design. Because only the tooth points of the saw blade do the actual cutting, with the rapidity of the cut determined by the pressure and penetration of each tooth on each cutting stroke, the increase in workpiece cross section or in the resistance to the length of blade contact results in a reduction of pressure per tooth or penetration.

These heavy-duty machines utilize "roll stroke" blade motion which progressively raises one end of the blade and lowers the other end as the blade is drawn through the work, thus developing the same rolling motion that is applied to a hand hacksaw. As this motion produces a cut slightly curved at the bottom, the blade is in contact with only a short section of the work at any one time. The required feed pressure is therefore greatly reduced, the resultant pressure per tooth is greatly increased for maximum and efficient tooth penetration, and more chip clearance is provided.

Large-capacity hacksaw is also efficient because low-cost, thin, high-speed cutting tools are available with proper beam strengths or blade widths and proper thicknesses of blade beam.

7.2.21 Work Setup

Work clamping (Fig. 7.49): The following are recommendations for clamping various types of work in conventional damping vises.

1. Thin strips should be flat, if at all possible.
2. Short ends of stock should be cut singly to assure accuracy and proper clamping.
3. Hexagonal stock should be laid on the plane of the bar; however, in the case of multiples almost any arrangement is satisfactory.
4. Round bar stock, when cut in multiples, should be nested in a square or rectangular bundle. The bars should be interlocked in layers with every other layer having one less bar than the first and last layers for more secure damping and, more importantly, for constant contact of the saw blade with the work without interruption from layer to layer. Equal number of bars in each layer causes the saw blade to "step out" of the cut after cutting each layer because of lack of support.

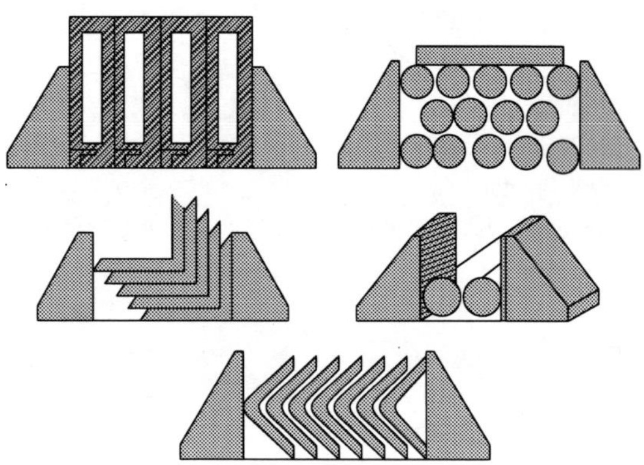

Fig. 7.49

5. Tubing should be clamped in the same manner as the round stock.
6. When rails are cut singly, lay the rail on its side, with the head of the rail against the stationary vice jaw. This method will give more cuts per blade and better cutting time.
7. Multiple or nested cutting of angles, channels, etc., can be done by loading the machine vice.
8. In most cases, an overhead guide bar should be used while clamping workpieces for multiple-cutting set-ups. When the workpieces are clamped with the movable vice jaw, the parts will readjust and automatically arrest themselves against the overhead guide bar. In this way the work is clamped from the side and top and held rigid for sawing.

Blade alignment and tensioning: Care should be taken in the alignment and tensioning of blades, as nearly all the machines in the field are of the draw-cut type. The following procedure is recommended:

1. It is important that the balance be mounted in the machine so that the teeth face the direction of the crank.
2. The blade pins should be near the size of the pinhole in the blade.
3. The slack should be taken up on the blade pins.
4. The blade clamp should be adjusted parallel to the blade holder.
5. The blade may now be securely clamped and then properly tensioned.

Insufficient blade tension is the most common cause of inaccurate and inefficient hacksawing. The correct amount of tension on each blade will vary according to the length, width, and thickness of the blade. When maximum tension is required for machines equipped with hexagonal nuts on their tension screws, the nuts should be tightened 8 flats on 14-inch blades, 12 flats on 18-inch blades, 14 flats on 24-inch blades and 12 flats on 18-inch blades after the blades are pretensioned. For machines which employ a compound-leverage double screw or a round narrow nut for tensioning, the optimum blade tension is best gained with the use of a stretch gague.

Methods by which the tension of the blade can be measured to a reasonably close degree include the use of: (1) a torque wrench on the blade-tensioning screws, (2) a pressure-indicating: device directly under the blade-tensioning screws and (3) a device for measuring the stretch of the blade.

Care should be taken in checking the crankshaft for proper rotation. The machine should begin to feed downward at approximately a 15° angle with the crankshaft at the beginning of the cutting stroke and lift up at the completion of the cutting stroke.

7.2.22 Operations

Starting a cut: When a cut is started on a sharp corner of the work, a fine-toothed blade should be used and applied to the work to engage as many teeth at one time as is possible. When starting on a corner is unavoidable, it is important that the cut be started with light pressure until the blade engages a larger cutting area, after which more pressure may be applied. In the case of thin sections of metal or light-wall tubing, light feed and higher speeds should be used. The following are safe rules for starting cuts in various types of sections:

1. **Small sections:** Bars, tubes, light angles, etc. Use more teeth per inch and light feed.
2. **Medium sections:** Squares, rounds, etc. Use medium-pitch blade and medium feed.
3. **Large sections:** Size greater than stroke of machine. Use coarse-tooth blade.
4. The first two rules are limited only to cutting bars or tubes singly. If they are cut in multiples, a coarser tooth and heavier feed can be used to increase production.

Cutting different materials: On thin sections of metal, light-walled tubing, and very soft metals, use light feed pressure. For heavier sections and harder metals, use heavier feed pressure. In power sawing, recommended speeds range from 50 to 160 strokes per minute, the higher speeds are used for cutting soft metals. For hard and tough materials, the lower speeds are advisable. This procedure will be found more efficient from the standpoint of production and blade economy.

Difficulties: The most common problem in the operation of a hacksawing machine is crooked cutting. The most probable causes of crooked cuts are: (1) misalignment or the work with the machine vice; (2) worn out blades; (3) excessive feed or speed; (4) insufficient blade tension; (5) a loose saw frame, saddle, or machine frame; and (6) improper cutting fluid or mixture.

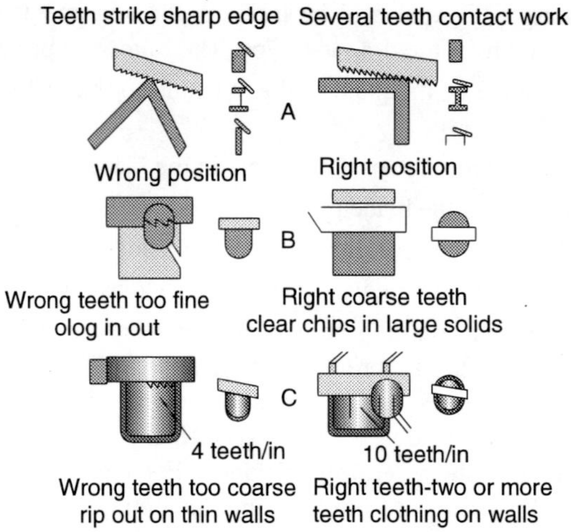

Fig. 7.50

Blade replacement: For maximum safety and the elimination of blade-breakage hazards, properly tensioned high-speed, welded-edge, shatter-proof hacksaw blades are recommended. If for any reason a blade is replaced in the process of a cut, the work should be turned over and a new cut started on the opposite side if at all possible. If this step is not feasible, the new blade should be entered into the kerf made by the old blade at a reduced power feed.

7.2.23 Power Band Sawing

The band saw machine has a long continuous band with many cutting edges or teeth travelling in one direction. The long band results in an uniform cutting action with an evenly distributed, low individual tooth load. The narrow blade is capable of following intricate contours and reduces very little material to chip waste, thus saving material and giving a low tool and machine operating cost per square inch of cut.

The two basic types of band saw machines, horizontal and vertical, are used for a wide variety of cutting operations including cutoff or straight sawing and contour sawing of nearly any form or outline. They are used for rough shaping to finish operations within ±0.003 in and are capable of closer tolerances when using continuous file bands and abrasive bands to finish work. Band machines can be used on practically all materials, which further increase their inherent versatility.

7.2.24 Vertical Machines

The vertical band saw machine, or "contour machine", although sometimes used for cutoff work, finds its greatest application in two other fields: shaping work by removal of unwanted material, both externally and internally, and slotting and splitting. The band runs vertically, and the work, carried on a horizontal table, is fed into the band. Machines are commonly available in a wide range of sizes from small tool room machines to large production models. Throat depths vary from 16 to 60 inch, band speeds from 2.5 to 15,000 fpm, and horsepower from 1 ½ to 15.

In some applications it is impossible or impractical to feed the work into the band, so machines in which the entire sawing head is fed into the stationary workpiece have been designed. On other applications, the work is fed manually over a fixed table or by power on a hydraulic table using simple fixtures, or where mass production warrants, more elaborate completely automatic machines and fixtures are recommended **(Fig. 7.51)**.

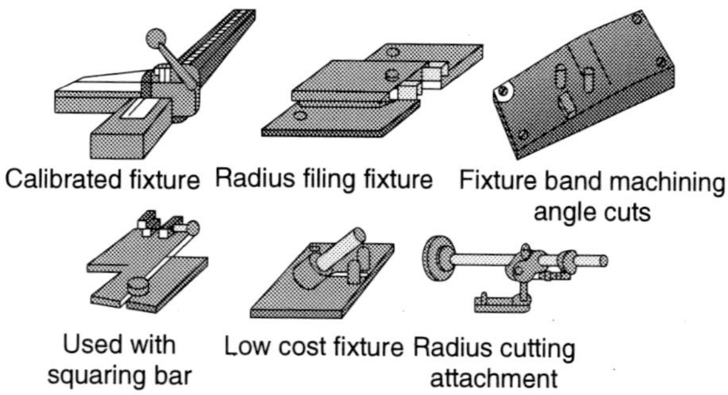

Fig. 7.51

The uniform downward cutting force provided by the band aids in holding the work, making elaborate, heavily constructed clamps or fixtures unnecessary, and it also allows soft, spongy or honeycomb materials to be cut without distortion. Machining time is low because the narrow band cuts directly to the layout line, which may be so intricate that other machining methods are impractical or impossible to use. The absence of reciprocating motion or exposed moving parts (only a small amount of the actual cutting portion of the blade is open and the downward cutting force makes the vertical band machine efficient and extremely safe to operate. Relatively unskilled workers can safely produce quantities of work to close tolerances with a minimum of training.

Various work tables, coolant facilities, blade welders and attachments for vertical band-saw machines are available. The simplest machines have fixed work tables which are generally inclinable for sawing compound angles. The gravity feed usually employed with this type of table has a control-cable arrangement to draw the work into the saw blade. This arrangement relieves the operator of pushing the work but still permits him to guide it along the layout line. Further advantages of power feeding are obtainable with hydraulically operated tables provided with speed and feed pressure controls, the foot-operated directional control leaves both the operator's hands free to load and unload work. Machines are available in which table tilt, blade-speed change, tool-post height, and band tension are hydraulically controlled. For sawing large cross sections, such as die blocks powerful variable-speed, feed tables with quick return motion are used. So that layout lines may be easily followed, air jets keep the work surface free of chips. Spray or flood coolants permit band saw machines to be operated to the fullest capacity of the saw blades.

7.2.25 Contour Sawing

The work can be easily guided manually or mechanically in contour sawing because of the uniform downward cutting force of the band. The blade beam strength greatly determines how fast the work can be fed into the blade and the wider the blade, the higher the beam strength. However, the width of the blade is also determined by the smallest radius to be cut. Internal sawing requires one drilled hole to start the blade, but additional holes may be drilled at sharp corners and small radii to permit the use of a wider blade. The saw can "file" the material to produce a corner within ± 0.002 inch. An even feed pressure is desirable for optimum results and blade life. Some machines with hydraulically fed tables have a servo control mechanism to assure a constant feed force against the blade. These machines sense the force and adjust the table travel (both direction and speed) to maintain the preset feed force when making straight or contour cuts. The servo control mechanism, coupled with a mechanical device to turn the work, enables difficult contours to be cut with high accuracy and little operator effort or experience.

7.2.26 Contour Sawing of Dies

Internal contour sawing is well suited to producing blanking dies and other straight-through die openings. It is the only process that permits both the die and the punch to be machined from the same piece of steel. This procedure is not complicated because every work step and related dimension is based on the thickness of the material.

1. The starting hole is drilled at an angle that varies with the thickness of the material. It is started inside the die layout line at the centre of the die block and emerges from the die block on the opposite side of the layout line.
2. With the table tilted at an angle slightly less than that of the starting hole, the slug is sawed out of the die block. The path of the saw on the surface is entirely inside the layout line, and the proper cutting angle at the bottom of the slug allows for excess material outside the die layout line.
3. With the slug or punch, section, removed from the die, is band-filed to the layout out line, and the slug is band-filed to fit the die. The filing operation removes all traces of the starting hole on the finished surfaces of both, the punch and the die.

7.2.27 Saw-blade Welding

Butt welding of the saw blade is an important part of contour work, particularly for internal sawing, because the blade must be threaded through the drilled hole and welded into a band before the sawing starts. Some machines have built-in butt-welding devices. If both the ends of the band are properly squared, properly aligned on the backs of the jaw clamps of the welder and carefully annealed by intermittent annealing for progressively shorter intervals until the weld is cool, no trouble should be experienced. A poor weld may be caused by misalignment, improper electrical setting, or wrong timing of the annealing period.

7.2.28 Production Tooling

In addition to the tool room uses of the vertical band sawing machine, such features as the hydraulic feeding table with foot control, high-speed steel cutting-edge saw blades, and coolant facilities make this type of saw suitable for high production. Because the cutting force is perpendicular to the table, and the feeding force is into the blade, the component force holds the work in place. Fixturing in many cases consists of a simple back stop or calibrated work fixture bolted to the table. The speed of set-up makes the vertical band machine suitable for short as well as long production runs. Parts with holes or circular openings can be dropped over a vertical mandrel. With fixtures of this type, the operator loads the machine with one hand, unloads with the other, and cycles the table with the foot control.

Slotting and splitting operations are well suited for the vertical band machine. Saw blades are obtainable to produce a kerf of the desired slot width.

7.2.29 Band Saw Finish

The surface finishes that can be expected from band sawing range between 125 and 500 μ in rms. The finish is dependent upon feed rate, saw pitch, saw velocity, and the material being sawed.

7.2.30 Horizontal Machines

The horizontal band sawing machine, or cutoff machine, should be considered when the following requirements are to be maintained in a cutoff operation:

1. **Speed.** Because the cutting is continuous and high-speed steel saw blades may be used, actual sawing time is short. Sawing rates up to 48 square inches/min for some steels, such as AISI 1018, are

attainable, with rapid-indexing work-feeding tables, fast cutting rates, and rapid set-up, production rates are high and cost per cut is low.

2. **Accuracy.** Horizontal band sawing machines will repetitively cut pieces to a length accuracy of ± 0.005 inch and with a flatness of cut within 0.002 inch/inch of work thickness or diameter.
3. **Low scraps losses.** The exceptionally narrow kerf of the saw blade, 1/16 inch, reduces to scrap less than half as much material as any other sawing method. The shorter the part being cut, the more important this consideration becomes, especially when expensive alloys are cut.
4. **Versatility.** Infinitely variable band speed and easily adjusted cutting feed permit most horizontal band machines to saw anything within their range and capacity with only minor adjustments. This is particularly important when a wide range of work sizes and/or work materials is encountered.

High-performance machines have infinitely variable saw blade velocities from 60 to 450 fpm with tachometers to permit accurate speed selection. In all cases the manufacturer's recommendations should be followed. To select the proper work feed and saw velocity, the work material must be considered.

The band blade is automatically tensioned hydraulically to 30,000 PSI, or approximately 1000 pound on each band strand. Feed pressure is uniform and unaffected by the force of gravity as the saw feeds downward because of a positive hydraulic feed compensator which automatically furnishes the proper hydraulic back pressure to the cylinder that supports the blade head. The machine stops automatically when it reaches the desired number of cuts or when its shuttle vice runs out of work.

7.2.31 Sawability of Work Materials

The characteristics of steel that affect sawing speeds and feeds are toughness, abrasiveness, free-machining qualities, and work-hardening qualities. Commercially annealed steels are in the best condition for sawing. Most low-carbon and low-alloy steels can be sawed at maximum cutting rates but require the use of a premium-grade, water-soluble coolant or lubricant to prevent chips from sticking to the saw teeth. To prolong band saw blade life, some machines are also provided with a power-driven wire brush to clean the teeth. The harder the steel is, the more detrimental its effect on tool life and cutting rates. Many steels can be cut at reduced rates even when hardness runs up to 450 Bhn. Hot-rolled or forged steels are generally more difficult to cut than cold-finished steels even though they are softer. Because the scales and oxides on hot-rolled, forged, or heat-treated bars have abrasive characteristics, the expense of removing the scale by pickling or turning may be repaid by increased tool life.

The work hardening of steels during sawing is often caused by dull teeth, excessive band speed or too light a feed pressure. Re-entering the same cut in which a band has failed will very likely lead to a second failure in a short time. Use of proper coolant and ample feed pressure is very important with work-hardening steels.

7.2.32 Band Saw Blades and Other Band Tools

Saw bands are differentiated by three characteristics: (1) metallurgy, (2) tooth shape and spacing, and (3) set.

Metallurgy: There are five types of saw blades based on the steel used and its heat-treatment

1. **Bimetalic high-speed steel blades:** These band saw blades have a high-speed steel cutting edge that is electron-beam welded to a fatigue resistant, high-tensile-strength, alloy steel back. The high-speed steel teeth are harder than other types and more abrasion-resistant, combining properties that give fast production and long tool life. The backing material is especially chosen for long life under frictions, weldability, and tensile strength, all of which contribute to high capability. Bimetallic blades are available in two cutting-edge harnesses: 64 to 65 Rc for sawing mild to tough materials and 68 to 69 Rc. for sawing the tough and abrasive difficult-to-machine materials, including special alloy, columbium-based, and titanium-based metals.

2. **High-speed-steel blades:** The high "red-heat" hardness of the teeth of these blades enables them to saw ferrous materials at elevated saw velocities to produce high cutting rates and long tool life. The teeth are hardened to 63 to 64 Rc, while the body, also composed of high-speed steel, remains flexible and springy.
3. **Intermediate-alloy steel blades:** These bands are made of a tough intermediate high speed-steel alloy with a tooth-tip hardness of 64 Rc. They are recommended for sawing pipe, tubing, structural-steel shapes, carbon-steel bars, and low-alloy steel bars. They have a high degree of toughness and high resistance to shock and breakage.
4. **Hard-edged, hard-backed carbon-steel blades:** These bands have a tooth-tip hardness of 63 Rc and spring-tempered backs with hardness of 44 to 48 Rc. These blades withstand high band tension that produces good suitability and ease in following a layout line. They are operated at low to medium saw velocities. The high tensile strength allows greater feeding force, resulting in faster cutting rates.
5. **Hard-edged, flexible-backed carbon-steel blades:** These blades have been traditionally used for all-around tool, die, and general machine-shop work, as well as for non-ferrous materials including woods and plastics. The teeth are hardened to 60 to 62 Rc and the back has hardness of not more than 74 on the RA scale. When sawing ferrous materials, the saw velocity should be adjusted from low to medium.

Tooth shape and spacing: Band saw blades regardless of metallurgical type employ one of the following three tooth styles:

1. **Conventional:** This is a closely spaced style with rounded gullets, the teeth having zero rake angles. This type is widely used for conventional sawing and for friction sawing **(Fig. 7.52).**
2. **Buttress or skip tooth:** This style also has zero rake angle, but the spacing is wide in order to provide elongated gullets. The increased gullet capacity accommodates the greater chip load required for cutting free-machining nonferrous alloys, woods, and plastics. This style in the hard-edged type is recommended for thick sections of ferrous metals.

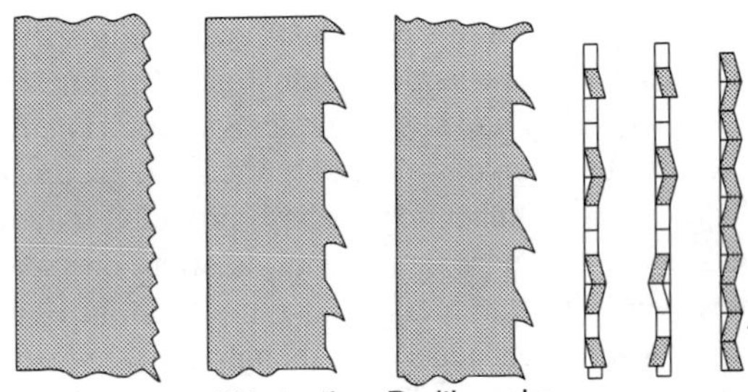

Fig. 7.52

3. **Hook tooth or positive rake:** This tooth style has a forward slanting tooth face with a correspondingly sharp cutting angle. It gives faster cutting rates and longer tool life but at a slight sacrifice to finish.

7.2.33 Tooth Set

Tooth set is the side bend given to saw teeth to provide clearance for the body of the saw band. This clearance prevents the blade from binding in the work and permits cutting of contours. Three patterns of set teeth are in general use.These are:

1. **Raker set:** This design has one unset tooth followed by a tooth set left and another set right. The rake or unset tooth acts as a stabilizer to compensate for the side pull of the set teeth and ensures accurate cutting, particularly under heavy cutting pressure. Raker set is used extensively in hard-edged metal saw blades **(Fig. 7.52d).**

2. **Wave set:** This design has teeth set to right or left in groups with a rake or unset booth between groups. It is used when the thickness of the material varies or for very thin materials that may catch or shock the teeth. With the teeth set in groups more than one tooth remains in the work at any given amount, thereby distributing the load and preventing shock **(Fig. 7.52e).**
3. **Straight set:** This style has teeth set symmetrically, one to the right followed by one to the left. This is the traditional set used primarily in spring-tempered blades **(Fig. 7.52f).**

Radius cutting controls the permissible band width, which in turn, affects the feed rate and pitch selected. Use the widest band applicable to job conditions, drilling holes where required avoiding small radii.

7.2.34 Other Band Tools

Bands other than saw blades with conventional teeth may also be used on vertical band sawing machines.

File Bands: Finish and cutting rate are achieved with principles of file-band selection comparable with those of saw-blade selection. Production filing with coarse-cutting bands can achieve filing rates approximately nine times faster than hand filing.

Diamond Bands: Hard, brittle materials such as stone and glass can be cut with diamond edged saw bands. Adequate coolant and accurate feed control are required.

Abrasive Bands: These bands are used instead of saw blades on the vertical band machine for polishing. A backup support is used as a platen to strengthen the band. A variety of grain sizes permits the choice of rapid stock removal of high polishing action.

Spiral Bands: For intricate sawing in any direction without turning the work, spiral bands are available which will cut in any direction. Two types are manufactured: spring-tempered type in various sizes for wood and plastics, and a hard-edged type for metal.

Special Bands: Special bands are produced for cutting cloth, meat, type metal, and many other materials. A saw specialist should be consulted for information on cutting any unusual material.

Saw Bands for Horizontal Machines: High-speed steel saw bands are widely used on horizontal band sawing machines because they can cut ten times faster and have a tool life up to thirty times greater than carbon-steel bands.

7.2.35 Circular Sawing

The circular-sawing machine method, as a cut-off method for all ferrous and non-ferrous metals and many plastics, has proved its efficiency. In sawing non-ferrous metals, this method has usually proved to be the most rapid, with excellent resultant finish, tolerance, and economy. The circular saw, compared with other types of cutoff saws, is a much heavier machine for similar capacity and contains many more moving parts. It is a more sophisticated machine tool and, generally, requires an operator with above-average skills.

The following selections, based on the workpiece type and material, must be made in order for circular sawing to be efficient:

1. The proper type of machine, whether horizontal or vertical
2. The correct saw blade
3. The proper grind of the blade
4. The correct speed and feed
5. The proper cutting lubricant

The circular sawing machine operates on the principle of a milling machine. Unlike the conventional milling machine, however the diameter of the blade used on a circular sawing machine is governed by the cutting capacity of the machine since the diameter of its spindle gear is usually smaller. The tool or blade must *be* firmly attached to the spindle especially when ferrous metals are being cut. Blade widths may range from ¼ to 1 inch. This relatively great tool thickness demands five to ten times more horsepower than that required by a modern hacksaw or band saw machine of comparable capacity. These factors call for a heavy, rugged design with fine workmanship to ensure a smooth, powerful drive to the saw blade. The gearing throughout the entire drive must be made with a high degree of accuracy to ensure a minimum of backlash in the blade.

The machine should have the necessary speeds and feeds to cut all the materials to be sawed. Generally, speed changes are accomplished through selective gearing, pick off gears, or sheave changes. Feeds can be mechanical, hydraulic, or pneumatic. In ferrous-metal cutting, the work is generally positively clamped mechanically or hydraulically, and the saw blade is fed either horizontally or vertically into the work. To facilitate feeding the work through the machine, some models are equipped with an automatic or mechanical lift roller to lift the work free of the vice and to enable the work to be moved freely in and out of the machine. The work may be clamped directly on rollers. Many smaller machines are furnished with complete automatic bar feeds.

Stock stops for short lengths are built into practically all circular sawing machines. Stock stops should be designed so that they can easily be moved out of the way to allow the cut pieces to fall clear, by actuating the stop with either a foot treadle or a cylinder. Machines with automatic bar feeds usually have the stop built into the feed assembly.

Chip cleaners or brushes are commonly used to keep the chips from clogging in the gullet. These cleaners are mounted at the rear of the head carrying the blade, enabling the blade to present a clean cutting edge to the work and eliminating pick up on the sides of the blade.

7.2.36 Applications

With proper selection of machine, blade, coolant, feed, speed, and blade grind, circular sawing can be most efficient for cutting bar diameters through 4 inch. The circular saw is not popular for sawing large diameters from 10 through 25 inch, primarily because of its high initial cost, exceedingly high blade investment, heavy chip or kerf loss, and tool cost. This inefficiency has been especially true when sophisticated metal tool steels, high-alloy steels, exotic alloys, and pre-hardened steels are sawed. However, for sawing non-ferrous materials such as copper, copper-nickel, brass, aluminium, and plastics, this method is most efficient, offers excellent economy, the fastest cutting cycle, and an excellent finish. When non-ferrous metals are sawed, the tool or blade does not have to be as thick or heavy as that used for sawing ferrous metals because of the greater machine ability of the non-ferrous metals. Kerf loss is therefore diminished, the horsepower required is reduced, and the massively heavy machine design is not needed.

For sawing the non-ferrous metals, circular saw machines are available in a wide range, from small sizes priced from a few thousand rupees to large sizes, with capacities as great as 47 × 18 inch. The more sophisticated circular saws are also available with fully automated rack-type bar feeds.

As an example, one circular saw will cut round, free-cutting brass to 5/8 inch $2^{1/8}$ inch lengths at an hourly production of 5,500 pieces. Blade life between grinds with this machine is 60,000 cuts, kerf or material loss is 0.055 in per cut, and finish is 50 μ inch rms. Blade regrinding tolerance to length is 0.001 inch, and tolerance for squareness is 0.001 inch. In slugging or cutting round bars averaging 5 to 6 inches in diameter, the average time per cut is 1 second. The same type of machine cutting bars as large as 10 inches in diameter is capable of making these cuts in an average time of 2 seconds per cut. Blades are operated at approximately 6000 fpm, and soluble oil and water are used as the coolant in a mixture of 25 parts water with 1 part oil. Blade life is approximately 5,00,000 in^2 between grinds.

Large copper cakes are cut by the circular sawing method, and for maximum efficiency some producers use double-headed saws with blades horizontally feeding or traversing towards each other. Just before the blades contact, one blade stops and retracts while the other blade passes. Through the centre to complete the cut. When sawing copper cakes as large as 27 × 16 inches, average time per cut is approximately 30 seconds.

7.2.37 Work Set-up and Operation

Whenever possible, the work is sawed singly so that simple bar storage tables and conveyors can be used, thus eliminating a great deal of material handling and saving the time used for assembling and bundling the work. The crop ends can be held to a minimum of scrap loss. Round bars are held in a 'V' block, and square or rectangular shapes are clamped directly on the bed or the supporting plate on the machine. At times it is possible and advantageous to saw two or more pieces in a single pass. **Figure 7.53** shows a number of work-clamping methods to facilitate gang cutting. When bars that are crooked or warped are sawed, whether singly or in multiples, care must be taken to ensure that they are properly seated in the 'V' block or in the vice. Special shapes can be sawed by holding them in special fixtures, or milling and slotting operations can be done by holding the work in fixtures and by using special blades.

7.2.38 Blade Selection and Application

Three types of saw blades are used in circular sawing: (1) solid blades. (2) inserted-tooth blades, and (3) segmental blades.

Solid blades: Solid blades are generally used in small diameters, not exceeding 14 or 16 inches. For blades 8 to 10 inches in diameter, to reduce kerf, obtain faster tooth penetration, and reduce horsepower, blades are available with reduced cutting thicknesses and increased flange thicknesses for rigidity. Blades as large as 8 inches in diameter can be obtained with a cutting thickness of 0.050 inch. Solid blades of 16 inches in diameter and larger are seldom used except for non-ferrous metals and plastics. Blades larger than 16 inches in diameter are generally of the inserted-tooth type.

Inserted-tooth blades: Inserted tooth blades have individual teeth mounted and held in place on the saw-blade disk by wedges. This arrangement makes individual tooth replacement easy, but because of the necessary backing required for each tooth, blades of this type are recommended only for extremely large-diameter work. Inserted tooth blades are difficult to arrange with close tooth-to-tooth centres or with a tooth pitch fine enough for prescribed work. These blades, ranging in diameters of up through 120 inches are seldom used except in steel mills and relatively heavy forging shops.

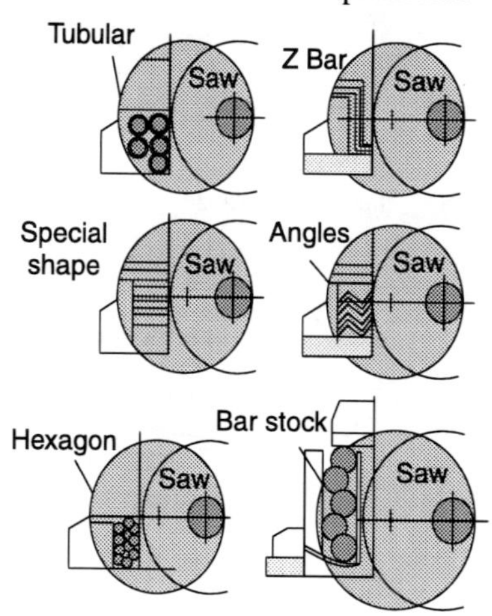

Fig. 7.53 Clamping various shaped material

Segmental blades: Segmental blades are constructed by a tongue-in-groove or tongue-in-lip arrangement of a series of segments mounted on a circular disk and riveted into place. This arrangement provides for a strong, rigid blade, and the segment construction allows for blade design into any specific pitch or tooth arrangement as required by the nature or size of the metal to be cut. The inserted tooth and segmental blade types offer tooth or segment replacement in the event of broken teeth or undue wear on an isolated tooth or teeth, thus helping to protect the initial large blade cost or investment.

Diameter and pitch: The saw blade chosen for specified work should have the correct diameter and pitch for the specific material to be sawed. For example, for sawing 9 inches diameter mild steel, the blade should have a diameter sufficient to clear through the 9 inches thickness in one pass, and a pitch should be selected to provide ample chip clearance in the tooth gullet when the blade is in the centre of the cut. Table 7 shows approximate pitch of blade to use for a given size of work.

When sawing thin or heavy walled tubing or structural shapes, a blade pitch slightly less than the thickness or width of the wall or section is recommended, regardless of the material. However, when sawing non-ferrous metals; with the blade revolving at far greater peripheral speed, a larger tooth pitch can be used.

Blade grinding: The proper grind of the circular saw blade is the most important single factor in efficient circular sawing. Through years of practice, the accepted grind is a cam generated tooth contour. This type of grind reduces the cutting friction and allows the chip to curl up, and a much smoother sawing action is obtained. Blades are generally ground with a high tooth and a low square tooth. The high tooth is 0.010 to 0.020 inch higher than the square tooth. The corners of the high tooth are bevelled at 45° so that one-third of the cutting edge is applied to the work. This roughing tooth takes out the centre of the cut or kerf. The square finishing tooth is the full width of the kerf and cleans up the corners of the cut. In this way, the chip is broken and the load distributed among the teeth. The roughing cutter or tooth does not receive the full load, and the load on the square tooth is divided and reduced. This tooth configuration is shown in **(Fig. 7.54).**

Table 7.1

Material	Stock size (inches)	Saw pitch
Mild steel	1-3	0.40-0.60
	4-7	1-3
	8-15	1-3
Alloy steel, annealed	1-3	0.25-0.50
	4-7	0.50-0.80
Alloy steel, as rolled	8-15	0.80-1.10
	1-3	4-7
	4-7	8-15
Cast iron	8-15	4-7
	1-3	8-15
	4-7	0.40-0.70
Soft brass copper	8-15	0.70-0.90
	1-3	0.90-1.10
	4-7	0.40-0.65
Hard brass bronze	8-15	0.65-0.80
	1-3	0.80-1.30
	4-7	0.40-0.70
Aluminum-magnesium alloys	8-15	0.70-0.80
	1-3	0.80-0.90
	4-7	0.50-0.80
	8-15	0.80-1.10

The proper grind of the circular saw blade tooth includes two important angles: the face or rake angle of the tooth and the clearance angle of the tooth. The rake angle can range from 12° to 25° and the clearance angle from 7° to 11°. For non-ferrous metals and soft ferrous metals, the angle should be ground at the upper limits-a 250° rake angle and an 11° clearance angle. For tough alloy steel the angles are reduced to the lower limits-12° for the rake angle and 7° for the clearance angle.

When large cross sections are to be cut, regardless of the material, and when a long chip per tooth is cut care must be taken to ensure sufficient space in the gullet of the tooth to accommodate the chip. The curling of the chip in the gullet is facilitated by maintaining an ample, well-rounded, smoothly finished gullet.

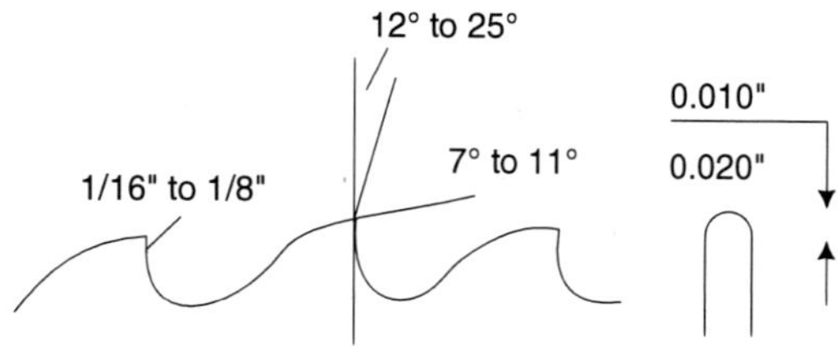

Fig. 7.54

Circular-saw blades with cemented carbide teeth have been used satisfactorily on non-ferrous metals and under certain adverse conditions on ferrous metals, but because of high blade cost, a high speed steel segmental or inserted-tooth blade should prove most efficient from cost and operating viewpoint for most work.

Feeds and Speeds: Recommended blade speeds for circular sawing of most commercially used round stock most efficient speed depends on the condition of the machine, the physical characteristics of the material sawed, and the type of finish, i.e., hot-rolled, forged, scaly, etc. Generally, the lowest speed that will produce the desired results should be used in order to extend blade life.

The feed rate for solid material should be set by determining the area in the section to be sawed.

For example, mild steel can be sawed at approximately 25 in^2/min. A 5-inch square would be 25 in^2, and the sawing time would be approximately 1 min. The rate of feed would then be 5 linear inches per minute (ipm). The travel of the saw-blade carriage in inches per minute can be computed from the following formula:

$$R = LT/A$$

where L = required length of travel to complete piece, in

A = area of work, inch square

T = work area traversed in unit time, in^2/min

R = feed rate, ipm.

In sawing steels, it is possible to estimate cutting time by means of a feed per tooth equation:

$$T = \frac{L \times 2P}{sfm \times 12 \times C}$$

where T = time, min

L = required length of travel to complete piece, in P = circular pitch of saw blade, inches,

C = feed per tooth, inches

sfm = surface feet per minute of sawing

For solid high-speed steel blades cutting material from 1/8 to 1½ inch in diameter, the feed per tooth should be between 0.001 and 0.0035 inch. For segmental or inserted-tooth high-speed steel blades cutting material from $1^{3/4}$ to 15 inches in diameter, the feed per tooth should be between 0.005 and 0.012 inches. The feed per tooth varies proportionally to the machinability of the material.

7.2.39 Friction Sawing

Friction sawing employs the friction created between a moving saw blade or band and a stationary workpiece to create heat to soften the workpiece near the blade and allow the saw to remove workpiece material. The blade must operate at a high enough speed and pressure to generate frictional heat faster than the workpiece can absorb or dissipate it. When the work reaches a high temperature below its melting point, its tensile strength is rapidly reduced, and the weakened surface is cut away by the moving blade. The removal of the initial surface exposes a succeeding surface to the same heating, weakening, and removal process.

To prevent the blade from reaching a high temperature and being weakened, the time a portion of the blade cutting edge is in contact with the work and being heated must be a small fraction of the time it is out of contact and being cooled. Therefore, the blade must have a much larger peripheral length than the length of the cut. For the reason only band sawing and circular sawing are suitable for friction cutting.

The key factors to proper friction sawing are: (1) heat dissipation from the blade rim at some point after emerging from contact, (2) freedom from side binding in the cut by proper blade alignment and adequate side grinding of the blade, (3) maintenance of a properly sharpened blade, (4) constant feed pressure, and (5) a drive motor suitable for maintaining constant blade rim speed.

7.2.40 Workpiece Materials

Hardness of a material does not limit the cutting ability of the saw. In fact, high-carbon steel, in either its hardened or its annealed condition, is more easily brought up to its weakening temperature by the saw than is low-carbon steel. The real criteria that determine whether a material can be satisfactorily friction sawed are its melting characteristics and its grain structure. The friction saw removes metal from the heated contact surface of any metal whose grain structure will not break down and which will lose sufficient tensile strength to break away from contact with the colder metal beneath without reaching a melting temperature.

Of the ferrous metals, all steels are readily severed in this manner. The grain structure of cast iron breaks down on the contact surface before it can be heated sufficiently to lose tensile strength. Therefore, cast iron is removed purely by an abrasive action without very satisfactory results. With the non-ferrous metals, such as; bronze, aluminium, or copper, the contact surfaces do not reduce in tensile strength sufficiently to break away from contact without melting. Consequently the metal being removed brazes to the warm steel blade rim before it can emerge from contact, and friction sawing of these metals is also unsatisfactory. Certain plastics and other materials may be friction-sawed if they are thermoplastic and can lose sufficient strength without melting.

7.2.41 Work Hardening

The depth of heat penetration in the material is very slight, even when the blade approaches the limiting length of contact are with proper side clearance, heat is concentrated on the contact surface immediately ahead of the blade, and side vice only at the blade rim corners. The depth of heat penetration ahead and to each side depends on the rate at which the blade rim advances; the faster it advances, the less the heat penetration. The large difference in temperature between the heated side surface near the blade rim and the cold metal behind the surface permits rapid cooling of each succeeding portion of the surface as the blade progresses through the cut. This sudden cooling will cause the sawed surface to harden if the steel has sufficient carbon or alloy content to be hardenable. However, this hardness is very shallow.

7.2.42 Friction Band Sawing

Band saw friction cutting is a separate method of vertical band sawing for the extremely fast cutting of ferrous materials. This work is usually performed on machines built for the purpose. The high band velocity provided by these machines is also desirable for sawing plastics, wood, and non-ferrous metals with standard bands. This frictional heat created is confined to an area immediately ahead of the saw, and only to a very small extent to the sides of the kerf. The resulting heat penetration is very shallow, seldom more than 0.002 inch deep. The selection of the proper band and sawing velocity is important in friction band sawing. If the teeth are too far apart, they will remove more material than has been softened; if they are too close, the band will have a tendency to load. The proper selection of band and speeds is important. If friction sawing of other than ferrous material is contemplated, a specialist should be consulted. The limitation of saw width to material thickness should be balanced.

Friction band sawing is particularly useful for cutting hardened steel, steel that has not been annealed, armour plate, stainless steels, welded sections, and similar materials that would be difficult or impossible to saw economically by other methods. Because the teeth are cutting away softened material in its plastic state, the force required to feed the work into the saw is relatively small. For the same reason, the pull down effect is so slight that the operators can in many cases hold the work manually when following complicated outlines or removing gates and risers from irregular castings. The friction band saw permits very rapid contour sawing of thin parts, usually limited only by the operator's dexterity in following the outline.

7.2.43 Circular Friction Sawing

Friction sawing can also be performed by a circular sawing machine with a smooth-rimmed blade because the rim becomes roughened on high-carbon steel. A blade with a smooth rim permits continuous production cutting of all high-carbon steels in their hardened or annealed condition whenever the carbon content ranges from 0.60 to 1.00 per cent. On low-carbon steel the rim does not roughen, and the blade periphery must be indented in order to produce satisfactory friction cutting. These teeth were cut with a topping and serration hob having 3 inches outside the diameter and single thread with 1/4 inch circular pitch. While cutting the indentations, the hob also cuts off the tops of the teeth concentric with the centre bore. The blade will readily sever any of the low-carbon steels with less energy than that required by the smooth rimmed blade when cutting high-carbon steels of the same sectional area and shape. On low-carbon steel the tops of the teeth do not roughen with continued cutting. With proper cooling, the rim will not adhere to foreign metal, and the energy required will remain nearly constant through several hours of cutting.

The blade may be redressed by the same hobbling operation used in forming the original teeth.

Burr: The burr is less when the length of the contact is short and the blade has suitable side clearance in the cut. If the same section is cut faster, the amount of burr will be less. If the corners of the blade periphery, as formed by the tops of the teeth and the sides of the rim, are sharp, the amount of burr will be minimum for that given cutting speed. If the machine has the capacity to generate heat on the maximum length of contact at a rate much faster than actually required, and if the total cutting time is very short and the blade is properly cooled, a minimum amount of burr can be maintained.

Speeds, feeds, and cutting lime: The efficiency of cutting for a given thickness of blade and condition of blade periphery is greater when the length of contact arc is shorter and the rate of feed faster, because the depth of heat penetration into the body of the material is less. The efficiency decreases as the corners of the blade periphery become rounded to a larger radius because this rounding increases the surface area on which heat is generated. The efficiency of friction sawing increases with the rim velocity for speeds up to about 22,000 sfpm but seems to remain nearly constant for higher values.

Cooling: Rim deformation and the adherence of foreign metal depend upon whether the rim becomes red hot before completion of the cut.

The cooling effect of air alone is not enough to prevent temperature rise on the circular blade periphery. If the ratio of the contact arc to the cooling is sufficiently high, and the duration of the cut is long enough, the entire exposed periphery will eventually display a visible red heat distinguishable from the sparks leaving the rim. This red heat will extend deeper into the rim with further duration of the cut but will disappear quickly the instant the cut is completed.

At such a high temperature, the original even rim surface will deform by rounding of the corners and widening into a mushroom-like shape. At these temperatures, metals removed will adhere to the rim of the saw.

If heat is dissipated from the blade rim by a coolant so that the rim cannot reach a red heat or show any indication of red heat on the periphery immediately ahead of contact for the entire duration of every cut, foreign metal will adhere to the rim regardless of the type of steel being cut. Furthermore, the deformation of the rim is reduced to such an extremely low value that successive cuts can be made without showing appreciable change in energy required.

Precautions

1. Hacksaw blade should be fitted on the frame carefully. It should neither be very tight nor very loose.
2. The job which is to be cut should not be held in a vice much higher otherwise there would be vibrations in it.
3. While cutting thin metal sheet, packing of wood, plastic or any other appropriate material should be used in its front and back.
4. Hacksaw blade should not be allowed to become slanting while using a hacksaw. If it becomes slanting, it may break.
5. If the blade starts cutting in a slanting manner, start a fresh cut. Operating a hacksaw in an old cut may break it.
6. If a blade gets broken while working, replace it with a blade which has been used almost for the same period rather than using a brand new blade. The new blade can break if it is operated in that old slot.
7. If necessary, water should be used as a coolant.
8. After using a hacksaw either its blade should be taken out or it must be made loose.

7.3 SCRAPING

7.3.1 Fundamentals of Scraping

Scraping is intended to remove high spots left after machining has been done on a job.

This is necessary with workpiece surfaces sliding on each other in order to achieve less friction and decent guidance and to obtain highest precision of the measuring surface.

Bearings are scraped so that it takes the same configuration as that of the journal. This is checked by smearing Prussian blue or red lead or any other marking media on the journal/mandrel. Marking media should be applied very thinly and uniformly all over to have the correct impression of the mating bearing surface. After introducing the journal into the bearing, it should be rotated one complete turn and withdrawn. The high spots on the bearing surface are scraped and the process is repeated till the impressions are uniform on the entire bearing surface. After completion of the scraping, the bearing should be cleaned thoroughly and the final assembly is to be made.

By this operation, still existing unevenness of workpiece surfaces is smoothened by filing is, planning or milling and high spots are removed as far as possible.

Such elevations are scraped and checked against master plates of known accuracy. After removal of smallest chips, by the cutter of the scraper many bearing points come into existence which are distributed over the entire surface of the workpiece densely. The more bearing points can be obtained on 1 cm 2, the better the quality of the scraped surface.

7.3.2 Action of the Scraper

The chisel with a cutting angle of less than 90° can deeply penetrate into the material and remove thick chips (**Fig. 7.55**). The scraper however has a larger cutting angle and thus penetrates only a little into the surface.

The removed chips have a thickness of about 1/100 mm and less. This is obtained by manipulating the scraper with a cutting angle (δ) greater than 90° under a slight pressure of the hand.

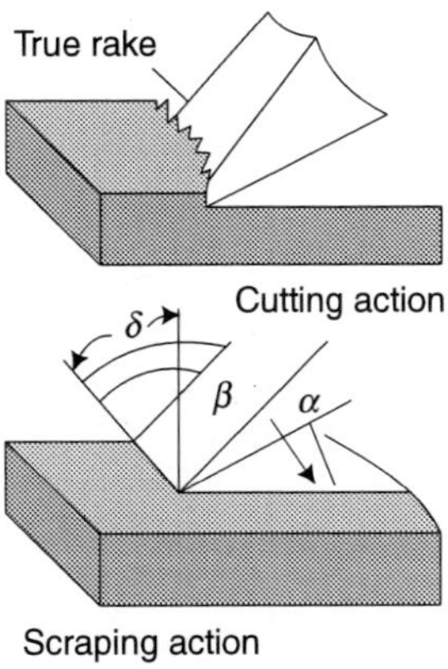

Fig. 7.55

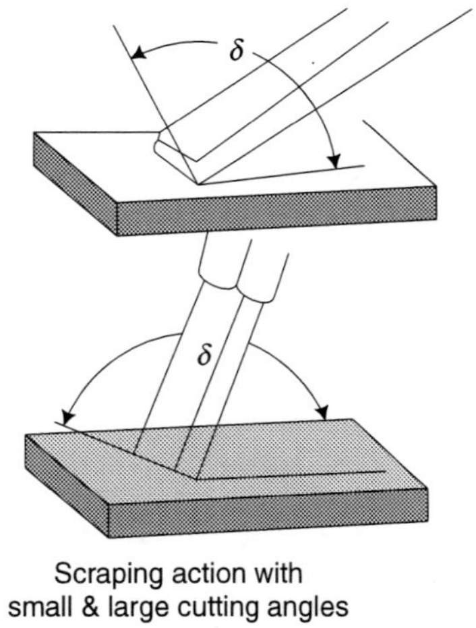

Fig. 7.56

In this working position of the scraper, the true rake of the tool is over the chip while it is below the chip in the cutting process.

The scraping action is the smaller that means smaller chips are removed. The greater the cutting angle (δ) the larger the cutting width of the scraper is **(Fig. 7.56)**

7.3.3 Kinds of Scrapers

Scrapers are a cutting hand tool. These are made of steel or alloy steel and their cutting edge is hardened and tempered. Material up to 0.002″ to 0.003″ should be scratched. It is different in shape and it is used like a file by using a handle in the tang. Square, triangular and round surfaces are scrapped with these scrapers. The length is from 100 mm to 250 mm and their thickness varies from 1 mm to 3 mm.

Flat Scraper (Fig. 7.57): Shape of this scraper is like a flat file. But its cutting edge is more in breadth than its body and is slightly rounded. Its edge is made at about 80° angle. It is used for scrapping high spots left on the plain surface of the material.

Triangular Scraper (Fig. 7.58): The shape is like a triangular file. But it is pointed towards the lower side. These are used to clean the corners, small diameter holes and deburring the edges of holes. The cross section of this is triangular. This has more number of cutting edges and the hollow portion between the cutting edges helps in resharpening easily.

Hook Scraper (Fig. 7.59): This is a special type of scraper. Its lower portion is bent and the rest is like a flat scraper. It is used to scrap the plain surface of a metal, where flat scraper do not work.

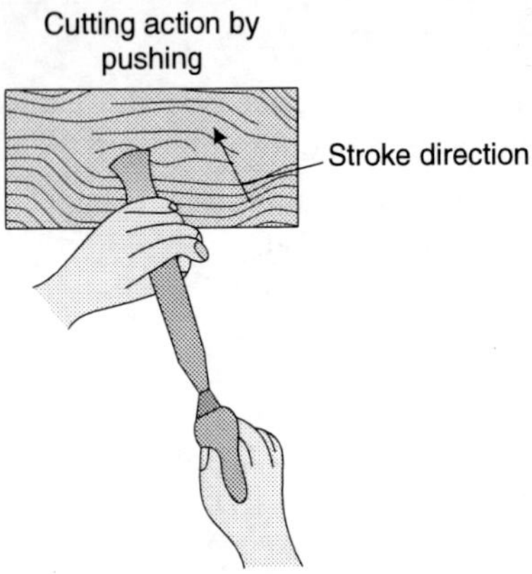

Fig. 7.57

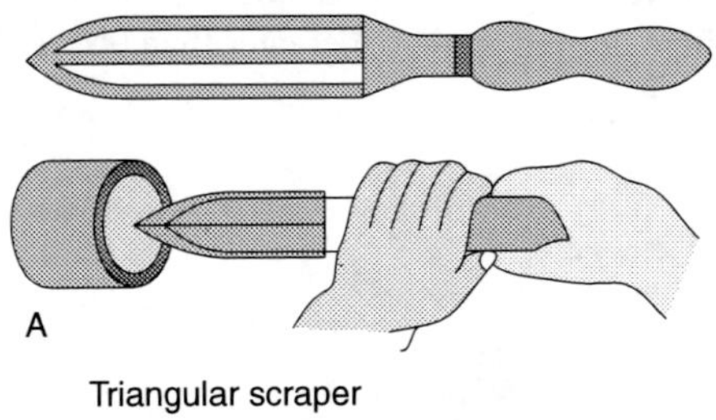

Fig. 7.58

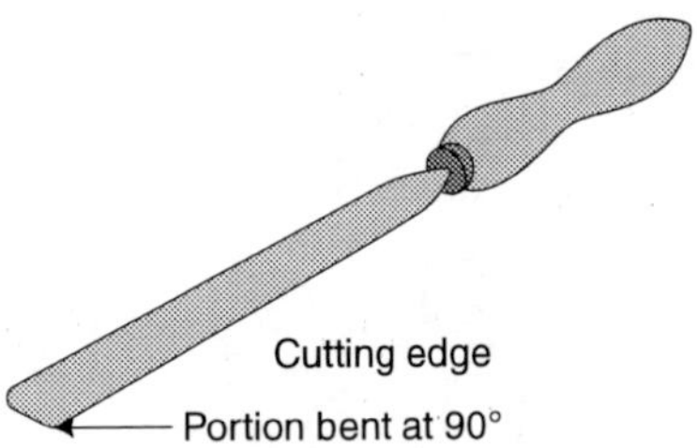

Fig. 7.59

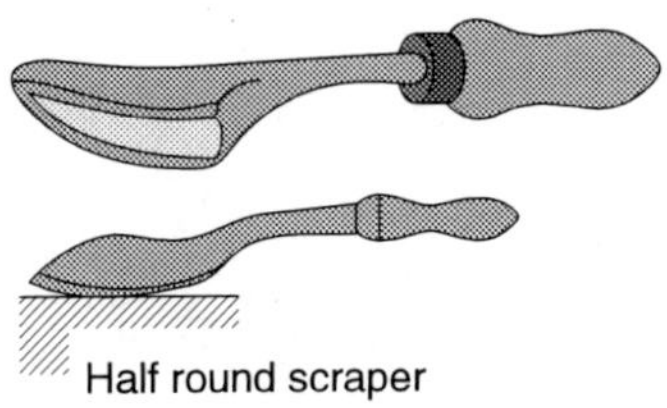

Fig. 7.60A

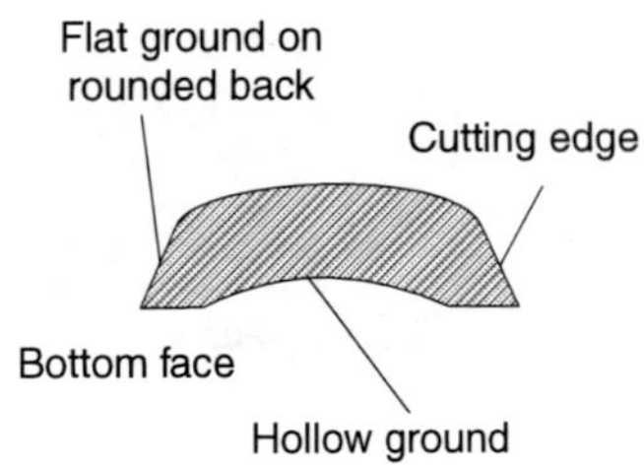

Fig. 7.60 B

Fig. 7.60

Half Round Scraper (Fig. 7.60): Its shape is like a half round file. But its body is hollow from inside. It is used for scraping the high spots between bearings and bushes made in two pieces. Cutting edge is ground on both the sides of the body. The cross section of the scraper is a segment and it tapers to a rounded point. The round bottom face is curved and is hollow in the middle.

The bottom facet and the flat surface are ground along the edges to form the cutting edge (**Fig. 7.60A).** The cutting angle is between 45° and 65° . The curvature at the cutting edge helps to make point contact while scraping, and also helps to remove small spots.

Bull Nose Scraper (Fig. 7.61): The bull nose scraper is used for scraping large bearings (flatter curves) by giving longitudinal strokes or sometimes with circumferential movements. The scraper has the cutting edge shaped into a flat circular disc. The cutting edge forms about two-thirds of a circle.

Double Handle Scraper: In this type of scrapers handles are fitted on both the ends of the scrapers. It is used for scraping broad surfaces like draw filing. This saves time.

Different types of scrapers which are in use are shown in **Fig. 7.62.**

7.3.4 Method of Scraping

Scraping is a process by which the high spots left on a job after the machining, are removed by using different scrapers through scratching. For the scraping process, Prussian blue or red lead paste is applied on the lathe surface plate equal to the surface of the job. Square jobs are rubbed on this paste.

Thus, only at the high spots the paste sticks. After ascertaining these high spots, these are removed by scraping.

The method of ascertaining the high spots on the square job and scraping is done (**Fig. 7.63**).

For scraping a bearing of two pieces of a shaft, the paste should be applied to the bearing surface. The bearing should be tied and revolved. By this process the paste will stick on the high spots of the bearing and then they are scratched with a scraper. Surface of a scraped job becomes completely accurate.

Fig. 7.61 Bull nose scriber

The following points should be borne in mind while using a scraper:

1. Hold the handle of the scraper in the right hand and its body in the left hand.
2. While using a scraper it should be bent at about 30° angle.
3. Do not keep a stroke of the scraper too long.
4. After scraping once, the scraper should be driven at right angle.
5. Prussian blue or red paste should be applied at appropriate place. It is most appropriate for metal of 0.002″ to 0.003″. For scraping of high spots, filing should be done again.
6. In order to find out high spots scraper should be rubbed on the surface plate and on shaft, etc., intermittently.
7. Scraping should be considered accurate only when the colour of the paste starts appearing on the entire surface.

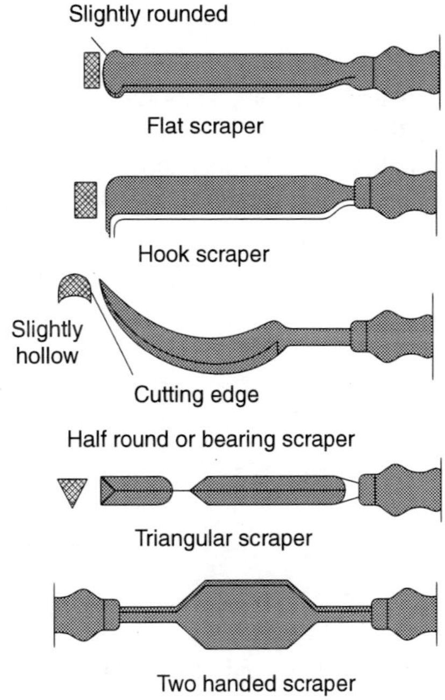

Fig. 7.62

For scraping even surfaces the square scraper is used. Its lip angle (ß) is 90° and the clearance angle about 30°. Thus these surfaces are generally scraped at a cutting angle of 120°.

The shape of the square scraper resembles that of a recut flat file of about 25 mm width and 5 mm thickness. The cutting part is offset at about 3 mm to avoid clasping. Scrapers are made from unalloyed and alloyed tool steel. The edge-holding property of a scraper made from alloyed tool steel is considerably higher. Old files are frequently used for the production of scrapers. Three-square, hollow ground and curved bearing scrapers are used for scraping rounded surfaces.

7.3.5 The Scraping Process

Handy work pieces are scraped in the vice.

Surfaces already treated by filing. Planning or milling is rough scraped to remove the still remaining more or less large channels and elevations from the bearing surface.

The still visible larger unevenness is equalized by means of a strong straighter which is moved over the workpiece with long strokes and under high pressure. In this operation large chips are removed. To remove the minor unevenness by removing smallest chips, the surface gets a finishing treatment by elastic scraping. The square scraper placed at a cutting angle of more than 90° is to be gripped in a way that the right hand guides the scraper while the left hand exerts the pressure during the forward motion.

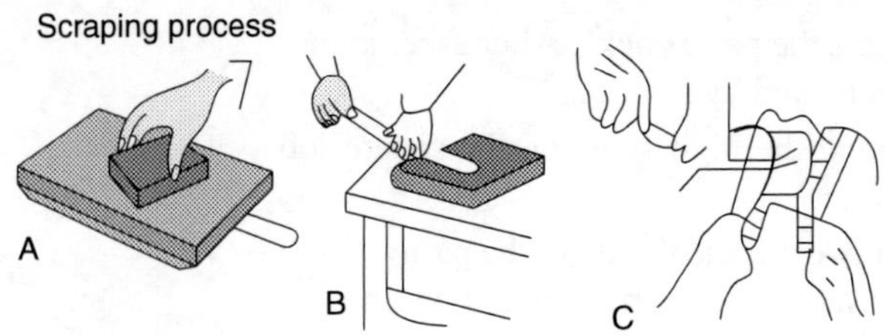

Fig. 7.63

The minor elevations on the surface of the workpiece remaining after completion of the rough-scraping process are no longer perceptible by the naked eye. This is achieved by scraping and checking against master plates of known accuracy. For this purpose a surface plate or a dressing table made from fine-grained grey cast iron showing a precisely even surface is used.

After having applied a fine coat of Chinese ink made of pine soot and oil on the surface by means of a small leather rag. The workpiece is slightly moved to and fro in a circular sense (**Fig. 7.64**) so that, a design of light and dark spots appears on the surface to be scraped. In case of large or heavy workpieces which cannot be placed on the plate, the same procedure is used inversely by moving a levelling straight edge over the workpiece.

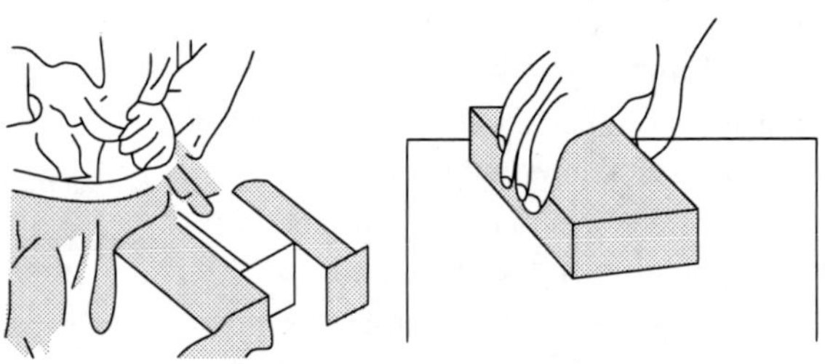

Fig. 7.64

All elevated and already bearing spots can be recognized by the spots that are dark coloured by the Chinese ink. Small chips are then removed from them by means of the scraper. A light scraper is used for finish-scraping. This scraper is moved in curves. It has a rounded cutting edge which avoids clenching and to removes narrower chips as well.

These procedures: scraping and checking against the master plates of known accuracy, also called dressing, are continued — continually changing the scraping sense until 3/4 of the surface of the workpiece will show bearing points, if ever possible (**Fig. 7.64**). During the treatment a scraping design will occur on the surface, this design occurs because of the repeated change of direction. It helps to determine the quality of the scraped surface.

The evenness of a scraped surface is checked with a bevelled steel straight edge according to the light gap process.

The bevelled steel straight edge is a slat with a hardened blade-like measuring edge. This measuring edge is placed on several points of the scraped surface. The smallest unevenness with deviations of more than 0.005 mm are visible as light gaps on the measuring edge of the bevelled steel straight edge.

7.3.6 Maintenance of Tools

The cutter of the scraper becomes blunt, particularly when scraping hard materials such as grey cast iron. It is to be resharpened on a water-cooled grind stone and subsequently strapped on an oil stone **(Fig. 7.67).** During this process, the burr remaining from grinding is removed. The edges of cutter are rounded to avoid the formation of channels when scraping.

As a precaution against corrosion the tools for scraping and checking against master plates of known accuracy should be greased.

After use they should be covered with a wooden board to protect them against damages by blows and shocks.

The bevelled steel straight edge should always be kept in wooden box. The scraper should also be kept in a way as to prevent the blade from getting damaged.

A slight grinding of the cutting edges on a finer silicon carbide wheel (120-150 grit size) can serve the purpose. The scraper blade movement should be in the form of an arc to have a convex cutting edge.

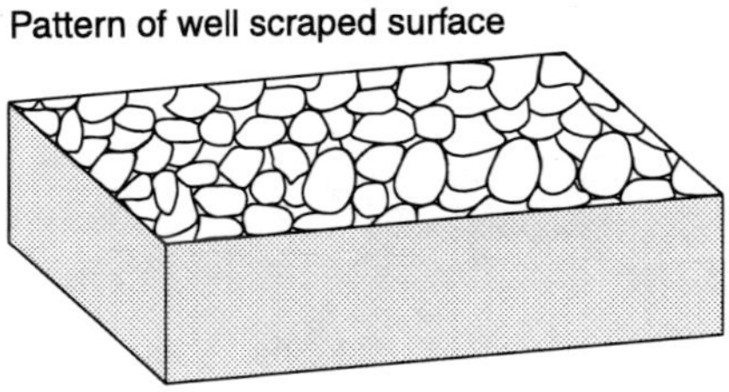

Fig. 7.65

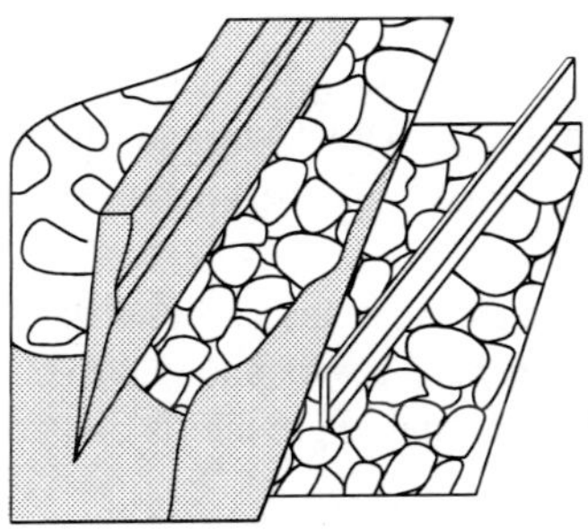

Fig. 7.66

After grinding, the cutting edge should be smoothened by gently rubbing on an oil stone, so that the cutting edges are completely honed.

The scraper should not be ground frequently as it reduces the cutting efficiency. Therefore, honing should be done for resharpening.

The cutting angle of a flat scraper should be between 85° and 90°.

Half round scrapers are ground on their side and flat bottom faces to form cutting edges (**Fig. 7.66).**

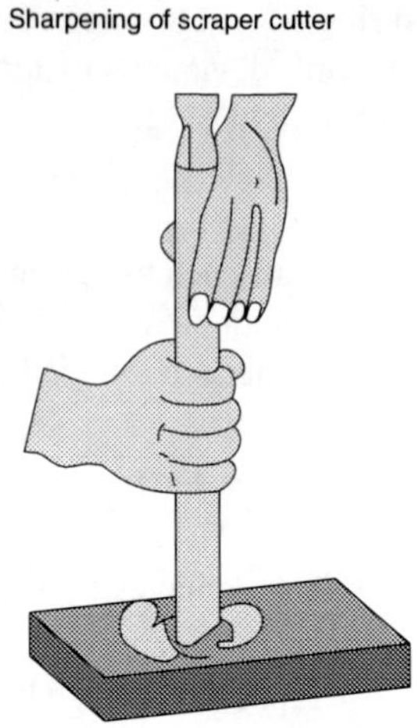

Fig. 7.67

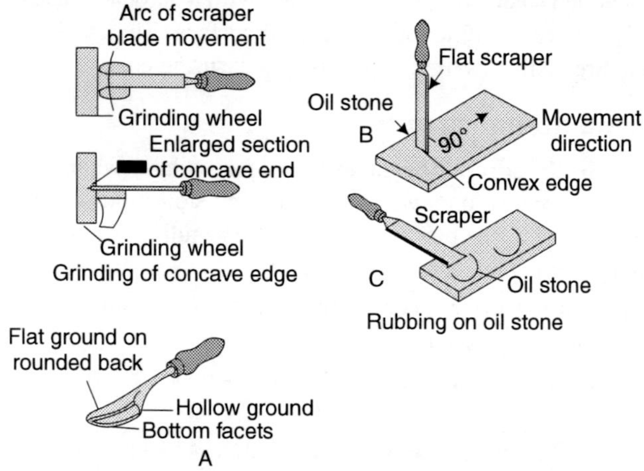

Fig. 7.68

The cutting angle should be between 45° and 60° and this angle is formed by grinding, and is followed by honing on an oilstone. The process of rubbing the bottom face along the oilstone surface is shown in (**Fig. 7.69**) with rock and roll movement. Similarly, flat ground on rounded back sides is honed on an oilstone to have smooth cutting edges.

Triangular surfaces are also ground flat on their curved edges, i.e., a facet of about 2 to 3 mm to form 55° to 60° cutting angle and is and followed by oilstone honing.

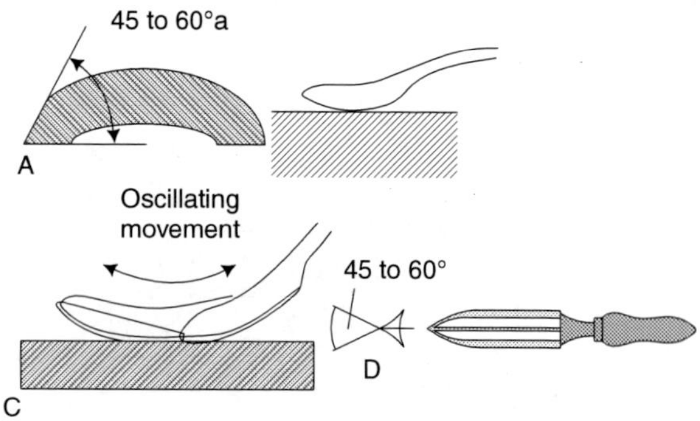

Fig. 7.69

Judgment on Scraping Work

Mistake	Cause
The surface scraped and checked against a master plate of known accuracy for the first time shows only few bearing spots.	The workpiece was not rough-scraped by long and powerful scraping strokes.
After scraping and checking against the master plate of known accuracy the entire surface is black.	During scraping and checking against the master plate the workpiece has been moved over the edge of the surface plate thus blurring the design. Too much Chinese ink was applied.
The cutting edge of the square scraper is becoming blunt after a few scraping strokes.	The material of the scraper is not hard enough for scraping the corresponding material.
The surface plate and the beveled steel straight edge show cracks, channels, or drag marks on their surfaces	The workpieces were only rough-worked before scraping and checking against the master plate, or the corresponding tools were not protected against damages.

Precautions

- Select the scraper according to the requirement of the job.
- Scraper should have proper edge.
- Hands should be free from any sticky substance while scraping.
- Scraper should not be used without a handle.
- Neither touch the scraper's edge nor move hands on its surface.
- Different places of surface plate should be used to obtain high spots.
- After scraping the surface should be properly cleaned.
- Scraper should be kept separate from other tools.

7.4 SHEARING

7.4.1 Fundamentals of Shearing

Shearing is a chipless working process. Sheet-metal strips or other shapes of sheet metal bands or plates as well as parts of bars or profiled slats are cut by shearing. The cutting edges of the plate shears are moved under pressure in close contact by hand or by mechanical means.

After the shearing resistance of the material has been overcome, the action produces the severing, the cut. Therefore this method is also called cutting or blanking.

By blanking one generally understands the separation (severing in the shearing method) of material from plates, bars or strips, e.g., cutting-off, cutting-out, cutting-in or punching. The cutting-out and punching methods were also called stamping-out. Nowadays punching only refers to reshaping by punch and die.

Cutting-off (Fig. 7.70): Cutting-off is the complete separation in non-closed lines with shears or cutting tools, e.g., cutting of sheet-metal strips, cutting of any desired length of bars, notching on L-, T, U profiles.

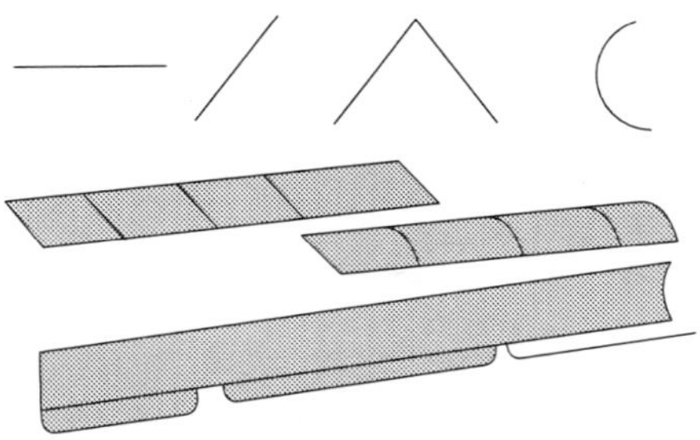

Fig. 7.70

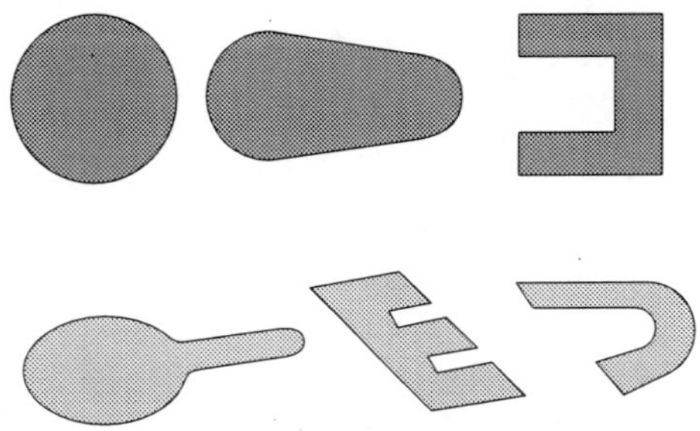

Fig. 7.71

Cutting-out (Fig. 7.71): Cutting-out is the complete separation of any desired shapes in a closed line by means of shells or cutting tools, e.g., cutting out with curve or circular shears.

Cutting-in (slitting) (Fig. 7.72): Cutting-in (slitting) is partial separation of the material by means of shears or cutting tools, e.g., cutting-in for subsequent bending.

7.4.2 Operation of Shearing

In shearing, two wedge-shaped cutter edges, arranged in one level, pass adjacently and penetrate in a uniform manner into the material under increasing pressure from both sides. During this process the material resists

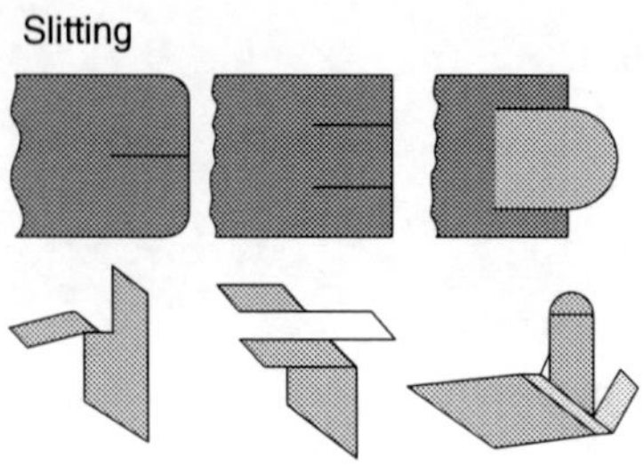

Fig. 7.72

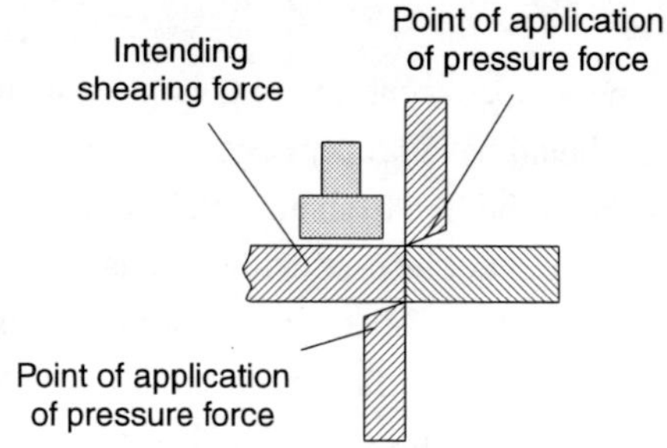

Fig. 7.73

at an increasing degree with regard to the depth of the penetrating cutter edge on the cutting point, and to the shearing resistance of the material.

At first a compression of the material structure occurs when the shear blades attack; the material is indented from two sides (**Fig. 7.73**). The pressure forces starting the shearing operation act on the centre of the indent surface. These forces deviate more and more from the cutting line to the outside when the indent surface is expanding. The initially unimportant twisting action of the cutters will distort the workpiece more and more from its horizontal level to a lateral one.

Mostly it is not possible to resist this effect of twisting of the material by hand. A blank holder placed on the shear's body, and adjustable to any thickness of the workpiece, secures the workpiece and prevents it from being distorted.

The further pressure in the sharing process shears and breaks the section weakened by the indents (**Figs. 7.73 and 7.74**). The point of the material where severance occurs after shearing is not even as this is the case when sawing, but is uneven, step-shaped, and requires subsequent treatment by filing. The individual steps of the shearing operation can be more or less clearly recognized at the severing point depending on the kind of material (**Fig. 7.75**).

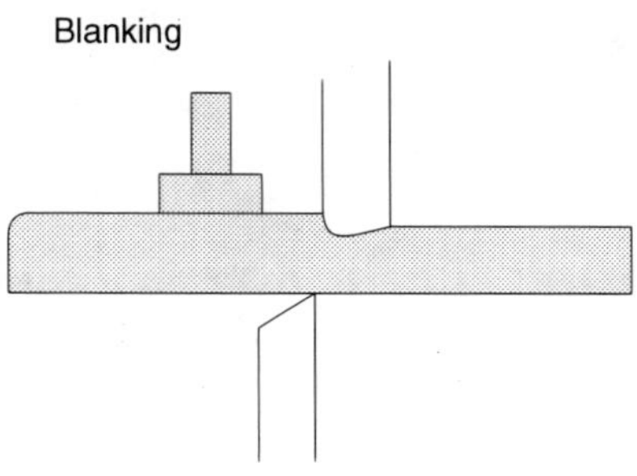

Fig. 7.74

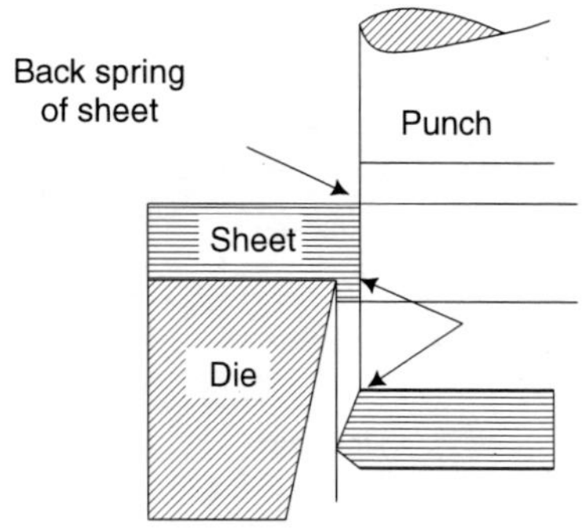

Fig. 7.75

7.4.3 Angle on the Cutter, Dynamic Effect and Clearance of Cutters

Owing to the varying stresses on the cutting tools, particularly on the shearing blades, it is necessary to adapt them to the material as well as to the shape and the point grinding of the shearing work (**Fig. 7.76**). As this is the case with all cutting tools aiming at good cutting results with a neat cutting face, the shear blades must have an equal angle and a precise clearance of the cutters. Hard and thick materials require a larger lip angle (β) than soft and thin materials. The shear blades of hand shears, usually made of good tool steel, have a lip angle of 75° to 85°, while the clearance angle and the side rake are at least 2° to 3° each.

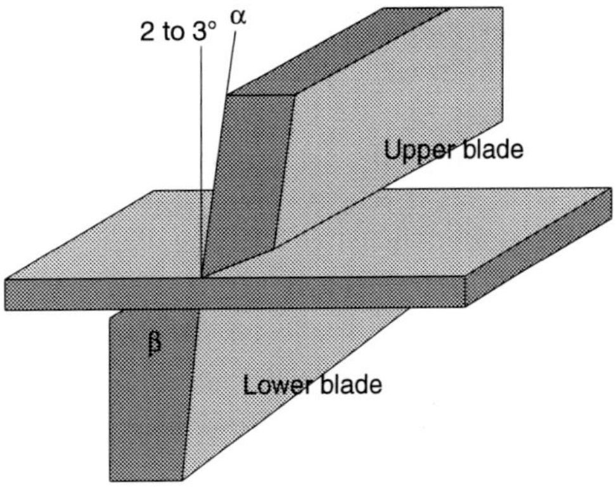

Fig. 7.76

It is not possible to point grind or resharpen wedge-shaped shear blades of hand or machine shears in every workshop. Sharpening of shear blades should therefore be done by an emery wheel to avoid the blades getting damaged.

To reduce the friction occurring on the blades during the shearing process, they have to pass each other at only a light clearance.The said clearance should be 0.015 to 0.3 mm when shearing workpieces from 0.25 to 0.5 mm thickness. If the clearance is too large, the material, particularly thin plates would skew between the shear blades.

7.4.4 The Lever Facilitates the Shearing Process

The cutting pressure needed for the shearing process is produced by lever action. The lever highly increases the applied force. This is clearly proved by the lever principle.

Force multiplied by power arm = load multiplied by load arm

$$P \times a = Q \times b$$

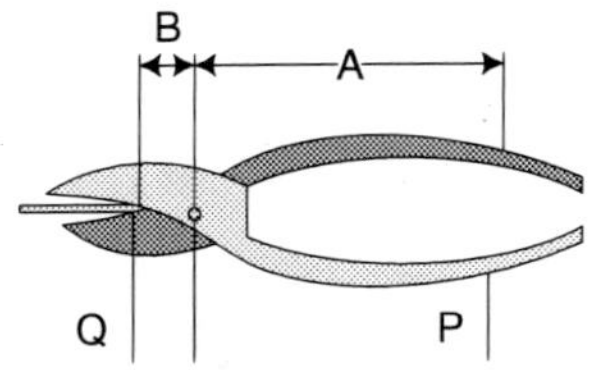

Applying a small force on the handle of the shears, the achieved pressure on the cutting edge will be the higher the longer the power arm is, or the shorter the load arm of this dissymmetrical lever is.

With hand shears: The shear blade and the handle form a two-side and dissymmetrical lever. A lever is two-sided if the supporting point or centre of motion is situated between the points of power and load. The lever of the shears is a rigid leg which can be moved around a centre of motion. The power arm (A) of the lever of hand shears is situated between the centre of motion and the point of impact of the shear blade. The load arm shifts continually during the shearing process. At the beginning of the cutting action, the load arm is short but the more the cutting point advances towards the point of the shear blade, the greater becomes the load arm. Thus, the force on the handle of the shears must continuously grow with the increasing displacement of the shearing point until finally the material is separated.

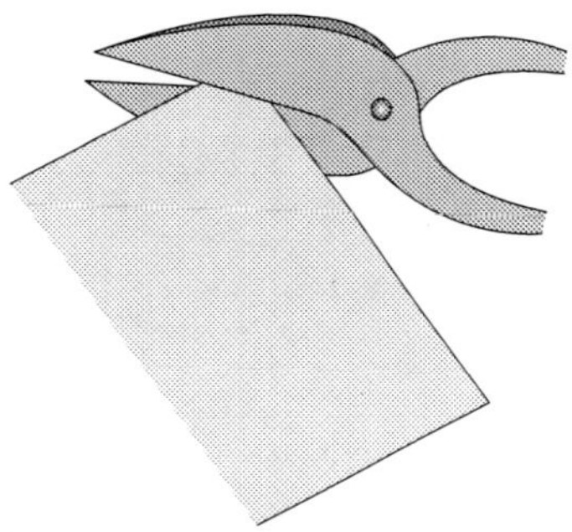

Fig. 7.77

Example: Separating a sheet metal strip with the hand shears requires an effort to overcome a resistance of Q = 90 kg.

How great must be the required force (P), if the distance of the severing point from the centre of motion (b) is 30 mm and the length of leg (a) is 180 mm?

$$P = \frac{Q \,.\, b}{.a} = \frac{90 \,.\, 30}{180} = 15 \text{ kg.}$$

Tin men's shears and lever shears have a unilateral lever. In this case the supporting point is located at the end of the lever while the force (P) and the load (Q) engage on one side of the centre of motion only. The lever principle valid for the bilateral and dissymmetrical lever is also applicable for the unilateral lever.

Shearing thin plates using hand shears: Thin plates of 0.1 mm to 1.5 mm thickness can be cut with the hand shears depending on the kind of material. The cutting faces of all types of hand shears are hollow ground to ensure burr-free blanking of the blades up to their point and to avoid elastic giving off.

If the blades of the hand shears are opened too far, the shears do not cut right away but push the plate forward. The thicker the plate is the larger is the shearing angle. The shears only cut at a blade opening of 15° to 25° without shoving the material.

According to the lever principle cutting is easier near the cutting point to the centre of motion of the handles of the shears. The plate should be placed as far as possible into the opening of the shears. In order to keep the plate in this position it is pressed against the shears in the direction of cutting.

Different kinds of hand shears **(Fig. 7.78),** are used according to shape and position of the cutting lines on the plates **(Fig. 7.79).**

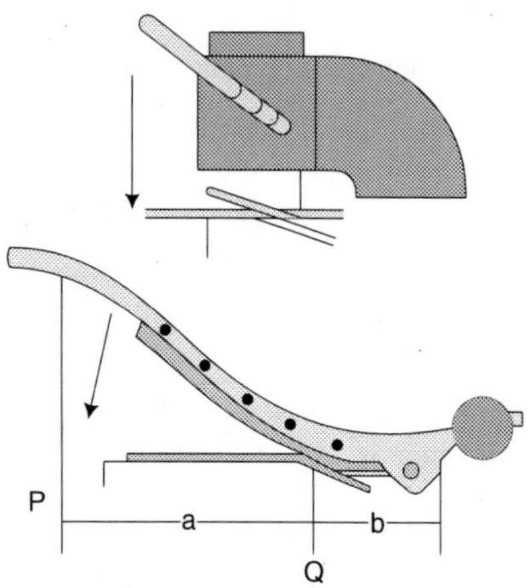

Fig. 7.78

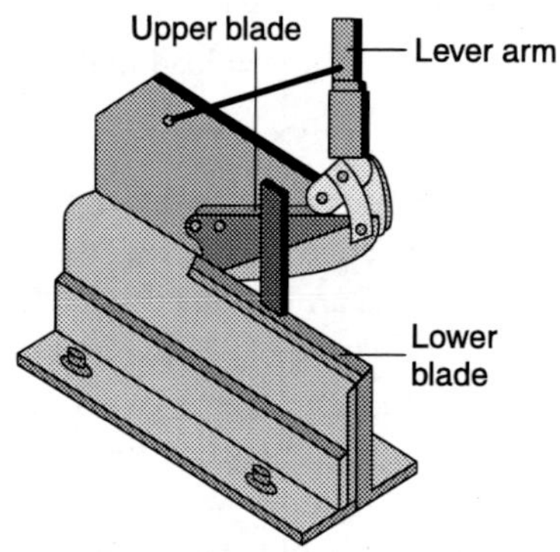

Fig. 7.79

Lever shears: Lever shears are used to cut sheets which cannot be cut with hand shears **(Fig. 7.79).**

The lever shear possesses a fixed lower blade and a moving upper blade. The sheet being cut is prevented from ting by a clamping device which can be adjusted to the thickness of the sheet. The knife-edge cutter of the upper blade is curved so that the opening angle at the point of cut remains constant.

Circle cutting and curve cutting machines: These machines are used to cut circles and curves of the desired shapes. When cutting curves, the sheet must be guided by the hand **(Fig. 7.80).**

7.4.5 Applications

(A) Hand shears are mostly used for straight cutting lines and for cutting of external curvatures. Generally, designed as right hand shears, i.e., seen in the cutting direction the lower die of the shears opening is arranged at the right-hand side. The left-hand shear part is only used for cutting points not accessible to the right hand shear part.

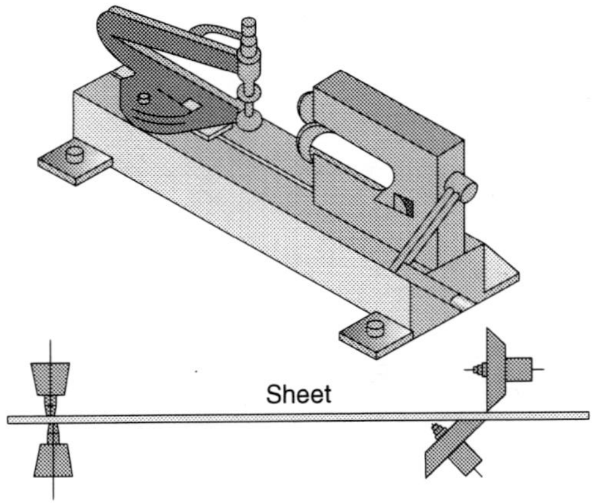

Fig. 7.80

(B) Especially for cutting points with straight cutting lines that are difficult to reach. Bent blade tinmans shears have laterally offset dies at 45°.

Tinmans shears with upward bent handles are used for the same purpose as bent blade tinmans shears. The dies of these shears are offset upwards.

(C) For inner curvatures. The blades are curved and pointed.

(D) For long and straight cutting lines, e.g., for cutting of strips and bands over the entire length of the plate.

Only the following thicknesses of plates can be cut with the hand shears.

Hand power does not suffice for cutting thicker sheet metal plates or other materials than those shown in the table. When chucking one leg of the hand shears in the vice, the full hand force can act on the other leg, thus allowing easier cutting.

It is also possible to apply greater force with the **bench shears (Fig. 7.79)**. Electrically operated hand shears, e.g., the **Bosch-shears** facilitate the cutting operation considerably. Often it is necessary to use different kinds of hand shears for the same workpiece.

Wire must not be cut with tinmans shears because the blades would break out or become notch. For this purpose, wire shears are used (hand shears with small opening).

When cutting narrow and short sheet metal strips, it sometimes happens that the ball of the thumb is pressed in between the extremities of the hand shears.

7.4.6 Shearing of Thick Plates and Bars with the Lever Shearing Machines

Sheets of steel and non-ferrous metals whose thickness exceeds 3 mm as well as profilated bars and wire are separated by means of **lever shears** or **machine shears**.

Contrary to hand shears where both shearing blades are moved, the lever and machine shears have a fixed lower blade, while the upper blade is movable and in most cases slightly rounded.

The lever shearing machine is used for cutting sheet steel up to 5 mm thickness; round steel up to a diameter of 13 mm, and flat steel up to 6 mm. Twisting of the material hardly ever occurs with these shears.

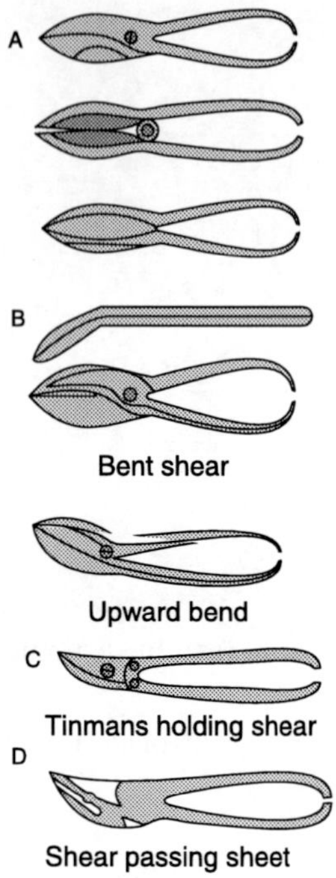

Fig. 7.81

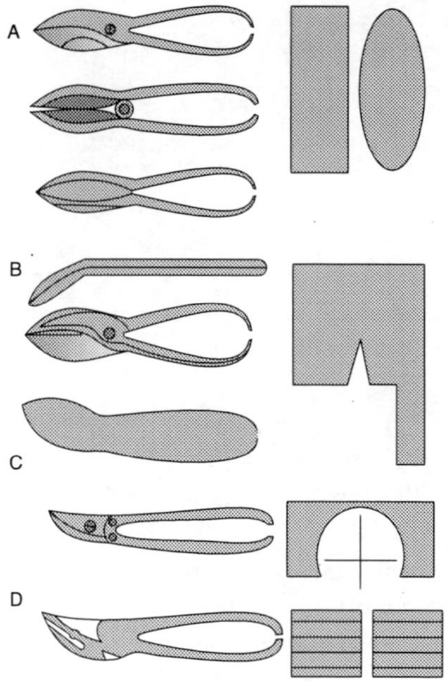

Fig. 7.82

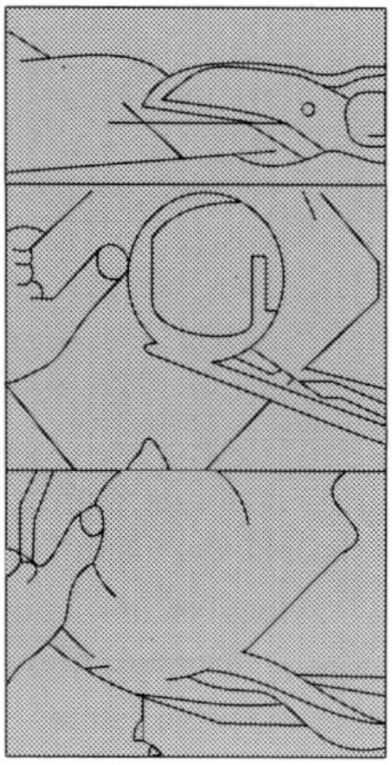

Fig. 7.83

Material		Thickness of plates
Steel up to 34 kg/mm^2		Up to 1.0 mm
	37 kg/mm^2	Up to 0.8 mm
	42 kg/mm^2	Up to 0.6 mm
	50 kg/mm^2	Up to 0.5 mm
Aluminium		Up to 2.5 mm
Aluminium alloy		Up to 10. mm
Lead		Up to 5.0 mm
Copper		Up to 1.2 mm
Brass		Up to 0.8 mm
Zinc		Up to 1.5 mm
Muscovite		Up to 3.0 mm
Hard paper		Up to 2.0 mm
Hard tissue		Up to 2.0 mm
Card board		Up to 6.0 mm

A long lever arm adapted to the stress of the shear blades, and which may be prolonged by placing on a tube, produces the cutting pressure needed for the shearing process.

The gate shears for sheet steel and sectional steel are lever shears with double transmission gearing and with high cutting capacity. They are able to cut sheet steel up to 12 mm thickness, round steel up to 36 mm diameter, and L- and T-steel up to 70 × 9 mm.

The shearing machine is equipped with a mechanically moved upper blade.

To prevent accidents, the lever of the shears is to be secured against falling down, suspending it with the hold-fast after use.

7.4.7 Special Shearing Work

The circular and curve shears are used to make circular discs, rounds, annulus parts, curved and also long cuts from metal sheets.

Two rotating circular shear blades arranged over each other produce a limited or endless cut by moving the hand lever. The cutters pass along each other in close contact, same as with all shear blades. The lip angle of the cutter, on an average of 80° to 85°, is formed by the truncated cone shape of the shear blades.

For cutting circular bottom parts and rings, it is advisable that the plate is chucked by the back centre of the adjustable bow placed on the circular shears driven through motor, and then guided between the shear blades in an arc of a circle.

The inclined position of the lower shear blade allows circular cutting with particularly small diameters, and reduces warping of the plate.

When cutting curves, the guiding bow of the circular shears is not employed. The plate is guided along the shear blades by hand following the marking line.

Judgment on Shearing Work

Mistakes	Causes
The cut shows deviation from the marking line to the right or to the left; cutting of curves of circles suppressed the marking line.	Ends of plates were not lifted to adapt them to the marking line. Marking was not visible when cutting because round cutting was not affected in clockwise direction.
Cutting points show considerable burr.	Ply between shear blades was too large. *Remedy:* reducing of ply by tightening the nut of the shear bolt, or by inserting thin plates behind the mounted cutters of the lever and machine shears.
Shear blades broke out or show notches.	Over stressing of blades either by too thick materials and placing a prolonging tube on the lever or by cutting of wire (spring steel) or hardened steel; cutting of bars by means of the tin men's shears.
Blades penetrating into the material only when applying excessive force.	Blades have become blunt because of strong wear. Resharpening of shearing blades by experts only.

7.4.8 Snips (Straight & Bent)

A snip, also called a hand shear, is used like a pair of scissors to cut thin, soft metal sheets. Snips are used to cut sheet metal up to 1.2 mm thickness **(Fig. 7.84).**

There are several types of snips available for making straight or circular cuts, the most common being straight snips and curved snips.

The choice of shears (snips) depends on the shape and type of the cut required.

Straight snips: These are used for making straight cuts and large external curves **(Fig. 7.86).**

Straight snips have thin blades which are only strong on a vertical plane. They are, therefore, only suitable for straight cuts and external curves when surplus waste has to be removed.

While cutting, the blade of the snips should not cover the marking **(Fig. 7.87).**

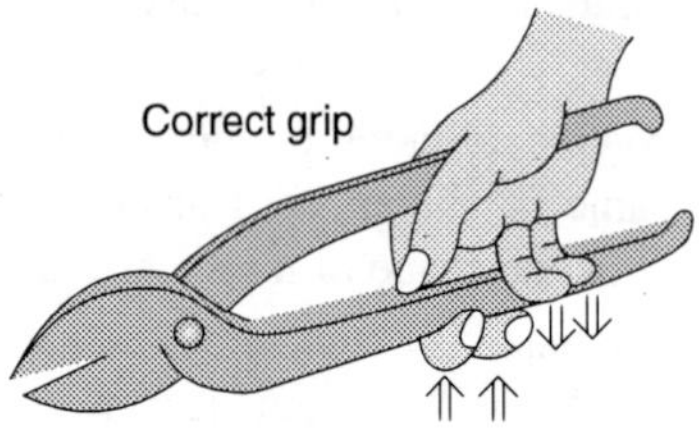

Fig. 7.84

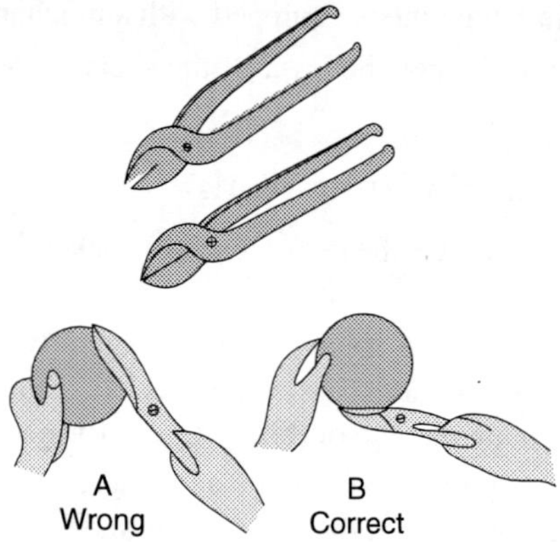

Fig. 7.85

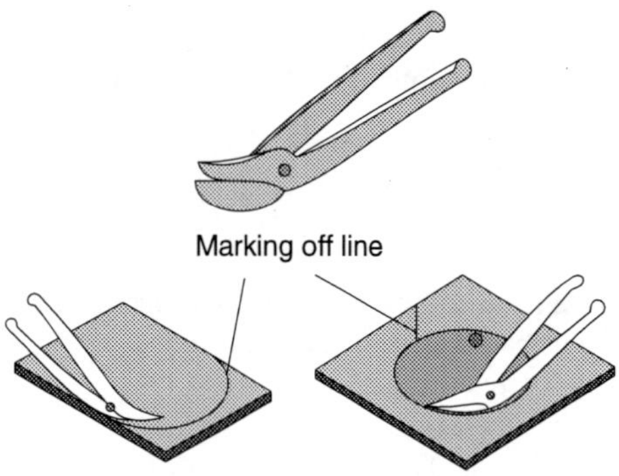

Fig. 7.86

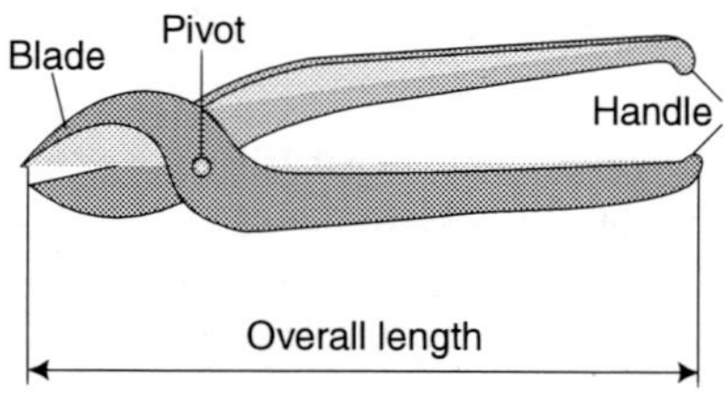

Fig. 7.87

Bent snips: These snips have curved blades for making circular cuts. They are also used for trimming cylindrical or conical work in sheet metal (**Fig. 7.86**). Snips are specified by the overall length and the shape of the blade (**Fig. 7.87**).

7·5 FILING

Filing is a method of machining or smoothing surfaces from which only a minimum amount of material is to be removed. In filing, numerous cutting ridges or edges, chiselled or ground into a hard-steel tool, act as the cutting medium on the workpiece surface. The presentation of numerous cutting edges has the advantage of requiring less time and labour than stock removal with a single cutting edge. The most common techniques of industrial filing are band filing and rotary filing or burring.

A file is used to cut, trim or finish a job of metal or wood, in order to give them a shape according to our needs such as making the job round, square or angular. File is a hard steel instrument with numerous cutting points. Metal is cut in small or tiny particles with the help of a file. Therefore, mostly a file is used for giving final finishing touches to a job of metal or wood. After chipping a job of metal generally becomes rough and a file is used to make it smooth. In case a machine part is larger in size than the required size, a file is used to make it suitable. In addition, a file is used where other cutting tools cannot be used. Files are also used for sharpening the edge of some other cutting tools.

7.5.1 Fundamentals of Filing

The surface quality of rough-treated workpieces is improved by filing. The teeth of a file remove small chips from the surface of a workpiece and produce smooth, neat and precise surfaces as well as an improved shape and dimensional finish of the workpieces.

Chip removal is effected by the cutting and scraping action of the file tooth.

By moving the file by hand, the cutter of the wedge-shaped tooth of the file penetrates into the material by the produced force. Chip removal takes place by a series of chisel-shaped teeth arranged one after the other, which are cut or milled into the surface of the file.

A good chip removal will be achieved by the shape of teeth (**Fig. 7.88**). This shape of teeth, however, cannot be obtained with the mostly used manufacturing process, the re-cutting, because the material, displaced by the pressure and the cut of the bolt chisel, would throw up the surface (**Fig. 7.89**).

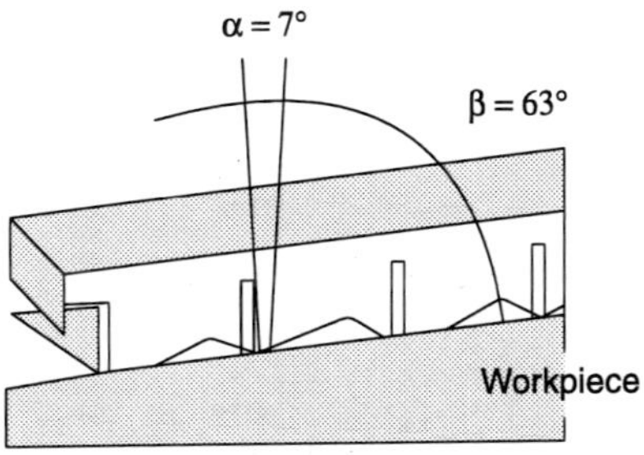

Fig. 7.88

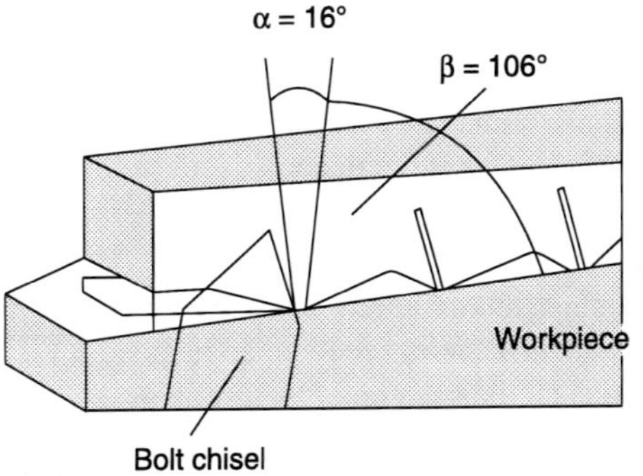

Fig. 7.89

Also the milled tooth of a file (**Fig. 7.90**), where the root is rounded to avoid clogging of the filings, has a shape which is only approximately the desired tooth shape. The angle of the cut of the re-cut tooth of a file is greater than 90°; the angle of rake is negative. Thus, re-cut files produce a scraping action whereas milled files produce a cutting action.

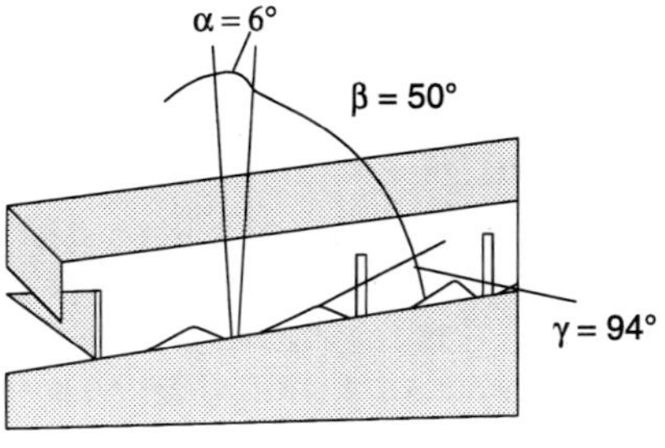

Fig. 7.90

7.5.2 The Cut of a File

By the cut of the bolt chisel on single-cut files, notches (cuts) are sunk over the entire width of the file surface; they are generally arranged at an angle of 71° diagonally to the cut direction to obtain a better cutting effect and chip removal (**Fig. 7.91**). Chips are removed to one side.

The simultaneous bite of several teeth over the entire width requires great effort, particularly when filing hard materials. Single-cut files are therefore only used for filing soft materials, e.g., tin, lead and aluminium.

Horn, synthetic and pressed materials are worked with special single-cut files.

For working hard materials, double-cut files are used. The over- or up-cut is worked in the form of a cross in the under-cut or first course (**Fig. 7.92**). Thereby the broad tooth of the single-cut file is divided in several small teeth which provide for a better chip removal.

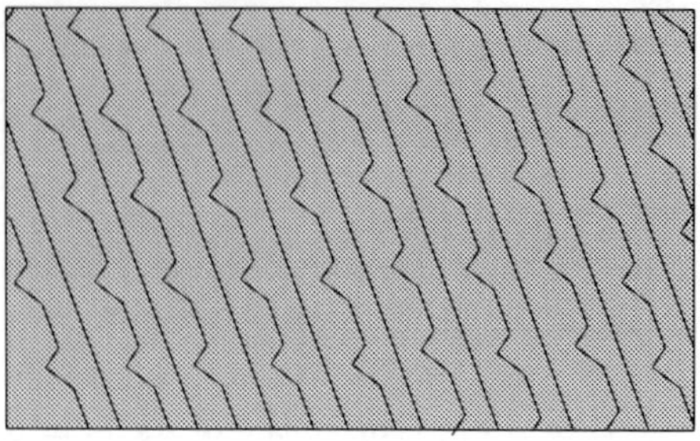

Fig. 7.91

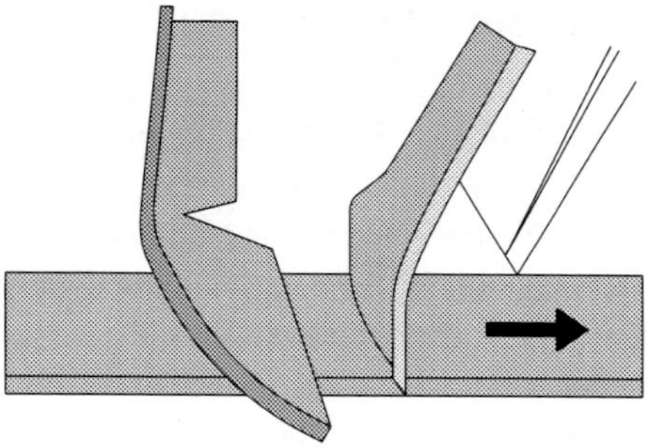

Fig. 7.92

Generally, the over-cut is sunk at an angle of 71° to the cut direction of the file over the under-cut whose angle is 54°. Thereby and by the unequal interval of the under- and the over-cut, the teeth are staggered so that a tooth will always cut those points of the workpiece surface which have only been partially filed by the preceding tooth. The cuts have the same angle arrangement and the teeth are placed exactly one after the other, as this is the case with a saw, thus producing considerable channels on the surface.

7.5.3 Kinds of Files

For obtaining the required surface quality of a rough-worked (∇), smoothened (∇∇) and super-finished (∇∇∇) surface, the interval of cuts, the spacing of teeth varies.

Spacing of teeth means the number of teeth arranged on 1 cm length. Spacing can have 4 to 120 cuts depending on the kind of the cut. The files are correspondingly classified (number of cuts from 00 to 10).

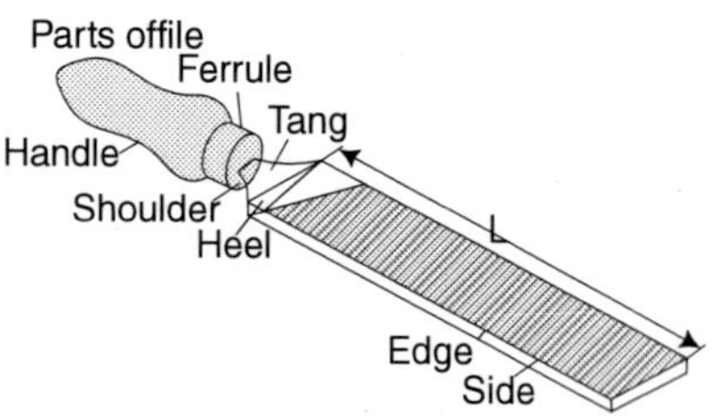

Fig. 7.93

Dimensions, shape and kind of cut are standardized according to DIN for the mostly used files; this means that these files have a uniform design.

File is made from high carbon steel. File is first made hard and then tempered except its tang part. It has the following main parts: (1) Tip or point, (2) Face or side, (3) Edge, (4) Heel, (5) Shoulder, (6) Tang, (7) Handle and (8) Ferrule (Fig. 7.93).

A wooden handle is fitted on its tang. Teeth are cut on its face and edge so that it can perform the job of cutting. The part having no teeth is called the heel. Thick side of file is called edge and its lower end is known as tip or point.

Tip or Point is the end opposite to tang.

Face or side is the broad part of the file with teeth cut on its surface.

Edge is the thin part of the file with a single row of parallel teeth.

Heel is the portion of the broad part without teeth.

Shoulder is the curved part of the file separating tang from the body.

Tang is the narrow and thin part of the file which fits into handle.

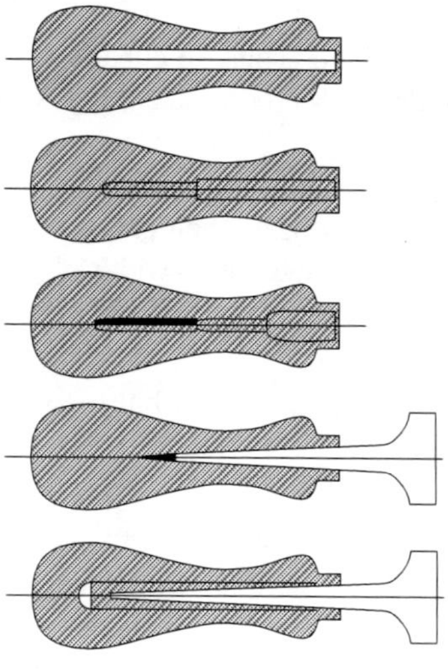

Fig. 7.94

Handle is the part fitted to the tang for holding the file (**Fig. 7.94**)

Ferrule is a protective metal ring to prevent cracking of the handle.

7.5.4 Classification of Files

Files of different types are used in various jobs for different purposes. These are specifically made for certain purposes. While making a selection of file for a specific job, some important points should be considered. In other words, we can say that files can be classified into the following categories:

1. On the basis of length
2. On the basis of shape
3. On the basis of grades
4. On the basis of cut

According to length: A file is measured according to its length. This length includes the whole file except its tang portion, i.e., length from heel to tip is the measurement of the file. Files of 100 mm to 450 mm length are used.

Length is the distance from the tip of a file to the heel (**Fig. 7.95**).

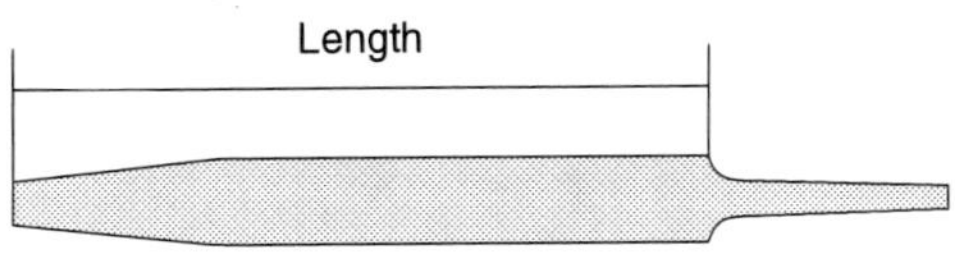

Fig. 7.95

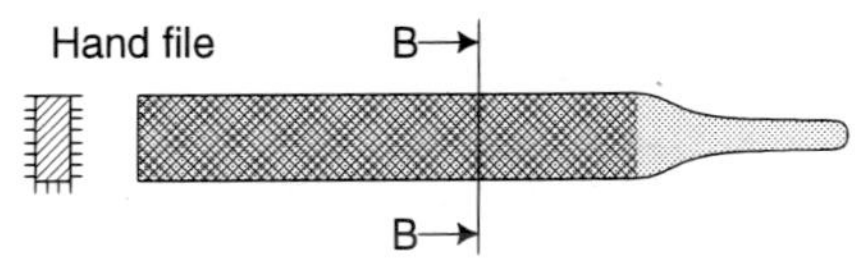

Fig. 7.96

According to shape: Files of different shapes are used for jobs of different shapes but in fitter trade, the following types of files are mostly used: (1) Flat file, (2) Round file, (3) Half-round file, (4) Triangular file, (5) Square file, (6) Hand file, and (7) Knife file.

Flat file: This file is of rectangular cross section in shape. It is made slightly tapered both in width and thickness. Two-thirds part of this file is parallel and the rest one-third is tapered. Double cut teeth are cut on the face and simple-cut teeth are cut on the edge. These are used to reduce flat surface by filing or finishing (**Fig. 7.96**).

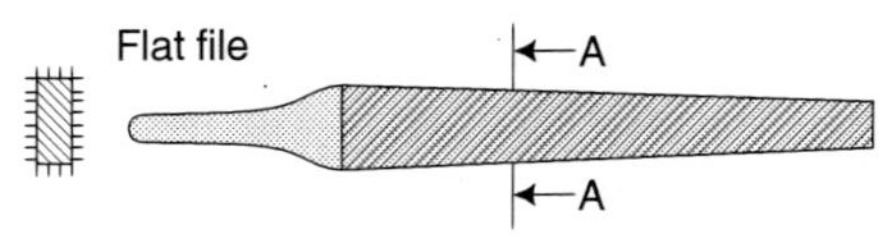

Fig. 7.96a

Hand files: These are similar to the flat files in their cross section. The edges along the width are parallel throughout the length. The faces are double cut. One edge is single cut whereas the other edge is safe edge. Because of the safe edge, they are useful for filing surfaces which are at right angles to surfaces already finished.

Round File: This type of file has a circular (round) cross section. Like flat file three-fourths of the file is parallel and the rest is tapered. It is used for enlarging the circular holes and filing profiles with fillets, rubbing or finishing key hole of small diameter (**Fig. 7.97**).

Half-Round File (Fig. 7.98): It is a file which is flat on one side and one-fourth curved on the other side. The shape is of a segment of a circle. The amount of convexity never equals a semicircle. Double cut dents are cut on it and like a round file it is also tapered. With this file the damaged hole can be set in order again.

Square File (Fig. 7.100): It is square in cross section and is somewhat tapered. Its measurement is taken from its square. It is used to file rectangular opening, square groove and corners, slot and key-way and splines.

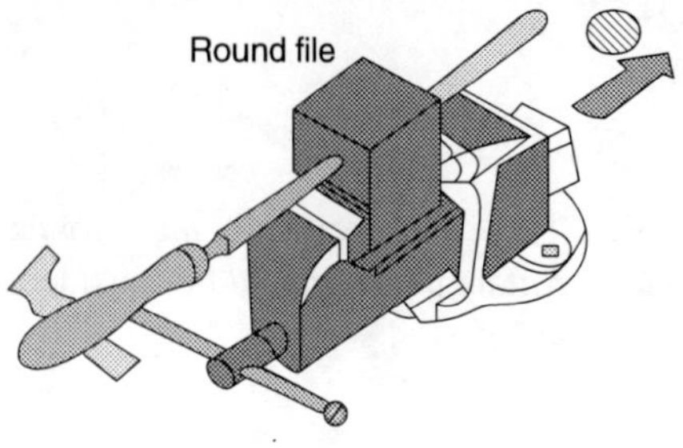

Fig. 7.97

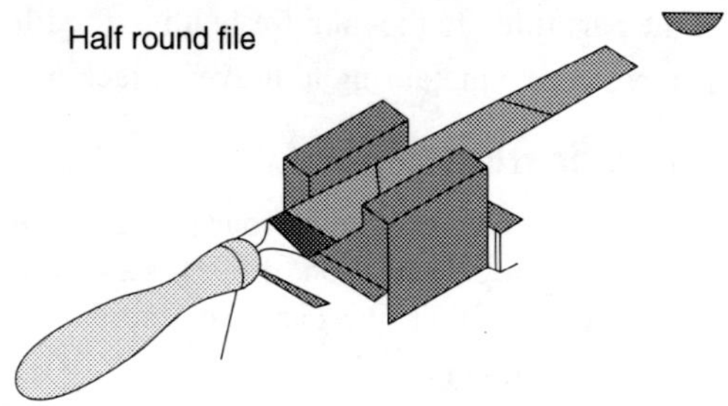

Fig. 7.98

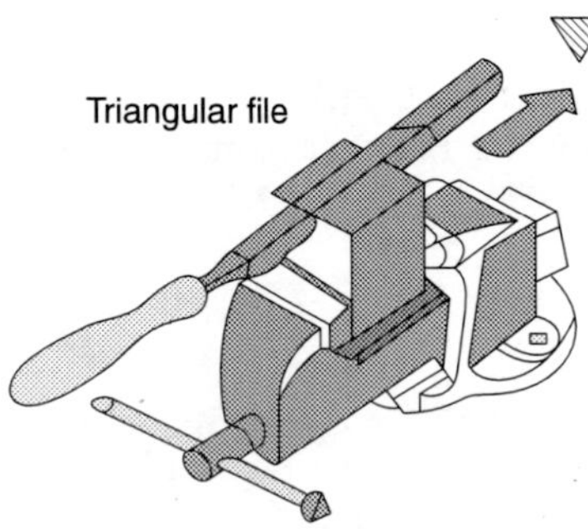

Fig. 7.99

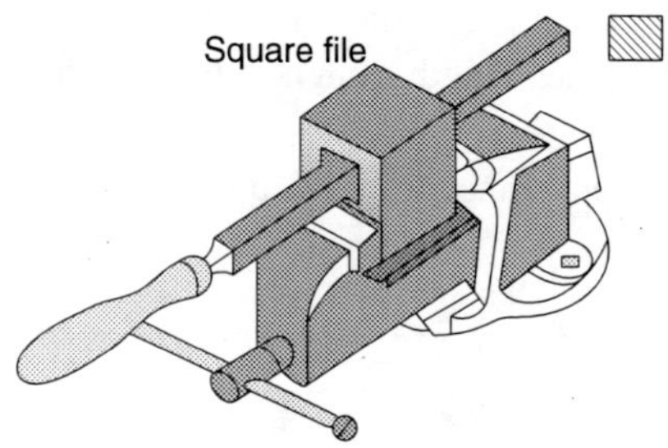

Fig. 7.100

Triangular File (Fig. 7.99): It is triangular in shape. Its angles are of 60°. It is also parallel in two-thirds part and is tapered in one-third part. Filing of "V" slot job, square and rectangular jobs of above 60° and below 90° angle is done with this file. It is also known as three square file. Teeth of the saw with which we cut wood can also be sharpened with this file.

Knife Edge (Fig. 7.101): It has a cross section of a sharp triangle. Its shape is like the edge of a knife. Its thin edge is at about 10° angle. Teeth are cut on its face as well as its edge. It is used for filing of small grooves and slots whose angle is less than 60°.

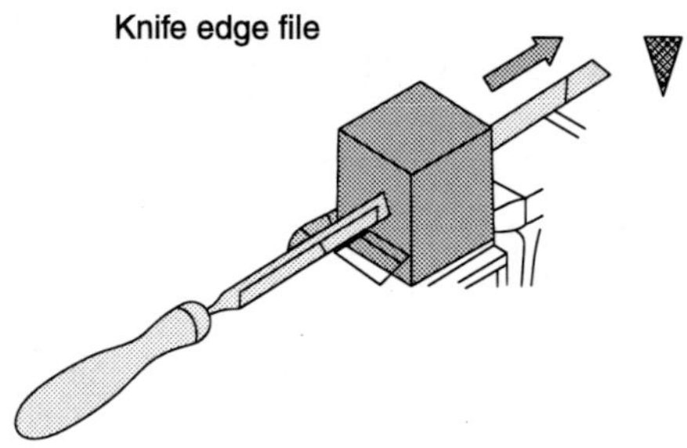

Fig. 7.101

It is generally used in lock industry for making keys.

Basically there are four types:

1. Single cut
2. Double cut
3. Rasp cut
4. Curved cut

Single cut file (Fig. 7.102): A single cut file has rows of teeth cut in one direction across its face. The teeth are at an angle of 60° to the centre line. It can cut chips as wide as the cut of the file. Files with this cut are useful for filing soft metals like brass, aluminium, bronze and copper.

Fig. 7.102 Single cut file

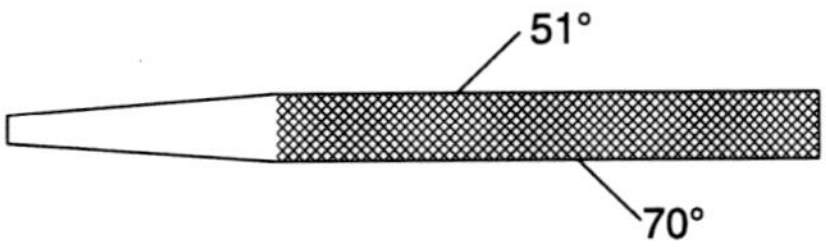

Fig. 7.103 Double cut file

Single cut files do not remove stock as fast as double cut files, but the surface finish obtained is much smoother.

Double Cut File (Fig, 7.103): A double cut file has two rows of teeth diagonal to each other. The first row of teeth is known as over cut and they are cut at an angle of 70°. The other cut, made diagonal to this, is known as up cut, and is at an angle of 51°. This removes stock faster than the single cut file.

Rasp Cut File (Fig. 7.104): The rasp cut has individual, sharp, pointed teeth in a line, and is useful for filing wood, leather and other soft materials. These files are available only in half round shape.

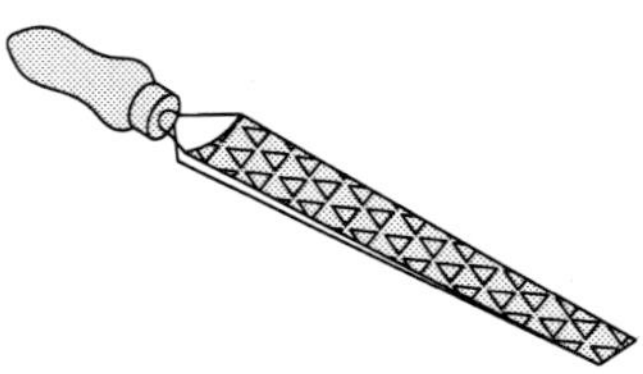

Fig. 7.104 Curved file

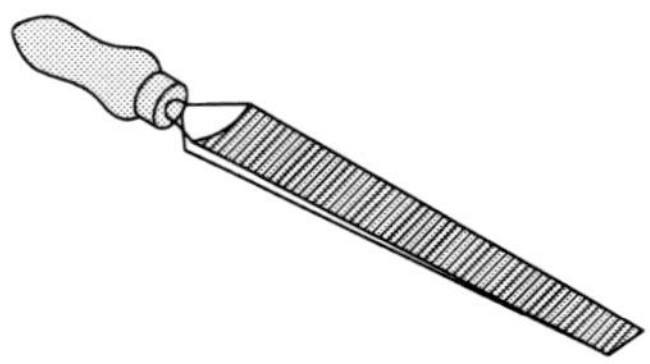

Fig. 7.105 Curved file

Curved Cut File (Fig. 7.105): These files have deeper cutting action and are useful for filing soft materials like aluminium, tin, copper and plastic. The curved cut files are available only in a flat shape.

According to grade: Grade of a file is determined on the basis of distance between the teeth of the file. Number of teeth per inch is determined on the face of the file. As the number of teeth per inch grows so the distance between the teeth keeps on narrowing, i.e., the number of teeth in the file increases. On the basis of these teeth their grades are specified. A surface rubbed by a file having lesser number of teeth is not that plain and fine as it is with that of a file having more number of teeth. The teeth of a file are formed by cuts made on its face. Files have cuts of different uses.

The teeth of a file are formed by cuts made on its face. Files have cuts of different types. Files with different cuts have different uses.

The selection of a file with a particular type of cut is based on the material to be filed. Single cut files are used for filing soft materials. But certain special files, for example, those used for sharpening saws, are also of single cut. Files are specified according to their length, grade, cut and shape.

Files are manufactured in different types and grades to meet the various needs.

Files are specified according to their length, grade, cut and shape.

File grades are determined by the spacing of the teeth. **(Fig. 7.106).**

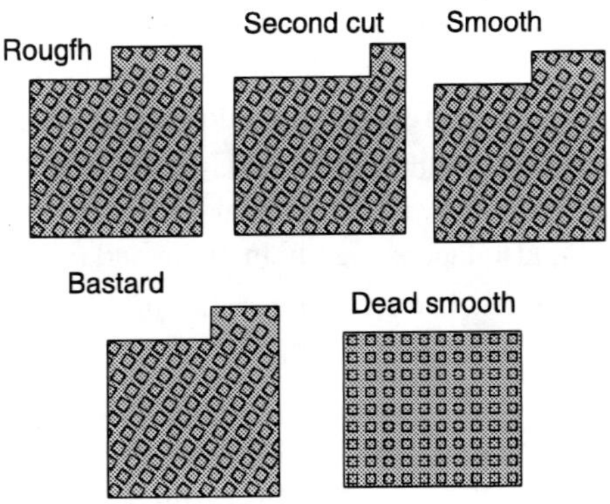

Fig. 7.106 Grades to files

Rough file: The teeth of big size and less in number are used for removing rapidly a larger quantity of metal. Its cutting is rough. It is mostly used for trimming the rough edges of soft metal castings. Filing done with rough file is coarse.

Bastard file: Teeth of this file are comparatively smaller than those of the rough file. It is used initially to shed metal (through filing) in large quantity. This file is used in cases where there is a heavy reduction of material.

Second cut file: This is used to give a good finish on metals. It is excellent to file hard metals. It is useful for bringing the jobs close to the finishing size. This is a file of medium grade and is mostly used to bring the job into the proper size. As compared to bastard file, filing done with this file makes the surface plain.

Smooth file: This file makes the surface quite plain, in addition to making the job of accurate size through filing. It removes small quantity of material and gives a good finish.

A dead smooth file: Its teeth are very close to each other and it rubs off the metal in very little quantity. It is used for bringing shine on the job after finishing and is used to bring to accurate size with a high degree of finish.

The most used grades of files are bastard, second cut, smooth and dead smooth. These are the grades recommended by the Bureau of Indian Standards. (BIS)

7.5.5 Grade of Files

Different sizes of files with the same grade will have varying sizes of teeth. In longer files, the teeth will be coarser.

Number of teeth on file: On files of different grades teeth are cut in the following manner:

Grade of file	**Number of teeth**
Rough	20 to 25 teeth per inch
Bastard	25 to 30 teeth per inch
Second cut	35 to 40 teeth per inch
Smooth file	40 to 65 teeth per inch
Dead smooth file	80 to 100 teeth per inch

Thus selection of a file is done after considering the following four points and the utility of the job:

1. Shape of file
2. Cut of the file-teeth
3. Grade of the file on the basis of the number of its teeth.

While taking a file the following points should be ascertained:

Convexity and taper in files: Most files have the faces slightly bellied lengthwise. This is known as convexity of a file. This should not be confused with the taper of a file. A flat file has faces which are convex and it also tapers slightly in width and thickness.

Purpose: If the file is parallel in thickness, all the teeth on the surface of the work will cut. This would require more downward pressure to make the file 'bite' and also more forward pressure to make the file to cut.

It is more difficult to control a file of uniform thickness.

To produce a flat surface with a file of parallel thickness, every stroke should be straight. But it is not possible due to the seesaw action of the hand.

If the file is made with parallel faces, while giving heat treatment, one face may warp and become concave, and the file will be useless for flat filing. Excessive chip removal at the front or rear workpiece edge is prevented and filing of the flat surface is made easier because of the convexity on the cutting faces.

There are three advantages of its being convex:

At the time of filing, the entire surface of the job does not come into contact. Therefore there is less exertion.

If the pressure is in the centre, both the ends of the job are equally rubbed. If the file is used in a flat manner it is difficult to keep equal pressure on both the ends.

An entirely flat file carries the risk of being bent when it is hardened.

Besides convexity, files are tapered towards the lower part breadth-wise. While filing, the taper of the front side can also go into a narrow place, e.g., round file can be used in less than half an inch because its taper is less than this.

1. Choose the right type of file, according to the job.
2. There should be a handle fitted on the tang of the file. In small files, a small handle should be fitted and a larger handle in larger file. Otherwise the balance of the file would not be proper and power would be unnecessarily wasted.

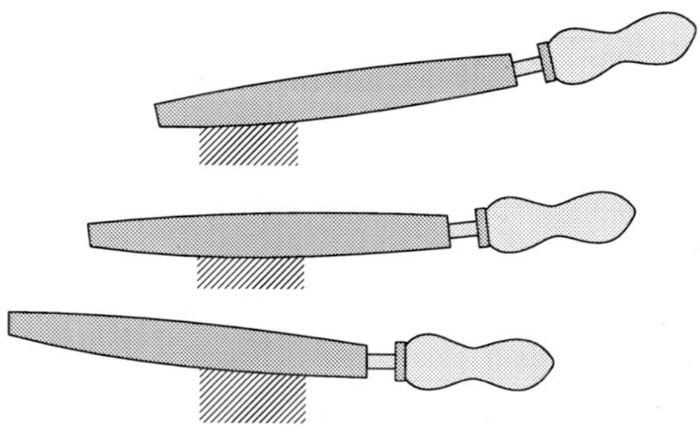

Fig. 7.107 Convexity of file

3. Hold the handle of the file with the right hand and hold the file end balanced with the left hand.

At the time of filing, left leg should be kept ahead and the right leg a little behind.

Pressure on the file should be exerted while moving it ahead, not while driving it back.

Only the hands should move with the file and not body. As far as possible, keep still.

Speed of moving the file should be at the rate of 30 to 35 strokes per minute.

Filing is done in three different positions.

1. Straight filing.
2. Cross filing.
3. Draw filing.

Straight filing: In this method one should stand almost in front of the vice and the file is moved in the straight direction of the vice. Through straight filing method the metal is quickly cut. But there is a great need for alertness to keep the balance of file. If a file of large size is being used, then the left hand should be kept as

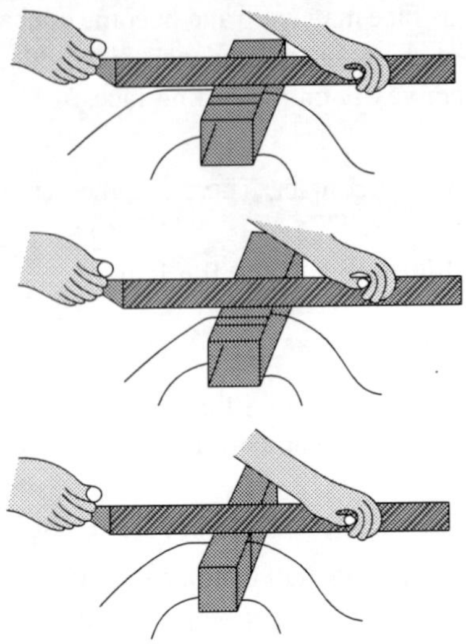

Fig. 7.108

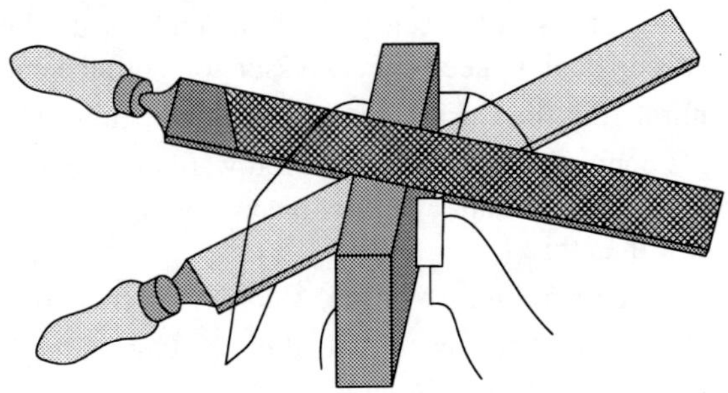

Fig. 7.109 Cross filling

shown in **Fig. 7.108 (A)**. If finishing work is being done then the file should be held in the left hand as shown in **Fig. 7.108 (B).**

Cross Filing (Fig. 7.109): By this method the file is moved from one end to the other end of the job in one stroke and in the second stroke filing is done on the opposite ends of the job. Through cross-filing the entire job becomes plain and smooth.

Draw Filing (Fig. 7.110): By this method the high spots of the job are easily filed. This filing is to bring finishing on the job which shines.

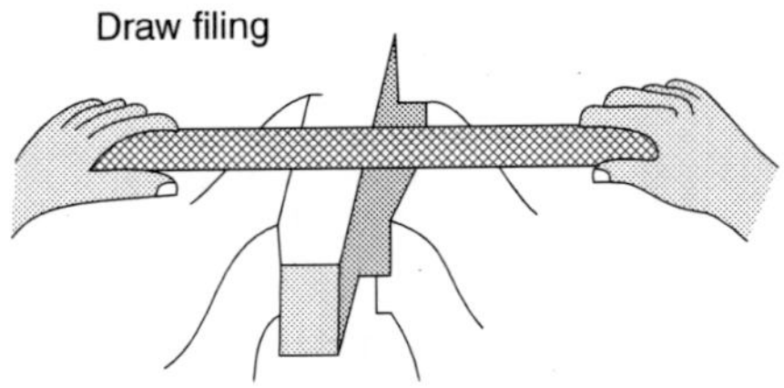

Fig. 7.110

While draw filing, the handle is removed. Chalk is rubbed on the face of the file so that the particles cut are not left in the file itself. Otherwise the file would leave some marks on the material.

7.5.6 Needle Files

These files are usually available in sets with assorted shapes. In one set there are 8, 10 or 12 files. These files are of different grades (bastard and smooth). Their length ranges from 100 mm to 200 mm. There is no need to add a handle separately to the files since their tang is made like a handle. These are very delicate, and are to be handled with care; there is risk of their being broken. These files are used by the watch makers, tool and die makers, and goldsmiths.

The common shapes of needle files are round edge, flat edge, flat taper, half round, triangular, square, round, knife, feather edge, crossing, barrette and marking.

7.5.7 Nomenclature of Needle Files

Length: These files are available in a nominal length of 120 mm to 180 mm.

Grade: The grades of cut may be identified by the cut number as bastard – cut 0, smooth – cut 2.

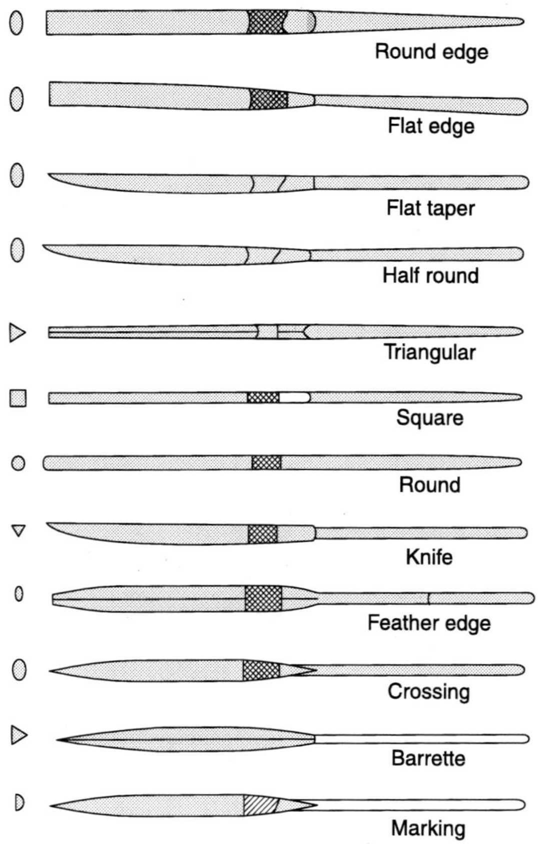

Fig. 7.111

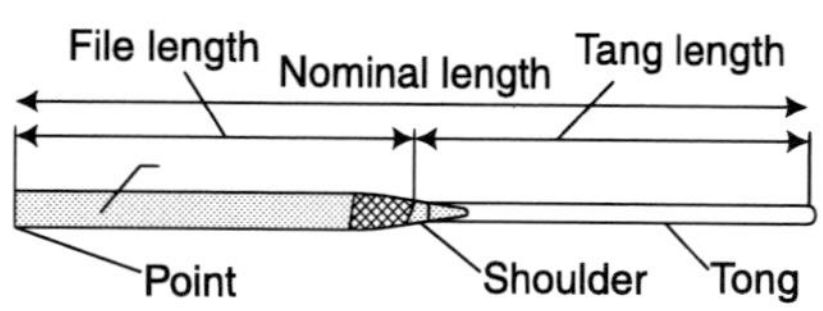

Fig. 7.112

7.5.8 Designation of Needle Files

The needle files are designated by their names, grade of cut, nominal length and BIS number.

Example: A flat edge needle file with grade of cut bastard, having a nominal length L = 160 mm shall be designated as flat edge needle file bastard, 160 IS 3152.

Special Files: In addition to the common type of files, files are also available in a variety of shapes for 'special' applications. These are as follows.

Riffler files (Fig. 7.113): These files are used for die-sinking, engraving and in silversmith's work. They are made in different shapes and sizes and are made with standard cuts of teeth.

Mill saw files (Fig. 7.114): Mill saw files are usually flat and have square or rounded edges. These are used for sharpening teeth of woodworking saws, and are available in single cut.

Mill saw files contain only single cut dents. It is also used for polishing of brass jobs and for filing on the lathe machines.

Mill saw files are of flat section with edges either round or flat.

Block file (Fig. 7.115): These files have rectangular cross sections with teeth cut at the bottom face, and have provisions for fixing of handles.

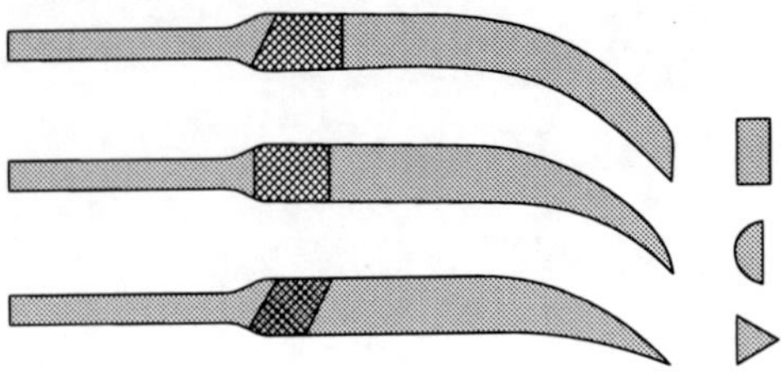

Fig. 7.113 Riffer files

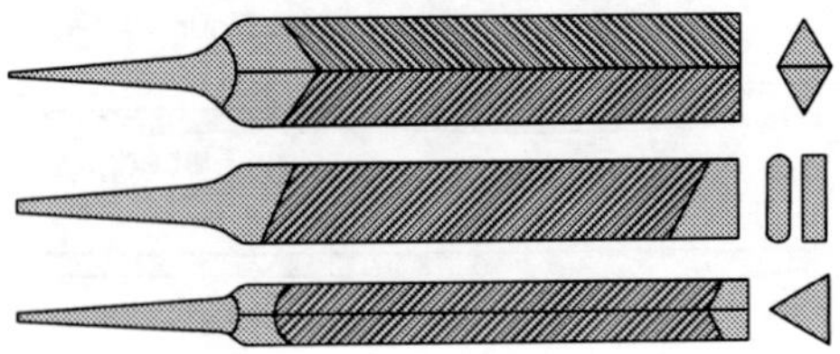

Fig. 7.114 Mill saw files

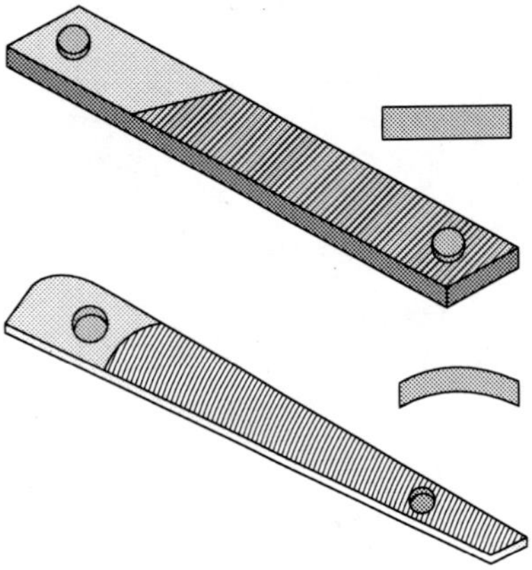

Fig. 7.115 Block files

Fig. 7.115a Use of mill saw file

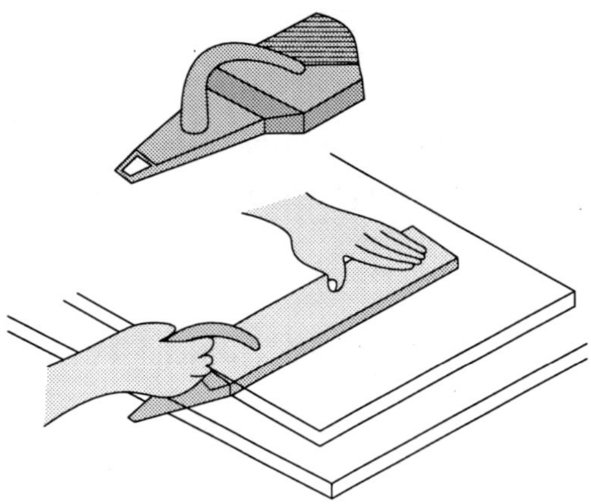

Fig. 7.115b Block file with handle

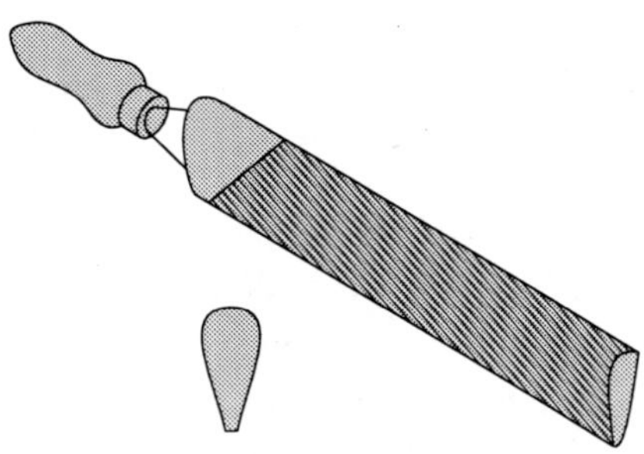

Fig. 7.116 Cross cut file

Block files are useful for filing surfaces. Large surfaces can also be filed using ordinary files fitted with special handles.

Cross-cut file (Fig. 7.116): These are used for sharpening cross-cut saws. The rounded edge of the file is useful for deepening the rounded gullets of saw teeth. The file is single cut.

Lathe files (Fig. 7.117): These are rapid, free cutting files designed to file jobs set on a lathe. This gives a smoother surface than the mill files. They are single cut and the teeth are longer. These files are also called long angle lathe files.

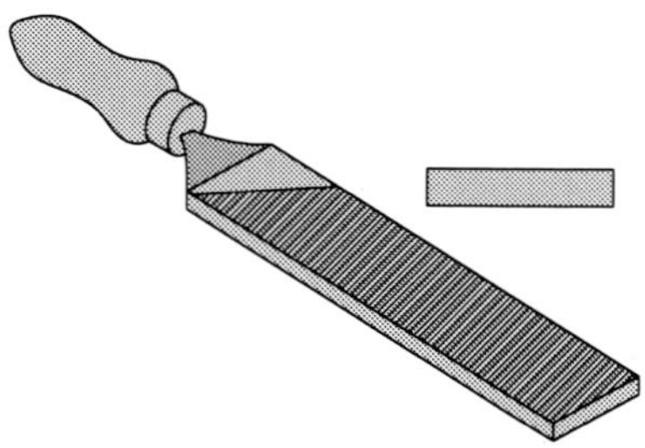

Fig. 7.117 Lathe file

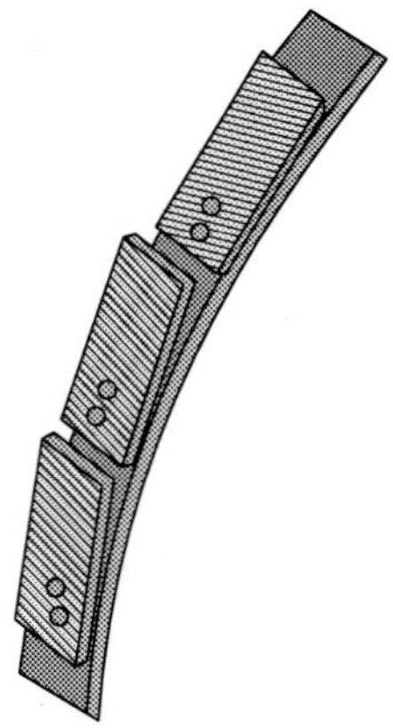

Fig. 7.118 Band file

Band files (Fig. 7.118): These are individual segments of files assembled on a band. The ends of the steel band are joined together to form a continuous loop.

These files are used on hand-saw machines for filing and finishing straight or contour forms according to the layout lines.

Crossing file (Fig. 7.119): This file is used in the place of a half round file. Each side of the file has different curves. It is also known as "fish back" file.

Fig. 7.119 Crossing file

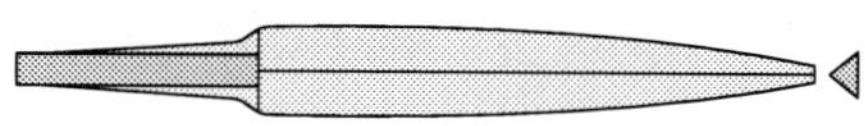

Fig. 7.120 Barret file

Barrette file (Fig. 7.120): This file has a flat, triangular face with teeth on the wide face only. It is used for finishing sharp corners.

Its two sides are short and one side is large. It is also called cantt file. It is used for enlarging the inside angle.

Tinker's file (Fig. 7.121): This file has a rectangular shape with teeth only at the bottom face. A handle is provided on the top. This file is used for finishing automobile bodies after tinkering.

Machine files (Fig. 7.122): These are of double cut, having holes or projections to fix to the holder of the filing machine. The length and shape will vary according to the machine capacity. These files are suitable for filing the inner and outer surfaces, and are ideal for die sinking and other tool-room work.

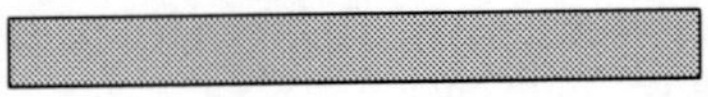

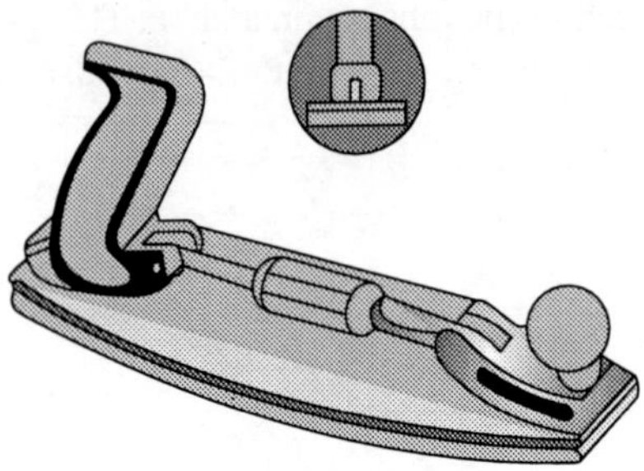

Fig. 7.121 Tinker's

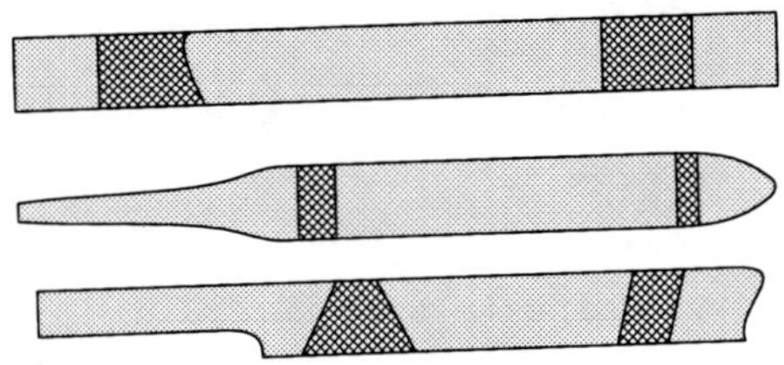

Fig. 7.122 Machine files for hand filling

Fig. 7.123 Warding files

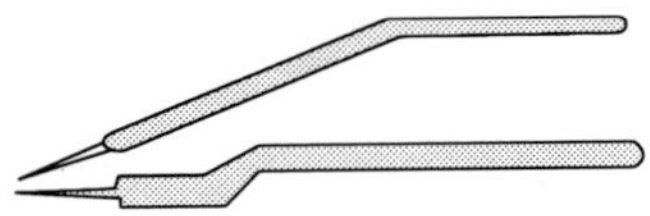

Fig. 7.124 Setting files

Warding File (Fig. 7.123): This is almost like a flat file. It is very thin, parallel in thickness and its tip is pointed. It is used for filing of jobs having narrow grooves. Particularly these are used by lock and key makers, and watch makers.

Pillar file: This file is similar to a hand file. Its breadth is parallel and thickness is tapered. It is used for filing of right angled corners of jobs. Besides this, where square files and hand files cannot be used a pillar file is used.

Swiss Pattern File: These files are more or less similar to warding files. Their tip is also pointed. But they have more dents cut per inch. The watch makers and other technicians doing precision work make use of these files.

Setting File (Fig. 7.124): This is a sort of flat file, which can be bent according to needs. With it the inside surface of a key way is filed.

Die sinkers file (Fig. 7.125): This is a special type of file, which is crooked and circular in shape. Dents are cut on both of its ends and the middle portion is made plain for holding it. With it filing can be done in deep spaces like die or mould.

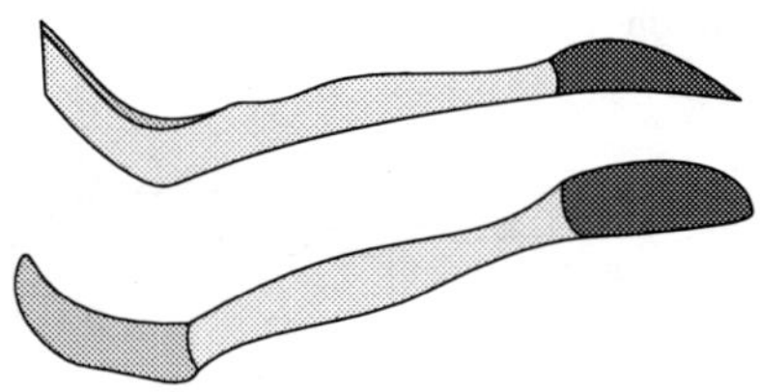

Fig. 7.125 Die sinker's files

7.5.9 Rotary Filing and Burring (Fig. 7.126)

These files are available with a round shank. They are driven by a special machine with a portable motor and flexible shaft. These are used in die sinking and mould making work. Rotary files and burrs are small cutters made of high-speed steel, solid carbide, or steel shanks tipped with carbide into which cutting edges have been ground or chiselled.

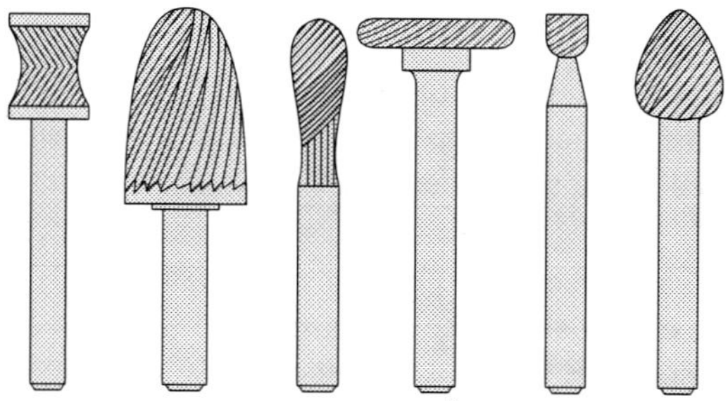

Fig. 7.126

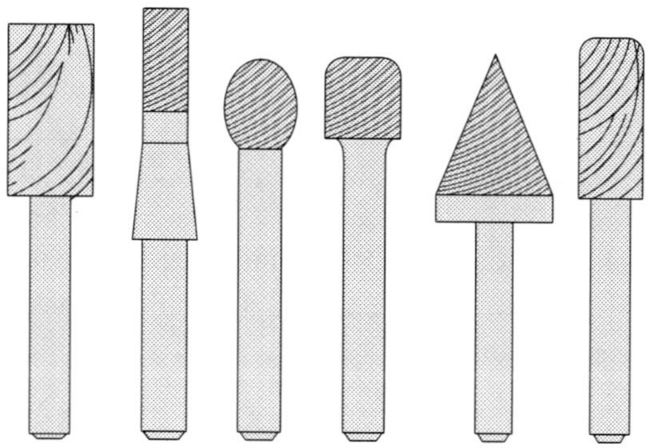

Fig. 7.126a

Rotary files and burrs are used on soft materials such as wood, plastics, leather, and fibre; commonly machined materials such as aluminium brass, and cold-rolled steel and hard or abrasive materials such as cast iron, stainless steel, high-speed steel, hard plastics, and ceramics. The cutters are used in portable tools powered by compressed air or electric motors, or through flexible shafts. Ground burrs are more efficient than hand-chiselled rotary files because the uniformity of the ground teeth and flutes prevents them from being easily clogged. Ground cutters may be re-sharpened at a fraction of the cost of the original tool.

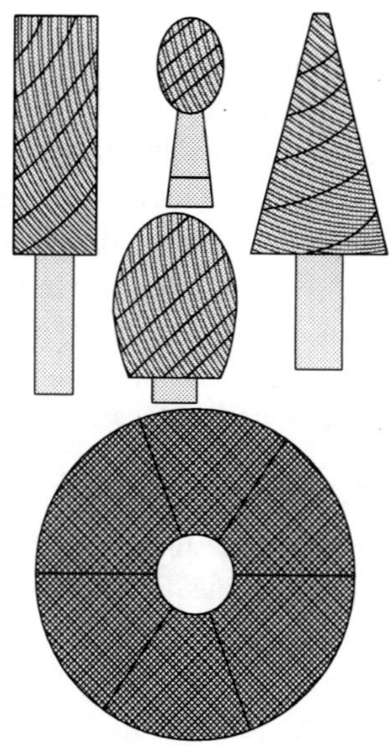

Fig. 7.126b Machine files

7.5.10 Cutter Shapes

Rotary files and burrs are manufactured in many shapes and sizes. A variety of tooth patterns is available to meet various requirements. Diamond fluting is designed for hardened-steel dies and forgings; herringbone fluting may be used where better finishes are required. Manufacturers usually supply the cutters with coarse, medium (standard), and fine teeth, most cutters are supplied with chip breakers to break up long chips. Tapered shapes ball and contour shapes, tools having front taper, back taper, or contours involving radii are generally arranged in five sections. Five teeth extend the full length of the cutter, with shorter teeth blending into the main teeth because of the difference in diameters.

Three types of teeth are used on rotary files and burrs: positive-rake, radial-rake, and negative-rake

7.5.11 Surface Finish

The surface finish obtained varies with the speed of the cutter. Higher speeds will result in better finish, but less material will be removed. Fine tools can be operated at lower speeds than coarse tools to obtain similar fine finishes. When constant speed power sources are used, the cutter diameters must be changed to vary the peripheral speed.

The application of these tools should be guided by many of the same rules used in most other machining and grinding operations. The file or burr should be fed at even rates and pressures to avoid irregularities in the workpiece surface. A short grip on the shank of the cutter is most ideal for accurate control.

7.5.12 Sawing and Filing Fluids

Sawing and filing are considered relatively easy operations compared with other metal removal processes. In some cases, sawing and filing may be performed without a cutting fluid, as in the sawing of high-carbon, high-chromium hardening and chlorinated high-viscosity cutting oil or synthetic cutting fluid should be used to:

1. Lubricate the saw blade and clean the tool by flushing away chips that might collect between the teeth.
2. Reduce the heat generated in the work and the tool thus contributing to longer tool life.
3. Eliminate or reduce the welding of metal chips to the cutting edges, thus extending tool life and improving workpiece surface finish.

7.5.13 Band Filing

Band filing is the fastest and most accurate method of filing. It is generally performed on vertical bands sawing machines and requires a speed range of 50 to 250 fpm. The file band is made up of file segments mounted on a flexible steel band. File bands are available in ¼ , $^3/_8$ and ½ inch widths; in flat, oval, and half-round shapes; in various numbers of teeth per inch; and with cut designs to suit industrial requirements. Bastard-cut files predominate for most steel and ferrous work. Short-angle or rasp-cut files, having deeply cut teeth and therefore ample chip clearance, are intended to cut soft metals and most nonmetals.

Coarse-cut files cut fast; bastard-cut files offer a compromise between cutting rate and finish; smooth-cut files give the highest degree of finish but at slower rate.

When band filing will fully load one machine, a machine with the limited speed range and medium or light pressure needed designed especially for this application should be used. Band filing, in addition to tool and die making, is used in many foundries and other manufacturing operations to remove flash and burrs and to smooth rough surfaces.

7.5.14 Pinning of Files

During filing, sometimes the metal chips (filings) clog between the teeth of files. This is known as ‘pinning’ of files.

Files which are pinned will produce scratches on the surface being filed, and also will not bite well.

Pinning of the files is removed by using a file brush also is called a file card.

Besides this, if dry chalk is rubbed on the file, there would be less pinning.

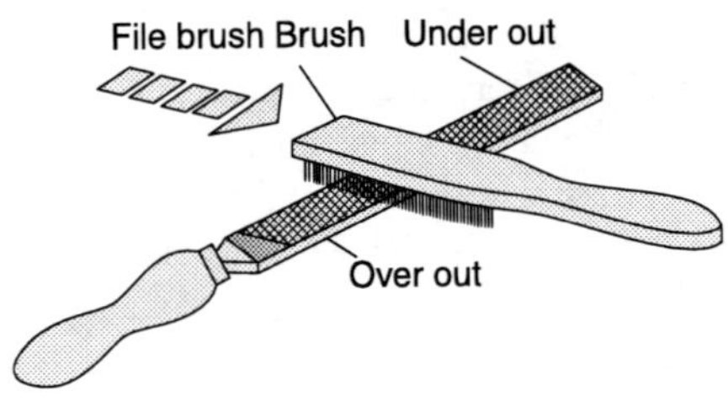

Fig. 7.127 Pining of files

7.5.15 Precautions

1. Select the file according to the job.
2. Do not use a file without handle.
3. According to the size, complete handle should be fixed on the file.
4. First, a new file should be used on a soft metal and then it can be used on solid metal.
5. If substantial material is to be cut then chipping should be done first and then filing.
6. For filing of sheet metal jobs, the job should be fixed in a vice by using wood on both its sides. It should also be ensured that the sheet should not be higher than the vice.
7. While filing a sheet, cut it in length.
8. Make use of chalk in order to save the job from pinning. If there are metal particles on the file, clean with a brush.
9. Keep the file away from other tools. They should not be heaped together.
10. File should not be heated.
11. Do not apply oil or grease on the file.

EXERCISE

1. Make a sketch of a chisel in the working position when chiselling a groove in hard steel. Name the angles by using Greek letters and state their approximate degrees.
2. Why does the cutter of a slim chisel easily break out when working on hard material?
3. Why is the chisel hardened on the cutting edge?
4. Why must the wedge hindering the hammer to come off be seated in an inclined position?
5. In which way can accidents be prevented during chiselling work?
6. Why does the saw often clasp during spot sawing?
7. Why preference is given to fine-toothed saw blades when sawing tubes and profiles?
8. How can choking of saw blades in the saw notch be avoided?
9. Why is sawing done more slowly in tool steel and grey cast iron than in structural steel?
10. Mention the reasons for using coarse-toothed saw blades for soft materials and long saw notches!
11. What must be observed when setting the saw blade and clamping the workpiece?
12. Why must the saw be released when the work is interrupted for longer periods?
13. What kind of accidents can occur when sawing by hand?
14. Why are surfaces scraped?
15. When is the scraping effect greater with a small or with a large cutting angle?
16. State the reasons for the rough-scraping process.
17. Which points must be scraped after checking against the master plate?
18. For what reason is the 'scraping sense changed?
19. How is a well scraped surface recognized?
20. In which way are the tools used for scraping and checking against the master plate protected against damages?

21. What do you understand by: (a) cutting-off, (b) cutting-out, (c) cutting-in?
22. Why does the material of high strength require much force when being cut?
23. What produces unevenness at the cutting point of the material during the shearing process?
24. Why are the blades moved along each other with little play?
25. Explain the lever action: a) with hand levers, b) with guillotine shears.
26. Why it is not allowed to elongate the lever arms of shears by means of a tube?
27. Why can generally only thin plates up to 1.2 mm thickness are cut with hand shears?
28. Why do shears generally only cut when reaching an opening of 15°?
29. Why the upper shear blade of lever and guillotine shears is slightly curved in most cases?
30. When will a workpiece throw off when cutting?
31. What influence has the setting of the blank holder on the neatness of the cut?
32. What accidents may occur when shearing and in what way can they be prevented?
33. What does the selection of files depend on?
34. Why is flute cut of a file transversal to file longitudinal axis?
35. What kinds of materials are worked with single-cut files?
36. Why have the under- and over-cut of file different angles to the longitudinal axis?
37. Why do workpieces that are clamped too high reduce the service life of the file?
38. How can the formation of channels be avoided when smoothening?
39. Describe the filing of a pivot in the shell wood.
40. In which way are plates clamped for filing?
41. What are the disadvantages of one-side clamping in the vice?
42. In what manner are file handles fixed?
43. What accidents may occur when doing filing work, and how can they be avoided?
44. Name the tool, used under the category of "cutting tools".
45. What is meant by hot chisel and cold chisel?
46. Draw a sketch of chisel showing its different parts.
47. What is the importance of chisel's cutting angle?' What should be the angle for cutting mild steel, cast iron, tool steel, brass, aluminium, and copper?
48. Describe chipping process and what precautions are necessary for this process?
49. What are the different kinds of files and on what basis files are classified?
50. What is the difference between double cut and second cut?
51. Why taper is given in a flat file?
52. Comment on the following:
    (i) Pinning.
    (ii) File -card.
    (iii) Draw file.
    (iv) Needle file.
    (v) Rasp cut file.

53. What is the function of a scraper? How many kinds of scrapers are there?
54. Draw a sketch of a scraper.
55. Write a brief note about the types of hacksaw frames.
56. What is the basis for classification of hacksaw blades?
57. What precautions are necessary for hacksaw?
58. What is meant by setting of hacksaw blade?

Drilling, Reaming, and Related Operations

Chapter 8

8.1 INTRODUCTION

The majority of all workpieces have holes, either through bores or blind holes.

Drilled holes serve many purposes, e.g., to take up rivets, screws, bolts, shafts, pistons, and furthermore to pass through gases, fluids, etc.

Drilling and boring are cutting operations and produce round holes and bores in metallic and non-metallic materials. The holes and bores are cut out of the material with a cutting tool. In most of the cases, drilling machines are used for this purpose. However, lathes, capstan lathes, and automatics are also used.

Holes in workpieces, besides by drilling and boring, can be made by other methods also, e.g., by punching, perforating, enlarging with a drift, gas cutting, casting, etc. This, as a rule, is cheaper than the drilling operation.

However, with none of these methods the diameter of holes, centre-to-centre distances and the surfaces can be obtained as accurately as by drilling and boring. Drilling and boring operations, therefore, is an important operation in all kinds of metal working industries.

Very often, drilled holes are finished by refining procedures, such as reaming, grinding and honing.

Movements while drilling on the drilling machine

The tool used for cutting holes in the full material is called twist drill, which is provided with two cutting edges. In order that the cutting edges can cut off chips, two movements are required simultaneously (**Fig. 8.1**).

The drill rotates. The rotary motion is called cutting or main motion.

In special cases, this action is effected by rotating the workpiece, e.g., while boring on the lathe.

As main motion, the cutting speed is measured in m/min. The highest cutting speed is on the periphery of the drill and it decreases towards the 'centre of the drill.

The drill is moved in a straight line towards the fixed workpiece. This movement is called feed and it controls the thickness of the chip.

The feed operation can also be effected by moving the fixed workpiece towards the rotating drill, as it is done with some small bench type drilling machines where the table is raised.

The feed is measured in mm/R. As the drill is provided with two cutting edges, the thickness of the chip is half the feed.

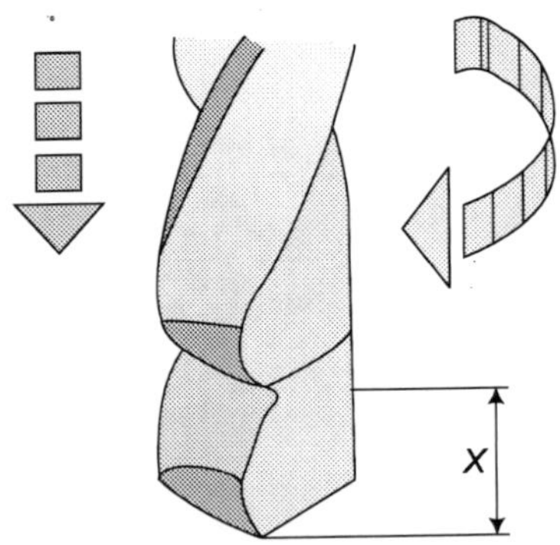

Fig. 8.1

By the simultaneous double action of the cutting or main motion and the feed, each cutting edge of the drill describes a spiral and thereby produces a constant flow of chips.

In drilling operations, we distinguish between drilling holes in the full material and enlarging the holes **(Figs. 8.2 and 8.3).**

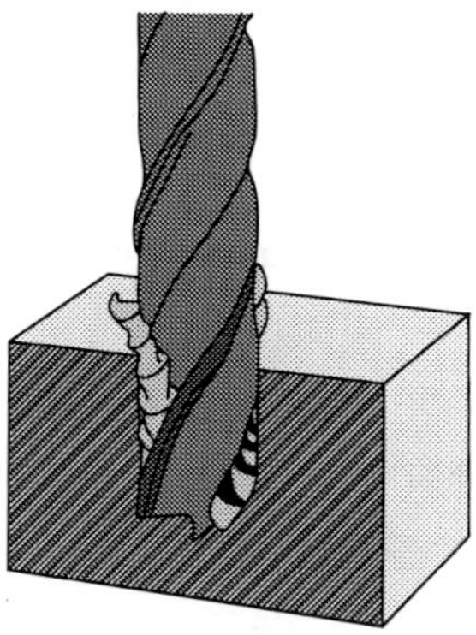

Fig. 8.2

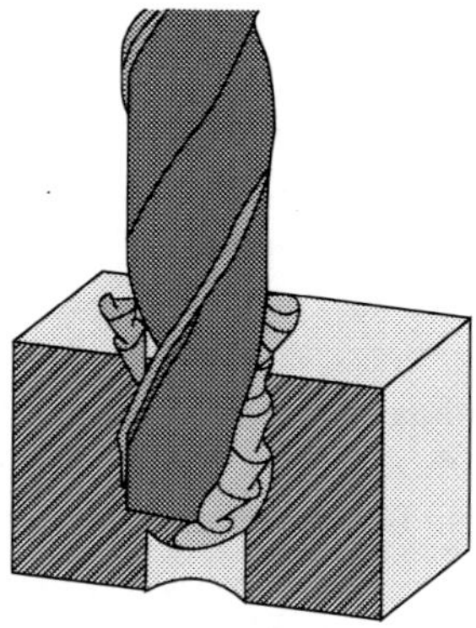

Fig. 8.3

For enlarging the holes on a drilling machine, next to the twist drill, very often the core drill, provided with three to four cutting edges, is used.

8.2 FUNDAMENTALS OF DRILLING

In order to bore round holes in different jobs, drill and drill machines are used. The tool used for boring hole is called drill. The machine with whose help the drill is moved is called the drill machine. The process of boring holes with drill and drill machine is called drilling.

8.3 DRILL

These are made of carbon steel; high speed steel and alloy steel. Drill is of three different shapes:

1. Flat drill
2. Straight fluted drill
3. Twist drill

Flat drill (Fig. 8.4): This type of drill is generally made by beating slender sticks of high carbon steel.

Its cutting edge is ground and then hardened and tempered. It can be made in any shape or size. It is not used for boring holes in iron. It is usually used for boring holes in carpentry work and soft metals. Cutting point is kept at 90° angle. But it has the following shortcomings:

1. Holes bored by this drill are neither round nor in one direction.
2. Holes are made of bigger diameter than the drill size.
3. For boring holes with flat drill much pressure and power is needed.
4. Chips cut by this drill do not come out of the holes themselves.

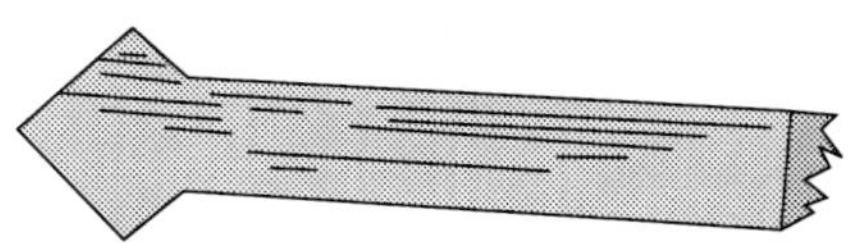

Fig. 8.4

5. Coolant is not able to reach up to the cutting edge. Therefore, soon it becomes hot and starts getting damaged.
6. By excessive grinding, the diameter of the cutting edge gets reduced.

Flat drill has some advantages also. These are:

1. On those solid metals where other drills get broken, this drill can be used after properly hardening and tempering.
2. These are made in different sizes after forging them.
3. These drills cost very less.

Straight fluted drill: In this type of drill two grooves from top to bottom are cut opposite to each other. This groove is known as flute. This is used to bore holes in soft metals such as copper, brass, aluminium. Their cutting angle is made differently according to the metal. As compared to flat drills, holes can be bored with it cleanly and in a better shape. Special drills are used of straight flute which is known as Eureka Drill.

Twist drill (Fig. 8.5): The most customary twist drills are standardized. Twist drills are used for boring holes quickly, clean and of accurate size. They are often made of slender sticks of high-speed steel and alloy steel. Then cutting edge is made on bottom end and a twisted groove is cut on its body from top to bottom which is known as flute; hence the name twist drill. The parts are:

1. Tang
2. Shank
3. Body
4. Point
5. Neck

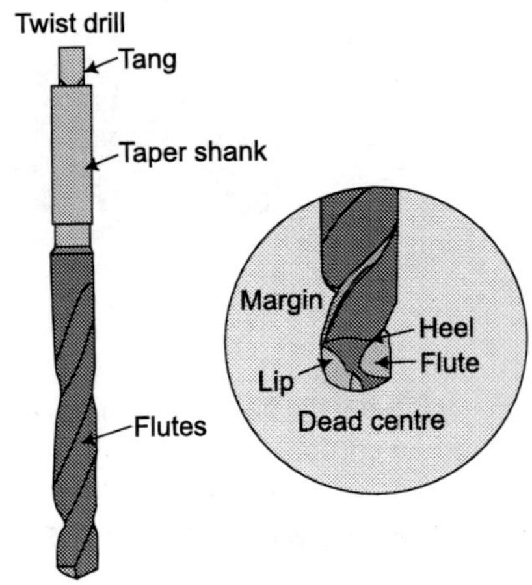

Fig. 8.5

The shank serves for fitting in the machine spindle or holding in the drill chuck. The shank can be cylindrical or tapered. The cutting part gets its basic form by two helical flutes.

The material profile, remaining between the flutes, is called centre web. The two main cutting edges (lips) are formed by grinding of the point. Between the two clearance faces, the chisel edge is located, which forms the angle of point with the two main cutting edges. The chisel edge does not cut; it only scrapes. It pushes the material out of the centre of the hole in front of the main cutting edges and hereby consumes about 40% of the feeding power. The lands guide the drill and prevent rubbing of the heel in the drill hole. To avoid damaging of the lands while drilling deep holes, the diameter of the drill is reduced towards the shank by about 0.05 mm in 100 mm length.

As on all other cutting tools, there are clearance, rake and wedge angles on twist drill. The angles appear on each of the two main cutting edges.

Tang: The uppermost part of the twist drill is called tang. It is small and flat in shape. It fits into the slot of the drilling machine spindle.

Shank: Shank is the portion of the drill by which it is fitted on to the machine. It is held in a socket drill chuck. Shanks are of two types.

Taper shank: Used for large size drills with a Morse taper. A drill with this type of shank can be easily fitted in the drilling machine. The risk of their being slipped is negligible because at the time of giving pressure, their hold in the spindle or socket is strengthened.

Straight shank (Fig. 8.6): This is generally used in drills of smaller diameter and gripped in drill chucks.

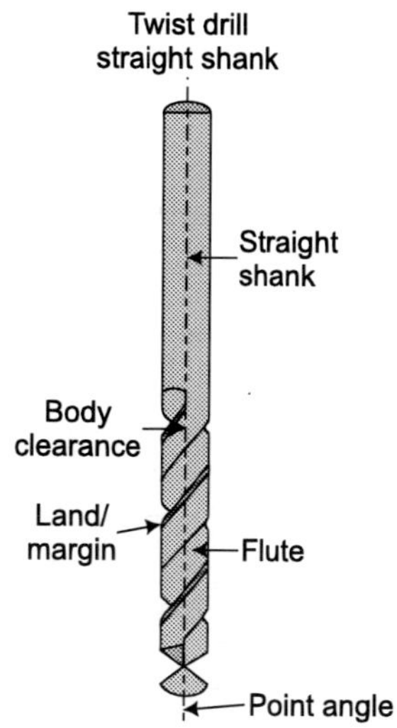

Fig. 8.6

Body: The main portion between the point and the shank is the body. It is the main part of the drill. It extends from the extreme cutting end to the beginning of the shank. Hole is bored with this part. Different types of spiral grooves are cut in the body. The parts of the body are discussed below.

Flute: The spiral grooves in the body of the drill are called flutes. Cutting edge is made on their border, these grooves act as an outlet passage for clearing the chips cut by the drill. Through these grooves the coolant reaches the point.

The flutes help to form the cutting edges. They cut the chips and allow these to come out. The coolant flows to the cutting edges through the flutes.

Land/margin: The land/margin is the narrow strip which extends to the entire length of the flute. It is the cylindrically ground surface on the leading edges of the drill-flutes. The width of the land is measured at right angles to the flute. On the verge of flutes a bright line is visible, which extends to the entire length. At the time of drilling a hole, this part comes into contact of the job. The diameter of the drill is measured according to this land/margin.

Body clearance: Body clearance is the part of the body which is reduced in diameter to reduce the friction between the drill and the hole being drilled. It saves the drill from getting hot and damaged.

Web (Fig. 8.7): The central portion of the drill situated between the "roots" of the flutes and extending from the point end towards the shank is known as web. The point end of the web forms the chisel edge. It gradually increases in thickness towards the shank.

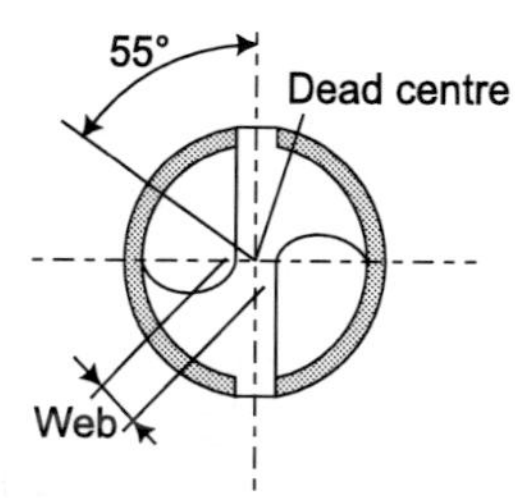

Fig. 8.7

Web is the metal column which separates the flutes. It gradually increases in thickness towards the shank.

Point: The lower portion of the drill is called point. It is made in the form of an angle. Different angles are made for different metals. Its point is generally ground at 118° angle. It consists of lips or cutting edges heel and dead centre.

Lip: The cutting edge of the drill is known as lip. This edge is there with each flute of the drill. The length and angle of the lip should be equal otherwise the hole would not be proper.

Lip clearance: The portion behind the cutting edge is ground at a special angle. This angle is known as lip clearance. This angle is small for boring holes in solid metal and more for holes in soft metals. Ordinarily, this angle is from 8° to 15°. This angle cuts the metal and the back portion does not come into contact.

Heel: The back portion of the cutting edge is known as heel.

Dead centre: The last pointed end of the drill is called dead centre. It should be exactly at the centre of the drill.

Drill angles: Like all cutting tools the drills are provided with certain angles for efficiency in drilling. They are listed below.

Point angle/cutting angle (Fig. 8.8): The point angle of a general purpose (standard) drill is 118°. This is the angle between the cutting edges (lips). The angle varies according to the hardness of the material to be drilled.

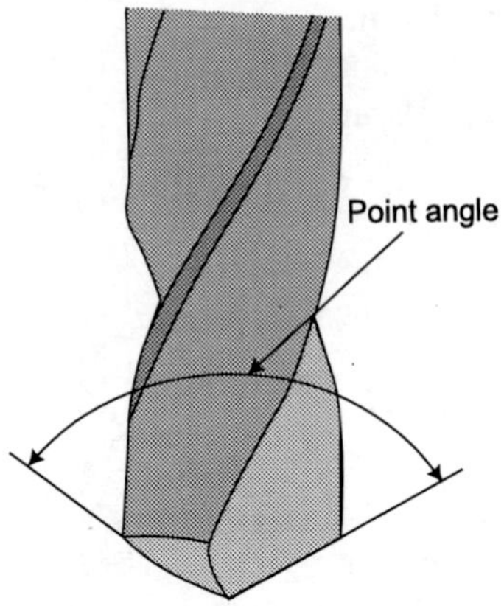

Fig. 8.8

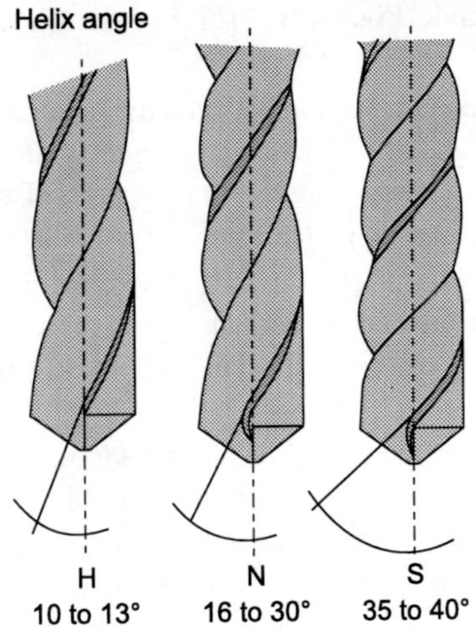

Fig. 8.9

Helix angle (Fig. 8.9): Twist drills are made with different helix angles. The helix angle determines the rake angle at the cutting edge of the twist drill.

The helix angles vary according to the material to be drilled. According to Indian Standards, three types of drills are used for drilling various materials.

Type N for normal low carbon steel.

Type H for hard and tenacious materials.

Type S for soft and tough materials.

Type N for general purpose drilling.

Rake angle (Fig. 8.10): Rake angle is the angle of flute (helix angle).

Clearance angle (Fig. 8.11): The clearance angle is to prevent the friction of the tool behind the cutting edge. This will help in the penetration of the cutting edges into the material. If the clearance angle is too much, the cutting edges will be weak, and if it is too small, the drill will not cut.

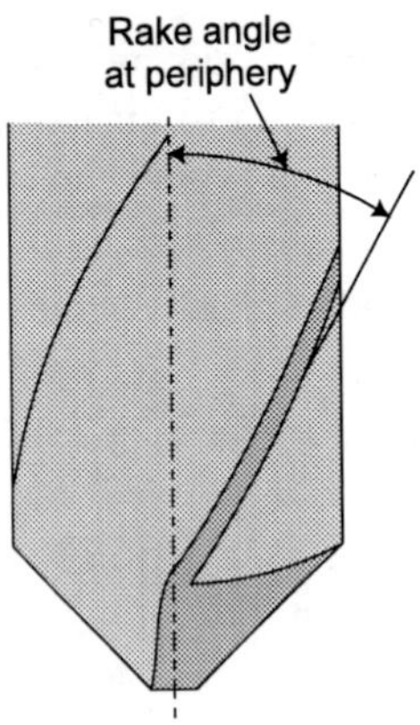

Fig. 8.10

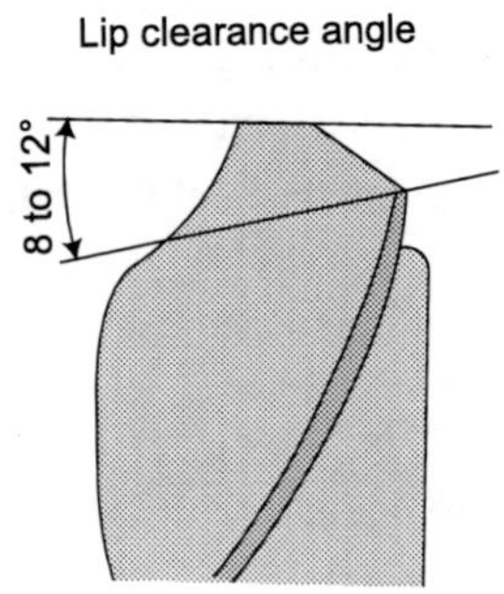

Fig. 8.11

Chisel edge angle web angle (Fig. 8.12): This is the angle between the chisel edge and the cutting lip.

Designation of drill:

- Diameter
- Tool type
- Material.

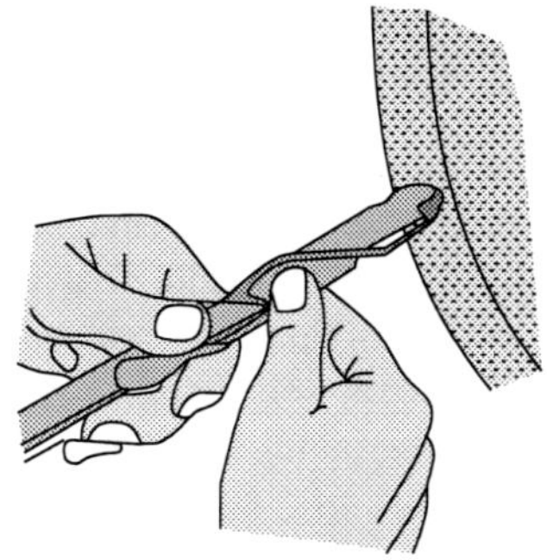

Fig. 8.12 Grinding by hand

Example: A twist drill of 9.50 mm diameter of tool type 'H' for right hand cutting and made from HSS is designated as

Twist drill: 9.5 – H –IS 5 101 – HS

9.5 — Diameter, H — Tool type, IS – Number – HS – Material.

If the tool type is not indicated in the designation, it should be taken as Type 'N' tool.

8.4 ACTION OF THE TWIST DRILL (FIG. 8.1)

8.4.1 Cutting Edge and Formation of Chips

By the circular cutting motion of the twist drill, the cuneiform (wedge-shaped) cutting edges detach chips and penetrate into the material by the force of the feed. The cutting effect of the twist drill depends mainly on the shape and the angles of the cutting edge, as this is the case with all cutting tools.

The drill must have an appropriate cross section to provide for a good formation and removal of chips, as well as a reliable guide. The inventor of the twist drill, Martignoni, gave it in 1863, a basic shape answering the mentioned requirements. The drill is made of round steel by cutting out two helical sloping flutes. The cutting edges are made by point grinding of this drill shape.

The centre line between cutting edges on the foremost point is formed by the centre piece of the drill, the centre or centre web, and by the relief-ground front rakes. Its most favourable position with regard to the main cutting edges is to be 55° with correctly pointed drills. The said centre line between the cutting edges does not act as a cutter but it presses the material laterally to the main cutting edges. Chipping work is thus rendered difficult. To facilitate the cutting action, the centre line (between the cutting edges) is shortened by tapering (internal grinding) in such a way that it still amounts to at least 1/10 of the drill diameter.

In case of larger diameters of drills, a smaller hole is at first drilled into the material to allow a free movement of the centre line between the cutting edges. The cuneiform main cutting edge is situated between the front rake and the flute. The said flute serves as true rake. The pitch of the flute is the hook (γ) (spiral angle). A changed spiral angle with an unchanged spiral angle also alters the hook and the lip angle of the cutting edge. Due to the fact that the lip angle of a cutting edge should be large for working hard material and small when treating soft materials, a drill with a large spiral angle is necessary for soft materials whereas a drill with a small spiral angle is needed for hard materials. The helical turned flutes form the cutting angle (δ) and they serve the chip removal. The role played by the flutes is to remove the chips from the hole, and to conduct the cooling liquid to the cutting edges.

The heels whose contours run parallel to the flutes, give the drill a firm guidance in the drill hole. The heels are produced by relief-milling of the outer surfaces remaining after the flutes have been made. The guide heels are edged with precision and true to dimensions of the drill diameter. This results in safely guiding the drill, less friction with the workpiece and neat hole walls. For drilling of deep holes, the diameter of the drill is reduced by up to 0.1 mm on 100 mm length for saving the heel.

8.4.2 Correct Sharpening of the Drill Point

Best drilling results are only obtained with well sharpened drills. Blunt drills produce a stuffed burr on the extremity of the hole. They are recognized by the worn out outer corners of the main cutting edges. It is possible to resharpen them on the usual grinding wheels, and in this operation the saddle-shaped front rakes of the drill are reground. To avoid drawing the temper of cutting edges when grinding, they should be cooled down several times. In case the drills have broken off or the temper has been drawn, one should first grind one surface at a right angle to the drill axis. Correct sharpening of the drill point can only be obtained with a twist drill grinding machine or with hand operated grinding devices.

Mistakes arise frequently when sharpening drills by hand:

Mistakes	*Select*
Drill point is outside the drill axis Cutting edges have different lengths	Drilled hole becoming too large Drill runs off centre
Different angles to drill axis Different lengths of cutting edges	Drilled holes not circular Unbalanced strain of drill
Drill point outside the drill axis	Drilled holes too large
Unequal angles	Drill runs off centre
Different lengths of cutting edges	Cutting edge gets rapidly blunt

The lip angle affects the drilling capacity as well as the accuracy to gauge and surface quality of the drill hole. The main cutting edges shall be ground sharp and straight. Cutting edges arched forward or backward, wear out quickly. If the main cutting edges are of uneven length, the drill hole will become too big; if they do not stand symmetrical to the drilling axis, only one cutting edge works, which then becomes blunt prematurely. A grinding gauge is used for testing of the lip angle. Blunt cutting edges cause a rough surface in the hole. The clearance angles have the correct size, when the angle of the chisel edge between the cutting edges is 55°. To reduce the unfavourable effect of the web, this is shortened by grinding on larger drills. Grinding of the web is not necessary when bigger holes are rough drilled.

8.4.3 Advantages of the Twist Drill

During regrinding, diameters and spiral angles (rake angles) are retained to the end. The chips are conveyed automatically out of the drill hole by the helical flutes. For that reason, the regrinding should be

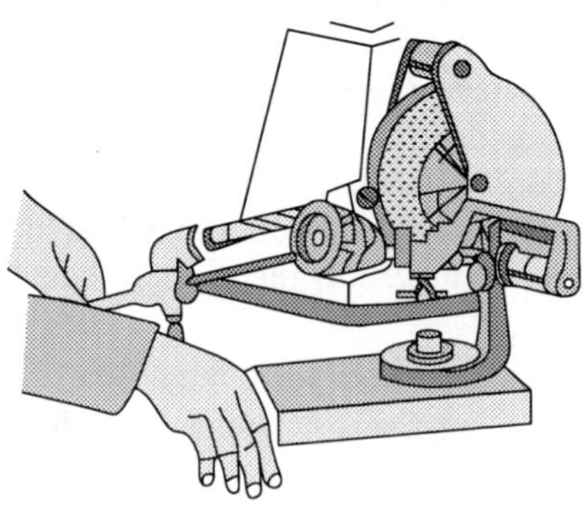

Fig. 8.13 Grinding with a device

performed on a tool grinding machine. Cooling is necessary to prevent the heating up of the cutting edges.

The grinding of a chamfer is advantageous for drills which are used for drilling cast iron. This breaks the chips and relieves the cutting edges. The result is that the cutting edges will have a longer life.

After use the drill has to be cleaned. Shank and cutting edges must be protected against damage. It is practical to store the drills in wooden blocks arranged by diameters. Thereby unnecessary searching for the required drill will be avoided.

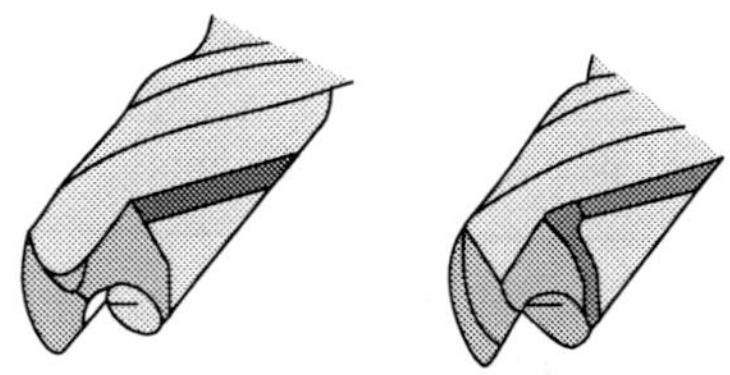

Fig. 8.14

Checking the drill points by means of the grinding gauge allows to recognize the mistakes and to almost eliminate them by hand grinding. Equal cutting edges and angles ensure precise drilling and a uniform wear of the drill. With unequal cutting edges the drill gets blunt too soon.

Kinds of drills

For drilling in material	*Angle of point in degrees*	*Pitch*
Steel	118°	
Grey cast iron	118°	Normal pitch – spiral angle 25° to 30°
Aluminium alloys, hardened (duraluminium)	118°	
Aluminium	140°	Small pitch – spiral angle 35° to 40°
Aluminium alloys – soft	140°	
Silumin	104°	
Electron and other magnesium alloys hard tissue	100°	
Copper	118 to 125°	
Brass	130°	Large pitch – spiral angle 10° to 160°
Moulded , laminated material	80° to 90°	
Non-laminated moulded material	50° to 60°	
Hard rubber, ebonite	30°	

Reference values for spiral angle

Diameter range-d	*Type W*	*Type H*	*Type N*
Up to 0.6	—	—	16°
More than 0.6 to 1	—	—	18°
More than 1 to 3.2	35°	10°	20°
More than 3.2 to 5	35°	12°	22°
More than 5 to 10	40°	13°	25°
More than 10	40°	13°	30°

Direction for application of tool types N, H, W

L Material to be machined	Type of tool	Lip angle
Steel, cast steel 40 – 70 kg /mm^2	N	118°
70-120 kg/ mm^2	N	130°
Cast iron, malleable cast iron	N	118°
Brass, up to Ms 5	H	
From Ms 60 on	N	118°
Copper up to 30 mm drill ?	W	140°
More than 30 mm drill ?	N	
Alum, alloy, cutting long chips	W	140°
Cutting short chips	N	
Moulded at thickness *s*	H	80°
Plastic at thickness *s*	W	
Laminated plastic, hard rubber	H	80°
Marble, slate, coal	H	80°

8.4.4 Letter Drills and Number Drills

The letter drill series consists of drill sizes from A to Z.

The letter 'A' drill is the smallest with 5.944 mm diameter, and the letter 'Z' is the largest, with a 10.490 mm diameter (Table 8.1).

In the number drill and the letter drill series, the correct diameter of the drill is gauged with the help of the respective drill gauges. A drill gauge is a rectangular or square shaped metal piece containing a number of different diameter holes. The size of the hole is stamped against each hole.

Number drills: The number drill series consists of drills numbered from 1 to 80. The No.1 drill is the largest, with 5.791 mm diameter, and the No. 80 drill is the smallest, with 0.35 mm diameter. There is no uniform variation in the drill diameters from number to number. To find the correct diameter of a number drill, refer to a drill size chart or a handbook. Number drill series are also known as 'wire gauge' series.

Letter Drills:

Table 8.1

Letter	Diameter inches	mm
A	0.234	5.944
B	0.238	6.045
C	0.242	6.147
D	0.246	6.248
E	0.25	6.35
F	0.257	6.528
G	0.281	6.629
H	0.266	6.758
I	0.272	6.909
J	0.277	7.036

K	0.281	7.137
L	0.29	7.366
M	0.295	7.493
N	0.302	7.671
O	0.316	8.026
P	0.323	8.204
Q	0.332	8.433
R	0.339	8.611
S	0.348	8.839
T	0.358	9.093
U	0.368	9.347
V	0.377	9.576
W	0.386	9.804
X	0.97	10.084
Y	0.404	10.262
Z	0.413	10.49

Number Drills

No.	*Inch*	*mm*	**No.**	*Inches*	*mm*	*No.*	*Inches*	*mm*	*No.*	*inches*	*mm*
1	0.228	5.791	21	0.159	4.039	41	0.096	2.438	61	0.0390	1.00
2	0.221	5.613	22	0.157	3.988	42	0.0935	2.375	62	0.0380	0.98
3	0.213	5.410	23	0.154	3.912	43	0.089	2.261	63	0.0370	0.95
4	0.209	5.309	24	0.152	3.861	44	0.086	2.184	64	0.0360	0.92
5	0.2055	5.220	25	0.1496	3.797	45	0.082	2.083	65	0.0350	0.90
6	0.204	5.182	26	0.147	3.734	46	0.081	2.057	66	0.0330	0.85
7	0.201	5.105	27	0.144	3.658	47	0.0785	1.994	67	0.0320	0.82
8	0.199	5.055	28	0.1405	3.569	48	0.076	1.930	68	0.0310	0.79
9	0.196	4.978	29	0.138	3.454	49	0.073	1.854	69	0.0292	0.75
10	0.1935	4.915	30	0.1285	3.264	50	0.070	1.778	70	0.0280	0.70
11	0.191	4.851	31	0.120	3.048	51	0.067	1.702	71	0.0260	0.65
12	0.189	4.801	32	0.116	2.948	52	0.0635	1.613	72	0.0240	0.65
13	0.185	4.699	33	0.113	2.870	53	0.0595	1.511	73	0.0240	0.60
14	0.182	4.623	34	0.111	2.819	54	0.055	1.395	74	0.0225	0.58
15	0.180	4.572	35	0.110	2.794	55	0.052	1.321	75	0.0210	0.52
16	0.177	4.496	36	0.1065	2.705	56	0.0485	1.181	76	0.0200	0.50
17	0.173	4.394	37	0.104	2.642	57	0.043	1.092	77	0.0180	0.45
18	0.1695	4.305	38	0.1015	2.578	58	0.042	1.067	78	0.0160	0.40
19	0.166	4.216	39	0.0995	2.527	59	0.041	1.041	79	0.0145	0.38
20	0.161	4.089	40	0.098	2.489	60	0.040	1.016	80	0.0135	0.35

8.5 THE DRILLING PROCESS

8.5.1 Preparing

The centre of the hole to be drilled is marked**.** Mostly the centre point is marked by two inter-crossing marking lines. They are obtained after the distance from the edge or from the hole has been determined by means of the steel straight edge or the parallel marker. The crossing point of the marking lines is centre punched to ensure correct guiding of the centre line between cutting edges and to avoid running off when drill points bite in. The centre mark should be large enough to allow the centre line (between cutting edges) of the drill to engage. For large holes it is necessary to drill a smaller hole first because it is not possible to make such large centre marks as necessary to provide for a correct guidance of the drill point.

In order to provide for coinciding of the centre punched point with that of the drill hole, a hole periphery is marked around the centre point by means of a divider, and test centre points are punched. For precision drilling, another control circle is marked around the periphery of hole permitting to determine immediately a running off centre of the drill point. For this purpose the drill is repeatedly lifted during spot drilling. If the drill has run off centre, centre punching must be repeated or a groove is cut before centre punching again.

8.5.2 Chucking of Drills

The common drill holding devices are drill chucks and sleeves and sockets.

While chucking, what matters above all, is that the drill runs true. Drills which are not running true, break off easily. Drills with taper shank are pushed into the tapered bore of the main spindle of the drilling machine. Thereby, the slotted hole receives the taper tang. Differences in size between taper shank and taper hole are compensated by standardized taper sleeves. During work, the drill is taken along by the friction on the surface of the taper. However, this is only possible when inside and outside tapers are undamaged and free from dirt. Foreign bodies between the taper walls are responsible for out of true running of the drill. Therefore, the tapers must be cleaned before putting them together. The taper tang is not used as catch, but for loosening of the drill with the drill drift. Before loosening the drill, a piece of wood may be placed underneath, so that the point will not get damaged by falling down.

For chucking of drills with cylindrical shafts, concentrically drill chucks with two or three jaws are used. Attention is to be paid that the drill is pushed up into the chuck as far as possible, thus it does not slip during the operation. Frequently, the chucks have two surfaces inside, to which the side faces of the upper part of the drill join to. Thereby, a safe rotation of the drill is obtained.

Quick change collet chucks permit the chucking of drills without disengaging the machine. They are particularly used in mass production.

For drilling the drill consisting of shank and cutting edges is chucked in the drill press. Generally, the drill is made of tool steel. If the material is cut at a high speed (cutting speed), drills made of high-speed cutting steel are used to increase the cutting capacity. For highest cutting effects, drills are used with cutting edges made of hard metal (e.g., Widia,) which have been hard soldered on the drill. The shank of the drill can be cylindrical, tapered or square. Drills of tool steel or high-speed cutting tool steel can be manufactured with diameters from 0.25 mm to 20 mm; drills with a tapered shank have generally diameters from 2 to 80 mm. Twist drills with a tapered square shank serve as tools for the retched brace and the bit brace. Accuracy of a drilled bore depends mainly on the precise action of the drill. It is therefore necessary to chuck the drill exactly centrically so that it does not hook. The cylindrical shank of the drill can be chucked in a two-jaw or in a three-jaw chuck, which will be fixed in the inside taper of the drill spindle with a drill socket.

The drill must be introduced up to the base of the chuck and fit tightly to avoid displacement and twisting when drilling. Drills which move in the chuck in case of a higher cutting resistance, will get the shank channelled and can no longer be chucked centrally.

8.5.3 Drill Chuck (Fig. 8.15)

Straight shank drills are held in drill chucks. For fixing and removing drills, the chucks are provided either with a pinion and key or a knurled ring.

The drill chucks are held on the machine spindle by means of an arbor fitted on the drill chuck

8.5.4 Taper Sleeves and Sockets (Fig. 8.16)

Taper shank drills have a Morse taper.

Sleeves and sockets are made with the same taper so that the taper shank of the drill, when engaged, will give a good wedging action. Due to this reason Morse tapers are called self-holding tapers.

Drills are provided with five different sizes of Morse tapers, and are numbered from MT 1 to MT 5

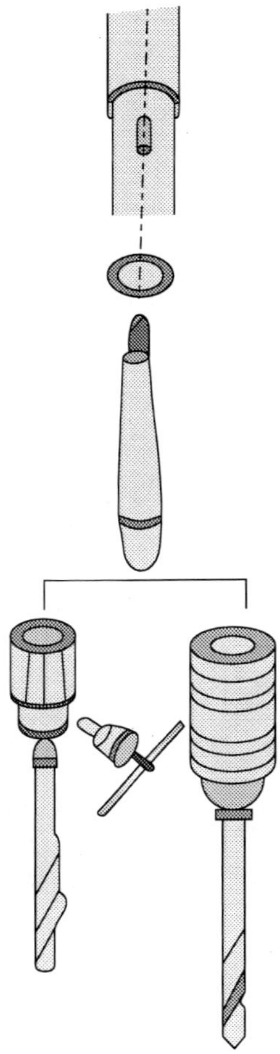

Fig. 8.15

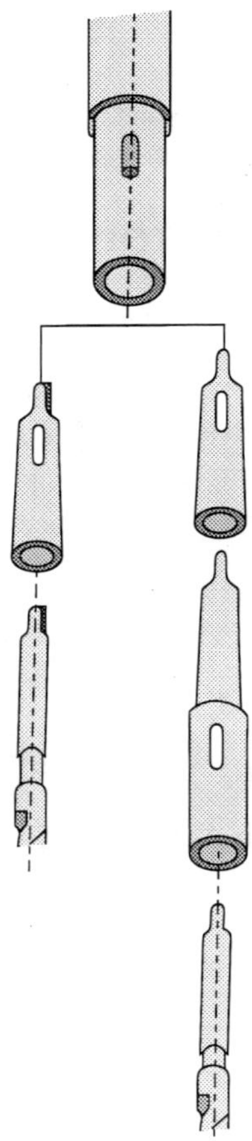

Fig. 8.16

In order to make up the difference in sizes between the shanks of the drills and the type of machine spindles, sleeves of different sizes are used. When the drill taper shank is bigger than the machine spindle, taper sockets are used (**Figs. 8.15 & 8.16**).

The tapered shank of the drill serves for clamping in the inside taper of the drill spindle. The shank of large drills fits exactly into the drill spindle. In case of small drills, one or two taper sleeves are placed onto the shank to adapt the conical shank to the inside taper of the drill spindle. When chucking the tool, the tapers of the drill and the drill spindle must be free from chips and oil because the drill is fixed to the taper wall by the friction and is turned with it.

8.5.5 The Drill Drift

The drill drift is used for pulling out the drill with tapered shank or the drill chuck. The drift is conical-shaped, and by a brief pressure with a lever or by light blows of a hammer on the tapered tang of the drill, the latter is released and will fall out. Drill and working table are to be protected against damages by a wooden plate.

Do not leave drill drift and square spanner in the tool because otherwise injuries to the hand may be caused when starting the drill spindle.

8.5.6 Quick-change Chucks

These provide for changing the drills with the drill spindle running. After lifting the outer sleeve, the drill with the adapter can be introduced into the chuck. By releasing and falling down of the sleeve the tool is locked in place: the drill is kept by the chuck and driven.

8.6 CLAMPING THE TOOLS

Clamping is done to secure workpieces and especially short ones against pulling around caused by the rotating effect of large drills.

8.6.1 Long and Handy Workpieces

These are generally on a base of wood or steel on the work table of the bench-type or column drilling machine and held by the hand. The seating surfaces of the workpiece, base and drilling table must be clean and without dust and burr. The bases serve for protecting the table against piercing.

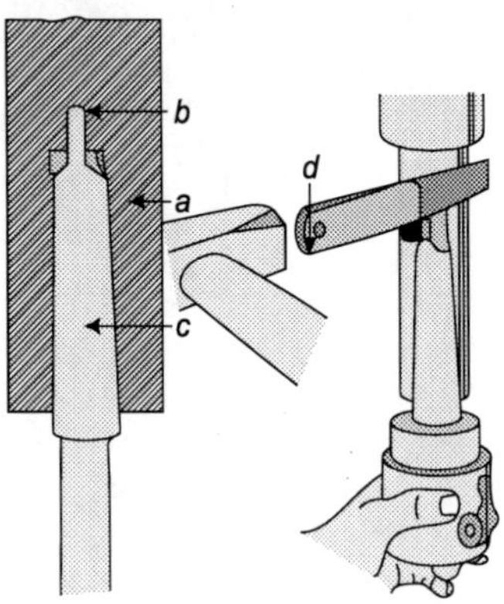

a-Drill spindle c-Drill Shank
b-Slotted hole d-Drill drift

Fig. 8.17

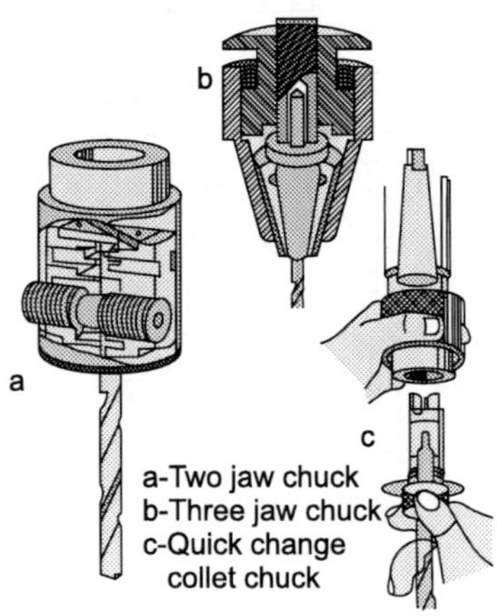

a-Two jaw chuck
b-Three jaw chuck
c-Quick change collet chuck

Fig. 8.18

8.6.2 Cylindrical Workpieces

These have the best support in the drilling V-block. Cylindrical, flat, square or hexagon bars must often be drilled on their front surface. They have a small bearing surface. It is advisable to centre the workpiece and to place it on a centre in the T-slot of the drilling machine table, the tool centre being exactly opposite the drill point.

To avoid pulling around, the bars are held with a hand vice.

Loose holding of small workpieces by means of pincers is not safe. There is a danger that they will swing around and cause injuries to the hand.

8.6.3 Small Workpieces

These are firmly and safely clamped in the clamping tool, the drill press vice or the drilling square; several parts of the same kind are clamped in the drilling fixture. Fastening is affected by thrust pieces and clamping screws. The drill is guided in hardened drill bushings. Drilling fixtures save time for marking and clamping.

Perfect clamping of workpiece and chucking of tool eliminates vibrations (oscillation, chattering) and yields accurate drilled holes with a neat hole wall.

8.7 THE DRILLING PROCESS

Working with the column drill press: All workpieces that can be clamped or placed on the drilling machine table in the workshop, are drilled with the column drill press.

Before starting, make sure that loose shirt arm sleeves or arms of jackets or long hair can be seized by the drill. Serious injuries could otherwise result. Have your hair cut short and wear tight fitting clothes and a cap!

The number of revolutions is determined by the diameter of drill.

The number of revolutions per minute (R.P.M.), the speed (n) of the drill, can be changed by the belt on the stepped belt pulleys. When the belt is placed on the smallest pulley over the drill spindle, the drill rotates at maximum speed.

Only small drills can withstand a high speed. Drills with large diameters must operate at a low speed.

With every revolution, the outer points of the cutting edges make a way corresponding with the length of the circumference U = diameter of drill (d) multiplied by 3.14 ($U = d\pi$).

The outer points of the cutting edges of a drill having a diameter (d) of 5 mm and a speed (n) of 100 revolutions cover a distance of $d\pi n = 5 \times 3.14 \times 100 = 1570$ mm or 1.570 m.

With a drill of a diameter (d) of 20 mm and a speed (n) of 100 revolutions they cover a distance of $d x \pi n = 20 \times 3.14 \times 100 = 6280$ mm or 6.280 m.

Thus, the small drill at 100 R.P.M. cuts the chips on a length of 1.57 m, whereas the large one rotating at the same speed cuts on a length of 6.28 m.

If the drill makes 100 revolutions per minute (n = 100 R.P.M.), the cutting speed (v) of the small drill is 1.57 m/min, and that of the large drill 6.28 m/min. This means that the cutting speed of the large drill is four times higher than that of the small drill. But to avoid a too high cutting speed and a too high stress because of the heat effects on the blade, the number of revolutions of large drills must only be one- fourth of that of small drills.

This becomes apparent when considering the equation for the cutting speed *v*:

$$V = \frac{d \times \pi \times n}{1000} \text{ m/min, i.e., Cutting speed in m/min} = \frac{\text{Diameter of drill mm} \times \pi \times \text{speed}}{1000}$$

This means: small drill diameters = high speed; large drill diameters = small speed.

The cutting speed in the drilling operation is determined by the material of the tool and of the workpiece. Empiric values have been found out for the cutting speed for the various materials and diameters; these values should not be surpassed. Converting the formula, it is possible to find out the number of revolutions for any drill diameter.

$$n = \frac{v \times 1000}{d \times \pi} \text{ i.e., number of revolution/min RPM} = \frac{\text{Cutting speed} \times 1000}{\text{Drill diameter} \times \pi}$$

The speed, whose vertical movement affects the penetration of the drill into the material, also highly influences the cutting capacity. The advance movement comes about by displacing the drill spindle by means of a rack and toothed wheel and by the sleeve movable on the spindle. In case of large advance, the cutting edges have to cut thick chips whereby they are often over-strained thus breaking out.

It is therefore necessary to adapt the feed, which is measured in millimetres per revolution, to the material of the workpiece and of the tool as well as to the drill diameter and to the speed. Just before drilling through the material, the feed should be reduced to avoid clasping of the drill and thus its splitting up.

The time for drilling a hole will be the shorter, the more exactly the given figures of cutting and feed speed are observed.

By the friction on the cutting edges which considerably increases at high cutting speeds and feed, the drill gets heated. The heating of a tool reduces its hardness. The cutting edges will rapidly get blunt if the warmth generated on the drill is not carried off by an appropriate cooling agent. This coolant runs over the flutes to the drill point, increases its cutting capacity and improves the quality of the hole wall. For this reason the cutting part of the drill should be repeatedly withdrawn from deep holes, cleaned from chips and cooled down.

As the flutes would clog when unsuitable coolants are used, the cooling agents stated in the table and provided for the most important materials to be drilled in the metal trade, should be used. When drilling electron, do never cool with water because otherwise the chips may easily catch fire.

Material	*Cooling agent*
Steel, copper, bronze, aluminium	Soap water
Brass, aluminium alloys	Soap water or petroleum
Electron, grey cast iron	Only dry drilling

8.7.1 Drilling of Simple Holes on the Drilling Machine

For drill holes, which are used for rivets and screws, no special requirements have been laid down in regard to accuracy in size and surface quality.

Example: The butt strap is to be provided with a through hole of 16 mm diameter (for a hexagon head screw M 14.

Operation/ process plan

Operations	*Tools*
Marking	Scriber, try square, divider, centre punch, hammer
Chucking of the drill	Twist drill 16 N HSS
Mounting of the workpiece	Machine vice
Drilling of the hole	Rose bit
Deburring	

8.7.2 Marking of Drill Holes

During drilling, the twist drill first cuts in the middle of the hole. Therefore, the location of the drill hole is always outlined from the centre and marked by two centre lines. The point of intersection is marked with centre punch.

For accurate work, indication of a drilling and test circle is necessary (**Fig. 8.19**).

If a large number of workpieces are to be drilled with identical holes, a template may be used.

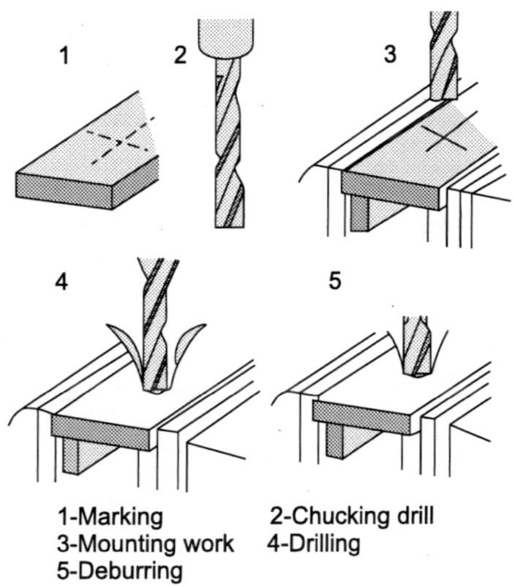

Fig. 8.19

8.7.3 Drilling of Hole

A twist drill of high-speed steel with a diameter of 16 mm, suitable for steel, is selected for drilling. A column drilling machine of medium size is suitable for the drilling work. For a cutting speed of 22 m/min. 475 RPM are required.

The feed will be 0.25 mm/R. Particular care is to be applied to the mounting of the workpiece and chucking of the drill.

8.7.4 Clamping of Workpieces on a Drilling Machine

The punched centre of the hole to be drilled must be lined exactly under the point of the drill. A vertical hole can only be obtained by a horizontal and proper seat of the workpiece. Therefore, the machine table must be

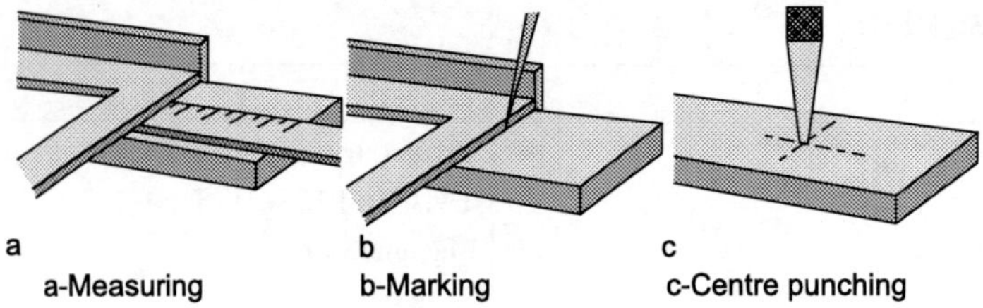

Fig. 8.20

free of chips and other particles. While drilling through holes, the machine table can get damaged by the drill. In order to prevent this, the drill must be able to pass through the chip hole of the table. If the machine table has no chip hole, a piece of wood or a steel parallel is to be placed underneath the workpiece.

During drilling torsion forces are produced which attempt to move the workpiece around. The torsion forces are particularly powerful, when the drill point emerges from the drill hole. The workpiece must be secured against moving around. Very large parts mostly lie firm by their own weight. The hand vice is suitable for holding of small parts. Occasionally, a bolt or angle plate, fastened in the T-slots of the drilling table can also be used as a stopper. In this way, the clamping of workpieces is safer. The safest way to clamp the workpieces is in the machine vice or directly on the machine table. Suitable 'T' bolts have to be selected for clamping. Round workpieces may be placed in a 'V' block. For drilling a large number of identical workpieces, drilling jigs are used. The workpiece is fastened in them. The drill gets its guidance by the drilling bush in the jig. Since marking and centre punching is not required, time is saved.

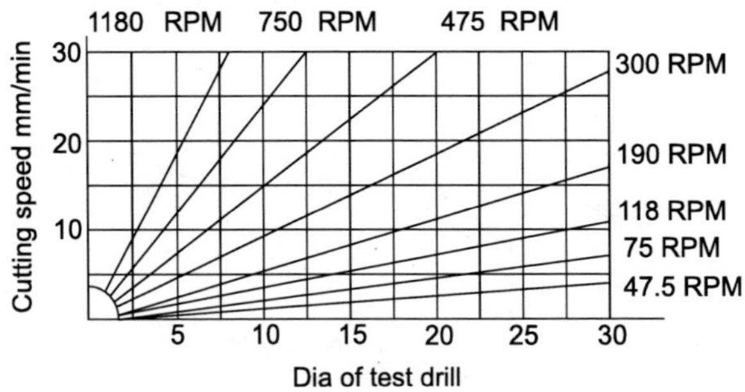

Fig. 8.21

8.8 WORK-HOLDING DEVICES

Workpieces to be drilled should be properly held or clamped to prevent from rotating along with the drill. Improperly secured work is not only a danger to the operator but can also cause inaccurate work, and breakage to the drill. Various devices are used to ensure proper holding.

8.8.1 The Machine Vice

Most of the drilling work can be held in a machine vice. Ensure that the drill does not drill through the vice after it has passed through the work. For this purpose, the work can be lifted up and secured on parallel blocks providing a gap between the work and the bottom of the vice (**Fig. 8.23**).

Workpieces which are not accurate may be supported by wooden pieces (**Fig. 8.24**).

8.8.2 Clamps and Bolts

Drilling machine tables are provided with T-slots for fitting bolt heads. Using clamps and bolts, the workpieces can be held rigidly. While using this method, the packing should be, as far as possible, of the same height as the work, and the bolt nearer to the work.

There are many types of clamps and it is necessary to determine the clamping method accordingly (**Figs. 8.22 to 8.27**).

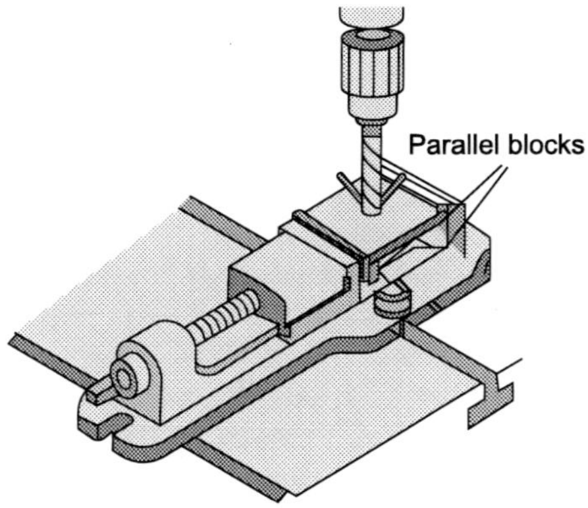

Fig. 8.22

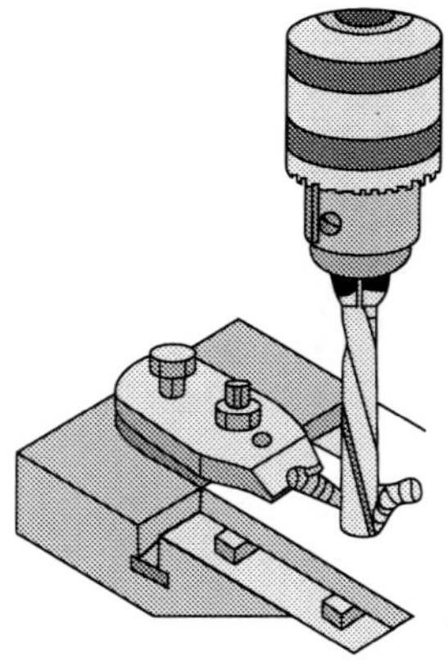
Fig. 8.23

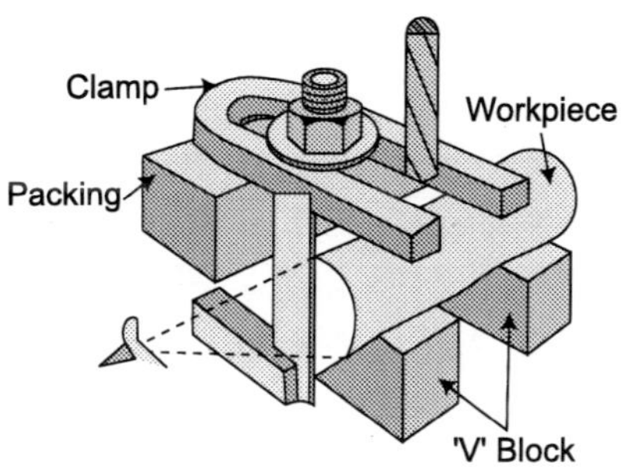

Fig. 8.24

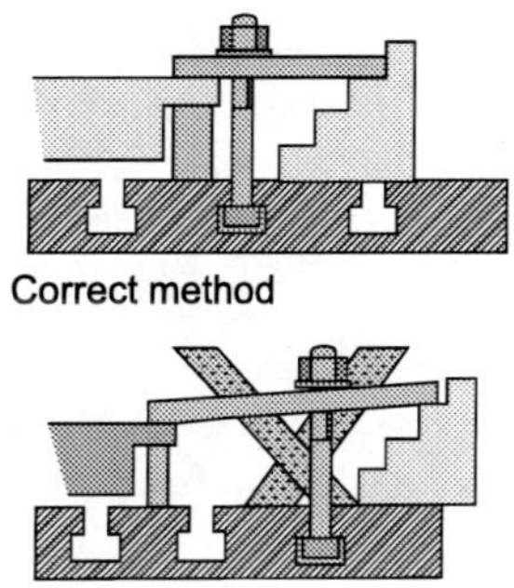

Fig. 8.25

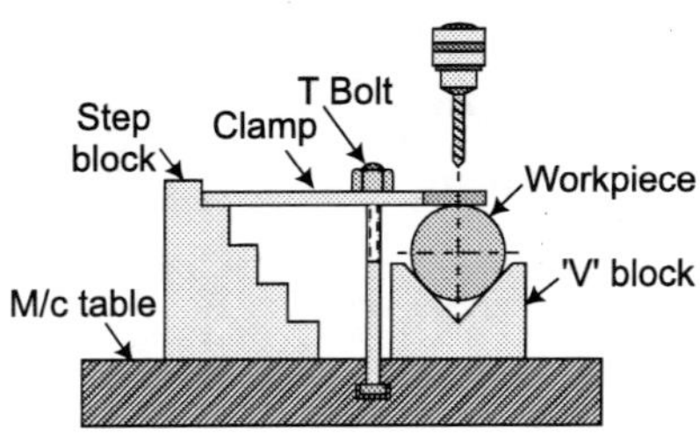

Fig. 8.26

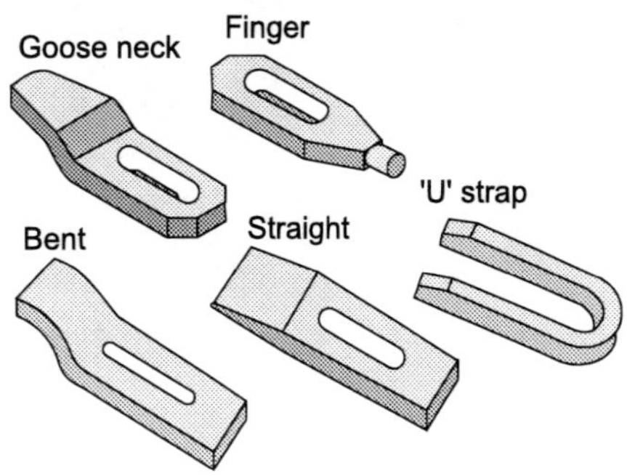

Fig. 8.27

8.8.3 Special Drilling and Boring Tools

The deep hole drill (tube bit) (Fig. 8.28): The deep drilling is suitable for drilling deep and accurate holes. It works with just one cutting edge. The centre bit is used when holes are flat at the base. A point serves for guidance.

The hollow drill (Fig. 8.28): This cuts a core out of the material. In most cases, it is used on special drilling machines.

Boring bars (Figs. 8.28, and 8.29): With inserted bits are used for boring of holes after pre-drilling. On the horizontal boring machine, self-supporting and guided boring bars are used. These are hardened and ground, thus they can run in guide bushes at any point.

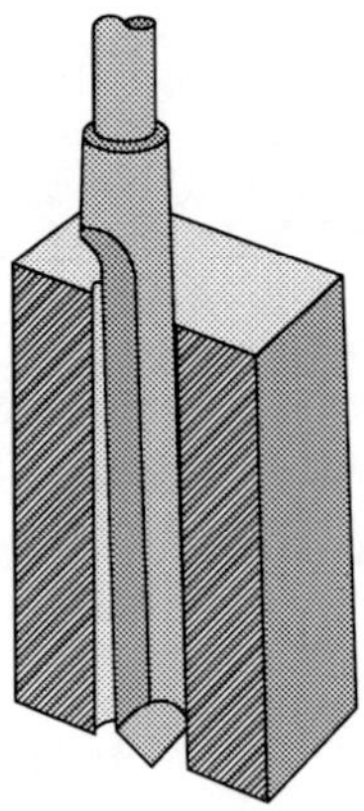

Fig. 8.28

The cutting out tool (Fig. 8.30): This may be used for cutting of sheet metal, etc.

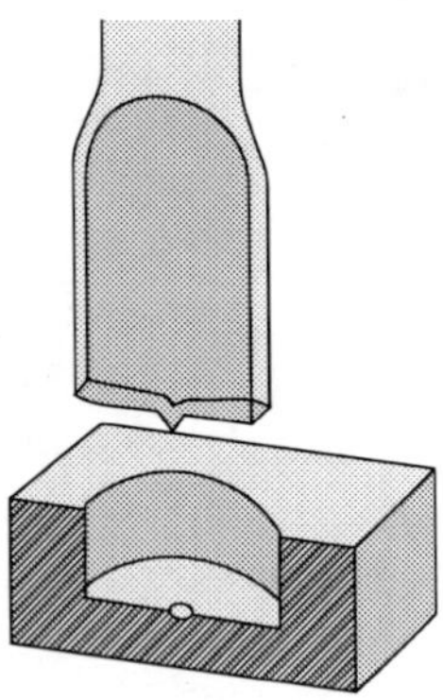

Fig. 8.29

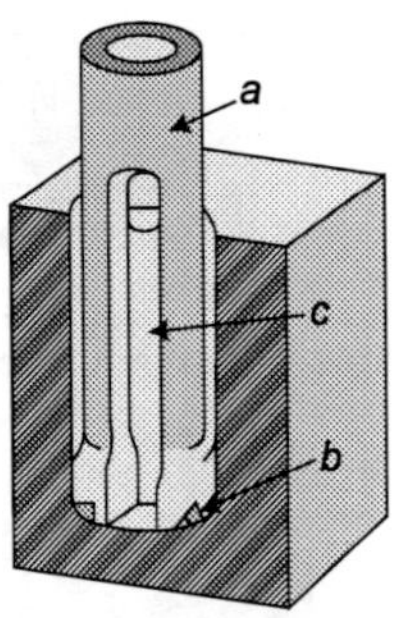

a-Hollow drill
b-Cutting edge
c-Core

Fig. 8.30

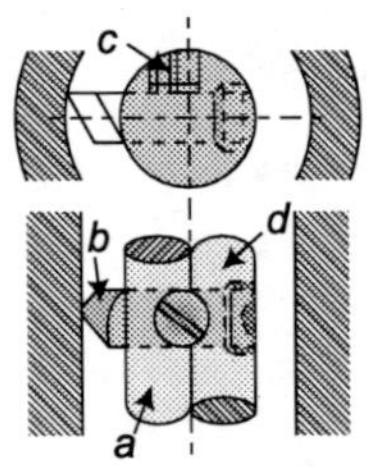

a-Hollow drill
b-Cutting edge
c-Core
d-Set screw

Fig. 8.31

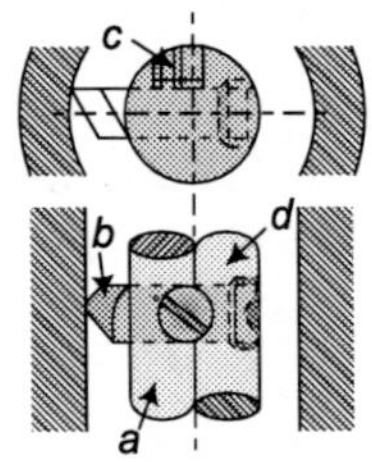

a-Hollow drill
b-Cutting edge
c-Core
d-Set screw

Fig. 8.32

Judgment on drilling work

Mistakes	Causes
Drilled bore is not neat	Drill was blunt or drill point was wrong; cooling agent was not sufficient or not appropriate; too large feed
Drilled hole is too large	Drill point was outside drill axis; angles at cutting edges were unequal; length of cutting edges was unequal, play or drill spindle was too large
Cutting edges of drill split up	Feed was too large; lip clearance was too strong
Drill does not cut	Drill was blunt; material was difficult to cut
The tang of the taper is distorted or broken.	The tapered shank of the drill had an inexact fit in the chuck due to damage or wearing out of the taper
Drill broke out when working soft and tough material	Flutes of the drill were obstructed by chips; material had hard or porous points.

8.9 CENTRE DRILLS

8.9.1 Centre Drilling

The combined operation of drilling and countersinking with one tool is known as a combination drill or centre drill. To prevent a drill from wandering off the centre, it is always a good practice to first spot the centre punch mark by centre drilling. The other purpose of centre drilling is to support the lengthy workpiece by the lathe centre for lathe operations. Jobs involving precision turning/ grinding will be supported between the centres.

8.9.2 Centre Drills

A centre drill is made of high-speed steel and it is cylindrical in shape. At both the ends it has a plain drill and countersink as integral parts. The centre drill is hardened and ground. It is available in standard sizes.

8.9.3 Classification as per Indian Standard

Indian Standard classifies centre drills into three types, Type A, Type B and Type R.

The difference lies in the formation of the countersink by each type.

8.9.4 Uses and Specifications

Type A centre drill **(Figs. 8.33 and 8.35)** is used to produce centre holes with the plain drilled portion and countersink. It is designated as centre drill A. 1.6 × 4.0 15:6708 which means that the centre drill is of Type A with the plain drill portion having a diameter of 1.6 mm and shank diameter of 4 mm.

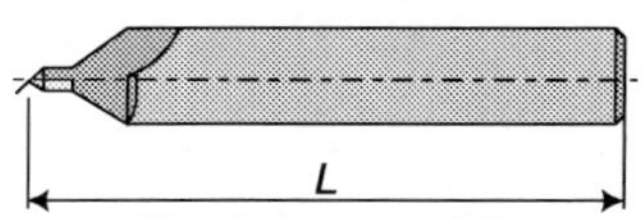

For dia less or = 0.8

Fig. 8.33

Type A centre drills with drill points 0.8 mm and below have the cutting edge on one side only. Centre drills with more than 1 mm drill point diameter have cutting edges on both ends.

Type B centre drill is used to produce a centre hole with plain drilled portion, with a countersink and has a further conical portion to form additional countersinking to protect the centre hole. The countersinking for providing the bearing surface for centres has an angle of 60° and the drill point has an angle of 120°.

This type is designated as Centre Drill B 1.6 × 6.3 IS: 6709 which means that the pilot diameter is 1.6 mm while the shank diameter is 6.3 mm (Table 8.2).

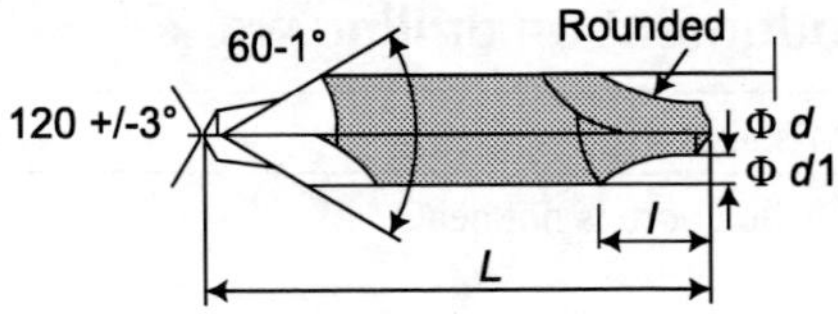

Fig. 8.34

The third type is designated as Centre Drill R 1.6×4.0 1.8.:6710. This also has provision to provide protected centre hole. This has an enlarged radius, machined along with the countersinking portion. IS: 6710.

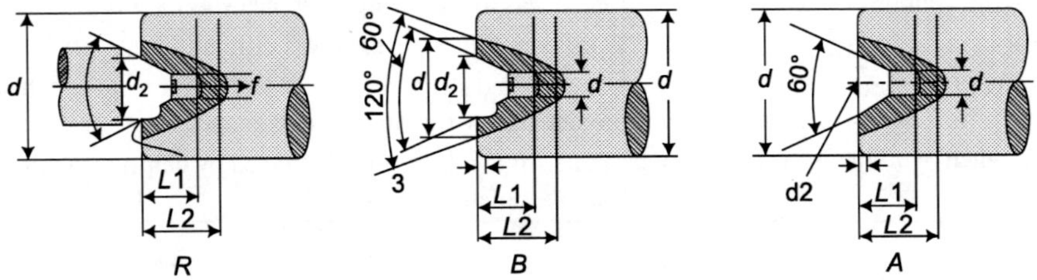

Fig. 8.35

Table 8.2

d	d_1	d	d_1	d_2	d	d_1
k 12	h 9	k 12	h 9	k 12	k 12	h 9
(0.5)	3.15	1.0	4.0	2.12	1.0	3.15
(0.63)	3.15	(1.25)	5.0	2.65	(1.25)	3.15
(0.8)	3.15	1.6	6.3	3.35	1.6	4.0
1.0	3.15	2.0	8.0	4.25	2.0	5.0
(1.25)	3.15	2.5	10.0	5.30	2.5	6.3
1.6	4.0	3.15	11.2	6.70	3.15	8.0
2.0	5.0	4.0	14.0	8.5	4.0	10.0
2.5	6.3	(5.0)	18.0	10.60	(5.0)	12.5
3.15	8.0	6.3	20.0	13.20	6.3	16.0
4.0	10.0	(8.0)	25.0	17.0	(8.0)	20.0
(5.0)	12.5	10.0	31.5	21.20	10.0	25.0
6.3	16.0					
(8.0)	20.0					
10.0	25.0					

8.9.5 Methods of Centre Drilling

Centre drilling can be done by a drilling machine, lathe or centring machine.

The selection of a particular method of centring depends on the size and shape of the component.

A drill chuck is used to hold the centre drill. Sometimes special holders/collets are also used. In mass production, the raw material ends are centred using a centre drilling machine.

Table 8.3

Diameter of workpiece	Type A	Without protective chamfer & straight running surface			Type B	With conical protective chamfer & with straight running surface				Type R	Protective chamfer with convex running		
	d_1	L_1	d_2	L_{2min}	L_1	b	d_2	d_3	$L_{2Min.}$	L_1	d_2	f	$L_{2Min.}$
Over 3 up to 9	–0.5	2	1.6	1.4	–	–	–	–	–				1.4
	–0.8	2.5	2	1.8	–	–	–	–	–	2	1.6	2	1.8
Over 9 up to 12	1	3	2.5	2.2	3.5	0.4	2.5	4	2.6	2.5	2	20.5	2.2
Over 12 up to 15	–1.25	4	3.2	2.8	4.5	0.5	3.2	4.75	3.3	3	2.5	3.15	2.6
Over 15 up to 20	1.6	5	4	3.5	5.5	0.5	4	5.6	4	4	3.2	4	3.5
Over 20 up to 30	2.5	6	5	4.5	6.6	0.6	5	7.1	5.1	33.2	4	5	4.5
Over 30 up to 40	2.5	7	6.3	5.5	8.3	0.8	6.3	9	6.3	46	5	6.3	5.5
Over 40 up to 60	–3.2	9	8	7	10	0.9	8	11.2	7.9	7	6.3	8	7
Over 60 up to 90	4	11	10	9	12.77	1.2	10	14	10.2	9	8	10	9
Over 90 up to 120	–5	14	12.5	11	15.6	1.6	12.5	18	12.6	11	10	12.5	11
Over 120 up to 180	6.3	18	16	14	20	1.8	16	22.4	15.8	14	12.5	16	14
Over 180	–8	22	20	18	–	–	–	–	–	18	16	20	18
	10	28	25	22	–	–	–	–	–	22	20	25	22
										25	25	31.5	

8.9.6 Centring in Two Operations

A straight or taper shank twist drill may be used to drill a hole to sufficient depth. Then the countersinking is done using a 60° countersink drill.

On a lathe, it may be done with the help of a lathe tool, and swivelling the compound slide through 300 and turn the tapered portion.

In lathe work, the size of the centre hole depends upon the diameter of the workpiece. Table IV gives the recommended sizes of centre holes.

8.9.7 Data for Centre Holes Type A, B and R

Bracketed sizes should be avoided wherever possible.

L1 = length cut off when the centre hole is not allowed to remain in the finished workpiece. If it is intended that the centre hole shall remain in the finished workpiece and shall also be protected against damage, then the centre holes which should preferably be used are Types B or R.

Refer Table 8.3

8.10 DRILLING THIN PLATES

Drilling holes on thin metals with the normal drill is difficult due to the following reasons.

The drill point will not have adequate material to support (**Fig. 8.36**).

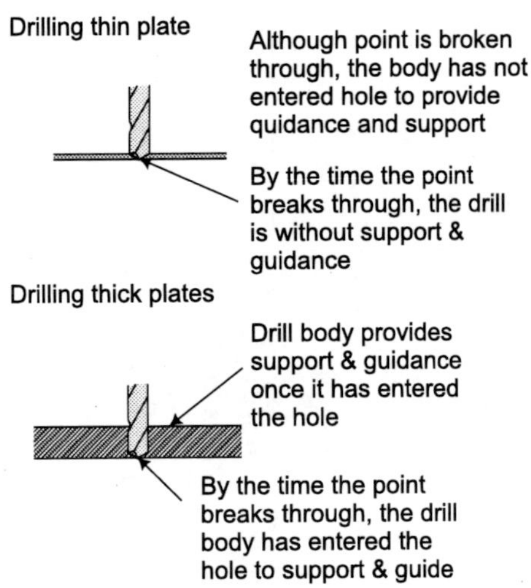

Fig. 8.36

8.10.1 Thin Plate Drilling (Fig. 8.36)

The drill point will break through the material and grab before the periphery of the drill enters in the hole.

The holes drilled will be out of shape.

When the drill grabs, there is a tendency to spin round and this is very dangerous and unsafe.

8.10.2 Spinning of Job (Fig. 8.37)

This can be overcome by resharpening the drill point as stated below.

8.10.3 Grinding the Drill Point with Increased Angle (Fig. 8.38)

This will reduce the length of the point and will permit the drill body to enter the metal before the point breaks through.

8.10.4 Thinning the Point of the Drill to Reduce the Web Thickness

This will improve the efficiency of the drill which has been ground with an increased cutting angle (**Fig. 8.37**).

8.10.5 Grinding the End of a Drill to form a 'W' Shape

This method is not suitable for very small drills (**Fig. 8.37**).

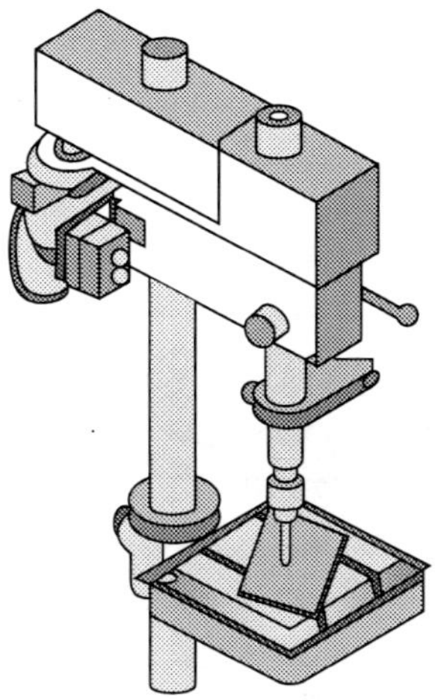

Fig. 8.37

Fig. 8.38

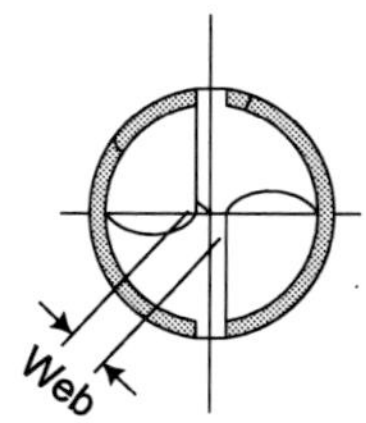

Fig. 8.39

This is sometimes called 'dead centre' drill. In this the centre point is slightly longer than the outer cutting edges. The hole cut will be clean and without any distortions. After cutting, a clean blank comes out.

For making bigger diameter holes on thin sheet metals trepanning types of tools are used.

8.11 TREPANNING TOOLS AND THEIR APPLICATIONS

8.11.1 Why Trepanning Cutters?

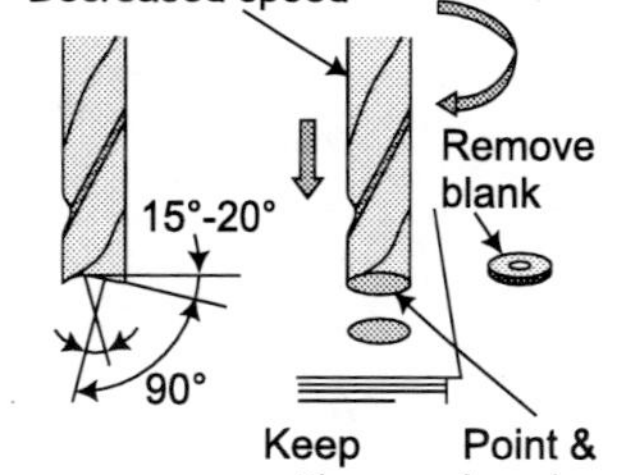

Fig. 8.40

If large holes are to be made in sheet metal with twist drills, the hole will be out of round and jagged. This is because the thickness of the sheet is insufficient to guide the drill point and to resist the cutting forces.

One way to overcome this problem is by using trepanning cutters. Instead of cutting away all the metal in the hole as swarf, the trepanning cutter merely removes a thin annulus of metal. This leaves a clean cut hole in the stock and a disc metal slightly smaller than the hole **(Fig. 8.41)**

The simplest type of a trepanning cutter is shown in Fig. 8.42. This is the traditional cutter. Another type of trepanning tool shown in **(Fig. 8.43)** is also used.

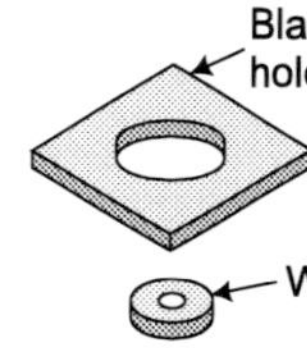

Fig. 8.41

For both cutters, first a small hole to suit the pilot is drilled for both cutters; first a small hole to suit the pilot is drilled at the centre of the required hole. The adjustable arm can be set so that the edge of the cutting tool produces the required size of hole. This may require a number of trials and adjustments before the correct size is reached. The pilot is located in the pilot hole and as the tool rotates it is fed through the work **(Fig. 8.44)**.

8.11.2 Hole Saw

Another type of trepanning cutter is a hole saw **(Fig. 8.45)**. It is used where a number of holes of the same size have to be cut.

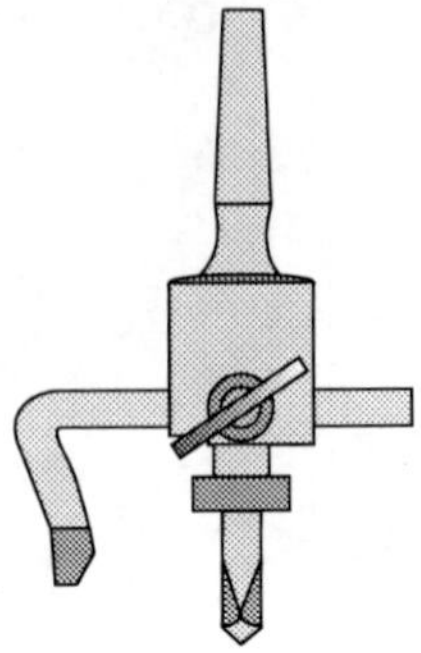

Fig. 8.42

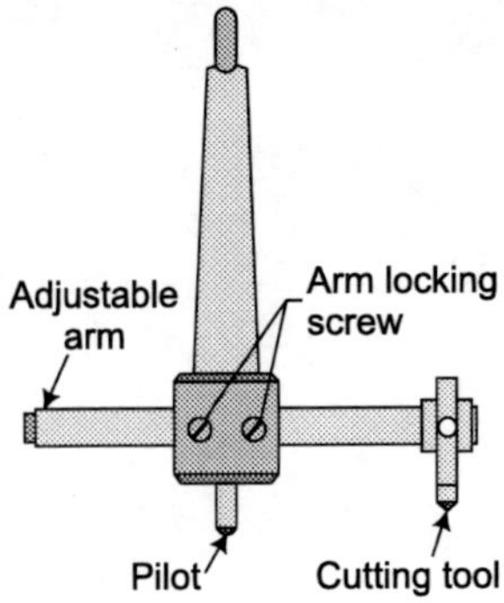

Fig. 8.43

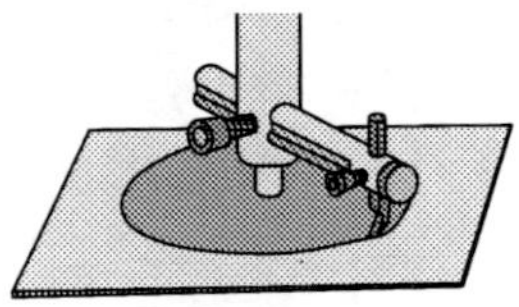

Fig. 8.44

Hole saws are of fixed size and are available in different steps of sizes. These cutters have teeth cut on the end with the normal rake and clearance angles. Body clearance is also provided to prevent rubbing as the tool progresses into the work.

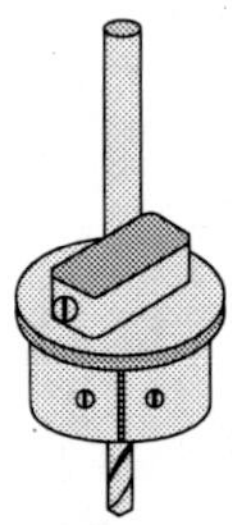

Fig. 8.45

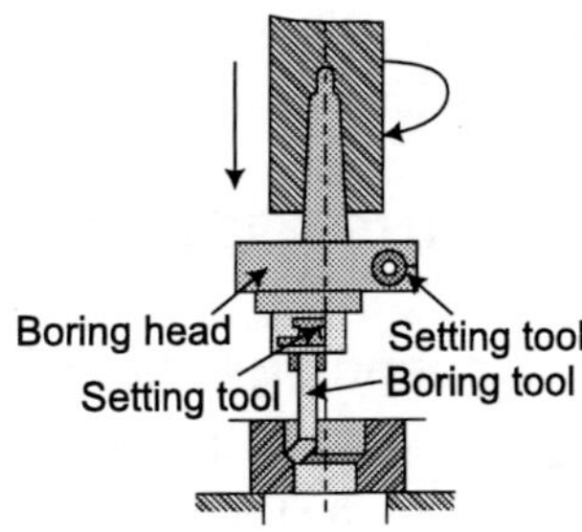

Fig. 8.46

8.11.3 Precautions

1. Cutting angle and clearance angle, etc., of the drill should be grounded according to the material.
2. Speed on the machine should be set according to the metal and the diameter of the drill.
3. Lips of the drill should be similar.
4. If a hole of large diameter is to be bored, first the hole should be bored with a small drill.
5. Sleeve or spindle should be cleaned before fitting the drill.
6. While drilling, a wooden socket should be placed below the job so that the drill below may not damage the machine bed by going across.
7. Drilling should not be done without properly clamping the job.
8. Drift should be used to separate drill from the machine.

At times, because of defective drilling either drill gets damaged quickly or the hole bored in the job is damaged.

The causes and remedies of such defects are given below.

Defects	*Causes*	*Remedies*
Breaking of cutting edge.	1. Deficient clearance angle. 2. Because of blow hole or solidity in the metal. 3. Because of drilling at a higher speed with blunt point. 4. Cutting angle may be of more degrees than it should be. 5. Drill may not be fitted properly. 6. Coolant may not be reaching the cutting edge.	1. Correct the clearance angle by grinding it again. 2. Before drilling the metal should be checked and drill with low feed speed. 3. Grind the drill again or reduce the speed. 4. Cutting angle should be suitably ground. 5. Fit the drill properly. 6. Supply coolant properly.
Drill gets broken from the front side.	1. Edge of the drill might have finished. 2. Spindle of the drill machine may be loose. 3. Feed of the drill or pressure or both may be less. 4. Job may not be properly clamped. 5 Chips might have blocked in the flutes	1. Grind the edge on the drill. 2. Repair the machine. 3. Give proper feed and pressure. 4. Clamp the job properly. 5. Clear the chips from the flutes.

Hole of the job defective

Hole of coarse surface	1. Edge of the drill may not be proper. 2. There might be excessive pressure on the drill. 3. Adequate coolant might not have been used. 4. Drill might have been fixed wrongly.	1. Grind the drill properly. 2. Reduce the pressure on the drill. 3. Use proper coolant. 4. Fix the drill properly.
When the drill bore and hole of larger diameter than needed.	1. Diameter of the drill might be larger. 2. There may be difference between length and angle of the lips. 3. Drills hold may not be proper. 4. Drill or machine spindle may be bent.	1. Take a drill of proper diameter.
Hole is not bored at the right point and deeper.	1. Careless drilling. 2. Job may not be set properly. 3. Drill is not ground properly. 4. Depth of drill may be set wrongly.	1. Drill carefully. 2. Set the job properly before drilling. 3. Grind the drill properly. 4. Set the drill depth correctly.

Oversized hole	1. Unequal length of cutting edge 2. Unequal angle of cutting edge 3. Unequal point thinning 4. Spindle running out of centre. 5. Drill point not in centre.	
Overheated drills.	1. Cutting speed is too high. 2. Feed rate is too high. 3. Clearance angle is incorrect. 4. Drill not sharp.	
Rough holes.	1. Feed rate is too much. 2. Drill cutting edges are not sharp. 3. Cooling is ineffective.	
Unequal flow of chips.	1. Cutting edges are not equal. 2. The point angle is not in the centre of the drill.	
Broken drill or split web.	1. Cutting speed is too high. 2. Feed rate is too high. 3. Work is not held rigidly. 4. Drill not held correctly. 5. Drill not sharp. 6. Point angle is incorrect. 7. Cooling is insufficient. 8. Flutes are clogged with chips.	

8.12 COUNTERSINKING

8.12.1 Definition

Countersinking is an operation of bevelling the end of a drilled hole. The tool used is called a countersink.

Countersinking is carried out for the following purposes: (a) to provide a recess for the head of a countersink screw, so that it is flush with the surface after fixing **(Fig. 8.47)** (b) to deburr a hole after drilling, (c) for accommodating countersink rivet heads, and (d) to chamfer the ends of holes for thread cutting and other machining processes.

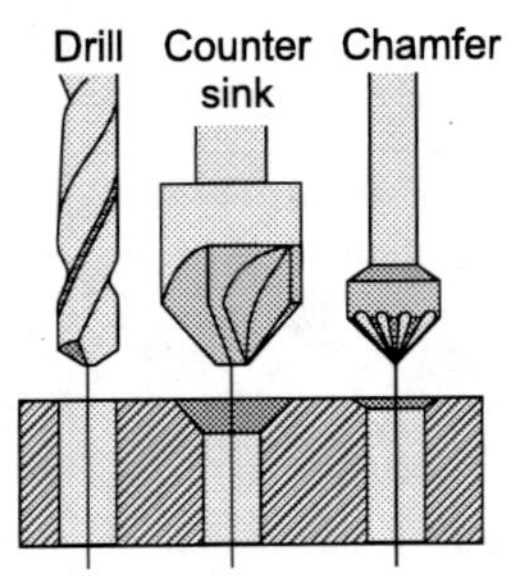

Fig. 8.47

8.12.2 Angles for Countersinking

Countersinks are available in different angles for different uses.

75° countersink riveting

80° countersink self-tapping screws

90° countersink head screws and deburring

120° chamfering ends of holes to be threaded or other machine processes.

8.12.3 Countersink

Countersink of different types are available.

The commonly used countersink have multiple cutting edges and are available in taper shank and straight shank (**Fig. 8.49**).

For countersinking small diameter holes special countersinks with two or one flute are available. This will reduce the vibration while cutting.

Countersinks with pilot (Fig. 8.48): For precision countersinking, needed for machine tool assembling and after machining process, countersinks with pilots are used.

They are particularly useful for heavy duty work.

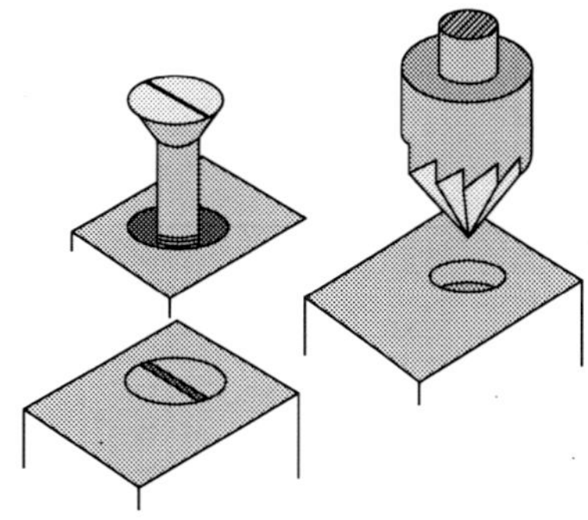

Fig. 8.48

The pilot is provided at the end for guiding the countersink concentric to the hole.

Countersinks with pilots are available with interchangeable and solid pilots.

Countersink hole sizes: The countersink holes according to Indian Standard IS 3406 (Part 1) 1986 are of four types, Type A, Type B, Type C and Type E. Type A is suitable for slotted countersink head screws, cross recessed and slotted raised countersink head screws.

These screws are available in two grades, i.e., medium and fine.

The dimensions of various features of the Type A countersink holes, and the method of designation are given in **(Fig. 8.50) and Table 8.4**.

Type B countersink holes are suitable for countersink head screws with hexagon socket.

The dimensions of the various features and the method of designation are given in Table 8.5.

Type C countersink holes are suitable for slotted raised countersink (oval) head tapping screws and for slotted countersink (flat) head tapping screws.

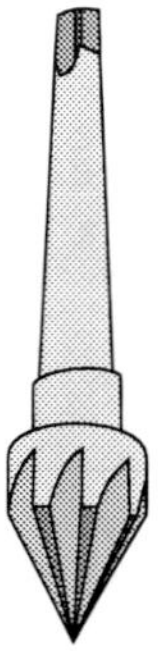

Fig. 8.49

The dimension of the various features and the method of designation are given in Table 8.6.

Type E countersinks are used for slotted countersink bolts used for steel structures.

The dimensions of the various features and the method of designation are given in Table 8.7.

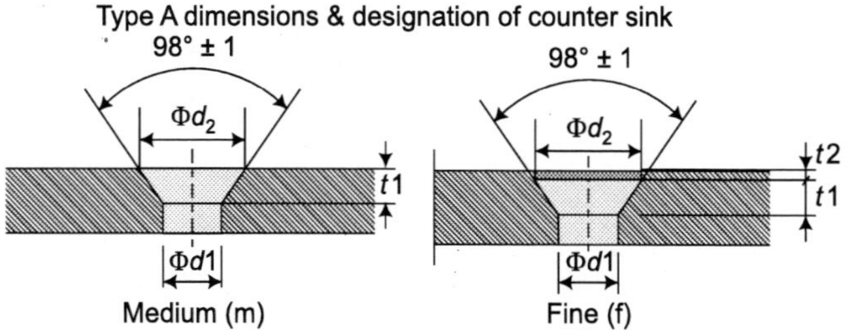

Fig. 8.50

Table 8.4

Nominal size		*1*	*1.2*	*(1.4)*	*1.6*	*(1.8)*	*2*	*2.5*	*3*	*3.5*	*4*	*(4.5)*
Medium series (m)	.d_1 H13	1.2	1.4	1.6	1.8	2.1	2.4	2.9	3.4	3.9	4.5	5
	d_2 H13	2.4	2.8	3.3	3.7	4.1	4.6	5.7	6.5	7.6	8.6	9.5
	t_1 =	0.6	0.7	0.8	0.9	1	1.1	1.4	1.6	1.9	2.1	2.3
	d_1 H12	1.1	1.3	1.5	1.7	2	2.2	2.7	3.2	3.7	4.3	4.8
Fine Series(f)	d_3 H12	2	2.5	2.8	3.3	3.8	4.3	5	6	7	8	9
	t_1 =	0.7	0.8	0.9	1	1.2	1.2	1.5	1.7	2	2.2	2.4
	t_2 + 0.1 0	0.2	0.15	0.15	0.2	0.2	0.15	0.35	0.25	0.3	0.3	0.3
Nominal size		*5*	*6*	*8*	*10*	*12*	*(14)*	*16*	*(18)*	*20*		
Medium Series (m)	d_1 H13	5.5	6.6	9	11	13.5	15.5	17.5	20	22		
	d_2 H13	10.4	12.4	16.4	20.4	23.9	26.9	31.9	36.4	40.4		
	t_1 =	2.5	2.9	3.7	4.7	5.2	5.7	7.2	8.2	9.2		
	d_1 H12	5.3	6.4	8.4	10.5	13	15	17	19	21		
Fine Series(f)	d_3 H12	10	11.5	15	19	23	26	30	34	37		
	t_1 =	2.6	3	4	5	5.7	6.2	7.7	8.7	9.7		
	T_2 + 0.1 0	0.2	0.45	0.7	0.7	0.7	0.7	1.2	1.2	1.7		

Note 1: Sizes shown in brackets are of second preference.

Note 2: Clearance hole d1 according to medium and fine series of IS :1821 dimensions for clearance holes for bolts and screws (second revision).

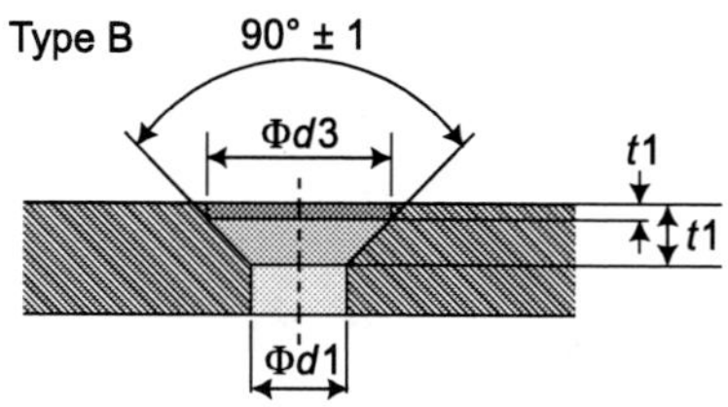

Fig. 8.51

Table 8.5 Dimensions and designation of countersink – Type B according to IS 3406 (Part 1) 1986.

Nominal size		*3*	*4*	*5*	*6*	*8*	*10*	*12*	*(14)*	*16*	*(18)*	*20*	*22*	*24*
Fine	d_1 H12	3.2	4.3	5.3	6.4	8.4	10.5	13	15	17	19	21	23	25
Series	d_2 H12	6.3	8.3	10.4	12.4	16.5	20.5	25	28	31	34	37	48.2	52
(f)	t_1 =	1.7	2.4	2.9	3.3	4.4	5.5	6.5	7	7.5	8	8.5	13.1	14
	t_2 + 0.1	0	0.2	0.3		0.4			0.5				1	

Note 1: Sizes shown in brackets are of second preference.

Note 2: Clearance hole of d1 according to medium and fine series of IS 1821 – 1982

Designation: A countersink type B with clearance hole of fine (f) series and having nominal size 10 shall be designated as – Countersink / 10 IS – 3406.

Dimension and designation of countersink – Type C according to IS 3406 (part1) 1986:

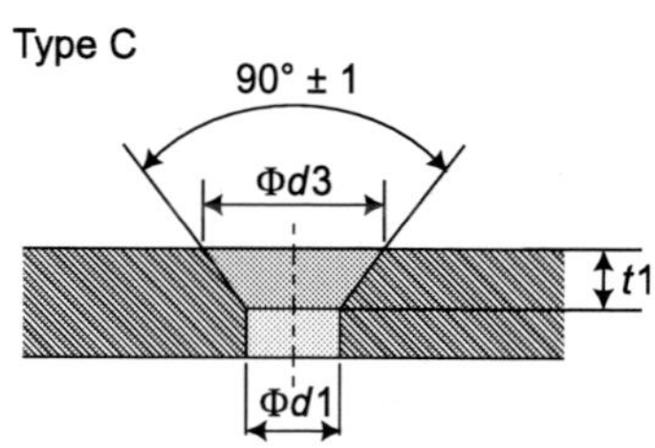

Fig. 8.52

Table 8.6

Screw size	*(0)*	*(1)*	*2*	*(3)*	*4*	*(5)*	*6*	*(7)*	*8*	*10*	*(12)*	*14*	*(16)*
d_1 H12	1.6	2	2.4	2.8	3.1	3.5	3.7	4.2	4.5	5.1	5.8	6.7	8.4
d_2 H12	3.1	3.8	4.6	5.2	5.9	6.6	7.2	8.1	8.7	10.1	11.4	13.2	16.6
t_1 =	0.9	1.1	1.3	1.5	1.7	1.9	2.1	2.3	2.6	3	3.4	3.9	4.9

Note: Sizes given in brackets are of second preference.

Designation: A countersink Type C for screws size 2 shall be designated as Countersink C 2 – IS 3406.

Dimension and designation of countersink – Type E according to IS 3406 (part 1) 1986:

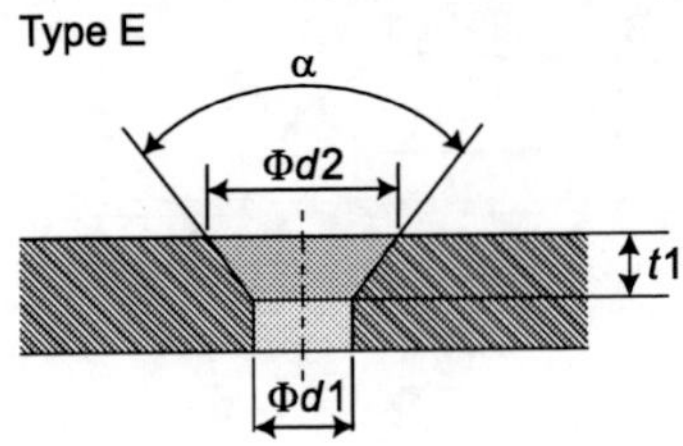

Fig. 8.53

Table 8.7

Nominal size	*10*	*12*	*16*	*20*	*22*	*24*
d_1 H12	10.5	13	17	21	23	25
d_2 H12	19	24	31	34	37	40
t_1 =	5.5	7	9	11.5	12	13
α ± 10	750	0		0		

Note: Clearance hole d1 according to fine series of IS 1821-1982

Designation: A countersink Type E for nominal size 10 shall be designated as Countersink E 10 to IS 3406.

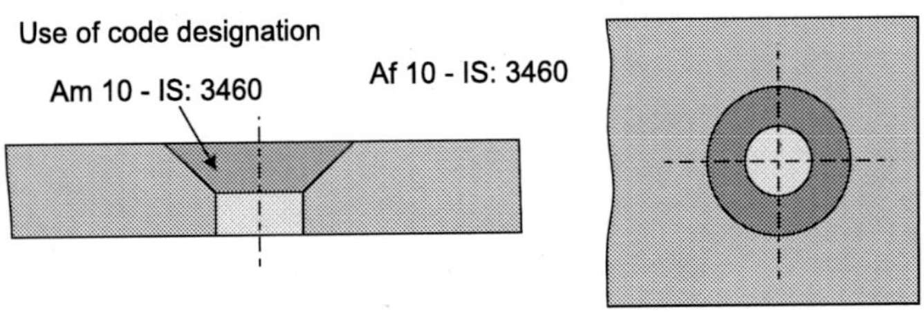

Fig. 8.54

Use of Dimension

The dimension of the countersink can be expressed by the diameter of the countersink and the depth of the countersink.

8.13 COUNTER BORING AND SPOT FACING

8.13.1 Counter Boring

Counter boring is an operation of enlarging a hole to a given depth, to house heads of socket heads or cap screws with the help of a counter bore tool **(Fig. 8.55)**.

8.13.2 Counter Bore (Tool) (Fig. 8.56)

The tool used for counter boring is called a counter bore. Counter bore will have two or more cutting edges.

At the cutting end, a pilot is provided to guide the tool concentric to the previously drilled hole. The pilot also helps to avoid chattering while counter boring (**Fig. 8.57**).

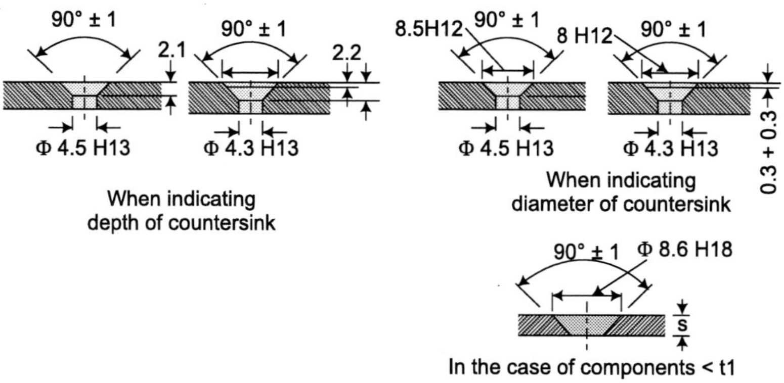

Fig. 8.55

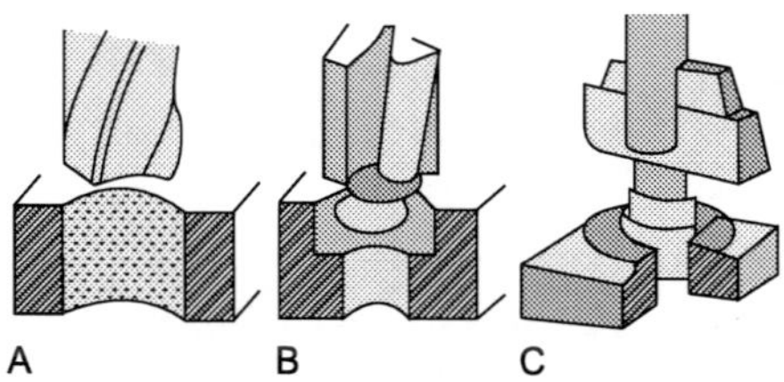

Fig. 8.56

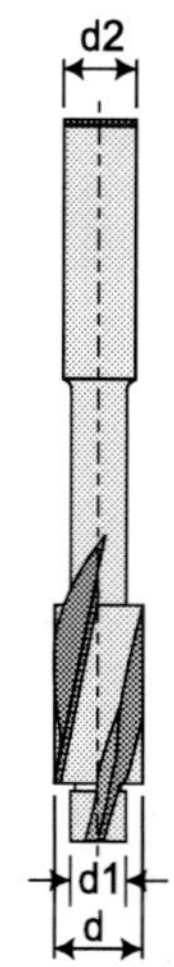

Fig. 8.57

Counter bores are available with solid pilots or with interchangeable pilots. The interchangeable pilot provides flexibility of counter boring on different diameters of holes.

8.14 SPOT FACING

Spot facing is a machining operation for producing a flat seat for bolt head, washer or nut at the opening of a drilled hole. The tool is called a spot facer or a spot facing tool. Spot facing is similar to counter boring, except that it is shallower. Tools that are used for counter boring can be used for spot facing as well (**Fig. 8.58**).

Spot facing is also done by fly cutters by end-cutting action. The cutter blade is inserted in the slot of the holder, which can be mounted on to the spindle (**Fig. 8.59**).

8.14.1 Counter Bore Sizes and Specifications

Counter bore sizes are standardized for each diameter of screws as per BIS.

There are two main types of counter bores. Type H and Type K.

The type H counter bores are used for assemblies with slotted cheese head, slotted pan head and cross recessed pan head screws. The type K counter bore is used in assemblies with hexagonal socket head cap screws.

For fitting different types of washers the counter bore standards are different in Type H and Type K.

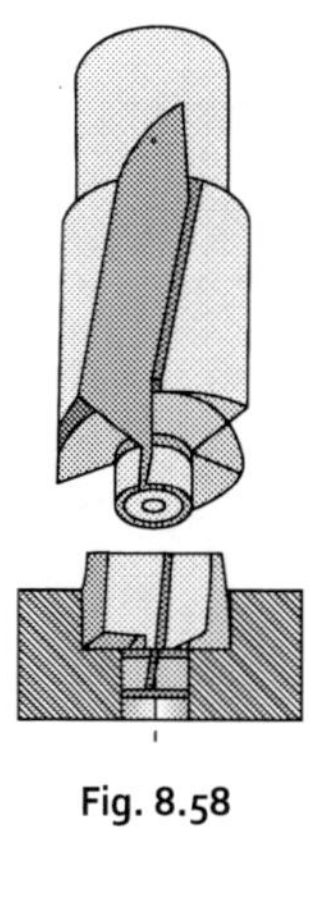

Fig. 8.58

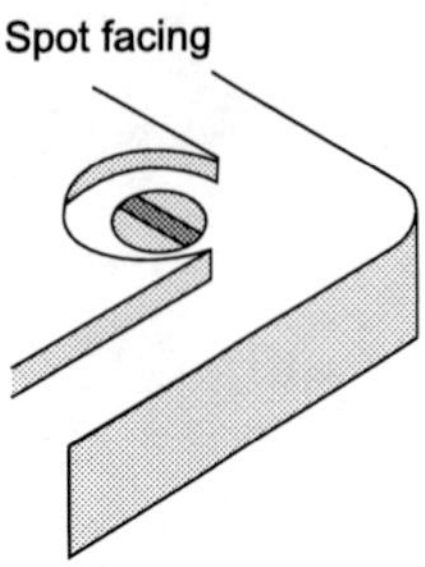

Fig. 8.59

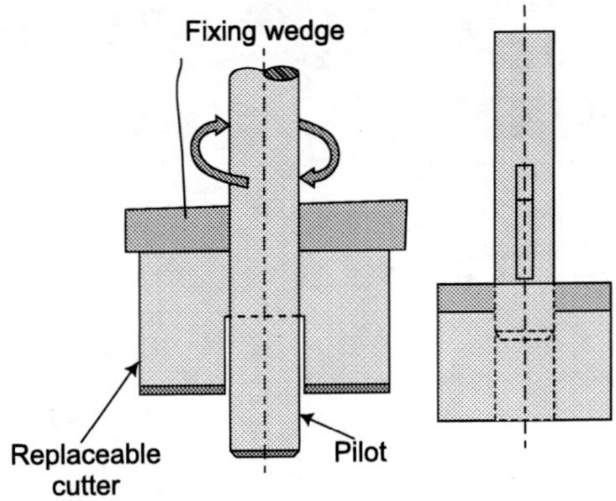

Fig. 8.60

The clearance hole d_1 is of two different grades, i.e., medium (m) and fine (f) and are finished to H13 and H12 dimensions.

The dimensions for type H and type K counter bores can be referred to IS 3406 (Part 2) **(Fig. 8.60)**.

Counter bore and clearance hole sizes for different sizes of screws.

While representing counter bore in drawings, counter bore can be indicated either by code designation or using the dimensions or using dimensions.

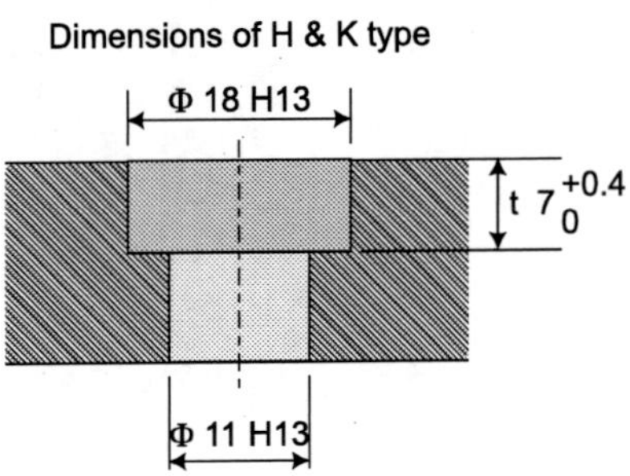

Fig. 8.61

Judgment on countersinking

Mistake	*Cause*
Countersinking is not "neat", the countersink is chattering.	Resharpening of cutting edges. Cutting edges are not uniform.
Countersinking is not true to dimensions and is offset towards the drill hole.	Hole was too much rough drilled or pilot guide was worn out. The tool was not centrically chucked.
Screw heads of cylindrical and conical countersinking project.	Countersinking depth was not checked. Countersink with cylindrical shank was sliding in the chuck.

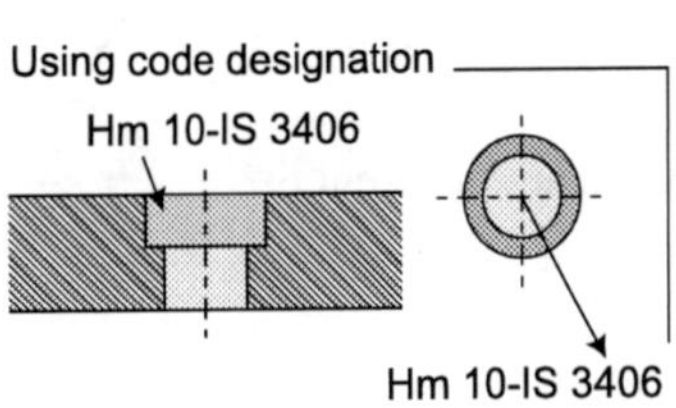

Fig. 8.62

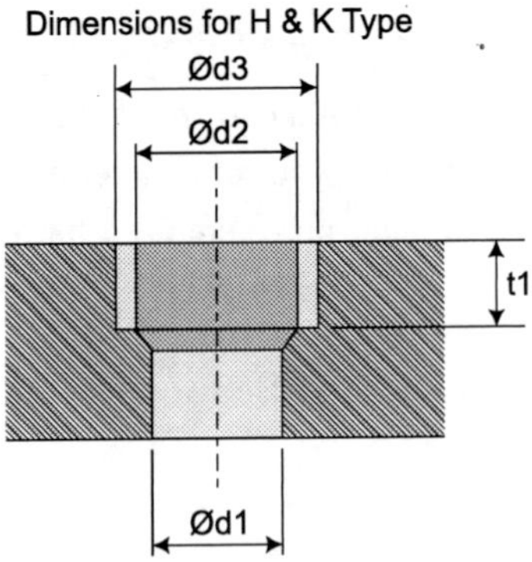

Fig. 8.63

8.15 DRILLING OF SMOOTH AND ACCURATE HOLES

Smooth and accurate holes mainly serve for fitting of bolts, shafts, bushes, etc. Certain requirements are set for the accuracy in size and surface quality. As a rule, the permissible tolerance is designated on the drawing by a fit and the surface quality by a symbol.

The holes drilled with a twist drill are neither smooth nor accurate in size and do not suffice for the requirements laid down for manufacturing accurate holes. A true bore with smooth walls can be produced by reaming. A reamer is used. This has teeth on the cylindrical part and will be inserted into the hole, after drilling and counter boring. During the rotation and feed motion, the teeth cut off fine chis. Reaming is a smooth finishing operation.

8.16 FUNDAMENTALS OF REAMING

Reaming allows to improve the wall and to increase the degree of accuracy of a hole made by a twist drill. Cylindrical pins and taper pins get a neat fit, and pivots are fitted into bushings in a sliding way and with little play.

The multipoint cutting edges of the reamer remove fine chips from the wall of the rough-drilled hole and produce an accurate and round boring with a smooth hole wall.

8.16.1 Action of the Reamer (Fig. 8.64)

Any bored hole which is to receive an accurate size and a neat wall must be drilled so much smaller that the rifles produced by drilling can be fully removed by reaming. For this reason holes up to 10 mm diameter are drilled by 0.1 mm or 0.2 mm smaller and hole of more than 10 to 20 mm diameter are drilled by 0.3 to 0.5 mm smaller than the finishing size of the bore hole.

Reamers are made of tool steel or high-speed steel. They are produced either with fixed cutting edges for a determined diameter or with adjustable blades allowing to adjust the reamer to different diameters of holes.

The unimportant chip removal of the reamer is effected by several cutting edges.

The front part of the cutters of the reamer, that is, the pilot point, is conically sharpened. It affects the actual cutting action whereas the other cutting edges only guide and smooth. Hand reamers are provided with a long pilot point to facilitate the penetration into the hole and to ensure good guidance. Machine reamers have a short pilot point for working of tough materials and for the production of blind holes; for working brittle material, the pilot point is longer.

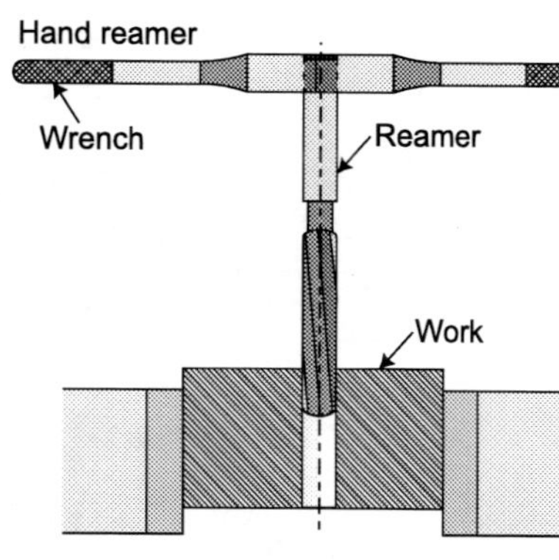

Fig. 8.64

Reaming by hand reamer is done manually for which great skill is needed.

Machine reamers are fitted on spindles of machine tools and rotated for reaming.

Machine reamers are provided with Morse taper shanks for holding on machine spindles.

Hand reamers have straight shanks with 'square' at the end, for holding with tap wrenches (**Fig. 8.64 and 8.65**).

8.16.2 Parts of a Hand Reamer

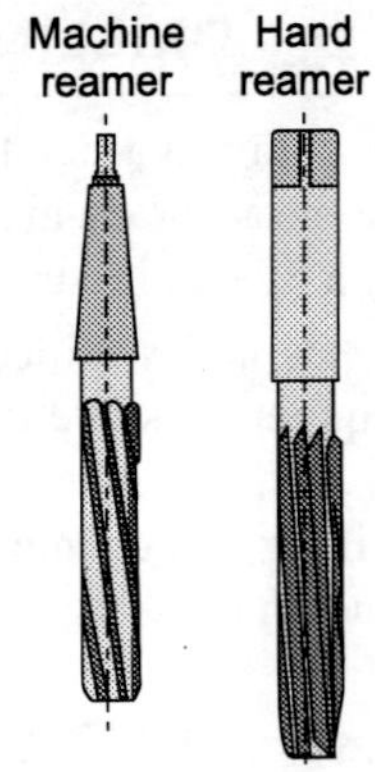

Fig. 8.65

The parts of a hand reamer are listed below (Fig. **8.66**).

Axis: Axis is the longitudinal centre line of the reamer.

Body: The body is the portion of the reamer extending from the entering end of the reamer to the commencement.

Recess: The portion of the· body, which is reduced in diameter below the cutting edges, pilot or guide diameters is referred to as recess.

Shank: The portion of the reamer which is held and driven is the shank. It can be parallel or taper.

Circular land: The cylindrically ground surface adjacent to the cutting edge on the leading edge of the land.

Bevel lead: The bevel lead is the cutting portion at the entering end of the reamer cutting its way into the hold. It is not provided with a circular land.

Taper lead: It is the tapered cutting portion at the entering end to facilitate cutting and finishing of the hole. It is not provided with a circular land.

Bevel lead angle: The angle formed by the cutting edges of the bevel lead and the reamer axis.

Taper lead angle: The angle formed by the cutting edges of the taper and the reamer axis.

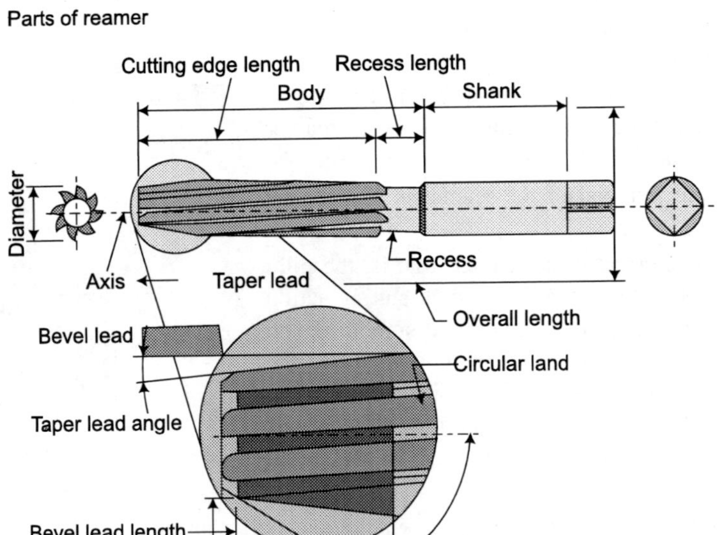

Fig. 8.66

8.17 TERMS RELATING TO CUTTING GEOMETRY FLUTES

The grooves in the body of the reamer, is to provide cutting edges, and to permit the removal of chips and to allow the cutting fluid to reach the cutting edges **(Fig. 8.66)**.

Heel: The edge formed by the intersection of the surface left by the provision of a secondary clearance and the flute **(Fig. 8.67)**.

Cutting edge: The edge formed by the intersection of the face and the circular land or the surface left by the provision of primary clearance **(Fig. 8.67)**.

Face: The portion of the flute surface adjacent to the cutting edge on which the chip impinges as it is cut from the work **(Fig. 8.67)**.

Rake angles: The angles in a diametral plane, formed by the face and a radial line from the cutting edge are the rake angles **(Fig. 8.68)**.

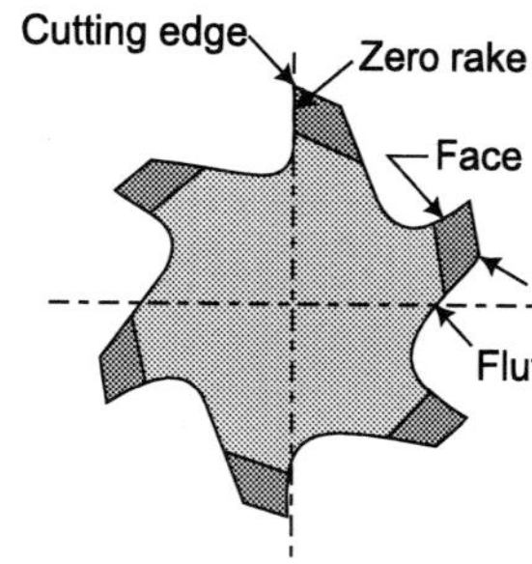

Fig. 8.67

Clearance angles: The angles formed by the primary or secondary clearances and the tangent to the periphery of the reamer at the cutting edge. They are called primary clearance angle and secondary clearance angle respectively **(Fig. 8.69)**.

Helix angle: The angle between the edge and the reamer axis is the helix angle **(Fig. 8.70)**.

Kinds of reamers:

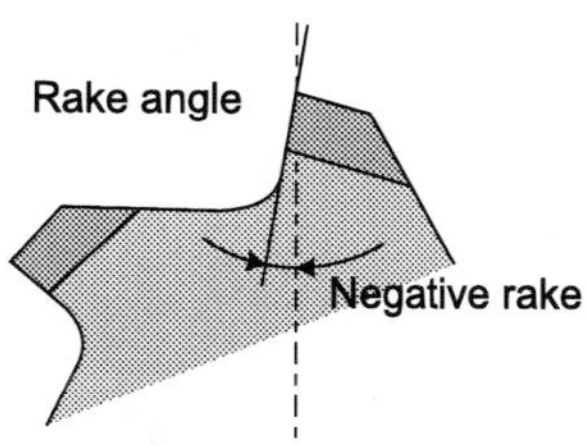

Fig. 8.68

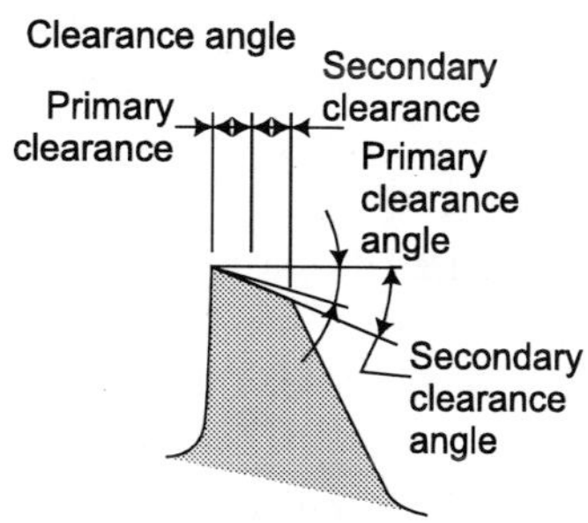

Fig. 8.69

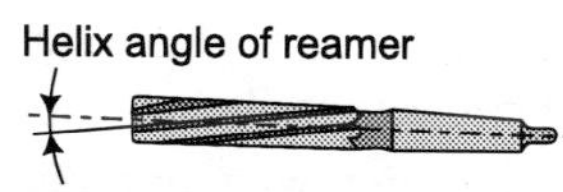

Fig. 8.70

8.18 GENERAL FEATURES OF HAND REAMERS (FIG. 8.70)

Hand reamers are used to ream holes manually using tap wrenches.

These reamers have a long taper lead. This allows starting the reamer straight and in alignment with the hole being reamed **(Fig. 8.71)**.

Most hand reamers are for right hand cutting.

Helical fluted hand reamers have left hand helix. The left hand helix will produce smooth cutting action and finish.

Most reamers, machine or hand, have uneven spacing of teeth. This feature of reamers helps to reduce chattering while reaming.

8.18.1 Types, Features and Functions

Hand reamers with different features are available for meeting different reaming conditions. The commonly used types are listed here under.

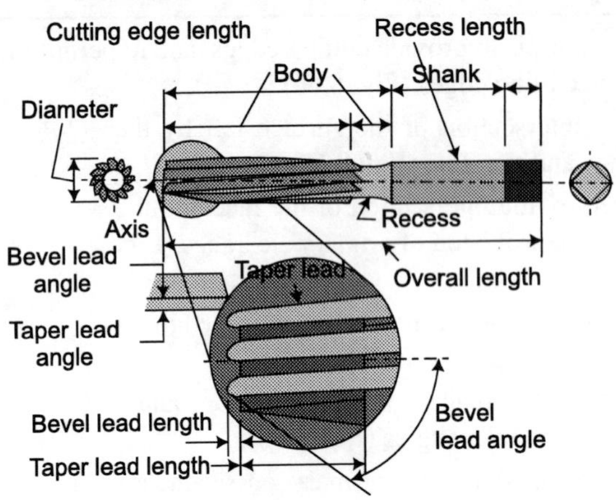

Fig. 8.71

Parallel hand reamer with parallel shank (Fig. 8.72): A reamer which has virtually parallel cutting edges with taper and bevel lead. The body of the reamer is integral with a shank. The shank has the nominal diameter of the cutting edges. One end of the shank is square shaped for tuning it with a tap wrench. Parallel reamers are available with straight and helical flutes. This is the commonly used hand reamer for reaming holes with parallel sides.

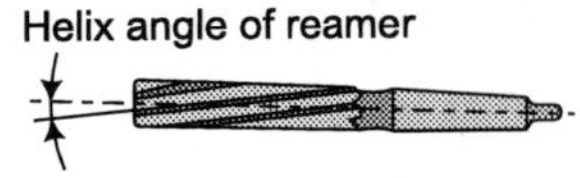

Fig. 8.72

Reamers commonly used in workshop produce H8 holes.

Hand reamer with pilot (Fig. 8.74): For this type of reamer, a portion of the body is cylindrically ground to form a pilot at the entering end. The pilot keeps the reamer concentric with the hole being reamed.

Socket reamer with parallel shank (Fig. 8.76A and 8.76B): This reamer has tapered cutting edges to suit metric Morse tapers. The shank is integral with the body, and is square shaped for driving. The flutes are either straight or helical.

The socket reamer is used for reaming internal Morse tapered holes.

Use of straight and helical fluted reamers (Fig. 8.75): Straight fluted reamers are useful for general reaming work. Helical fluted reamers are particularly suitable for reaming holes with key way grooves or special lines cut into them. The helical flutes will bridge the gap and reduce binding and chattering.

Taper pin hand reamer (Fig. 8.76C): This reamer has tapered cutting edges for reaming taper holes to suit taper pins. A taper pin reamer is made with a taper pin of 1 in 50. These reamers are available with straight or helical flutes.

Parallel hand reamer with parallel shank

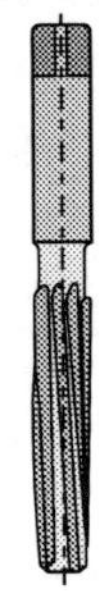

Fig. 8.73

8.18.2 Material of Hand Reamers

When the reamers are made as a one-piece construction, high-speed steel is used. When they are made as two-piece construction then the cutting portion is made of high-speed steel while the shank portion is made of carbon steel. They are butt-welded together before manufacturing.

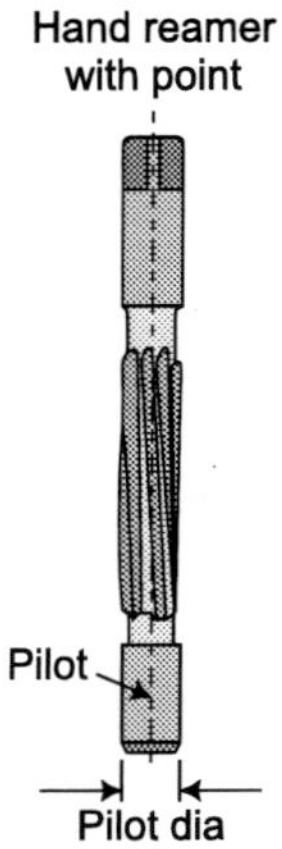

Fig. 8.74

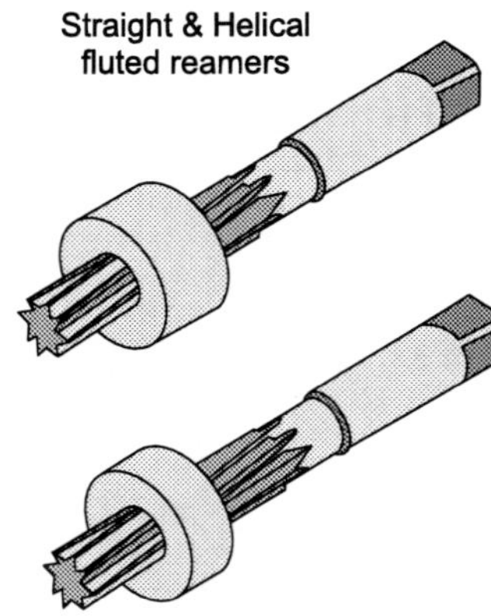

Fig. 8.75

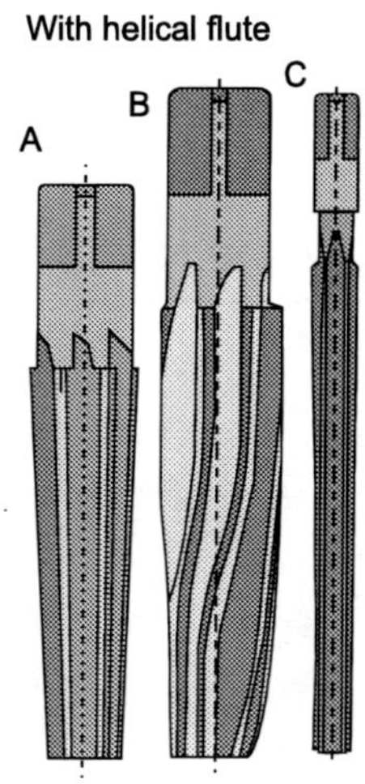

Fig. 8.76

8.18.3 Specifications of a Reamer

To specify a reamer the following data is to be given.

Type: Flute shank end size.

Example: Hand reamer, straight flute, and parallel shank of 0.20 mm.

The direction of the mist: The direction of mist be opposite to the cutting direction to avoid setting of the reamer into the boring hole.

Cylindrical reamers: Cylindrical reamers with fixed cutting edges will wear rapidly thus losing their size. This is especially disadvantageous with reamers for finish or precise drilling work. Then the tool can only be used as a roughing reamer.

Adjustable reamers: Adjustable reamers offer more advantages. They can be adjusted to several diameters within the range stated on the shank of the tool (**Fig. 8.76**). In that case the blades inserted in ascending grooves are moved by adjusting nuts.

The diameter can be increased or decreased in size according to range. Threads are cut in its body and grooves are tapered in which blades are fixed. These blades are moved up and down through the adjusting nuts and thus the size of the reamer can be increased or decreased. The size can be increased or decreased up to 0.5 mm. Gradually, the size can be increased and metal can be cut easily. If its edge becomes blunt, the blades can be taken out and their edge can be sharpened.

Expansion reamer: This reamer is similar to the adjustable reamer but it does not act like the adjustable reamer. This reamer admits a limited amount of size. Such adjustment is usually through wedge action controlled by a screw. In one inch diameter it can be increased or decreased only up to 0.005″. A taper hole is provided in its body. A taper plug having threads is fitted on it. Head of the plug is square which can be moved by a wrench or a spanner. Size is increased or decreased by this action. It is mostly used for surface finishing.

Taper reamer: This reamer is used for drilling a hole in a job making this hole a tapered one. There are two reamers in a set. One performs the rough job, the other one finishing. These can be used by hand or a machine because they have straight or taper shank.

Pilot reamer: This reamer is of a cylindrical shape in its inner part spiral teeth are cut and on both the sides plain shanks are built. On one side, a square tang is made on the shank in which, a handle is fixed and it is moved. It is used to make two holes opposite to each other of the equal size.

8.19 TERMINOLOGY AND DEFINITIONS

Reamer: A rotary cutting tool with one or more cutting elements used for enlarging to size and contour a previously formed hole. Its principal support during the cutting action is obtained from the workpiece (**Fig. 8.77**).

Actual size: The actual measured diameter of a reamer, usually is slightly larger than the nominal size to allow for wear.

Angle of taper: Angle of taper is the included angle of taper on a taper tool or taper shank.

Arbor hole: Arbor hole is the central mounting hole in a shell reamer.

Axis: Axis is the imaginary straight line which forms the longitudinal centre line of a reamer, usually established by rotating the reamer between centres.

Back taper: Back taper is a slight decrease in diameter from front to back in the flute length of reamers.

Bevel: Bevel is an unrelieved angular surface of revolution (not to be confused with chamfer).

Body: Body is the fluted full-diameter portion of a reamer inclusive of the chamfer, starting taper and bevel.

Chamfer: Chamfer is the angular cutting portion at the entering end of a reamer (see also secondary chamfer).

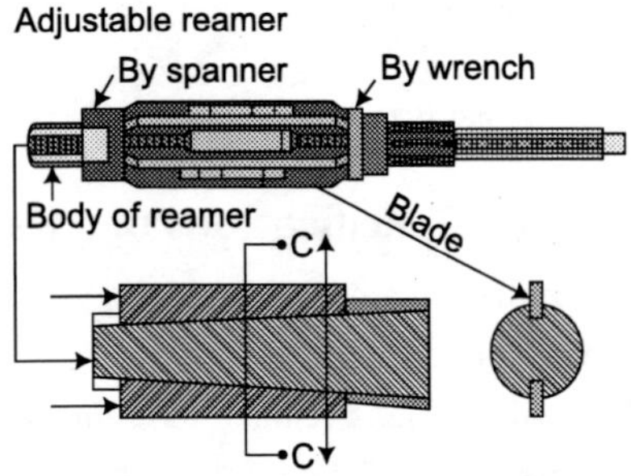

Fig. 8.77

Chamfer angle: Chamfer angle is the angle between the axis and the cutting edge of the chamfer measured in an axial plane at the cutting edge.

Chamfer length: Chamfer length is the length of the chamfer measured parallel to the axis at the cutting edge.

Chip breakers: Chip breakers are notches or grooves in the cutting edges of some taper reamers designed to break the continuity of the chips.

Clearance: Clearance is the space created by the relief behind the cutting edge or margin of a reamer.

Core: Core is the central portion of a reamer below the flutes which joins the lands.

Core diameter: Core diameter is the diameter at a given point along the axis of the largest circle which does not project into the flutes.

Cutter sweep: Cutter sweep is the section removed by the milling cutter or grinding wheel in entering or leaving a flute.

Cutting edge: Cutting edge is the leading edge of the land in the direction of rotation for cutting.

Cutting face: Cutting face is the leading side of the land in the direction of rotation for cutting on which the chip impinges.

External centre: External centre is the pointed end of a reamer. The included angle varies with manufacturing practice.

Flutes: Flutes are longitudinal channels formed in the body of the reamer to provide cutting edges, permit passage of chips, and allow cutting fluid to reach the cutting edges.

Angular flute: Angular flute is a flute which forms a cutting face lying in a plane intersecting the reamer axis at an angle. It is unlike a helical flute in that it forms a cutting face which lies in a single plane.

Helical flute: Helical flute is (sometimes called "spiral flute") a flute which is formed in a helical path around the axis of a reamer.

Spiral flute: On a taper reamer a flute of constant lead. In reference to a straight reamer, see the preferred term helical flute.

Straight flute: Straight flute is a flute which forms a cutting edge lying in an axial plane. Flute length is the length of the flutes not including the cutter sweep.

Guide: Guide is a cylindrical portion following the flutes of a reamer to maintain alignment.

Heel: Heel is the trailing edge of the land in the direction of rotation for cutting.

Helix angle: Helix angle is the angle which a helical cutting edge at a given point makes with an axial plane through the same point.

Internal centre: Internal centre is a 60° countersink with clearance at the bottom, in one or both ends of a tool, which establishes the tool axis.

Irregular spacing: Irregular spacing is a deliberate variation from uniform spacing of the reamer cutting edges.

Land: Land is the section of the reamer between adjacent flutes.

Land width: Land width is the distance between the leading edge of the land and the heel measured at a right angle to the leading edge.

Lead of flute: Lead of flute is the axial advance of a helical or spiral cutting edge in one turn around the reamer axis.

Length: Length is the dimension of any reamer element measured parallel to the reamer axis.

Limits: The maximum and minimum values designated for a specific element.

Margin: Margin is the unrelieved part of the periphery of the land adjacent to the cutting edge.

Neck: Neck is the section of reduced diameter connecting shank to body, or connecting other portions of the reamer.

Nominal size: Nominal size is the designated basic size of a reamer.

Overall length: Overall length is the extreme length of the complete reamer from end to end, but not including external centres or expansion screws.

Periphery: Periphery is the outside circumference of a reamer.

Pilot: Pilot is a cylindrical portion preceding the entering end of the reamer body to maintain alignment.

Rake: Rake is the angular relationship between the cutting face, or a tangent to the cutting face at a given point and a given reference plane or line.

Axial rake:. Axial rake applies to angular (not helical or spiral) cutting faces. It is the angle between a plane containing the cutting face, or tangent to the cutting face at a given point, and the reamer axis.

Helical rake: Helical rake applies to helical and spiral cutting faces only (not angular). It is the angle between a plane tangent to the cutting face at a given point on the cutting edge and the reamer axis.

Hook: Hook is a concave condition of a cutting face. The rake of a hooked cutting face must be determined at a given point.

Negative rake: Negative rake describes a cutting face in rotation whose cutting edge lags the surface of the cutting face.

Positive rake: Positive rake describes a cutting face in rotation whose cutting edge leads the surface of the cutting face.

Radial rake angle: Radial rake angle is the angle in a transverse plane between a straight cutting face and a radial line passing through the cutting edge.

Relief: Relief is the result of the removal of tool material behind or adjacent to the cutting edge to provide clearance and prevent rubbing (heel drag).

Axial relief: Axial relief is the relief measured in the axial direction between a plane perpendicular to the axis and the relieved surface. It can be measured by the amount of indicator drop at a given radius in a given amount of angular rotation.

Cam relief: Cam relief is the relief from the cutting edge to the heel of the land produced by a cam action.

Chamfer relief: Chamfer relief is the axial relief on the chamfer of the reamers.

Chamfer relief angle: Chamfer relief angle is the axial relief angle at the outer corner of the chamfer. It is measured by protection into a plane tangent to the periphery at the outer corner of the chamfer.

Eccentric relief: Eccentric relief is a convex relieved surface behind the cutting edge.

Flat relief: Flat relief is relieved surface behind the cutting edge which is essentially flat.

Primary relief: Primary relief is the relief immediately behind the cutting edge or margin, properly called relief.

Radial relief: Radial relief is relief in a radial direction measured in the plane of rotation. It can be measured by the amount of indicator drop at a given radius in a given amount of angular rotation.

Relief angle: Relief angle is the angle measured in a transverse plane between the relieved surface and a plane tangent to the periphery at the cutting edge.

Secondary relief: Secondary relief is an additional relief behind the primary relief.

Secondary chamfer: Secondary chamfer is a slightly relieved chamfer adjacent to and following the initial chamfer on a reamer.

Shank: Shank is the portion of the reamer by which it is held and driven.

Squared shank: Squared shank is a cylindrical shank having a driving square on the back end.

Starting radius: Starting radius is a relieved radius at the entering end of a reamer in place of a chamfer.

Starting taper: Starting taper is slightly relieved taper on the front end of a reamer.

Straight shank: Straight shank is a cylindrical shank.

Tang: Tang is the flattened end of a taper shank which fits a slot in the socket.

Taper per foot: Taper per foot is the difference in diameter between two points 12 inches apart measured along the axis.

Taper shank: Taper shank is a shank made to fit a specified (conical) taper socket.

Clearance and blade angles: Practically all cutting done by a reamer (except a hand reamer or taper reamer) is accomplished by the chamfer **(Fig. 8.77)** which forms a truncated cone on the end of the reamer and carries the cutting edges. The chamfer is shaped so that the reamer will start properly in the hole and the major portion of the cutting is done by the chamfer length. The longitudinal cutting edges should do little or no cutting. The clearance angles in the cutting portions of the reamer are employed to obtain cutting action in the proper places, prevent friction, and clear the chips. Reamers usually embody three types of clearance: chamfer relief, peripheral relief, and longitudinal relief.

Chamfer relief (Fig. 8.77) is obtained by the clearance angle ground on the cutting ends of the reamer blades to enable them to penetrate more easily beneath the workpiece surface and produce chips.

Peripheral relief is obtained by the clearance angle ground on the lands back of the cutting edges to relieve the radial pressure of the reamer flutes against the walls of the hole. This tends to reduce friction and provides a cutting edge that functions when the peripheral corner of the chamfer becomes rounded and dull.

Longitudinal relief is the slight taper embodied in standard reamers, which results in a smaller diameter at the shank end than at the cutting end, to prevent the reamer blade from causing side-cutting action due to lack of proper alignment or to dullness of the reamer, caused by wear. It is important that reamers used in angular floating holders should embody enough back taper to prevent interference resulting from improper alignment caused by these types of holders.

Chamfer angle usually has a magnitude of 45°, but for particular applications and after suitable trials in the workpiece material it may be as great as 50° or as small as 40°. As the greatest dulling of reamer blades occurs at the peripheral corners of the chamfer, the maximum effective cutting action will continue only as long as the original sharp chamfer angle is maintained.

Secondary chamfer (Fig. 8.78): Sometimes referred to as "starting taper" or "lead," consists of a chamfer ranging from 1° to 10° which is located immediately in the back of the first chamfer.

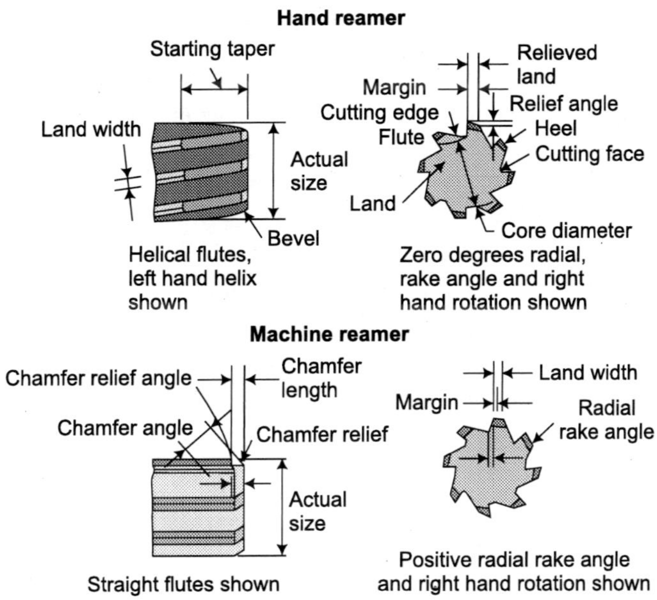

Fig. 8.78

Clearance should be provided for this second taper in such a manner as to bring it to a sharp edge. It should not be confused either with the primary chamfer-relief angle or with a similar expression that refers to the starting taper or lead of a hand reamer. The secondary chamfer functions to affect a scraping cut that helps to size and smooth the walls of the hole.

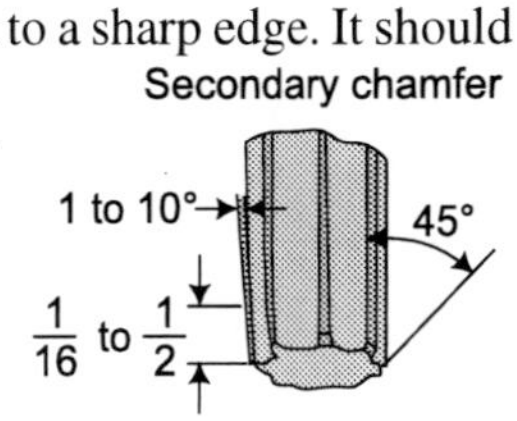

Fig. 8.79

Chamfer on hand reamers for light cutting should be approximately 45° on the entering teeth (blades). A hand reamer should cut on the lead or starting taper located in the back of the chamfer. The length of the hand-reamer lead section or starting taper customarily is equal to approximately the reamer diameter for sizes of 3/4 inch and smaller; it decreases to approximately one-half the reamer diameter for sizes ranging from 3/4 to 3 inch in diameter. The diameter of the lead section of the entering end should range from 0.010 to 0.015 inch under the nominal reamer diameter.

Hand of Spiral and Hand of Cut: The hand of a reamer cut refers to the direction of rotation (right or left) and may be determined by inspection of the chamfer end of the reamer when mounted so as to make a cut. If the rotary motion of the reamer is counterclockwise, cut is right-hand; if the rotary motion is clockwise, cut is left-hand.

The hand of the spiral or helical reamer may be determined by viewing the reamer from either end. If the flutes twist away from the observer in a clockwise direction, the spiral is right-hand; if the flutes twist away from the observer in a counterclockwise direction, the spiral is left-hand.

Most reamers **(Fig. 8.80)** are right-hand cutting and may have straight flutes: (a) left-hand spiral flutes, (b) or right-hand spiral flutes and (c) straight flutes cost less originally and are easiest to sharpen or recondition. This type of reamer has proved satisfactory for a great many applications. When selecting reamers for machine tools with play in the gears or with spindles not in the best of condition, it is usually advisable to choose those with left-hand spiral flutes. This type of reamer has a negative axial rake and therefore requires more pressure to feed the tool into the work. The additional pressure and the hand of the spiral opposite to the hand of the cut reduce the tendency for the reamer to jump ahead when easier cutting is encountered because the spindle is under tension and the gears are held in contact. For machine tools in very good condition, right-hand spiral-fluted reamers are satisfactory. They are inherently free-cutting and require less power; however, when used on a worn machine, they tend to dig into the workpiece, cause chatter, and produce a rough hole.

Marginal Widths as Related to Peripheral Relief Angles: The purpose of the margin of a reamer land is to size the hole by scraping away roughness caused by the dulling of the peripheral corners of the chamfer and to reduce friction caused by the reamer's rotating in the hole. When a reamer is ground down to a diameter smaller than the original size, the land should be relieved to a circular margin to prevent the reamer from damaging the walls of the hole. The marginal width and magnitude of a relief angle vary somewhat for particular applications, but the values in Table 8.8 be used as a basis from which to determine by actual trial the correct values to be used. By properly changing the relief angle, a reamer may be made suitable for use under a great variety of operating conditions.

Table 8.8 Suggestion width of margin and magnitude of relief angles for various sizes of reamers

Diameter of reamers	*Margin width*	*Magnitude of relief angle*
Under 3	0.02 – 0.0153	
3 – 6	0.153 – 0.254	20 – 25
6 – 12	0.23 – 0.254	15 – 20
12 – 20	0.254 – 0.381	11 – 14
20 – 25	0.308 – 0.432	9 – 10
20 – 32	0.305 – 0.457	8 – 9
32 – 50	0.406 – 0.59	7 – 8
Larger than 50	0.437 – 0.635	7 – 8

Kinds of Reamers

Reamers are made in many different forms **(Fig. 8.80)** including solid and inserted-blade types, adjustable and nonadjustable, and are available for either manual operation (hand reamers) or for machine use (chucking

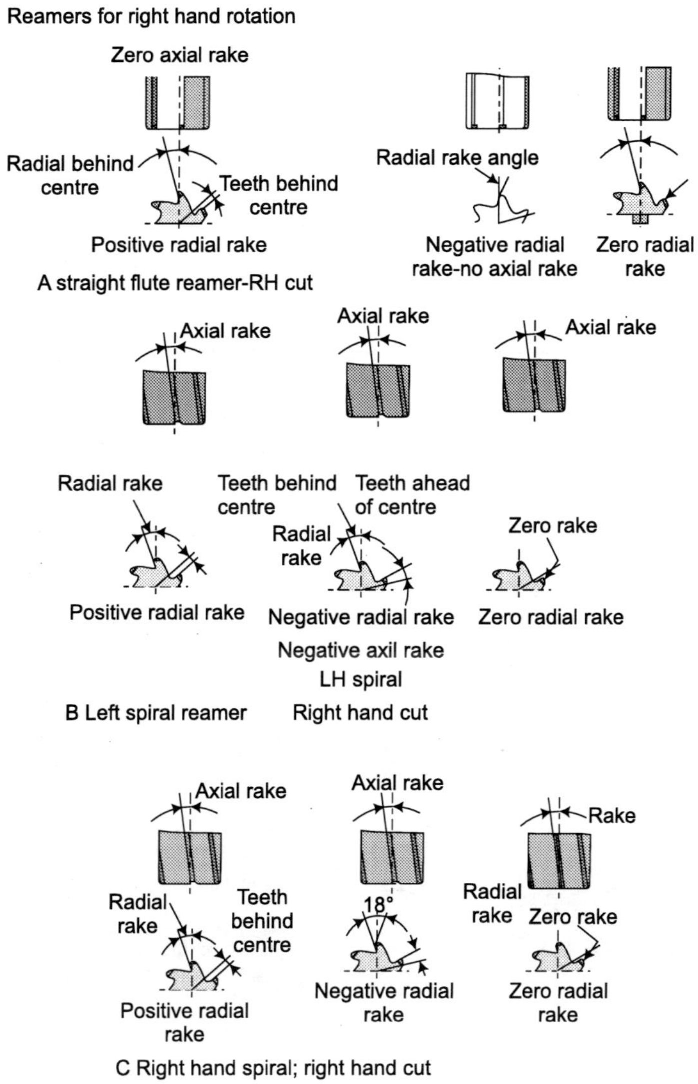

Fig. 8.80

reamers). Materials from which cutting elements of reamers are made include high-carbon steel, high-speed steel, cast alloys, and cemented carbides. It is advisable to consult the manufacturers of reamers regarding particular applications of these tools.

8.20 HAND REAMERS

A hand reamer is generally used to finish or accurately size a previously earned or bored hole, or for fitting bushings or bearings when only a small amount of metal is removed during the finishing process. Hand reamers are available in either the solid or the expansion type and may have either straight or left-hand spiral flutes. The solid type is made from either carbon or high-speed tool steel. The expansion type is regularly made from carbon tool steel as this material is better adapted to being expanded over a long section. Hand reamer blades are usually spaced irregularly around the reamer body.

Limit of expansion: The amount of expansion possible in an expansion reamer is rather limited and varies with different makes. Breakage of these tools usually is caused by the users attempting to expand them beyond the intended limits. One reamer manufacturer suggests the limits of expansion in as given below.

Recommended expansion limits for expansion reamers:

Size range	*Limit of expansion*
6 – 12	01524
12 – 25	0.254
25 – 32	0.33
40 – 50	0.381

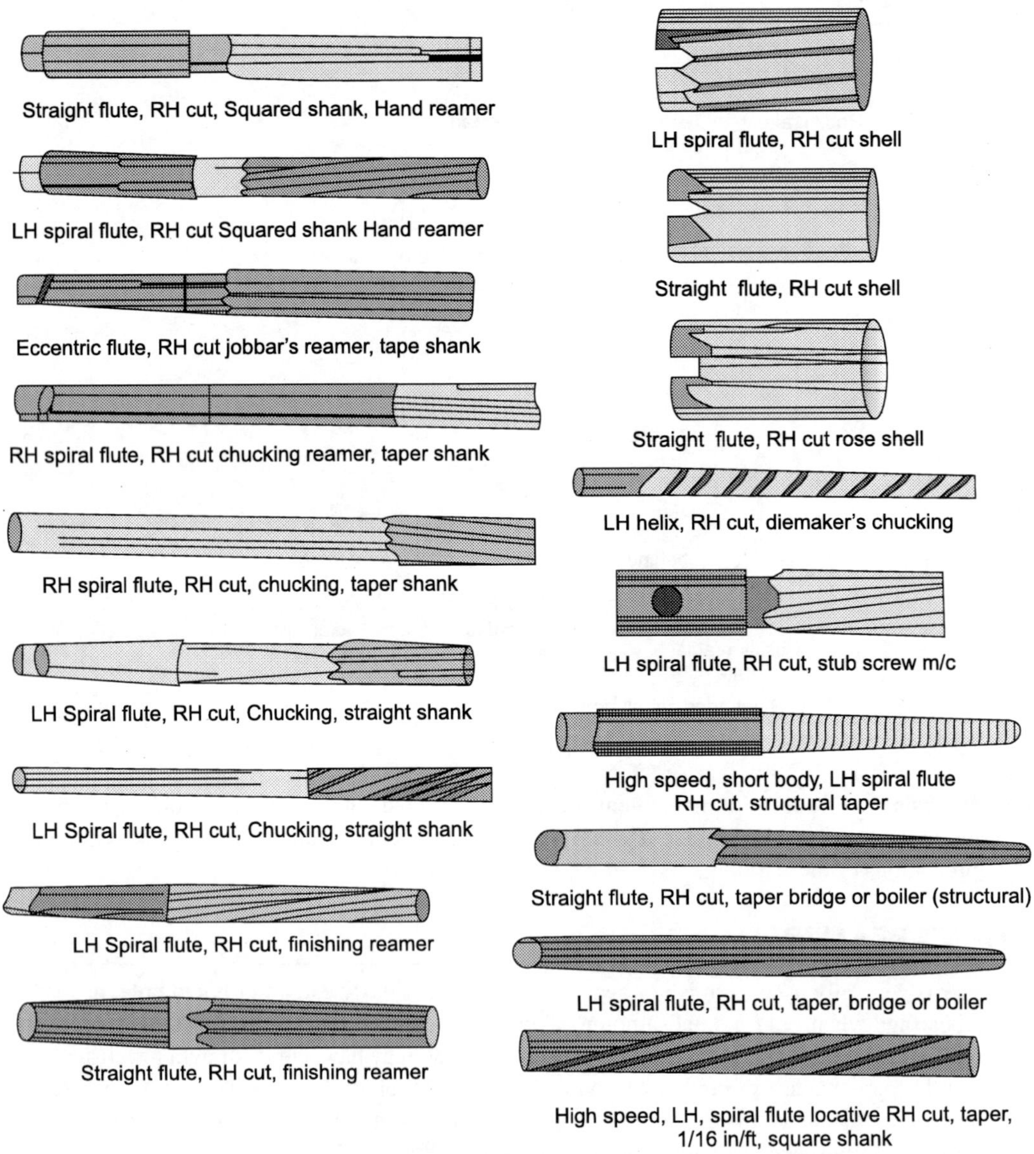

Fig. 8.81

8.21 CHUCKING REAMERS

Chucking reamers were designed primarily for application in turret lathes and screw machines where in many instances the tool is stationary and the work revolves. They are available in either straight-fluted, right- or left-hand spiral-fluted forms and may be either of the fluted chucking type **(Fig. 8.82)** or of the rose chucking type the essential difference between the rose chucking reamer and the fluted chucking reamer is that the rose type is not relieved on the periphery.

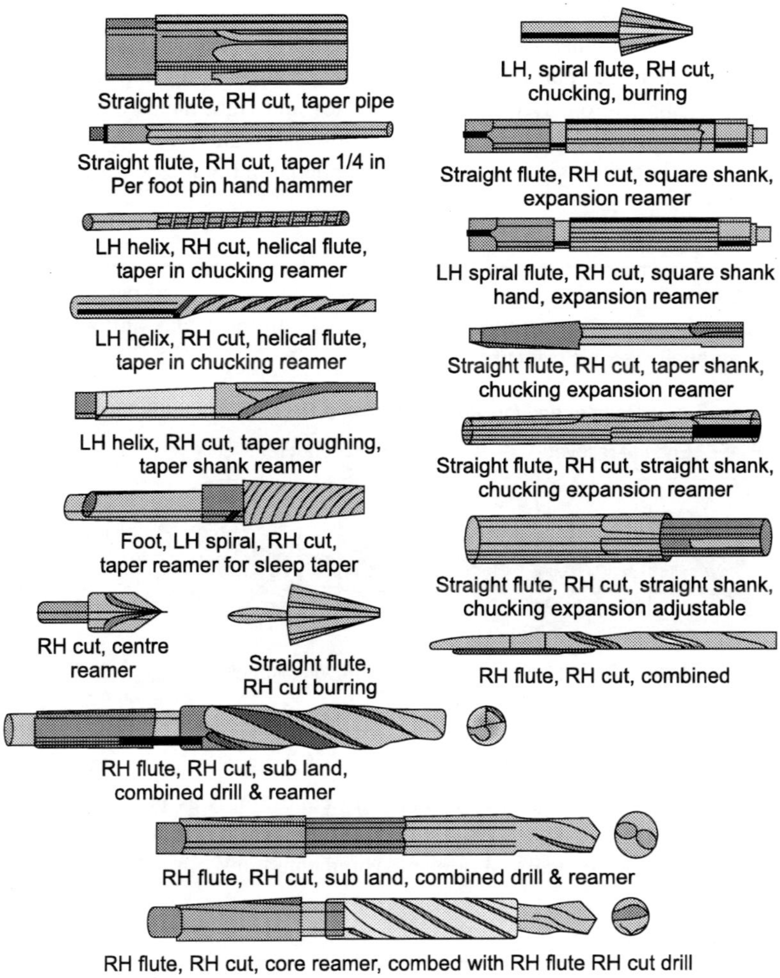

Fig. 8.82

8.22 STRAIGHT-FLUTED CHUCKING REAMERS

Straight-fluted chucking reamers usually embody staggered flutes, i.e., the flutes are irregularly spaced around the circumference of the reamer body. This is done to reduce the reamer's tendency to chatter because of slippage or torsion deflection. This is true of most straight fluted reamers including jobbers, shell, and hand reamers. Chatter may sometimes be minimized by using a reamer having a hand of spiral opposed to the hand of cut.

8.23 STUB SCREW-MACHINE REAMERS

Reamers used in automatic screw machines are required to function in a limited space, hence their short length. Stub screw-machine reamers are particularly adapted to use in floating holders, and a pinhole is provided in the shank for this application. Left-hand spiral flutes and right-hand rotation for cutting are considered standard; earners requiring left-hand rotation for cutting and reamers with right-hand spiral flutes are considered special types. Stub screw-machine reamers are available in numbered series.

8.24 END-CUTTING REAMERS

All reamers are considered to be end-cutting, but it is often necessary to produce a special angle in a hole or to use the reamer as a counter bore or as a seating tool in blind hole. In these and similar cases, the reamer end must be shaped to accomplish the purpose (**Fig. 8.83**).

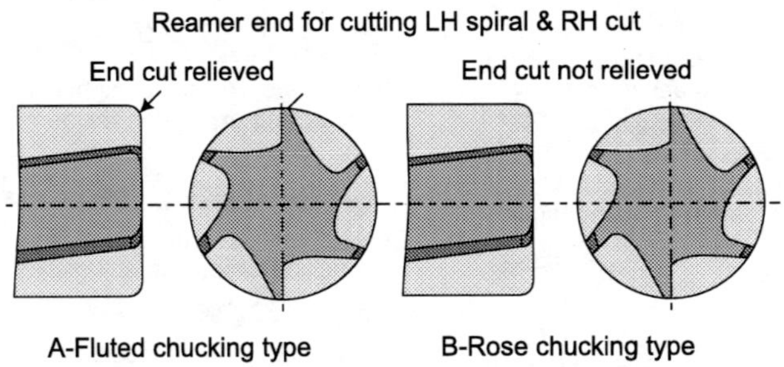

Fig. 8.83

When reamers are cut to centre, right-hand view, care should be taken to avoid era flute depth at centre, in which case no chips could be removed at or near centre. This condition is avoided, in the design shown, by having one land extend about 41/64 inch past centre with sufficient depth of flute to permit chip escape at centre. The mating land is thereby automatically shortened, as indicated in section A-A.

8.25 JOBBERS' REAMERS

Tapered-shank machine or jobbers' reamers usually have irregularly laced straight flutes; their lengths are approximately the same as hand reamers. Some applications require right- or left-hand spiral styles, but these are considered to be special and manufacturers do not ordinarily stock them.

8.26 SHELL REAMERS

Shell reamers, sometimes referred to as "hollow" reamers, are used on arbors with driving lugs. The holes in shell reamers are ground with a taper to assist in firmly seating the tool on the arbor. Usually, sufficient space is left between the end of the reamer and the shoulder on the arbor so the reamer can be removed without damaging it. Shell reamers are used in turret machines for line reaming of bearings and other similar applications.

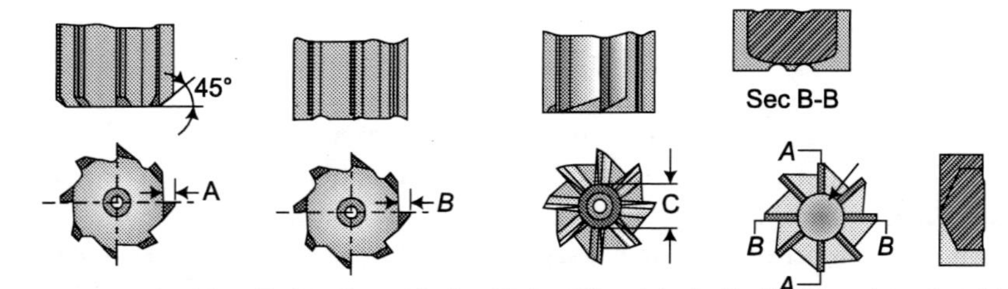

Regular standard A reamer | End cutting style B is same as A | End cutting style C specify dia | End cutting style cut to centre | Sec AA

Fig. 8.84

They are used where the tool is held stationary as the work rotates, where the work is stationary and the tool rotates, and where both tool and work are in motion. Because shell reamers are two-piece assemblies, they lack the rigidity of solid reamers; ordinarily they may not be used for the removal of excessive amounts of stock.

8.27 ADJUSTABLE REAMERS

An adjustable reamer may be used to obtain holes of varying diameters over a limited range. The adjustment should always be greater than the nominal size of the reamer. These reamers lack the rigidity of the conventional solid types and must be used accordingly; removal of excessive amounts of stock with them results in poorly finished holes (**Fig. 8.85**).

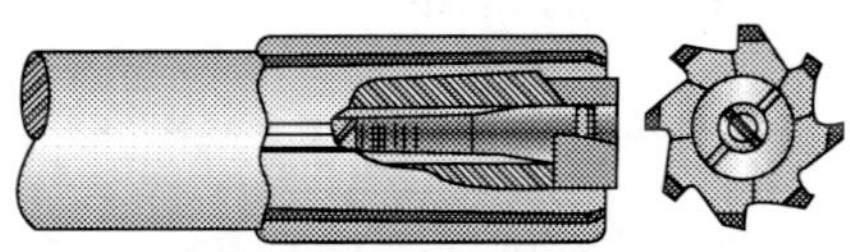

Fig. 8.85 Adjustable reamer

8.28 COMBINED DRILLS AND REAMERS

Drilling and reaming are usually considered to be separate operations, but in certain applications they may be combined and performed with one single tool with accompanying savings in time and labour. The combined drill and reamer is particularly adapted to drilling and reaming through holes. In order for the reamer to perform properly, the drill must pass completely through the work before the reamer flutes engage the hole. The length of such holes should not exceed the diameter of the drill. Blind holes cannot be satisfactorily processed with this type of tool for two reasons: 1) the reamer should feed faster than the drill feed will permit, and 2) the possible eccentric action of the drill would destroy the accuracy of the reamed hole. The drill portion is made from 1 % to 2 times as long as its diameter. The guide bushing should be of proper size to fit the reamer portion and long enough to guide the tool properly when the drill begins to cut. These tools are available in a number of combinations of hand and flute types of both the drill and reamer sections (**Fig. 8.82**).

The most common combined drill and reamer is made with the margins of the flutes of the drill section and reamer section coinciding, making it necessary to re-grind the tool completely or discard it when the original length of the drill section has been ground away.

The subland combined drill and reamer has a constant-diameter drill margin for the full length of the flute. The reamer flutes, larger in diameter than the drill are positioned behind the drill margin **(Fig. 8.82)**. When the reamer flutes are re-ground, the drill margins are not touched. This tool is more expensive to make than the more common combined drill and reamer, but it has much longer life and, in the larger sizes, is more suitable for production work.

Reamers for tapered holes The taper reamer, besides being a finishing tool, is a tool for heavy stock removal. At the end of the operation, the cutting edges are enaged in the cut through most of their length. Taper reamers must therefore be constructed as sturdily as possible, cutting edges must be adequately backed up, and flutes must be sufficiently large and wave well-rounded bottoms. For machine reaming it has been found that a reamer with few flutes and with cutting edges forming a steep left-hand spiral angle with the axis is the most practical. The number of flutes may vary from two to five in the ordinary range of reamer sizes. The left-hand spiral should be at an angle of approximately 45°, A spiral angle of greater magnitude than this does not tend to improve the quality of the cut but materially creases the end pressure required to feed the reamer into the work.

Taper reamers with steep spiral angles cannot be used successfully for hand operations because of the end pressure required; but the smoothness of operation in machine reaming outweighs the disadvantages of the necessary end pressure; the quality of the work is better and longer reamer life is obtained.

For hand operation, taper reamers with straight flutes or with not more than 10° of left hand spiral angle have proved to be best under most circumstances.

It is important that a taper reamer has sharp edges if it is to cut with the least resistance, even at a slight sacrifice of the taper accuracy. However, a 0.002- to 0.010-inch circular land or margin has proved beneficial for accuracy and finish.

8.29 TAPER-PIN REAMERS

Reaming holes for taper pins present special problems because the holes usually are of small diameters and comparatively deep. For hand reaming of taper-pin holes, the straight-flute or slow-spiral-flute type of the reamer has found wide acceptance.

For machine reaming of taper-pin holes, the helical type of reamer has proved to be more successful than other types; it is more substantial and provides an advantageous cutting action. Such reamers are made with left-hand spiral or helix of approximately 45° with the axis.

The spiral or helix is of the opposite hand to the direction of rotation for cutting, and as a result the rotary pressure (torque) is less, although more end pressure is required. The cutting action somewhat resembles that of a conventional reamer combined with that of a broach. The steep spiral of the helical-type reamer renders it unsuitable for hand operations because of the required high end pressures.

8.30 STRUCTURAL REAMERS

Structural reamers are frequently referred to by the names of their particular applications, such as bridge, car, or locomotive reamers.

The holes for rivets and bolts in structural steel members, such as I beams, angles, channels, and plates, usually are punched or drilled before the material leaves the shop and a certain amount of incorrect alignment usually is encountered when the holes are matched up to insert rivets and bolts as the material is being erected.

The structural or bridge reamer has a sharp taper at the front or entering end to enable it to enter the holes that are not fully aligned. The tapered portion functions to enlarge the holes until the entire body of the reamer can enter and enlarge them to the required size. Although some of this work is accomplished by drill presses in the shop, it more often is done with electric or pneumatic drilling machines at the erection site. This is a severe service as the reamer must perform the combined functions of a reamer and crowbar. Several types of structural reamers of straight and fluted styles (**Fig. 8.82**) are in use, each suited to a particular kind of work.

8.31 REAMER FEEDS AND SPEEDS

Reamers are operated at slower speeds and higher feed rates than drills of the corresponding diameter. Generally speaking, the speed of reamers may be 65 to 70 per cent of those for drilling the same material, and the feeds are often two to three times those for drilling. Table below gives conservative speeds for various materials when using high-speed-steel reamers. Solid and carbide-tipped reamers may be operated at speeds up to 150 per cent that of high-speed steel reamers, but their real efficiency is achieved through their resistance to abrasive wear.

Material	*Speed, sfpm*	*Feed, ipr*	*Material*	*Speed, sfpm*	*Feed, ipr*
Aluminium	150	A	Steel, 130 Bhn.	96	A
Bakelite	48	A	Steel, 150 Bhn.	81	B
Brass, leaded	180	A	Steel. 170 Bhn.	66	B
Brass, red or yellow	150	A	Steel, 200 Bhn.	57	B
Bronze, cast	150	B	Steel. 230 Bhn	48	C
Bronze, soft	180	A	Steel. 260 Bhn.	39	C
Copper	42	C	Steel, 300 Bhn 1	30	C
Duralumin	120	A	Steel, 330 bhn.	21	C
Everdure	36	B	Steel. 360 Bhn.	15	D
Glass	9	D	Steel, 400 Bhn.	9	D
Cast iron, chilled	9	D	Steel, cast	24	C
Cast iron, hard	48	C	Steel. forged alloy	30	C
cast iron, medium	66	A	Steel. Forged carbon.	36	C
Cast iron pear lite	60	A	Steel. Low-carbon.	72	B
Cast iron, soft	90	A	Steel, medium-carbon.	66	B or C
Malleable iron	66	B	Steel. High-carbon.	45	D
Monel	30	C	Steel, stainless	12	C
Nickel	42	C	Steel tool	36	D
Plastic	60	C	Titanium!	40	A
Rubber, hard	60	B	Wood. hard	300	A
			Zinc alloy.	150	A

Reamer feeds depend upon the type of material being reamed and the size or strength of the reamer. When reaming soft materials such as aluminium or brass the feeds may vary from 0.025 mm in per flute per revolution for small reamers to 0.123 in per flute per revolution for large reamers. For hard materials such as hard cast iron, stainless steel manganese steel and steel with a 330 Bhn and up, the feed per flute per revolution ranges from 0.010 to 0.038 mm to 0.015 in.

The cutting speed should be adjusted to eliminate any chatter in the reamer. The speeds and feeds of reamers will vary sometimes slightly and sometimes a great deal although reaming the same material. These variations may be due to the finish required, differences in the machine setup, and the design of the part.

Speeds that are too slow adversely affect the productivity of reamers without any particular increase in tool life; on the other hand, speeds that are too high may cause the material to cling to the edges and lands of the reamer, resulting in premature dulling of the cutting edges and rough holes. In reaming hard materials, a reamer may be quickly ruined if run at too high a speed.

Reamer feeds depend mostly on the type of reamer, the amount of material to be removed, and the finish desired. The more stock to be removed, the lower the feeds should be. Too coarse feeds tend to produce spiral marks or wavy finish; too fine feeds let the reamer idle in the cut and cause excessive wear.

Reamer, mm	*A*	*B*	*C*	*D*
3	0.152	0.127	0.101	0.07
12	0.304	0.254	0.18	0.127
25	0.51	0.41	0.305	0.203
50	0.813	0.66	0.51	0.33
60	1.1	0.9	0.72	0.45
75	1.42	0.14	0.9	0.58

8.32 STOCK-REMOVAL ALLOWANCE

Reaming is a sizing or scraping operation in which the tool cuts slightly larger than its own diameter, usually in direct proportion to the amount of stock to be removed. For efficient operation, reamers must be cutting at all times, which is possible only when they are being used in properly drilled holes. For semi-finish reaming in most metals, 0.38 mm should be allowed on holes less than 12 mm in diameter. For holes 12 mm inch in diameter and larger, a minimum of 0.76 mm stock allowance is recommended. For finish reaming, the allowances are 12 mm inch for holes less than 12 mm in diameter and 0.38 mm and up for larger holes.

Removal of too much stock by reaming often causes oversize and rough holes. Oversize holes, walls of holes roughened with grooves at or beyond the finished diameter, bell mouthed holes, or out-of-round holes are common causes of reamer failure. In improperly prepared holes the reamer has a tendency to wedge in the hole rather than machine it. The result is severe reamer wear and possible breakage.

8.33 REAMER GRINDING

Reamers are often required to produce holes that are in parallel alignment, at exact distances from location points on the work, and free from bell mouth at either end. The jigs and fixtures needed for such accurate operations must hold the work securely and incorporate bushings to guide the reamers properly. For long holes it is preferable to guide the reamer at both ends; for short holes usually the reamer is guided at one end, the entering side of the hole **(Fig. 8.86)**.

When the misalignment between the reamer and the hole to be machined amounts to about 30 per cent or more of the stock to be removed, there is often wear on the reamer in the form of a thread. The lead is equal to the feed, and depth is equal to the misalignment. The thread appears to be more readily produced when reaming tough and abrasive materials. When reaming other materials, only excessive marginal wear may be noticed.

8.34 CHATTER

It is usually possible to eliminate chatter by reducing the reamer speed increasing the feed, or reducing the lead-angle clearance. In some cases it has been found helpful to use a double clearance angle with minimum clearance on the first angle directly behind the cutting edge.

Reamer chatter is frequently caused by using a straight-fluted reamer when a helical (spiral) fluted reamer should be employed. Straight-fluted reamers should have unevenly spaced cutting edges to prevent any tendency to synchronize the slipping around from the mark of one tooth to that of the next. Chatter is sometimes caused by cutting edges biting into the work because of too much clearance on the chamfer or the reamer periphery.

8.35 COOLANTS

Cast iron is reamed dry, frequently with an air blast far cooling and blowing away the chips, but nearly all other materials require a suitable coolant. Water-soluble oil is satisfactory for most steels, malleable iron, brass, bronze, and aluminium. For fine finishes and close size control, a sulphurized and chlorinated oil-base cutting fluid is preferable. For heat treated steels, the harder stainless steels, and the more difficult to machine of the nonferrous alloys, chlorinated and sulphurized cutting oil may be required. These are oil-based cutting fluids and are consequently more expensive than the water-based soluble oils, but their usage is sometimes necessitated by the characteristics of the metal being reamed.

8.36 REAMER DRIVING AND HOLDING METHODS

The most usual means for holding and driving reamers is the three-jaw chuck. Another is the straight-sleeve and setscrew method. Taper shanks are sometimes flattened on one side and used in what is known as a "use-them-up" sleeve or socket. Reamers with adapters for quick-change chucks are used on production setups.

When reamers must guide themselves into previously made holes, they require floating holders to maintain alignment and prevent tapered, out-of-round and bell mouth holes. There are several types of floating holders, some of which permit an angular float; others permit a parallel float; still others permit both features.

8.37 REAMER SELECTION AND USE

The physical characteristics of the workpiece must be considered when a reamer is chosen for a particular application. Reamers for nonferrous metals, such as brass and aluminium, are cleared to a sharp edge (apex of relief angle coincides with cutting edge); reamers for steel are cleared to within 0.008 to 0.015 inch of the cutting edge (clearance-angle apex is back of cutting edge), leaving a cylindrical surface that aids materially in prolonging the life of the reamer.

8.38 REAMERS FOR MAGNESIUM ALLOYS, DIE-CAST METALS, ALUMINIUM, AND DOW METAL

For these metals, reamers with left-hand helical (spiral) flutes with no cylindrical margins and radial relief of .5 to 6° have proved successful. The grooves should be highly polished and the cutting edges sharp. A rake angle of 5° promotes shear cutting, which is desirable for aluminium.

8.39 REAMERS FOR MONEL, METAL, NICKEL-CHROMIUM ALLOYS, AND STAINLESS STEELS

For these materials, reamers with right-hand helical (spiral) flutes and radial cutting face and narrow margins have proved most successful.

For all applications it is essential that reamers are kept sharp.

8.40 PRODUCT DESIGN AS RELATED TO REAMING

Better reaming usually results when the product is designed to facilitate that operation. Where possible, provision should be made for the reamer to pass through the workpiece. This avoids the necessity for reaming a blind hole; when this is unavoidable, the depth of the cut should be controlled to prevent bottoming and probably damaging the reamer. It is advisable to avoid operations requiring multi-diameter reamers because the normally different cutting speeds of the different diameters make it difficult to produce true holes and good surfaces. A reamer should enter a hole at right angle to the work surface to permit all teeth to engage for a good start because reaming at an angle makes it difficult and sometimes impossible to turn out good work. Concentricity and alignment require line reaming, which requires that the holes be of equal or progressively smaller diameters to permit entrance and withdrawal of the reamers. Provision must be made for guiding the reamer bar or arbor at both ends.

8.41 CARBIDE-TIPPED REAMERS

Carbide-tipped reamers are basically the same in design as high-speed steel reamers, with the exception that the cutting blades have carbide tips. They may be used on a wide range of materials, including all the common industrial metals and the machineable plastics. The extreme hardness of cemented carbides enables them to withstand wear and abrasion; however, these materials are relatively brittle, and the reamer body, the flutes or blades, and the carbide tips themselves must definitely be shaped and dimensioned to resist deflection. Nearly all carbide-tipped reamers are intended for machine reaming.

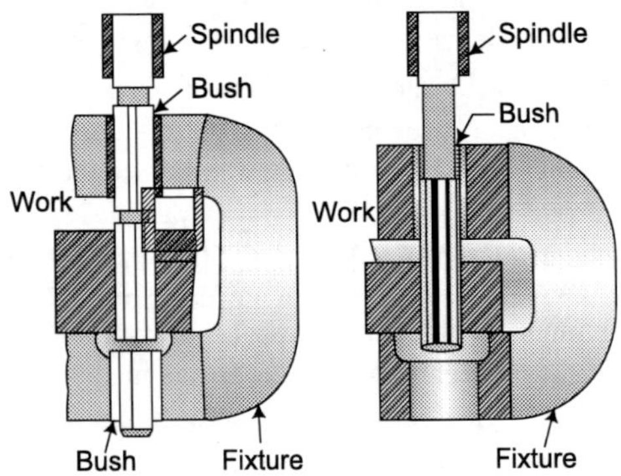

Fig. 8.86 Fixture to grind reamers

The terminology and definitions for reamers adopted and published by the ANS I21 are generally applicable to carbide-tipped reamers.

Certain angles and clearances are embodied in the cutting elements of carbide-tipped reamers to obtain cutting action in the proper places, prevent friction and clear the chips.

Practically all the cutting is done on the chamfer, which is in the form of a truncated cone on the entering end of the reamer. This chamfer is shaped to start the reamer properly in the hole.

A clearance on the chamfer of 6 to 80 on the carbide tip and a secondary clearance of 10 to 150 on the steel body have proved to be satisfactory for most workpiece materials. Under good working conditions and proper alignment of the reamer and the work a back taper of 0.076 mm/2.5 mm of flute length is adequate. Back taper should be increased to prevent binding of only slightly dull tools when reaming very abrasive materials. In general, circular land widths may range from 0.178 mm on 356 mm diameter to 0.51 mm on 300 mm-diameter reamers. Primary relief angles on the carbide tip range from 15° on X-inch-diameter to 7° on 300 mm-diameter reamers (**Fig. 8.87**).

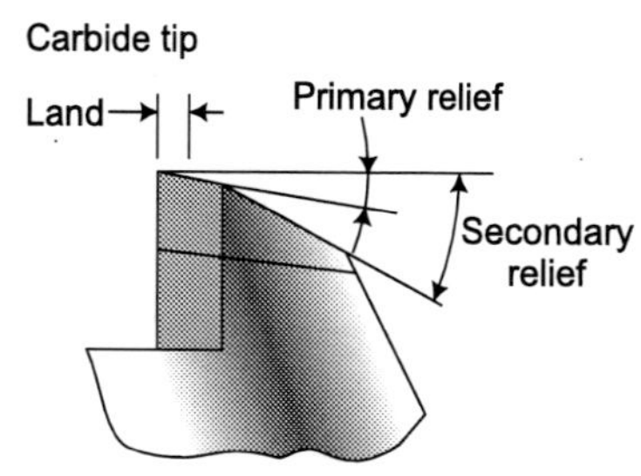

Fig. 8.87

Solid carbide reamers are available in diameters from 1 to 3 mm, and larger on special order. Since sintered carbide is a rigid material, a solid carbide reamer will finish a hole within close tolerances. If a hole is drilled slightly out of round or out of parallel a solid carbide reamer will usually not follow the irregularities but will finish a true hole around its own axis. Rough and finish-reaming operations often can be combined into one through the use of solid carbide reamers.

8.42 NUMBER OF FLUTES

Carbide-tipped roughing reamers may have either an even or an odd number of flutes depending upon the reamer size and the application. Standard and finishing types usually have an even number of flutes. The spacing of the flutes around the reamer body is sometimes uneven to minimize the possibility of chatter.

8.43 KINDS OF CARBIDE-TIPPED REAMERS

Carbide-tipped reamers may be made for either right or left-hand flutes, or with straight flutes, and may incorporate one, two, or more sections of different diameters for cutting two or more diameters in one operation (**Fig. 8.88**).

Carbide-tipped reamers in straight-flute design, with strips of tungsten carbide the full length of the flutes, are ideal for precision reaming to close tolerances. The wear on the diameter from bushings is eliminated and back taper can be reduced to a minimum, as the danger of pick-up on the diameter is reduced.

Solid carbide reamers are more practical in small sizes because of their rigidity, heat resistance, and strength. Tipped reamers in small sizes below 1/4 inch are fragile because of the removal of steel necessary to install the tip. Also, the rapid absorption of heat causes brazing failures in dry operation.

For best efficiency, carbide-tipped reamers should be used only on machines that are in good condition. Loose or misaligned spindles can cause serious damage to the reamer, and the resulting cut will not conform to the desired size or finish.

The magnitude of lead and rake angles, width of land, amount of clearance, and whether straight or spiral flutes should be employed are all related to the kind and characteristics of the material in the workpiece.

Carbide-tipped reamers are generally covered with a plastic coating over the tips for the protection of the cutting edges. This coating may be removed in such a manner that it can be replaced when the tool is not in use. The coolant or lubricant, when used, should be in abundant supply and should wash over the tool at all times to avoid cracking of the carbide from overheating and quenching.

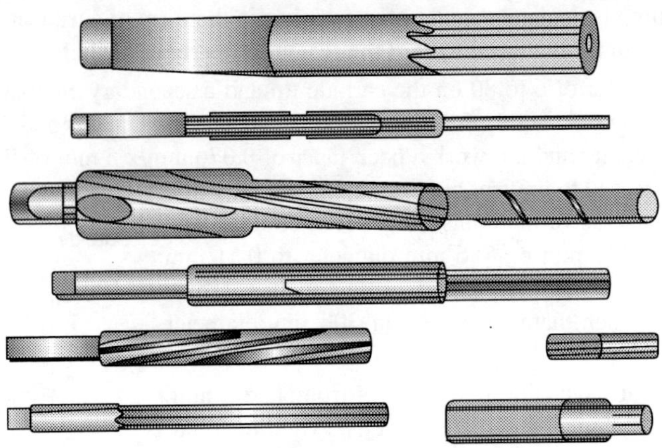

Fig. 8.88 Carbide tipped reamers

8.44 GRINDING SUGGESTIONS

Reamers should be sharpened as soon as they become dull enough to prevent proper cutting. It usually is necessary to sharpen only the cutting edge or the chamfer, but if it is necessary to regrind the diameter, the correct circular-land width must be maintained, as well as the back taper.

Carbide-tipped reamers may be reground on a standard cutter grinder with a 220-grit resinoid-bonded diamond cup wheel. The grinding wheel should be rotated in a direction to cut from the face of the tip towards its back. The use of a lubricant during grinding lengthens grinding-wheel's life and lessens the danger of surface cracking or "crazing" of the carbide from overheating. Do not quench carbide tools in water.

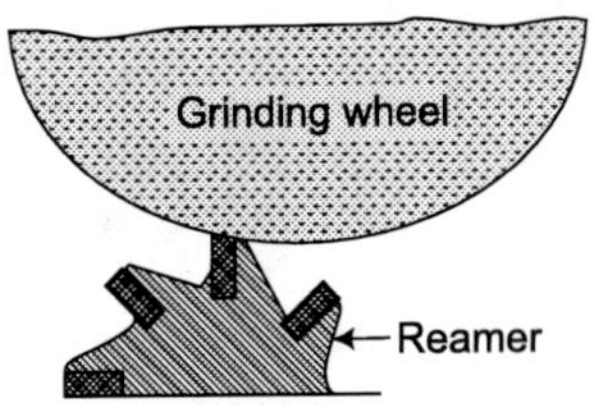

Fig. 8.89

8.45 INSERTED BLADE REAMERS

Inserted-blade reamers have been developed primarily because of economic considerations. Probably no inserted blade tool can be quite so rigid, accurate, and dependable as one constructed from a single piece of metal. However, the advantages inherent in solid construction soon are outweighed by the costs involved as sizes increase, and a point is eventually reached where the construction and use of solid tools are economically prohibitive and built-up construction is required.

Figure 8.90 shows the essential cutting elements of an inserted-blade reamer. The lead A, sometimes referred to as the chamfer, removes excess stock from the walls of the hole, and the cutting point B sizes the hole to the desired diameter.

To prevent the blade from binding or seizing in the hole, clearance C is provided by tapering the blade behind the cutting point.

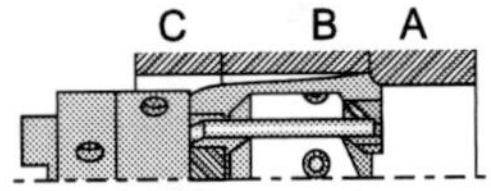

Fig. 8.90 Inserted blade reamer

8.45.1 Types and Construction

Inserted-blade reamers are commercially available in chucking reamers, shell reamers, and hand reamers. Chucking reamers are regularly furnished in sizes from 5/8 to 3 inch diameter, shell reamers from 1¼ to 6 inch diameter and hand reamers from 1 to 3 inch diameter. Arbors may be furnished with either.

Straight or taper shanks for use with shell reamers are available with either straight or taper holes. Straight-hole shell reamers usually are employed for reaming a series of holes in line, for which purpose they are mounted in tandem on a bar. An example is the line reaming of bearings.

An inserted-blade hand reamer ordinarily has a squared shank and a pilot sleeve following the blades. Blades in inserted-blade hand reamers usually are furnished without a spiral.

Chucking reamers and hand reamers are available in both right- and left-hand spiral. As a general rule, with conventional right hand cutting, right hand spiral blades are preferred for roughing and left hand spiral blades for finishing. However, each of these generalities is subject to modification and depends to a great extent on the condition prevailing during use. Right hand spiral has more of a tendency to dig and pull into the work, especially if conditions are not favourable to extreme rigidity, but may cut more freely and provide a better finish if conditions are favourable. Left hand spiral tends to act as a brake on the reamer and often allows work to be satisfactorily completed on setup which would be ordinarily impractical.

The conventional construction of an inserted-blade reamer consists of a body of comparatively soft, tough, and workable steel in which are positioned blades of a hardened material. Inserted-blade reamers are comparatively expensive but work efficiently. The body is usually constructed from alloy steel, heat-treated to the optimum of hardness, toughness, machineability, and the strength necessary to resist both the stresses generated by the cutting action and forces necessary to lock the blades in the body.

Reamer blades are available in carbon tool steel, high-speed steel, cast alloy, and the various cemented carbides, the latter being furnished either solid or brazed onto steel bodies.

Carbon tool steels are nearly obsolete but are sometimes used in specialized cases, usually in hand or line reamers on soft but "greasy" metals, such as some of the bronzes and babbitt. In this case the extreme sharpness to which carbon steel may be ground and honed is the determining factor in its selection.

The majority of reamer blades are made from high-speed steel. The cast alloys are used where a material with a greater production potential than the high-speed steels is required. They are usually classified midway between the high-speed steels and the carbides. The carbides are commonly used in the machining of nonferrous and cast-iron materials.

8.45.2 Methods for Holding Blades

The assembly must be rigidly locked in place, as any displacement of the blade would spoil the work. Of the various locking devices which have been developed, most are some modification of the wedge or screw principle.

8.45.3 Blade Sharpening

Three distinct grinding operations are involved when a reamer is sharpened (1) circular grinding, (2) backing off, and (3) grinding the lead. The reamer should be so mounted on centres that the grinding wheel will revolve against the cutting edge (not with it) to avoid burning and feathering. A cup or disk wheel should be employed to avoid hollow grinding.

To obtain accurately reamed holes within close limits, the reamer should be re-sharpened following any change in blade position (adjustment), as inserted-blade reamers are not considered adjustable within the meaning usually applied to hand tools.

Leads may be reground without changing blade position at more frequent intervals than radial grinding. For average work conditions and also for rough reaming of steel and tough bronze, the reamer should be ground as follows (**Fig. 8.92**).

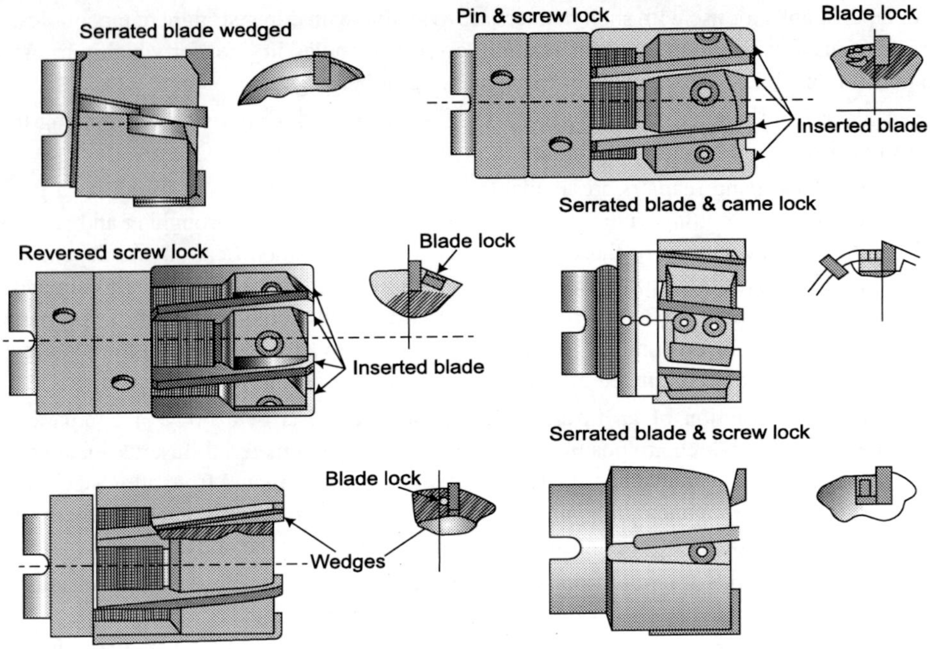

Fig. 8.91

Back-taper diameter at D should be 0.025 mm less than the cutting diameter for each 25 mm of blade length. Diameter across the heel of blade Y should be 0.076 to 0.127 mm less than the cutting diameter. Length of C should be 1.6 mm. Angle of lead A should be 45° and backed off with clearance.

The peripheral cutting edge should be backed off close to the edge at Z, leaving a 0.127 to 0.38 mm land.

When grinding a reamer with two lead angles, the main lead angle A as shown in **(Fig. 8.92)** remains constant at 45°, as in the general-purpose grind but it is necessary to vary the second lead angle B to obtain the best cutting conditions in various metals. For cast iron, bronze, aluminium, and most nonferrous alloys, angle B should be 3°, and its length C should be from 5 to 3 mm.

For free cutting, both lead angles should be backed off. Back taper, diameter across heel, and width of land should be nearly the same as used in the general-purpose grind, with the elements varying as conditions indicate. A maximum of back taper should accompany a relatively wide land to assure sufficient hole clearance. When chips adhere to the land and cause a rough hole to be produced it is an indication that the land is too wide.

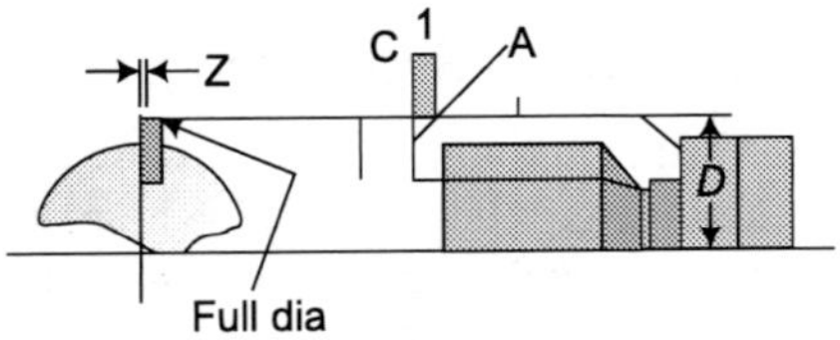

Fig. 8.92

8.46 CUTTING RAKE

A positive rake assists in reaming soft, stringy materials such as steel tubing and pipe, copper, soft aluminium, and the stringier of the nonferrous alloys. One side of the rake angle X **(Fig. 8.93)** should blend into a radius at the body of the reamer to force the chip to curl. The rake angle should increase directly as the softness of the material to be reamed. For steel tubing and pipe, angle X should be about 5°, varying to as much as 10° for copper and aluminium.

When the cutting edge of the reamer blade picks up the material, or when the hole produced is torn and rough, it may be advisable to increase the rake angle. Careful experimentation is torn and rough; it is advisable to increase the rake angle. Careful experimentation is sometimes necessary to develop the rake angle best suited to the particular application.

Rake angle & radius applied

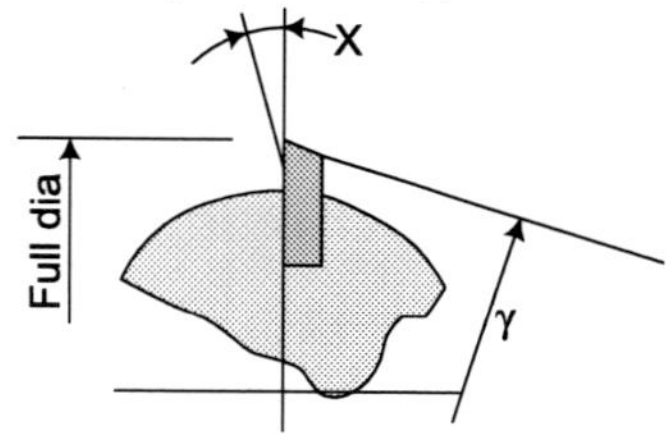

Fig. 8.93

8.47 CARE IN STORAGE

As with all edged cutting tools, inserted-blade reamers should be carefully protected in storage, especially from each other. When stored in metal racks or shelves nonmetallic dividers or separators should be provided, and the sharpened blades should be kept from direct contact with the shelves or each other. The use of thermoplastic materials, which melt at a comparatively low temperature and solidify to form a protective mass around a tool dipped into them, is one of the most satisfactory solutions to the problem of storing cutting tools.

8.48 BLOCK REAMERS (INSERTED-BLADE)

Adjustable block reamers are flat blocks that carry two or more cutting blades and are fitted to slots in suitable bars. The blades may be made of high-speed steel, cast alloy, or cemented carbide **(Fig. 8.94).**

Block-type equipment may be used generally in the field covered by ordinary reamers and in addition is frequently used to rough-bore, semi-finish bore, and even face when conditions are favourable.

For general-purpose use, taper-shank bars for drill presses and straight-shank bars for turret lathes are commercially available. A series of bars may be used in the stations of a turret lathe as needed to perform the required operations on a workpiece. For example, a hole in a cored casting may be rough-bored, semi-finish-bored, and reamed, using three bars in three stations of the lathe turret. Where not enough stations are available, it is possible to change blocks in one or more of the bars between cuts.

Typical block reamer

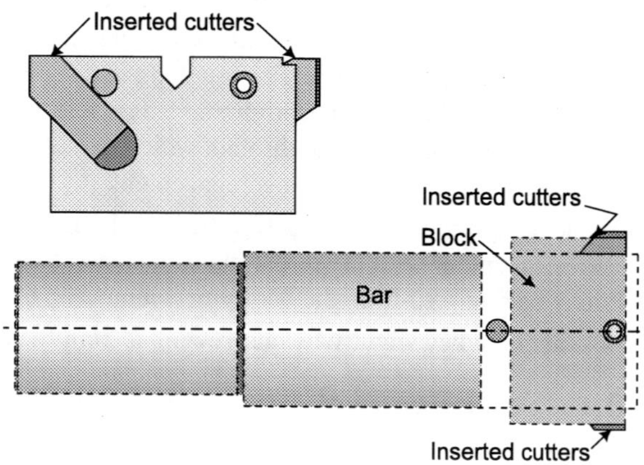

Fig. 8.94

Block boring and reaming are particularly adaptable to special bar constructions for use on horizontal boring mills.

The inserted-blade block construction adapts itself to all sorts of multiple-diameter in-line boring, counter-boring, reaming, and facing operations.

Large complicated castings with numerous bores may be handled efficiently with this type of equipment.

8.48.1 Sizes of Blocks

Inserted-blade boring and reaming blocks are commercially available for machining holes from 1 in up to and including 610 mm in diameter. Larger tools may be constructed on special order.

The blocks range in thickness from 8 to 32 mm. Each range of blocks considerably overlaps the sizes of those immediately preceding and following it. The thinner blocks are intended for the smaller bore sizes and are mounted on small-diameter bars; larger bars are employed for the thicker blocks.

8.48.2 Block-reamer Construction

The block bodies are usually made of alloy steel, heat-treated for strength and wearing qualities. Blades incorporate substantially dovetail cross sections and are retained by locking devices similar to those used for inserted-blade machine reamers.

Adjustment screws are provided for advancing the blades for resharpening, and to take the thrust when the reamer is cutting. A key or locking member intended to hold the block in position may also serve to centralize the cutting edges of the blade, and hold them equidistant from the bar axis so that both blades will cut equally and produce an accurate bore diameter.

A thrust member behind the block, normally retained by a ball and spring dented may be pushed out of the bar by finger pressure when the lock key is loosened slightly. This allows sufficient block movement to clear the lock key for removal from the bar slot. The V-shaped wedge portion of the key enables accurate centralizing when one side of the V is precision-ground until a master block indicates dead central when locked in the bar and is dial-indicated on centers.

Withdrawing or loosening the key slightly will permit the block to "float." This is desirable in most finish-reaming operations with this type of reamer. It assists in causing the block to cut to exact cutting diameter. However, the block should never float when there are ports or other irregular openings in the bore. Both cutting blades of a floated-block reamer must cut simultaneously, a condition that cannot exist when irregular openings are present in the bore. For roughing and semi-finishing cuts, the block should be firmly locked in position. A block is held with centralizing ledges and spring retainers. Another method of holding the block in the bar utilizes a tapered hole through which passes a cone-shaped locking key threaded on the large end. The key is screwed into the locking engagement.

This device provides for contact of the key on one side of the tapered hole in the block, forcing the block into the engagement with the thrust plates fastened to each side of the bar at the back end of the slot. The block may be removed from the bar by first removing the key.

Loosening the key slightly by unscrewing it from the locked position will provide a slight "float."

8.48.3 Stock Removal, Speed, and Feed for Block-type Reamers

The amount of stock left for finish reaming should seldom exceed 0.015 inch on the diameter, but enough stock should be left to ensure a definite cut and avoid rubbing.

More stock may be removed in roughing operations, but it becomes necessary to make certain that the amount of stock being removed is not beyond the reasonable capacity of the block. Because of their built-up construction, block reamers are necessarily more fragile than solid members.

Cutting speeds should be relatively low. When attempting to establish the proper conditions for a specific job, it should be kept in mind that the feed may be more safely increased than the peripheral speed.

For boring and/or reaming with high-speed-steel block tools, a safe starting speed of from 25 to 50 sfpm should be used, keeping the speed low for the harder materials.

A safe starting feed will be 0.008 to 0.012 ipm. If satisfactory size and finish are obtained with no apparent overloading of the tools or machine, increase the feed until it is the maximum obtainable without sacrificing any desirable results.

Larger production per unit of operating time is obtained with cast-alloy and cemented carbide cutting materials because they can be run at higher speeds than high-speed steel. Such increased speeds make it imperative that conditions suited to the hard cutting materials are maintained, and particularly that machine tools be in good alignment and without vibration.

In addition to increased output per machine-hour due to higher speeds, the hard metals make possible longer tool life between the grinds. To gain this advantage fully, greater care and more time must be given to tool sharpening.

The conditions required for the best performance of the cast alloys and cemented carbides vary too greatly to permit general recommendations similar to those given for high-speed steel. Each individual job and the conditions it presents require individual study and planning, and the makers of the particular hard cutting-tool material should be consulted regarding particular applications.

8.49 THE REAMING PROCESS

8.49.1 Preparing

Before reaming the cored hole it has to be determined whether the required diameter and the neatness of the finished face can be obtained. The reamer is selected according to the kind of bore and to the working method (by hand or by mechanical means); then the tool is chucked and the workpiece clamped.

When reaming by hand, the squared shank of the reamer is firmly chucked into the mostly adjustable tap wrench in order to provide for safe and straight guidance of the tool. The workpiece is clamped vertically into the vice.

8.49.2 The Process

Reaming requires sensitive working. The reamer is introduced by hand and in an exactly vertical position into the cored and well burred hole. Reamers that have been introduced in an inclined position and whose cutting edges attack in a tilted angle produce untrue holes and chatter marks involving the risk that the cutting edges break out. By right-hand rotation and with a slight and uniform pressure – simultaneously lubricating with oil or boring liquid – the reamer penetrates into the material.

The reaming process must often be interrupted because of hard points or chips between the material and the back of the cutter of the reamer which will jam the reamer or only allow jerky rotation. The reamer must then be carefully released and taken out by turning it in the cutting direction. In most cases these obstacles are overcome after cleaning the tool and starting the reaming action anew under slight pressure.

Guidance of thin workpieces is improved by mounting a thick plate or by reaming with a machine. In this case the reamer operates at lower speed.

Conical bores for receiving taper pins or machine tapers are reamed by means of a taper pin reamer. For this purpose a smaller hole is at first drilled mostly cylindrically according to the smallest diameter of the taper pin. For such rough drilling of holes for taper pins in thick workpieces, a taper borer can be used. Conical holes for machine tapers are visibly rough drilled in the form of steps.

Taper pin reamers for holes of pins get an amount of conicity of 1:50, but reamer with a conicity of 1:10 is also manufactured. For these taper holes causing a great chip removal of the material, one set of reamers (3 pieces) is used. The roughing reamer with step cutting edges serves for rough working. The conical shape is reamed with the roughing reamer provided with breaking flutes, while the finishing reamer with straight teeth is intended for finish smoothing of the bore hole. Work done with blunt reamers requires more time and does not achieve the required surface quality. We do our utmost to conserve the cutting property of reamers as long as possible. For this purpose they are cleaned after completion of the reaming work with a cloth to remove the remaining chips, thereafter oiled and stored in wooden boxes.

It is impossible to resharpen reamers by hand. The cutting edges can only be sharpened and then honed by means of special tool grinding machine.

8.50 MORE INFORMATION ON REAMING

Reamers are used to finish previously drilled holes accurately to smooth finish/or to enlarge a drilled or bored hole to produce a perfectly round hole with an accurate diameter and a straight and smooth surface. This can be done either by hand or on machine. Reamers used on machines are called machine reamers. The basic difference between a hand reamer and machine reamer is the lead angle at the cutting end **(Fig. 8.94)**. Hand reamers have long taper lead while machine reamers a short bevel lead.

The shank end of the hand reamers is square to facilitate reaming using tap wrenches. Machine reamers of small diameters have parallel shank and the larger reamers are provided with taper shanks.

8.51 TYPES OF MACHINE REAMERS

8.51.1 Machine jig Reamers (Fig. 8.95)

When reaming is done using jigs, the bushings of the jigs can be used to guide the reamer. Machine jig reamers are provided with special, long guide surfaces of standard diameters according to the diameter of the reamer. Reamers of this type are available with rear guides only or with front and rear guides **(Fig. 8.96a).** They can produce accurate holes if the spindle and bushes are aligned accurately. While reaming deep holes it is better to select the jig reamers with guides on both ends.

When reaming is carried out immediately after drilling in the same setting, renewable bushes are used on the jig **(Fig. 8.96b)**.

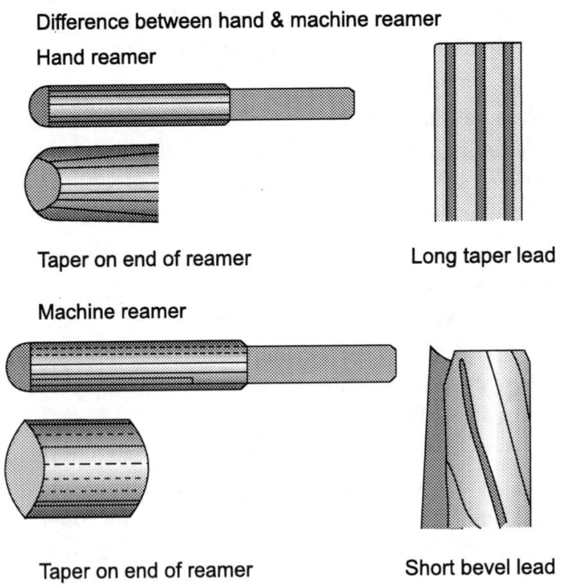

Fig. 8.95

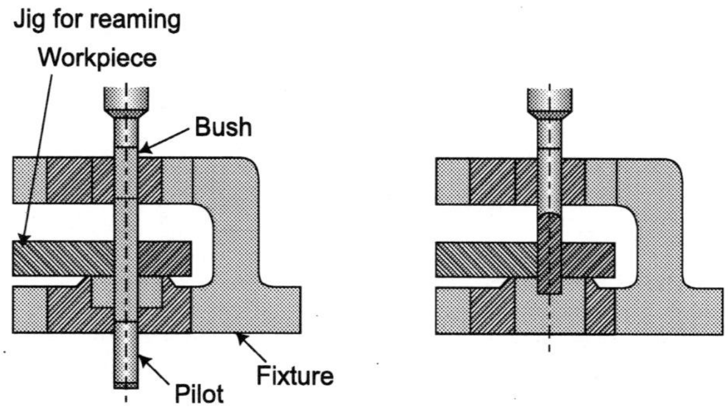

Fig. 8.96a

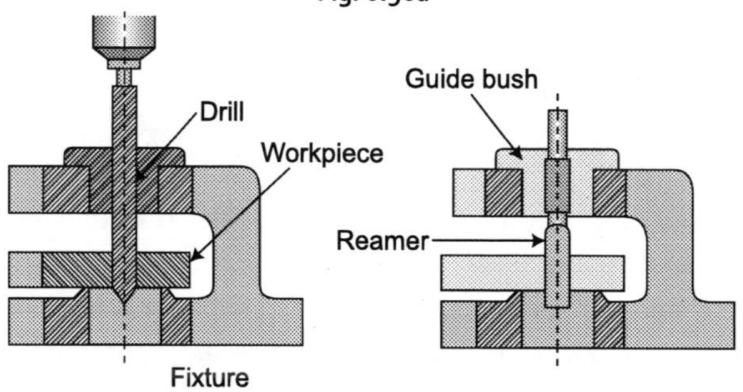

Fig. 8.96b

8.51.2 Solid Fluted Machine Reamer (Jobber Reamer) (Fig. 8.97)

These are identical to a hand reamer. These reamers are either straight fluted or with left hand helix to prevent the tendency of 'cork screwing' when rotated clockwise for reaming.

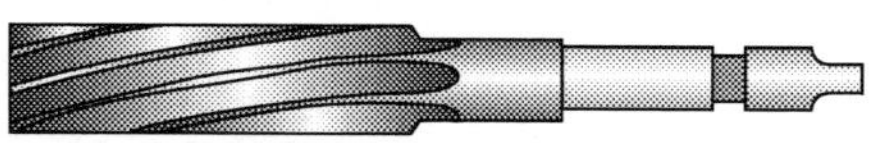

Fig. 8.97 Jobber reamer

8.51.3 Chucking Reamers (Fig. 8.98)

These reamers are similar to jobber's reamers but have shorter and deeper flutes, and are available as straight or helical fluted. This is a side cutting reamer and cuts along the full length of the land and produces smooth and accurately sized holes. The ends of these reamers are slightly chamfered to initiate the cutting action.

8.51.4 Rose Reamer/rose Chucking Reamer (Fig. 8.98 & 8.99)

This reamer is designed to cut on its end. The flutes help in chip clearance and act as guides while cutting. This is used when a considerable amount of metal is to be removed, and the finish is not very critical. This is sometimes used as a roughing tool. Final finishing is done with other finishing reamers.

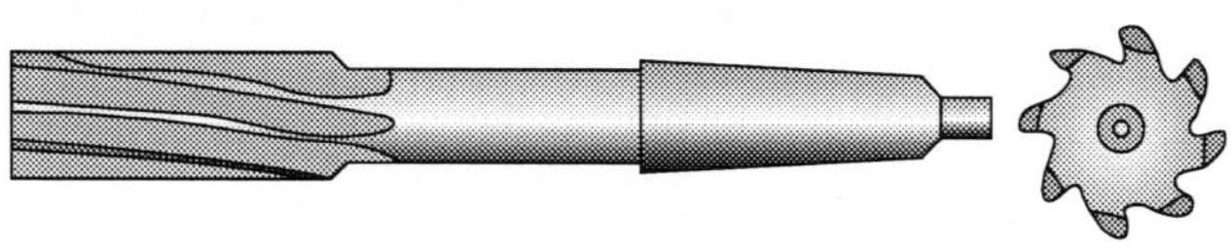

Fig. 8.98 Chucking reamer

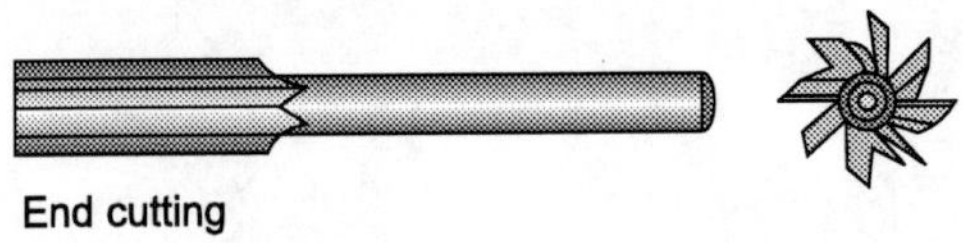

Fig. 8.99 Rose reamer

8.51.5 Adjustable Machine Reamer (Fig. 8.100)

These reamers are easy to adjust when worn out. They can be resharpened and adjusted back to the correct size. As such these reamers have longer working life than ordinary reamers. Adjustable reamers have adjustable insert blades. When worn out or damaged, the blades can be easily replaced. These reamers are not meant to produce holes of different sizes. However, they can be used to increase the holes slightly.

8.51.6 Shell Reamer (Fig. 8.101)

The shell reamer is an independent reaming unit which has a slightly tapered hole through the centre that permits the reamer to be held on a separate shank or arbor that has driving lugs. Several sizes of reamers may be used with one shank. Shell reamers are made with either fluted teeth having clearance, or the rose chucking type which cuts on the end only.

Note: Similarly, adjustable reamers are also available for hand reaming (**Fig. 8.102**).

The size of the reamer can be adjusted by moving the blades in the tapered slots using the nuts provided on either end of the blade.

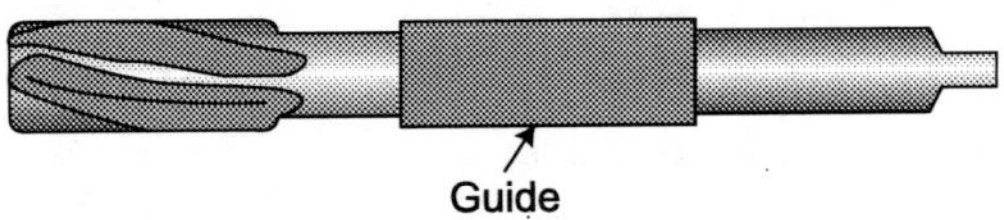

Fig. 8.100 Machine reamer

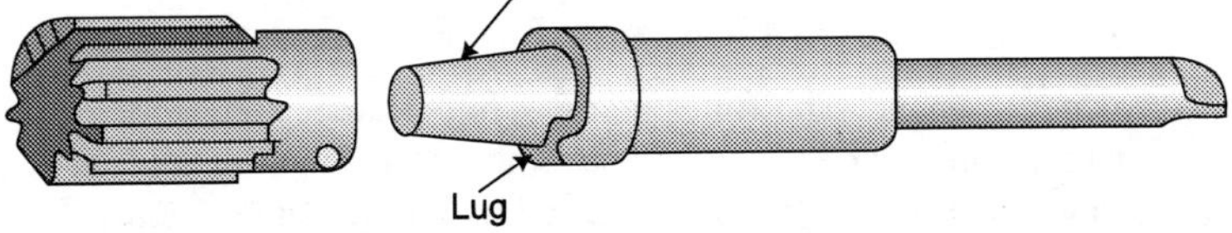

Fig. 8.101 Shell reamer

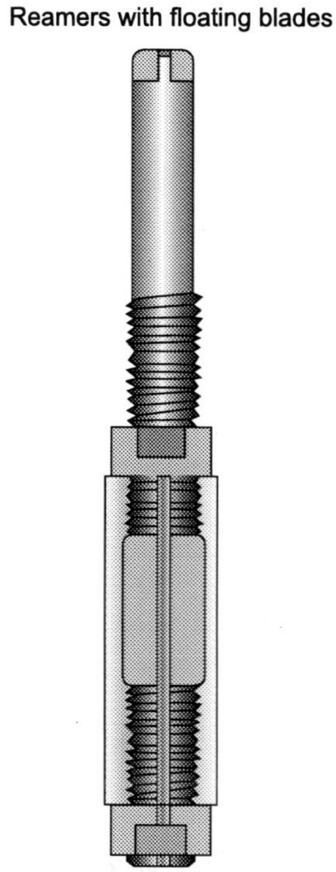

Fig. 8.102

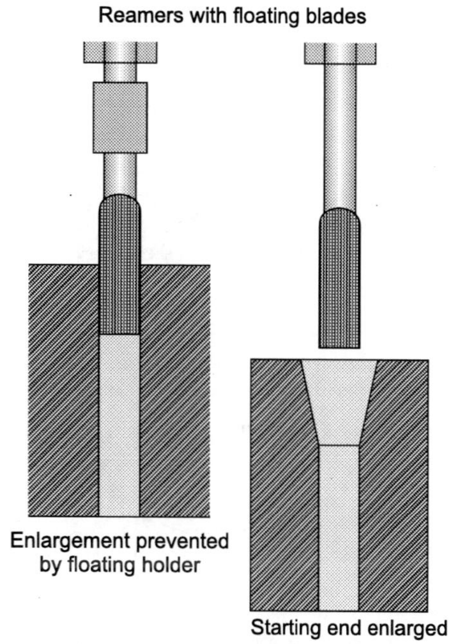

Fig. 8.103

8.51.7 Reamers with Floating Holders (Fig. 8.103)

While machine reaming, the taper lead at the reamer end guides into the hole being reamed. In the event of any misalignment, the hole being reamed can enlarge at the starting end. This can be avoided by the use of reamers with floating holders. Floating holders compensate minor discrepancies in the axis alignment. Floating holders are available with angular floats and parallel floats (**Fig. 8.104**).

8.51.8 Taper Reamers

Taper machine reamers are manufactured in all standard tapers and with tapered shanks. They can be mounted directly in the spindle of the machine.

For taper reaming the diameter on the hole drilled is slightly smaller than the finished diameter of the small end of the taper.

While reaming, the taper reamer will have to remove more material at the big end and less material at the small end (**Fig. 8.105**). While cutting, the entire length of the reamer will be in contact with the workpiece. This can cause chatter marks and poor finish.

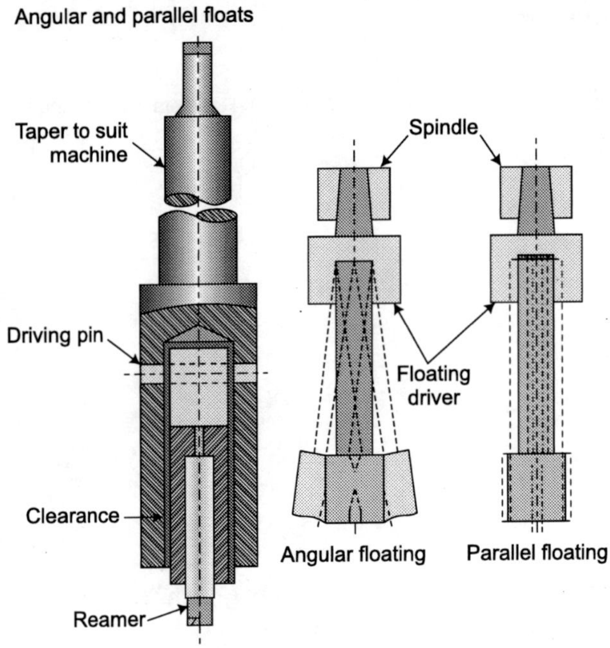

Fig. 8.104

For better results use a roughing reamer first and then finish with a finishing reamer. Step drilling the hole will help to reduce the strain on taper reamers (**Fig. 8.106**).

These reamers are used for reaming taper holes needed for fitting taper pins.

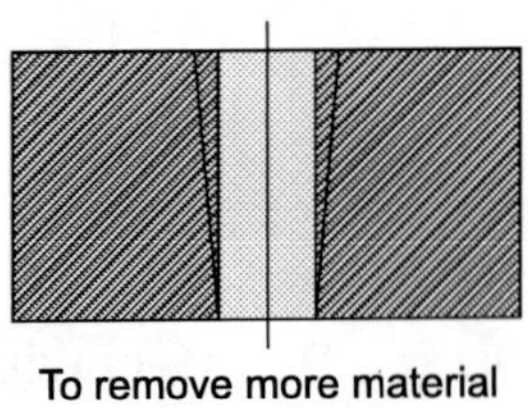

Fig. 8.105

8.52 REAMING ALLOWANCES

The method by which a hole has been produced in the workpiece determines the amount of material left for reaming. Rough drilled or punched holes should be rough-reamed before a finishing reamer is used. The amount of material left in a hole for reaming purposes depends on factors such as the condition of the machine, the type of material, the finish desired, the speed and feed of the reamer, and the use of cutting fluid.

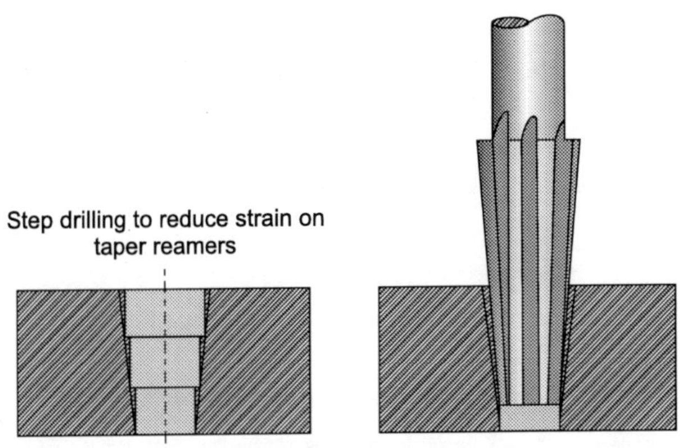

Fig. 8.106

8.53 REAMING SPEEDS AND FEEDS

Reaming speed is affected by the following factors.

The reaming speed is about one-third the drilling speeds, for similar material. High reaming speeds are used on rigid set-ups. Less rigid set-ups require slower speeds. Holes requiring a fine finish, and close tolerance require slower speeds. The use of cutting fluids helps to produce better surface finishes.

The chart below shows the reaming speed for high-speed steel reamers. Carbide-tipped reamers can be used at twice the cutting speed. If H.S.S. reamers produce chatter while cutting at any of these speeds, reduce the speed until the chatter is eliminated.

Recommended reaming speeds for H.S.S reamers

Material	*Speed m/min*
Aluminium	25-40
Brass	25-35
Bronze	10-20
Cast iron	10-15
Mild steel	10-15
Stainless steel	10-15
Magnesium	35-50

Feeds: The feed used for reaming is generally about twice that for drilling. If the feed is too slow, it produces a glazed surface, excessive reamer wear, and chatter. A fast feed reduces the accuracy of holes and produces a poor surface finish. In the case of taper reaming, feed the reamer slowly, since the reamer is cutting along the entire length.

8.54 DRILL SIZE FOR REAMING

For reaming with a hand or a machine reamer, the hole drilled should be smaller than the reamer size.

The drilled hole should have sufficient metal for finishing with the reamer. Excessive metal will impose a strain on the cutting edge of the reamer and damage it.

8.55 CALCULATING DRILL SIZE FOR REAMER

A method generally practised in workshop is by applying the following formula.

$$\text{Drill size} = \text{Reamed size} - (\text{Undersize} + \text{Oversize})$$

Finished size: Finished size is the diameter of the reamer.

Undersize: Undersize is the recommended reduction in size for different ranges of drill diameter (see Table below).

Undersize for reaming:

Diameter of ready reamed hole mm	*Undersize of rough bored hole mm*
Under 5	0.1 to 0.2
5 to 20	0.2 to 0.3
21 to 50	0.3 to 0.5
Over 50	0.5 to 1

Oversize: It is generally considered that a twist drill will make a hole larger than its diameter. The oversize for calculation purposes is taken as 0.05 mm for all diameters of drills.

For light metals the undersize will be chosen 50% larger.

A hole is to be reamed on mild steel with a 10 mm reamer. What will be the diameter of the drill for drilling the hole before reaming?

Drill size = Reamed size - (Undersize + Oversize)

Finished size = 10 mm

Under size as per table = 0.2 mm

Over size = 0.05 mm

Drill size = 10 mm — 0.25 mm = 9.75 mm

Note: If the reamed hole is undersize, it is because the reamer is worn out.

Always inspect the condition of the reamer before commencing reaming.

For obtaining good surface finish:

- Use a coolant while reaming. Remove metal chips from the reamer frequently. Advance the reamer slowly into the work.

8.56 DEFECTS IN REAMING — CAUSES AND REMEDIES

- Reamed hole undersize
- If a worn out reamer is used, it may result in the reamed hole bearing undersize. Do not use such reamers.
- Always inspect the condition of the reamer before using
- Surface finish rough.
- The causes may be any one of the following or a combination thereof.
- Incorrect application.
- Swarf accumulated in reamer flutes inadequate flow of coolant
- Feed rate too fast.
- While reaming, apply a steady and slow feed-rate.
- Ensure a copious supply of the coolant.
- Do not turn the reamer in the reverse direction.

Judgment on reaming

Mistakes	*Causes*
Boring hole is not neat.	The cutting edges of the reamer are blunt or have notches.
Teeth of the reamer have broken out.	Reamer was turned in the wrong sense after jamming.
The bored hole is too large and shows chatter marks.	Reamer was placed in an inclined position and incorrectly guided, jerky operation.

8.57 SCREW THREAD

A screw thread is a ridge, usually of uniform section and produced by forming a helical groove on the external or internal surface of a cylinder, or by forming conical spiral on the external or internal surface of a cone or frustrum of a cone. Screw thread formed on a cylinder is known as a straight or parallel thread, to distinguish it from a taper screw thread, which is formed on a cone or frustrum.

Thread: A thread is a portion of a screw thread encompassed by one linear unit of measure. One single start thread is equal to one turn **(Fig. 8.107).**

An external screw thread is formed on the outer surface of a cylindrical part. **Examples:** bolts, screws, studs, threaded spindles, etc.

An internal thread is formed on the inner surface of a hollow cylindrical part. **Examples:** nuts, threaded, lids, etc.

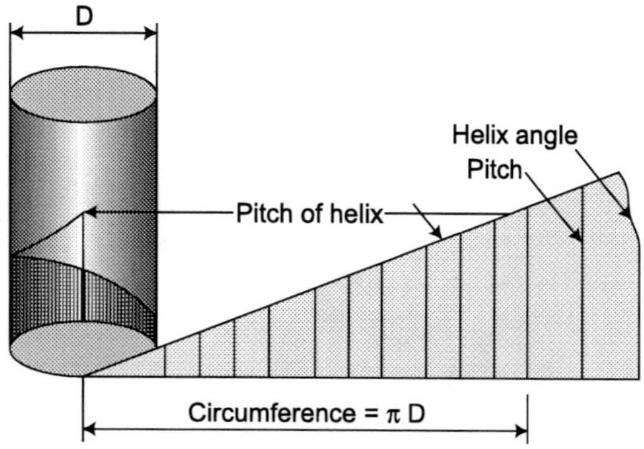

Fig. 8.107

External threads and internal threads are assembled together for different engineering uses.

8.58 USES OF SCREW THREADS

Screw threads are used as fasteners to hold together and dismantle components when needed, to transmit motion on machines from one unit to another, to make accurate measurements, to apply pressure and to make adjustments **(Fig. 8.108 to 8.112).**

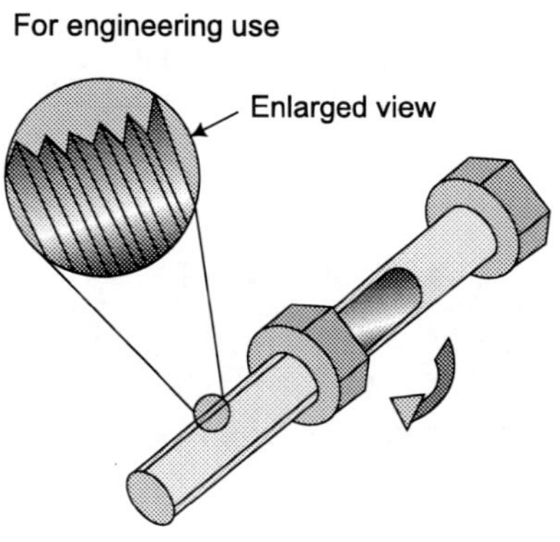

Fig. 8.108

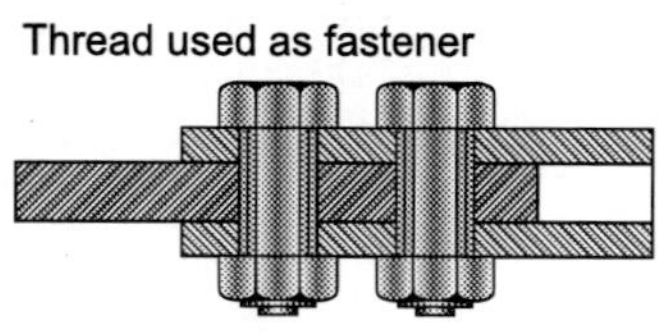

Fig. 8.109

Fig. 8.110

8.59 PARTS OF A SCREW THREAD (FIG. 8.113)

Crest is the top surface joining the two sides of a thread.

Root is the bottom surface joining the two sides of adjacent threads.

Flank is the surface joining the crest and the root.

Thread angle is the included angle between the flanks of adjacent threads.

Depth is the perpendicular distance between the roots and crest of the thread.

Major diameter in the case of external threads it is the diameter of the blank on which the threads are cut and in the case of internal threads it is the largest diameter after the threads are cut that are known as the major diameter. **Fig. 8.114)** This is the diameter by which the sizes of screws are stated.

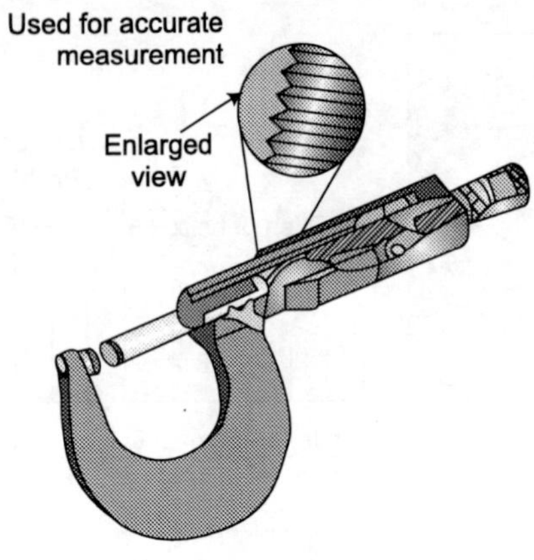

Fig. 8.111

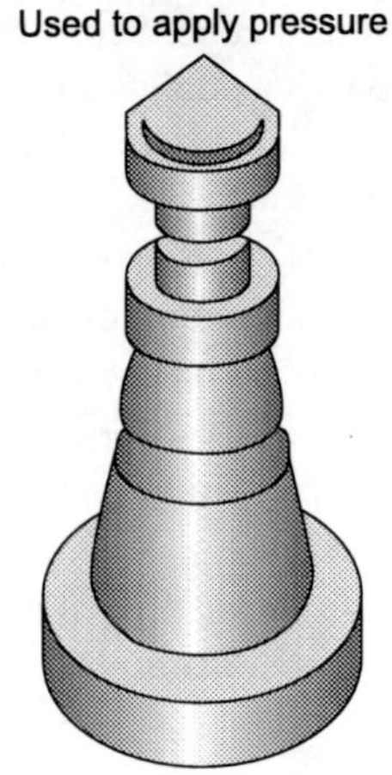

Fig. 8.112

Minor diameter for external threads, the minor diameter is the smallest diameter after cutting the full thread. In the case of internal threads, it is the diameter of the hole drilled for forming the thread which is the minor diameter.

Pitch diameter (effective diameter) is the diameter of the thread at which the thread thickness is equal to one half of the pitch.

Pitch is the distance from a point on one thread to a corresponding point on the adjacent thread measured parallel to the axis.

Lead is the distance a threaded component moves along the matching component during one complete revolution. For a single start thread the lead is equal to the pitch.

Helix Angle is the angle of inclination of the thread to the imaginary perpendicular line.

Hand is the direction in which the thread is turned to advance. A right hand thread is turned clockwise to advance, while a left hand thread is turned anticlockwise **(Fig. 8.115).**

Example: The shaft of a right pedal of a bicycle is screwed in with the left-hand thread to avoid unscrewing by riding the bicycle.

A left-hand thread also avoids troubles caused by mixing up screw parts.

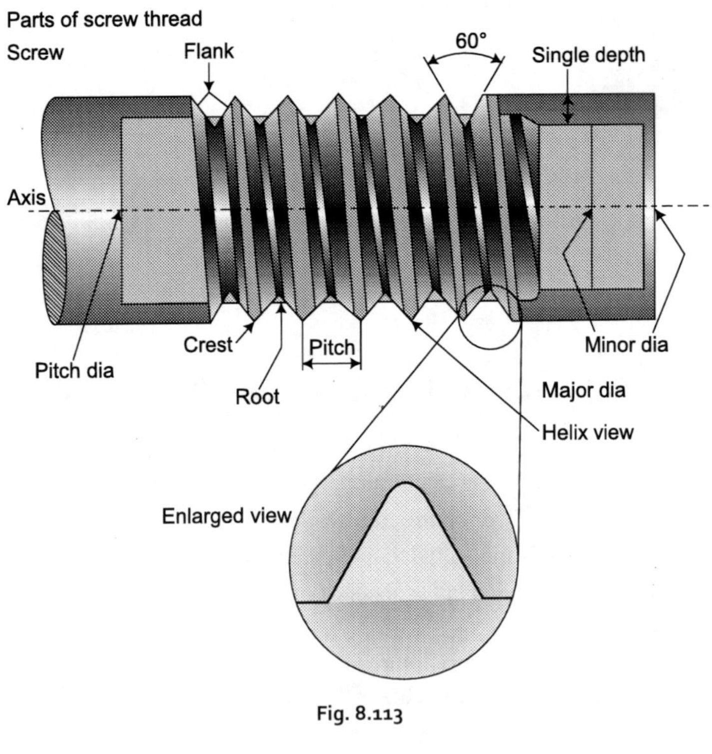

Fig. 8.113

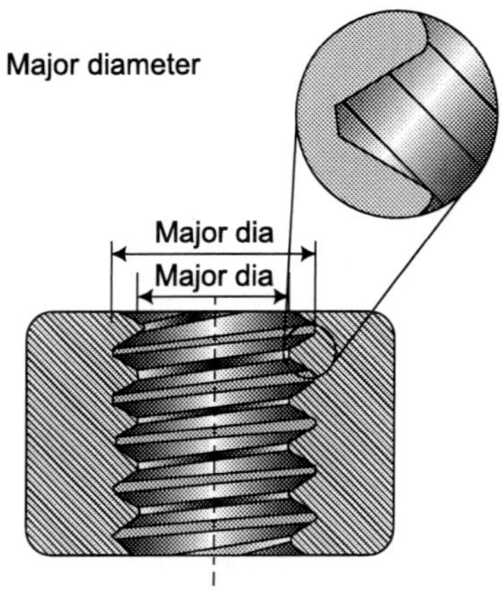

Fig. 8.114

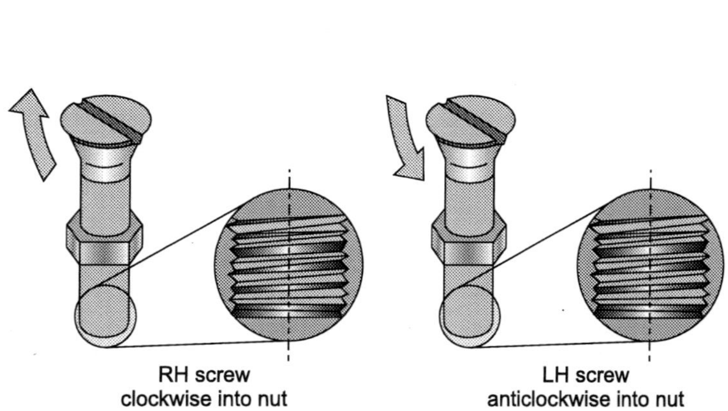

Fig. 8.115

Example: The valve of a hydrogen cylinder has a left-hand thread in order to avoid confusion with the valve of an oxygen cylinder (right-hand thread).

Last but not the least, the clamping nut of a turn-buckle has a left-hand thread on the one side and a right-hand thread on the other side in order to draw together two opposite structural steel parts.

8.60 BASIC FORMS OF SCREW THREADS (FIG. 8.116)

Screw threads of different forms are available for meeting the various requirements. The basic forms of screw threads are:

1. 'V' threads
2. Square threads
3. Trapezoidal threads.

8.60.1 'Vee' Threads (Fig. 8.116 A)

These threads are of 'V' shape. Vee threads of different types are available. Vee thread is the most commonly used form of threads, and is used for domestic and industrial applications like bolts, nuts and spindles for micrometer, etc.

The turns of the thread have a tree angular shape of section. The crest of the thread are truncated or rounded. The root of the thread of the bolt is always rounded to reduce the notch effect.

These can be cut by hand, whereas all other kinds of threads, such as trapezoidal threads, round threads are cut by machines because more material has to be detached.

8.60.2 Types of 'Vee' Threads

'Vee' threads are available in different forms and standards. The types of 'Vee' threads used for general engineering threaded fasteners are:

1. ISO metric thread
2. ISO inch (unified thread)

3. British Standard Whitworth thread
4. British Standard fine thread.

Square threads (Fig. 8.116A): The cross section of these threads is square in shape. These threads are very strong, and are used on fly presses, screw jacks, and vices and spindles etc.

Knuckles thread (Fig. 8.116B): This is a modified form of square thread. In this case the crest and root are rounded to form a semi-circle. An example of the use of this thread is the one used for railway wagon couplings.

Trapezoidal threads (Fig. 8.116C): This is also a modified square thread. The angle between the flanks is 30°. These threads are also strong and are used for easy engagement and disengagement in transmission.

Lead screws of machine tools are made of trapezoidal threads. Acme thread is another type of trapezoidal thread with 29° angle between flanks.

Saw-tooth threads (Fig. 8.116D): This form of thread has an inclination of 3° on one side and 30° on the other side. These threads are used in places where sliding forces act in one direction.

Spindles of carpenter's vices and quick release presses have saw-tooth threads. Buttress thread is another thread which is similar to saw-tooth thread. This thread has one flank vertical, and the other at 45°.

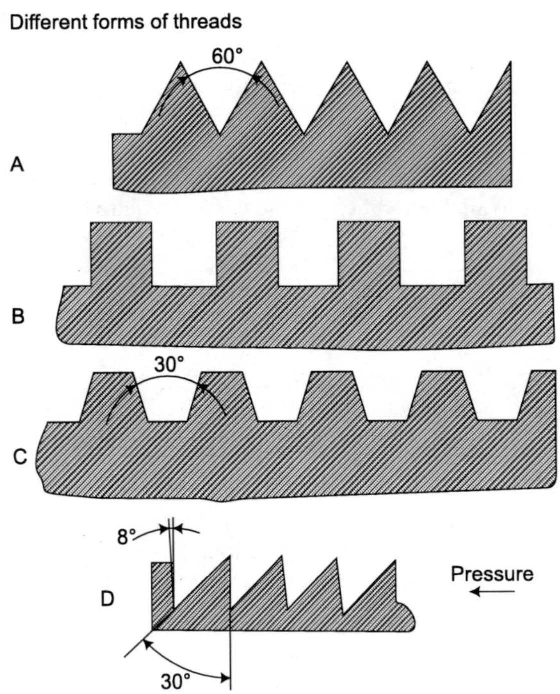

Fig. 8.116

The thread angle is 60°. The root of the external thread is rounded. The crest of the external thread is flat, but sometimes it is rounded depending on the type of the manufacturing process. The root of the internal thread is cleared beyond the width equal to one eighth of the pitch, and is rounded. The crest of the internal threads is left flat.

ISO metric thread (Fig. 8.117): This is the form of thread approved by B.I.S. for threaded fastening. Two series of threads have been identified by them. They are ISO Metric (coarse) and ISO Metric (fine).

Designation of ISO Metric Thread: ISO metric coarse thread is designated as: **M 12 × 1.25:** The symbol M indicates that it is ISO metric thread and 12 is the diameter of the thread. For the coarse series the pitch of the threads is standardized for each diameter.

ISO metric fine threads are designated as: M 12 × 1.25

The addition of 1.25 in this case indicates the pitch of the thread.

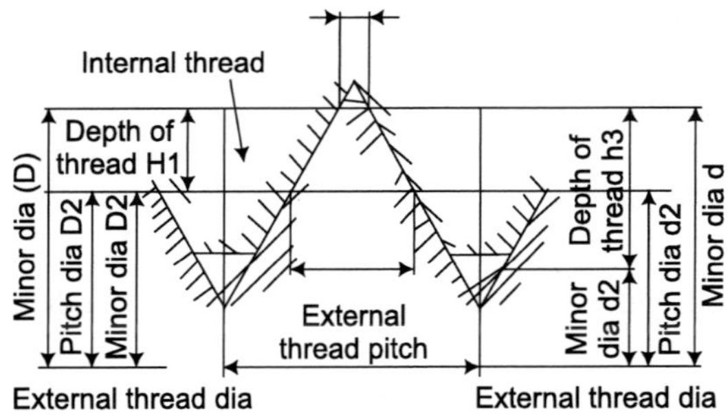

Fig. 8.117 ISO metric thread

ISO Inch (Unified) thread: The ISO inch system (unified) is a recognized standard for interchangeability with the American National Thread.

These threads are used for general purpose engineering threaded fastening and are of two types:

- Unified coarse (UNC)
- Unified fine (UNF)

For unified threads the angle is 60°. The thread profile is similar to that of the ISO metric thread.

Designation of ISO inch (unified) threads:

Examples

(a) 1/4 20 UNC

(b) 1/428 UN

(**a**) indicates that the diameter of the thread is 1/ 4″, it has 20 threads per inch (TPI) and the ISO thread series is UNC (unified course). (b) has 28 TPI and is of UNF series.

British Standard Whitworth (BSW) Thread (Fig. 8.118):

This thread is being replaced by ISO metric thread. However, the application of this thread is still being continued in a limited manner, particularly in the production of spare parts and repair works.

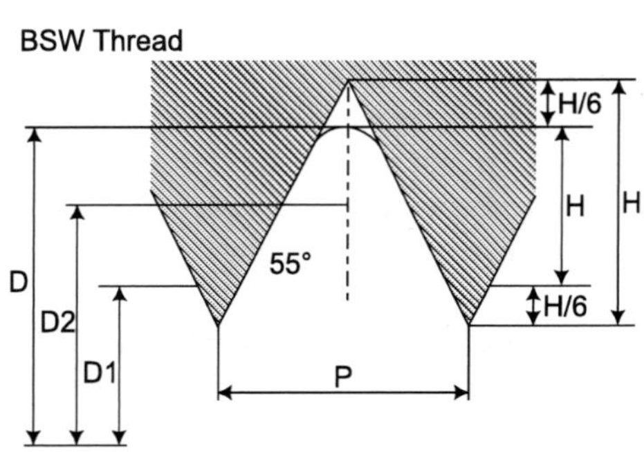

Fig. 8.118

8.61 FUNDAMENTALS OF THREAD CUTTING BY HAND

With thread cutting, grooves of spiral lines are cut on the periphery of cylindrical workpieces. These recessed helical lines together with the remaining material from the winding or the screw thread. It results the screw bolt or the screw spindle with male thread.

If the screw windings are cut into the inner wall of a hollow cylinder, the screw nuts with female thread results.

The observance of standardized sizes is very important for making perfect angular threads. This particularly applies to the threaded parts which are to be exchangeable. In this connection, five different sizes are to be considered: the outer diameter, the core diameter, the flak diameter, the angle of thread and the pitch.

The outer diameter is the nominal diameter which is measured above the crest of thread of a screw bolt.

The core diameter is the inner diameter of a screw bolt.

Outer diameter and core diameter are the secondary sizes of a thread. It is only necessary that their sizes are smaller with screwed bolts than with threaded nuts.

The three main elements of a thread are:

(a) The flank diameter is the intermediate size between the outer and the core diameter. It determines the size of the bolt and the nut.

(b) The angle of thread indicates the oblique position of the thread flanks towards each other. It has great influence on the quality of a thread.

Too small or too large an angle of thread will result in unequal flank contacts and is considered the worst defect of a thread.

(c) The pitch is the vertical distance between the crests of the thread.

The slightest deviation from the prescribed size will result in the fact that bolt and nut can only be screwed into each other with a few turns depending on the gravity of the mistake.

8.62 THE WHITWORTH THREAD AND THE METRIC THREAD

The Whitworth thread and the metric thread was developed by the Englishman Whitworth in 1841; owing to the exports of British machined it became widely known in Germany and other countries

The metric thread resulted from a SI (System Internationale = international system) according to an arrangement made between Germany, France and Switzerland in 1898. For standardization this thread type has been accepted and used.

With the Whitworth as well as with the metric threads, which are also called normal or standard screw threads, the pitch increases proportionally with the diameter. In both the systems there are fine threads. These fine threads are threads with a small pitch. They are used for thin walled workpiece, e.g., for threaded bushings and tubes where it is not possible to cut a deep thread; besides they are used for threaded parts which must not loosen when vibrating.

8.63 KINDS OF THREAD-CUTTING TOOLS

8.63.1 Threading Die (Fig. 8.119)

The threading die is used to cut male threads. The threading die is a round and hardened steel plate with a female thread of tool- or high-speed steel. For the formation of the threading chasers, the turns of thread are interrupted by two to five transversal slots, according to the size of the thread chamfers; the said slots at the same time serve for chip removal and lubrication.

Types of dies

- Circular split die/button die
- Half die
- Adjustable die.

The thread is finish-cut by a single turning up the threading die which is done by hand. This means that the sectional shape of the angular thread is made in one cutting operation and with large chip removal.

The actual chip removal is effected by the taper chamfer, while the other chasers only perform a smoothing action. The chamfer passes over the two first turns of the thread and is produced by spot facing below 60°. In order to achieve a better cutting effect, the chasers situated in the chamfer are back-tapered. The chamfer is arranged on both sides of the threading die, so it does not make any difference what side of the threading die points upwards when chucked in the die stock. Split and non-split dies can be chucked in the stock either direct or with collets. The contour of the threading die is mostly conically spot drilled on four points for chucking in the stock or in the collet to ensure engagement of the locking screws (**Figs. 8.119 and 8.120**).

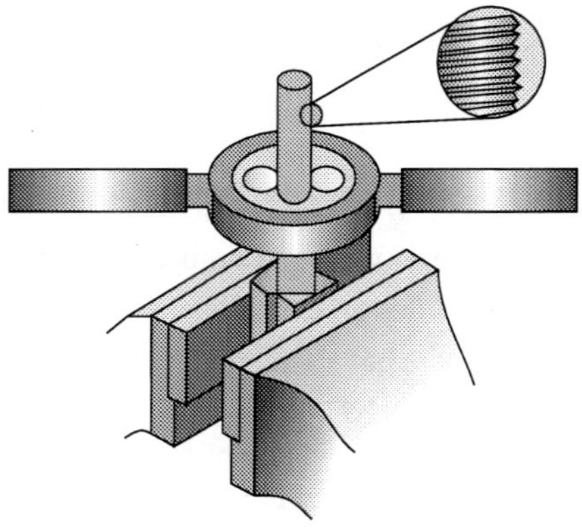

Fig. 8.119 Threading die external

The threading die is pre-split by a notch situated between two taper holes.

Non-split and new threading dies cut bolts of soft material are mostly too short. After grinding off the ridge remaining on the notch, the now split threading die can be opened a little by tightening the expanding screw.

The thrust and locking screws serve for fastening. While in this case the thrust and locking screws on the left and on the right of the expanding screw are fastened at first, all four screws of non-split threading dies are tightened for chucking in a uniform manner.

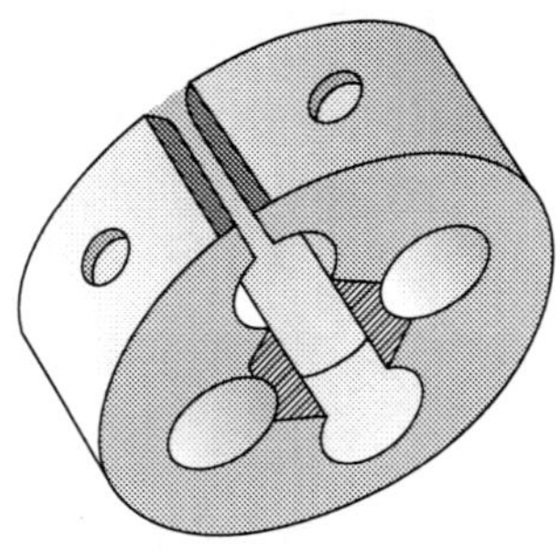

Fig. 8.120 Split die

It is not possible to cut all sizes of threads on all kinds of material perfectly. To make a thread in one single cutting action requires important force due to the considerable chip removal, especially when cutting steel; the thread becomes rough.

8.63.2 The Stock and Die (Fig. 8.121)

It is used for cutting large bolt and pipe threads. The stock and die are opened at the beginning of the cutting action in a way that only the tips of the cutting dies cut slightly. By repeated turning up and adjusting of the cutting dies, the thread will be cut true to dimensions. The stock and die can be lifted after each cutting action. Pipe threads are cut by means of the pipe stock and die. Pipe threads of the same pitch but with different diameters can be cut with the same set of cutting dies.

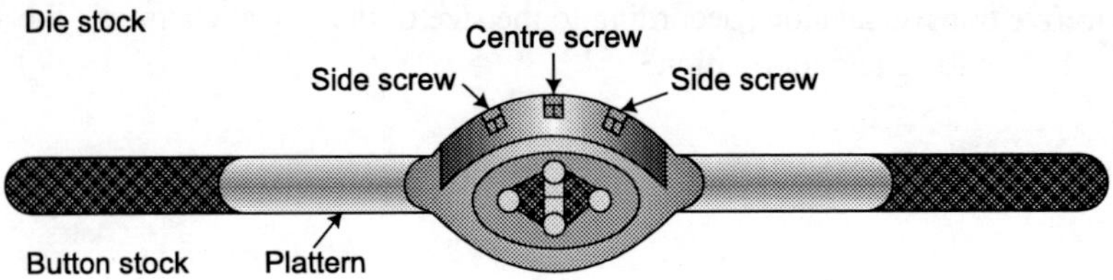

Fig. 8.121

8.63.3 Half Die

Half dies are stronger in construction.

Adjustments can be made easily to increase or decrease the depth of cut.

These dies are available in matching pairs and should be used together.

By adjusting the screw of the diestock, the die pieces can be brought closer together or can be moved apart.

They need a special die holder **(Fig. 8.122).**

8.63.4 Adjustable Screw Plate Die (Fig. 8.123)

This is another type of a two-piece die similar to the half die.

This provides greater adjustment than the split die.

The two die halves are held securely in a collar by means of a threaded plate (guide plate) which also acts as a guide while threading.

When the guide plate is tightened after placing the die pieces in the collar, the die pieces are correctly located and rigidly held.

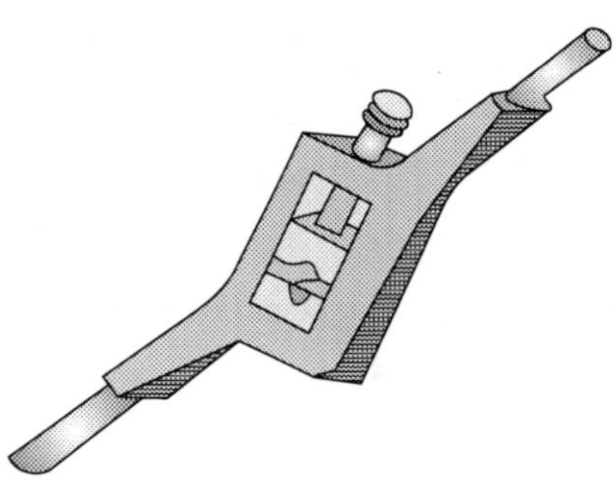

Fig. 8.122 Die holder for half die

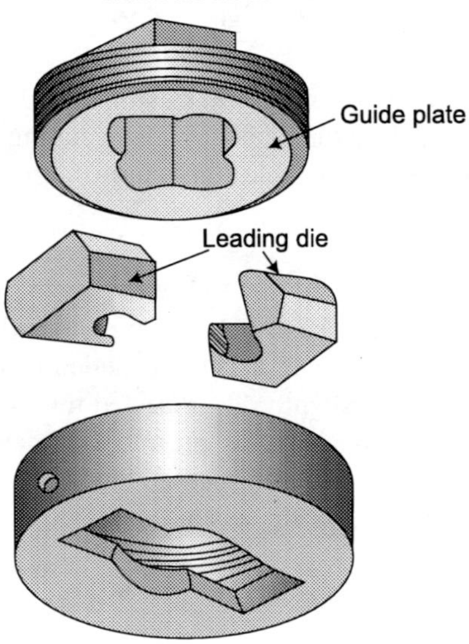

Fig. 8.123 Adjustable screw plate die

The die pieces can be adjusted, using the adjusting screws on the collar.

This type of die stock is called quick cut diestock **(Fig. 8.124).**

The bottom of the die halves is tapered to provide the lead for starting the thread. On one side of each die head, the serial number is stamped.

Both pieces should have the same serial numbers.

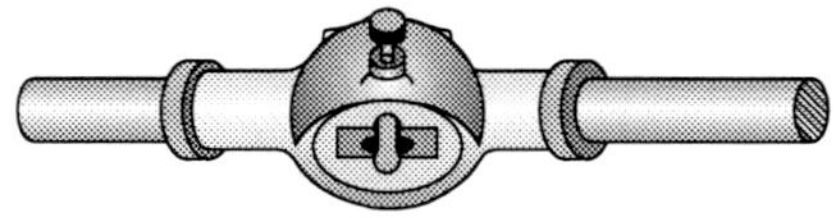

Fig. 8.124 Quick cut die stock

8.63.5 Die Nut (Solid Die) (Fig. 8.125)

The die nut is used for chasing or reconditioning the damaged threads.

Die nuts should not be used for cutting new threads.

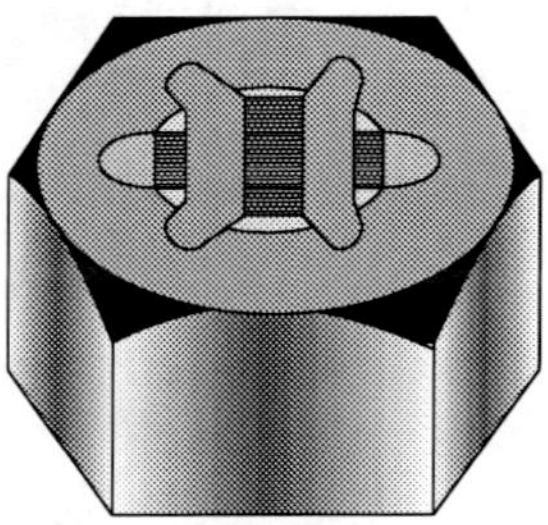

Fig. 8.125

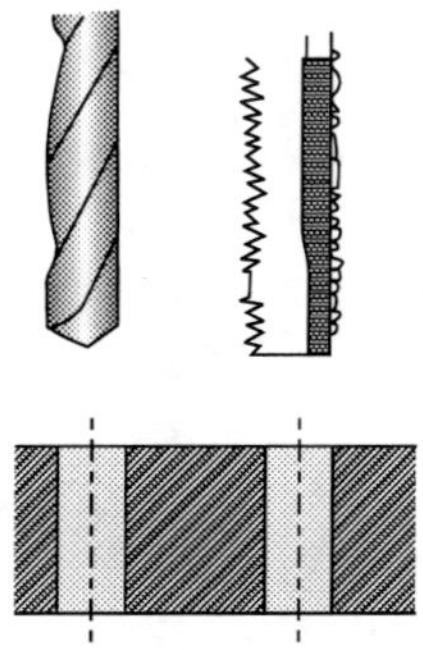

Fig. 8.126

The die nuts are available for different standards and sizes of threads.

The die nut is turned with a spanner.

8.64 HAND TAP (FIG. 8.127)

The tap is used for cutting inner or female threads. For producing the cutting dies, three or four half round longitudinal slots are milled into the cylindrical part of the prefabricated rough forging of the tap. These slots of the finished tap receive and remove the chips.

After incorporating the said slots there will only remain a small core which is not able to withstand a high cutting stress. The internal thread, therefore, cannot be made in a single cutting action. For this reason a set of taps is used for cutting inner threads by hand, and the cutting work is distributed as follows: with the set to three taps (the taper tap, the plug tap and the bottom tap) and with the B-set to two taps (the taper- and the bottom-tap).

The tap, same as the threading die, is provided with a taper chamfer which effects the actual chip removal. The chamfer is long for through holes and short for blind holes and brittle materials.

The end of the tap shank is formed as a square. On this square a suitable tap wrench is put on or fastened; by means of this tap wrench the tap is turned in the core hole. Tap wrenches whose dies are adjustable can be used for different taps.

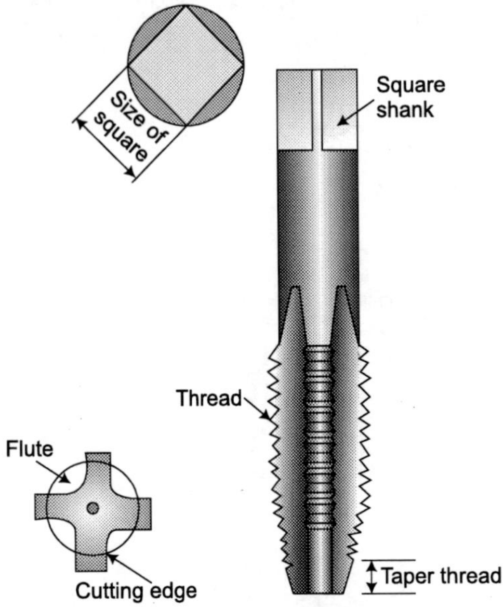

Fig. 8.127 Hand tap

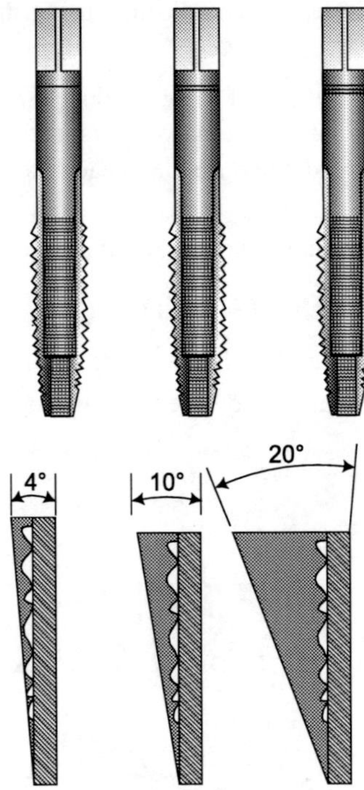

Fig. 8.128 Hand tap set

8.64.1 Solid Taps

Tap terms and definitions (Fig. 8.129 and 8.130): The following terms and definitions are adapted in part from the Metal Cutting Tool Institute, and from American National Standard ANSI B 94.9

The **major diameter** is the largest diameter of a straight thread and the largest diameter of a taper thread in any given plane normal to the axis.

The **minor diameter** is the smallest diameter of a straight thread and the smallest diameter of a taper thread in any given plane normal to the axis.

Pitch diameter on a straight screw thread is the diameter of an imaginary cylinder the surface of which would pass through the threads at such points as to make equal the width of the thread ridge and the width of the groove cut by the surface of the cylinder. The pitch diameter of a straight tap is measured at the first full thread back of the chamfer. On a taper screw thread it is the diameter, at a given distance from a reference plane perpendicular to the axis of an imaginary cone, the surface of which would pass through the threads at such points as to make equal the width of the thread ridge and the width or the groove cut by the surface of the cone.

The **cutting edge** is the leading edge of the land in the direction of rotation for cutting and which does the actual cutting.

The **cutting face** is the leading side of the land in the direction of rotation for cutting; on which the chip impinges.

Single thread is a thread in which the lead is equal to the pitch.

Two-start (double) thread is a thread in which the lead is equal to twice the pitch.

Three-start (triple) thread is a thread in which the lead is equal to three times the pitch.

Four-start (quadruple) thread is a thread in which the lead is equal to four times the pitch.

An interrupted thread tap is one having an odd number of flutes, with every other too along the thread helix removed.

Relief is the condition whereby metal is removed from behind the cutting edge to provide clearance between the part being threaded and a portion of the tap land.

Centre relief is the clearance produced on a tap land by reducing the diameter of the entire thread form between the cutting edge and the heel.

Coneccentric relief is the radial relief in the thread form starting back of a concentric margin.

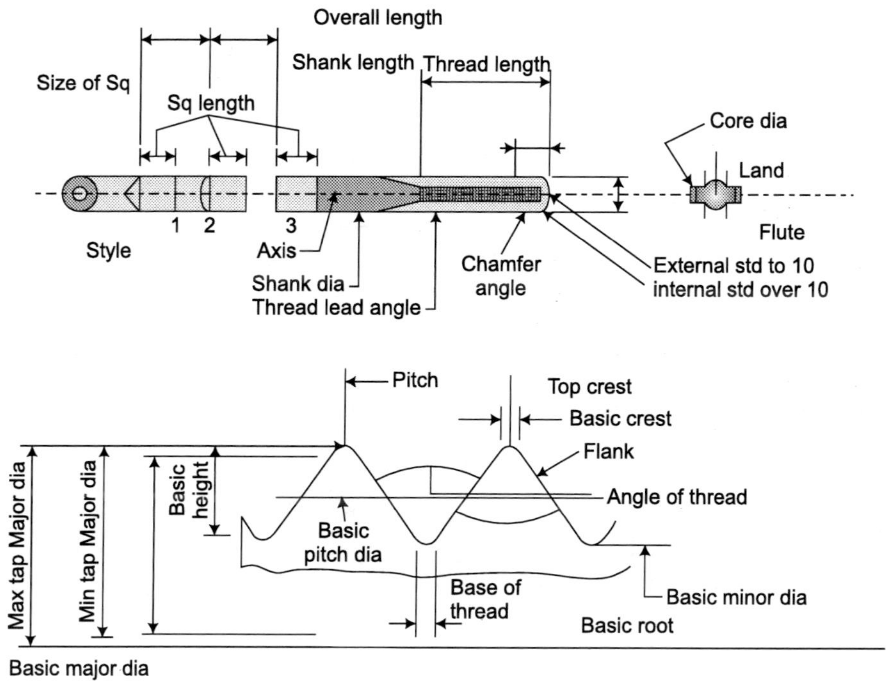

Fig. 8.129

Figures (8.129 & 8.130) shows a comparison of styles 1, 2, 3 centres and the points of measurement for overall length of square for each style. Style 1 is standard for machine screw sizes to and including No. 12 size. Style 2 is standard for sizes larger than no. 12 and including 3/8 inch size. Style 3 is standard for sizes larger than 3/8 inch.

Double eccentric relief is the combination of a slight radial relief in the thread form starting at the cutting edge and continuing for a portion of the land width, and a greater radial relief for the remainder of the land.

Eccentric relief is the radial relief in the thread form starting at the cutting edge and continuing to the heel.

Flattened land relief is the clearance produced on a tap land by truncating the thread between the cutting edge and the heel.

Grooved land relief is the clearance produced on a tap land by forming a longitudinal groove in the centre of the land.

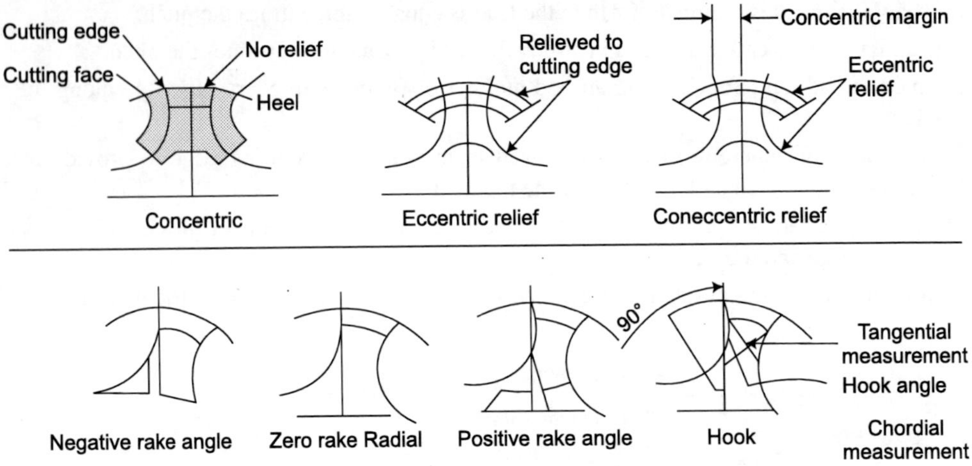

Fig. 8.130

8.65 TYPES OF TAPS

Hand taps for a particular thread are available as a set consisting of three pieces. These are:

- First tap or taper tap
- Second tap or intermediate tap
- Plug or bottoming tap.

These taps are identical in all features except in the taper lead.

The taper tap is to start the thread. It is possible to form full threads by the taper tap in through holes which are not deep.

The bottoming tap (plug) is used to finish the threads of a blind hole to the correct depth.

For identifying the type of taps quickly. The taps are either numbered as 1, 2 and 3 or rings are marked on the shank.

The taper tap has one ring, the intermediate tap has two rings and the bottoming tap has three rings.

8.66 TAP WRENCHES

Tap wrenches are used to align and drive the hand taps correctly into the hole to be threaded.

Tap wrenches are of different types:

- Double ended adjustable wrench,
- T-handle tap wrench, solid type tap wrench.

8.66.1 Double-ended Adjustable Tap Wrench or Bar Type Tap Wrench

This is the most commonly used tap wrench. It is available in various sizes. These tap wrenches are more suitable for large diameter taps, and can be used in open places where there is no obstruction to turn the tap. It is important to select the correct size of a wrench.

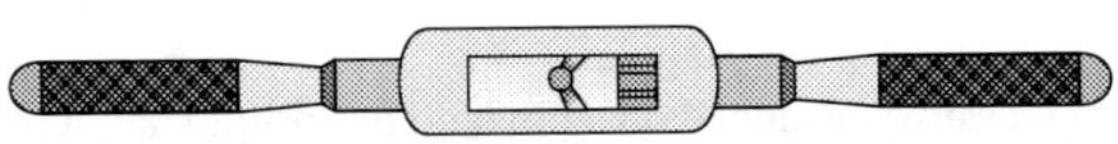

Fig. 8.131 Double ended adjustable tap wrench

8.66.2 T-Handle Tap Wrench (Fig. 8.132)

These are small adjustable chucks with two jaws and a handle to turn the wrench.

This tap wrench is useful to work in restricted places, and is turned with one hand only.

This wrench is not available for holding large diameter taps.

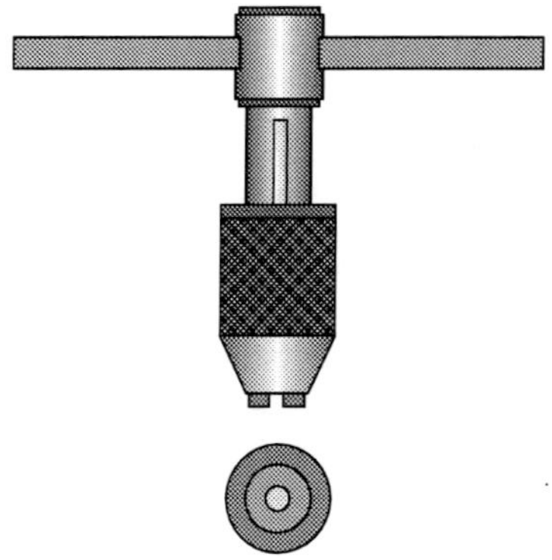

Fig. 8.132 'T' handle tap wrench

8.66.3 Solid Type Tap Wrench (Fig. 8.133)

These wrenches are not adjustable.

They can take only certain sizes of taps. This eliminates the use of wrong length of the tap wrenches, and thus prevents damage.

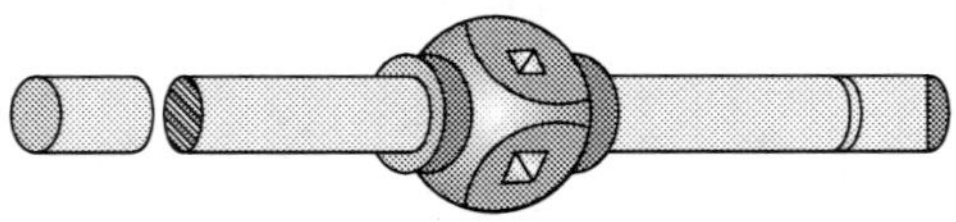

Fig. 8.133 Solid tap wrench

8.67 TAP DRILL SIZE

Before a tap is used for cutting internal thread, a hole is to be drilled. The diameter of the hole should be such that it should have sufficient material in the hole for the tap to cut the thread.

Tap Drill Sizes for Different Threads:

ISO metric thread

Tapping drill size for M 10 × 1.5 thread

Minor diameter = Major diameter - 2 × depth

Depth of thread = 0.6134 × pitch of a screw

2 depth of thread = 0.6134 × 2 × pitch = 1.226 × 1.5 mm = 1.839 mm

Minor diameter (D1) =1 0 mm - 1.839 mm = 8.161 mm or 8.2 mm

This tap drill will produce 100% thread because this is equal to the minor diameter of the thread. For most fastening purposes a 100% formed thread is not required.

A standard nut with 60% thread is strong enough to be tightened until the bolt breaks without stripping the thread. Further, it also requires a greater force for turning the tap if a higher percentage formation of thread is required.

Considering this aspect, a more practical approach for determining the tap drill sizes is:

$$\text{Tap drill size} = \text{Major diameter} - \text{pitch} = 10\text{ mm} - 1.5\text{ mm}$$

$$= 8.5\text{ mm.}$$

Compare this with the table of tap drill sizes for ISO metric threads.

ISO inch (unified) threads formula:

$$\text{Tap drill size} = \text{major diameter} - \frac{1}{\text{No. of threads per inch}}$$

For calculating the tap drill size for 5/8″ UNC thread

$$\text{Tap drill size} = 5/8'' - 1/11'' = 0.625'' - 0.091'' = 0.534''$$

The next drill size is 17/32″ (0.531 inches)

Compare this with the table of drill sizes for unified inch threads.

8.68 BLANK SIZE FOR EXTERNAL THREADING

It has been observed that the threaded diameters of steel blanks show a slight increase in diameter. Such increase in the diameter will make the assembly of external and internal threaded components very difficult. To overcome this, the diameter of the blank is slightly reduced before commencing the threading.

The diameter of the blank should be less by 1/10th of the pitch of the thread.

Commercial tap drill sizes ISO inch (Unified) thread

NC (National Coarse)			*NF (National Fine)*		
Tap size	*Threads per inch*	*Tap drill size*	*Tap size*	*Threads per inch*	*Tap drill size*
5	40	38	5	44	37
6	32	36	6	40	33
8	32	29	8	36	29
10	24	25	10	32	21
12	24	16	12	28	14
1/4	20	7	1/4		3
5/16	18	F	5/16	24	1
3/8	16	5/16	3/8	24	0
7/16	14	U	7/16	20	25/64
1/2	13	27/64	1/2	20	29/64
9/16	12	31/64	9/16	18	33/64
5/8	11	17/32	5/8	18	37/64
3/4	10	21/32	3/4	16	11/16
		....			
7/8	9	49/64	7/8	14	13/16
1	8	7/8	1	14	15/16
1 1/8	7	63/64	1 1/8	12	1 3/6
1 ¼	7	17/64	1 1/4	12	1 11/6
1 3/8	6	17/32	1 3/8	12	1 19/64
1 ½	6	1 11/32	1 1/2	12	1 27/64
1 ¾	5	1 9/16			
2	41/2	1 25/32			
1/8	27	11/32	1	11 1/2	1 5/32
1/4	18	7/16	1 1/4	11 1/4	1 1/2
3/8	18	19/32	1 1/2	11 1/2	1 23/32
1/2	14	23/32	2	11 1/2	223/16
3/4	14	15/16	21/2	8	2 5/8

Example: For cutting the thread of M 12 with 1.75 mm pitch the diameter of the blank is 11.80.

Formula, D = d-p/10

12 mm – 0.175 mm = 11.825 or 11.8 mm.

where, d = diameter of bolt

D = the blank diameter

p = pitch of thread

8.69 LUBRICANTS USED FOR THREAD CUTTING

Good and appropriate lubricants considerably reduce the friction on the cutting point and also reduce the risk of plucking out of the turns of the thread.

The friction occurring when cutting varies with the different materials; therefore the lubricant should be chosen according to the material. Threads on workpieces of grey cast iron and electron are cut without using lubricants because the cutting oil would form a paste in combination with the fine chips of the material. Machine oil is not suitable at all because it would produce rough flanks.

Materials	*Lubricants*
Steel, unalloyed	Cutting oil or coal oil
Steel, alloyed	Cutting oil, turpentine
Grey cast iron	(dry)
Brass	Cutting oil or coal oil
Bronze	Cutting oil or coal oil
Aluminium alloys	Petroleum
Electron	(dry)

8.70 THREAD CUTTING PROCESS

8.70.1 Cutting of Male and Female Threads

Cutting of male threads is by means of the threading die: The end of the bolt receives a chamfer to ensure a good cutting start of the cutting tool; to achieve this spot facing, the threading die is placed straight onto the bolt and turned under a slight pressure so that the cutting edges can easily attack. Thereafter cutting is done without pressure. Threading dies that are placed skew-angled to the tap axis will produce slanting and drifted threads. The threading die is to be turned back periodically several times to allow the lubricant to penetrate to the cutting spot. Before placing the die again, the flute must be cleaned from chips.

Cutting of male threads is by means of stock and die: The stock and die are pushed over the bolt with open dies. Then the die stock dies are tightened by means of an adjusting screw of that the slightly turned stock and die will cut a slight thread groove into the bolt. The thread is finished-cut in several operations by repeated adjusting of the cutting dies. The turns of the thread will thus become neater and more accurate than those cut with the threading die. After completion of the cutting work, the stock and die are simply lifted.

Cutting of female threads by means of taps: The tap has a good cutting ability when the core holes have been spot faced. When placing the tap, the vertical position should be verified with a square. The taps are to be turned in the core hole in a uniform manner and not by jerks, the correct sequence being: taper tap, plug tap and bottom tap. If it is difficult to get the chip loose, the tap should be slightly moved up and down. Failure to work with the taper tap or the plug tap will produce poor and drifted threads.

Most defects when making threads are caused by worn taps and threading dies as well as by the absence of appropriate checking means.

Even threads that have been well cut with a threading die show undulating flanks. They cannot be avoided when cutting with a threading die and are perceptible when magnified.

Threads cut in such a way have only the crests as supporting parts whereas the flanks do not contact. When tightening such threads, the crests of the thread will bend until the flanks contact.

But as the force should be received by the flanks when tightening nut and bolt, the utmost must be done to obtain a good contact of the flanks.

Only gauging instruments allow perfect verification of the quality of threads.

Judgment on thread cutting work

Mistake	*Cause*
The male thread has become skew in the cutting process.	The threading die was chucked in the collets or in the die stock in a slanting position. The cutting tools were placed in a slanting position and were not guided. One-side pressure was exerted when cutting .
The turns of thread are rifting.	Bolt diameter is too large. Insufficient lubrication . Bolt diameter was too small.
Thread has not been cut out sharp enough.	The thread core holes were not spot faced.
The first turns of a female thread are rifting.	The flutes of the tap are clogged. Placing the tap in a slanting position.
Further turns of thread are rifting.	Working by jerks with unsuitable tap wrench. Placing the tap on the bottom of the hole when cutting blind holes.
The tap is broken.	The tap was loosened with violence after skewing.

8.71 MORE ON DIES AND TAPS

8.71.1 Die Heads and Collapsing Taps

Theoretical horsepower requirements for self-opening die heads and collapsing taps vary directly with the volume of material per unit time and the cutting speed. They will also vary for any given pitch, material hardness, chamfer angle, cutting angle, and chaser dullness.

For practical purposes, these values are straight-line values and may be extended to higher cutting speeds. Factors for applying the values to other materials are also given.

Special cutting angles or short chamfers will require more horsepower. The same pitch at the same cutting speed requires the same amount of horsepower. The number of chasers used for cutting a given pitch will have some effect on the horsepower requirements, but not enough to consider, when using the table. High tangential loads are encountered when cutting coarse threads and will be higher when cutting Acme threads of the same pitch.

Adequate provision for work holding must be made when chucking parts for cutting coarse threads to resist the high torque involved. Some chuck-jaw penetration of the workpiece is needed to hold it, because

Table 8.9 Net horsepower required for cutting threads in cold-rolled steel and cast iron*

Pitch	Area cut, sq in.	Tang-ential cut load, lb	Cutting speed, sfpm										
			16	18	20	25	30	35	40	45	50	60	70
4	0.0203	4510	2.19	2.40	2.73	3.41	4.10	4.78	5.46	6.16	6.83	8.22	9.60
4.5	0.01603	3560	1.73	1.95	2.17	2.72	3.25	3.79	4.33	4.88	5.40	6.50	7.60
5	0.01299	2890	1.40	1.58	1.75	2.19	2.63	3.07	3.51	3.96	4.38	5.26	6.14
6	0.00901	2020	0.972	1.09	1.21	1.52	1.81	2.12	2.42	2.72	3.03	3.63	4.23
7	0.00663	1470	0.717	0.805	0.894	1.12	1.34	1.57	1.79	2.01	2.24	2.68	3.13
8	0.00507	1130	0.547	0.616	0.683	0.855	1.03	1.19	1.37	1.54	1.71	2.05	2.49
9	0.00401	890	0.434	0.488	0.543	0.678	0.818	0.950	1.09	1.22	1.35	1.63	1.90
10	0.00325	722	0.350	0.394	0.437	0.546	0.656	0.766	0.878	0.982	1.09	1.31	1.53
11	0.002682	596	0.290	0.326	0.362	0.452	0.543	0.633	0.728	0.816	0.905	1.09	1.27
12	0.002255	504	0.242	0.273	0.303	0.379	0.456	0.530	0.612	0.684	0.755	0.910	1.06
13	0.00192	427	0.207	0.232	0.259	0.324	0.388	0.454	0.517	0.584	0.650	0.745	0.870
14	0.00166	368	0.180	0.201	0.224	0.279	0.355	0.391	0.446	0.502	0.561	0.672	0.784
16	0.00127	282	0.136	0.154	0.171	0.214	0.257	0.299	0.342	0.385	0.427	0.513	0.599
18	0.000999	222	0.108	0.121	0.134	0.168	0.202	0.235	0.279	0.303	0.336	0.403	0.470
20	0.000813	180	0.088	0.099	0.110	0.138	0.165	0.193	0.220	0.248	0.275	0.330	0.385
24	0.000562	125	0.061	0.068	0.076	0.094	0.113	0.132	0.151	0.170	0.189	0.227	0.265
28	0.00414	92	0.046	0.051	0.057	0.071	0.085	0.099	0.103	0.128	0.143	0.170	0.198
32	0.000317	70.5	0.035	0.039	0.043	0.054	0.065	0.076	0.086	0.098	0.108	0.130	0.152
36	0.000250	55.5	0.028	0.031	0.034	0.043	0.052	0.061	0.069	0.078	0.086	0.103	0.120
40	0.000203	45	0.022	0.025	0.028	0.35	0.041	0.049	0.055	0.063	0.069	0.083	0.097

*For other materials, use the following factors: alloy steel 1.3; carbon tool steel and annealed high-speed steel 1.7; allloy steel, 30 to 32 R_C, 2.3.

friction alone is usually insufficient. A key or similar driving device on the workpiece can be used on long work of high accuracy; long-travel threading machines should be used. When serrated chuck jaws penetrate long threaded workpieces that must be regripped, the marks are sometimes objectionable. Also, there may be some lead error caused by regripping.

8.71.2 Self-opening Die Heads and Collapsing Taps

Both stationary and rotary types of self-opening die heads are external thread-cutting tools in which the chasers automatically withdraw from the work at the completion of the thread, making reversal of the machine unnecessary. Tools of this description in addition to saving considerable time have many other outstanding advantages over solid threading dies. The chasers made in sets of four or more can be easily removed for resharpening and in addition can be quickly and accurately adjusted to size. These tools with suitable chasers will cut either right- or left-hand threads, straight or taper, in a range of diameters within the rated capacity of the tool.

Die heads are designed for a variety of applications. They are selected on the basis of the job to be done and the machine upon which it is to be used. All self-opening die heads must position the chaser properly for correct cutting action and open automatically out of the thread, thus permitting the die to be retracted rapidly from the work.

Selection of the die-head style and size depends upon whether the die head is stationary or revolves, and the manner in which it is opened or closed. Die-head size is selected by size and clearances on the machine the head is to be used on, and the size of threads to be cut.

A small head is more suitable for cutting relatively fine pitches on short threads because it is light in weight and adjusts itself readily to slight misalignment. Coarse threads, long threads, and tough materials require large-capacity die heads if they will clear the machine. Die heads are not standard as to size; each manufacturer has his own standard sizes.

8.71.3 Stationary Die Heads

These are self-opening die heads which are particularly suited for hand-operated turret machines, single-spindle automatics, or other similar applications where the die head does not rotate. A closing handle or other means is provided for resetting the tool.

Pull-off trip (Fig. 8.134A): This tool is especially suited for the general run of turret machine or single-spindle automatic work when the thread is not too short. Roughing and finishing attachments are frequently furnished for heavy threading.

In operation, this head opens after its forward travel has been stopped, and the slightly continued cutting action of the chasers advances the front or body part of the tool sufficiently to release the locking mechanism, allowing the chasers to be withdrawn from the work.

Internal trip (Fig. 8.134B): The opening action of this type is effected by the work contacting an internal adjustable stop which is set to produce the required length of thread. It is very desirable in cases where the work to be threaded varies in length, such as in pipe threads, but regardless of this condition, the thread length must be accurately maintained.

Outside trip (Fig. 8.134C): This die head is intended for short-length threads, especially those of fine pitch. It is constructed to reduce the stress on the threads when the head is tripping.

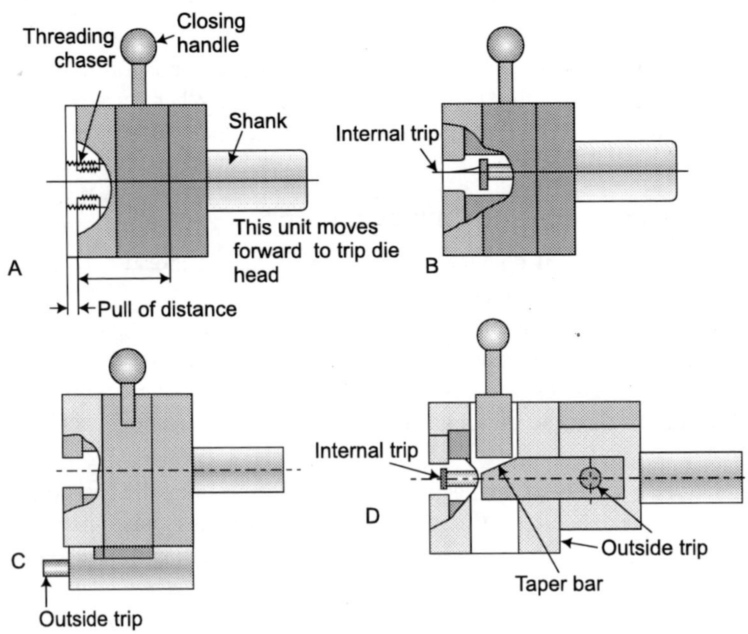

Fig. 8.134

Stationary taper die heads (Fig. 8.134D): The chasers in a taper die head gradually recede as they advance while cutting the thread. When the desired length has been threaded, the head opens automatically, releasing the chasers from the work. A closing handle or equivalent means is provided for resetting the tool. This method eliminates the objectionable stop lines sometimes left on the work, indicating when the chasers stop cutting when the head opens. Taper threads can be cut with conventional die heads using taper chasers, but the length of the thread is limited by the length of the chasers and there is also the possibility that stop lines will be left on the work. This method also requires considerably more power.

Internal trip (Fig. 8.134): The taper action of this tool becomes effective when the work contacts an internal adjustable stop which permits the chasers to recede gradually as the thread is being cut. When the desired length has been produced, the chasers automatically withdraw from the work.

Outside trip (Fig. 8.134D): The taper action by this method takes place when an outwardly extending pin contacts a stop on the machine, permitting the chasers to recede gradually as the thread is being cut.

8.71.4 Rotary Die Head (Fig. 8.135)

These heads are particularly designed for use on multiple-spindle automatics. Drilling machines, and regular threading machines, or where it is necessary to open and close the die head automatically.

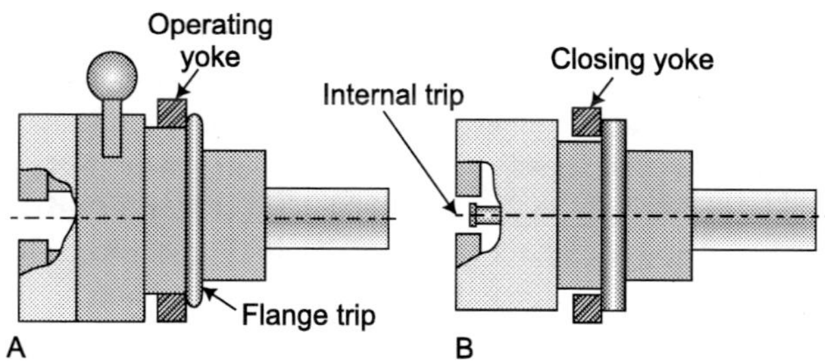

Fig. 8.135 Rotary die head

8.71.5 Flange Trip (Fig. 8.135A)

This tool is tripped and reset by means of a yoke operating in an outer groove of the die head. When the spindle carrying the die head advances and the forward motion of the yoke is stopped, the head opens. When the spindle returns and the yoke is again stopped, the head is dosed.

8.71.6 Internal Trip (Fig. 8.135B)

Although this die head is closed by a yoke, the opening action is effected by the work contacting an internal adjustable stop which is set to produce the required length of the thread. It is particularly desirable in cases where the work to be threaded varies in length, such as pipe threads, but regardless of this condition, accurate thread length must be maintained.

8.72 ROTARY TAPER DIE HEADS (FIG. 8.135)

The chasers in a rotary taper die head gradually recede while cutting the thread. When the desired length has been cut, they are automatically withdrawn from the work. A closing yoke or equivalent means is provided for tripping as well as resetting the tool. This method eliminates the objectionable stop lines sometimes left on the work indicating where the chasers slop cutting when the head opens. Taper threads can be cut with conventional die heads using taper chasers, but the length of the thread is limited by the number of threads on the chasers, and there is also the possibility of stop lines being left on the work. This method also requires considerably more power.

When reverse taper threading, the chasers start cutting at the large end of the work and gradually converge while advancing towards the small end.

8.73 SOLID ADJUSTABLE DIE HEADS

The chasers of solid adjustable die heads do not automatically withdraw from the work at the conclusion of the threading operation. They can be easily removed for resharpening and accurately adjusted for size, giving this tool a definite advantage over solid dies.

Solid adjustable die heads are used on multiple-spindle automatic screw machines and other similar applications where tool space is limited and self-opening die heads cannot be used. Means must be provided on the machine for reversing the die head or for reducing the spindle speed of the die head with respect to the workpiece. The type of shank will vary with the application.

8.74 SPECIAL DIE HEADS (FIG. 8.136A)

Included are modifications of standard tools frequently designed to meet a wide variety of special requirements. Guide bushings are added to the face of die heads to **(Fig. 8.136B)** guide the work and ensure concentricity of the thread. In other cases, they are inserted in the bore of the head **(Fig. 8.136 B),** when concentricity must be held with a previously turned portion ahead of the thread. Pilots **(Fig. 8.136 C)** like guide bushings, serve the same purpose when threading must be accurate in respect to a finished hole in the work. A tap or reamer can be used in the same manner to combine two operations.

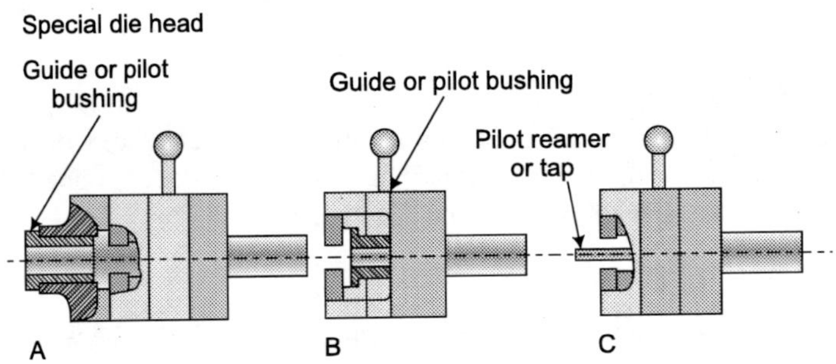

Fig. 8.136

8.75 DIE-HEAD CHASERS (FIG. 8.137)

Three types of chasers are in use: tangent, circular, and radial. Radial chasers may be of the inserted type or one piece, and may be either milled, hobbed, or ground.

Each style of chaser has certain characteristics, the value of which is determined by the job to be accomplished.

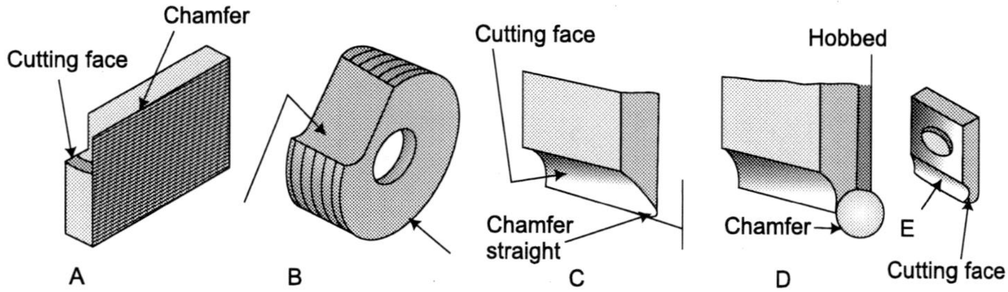

Fig. 8.137

8.75.1 Tangent Chasers (Fig. 8.137 A)

Tangent chasers are flat-type blades threaded on one side and mounted in holders so that this side, at its inner end, contacts the work tangentially to form the cutting edge. These chasers are resharpened on the end only. The chamfer angle remains constant throughout the life of the chaser, or it may be decreased at any time.

There are two types of tangent chasers: one has the helix angle incorporated in the chaser; the other has straight threads ground in its face, and the proper helix angle is obtained by tipping the chaser in the die head. The type used depends upon the make of die head.

8.75.2 Circular Chasers (Fig. 8. 137B)

Circular chasers have a single flute which provides the cutting edge and also the necessary chip space. These chasers are mounted on holders and resharpened on the cutting face only. The chamfer angle remains constant throughout the life of the chaser, or the chamfer angle may be decreased at any time.

8.75.3 Milled-type Radial Chasers (Fig. 8.137 C)

Milled-type radial chasers fit into chaser slots machined in the body of the chaser head. Chasers of this description have the threaded end formed with a positive rake angle. They are furnished with either cut or ground threads and are usually resharpened on the chamfer. Occasionally they are lightly reground on the cutting face. The helix angle is incorporated in the chaser.

8.75.4 Hobbed-type Radial Chasers (Fig. 8.137 D)

Hobbed-type radial chasers have the threaded end formed concave as produced by a hob or tap. General practice is to sharpen these on the cutting face. Occasionally they are lightly ground on the chamfer. The helix angle is incorporated in the chaser.

8.75.5 Inserted Radial Chasers (Fig. 8.137 E)

Inserted radial chasers are blade-type chasers that are locked in the die head carriers. They are used to advantage on short-run jobs because of low initial cost, wide threading ranges, and minimum downtime when changing from one diameter/pitch combination to another. They are also used on long-run jobs where chasers tend to wear rapidly or where machine idle time is expensive, since dull chasers can be replaced rapidly.

8.76 COLLAPSING TAPS

These taps are internal-thread cutting tools in which the chasers automatically withdraw from the work at the completion of the thread, making reversal of the machine unnecessary. These tools save the reversal time of solid taps. The chasers, made in sets of three or more can be easily removed for resharpening and can also be quickly and accurately adjusted to size. These taps with suitable chasers will cut either right- or left-hand threads, straight or tapered in a range of diameters within the rated capacity of the tool.

8.77 STATIONARY TAPS (FIG. 8.138)

Stationary taps are collapsing taps which are particularly suited for hand-operated machines or other similar applications where the tap does not rotate. A closing handle or other means is provided for resetting the tool. The tap bodies are provided with either a hand trip or a trip plate. The taps provided with a trip plate are adjustable to govern the length of the thread to be cut. Hand trips are suitable where the construction of the part is such that it is desirable to both trip and reset the tap by hand. They may also be used as a lever trip by having the closing handle contact a stop on the machine.

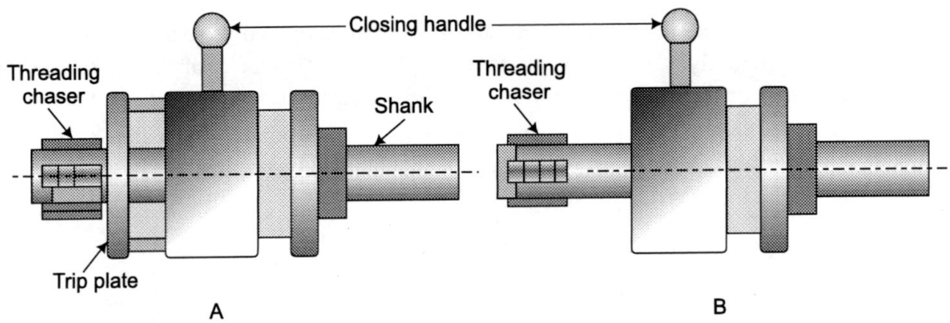

Fig. 8.138 Stationary collapsible taps

8.78 ROTARY TAPS

Rotary taps are especially designed for use on multiple-spindle automatics and drilling machines or where it is necessary to open and close the tap automatically. The taps are opened automatically by either a plate trip or a sleeve trip.

Taps provided with a trip plate **Fig. (8.139A)** are adjustable to govern the length of the thread to be cut. Resetting requires a yoke arranged to contact the rear face of the movable outer sleeve of the tap.

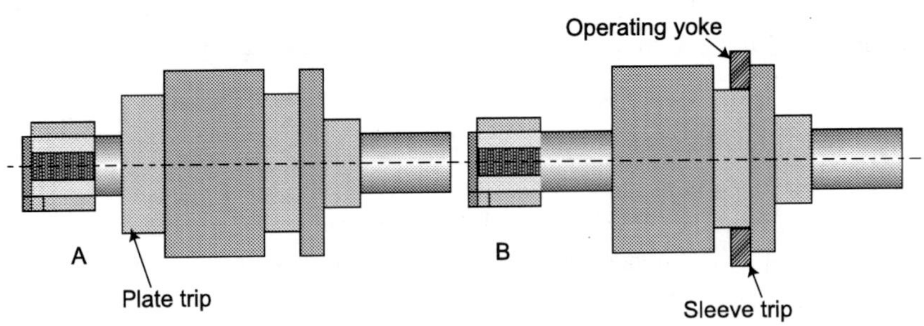

Fig. 8.139 Rotary collapsible taps

8.79 TRIPPING BY THE SLEEVE TRIP

Tripping by the sleeve trip is particularly desirable for multiple-spindle automatics, threading machines, and similar applications. Both tripping and resetting are by means of a yoke operating in the groove of the outer sleeve of the tap **(Fig. 8.139B)**.

8.80 SOLID ADJUSTABLE TAPS

The chasers of solid adjustable taps do not automatically withdraw from the work at the conclusion of the threading operation. They can be easily removed for resharpening and accurately adjusted for size, giving this tool a definite advantage over solid taps.

Solid adjustable taps are used on multiple-spindle automatics, automatic screw machines, and other similar applications where tool space is limited and collapsing taps cannot be used. Means must be provided on the machine for reversing the tap or reducing the spindle speed with respect to the workpiece. The type of shank will vary with the application.

It is not uncommon to use solid adjustable taps for sizing purposes in cases where, for various reasons, it is desirable to finish a thread by hand.

8.81 SPECIAL TAPS

Included are the modifications of standard tools which can be designed to meet a wide variety of unusual requirements. Pilots in lengths to meet specific requirements can be added to the front end of the tap in place of the usual cap to ensure concentricity with previously bored holes.

Guides can be placed on the body of the tap behind the chasers for alignment with a previously finished hole in the work or a guide bushing in the jig or fixture.

Taps can also be provided with reamers. In some applications the reamer is attached to the front end of the tap in place of the cap; in other instances reamer blades are inserted between the chasers, which collapse during the reaming operation. Taps arranged in this particular manner are not suitable for heavy cuts.

For tapping two threads of the same pitch but of different diameters, it is sometimes possible to combine two solid adjustable taps in a tandem manner so that both threads can be tapped simultaneously.

8.82 TAP CHASER

Three types of chasers are in use: regular, overhanging and circular.

8.82.1 Regular Chasers (Fig. 8.140 A)

Regular chasers do not extend through the front cap of tap body and are furnished when tapping is not close to a shoulder or bottom. They can be sharpened on both the chamfer and the cutting face.

8.82.2 Overhanging Chasers

Overhanging chasers extend through the front cap of the tap and are required for bottoming and close to shoulder work. Practically, all the resharpening is done on the cutting face. Occasionally they are lightly reground on the chamfer.

8.82.3 Circular Chasers (Fig. 8.140 C)

Circular chasers have a single flute which provides the cutting edge and also the necessary chip space. These chasers are resharpened on the cuttibg face only.

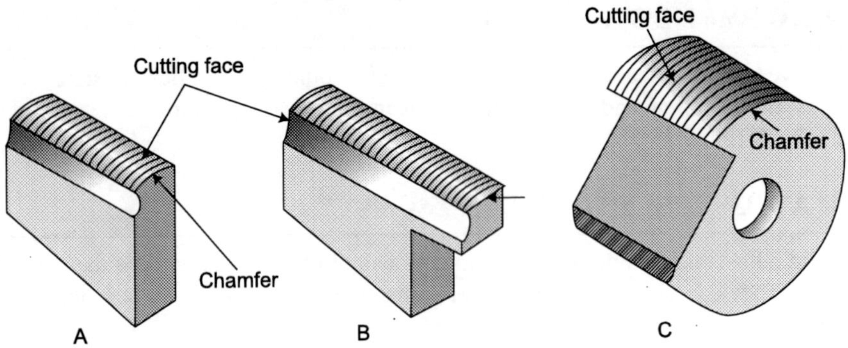

Fig. 8.140 Tap chasers

8.83 SEMIAUTOMATIC SINGLE-POINT THREADING

The semiautomatic high-speed threading lathes are designed to utilize carbide tools. Operating in the range of 300 sfpm and upward, they produce threads of high-quality surface finish on all machineable materials including those steels which, because of alloy constituents and/or hardness, do not readily lend themselves to high-speed steel tools.

The cutting action is generally analogous to a conventional screw-cutting lathe in that the tool is fed into the workpiece (penetration), traversed longitudinally to produce the required thread lead (feed), withdrawn from the work at the end of the thread length (recoil), and rapidly traversed to the starting point (return). This sequence is automatically repeated at successive increments of depth until full thread depth is reached, at which time the machine automatically stops and resets itself to repeat the cycle on the next workpiece.

Parts may be held by manual or air chucks, collets, between cinders, or by virtually any method applicable to lathes.

The feed and return motions are obtained through a drum-type cam or by a reversing lead screw, depending on the make of the machine. In the former case, a drum-type cam permits rapid cycling by combining the functions of a conventional lathe's lead screw, half nut, and manual return of the carriage.

An additional cam provides the penetration and recoil movements, and a third cam system, whose action is superimposed on the penetration and recoil motion, provides the increments of in-feed per pass, which may be either equal or diminishing in nature.

Spindle speed, number of passes, and in feed increments are quickly determined and set by simple dial adjustments.

The machine design is such that virtually any thread may be cut, right- or left-hand, straight or taper, internal or external single or multiple-start. Tooling costs are low compared with thread milling, grinding, and die-head threading, since it is only necessary to produce the thread form (V, Acme, Whitworth, API, etc.) on a single-point carbide tool bit. With suitable attachments, secondary operations such as turning, boring, and facing may be incorporated. Ease of setup, low initial tooling costs, and inherent speed and accuracy make the application of this threading method practical both for jobbing runs and for sustained high production.

Recommended practice is to feed the tool straight into the workpiece, i.e., perpendicular to the workpiece axis, thus providing an even, equal and sufficient chip load on both sides of the threading-tool form. The chip formation, as shown consists of a single chip whose cross section approximates that of the thread form. Tool wear and tool life are commensurate with other single-point carbide-tool applications.

8.84 THREAD TAPPING

To tap holes efficiently and economically, the proper style of tap should be used for the class of material under consideration. One of the first things to be considered is the style of flute to dispose of the chips while four-flute laps may dispose of the chips in shallow or large-diameter holes, or when tapping materials such as cast iron where the chip breaks up, they will not dispose of stringy steel chips for deep or small-diameter holes. Two or three flute taps may also clog with chips, necessitating the use of spiral-point or spiral-flute taps or cold-forming taps to dispose of the chips.

Standard taps are available in spiral-point styles which are designed to push the chip ahead of the tap in through holes. Spiral-flute and high spiral-flute taps pull the chips out of the tapped hole and may be used in either through or blind holes. Cold-forming taps form the thread and therefore produce no chips. They are limited to use in ductile materials and for applications where a seam in the crest of the thread at the minor diameter is not objectionable.

In most cases standard taps in one of the many styles and size limits are the most economical choice. Even though standard taps cost much less than special taps, modem production methods sometimes dictate the use of special taps for their greater economy in the cost per tapped bole.

When special-design taps are required, there are many other factors to consider besides their pitch and size. These factors are the style of flute, the width of land, the overall length (which may be shorter or longer), the amount of hook and the chamfer (which may be special), or the pitch diameter of the tap itself. The material being tapped will affect the pitch diameter of the tap, and certain materials may require eccentric relief.

8.85 GROUND-THREAD VS. CUT-THREAD TAPS

Although high-speed ground-thread taps cost more than cut-thread taps, their cost per tapped hole is usually much less than a cut-thread taps. When tapping at high speeds, a ground-thread tap will usually produce six times as many holes as a cut-thread tap. Since the cost of a high-speed ground tap is but a little higher than for a cut-thread tap, the adoption of ground-thread taps for all work will show considerable savings plus finer work, and with less strain on the tapping machine.

A ground-thread tap requires considerably less power to operate than a cut-thread tap. All conditions being equal, the ground-thread tap requires about one-fifth the number of sharpening for the same number of tapped holes as a cut-thread tap.

Where a go-not-go gauge is used to check tapped holes, a ground-thread tap of the proper pitch diameter will produce greater accuracy than a cut-thread tap, not only on one tapped hole but on an infinite number of holes. While cut-thread taps can be used on some classes of work, such as lB or 2B class of tapped holes, it is still more economical to use ground-thread taps because of their longer life.

8.86 CAUSES OF VARIATION IN TAPPED HOLES

The most important of all requirements for precision tapping is a true-running spindle; also of major importance are true-running taps. A tap may not run true just because the machine spindle does. Sleeves and collets that hold the taps must be equally accurate. Tap shanks should fit the holes in the sleeves or collets with a good slip fit, neither loose nor tight. Holes cannot be uniformly tapped to close tolerance unless the pitch diameter of the tapped hole and the lead and angle of the thread are all carefully held to specification.

It is of utmost importance that the chamfer on the tap be accurately ground on every land. Even though the tap runs true, if the chamfer is not accurately ground, there will be considerable variation in tapped holes.

A tap mounted in a machine having a loose spindle will not consistently produce accurately tapped holes. When pressure is applied to a loose spindle during a tapping operation; the tap is forced out of alignment with the hole, and an unnecessary strain is placed on the tap. As the tap enters the hole and starts to cut, there is a tendency for the tap to wobble and shave a slight amount of metal off the side of the thread. When the tap

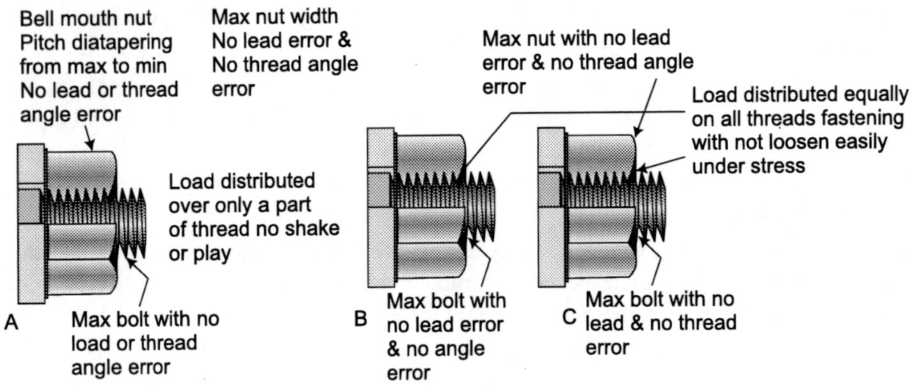

Fig. 8.141 Quality of tapper threads

is reversing, the tapped hole, being much larger than the tap itself, gives no support to the tap, and the tap, if relieved, cuts while running backward and causes rapid dulling and early breakage.

A comparison is shown **(Fig. 8.141),** between the quality of threaded connections obtained under various conditions of tap trueness and pitch tolerances on a nut and bolts. These conditions may be summarized as follows:

1. **Poor condition:** Bell mouthed tapped hole **(Fig. 8.141 A)** condition resulting when the tap does not run true, when too much pressure is applied to the tap, when tap lands are too narrow, or when used in some floating chucks. Narrow lands will permit the tap to shave material from the sides of the first threads.
2. **Good condition:** Tapped with maximum bolt and maximum nut **(Fig. 8.141 B)** no lead or thread-angle errors. Metal-to-metal contact is obtained over the entire length of engagement.
3. **Ideal condition:** Tapped with maximum bolt and minimum nut **(Fig. 8.141 C)**); no lead or thread-angle errors. Metal-to-metal contact is maximum over the entire length of engagement.

8.87 CHIP-ROOM PROVISION

Adequate chip room in the flute of the tap and in the work is an important consideration. When an improperly designed tap is used, especially for tapping steel and other materials that produce string chips, the chips are crowded into the flutes. This crowding increases the amount of power needed to drive the tap, and causes excessive tap breakage.

If the flutes are too shallow and too narrow, the chips will not curl properly, and the amount of power required to break them is considerably increased. The tap must do double duty: it must cut the metal out and crowd the chip into the flute as well as break it up.

The physical limitations of the smaller taps do not permit sufficient flute room to get rid of the chips readily. It is therefore necessary to use fewer flutes and narrower lands on the tap so that it will cut more freely and allow chip room. Spiral-point taps, which push the chips ahead of the tap, or high-spiral-flute taps may also be used to remove chips from smaller holes. Therefore, on taps smaller than 5/16 inch in diameter, the use of four flutes is not advisable on unified coarse (UNC) threads, as there is not sufficient chip room in the flutes and it is difficult to tap holes commercially, especially in steel. It is possible to use a four-flute tap in cast iron or certain grades of brass where the chip breaks up into a granulated form, but there is not sufficient chip space for four-flute taps, and they are usually unsuitable for production tapping of materials that produce stringy chips.

In low-carbon steels, especially cold-headed or upset steels, even a three-flute tap does not allow sufficient chip room, as the metal is soft and gummy. Steel of this stringy character should be permitted to curl out while tapping, for which service a two-flute tap is best, e.g., a tap with two lands and two straight flutes. A two-flute tap with very wide flutes and narrow lands provides more flute area in which the chips may curl. A high-spiral-flute tap may have more flutes, since it draws the chips out of the hole. Except in blind holes, a spiral-pointed tap works well. The flute area need not be so great because the chips move ahead of the tap.

The crowding of chips causes trouble in holding size, especially where the chips pack, as it tends to force the tap to cut oversize, and there is more danger during tap reversal that the chips will reverse with the tap and cause welding or loading. There are other conditions such as blind-hole tapping, where sometimes the tap must go to the bottom, again creating chip disposal problems. It is impractical to tap holes without leaving sufficient room for chips and in the smaller sizes, if a tap were made with narrow lands and deep flutes, it would lack sufficient mechanical strength for tapping. There is a limit to the depth of flute and the width of land that can be built into any tap, depending upon its diameter and pitch.

Three- and four-flute taps are usually furnished as standard. Two-flute taps in some sizes are special, but wherever they can be used to advantage, regardless of their cost, they provide the most economical production. The three-flute, 1/4 X 20 tap on ordinary cold-rolled steel will outperform a four-flute tap at least twenty to one.

8.88 LOADING OR PICK-UP OF METAL

A serious problem of variation in tapped holes is the loading or pick-up of metal on the tooth flanks of the tap. This loading has been an endless source of trouble and is the cause of most tap breakage and the variation in size of tapped holes. This pick-up of metal is critical in different materials. It is more critical in certain grades of materials when the tap reverses, causing a very slight pick-up of metal. With every tapped hole a little more metal is picked up until finally it is above the surface of the tap. If the tap continues to operate and pick up metal, it will gradually weld and break.

When a tap loads, there is a noticeable increase in size of the tapped hole, which gradually increases as more metal is picked up so that in a short time the go-not-go gauge will enter the hole. The threads also become slightly roughened, and this further contributes to the pick-up of metal.

Loading or pick-up of metal by a tap is a form of welding and welding requires three simultaneous conditions to be present: (1) heat, (2) pressure, and (3) two clean surfaces (free of oil or tapping fluid). Elimination of anyone condition can prevent welding.

When a hole is tapped, there is always a certain amount of expansion caused by the cutting action of the tap. There is another expansion that takes place from the heat created by friction. When a tap starts, there is a slight hole expansion; as it goes farther into the hole, there is an increase in heat and the tapped hole has expanded a slight amount. When the tap reaches the end and is ready for reversal, the pressure of the cutting action of the tap is removed and a slight shrinkage takes place while the tap is reversing. If the shrinkage is very great, it wipes out most of the cutting oil, and a slight pick-up of metal takes place. Many materials are very sensitive to this pick-up of metal, while others have no affinity for the tap. Every type of material will pick up at some time or other if improper cutting oil is used, if an incorrect style of tap with too wide a land is used, if there is crowding of chips in the flute, if the tap is dull, or if the work shifts while tapping.

Most of the pick-up on the tap due to the material can be eliminated, although certain materials, particularly low-carbon steels that are soft and gummy, cause trouble since they have a tendency to expand and shrink very rapidly under pressure. This is noticeable in upset or cold-headed materials. Seamless steel tubing is another material that causes metal pick-up to a great extent. Sometimes even in free-cutting screw stock one lot of steel will tap very freely and cause no trouble whatever in pick-up, yet another lot of the same material will cause trouble in metal pick-up.

As a general rule, when metal pick-up occurs, the first thing to look for in order to eliminate this condition is the cutting lubricant; then examine the lands of the tap, as they are generally too wide and create too much friction, which tends to increase pick-up of metal. The use of taps with narrower lands sometimes eliminates this trouble. Parts to be tapped often have thin walls. In these cases there is not sufficient support for the cutting action of the tap; therefore, too great an expansion takes place. In this case the shrinkage after tapping is tremendous, but it can be eliminated by narrowing the lands of the tap so as to have less frictional contact. The outside surfaces of swaged, upset or special drawn materials that have certain strains set up in the drawing or upsetting will in many cases be work-hardened; the interior will be soft, and when a tap starts to cut, it releases some of these strains and causes metal pick-up.

In general, most metal pick-up is due to the quality of the cutting lubricant used, although the quantity of oil per tapped hole has some effect on pick-up. If oil is to be used at all, there must be sufficient quantity to lubricate the tap and the oil must be clean.

8.89 EFFECT OF MACHINE RANGE ON TAPPING

The range of thread sizes for which a tapping machine is designed is sometimes responsible for the variation in tapped holes. Many machines tap entirely too great a range. A machine that is efficient and sturdy enough to tap ½ -inch holes is hardly light or sensitive enough for tapping No. 2, or even No. 10 holes. Somewhere within its range a machine will be most efficient. If it is suitable for tapping with a No. 2 or No. 10 tap, it is hardly rugged enough for a ½ -inch tap.

If a small tap is put in a large machine the possibility of producing low-quality threads is multiplied many times by the excessive weight and inertia and the lack of sensitivity of the smaller machine. On the other hand, if a large tap is used and the capacity of the machine is crowded, there is the problem of the extreme pressures required to start the tap. Usually the weight of a chuck alone is detrimental in tapping over a large range, since a chuck that is strong enough to hold a ½ -inch tap even if there are no spindles or clutches with which to contend, would be entirely too heavy for a No. 2 to a No. 10 tap. For real efficiency, it is necessary to have tapping machines that have less rather than more range in their capacity.

Tapping on a screw machine involves the same problems as on a tapping machine, and tap holders are usually designed so that they carry the maximum size of tap for which the machine is rated. This, together with the fact that they have provision for float, makes it a rather difficult job to hold tolerances on the screw machine when heavy holders are used for light taps. Holders for the screw machine should be designed according to the taps rather than the capacity of the machine.

8.90 CUTTING FLUIDS

In high-speed work, the cutting fluid used is the most important single factor for successful tapping on a commercial basis.

No one cutting fluid is efficient and economical for all tapping operations; several grades are necessary for tapping different kinds of material.

In operation, once the tap is buried in its work, its feed cannot be varied as in other metal cutting operations. The limitations are chip space in the tap itself, the small number of cutting teeth that do all the work, and the physical limitations of the tool, especially in taps under ½ inch in diameter.

On the average, less than 10 per cent of the useful life of taps is realized. Very rarely do they wear out in actual service. In most cases taps will load, eventually weld, and then break, often because of the cutting fluid used.

The manner in which fluid is applied to a tap is very important. It is not enough to have a stream of fluid running onto the tap; it must be applied properly. In horizontal tapping, it is necessary that the fluid be applied to the work horizontally rather than vertically. It must fill the hole being tapped and assist in washing the chips away. There must be a sufficient volume of fluid; merely allowing a small stream to flow on the tap is not enough for high-speed work.

8.91 PERCENTAGE OF THREAD IN TAPPED HOLES

The percentage of depth of thread being tapped is important to efficient and economical ·tapping. Too great a percentage throws an unnecessary strain on the teeth of the tap and serves no useful purpose. The greater percentage, the more power required to tap, the more difficult to hold size, and the greater the amount of tap breakage, because more metal must be removed in a given time. Many difficulties in tapping are direct results of attempting to tap too great a percentage of thread.

Even the generally accepted 70 to 75 per cent depth of thread is rather difficult to produce in certain small-sized taps of the UNC standard, since the smaller taps in the coarse pitches have to remove too great a percentage of metal in relation to their dimensions and strength.

As a general rule, the tougher the material, the less should be the percentage of the thread. It is difficult to tap even 60 per cent stainless steel: monel, copper, and some of the heat-treated alloy steels. Manufacturers who have studied this problem have reduced the percentage of thread in these materials, in many cases, by as much as 50 per cent without sacrificing the usefulness of the tapped hole.

In tapping to close tolerances, such as class 3 or class 4 fit, the difficulties of maintaining size are multiplied many times when too great a percentage of depth of thread is ·tapped. In rare cases where a greater percentage is necessary and close tolerances must be maintained it is more practical and economical to tap in two operations with an undersize tap for the first operation and the second tap merely to size the hole.

Standard drills do not always produce the percentage of thread desired. Sometimes a size is entirely too large while the next size smaller is entirely too small. Millimetre drills in many instances fit in between the standard drills and give the percentage desired. A good example is a ½ × 20 tap. If an 80 per cent thread is required, there is no standard or metric drill that will give exactly this percentage. The nearest would be a 7/16 inch drill giving 96 per cent depth of thread; the next size larger would be a 29/64-inch drill giving 72 per cent depth of thread. A millimetre drill in between would be 11.5 mm, which would give a 73 per cent depth; therefore an 80 per cent thread cannot be obtained with standard tools.

Drills can be commercially supplied in many sizes in between the regular sizes, and it would be well to use some of them, since the cost of a drill is generally much less than the tap; besides, there is considerable saving from work that is rejected as having poor threads. The additional burden of removing extra metal causes many difficulties: rough threads, tap loading, and bell mouthed holes.

If small-diameter taps of coarse pitches are closely examined, it is found that, after the amount of metal removed to form the flute of the tap and the metal removed to form the thread are deducted, more than 50 per cent of the total cross-sectional area of the tap has been cut away. This is one of the reasons why it is difficult to remove a greater percentage of thread in a tapped hole. The core diameter of such a tap is very small in relation to its outside diameter, and the tap is weakened to such an extent that, if it is attempted to tap 70 to 90 per cent depth of thread, the tap breaks. On the larger sizes there is an entirely different condition, since the amount of metal cut away for the flute and thread provides sufficient core diameter so that the tap is capable of producing a large percentage of thread without breaking.

Definite limits for minor diameter on nut threads have been set for both the UNC and the UNF series. They provide a depth of thread not less than 62 to 75 percent (depending on the size) and not more than 83-1/3 per cent of the basic thread depth. These percentages have ample margin for safety so that, even with the maximum size of nut and the minimum size of screw giving the loosest possible condition for each class of fit, the screw will break before the thread will strip in the tapped hole. There is no reason why these percentages should be exceeded in the average run of commercial work. In fact, in several of the sizes such as 4 × 36, 6 × 32, 8 × 32, 10 × 24, and ¼ × 20, it is rather difficult to tap even the minimum of 62 per cent, particularly in alloy steels or where the depth of the tapped hole exceeds the diameter of the tap.

8.92 EFFECT OF TAP DIAMETER AND PITCH

The percentage of thread in a tapped hole is, in order of importance, governed by: (1) the diameter and pitch of the tap, (2) the hardness and toughness of the material being tapped, (3) the depth of tapped hole, and (4) the kind of hole, whether blind or through. As an example, the factor of tap diameter and pitch would make it difficult to tap an 8 × 32 hole in tool steel with 70 per cent depth of thread, since an 8 × 32 has a small diameter in proportion to its pitch. Yet, the same pitch on a 14-inch tap would be entirely practical, since a ¼ × 32 tap has sufficient strength to tap even 100 per cent of thread. It can therefore be seen that the smaller the diameter and the coarser the pitch of the tap, the less percentage of thread should be tapped. This depend on the material, but in no case should it ever exceed 83 per cent.

8.93 EFFECT OF MATERIAL HARDNESS AND TOUGHNESS

The hardness and toughness of the material to a great extent govern the amount of material that a tap will remove. Generally, the harder and tougher the material, the less the percentage of thread. For example, on 3140 nickel steel it would be difficult commercially to tap 70 per cent depth of thread yet on a high-grade free-cutting; screw stock it would be entirely practical. There are many other tough alloy metals, such as copper, monel, nickel, bronze, and the various grades of alloy steels, in which it is difficult to tap a large percentage of thread. The minimum percentage should be adopted wherever possible.

8.94 EFFECT OF TAPPED-HOLE DEPTH

The depth of the tapped hole is the third important factor governing the percentage of thread. Calculations should be based on the length of engagement, i.e., the length of contact between a screw and a tapped hole measured axially; this length of engagement should equal the basic major diameter. It would be possible to tap greater percentages of threads if the depth of the tapped holes were less than their basic major diameter; therefore the percentage of thread should be reduced, wherever the tapped hole exceeds the basic major diameter. This is particularly true in tapping blind holes, especially with the smaller taps and the coarser pitches where there is difficulty in finding room for the chip.

When too great a percentage of thread is being tapped, the hole is being reamed instead of being threaded. This is due to the excessive pressure required. Before the tap can take hold, the chamfer starts reaming the hole instead of cutting the thread. With the proper percentage of thread very little pressure is required to start a tap. When too much pressure is applied, the tap lacks support in the thread and cuts away each succeeding thread as the tap revolves. This quite frequently happens in automatic screw machines when the cams are not correct for the lead of the tap. The cams may be too slow or too fast.

8.95 SPEEDS FOR TAPPING

A high-speed ground-thread tap should be able to operate at the same speeds as a drill, milling cutter, or any other metal-cutting tool, since the same steel and heat-treatment are used in a tap. It is true that, if the wrong style of tap is used and permitted to float or wobble in all directions, the speed must be cut down, but even at slow speeds, tapping is still unsatisfactory. A tap will work at the same surface feet per minute as any other metal-cutting tool if the proper style is used and is held rigidly and accurately as in other metal-cutting tools; in fact, a high speed ground-thread tap is relatively short-lived if operated at carbon-steel speed. Although this critical speed varies with the material, depth of hole, and pitch, a little care and study of each job will soon establish the maximum speed of each size of tap for each grade of material.

8.96 FACTORS GOVERNING TAPPING SPEED

In tapping, the feed cannot be governed, since it is entirely dependent upon the pitch, which determines the feed per revolution of the tap. For this reason no hard-and-fast rule can be laid down, nor can any set formula be applied to the peripheral speed of a tap, since the problem is greatly complicated by the many pitches and diameters involved. Conditions that govern the speed of a tap are:

1. Outside diameter of the tap
2. Pitch of the tap
3. Material being tapped
4. Depth of tapped hole
5. Percentage of thread in tapped hole

Example: To operate a ½ inch × 20 tap at 150 sfpm in free-cutting screw stock, the tap would have to revolve at 1146 rpm. However, the pitch of the tap enters into this to a large extent, as the tap must cut the thread in one pass. When tapping a ¼ inch hole with a ½ × 20 tap in screw stock, it is safe to operate this tap at 150 sfpm. Using a ½ × 13 tap instead of the ½ × 20 tap and operating the coarser-pitch tap at the same speed of 1146 rpm, considerably more metal would be removed per revolution and in fewer revolutions. In other words, with a ½ × 20 tap, five revolutions are required to tap a ¼ inch hole, whereas with the ½ × 13 taps only 3¼ revolutions are required. Since the depth is much coarser and more metal must be removed in less time, it is obvious that we should not operate the ½ × 13 tap as fast as the ½ × 20. Metal removed is excessive with the coarse pitch in comparison with the fine. Therefore, the speed must be slowed down and the amount of metal removed in a given time remains approximately the same.

If, instead of the screw stock cited above, tapping is to be done in the tougher and stringier nickel steel, more power will be required. Tapping ¼ inch holes with a ½ × 20 tap, the speed would have to be reduced to approximately 80 sfpm or 600 rpm. Again, if it is desired to change the tap to ½ × 13, the speed must be reduced to offset the increased metal removed.

A 1 inch deep hole cannot be tapped at the same speed used for a hole ¼ inch deep. Experience has proved that the speed must be reduced approximately 5 per cent every time the depth of a tapped hole is doubled. In other words, if a hole ¼ inch deep is being tapped at a certain speed and is increased to ¼ in deep, the speed must be reduced by 5 per cent. The speed must also be reduced in relation to the coarser pitch and to the material being tapped.

The percentage of the thread has a direct relation to the speed at which a tap may be operated. Obviously a 50 per cent thread, regardless of the material, may be tapped at a much higher speed than an 80 or 100 per cent thread. This again is governed by the amount of material removed. The more material removed per revolution, the more heat is built up in the tap per revolution.

8.97 MECHANICAL FACTORS AFFECTING TAPPING SPEED

The rigidity of the spindle, condition of the machine, running trueness of the tap, cutting fluid used, type of fixture, style of tap, and width of lands in the tap all have some relation to the speed of the tap. The rigidity of the spindle is very important. If a tool is permitted to wobble and float, the cutting tool cannot be operated at its maximum efficiency. The condition of the machine has some bearing on the speed at which a tap may be operated. Loose slides or turrets cause trouble, since after power is applied and the tap has entered the hole and has started to cut, any movement causes loading. Again, if the tap does not run true with the axis of the spindle, it will start tapping straight and then have a tendency to move at a slight angle. This also has some relation to the speed of the tap.

Sometimes the fixture used does not support the work properly while the tap is working either tapping or reversing, and it will cause trouble at high speed. To tap efficiently at high speed, the fixture must not only support the work but keep it from shifting.

Tapping cannot be done as fast horizontally as vertically. The principal reason for this is that the chips cannot be disposed of as rapidly, and there is always the danger at high speed of the chips reversing themselves with the tap and loading it rapidly. When tapping is done vertically, this problem is not so serious if the proper style of the tap is used.

When all conditions are correct for efficient tapping, the life of a tap at high speed is even greater than at slow speed, since a tap that is operated at its critical speed operates at its maximum efficiency and cuts more tapped holes between the grinds.

It is always advisable to increase the speed on the tap reversal to at least double that of the tapping speed, for example, tapping in at a speed of 1000 rpm and tapping out at 2000 rpm. If a tap is reversed at high speed and is not held rigidly, it has a tendency to cut backward and will soon be ruined.

In general, the harder the material, the slower is the tapping speed. When there is a combination of hardness and toughness such as is usually found in heat-treated nickel-steel alloys, it is highly advisable to tap in two separate operations, roughing and finishing.

8.98 FIXTURES FOR TAPPING

The first consideration in designing fixtures for tapping is safety, because many fixtures are fast-acting and hand-fed. Since many air-operated automatic or semiautomatic fixtures will clamp and index in less than 1/10 sec the importance of safety to the operator cannot be stressed too strongly.

The second consideration is the disposal of chips. The fixture should be designed to eliminate the necessity of cleaning it prior to locating the next blank. The angle of oil flow should be such as to wash out the fixture thoroughly, eliminating a cleaning operation.

The third consideration is the design of the fixture itself. It would be uneconomical to design an automatic indexing dial fixture to tap 25,000 pieces, when the job will never be run again; instead, a hand-fed simple post-type fixture should be used. Many hand-fed jobs in post-type fixtures run as high as 3000 tapped holes per hour.

The fourth consideration is the height of the fixture and the angle of feeding. No hand-fed fixture should be less than 4½ inch high, and they should be high enough to allow use of a chute if the blanks are automatically ejected.

Operators prefer to feed blanks from the right-hand side of the fixture rather than from the front.

The standardization of master base fixtures, with simple nesting blocks or clamping blocks, can simplify the design of tooling. When fewer than 10,000 pieces are to be tapped, the nesting blocks need not be hardened. Another feature is that spare parts, blank-nesting blocks, locating plates, etc., can be carried in stock and stored in the tool crib in much less space than the completed fixture unit. Not only do standardized master fixtures permit simpler and cheaper tooling, but they are generally a better quality than fixtures designed especially for single-purpose applications.

8.99 DESIGN HINTS

Too much clearance in the nesting block often causes considerable trouble in holding close tolerances; too little clearance slows up the feeding of the blank in its proper position.

Thin work must always be supported under the tap so that parts do not bend.

Sufficient chip clearance must be maintained so that chips will fall freely outside the fixture.

Some parts can be tapped without clamping by using a thin stripper plate to hold the part in the nesting block. When parts must be clamped, it is important that the fixture be self-entering. Sometimes at the point of clamping, the variation in different diameters is too great and causes tap damage, especially when the tap is reversed.

8.100 TAPPING IN TWO OPERATIONS

Although it is more economical to tap in one operation whenever possible, there are two factors that sometimes make it more economical to do it in two operations. The first factor is the depth of the hole and the closeness to the bottom, where these considerations may not permit sufficient chip room either in the hole itself or in the flute of the tap. Second, some of the various grades of heat-treated alloy steels are very hard and brittle and highly abrasive; others are very tough. All these alloy steels dull the tap quickly, and it is difficult to maintain an accurate tapped hole unless the work is done in two operations.

It is better to use a special undersized tap for the roughing operation and a standard tap for the second operation. To do this economically and efficiently, the first tap should be at least 0.010 inch undersize to and including 20 threads per inch and at least 0.015 inch undersize over 20 threads per inch. This gives the finishing tap a chance to take hold and cut metal instead of merely rubbing the previous threads out. This also gives the finishing tap a chance to correct any errors that are left by the roughing tap. If there is not sufficient metal left for the finishing tap; it will merely follow the first tapped hole and will not correct any of the errors such as lead, Bell-mouth and poor threads.

It is also important when tapping in two operations that the first operation with an undersized tap be just as accurate as the finishing operation.

Many jobs that are done in two operations could be trapped in one operation if the drilled hole were deeper. This is a common fault, and in many cases there is sufficient room to drill a deeper hole. The cost of drilling a hole slightly deeper is nothing in comparison with the saving in tap troubles. Trouble is often encountered when tapping deep, blind holes of some pitches if a coarser-pitch tap is used. This is due to the amount of material removed for the same diameter of the tap. For instance, a much deeper hole can be tapped with a ¼ × 28 tap in one operation than with a 1/4 × 20 tap. Also, a much deeper hole can he tapped when cutting only 50 per cent depth of thread than with SO per cent depth This of course, is due to the cubic inches of metal removed, which must be confined either in the flute of the lap or ahead of the tap, unless pulled out of the flute by a spiral- or-high-spiral-flute tap.

The grade of steel is an important factor in causing tap trouble, since many of our heat treated alloys dull a tap quickly even though there is sufficient room for chips. When close tolerances must be maintained

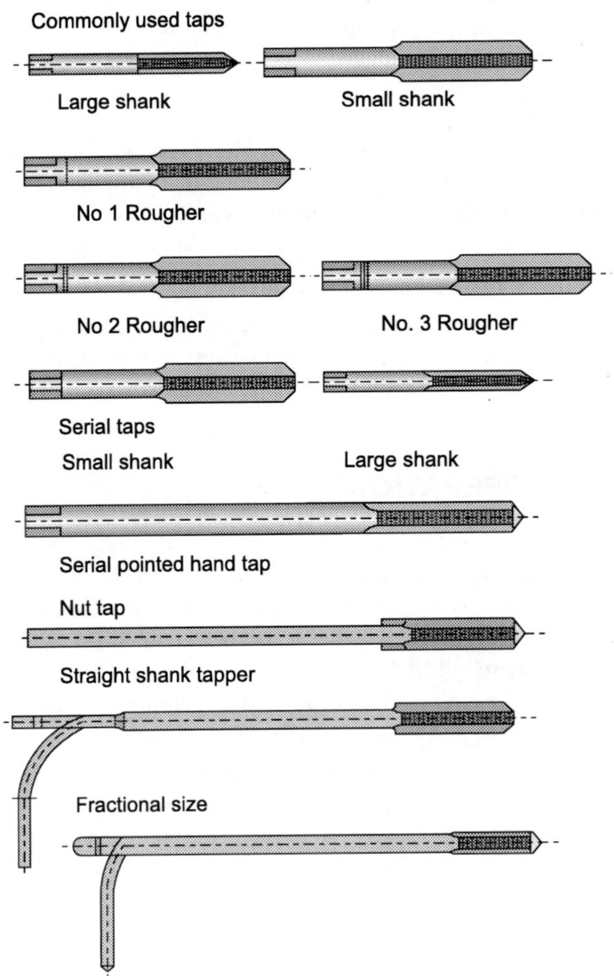

Fig. 8.142 Machine screw taps

on these steels, it is far better to use an undersized tap for roughing and a standard tap for finishing, as it is difficult to maintain size if the tap is even slightly dulled.

When blind holes are tapped in two operations, it is important that the first operation be at least one-half to one thread deeper than the second operation, so that the finishing tap need not remove a great amount of metal when it reaches the bottom of the tapped hole. Another point of importance is that the finishing tap should have a shorter chamfer than the roughing tap. Confining the cutting action on a few teeth for finishing makes it easier to maintain size and resharpen the tap.

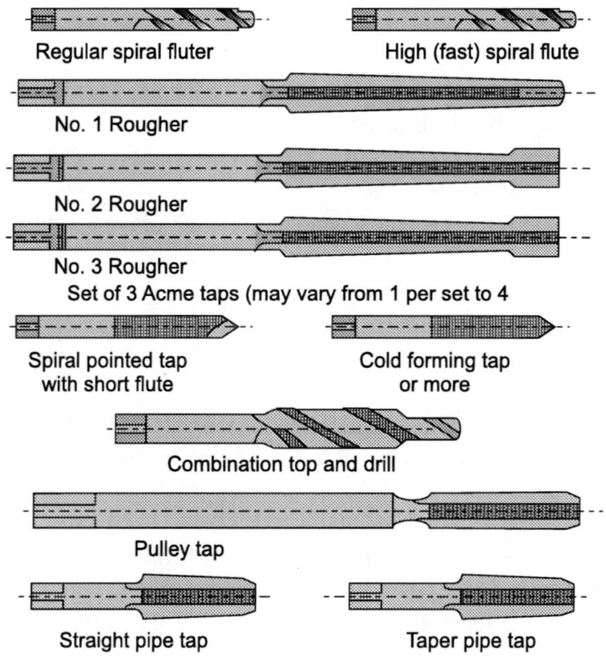

Fig. 8.142(a)

Solid Taps

Tap terms and definitions:

Styles of taps: Only the most commonly used styles of taps (**Fig. 8.142(a)** and their principal uses, together with the outstanding features of each, are given. Reference should be made to tap manufacturers' catalogues to ascertain the standard features of stock taps for specific dimensions and thread limits.

Regular hand taps: The plug hand tap is most widely used for production tapping because of its simplicity and availability. It is used for both hand and machine tapping. When intended for hand tapping, taps are usually furnished in sets of three of a size, namely, taper, plug, and bottoming. The dimensions of the three types are identical, their only difference being in the length of the chamfer at the point. The chamfer on the taper hand tap has a length of eight to ten threads; the plug hand tap has three to five threads chamfer, and the bottoming hand tap one to one and one-half threads chamfer.

For hand tapping, the taper hand tap should be used in through holes with coarse-pitch threads, as it assures straighter starting. It is also recommended for the harder metals. Plug hand taps can be used for tapping through holes in soft metals or fine-pitch threads.

Hand taps are available in fractional sizes from 1/16 to 4 inch diameter.

Serial hand taps: These are used for tapping deep holes, open or blind, by hand, especially in tough metals, because they cut much more easily and produce smooth threads. They are usually made in sets of three, numbered, respectively, 1, 2, and 3, with one, two, or three rings, respectively, on the shank near the square.

The No. 1 tap makes the roughing cut and is smaller both in major and pitch diameter. The No. 2 tap cuts a little fuller thread, and the No. 3 finishes the thread. These taps are available in the same sizes as regular hand taps.

Spiral-pointed Taps. These taps are usually used for production tapping in machines, not for hand tapping. A spiral point causes the tap to cut with a shearing action and throw the chips ahead of the tap, making it most suitable for through holes, but it may be used for bottoming on special applications. Except for the flute depth and the spiral point, the physical dimensions are identical to the regular hand tap.

This tap has fewer and shallower flutes than most hand taps, as they are not required for chip passage. The taps can be run at a higher speed and they require less power to drive than regular hand taps.

Machine-screw taps: These are small-sized hand taps, except that they are designated by Nos. 0 to 14, instead of fractional sizes. Machine-screw sizes for tapped holes below ¼ inch diameter should be specified, because screws in machine-screw sizes are more readily obtainable than fractional-size screws below ¼ inch.

Spiral-fluted hand taps: These taps are made with spiral instead of straight flutes. They are used to draw the chips out of a blind hole in stringy materials such as aluminium, magnesium, brass, copper, and die-cast metal. Spiral-fluted taps are available in ground-thread plug and bottoming styles, Nos. 3 to 12 machine-screw sizes, and ¼ to ½-inch fractional sizes. High spiral styles are most effective in drawing chips out.

Nut taps: These are used in nut-threading machines, bolt cutters, drill presses, etc., where the nature of the work requires a tap with a long shank, also for small quantity production of nuts, especially in tough materials. The threaded portion has a long chamfer to distribute the cutting over many threads, also a point small enough to facilitate easy starting. The shank is long and smaller than the minor diameter of the threads, permitting several nuts to accumulate on the shank after tapping. When the shank is filled, the tap is taken out of the machine and the nuts removed. They are available with cut threads and ground threads in sizes from 3/16 to 2 *Yz* inch diameters.

Straight-shank taper taps: These taps are used for production tapping of nuts. The shank lengths are longer and the thread lengths are shorter than nut taps. The chamfer is 11 to 12 threads for coarse-pitch taps and 15 to 17 threads for fine-pitch taps.

Bent-shank taper taps: This style of tap is used in automatic tapping machines for production tapping of nuts. The chamfer and the length of thread differ depending on the application. This style of tap is also made with a reusable shank, and only the threaded-tapping end (tap nib) is replaced when worn.

Pulley taps: These are used for tapping oil-cup and setscrew holes in pulley hubs. The shank is made to the basic major diameter of the tap to act as a guide through a hole in the pulley rim. There is more than one overall length of tap for each size because of the varying diameters of pulleys. Pulley taps are frequently used for applications where a long shank is required. The chamfer on the thread is plug style. Fractional sizes from ¼ to ¾ inch diameter are available.

Stove bolt taps: These are used for tapping nuts for stove bolts made to hold stoves together. Two standards are recognized today, the old or "manufacturers' standard" and the comparatively new "American standard." All sizes of the latter conform to the unified regular thread specifications so that regular taps can be used. None of the manufacturers' standard threads are regular screw-thread sizes; they are a distinct, separate standard. The sizes range from 1/3 to 1/2 inch diameter with a plug-style chamfer.

Taper pipe taps: These are used for tapping pipe fittings on steam, fluid, chemical, and compressed gas systems. Threads are tapered ¾ inch per foot to make a tight joint. When used on tough, stringy material, the interrupted thread tap prevents tearing because on each land the threads cut in alternate grooves, leaving the whole thread section to support cutting on only one side, whereas full-form, threads cut on both sides. Taper-thread taps cut on the entire length of contact in comparison with straight taps, which cut only on the chamfer.

When it is desirable to tap fittings which can be assembled without the use of compound, ground-thread pipe taps having the American standard dry seal pipe-thread form are used. The sizes available range are from 1/16 to 4-inch nominal pipe size.

Straight pipe taps: These are used for tapping lock nuts and couplings for taper-thread assemblies and have the same cutting action as any regular hand tap. The thread dimensions and threads per inch are governed by the corresponding taper-pipe-thread size, the size at the gauge notch being the basic dimension for straight pipe threads.

Combined tap and drill: These are used for drilling and tapping threads in one operation where the drill is through cutting before the tapping begins; also used for tapping cored holes. Water and gas companies use these tools together with a special device for connecting a service pipe to mains under pressure.

Acme taps: These are used to tap the nuts for translating, pressure, and clamping screws used in many varieties of applications.

Each application of acme threads presents its own problems. The same is true with acme taps, each tap or set of taps being designed for its specific job. The design of these taps is governed by diameter, pitch, lead, length of nut, material, through or blind holes, and method of use.

Full working strength of the screw is matched by a nut 1 *Yz* to 2 screw diameters long. Errors of lead increase with the length of the gut and make the added length of little value.

Choice of Taps

Tap grades: Taps may be regarded as belonging to one of two grades: cut-thread, or ground-thread.

Cut-thread taps are threaded to size before hardening; therefore, they are inspected to wider tolerances that must allow for changes in the steel due to heat treatment. They are used for tapping short nuts used for fastening and many other places where loose fits and moderate pressures are encountered.

Ground-thread taps are finish-threaded after hardening and are inspected to much closer tolerance for lead and pitch diameter.

Lead tolerance is ±0.0030 inch/inch for ungrounded thread taps and ±0.0005 inch/inch for ground thread taps. Pitch-diameter tolerance on ungrounded thread taps is from 0.001 to 0.0055 inch/inch, whereas ground threads vary in pitch-diameter tolerance from 0.0005 to 0.0025 inch/inch depending upon the size. For example, the pitch diameter on ungrounded thread taps, size 74-20 NC, is 0.002 inch and on the 1-8 NC is 0.003 inch, whereas ground thread taps over this size range have a tolerance of 0.0005 inch/inch.

Screw-thread assemblies that must carry heavy loads or be closely fitted require screws and nuts with accurate lead; otherwise off-lead condition will impair the working of the whole assembly. Ground thread taps will produce a nut to close lead tolerances.

It is further necessary for the screw to be threaded with as close a lead as the nut to avoid conditions of lead, even though ground thread taps are used to thread the nut.

Carbon-steel taps, being cut-thread, are further limited to materials and speeds where carbon-steel tools prove economical. Holes tapped in brass and other materials of a non-abrasive nature can be produced with carbon-steel taps. They are also used in machines that do not have the power to drive high-speed steel taps to their best efficiency.

High-speed steel taps are more generally used than carbon-steel taps. High-speed steel tools can be used where higher cutting temperatures are generated without danger of losing their temper. Abrasive material and tough alloy steels require high-speed steel taps. Ground thread high-speed taps are used mostly because of their greater accuracy and ability to produce tapped holes for the closer fits. Ground-thread taps have much better cutting surfaces, because any scale or surface imperfections are removed after heat-treatment.

Cut thread taps may be used under normal conditions and in average materials for producing tapped holes to this classification.

8.101 METHODS OF PRODUCING THREADS

Production of threads depends on the following factors:

- Type and number of components required,
- The accuracy of the thread and its surface finish,
- Availability of machine tools,
- Skill of the operator, etc.

Methods of producing threads are by:

- Using hand tools like taps and dies
- Using single point cutting tool on lathe
- Using multi-point cutting tool called chasers
- Using coventry die-head and collapsible taps in
- Production lathes
- Thread rolling
- Thread milling
- Thread grinding
- Thread casting (die casting or moulding)

8.101.1 Using Taps and Dies (Fig. 8.143)

These are commonly used for general purpose bolts and nuts. Taps are used to produce internal threads. Dies are used to produce external threads.

Taps and dies can be used only for producing standard 'V' threads for both coarse and fine pitches. Machine taps are also available for cutting threads.

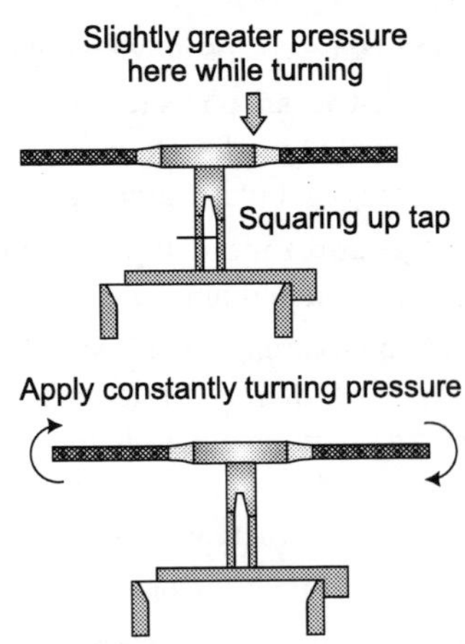

Fig. 8.143

8.101.2 Using Single Point Cutting Tools on Lathe (Fig. 8.145)

Both internal and external right hand and left hand threads can be cut by this method. Any form of thread to a required pitch can be cut or produced by using the corresponding tools. Accuracy of the thread depends on the skill of the operator.

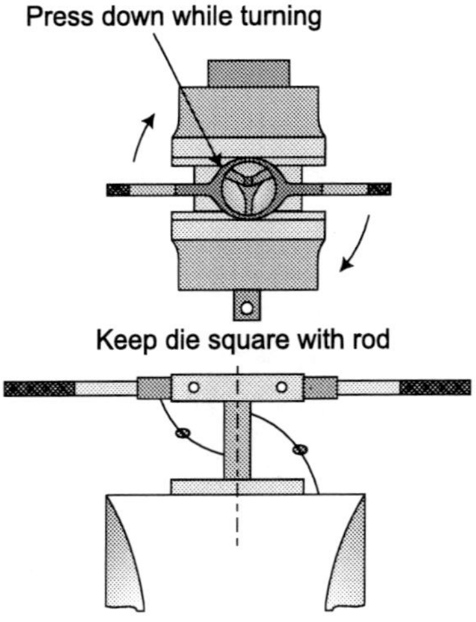

Fig. 8.144

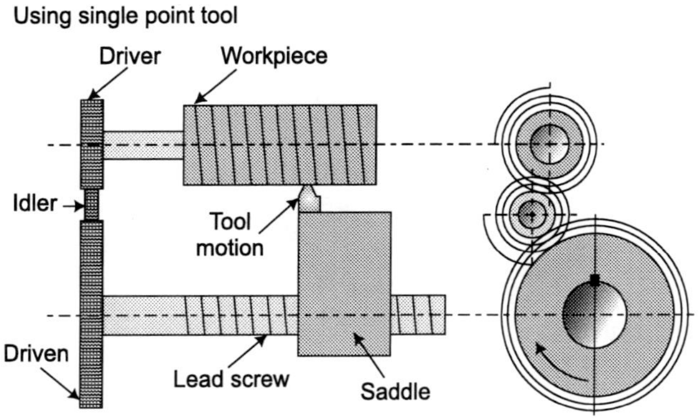

Fig. 8.145 Using single point tool

8.101.3 Using Chasers (Fig. 8.146)

Chasers are multi-point cutting tools to produce external 'V' threads. There are different types of chasers, each one having its own special characteristics.

Usually hand chasers are used to finish the thread, and machine chasers are used for producing the threads. Machine chasers are used in conjunction with die-heads.

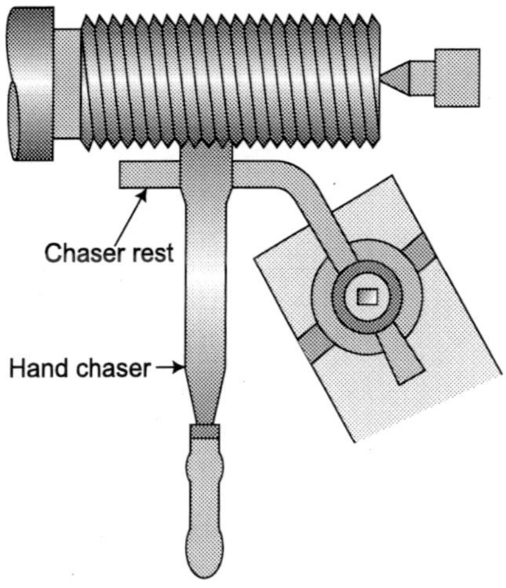

Fig. 8.146 Using chaser

8.101.4 Using Coventry Die-head and Collapsible Taps (Fig. 8.147)

It is used in mass production to produce threads on Capstan, Turret lathes and automats. A highly skilled operator is not essential for cutting threads, but for setting, skilled setters are required. This method is limited to producing 'V' threads.

Collapsible taps are used with floating holders for producing internal threads.

8.101 5 Thread Rolling (Fig. 8.148)

It is used in mass production. In this method, the thread is not cut and chips will not be produced. Thread is produced by rolling. Rolling is a process of cold forging resulting in a plastic deformation. The job is turned to the pitch diameter before rolling. The rollers may be flat or of disc type. Threads produced by rolling have better strength and a good finish.

8.101.6 Thread Milling (Fig. 8.149)

In this method, threads are produced by thread milling cutters. The operation is performed on special thread milling machine to produce accurate threads in small or large quantities. In this operation, there are three

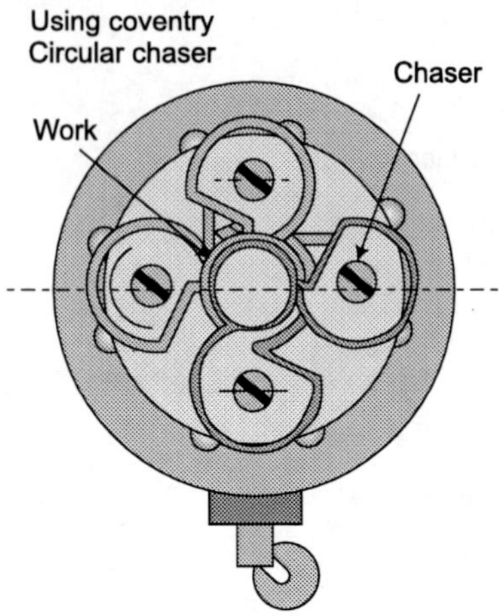

Fig. 8.147(a)

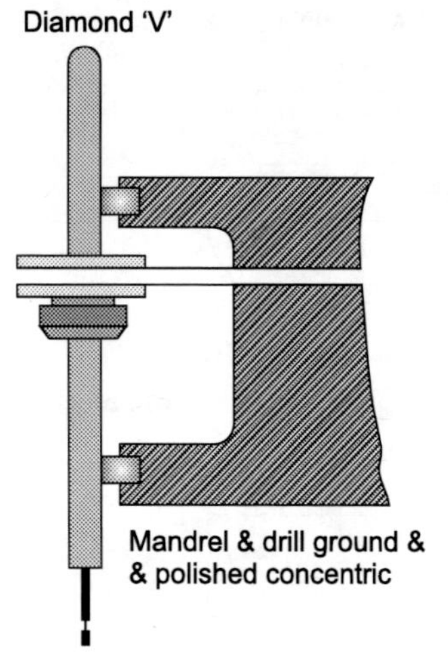

Fig. 8.147(b)

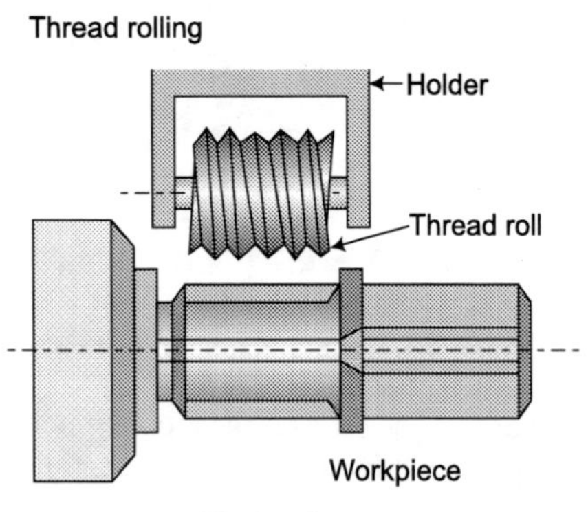

Fig. 8.148

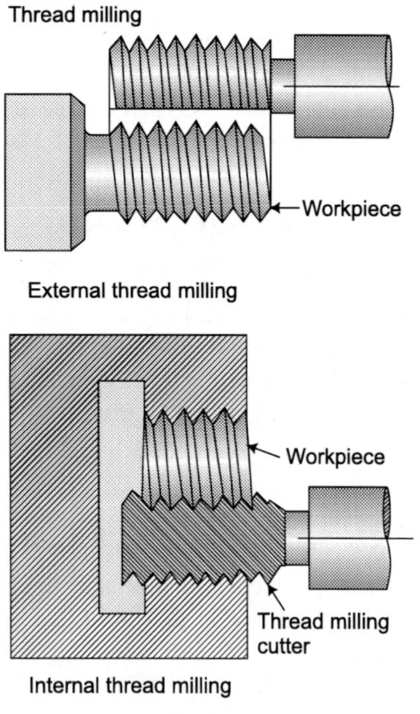

Fig. 8.149

driving motions, that is, for cutter, work, and longitudinal movement of the cutter. The thread is completed in one cut by setting the cutter to the full depth of the thread, and then feeding it along the entire length of the workpiece. Internal thread milling can also be performed to produce accurate internal threads.

8.101.7 Thread Grinding (Fig. 8.150)

Special grinding machines are required for thread grinding. External and internal threads can both be ground.

There are two main types of thread grinding, i.e., single rib and multiple ribs.

Single rib grinding involves the use of a narrow grinding wheel which is shaped to the required form of thread.

Multiple rib grinding is done with a grinding wheel with many grooves on its face. This type of wheel grinds many threads at one time **(Fig. 8.151)**

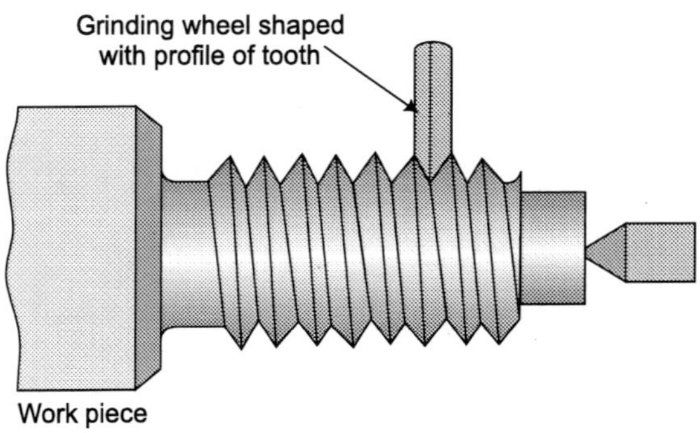

Fig. 8.150 Thread grinding

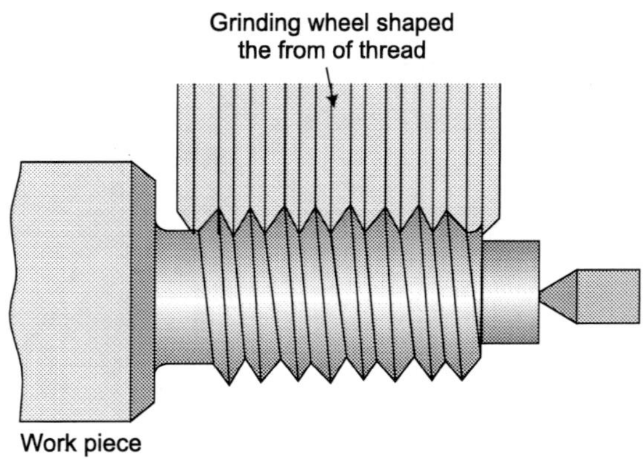

Fig. 8.151 Multiple rib grinding

8.101.8 Thread Casting

Casting can be used to produce crude threads. These threads are produced either by die casting or mould for plastics. Usually threads on non-metals are produced by this method.

8.102 MICRO DRILLING

Micro drilling is the accurate placement and production of holes of 0.06 mm (0.020 inch) diameter and smaller. Recent advances in the field have led to such micro drilling to describe the production of holes smaller than 0.06 mm, and the present state of the art defines the smallest as 0.000635 mm diameter. Pivot- or spade-type drills are used, and the drills may be made from steel, tungsten carbide, ceramics, refractory material, or diamond.

To understand micro drilling technology, two general misconceptions need first to be corrected. The notion that drill speed (in rpm) should increase as the drill diameter grows smaller is not correct, and is based on an assumption that surface feet per second at the drill should remain constant for any given material. The second erroneous concept is that the working life decreases in direct proportion to the decreasing sizes of

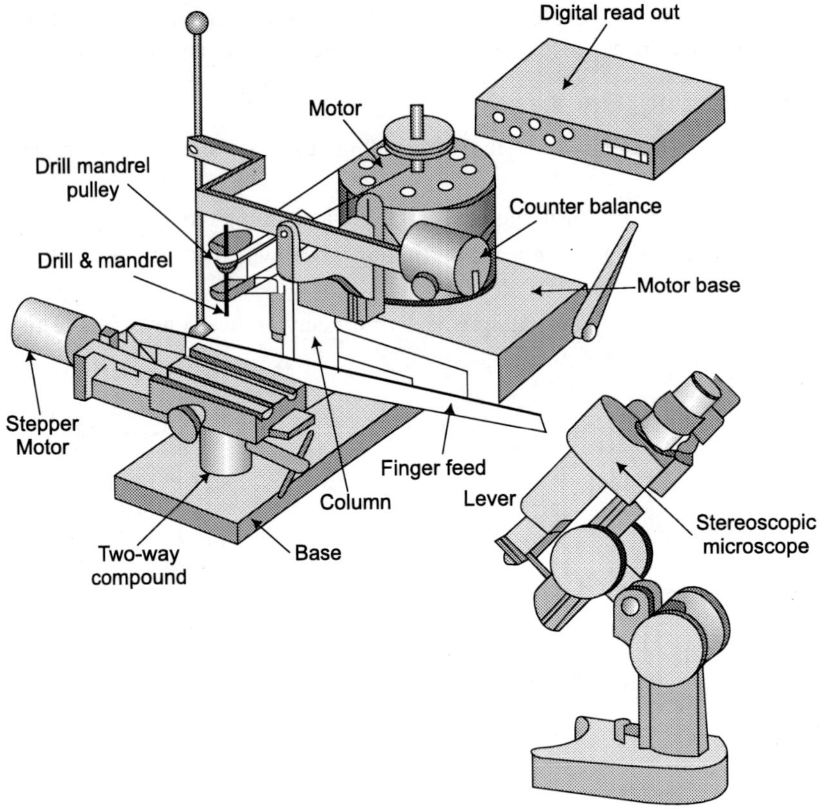

Fig. 8.152 Micro drill equipment

drills. Micro drills operate optimal at 1200 to 3500 rpm, and a 0.1524 mm precision micro drill can cut more linear working feet of material in its life than can a ¼-inch twist drill.

Most basic micro drilling machines are simple in design, and resemble the machine shown in **(Fig. 8.152)**. The normal techniques of gripping and rotating a drill (chucks, collets, tapers, direct drive, etc.) are themselves much too inaccurate for micro drilling, which requires absolutely no rotational eccentricity at the drill tip. The tool is held in place against four diamond bearings by the tension of the elastic belt **(Fig. 8.153)**. Only the diamond bearings, inset into the 450 jaws of the tool holder, contact the surface of the mandrel. These bearings must be dressed and ground with the greatest care, as tolerances are extremely close.

Hand motions of the micro drill operator are translated by a feed-control mechanism at a leverage ratio of 20:1, giving the operator greatly amplified sensitivity control. Work progress is usually viewed through a 20-power microscope. Working with identical magnification and mechanical advantages allows the operator to do micro drilling in what appears to him as a real-world scale. Micrometer stops limit the drill travel; adjustment increments are 1 μ each, and the adjustment can be made manually or by a stepper motor.

The choice of tool material for micro drills depends upon specific characteristics of the material being micro drilled. Quality tool steel has been used for all sizes of drills down to the smallest; the limitation on size of steel micro drills is the grain

Structure of the metal. Drills are also available in tungsten carbide, special alloy tool steel, ceramics, refractory materials, and diamond. Diamond is the only material able to drill certain of the ceramics and refractory.

Because rotational eccentricity is zero, mandrels and drills can be interchanged, and the centre line of each successive tool will be correct and remains entered. The drill mandrel, which bears against the diamond bearings, is made of a carefully machined, lapped, and hardened alloy that will ride on the bearings without scoring or scribing. The mandrel must show no thermal expansion over the range of operating temperatures (50 to 1500 F). Each mandrel contains, or is equipped with, a pulley to accept the belt drive. The pulley can be attached to the mandrel or can be removed for use on many mandrels through a collet. The concentricity of the pulley with the axis of the mandrel and drill is not critical.

Drills supplied by the manufacturer are well within the tolerances of micro drilling for the given drill size and are so graded. Variance between two drills of the same size should be controlled to within a thousandth of 1 per cent.

8.103 MICRO MACHINING

Because micro drilling alone is almost never a final operation on any hole and further finishing work is needed, the development of a micro machining tool changer was an almost necessary next step. A drilled hole usually requires reaming, broaching, lapping, and polishing by tools that must operate on the exact axis of the original micro drill and must be in perfect concentricity to the micro drilled hole. The mandrels, bearings, and tool mounting as shown on the tool changer in **(Fig. 8.152)** are identical to those of the micro drill in **(Fig. 8.153)**. The difference between the machines lies in the ease of changing tools. With the micro machining tool changer, only a few seconds are required to remove and replace a tool.

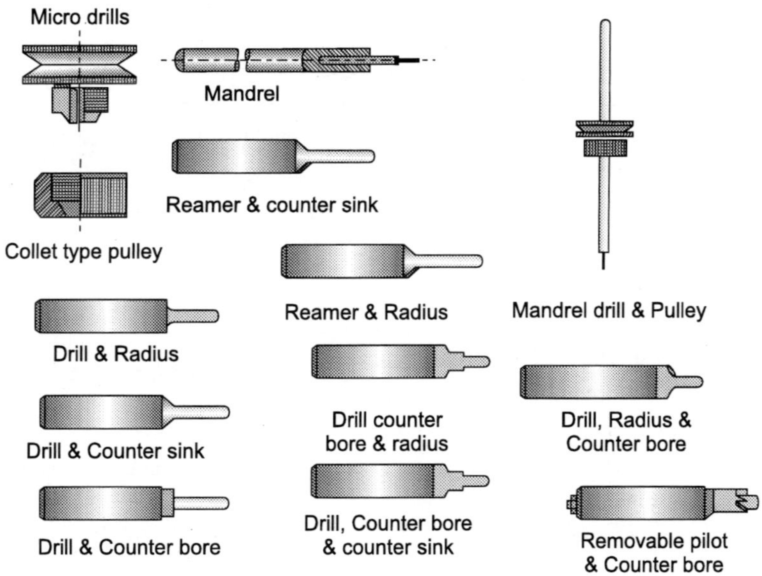

Fig. 8.153

The tools need not be limited to those requiring rotation. Micro burrs, punches, end mills, shavers, trepans, lathe tools, box tools, broaches, reamers, and lapping or polishing tools can be used in addition to the micro drills. The tool changer will also accept lasers and ultrasonic transducers. The initial holes made by lasers almost always require additional micro machining.

Both micro drilling and micro machining can be fully automated, or automated to the extent acceptable to production economics. Where work is defined and long-term, such as continuing production run on micro drilling nozzles, total automation, using stepper motors, for example, can regulate an ongoing work procedure more accurately than can an operator.

Like the micro drill, the micro machining tool changer has mechanical stops to check the motions of the tool and to supply necessary motions to the workpiece. Adjustment, either manual or automated, of these stops allows in feeding in increments of 1 **μ** or less.

Tolerances on the drills and tools used are given in three classes: (1) +0.0000 inch, –0.0000 inch; (2) +0.0001 inch, –0.0001 inch; and (3) +0.0002 inch, –0.0002 inch. The point angle on drills for steel and general use is 1350 while the angle for softer materials is 1180. The clearance angle is 80 for hard materials and 150 for soft materials. For optimum results in micro drilling and micro machining a particular material, the tool manufacturer should be consulted.

8.104 MICRO DRILLING SPEEDS

Optimal micro drilling is done at speeds of from 1300 to 3500 rpm, with the faster speeds used for drills larger than 0.020 inch. Diamond and refractory material drills have special speed requirements based upon their ability to withstand torque and the problems of stock removal. An experienced and skilled operator watching the drilling can usually select the best speed for a given drill and material. However, consultation with the tool manufacturer is the best and wisest procedure. Micro machining, because of the many operations possible with the 1001 changer, requires speeds ranging upto 5,00,000 rpm.

EXERCISE

1. What does good formation of chips with the twist drill depend on?
2. What does the lip clearance of the cutting lips serve for?
3. What influence has the centre line between cutting edges on the chip removal?
4. Why should greater holes be spot-drilled?
5. What role do the helical sloping flutes of the drill play?
6. What materials are not cooled when drilled?
7. Why are drills with a small pitch used for drilling soft materials?
8. How does the twist drill obtain a safe guidance?
9. What kind of errors may occur when grinding the drill by hand?
10. Which are the means used for measuring the angle of point, the lip clearance and the position of centre line between the cutting edges?
11. How is it possible to recognize whether the twist drill is made of tool steel or high-speed cutting steel?
12. What influence has the chucking of the drill on the accuracy of a drilled hole?
13. What must be observed when clamping twist drills with a cylindrical shank and by tapered shank?

14. Why must small workpieces always be firmly and safely clamped?
15. Describe the possibilities of clamping workpieces.
16. What are the different types of drill?
17. Draw a sketch of twist drill explain its parts.
18. What are the causes of defects drills ?
19. Give a summarized account of cut angle of drill for different materials.
20. What precautions are necessary for drill grinding?
21. What is meant by drill's feed speed? What purpose these serve for different materials?
22. What causes defects in a drill ?
23. Which are the different types of drills?
24. What are the special features of radial drill machine? Draw a sketch of radial drill and explain its work process.
25. What is the difference between gang drill machine and multi spindle drilling machine?
26. What functions does a drill machine perform?
27. Comment on the following:
    (a) Drill chuck
    (b) Drill sleeve
    (c) Drill drift
28. What are the functions of tap die?
29. Draw a sketch of tap and show its parts.
30. Why has the spiral countersink a better guidance than the twist drill?
31. Why are countersinking for countersunk not made with a countersink?
32. Why is the pilot guide hardened?
33. What are the advantages of exchangeable pilot guides?
34. Why do the heads of countersunk rivets project a little?
35. How do too deep countersinking come about?
36. What is to be observed while resharpening countersink?
37. What is the purpose of the pilot point of a reamer?
38. What is the purpose of chamfers?
39. What is the advantage of spiral-fluted reamers?
40. For what kind of work are adjustable reamers used?
41. How is it to be avoided that reamers are placed in an inclined position?
42. Why must reamers only be moved in cutting direction?
43. Explain how a conical boring hole is to be made and do it yourself in your workshop.
44. What are the uses of a reamer?
45. Draw a sketch of a reamer, showing all its parts.
46. What are the different types of a reamer?
47. What is the difference between a hand reamer and a machine reamer?
48. Write about the uses of an adjustable reamer.

49. What is the function of a screw extractor?
50. What is the purpose of the pilot point of a reamer?
51. Explain how a conical boring hole is to be made.
52. Produce the helical line on drawing paper.
53. Name the five most important sizes of a thread.
54. What forms of thread are used for spindles?
55. For what purposes are angular threads used?
56. What are the characteristics of a) the metric thread and b) the Whitworth thread?
57. How are pitch errors to be recognized?
58. Why are thin-walled workpieces fitted with fine threads?
59. What is the difference between a) Whitworth fine thread and a Whitworth pipe thread?
60. How is it possible to recognize a) a right-hand thread and b) a left-hand thread?
61. What are the effects of the difference of a threading die and a stock and die?
62. What must be observed when chucking split threading dies into the die stock?
63. How are threading dies adjusted?
64. What function has the chamfer of the threading die?
65. In which way is it possible to obtain easily and correctly spot facing by means of the threading die?
66. Why is it necessary that mostly three or two different taps cut into a rough hole when making female threads?
67. Describe the manufacture of a threaded hole.
68. Why is it necessary to rough tap blind holes deeper than the depth of thread?
69. What must be observed when selecting the tap wrench?
70. How can rough threads be avoided?
71. What are the means necessary for measuring the outer and the core diameters of a thread?
72. What are the means for measuring the pitch of a thread?
73. How is it possible to establish the dimensional accuracy of the flank diameter?
74. Produce the helical line on drawing paper.
75. Name the five most important sizes of a thread.
76. What forms of thread are used for spindles?
77. For what purposes are angular threads used?
78. What are the characteristics of: a) the metric thread and b) the Whitworth thread?
79. How are pitch errors recognized?
80. Why are thin-walled workpieces fitted with fine threads?
81. What is the difference between a Whitworth fine thread and a Whitworth pipe thread?
82. How is it possible to recognize: a) a right-hand thread and b) a left-hand thread?
83. What are the effects of the difference of a threading die and a stock and die?
84. What must be observed when chucking split threading dies into the die stock?
85. How are threading dies adjusted?

86. What function has the chamfer of the threading die?
87. In which way is it possible to obtain easily and correctly spot facing by means of the threading die?
88. Why is it necessary that mostly three or two different taps cut into a rough hole when making female threads?
89. Describe the manufacture of a threaded hole.
90. Why is it necessary to rough tap blind holes deeper than the depth of thread?
91. What must be observed when selecting the tap wrench?
92. How can rough threads be avoided?
93. What are the means necessary for measuring the outer and the core diameters of a thread?
94. What are the means for measuring the pitch of a thread?
95. Define threads and their advantages.
96. What is the difference between left-hand threads and right-hand threads ?
97. Comment on the following:
    (a) Pitch of threads, (b) Pitch diameter, (c) Minor diameter, (d) Major diameter,
    (e) Depth of thread. Show the above by drawing their figures.
98. What do you understand by pitch of the thread and lead of thread?
99. What are the different standards thread?
100. Draw a figure of acme thread. Mention its uses.
101. What is the function of worm thread.
102. Comment on the following:
    (a) Crest (b) Root (c) Thread angle (d) Helix angle (e) Number of threads.
103. Explain in brief the main sources of thread making.

Hand Tools

Chapter 9

In this chapter we shall discuss handy tools which are frequently used at the workplace. Jobs like assembling, dismantling and maintenance require various kinds of tools. In addition, other important tools used in workshops have also been described in detail. These tools are:

- Vice
- Hammer
- Screw drivers
- Pliers
- Spanners
- Wrenches
- Clamps

9.1 VICE

In a workshop, it is absolutely essential to hold a job strongly for the purpose of carrying out filing, drilling, chipping, milling, etc. The mechanical contrivance used for holding jobs is called a vice. For different jobs, different kinds of vices are used. They are:

1. Bench vice
2. Machine vice
3. Pipe vice
4. Leg vice
5. Carpenter vice
6. Hand vice
7. Pin vice
8. Tool maker's vice
9. Drill vice

9.1.1 Bench Vice (Figs. 9.1 and 9.2)

The body of this vice is made of cast iron. Both of its jaws are made of tool steel. Spindle handle is of mild steel. This is the most widely used vice in a workshop. It contains the following parts:

1. Base
2. Spindle

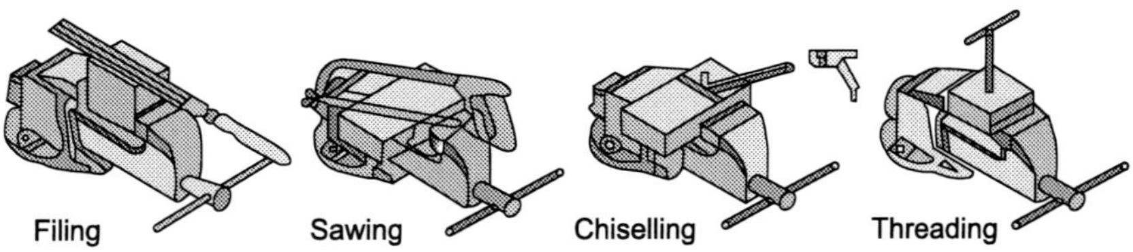

Fig. 9.1 Uses of bench vice

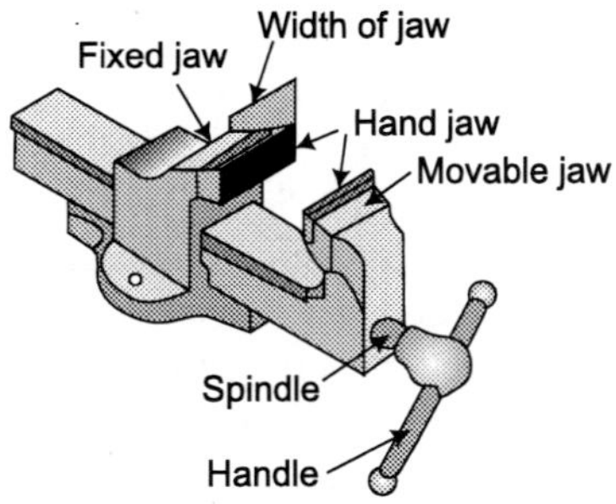

Fig. 9.2 Parts of vice

3. Handle
4. Fixed Jaws
5. Movable Jaw
6. Jaw Plate

Vice clamps or soft jaws (Fig. 9.3): To hold a finished work use soft jaws (vice clamps) made of aluminium over the regular hard jaws. This will protect the work surface from damage.

Do not over-tighten the vice as, otherwise, the spindle may be damaged.

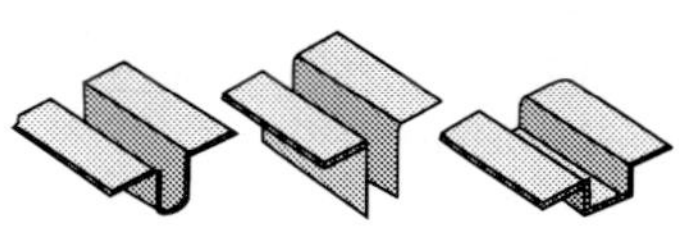

Fig. 9.3 Clamps for soft jaws

Construction of bench vice: Fixed jaw is cast along with the vice. movable jaw is fixed with it. In both the jaws, tool steels plate are fixed with screws, which have dents. It is because of these dents that the grip of the job is strong. When the jaw plate is damaged it can be replaced.

Spindle is fitted opposite to the movable jaw, which is fixed with the help of a pin and a spring. This spindle sets in the threaded box-nut tightened in the fixed jaw. In the spindle and the box nut square threads are cut. Handle of the round rod is fixed in a hole made in the part of the vice. By revolving this handle the movable jaw moves ahead or backwards. Job is held between the two jaws. The size of the vice is taken from its breadth. Vice is tightened on the work table with two or three bolts.

According to the utility and performance, the bench vice is of the following types:

1. Simple parallel jaw bench vice
2. Swivel base bench vice
3. Quick release bench vice
4. Combination bench vice

5. Machine vice
6. Leg vice
7. Pipe vice
8. Hand vice
9. Tool maker's vice
10. Pin vice

Simple bench vice: The movable front jaw is moved in a circular way against the fixed jaw thus only chucking on some points. The jaws are beaten up and hardened.

Swivel base bench vice: This vice is a simple bench vice. Its lower base is constructed in such a way that it could be swivelled. For swivelling the tightening lever fixed at the base is loosened and after swivelling, again the lever is tightened. One advantage of using this vice is that having fastened the job once, it can revolve the job in any direction for doing different operations like filing, chipping, etc. It saves time.

Quick release vice (Fig. 9.4): In this type of vice, buttress threads are cut on the spindle and a half nut is used with it. On one side of the spindle a release lever is fixed, which can be pressed to pull out movable job without revolving the spindle. When the lever is pressed the half nut is pressed down and the contact between the spindle and the nut is disconnected. As soon as the pressure from the lever is released the nut comes to its previous position with the help of a spring and at this time this vice works like an ordinary vice. The main facility provided by this vice is that for carrying out various operations like filing, chipping, scrapping, etc. The job can be frequently fastened or removed.

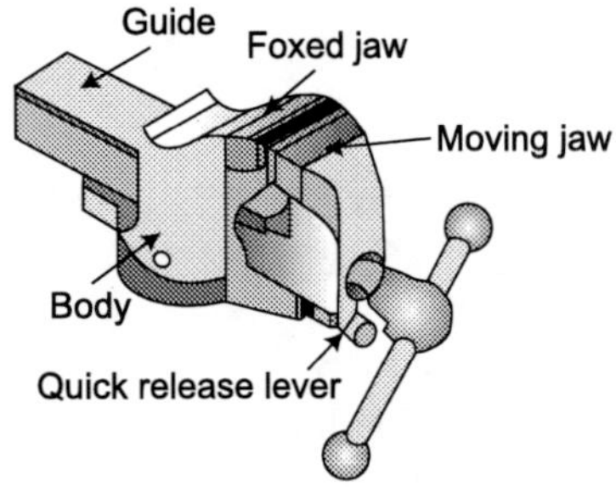

Fig. 9.4 Quick releasing vice

If the trigger at the front of the movable jaw is pressed, the nut disengages the screw and the moveable jaw can be set in the desired place quickly.

Combination vice: Under both the parallel jaws, "V" shaped jaws are also made. Thus, along with flat job, round pipes and rods, etc. can be easily held.

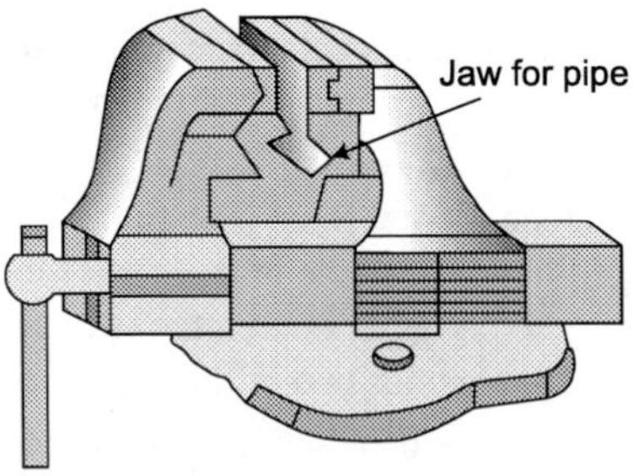

Fig. 9.5 Combination vice

9.1.2 Machine Vice (Fig. 9.6)

It is necessary to hold a job strongly for carrying out drilling, milling, slotting or shaping. For holding the job for these operations a machine vice is used. This machine vice is fastened on the table. For fastening it on the table "T" bolts are used, which fasten the vice firmly and easily in the "T" slot made in the table. Machine vices are of the following types:

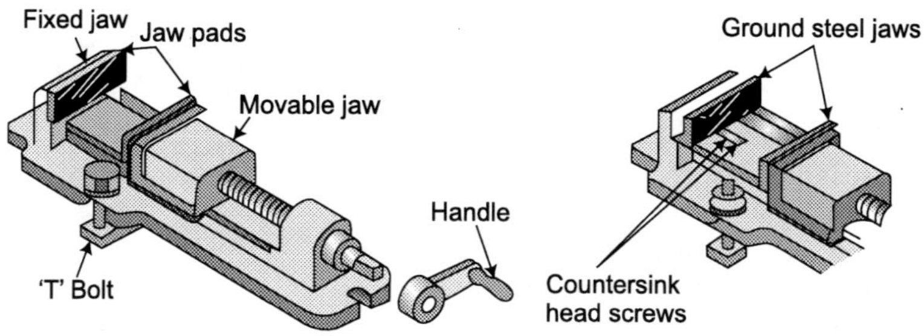

Fig. 9.6 Machine vice

Machine vices are used on drilling, shaping, and milling machine tables to hold workpieces with parallel surfaces.

The main parts of a machine vice are:

- Base
- Fixed jaw
- Movable/sliding jaw
- Jaw pieces/jaw pads
- Handle
- Spindle.

The base is bolted directly to the machine table by means of tee slot bolts

Fixed jaw is an integral part of the base.

Movable jaw is mounted on an accurately machined guide.

This is operated by means of a handle.

Jaw pads are hardened and ground. They are held in position by means of countersink head screws or socket head screws.

Machine vices are available in a wide range of sizes and types.

1. Plain or fixed vice
2. Flanged machine vice
3. Swivel machine vice
4. Universal machine vice
5. Vertical machine vice

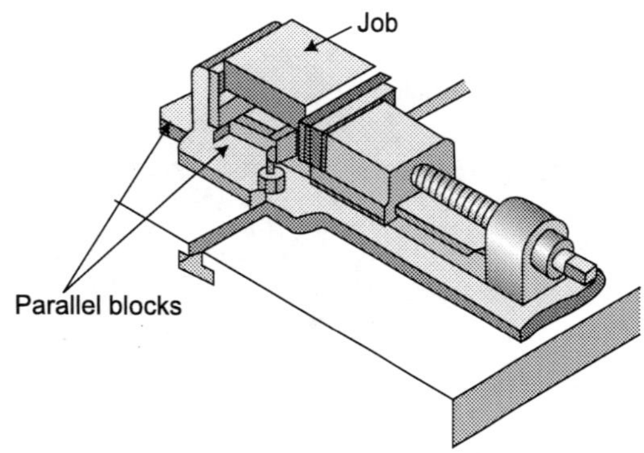

Fig. 9.7 Fixed or plain vice

Plain machine vice (Fig. 9.7): This machine vice is of stronger construction and used for general purpose heavy duty work. Its fixed jaw is long in size. The movable jaw moves forward or backward on revolving the spindle.

Flange machine vice: This is also called heavy duty machine vice. This vice is used for fastening such jobs which get frequent shocks at the time of machining. On one end of the spindle a square end is made in which a box type handle is fixed to revolve it.

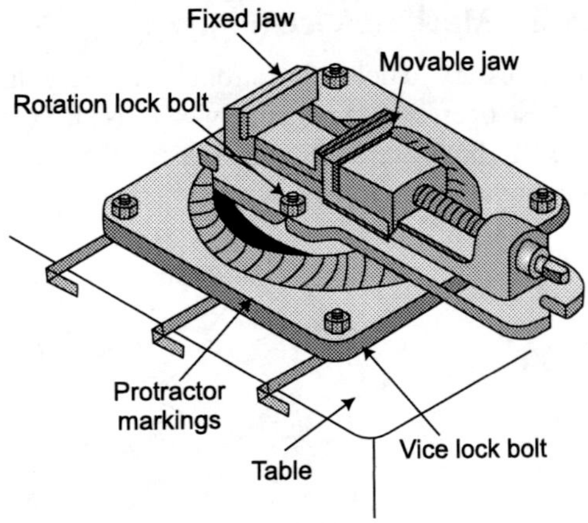

Fig. 9.8 Swivel vice

Swivel machine vice (Fig. 9.8)

In construction and shape this vice is similar to flange machine vice but its base is different. With the help of this base the vice can be revolved at any angle. Its advantage is that without separating the job from the vice straight or angular machining can be done. This can be set at any desired angle in the horizontal plane with the help of a graduated disc.

Universal machine vice (Fig. 9.9)

This is a highly useful machine vice. The speciality of this vice is that with it the job can be held in vertical, horizontal or angular position. It is mostly used in tool-rooms. It can be set at any desired angle in both horizontal and vertical planes.

Another position of universal machine vice is shown in **Fig. 9.10.**

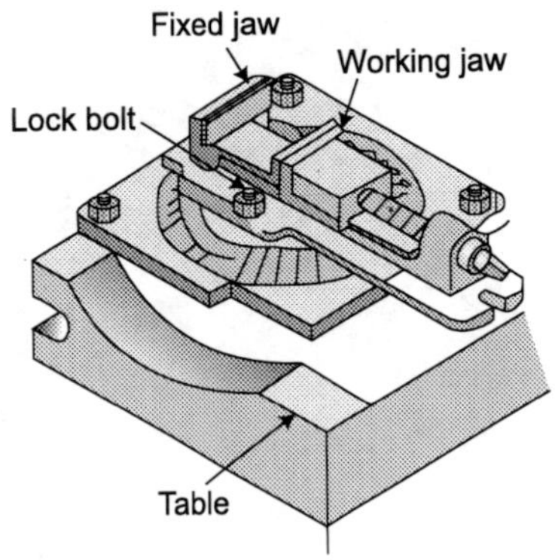

Fig. 9.9 Universal machine vice

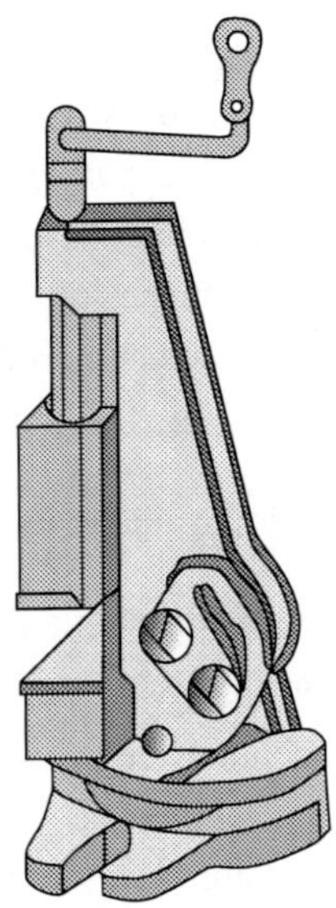

Fig. 9.10 Another position of universal vice

Vertical machine vice (Fig. 9.11)

This vice is different from all other types of vices because in this vice jobs are held in vertical manner, instead of jaws there are flanges. With this vice job can be easily held for machining.

For accurate alignment of the vice with the machine tool table, tenons (locating pieces) **(Fig. 9.12)** are also provided at the base of the vice.

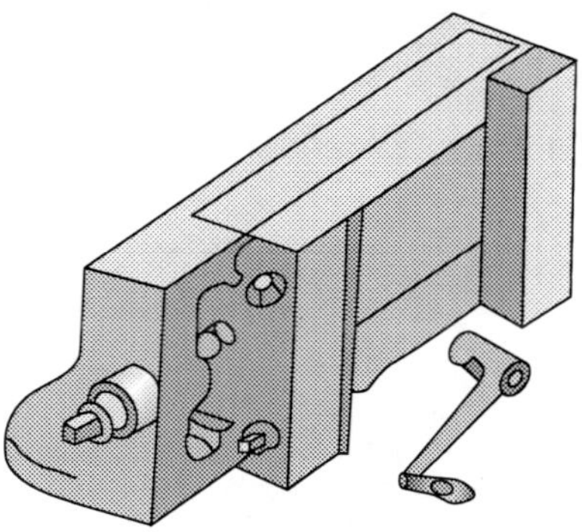

Fig. 9.11 Vertical machine vice

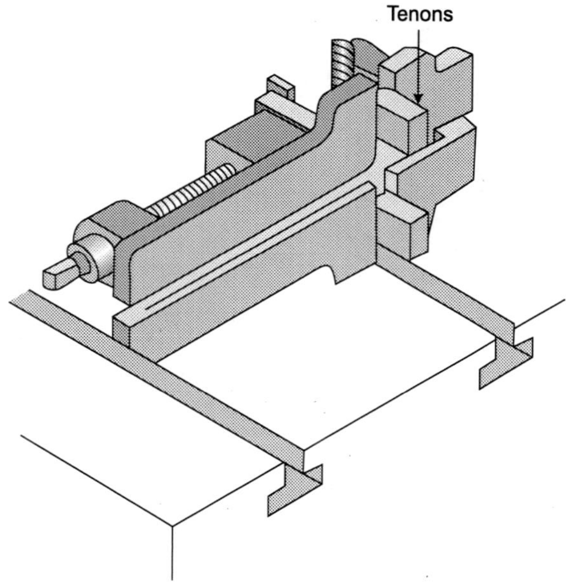

Fig. 9.12 Machine vice with tenons

9.1.3 Pipe Vice (Fig. 9.13 and 9.13A)

Spindle and jaw are fitted vertically in this vice. Its jaws are of "V" shape and dents are cut in the jaws. That's why this vice performs the job of holding the pipe and round jobs easily and firmly. In its jaw "V" is made at 90° angle.

The pipe vice grips the work at four points on its surface.

Portable folding pipe vice (Fig. 9.13B): This vice can be folded and carried easily to any working place. This is similar to the quick-releasing type pipe vice.

Chain pipe vice (Fig. 9.13C): This vice is used to hold larger diameter pipes up to 200 mm diameter. The pipe is gripped by means of a chain and the serrations are provided on the vice jaws.

9.1.4 Hand Vice (Fig. 9.14)

This vice is almost similar to the leg vice with a long leg and opens like that vice. Only small jobs can be held in this vice.

Hand vices are used for gripping screws, rivets, keys, small drills and other similar objects which are too small to be conveniently held in the bench vice. A hand vice is made in various shapes and sizes. The length varies from 125 to 150 mm and the jaw width from 40 to 44 mm The jaws can be opened and closed using the wing nut on the screw that is fastened to one leg, and passes through the other.

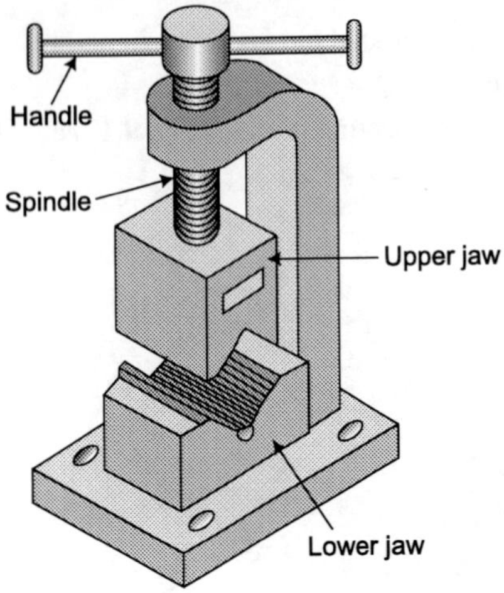

Fig. 9.13 Pipe vice

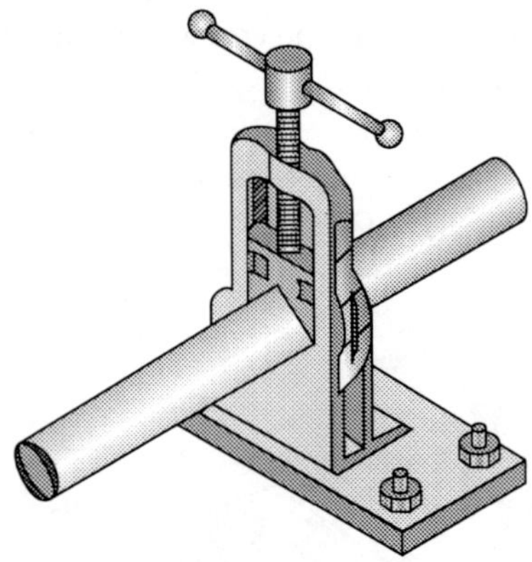

Fig. 9.13(a) Pipe vice in action

Fig. 9.13(b) Portable folding pipe vice

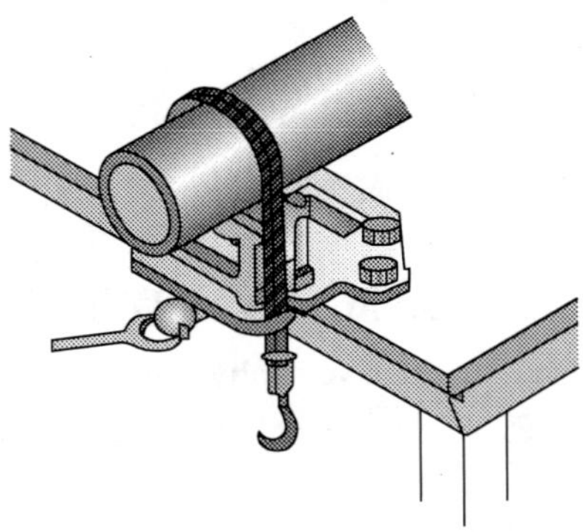

Fig. 9.13(c) Chain pipe vice

As per the name it is held by hand. Instead of handle a wing nut is used which is attached to the spindle fixed between the two jaws. Between the two legs, a spring leaf is fixed to open easily.

9.1.5 Leg vice (Fig. 9.15)

A leg vice is a holding device generally used in a forge shop for bending and forging work. It is made of mild steel to avoid breakage while hammering.

The leg vice is generally used in smithy shops. Hot job is fastened in it and operations like forging, bending, etc. are done. It is fitted in the work bench in a hanging manner. It contains one leg which is straightaway placed on wood. Movable jaw and fixed jaw are joined with each other through a hinge. Between these two jaws a strong spring leaf is fixed, with whose help the movable jaw handle opens easily when revolved. On

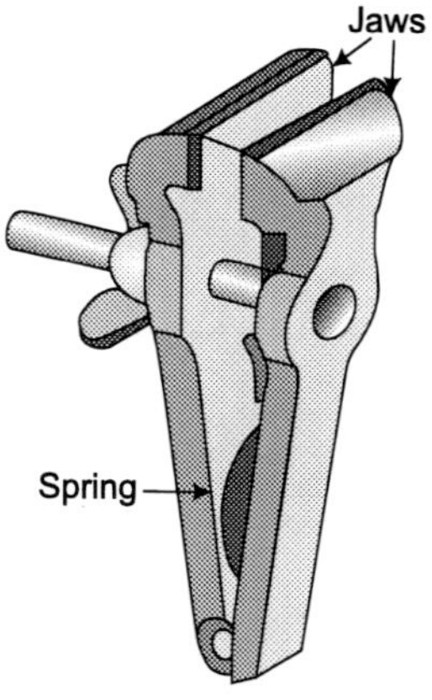

Fig. 9.14 Hand vice

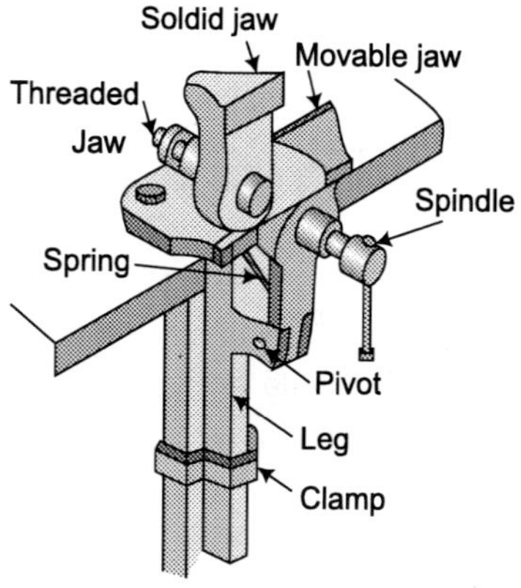

Fig. 9.15(a) Leg vice

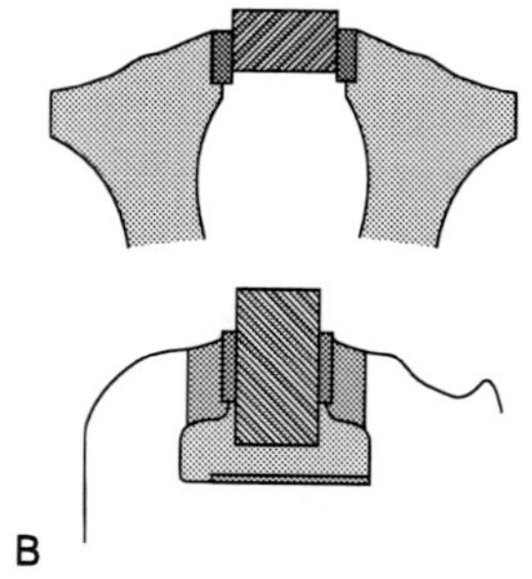

Fig. 9.15(b)

the upper side of both the jaws a spindle is fitted in which square threads are cut. An ox nut is fitted in the fixed jaw because of this when the handle is revolved, the movable jaw opens towards the outside.

Its entire body is made of wrought iron. That is why when we strike hammer blows it does not have any bad effect. It is a holding device used in forge shop for bending and forging work.

Since the hinged jaw moves in a radial path, the job held in this vice is not gripped properly because of the line contact **(Fig. 9.15 B)** Hence, a work which can be carried out on a bench vice is not held on a leg vice. Jobs which require hammering only are held on a leg vice.

The following are the main parts of a leg vice.

1. Solid jaw
2. Movable jaw
3. Threaded jaw
4. Spindle
5. Spring
6. Pivot
7. Leg
8. Clamp

9.1.6 Pin Vice (Fig. 9.16)

This vice functions as a small drill chuck. There are three jaws like a chuck. The upper part is thin and long which is either knurled or is made of six or eight panes. There is a hole across its body, in which long wires or pins can be easily held. It can be opened or tightened by holding it between the small finger and the thumb of hand. It is used to hold small jobs like screws, pins, etc., by the watch makers or instrument mechanics.

The pin vice is used for holding small diameter jobs. It consists of a handle and a small collet chuck at one end. The chuck carries a set of jaws which are operated by turning the handle.

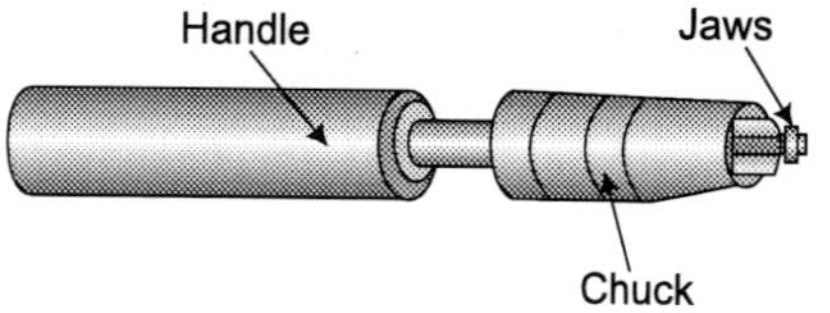

Fig. 9.16 Pin vice

9.1.7 Tool-maker's vice (Fig. 9.17)

It is generally for holding nonferrous metal jobs and other small jobs by the tool makers. Its jaws are made plain, which avoid bruising of the job. There is no need to fix this vice on the table or other place. It can be held in hand or simply placed on a table for filing, etc.

The tool-maker's vice is used for holding small works which require filing or drilling and for marking of small jobs on the surface plate. This vice is made of mild steel.

Tool-maker's vice is accurately machined.

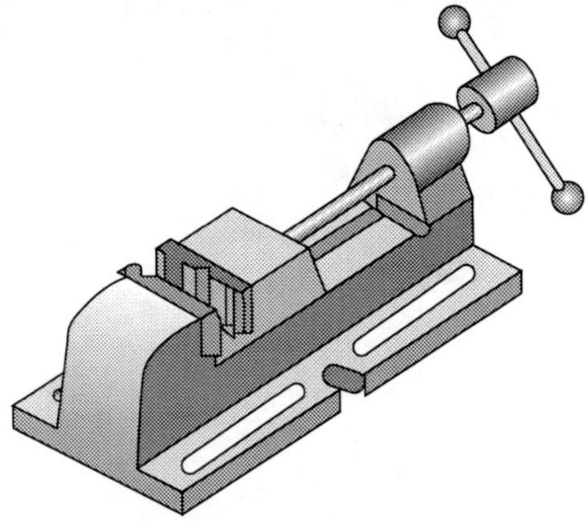

Fig. 9.17 Tool maker's vice

9.1.8 Drill Vice (Fig. 9.18)

It is a special kind of machine vice. It is used for holding the job while boring holes with the drilling machine. It is made of steel. A handle is attached to it. By lifting it up the jaws can be separated. A threaded spindle is also provided for setting its strokes after lifting the handle. There is no need to tighten this machine drill on the table. It is kept on the table in the normal way and operated with the handle.

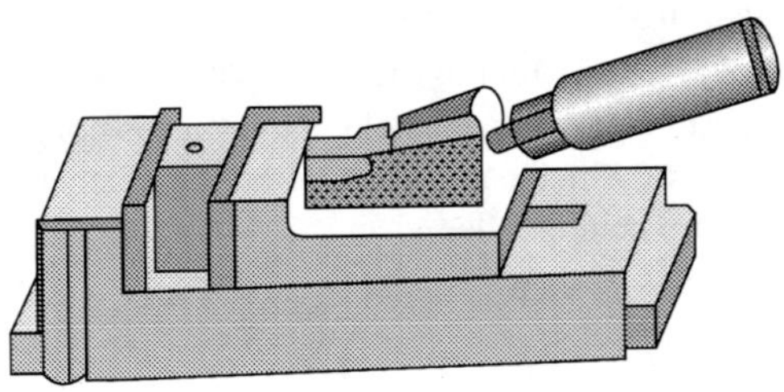

Fig. 9.18 Drill vice

9.1.9 Carpenter Vice

This vice is used only for woodworks. Its spindles are made of iron but jaws are made of wood. It is used for holding wood and carrying out different operations on it. Buttress threads are cut on its spindle and a spring is fitted in it. This vice also functions almost like the quick release vice.

9.2 CLAMPS

Clamps of different types are used for preventing the movement of work.

Requirements of clamping devices:

1. Clamps should be able to manipulate for easy loading.
2. Should provide the required clamping force.
3. Should be capable of locking with minimum movement.
4. Should accommodate a range of sizes of jobs.

Typical clamping devices, employing a screw and nut to provide the clamping force.

9.2.1 Vice Packing Clamp

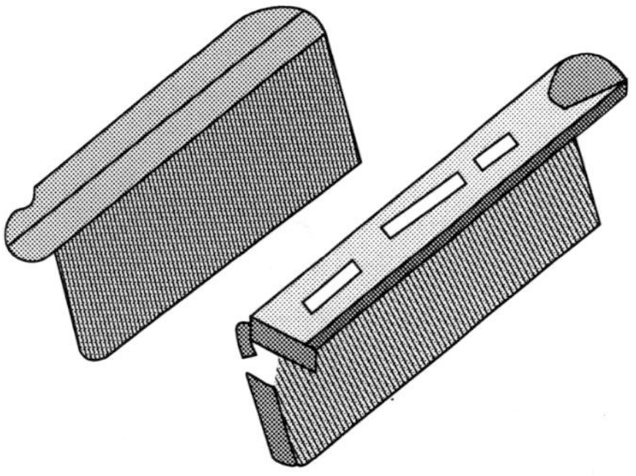

Fig. 9.19 Plain packing clamps

In ordinary bench vice parallel jaws are made. These are of hard metal like tool steel. Slanting dents are cut on it so that the job could be held firmly. But because of the teeth, brushing marks are left on the soft jobs. In order to prevent bruising vice packing clamp couple is used. The shape of the job can be different; therefore these are made in four different designs.

Plain Packing Clamp (Fig. 9.19): These types of clamps are made of copper, brass or aluminium sheets by bending it in the shape of a jaw from the upper side.

It is used for holding square or flat jobs.

Half Round Packing Clamp (Fig. 9.20): These clamps are made by soft metal casting. Semicircular grooves are cut. It is used for holding small rods, pipes, etc., for cutting threads, etc., even when it contains parallel jaws.

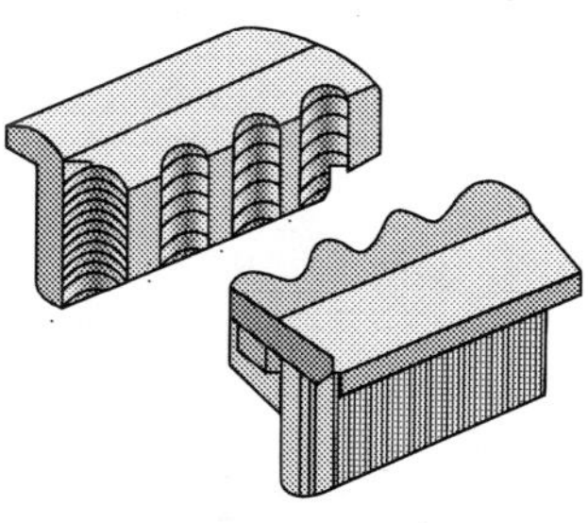

Fig. 9.20 Half round packing clamp

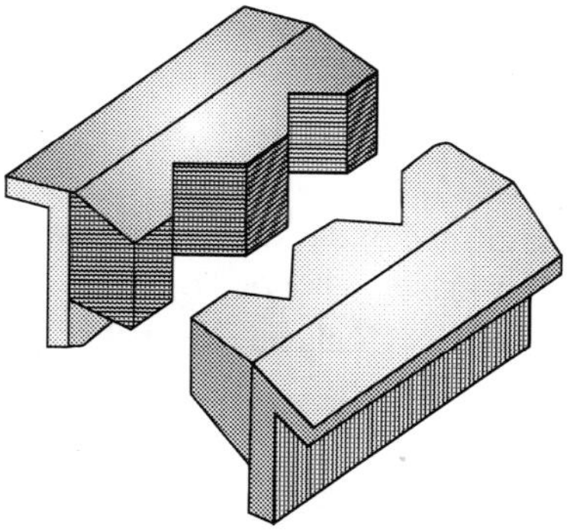

Fig. 9.21 'V' packing clamp

"V" Packing Clamp (Fig. 9.21): This type of clamp is also made in the same process as a half round clamp.

In place of semicircular grooves "V" shaped grooves are cut. With this clamp pipes are firmly held in the vice with parallel jaws.

9.2.2 "U" Packing Clamp (Fig. 9.22)

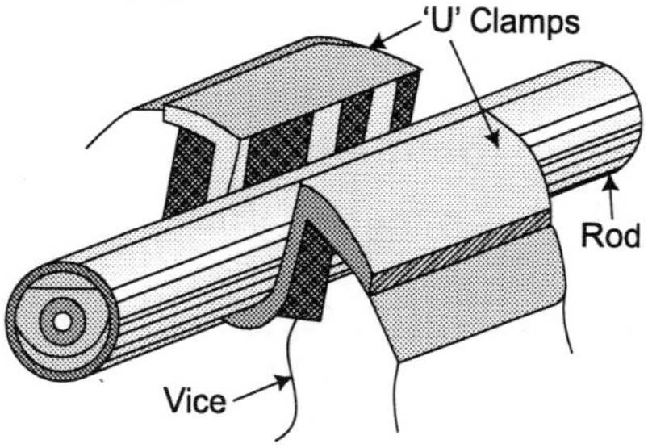

Fig. 9.22 'U' clampling

This clamp is made of the shape of letter "U". It is fitted in the parallel jaw vice. With this clamp round rods and pipes can be held lengthwise. This clamp is needed for making key way in a shaft.

9.2.3 "C" Clamps (Fig. 9.25)

These clamps are in the shape of a "C". The "C" clamp has its body forged or cast. One end of the clamp is machined flat. The other end is drilled and threaded to accommodate a screw-rod

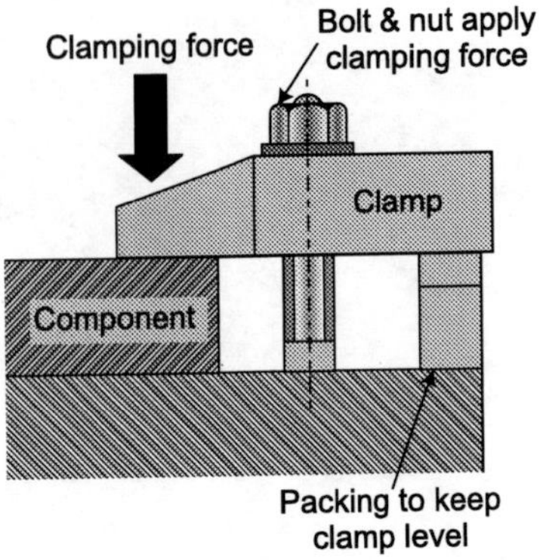

Fig. 9.23

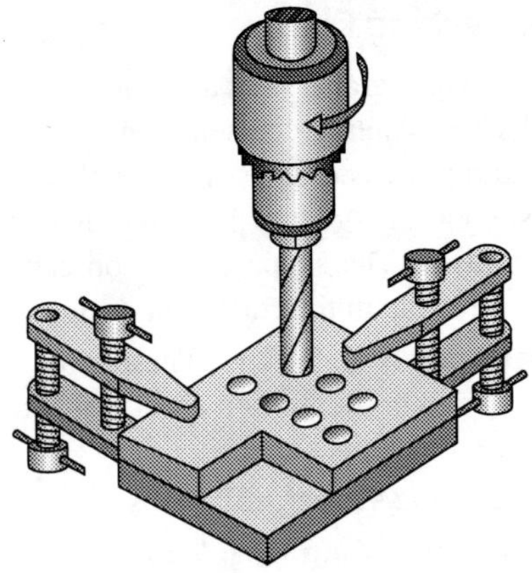

Fig. 9.24 Example of clamping

which is operated by a handle. The screw-rod carries a swivel pad which is free to revolve. The clamp is hardened and the face is serrated.

These clamps are used to hold work, an angle plate or a drill press table, and also, for holding two or more workpieces together.

The swivel pad on the end of the clamping screw helps in clamping surfaces which are not parallel. "C" clamps are available for light and heavy duty work.

An example of the use of "C" clamp **(Fig. 9.25 A)**

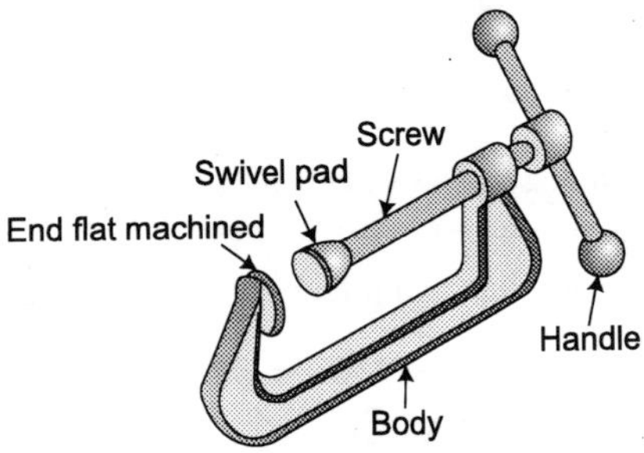

Fig. 9.25

Fig. 9.25(a) Example "C" clamp

9.2.4 Tool-maker's clamps (Fig. 9.26)

This clamp is the most commonly used by tool-makers for holding small, machined, flat pieces for further operations. They have two rectangular pieces of steel perfectly machined. The inner faces which come in contact with the workpiece are perfectly parallel. They are assembled by means of two threaded rods. The screw-rod (A) is rotated in one direction to adjust the gap between the two holding faces. The other screw (B) when tightened maintains the required pressure.

The head of the screw-rod (B) is provided with a hole through which a cylindrical pin may be passed for tightening purposes. The tool-maker's clamps are for holding a previously machined work which is flat and parallel.

The tool-maker's clamp is not suitable for doing heavy operations on the workpiece since the contacting and holding area of the clamp is limited. It is meant for holding light jobs.

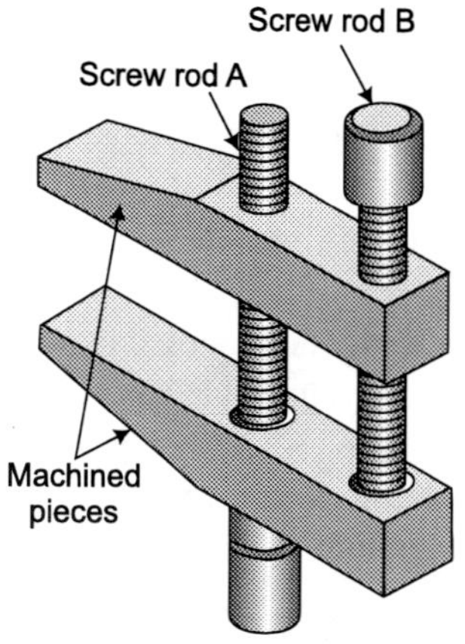

Fig. 9.26

9.2.5 'U' Clamps (Fig. 9.27)

These are clamps used along with 'Y' blocks as an accessory. These clamps serve the purpose of holding the round work securely in the 'Y' groove for layout operations as well as for machining operations.

Using clamps of different types along with bolts the workpieces can be held very rigidly (**Fig. 9.29).**

For holding small jobs or two-three machine parts other tools like vice are also used which are called clamps. Mainly the clamps are of the following two types: 1. Tool maker's clamps (Fig. 9.26) 2. 'C' clamps (Fig. 9.25)

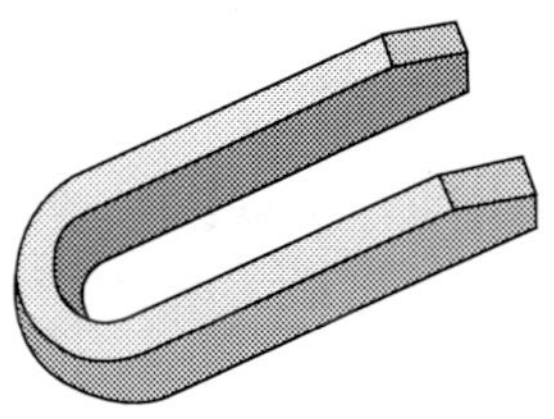
Fig. 9.27 'U' clamps

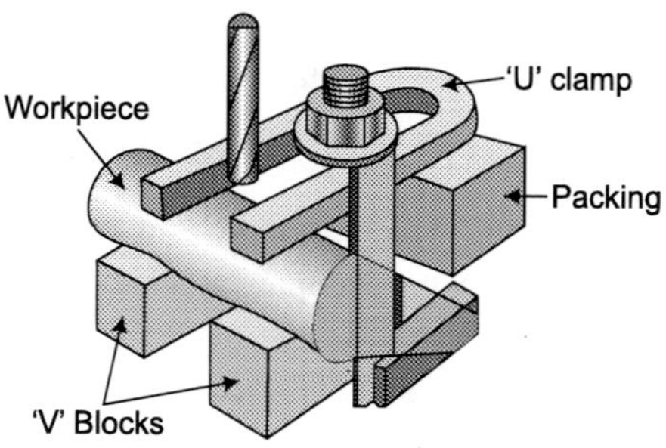

Fig. 9.28 Use of 'U' clamps

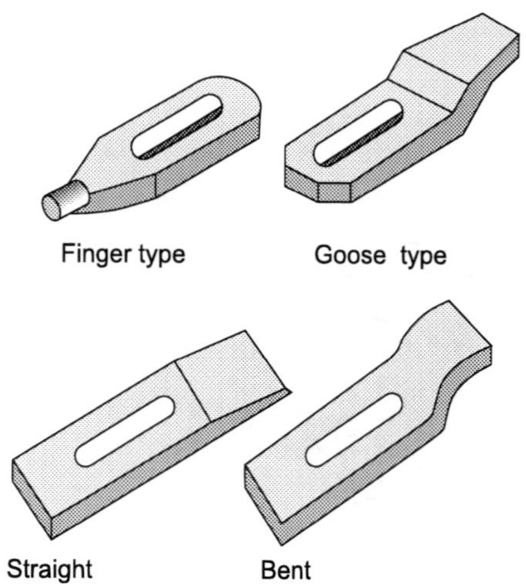

Fig. 9.29 Type of clamps

Points to be noted while using machine vices:

1. Ensure that the base of the vice and the top of the table are free from burrs.
2. Provide more contacting area between the work and jaws for better grip.
3. Use parallels to elevate the workpiece. Elevate it to accomplish the required machining.
4. As far as possible avoid applying cutting pressure against the movable jaw.

Workpieces are set up for drilling and other operations depending on the shape of the work and the facilities available on the machine for holding.

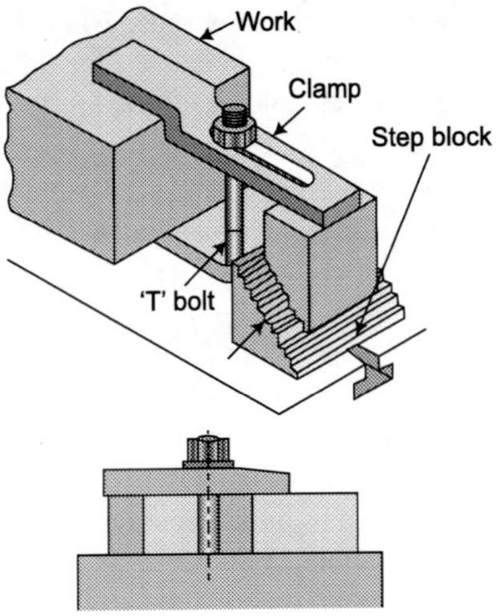

Fig. 9.30 Uses of step blocks

9.2.6 Use of Step Blocks (Fig. 9.30)

Heavy workpieces can be directly held on the machine table using a pair of step blocks. The height of the clamp can be lifted to the required level by adjusting the height of the step blocks. This permits clamping parallel to the contact surface of the work.

As far as possible the bolt should be placed near the workpiece for rigid holding.

Large workpieces which are to be lifted from the machine surfaces can be held using parallel blocks, place clamps and step blocks.

The clamps may be placed in the opposite direction for preventing the workpiece from rotating while drilling. The parallel blocks will assist in holding the work parallel to the work table.

Step blocks placed below the clamping point on the work will also be helpful for lifting the work parallel to the work table ensuring good holding.

When a number of clamps are to be used, ensure even tightening of the bolts.

9.2.7 Use of Vee blocks (Fig. 9.31)

Cylindrical workpieces can be conveniently held on Vee blocks. Vee blocks can be correctly aligned on the machine tables using a straight edge in the T-slot.

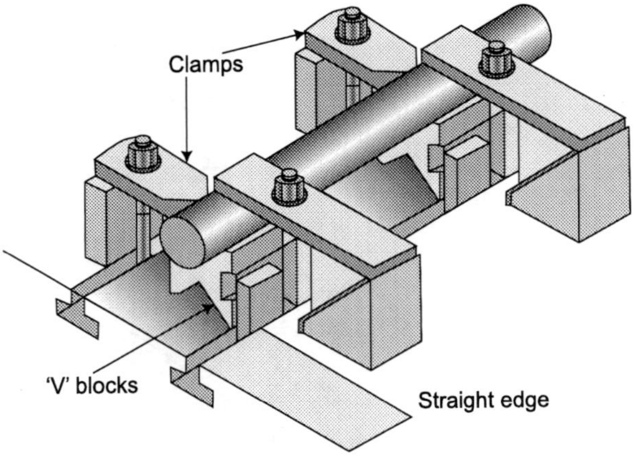

Fig. 9.31

Long workpieces would need two Vee blocks for proper support.

9.2.8 Use of round rods (Fig. 9.31)

For ensuring a good grip on a machine vice round rods are used for providing a line contact.

Slight surface angularity on one gripping face will not affect the angularity of the surface being machined. The surface can be finished with respect to one face machined correctly.

Be sure that the work has a good contact with the parallels and on the opposite side of the round.

Points to be noted while using machine vices:

Ensure that the base of the vice and the top of the table are free from burrs.

Provide more contacting area between the work and jaws for better grip.

Use parallels to elevate the workpiece. Elevate it just enough to accomplish the required machining.

As far as possible avoid applying cutting pressure against the movable jaw.

9.3 PRECAUTIONS

1. Job should be fastened nearly at the centre and it should not be very high from the surface of the vice.
2. While chipping, care should be taken to ensure that the blow of hammer should not hit the vice.
3. Job should be tightened in the jaw only by moving the handle manually. A hammer should not be used for this purpose.
4. This job should be held in the jaw with wood packing otherwise at the time of filing, etc., chattering takes place.
5. Jobs which have already been finished should be fastened on the vice without packing the damp.
6. The height of the vice should not be more than the knee of the worker. If the height is less then the wooden blocks should be used below the vice.
7. Lubrication must be done in the spindle of the vice once a week.
8. After using the vice it should be cleaned. While cleaning it should be ensured that the tiny particles of metal do not go into the spindle.

9.4 HAMMERS

An engineer's hammer is a hand tool used for striking purposes while:

1. Punching
2. Bending
3. Straightening
4. Chipping
5. Forging
6. Riveting

It is used to beat or strike blows on jobs or metals or for driving nails, etc. It is used to straighten or bend a job made of metal. In addition, it is also used for riveting, chipping and forging jobs.

Classification of hammers is done according to their pein, face, weight and shape.

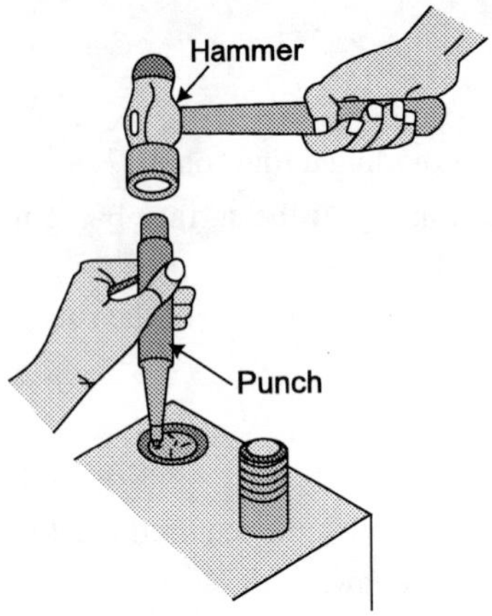

Fig. 9.32 Use of hammer

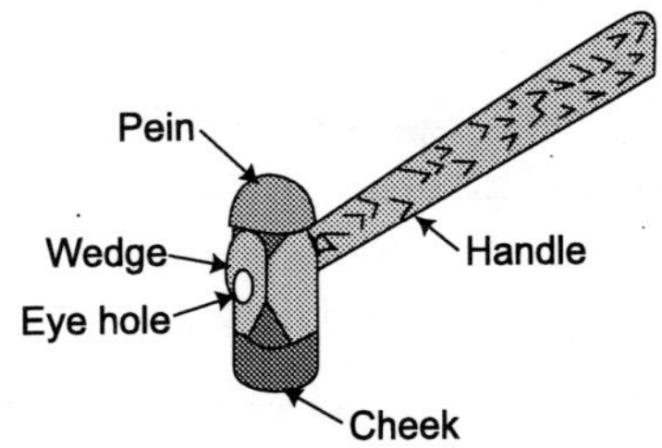

Fig. 9.32(a) Part of hammer

The major parts of a hammer are its head and handle.

The head is made of drop-forged carbon steel, and a wooden handle which must be capable of absorbing shock.

The parts of a hammerhead are **(Fig. 9.32a):**

1. Face
2. Pein
3. Cheek
4. Eye hole.

Face: The face is the striking portion. Slight convexity is given to it to avoid digging of the edge. The face of the hammer is used for general work, such as striking chisels and punches and levelling and working over joints.

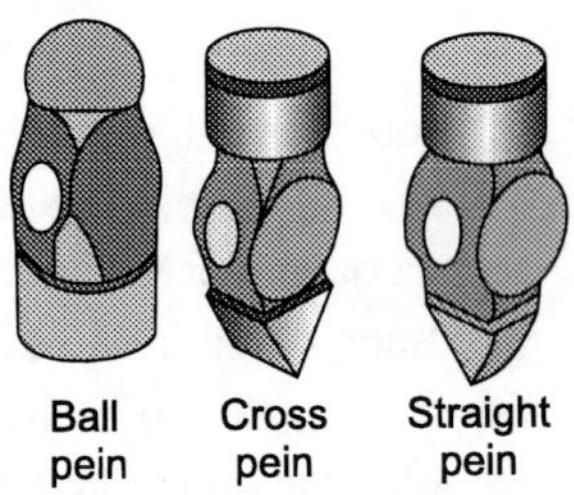

Fig. 9.33 Faces of pein hammer

Pein (Fig. 9.33): The pein is the other end of the head. It is used for shaping and forming work like riveting and bending. The pein is of different shapes like the ball pein, cross pein or straight pein.

The face and the pein are hardened.

Cheek: The cheek is the middle portion of the hammerhead. The weight of the hammer is stamped here. This portion of the hammerhead is left soft.

Eye hole (Fig. 9.34): An eye hole is meant for fixing the handle. It is shaped to fit the handle rigidly. The wedges fix the handle in the eye hole.

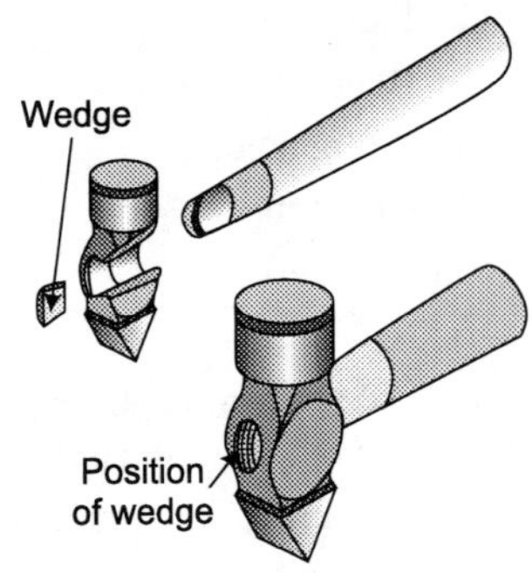

Fig. 9.34 Eye hole of hammer

These hammers are made of cast steel or carbon steel. Their pein and face are hardened and tempered. The middle body is kept soft. On one end of the body, face and pein are made. An oval-shaped hole is made in the body in which the handle is fitted by using wedge. Because of the wedge the hole is somewhat enlarged and there is no risk of handle becoming loose and coming out. The length of the hammer depends on its weight. On heavy hammers long handles are used. Contrarily on the light weight hammers shorter handles are used. In hand hammers handles of 8 to 12 inches are used. According to IS 841-1957, the handles used for marking, etc., have 100 to 300 gm weight. Heavy hammers are of 900 gm in weight.

9.5 KINDS OF HAND HAMMER

Based on their utility hand hammers are of several types. Their classification is done on the basis of their panes and faces. According to the face and pein the hammers are of the following types:

- Ball pein hammer
- Cross pein hammer
- Straight pein hammer
- Double face hammer

Ball pein hammer (Fig. 9.35): A ball pein head is used to spread metal in all directions.

This hammer has a semi-spherical pein suitable for riveting. It is used for shaping the cylindrical end of a metal rivet to form a rivet head.

This type or hammer is widely used in the workshops. On one end of this ball pein hammer a plain face is made and on the other end a ¾ inch round like a ball is made. It is used for setting rivet. It is also used to strike blows on metal sheet up to certain depth. Its weight is from 100 to 900 gm.

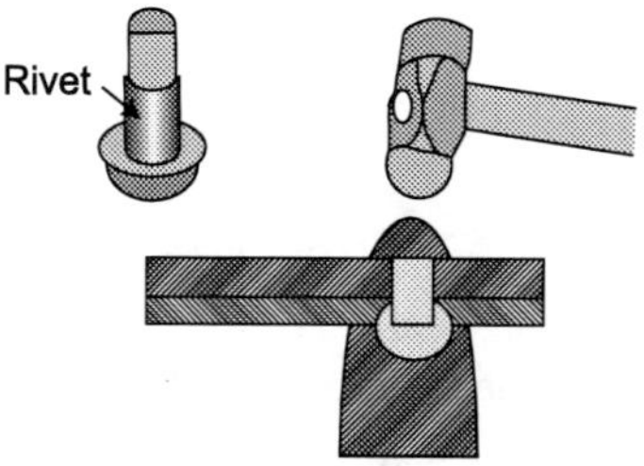

Fig. 9.35 Uses of ball pein hammer

Cross pein hammer (Fig. 9.36): This hammer has a face on the one end and a pein at the other. The handle is fixed at right angle. It is used for making channel, "V" grooves, inside corner of the job and for expanding metal sheets. Its weight is from 200 to 900 gm.

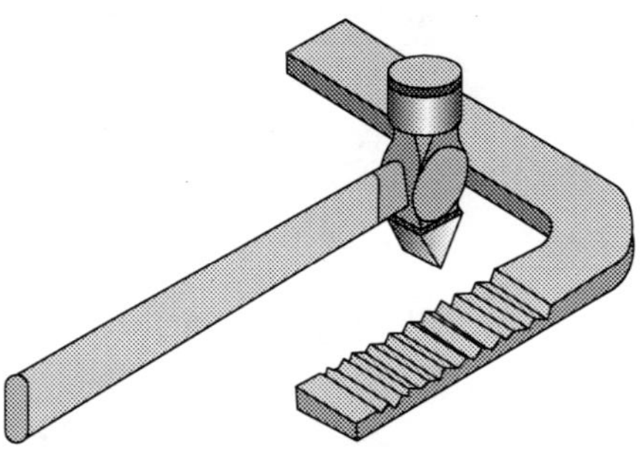

Fig. 9.36 Uses of cross pein hammer

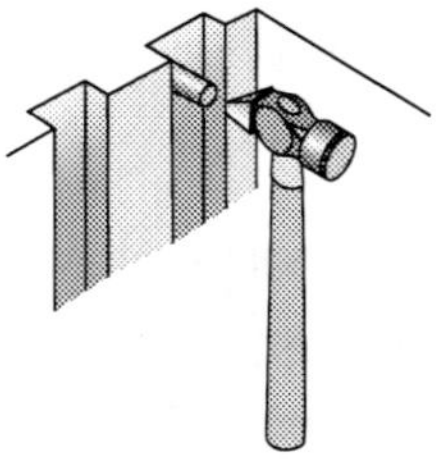

Fig. 9.37 Uses of straight pein hammer

A cross pein head is used to spread metal in one direction in the line of striking.

Straight pein hammer (Fig. 9.37): This hammer is almost similar to cross pein hammer but its pein is in straight line to the hammer handle. It is also used like a cross pein hammer for making channel, "V" grooves and expanding the metal sheets. These are of light weight as well, 100 to 900 gm.

A cross pein head is generally used to spread metal in one direction in the line of striking.

Double face hammer (Fig. 9.38): This type of hammer is mostly used for blacksmithy jobs. On both its ends similar pein faces are made.

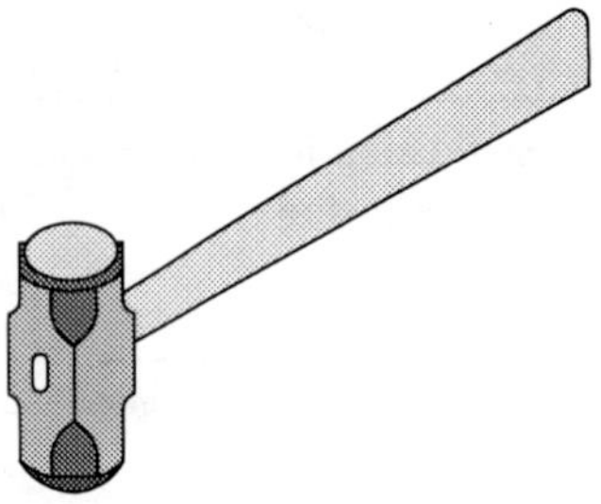

Fig. 9.38 Double face hammer

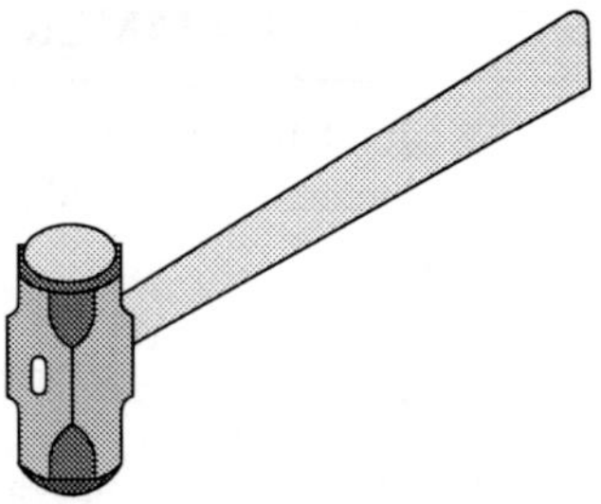

Fig. 9.39 Sledge hammer

Sledge hammer (Fig. 9.39): These are mostly used in blacksmithy for straightening round rods, iron bars, angle iron, channel or flat iron. Its shape is similar to double face hammer. As these are used for heavy jobs they are heavy, from 4 to 10 kg.

A sledge hammer is a heavy hammer with a long handle. It may weigh from 4 to 16 kg. These are used for heavy work. The length of the handle increases with the weight.

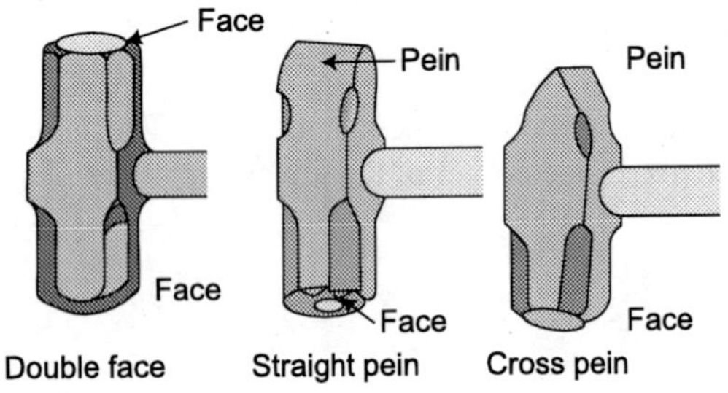

Fig. 9.40 Sledge hammers

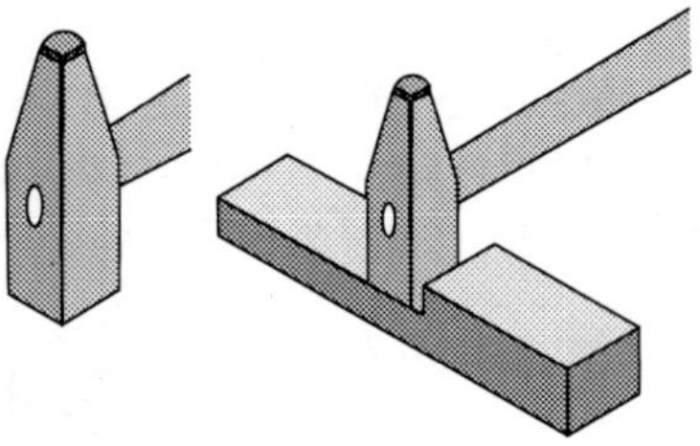

Fig. 9.41 Set hammer

Set hammer (Fig. 9.41): A set hammer is used to set down at right angles (shouldering). Generally, hammers will not produce sharply defined border lines.

A set hammer is made of 0.8% carbon steel.

Wooden or steel-wired handles are provided for holding these hammers.

Claw hammer (Fig. 9.42): This is a special type of hammer. On one end of this hammer a round face is made and on the other end its pein is inclined towards the handle. In its centre a slot is cut, with whose help nails, etc., can be easily extracted (taken out). This hammer is generally used by the carpenters.

Soft hammer: On the jobs which are likely to get scars when hammered, soft hammers are used.

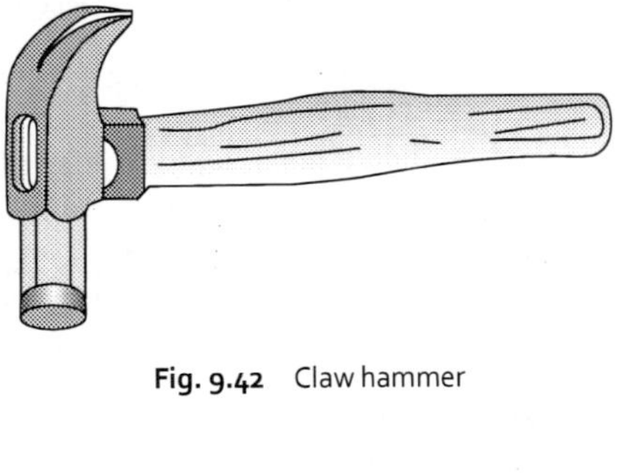

Fig. 9.42 Claw hammer

The engineer's hammer or other steel hammers cannot be used for all types of shop work.

Soft hammers made of different materials are used for preventing damage to the workpieces while working

Wooden and rubber mallets **(Fig. 9.43)**

These mallets are useful for working on sheet metal, wood, etc.

Mallets are soft hammers and are made of raw hide, hard rubber, copper, brass, lead or wood, and are used to strike a soft and light blow on the metal.

Fig. 9.43 Wooden mallet

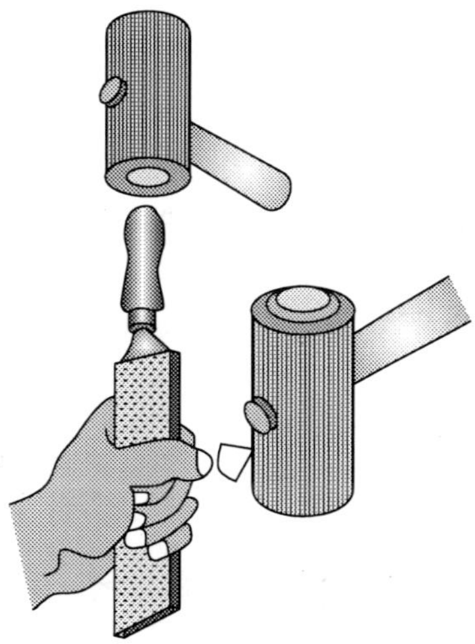

Fig. 9.44 Rubber mallet

Standard wooden mallets are used for general purpose work like flattening, bending, etc. **(Fig. 9.45).**

Bossing mallets are used for hollowing panel beatings **(Fig. 9.45).**

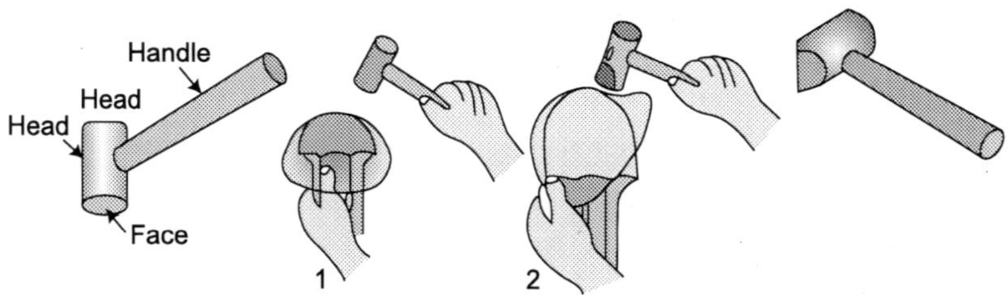

Fig. 9.45 Wooden hammer

An end-faked mallet is used for stretching, hammering, etc. **(Fig. 9.45A).**

Copper hammer (Fig. 9.46)

These hammers with copper inserts on the head are useful for setting jobs on machines and dismantling and assembling of components.

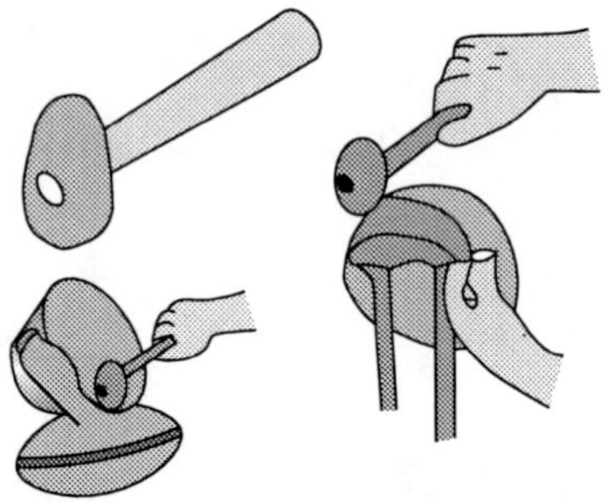

Fig. 9.45(a) Shaping hemisphere bowl by raising

Fig. 9.46 Uses of copper hammer

Rawhide mallets (Fig. 9.47)

These hammers with leather inserts are used for tapping machined components for assembling or setting purposes.

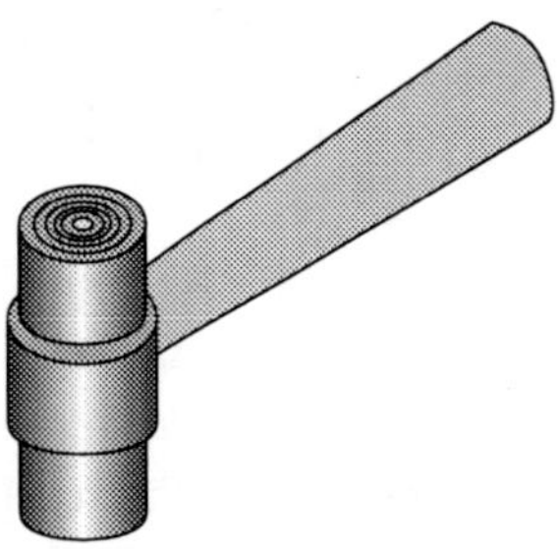

Fig. 9.47 Raw hide hammer

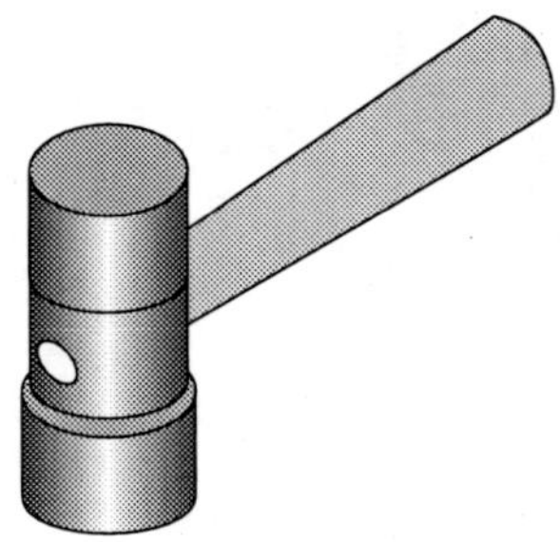

Fig. 9.48 Plastic hammer

Plastic hammers (Fig. 9.48)

Plastic hammers are available with solid and interchangeable heads. These are quite useful for handling light assembly work.

Lead hammer (Fig. 9.50)

For giving heavy blows for certain stubborn components in restricted places of assembly, lead hammers are sometimes used.

Lead being heavy does not bounce and at the same time it protects the surface.

Power hammer (Fig. 9.51): For heavy jobs of forging, bending, etc., in blacksmithy, power hammers are used. These are power driven machines. Electric or pneumatic power is

Plastic hamer

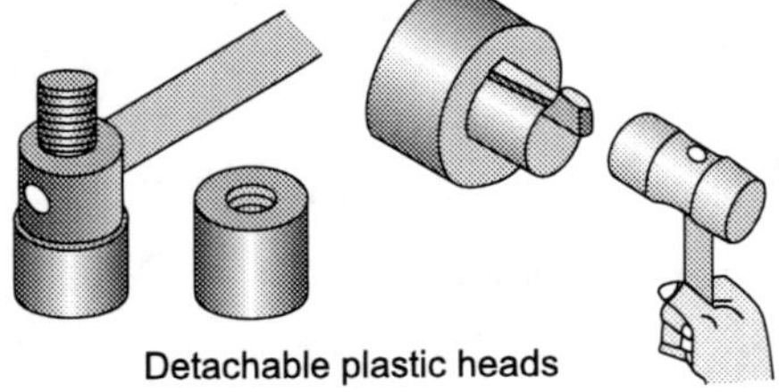

Detachable plastic heads

Fig. 9.49

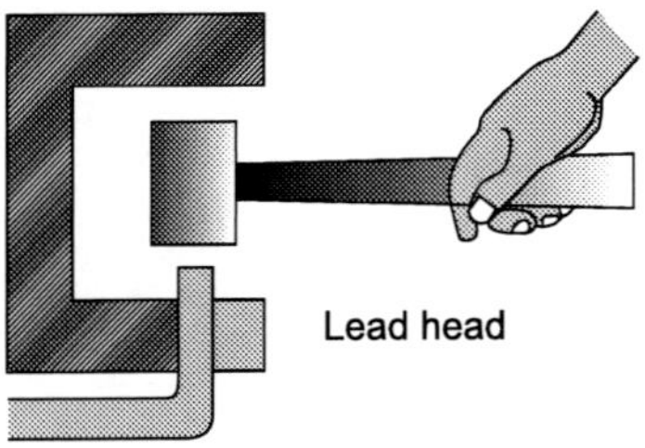

Fig. 9.50

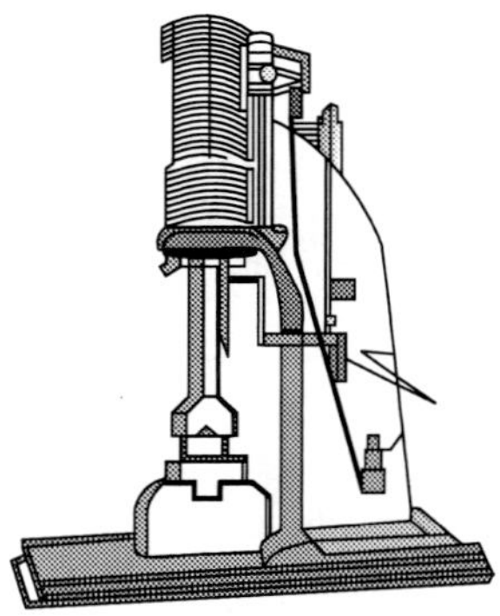

Fig. 9.51 Power hammer

used. Power hammers were driven by steam in olden days. Little effort is required for working on these and they save time also.

9.6 PRECAUTIONS

1. Do not use oil, grease or any other oily substance on the handle of the hammer and if it is there, it should be properly cleaned.
2. Hammer handle should be plain otherwise it can cause boils on the palm.
3. Hammer handle should be kept tight otherwise the hammer would come out and the handle and may cause an accident.
4. Due to excessive use the face of the hammer becomes a mushroom, it should be ground properly.
5. Hold the handle always at its end, it saves energy.
6. Choose the appropriate and suitable hammer according to the nature of the job.

9.7 SCREWDRIVER (FIG. 9.52)

Screwdriver is used for tightening or loosening screws and is available in various lengths. A round rod with flat end, having a handle at the other end is called a screwdriver. A screw driver has three main parts:

(1) Handle, (2) shank and (3) blade.

Handle: Generally, wooden or plastic handles are fixed to the screwdrivers. In order to save the wooden handle from being broken, an iron ring is fixed.

Shank: The middle part between the blade and the handle is called shank. On its lower part tip or blade is made. On the upper part either there is a hole through which a wire is passed or this end is made flat so that when this part is fixed to the handle it does not allow the shank to move separately. These are generally made of carbon steel.

Blade: Blade is the lower part of the shank. It is also known as tip. This part is hardened and tempered.

Fig. 9.52

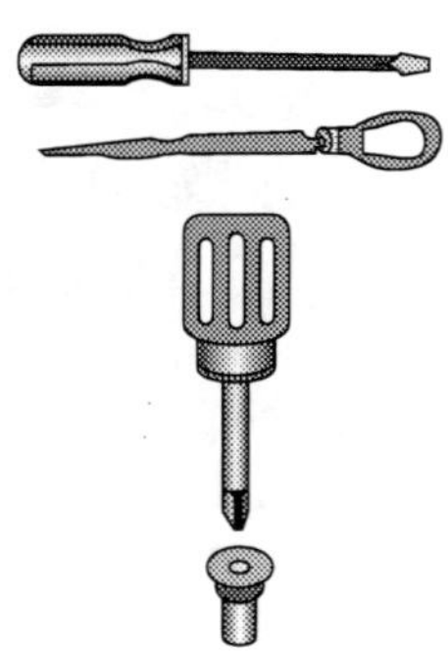

Fig. 9.52(a)

Length of a screwdriver is measured excluding its handle. The length of the screwdriver varies from 50 to 450 mm.

9.8 KINDS OF SCREWDRIVERS

1. Ordinary flat screwdriver
2. Offset screwdriver
3. Ratchet screwdriver
4. Philips screwdriver
5. Watch maker screwdriver
6. Carpenter screwdriver

Ordinary flat screwdriver: This screwdriver is used for tightening or loosening different types of screws. These are available in small as well as larger sizes. These screwdrivers have a handle of wood or plastic fixed to them. According to needs there are light duty **(Fig. 9.52)** or heavy duty screwdrivers **(Fig. 9.53)**. They have a flat tip. **(Fig. 9.53)**

The blade of heavy duty screwdriver is square and the shank is also square section for applying extra twisting force with the end of a spanner **(Fig. 9.54).**

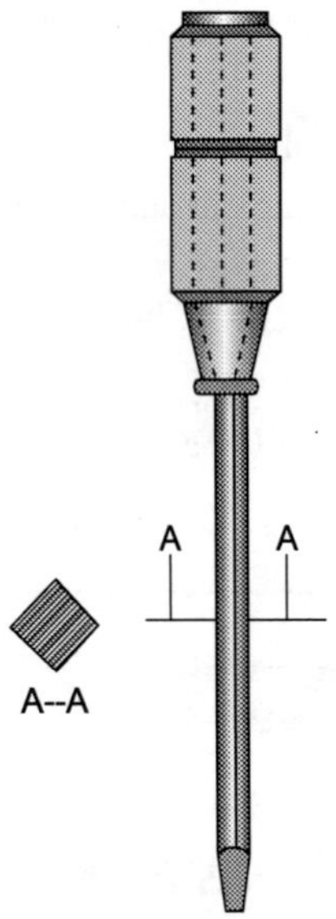

Fig. 9.53 Heavy duty screwdriver

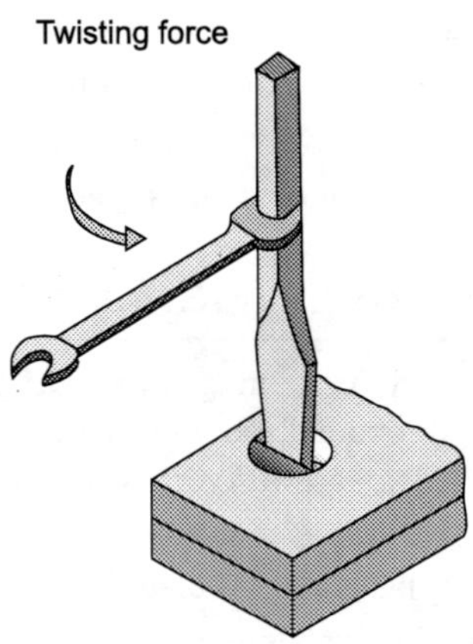

Fig. 9.54

Offset screwdriver (Fig. 9.55): These screwdrivers also have flat blade. But no handle is used in these screwdrivers. Their shank is turned at 90° angle at both the ends and flats are made on both the ends. In case of narrow or slanting spaces, where ordinary flat screwdriver cannot be used, offset screwdriver is used.

These are useful in some specific situations **(Fig. 9.56).** Where a normal screwdriver cannot be used because of the length of the handle, the offset screwdriver is preferable. They are also useful for applying greater turning force.

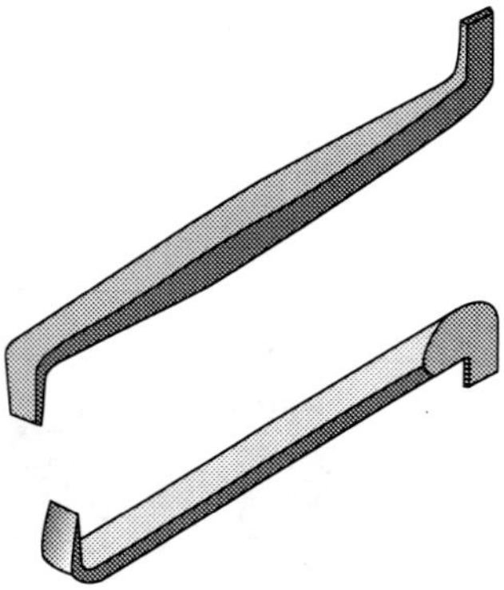

Fig. 9.55 Offset screwdriver

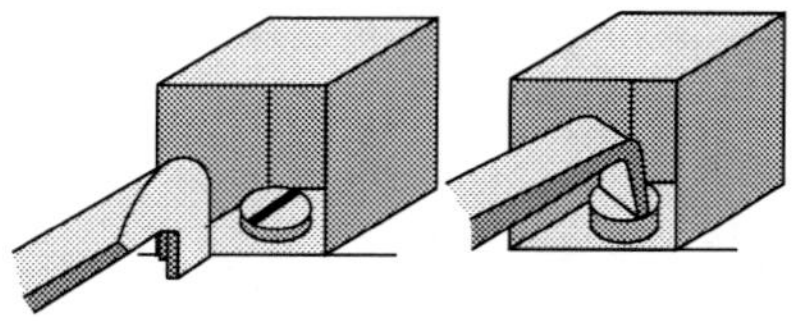

Fig. 9.56 Uses of offset screwdriver

For quicker application ratchet offset screwdrivers are also available with renewable tips.

Ratchet screwdriver (Fig. 9.57): In this type of screwdriver a spring or ratchet is fitted in the handle. There is a button at its handle, when this button is pressed the shank of the ratchet screw starts revolving through the spring. When the button is left free again the shank comes to its original position. With the help of this screwdriver the screw can be quickly tightened or loosened. It is a hand tool used for turning screws, usually having a handle of wood, plastic, etc., and a steel shank with a flattened square-cut tip that fits into a slot in the head of the screw in which the blade turns in only one direction and a spiral in the handle (so the blade rotates) with downward pressure on the handle.

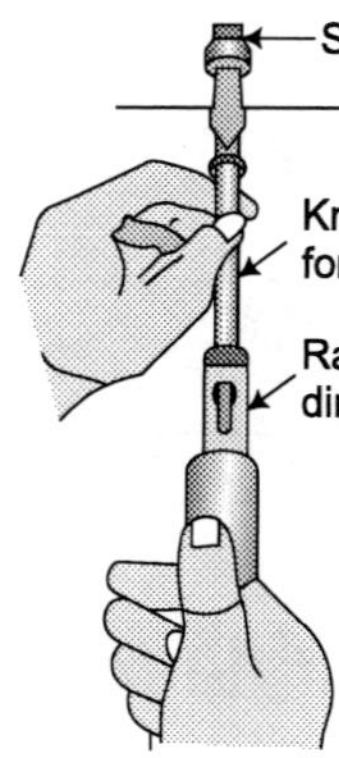

Fig. 9.57 Ratchet screwdriver

For using normal screwdriver the following actions are required:

Position the flat tip of the screwdriver into the slot on the head of the screw, turn a half turn disengage the tip, move, i.e., turn the screwdriver, position the tip into the slot and half turn the screwdriver. Repeat this till the screw is tightened fully.

When the ratchet screwdriver is used, position the tip of the screwdriver into the slot of the screw, push the handle of the ratchet screwdriver, and repeat pushing till the screw is tightened.

This is a advantageous since lot of time is saved.

Philips screwdriver (Fig. 9.59): Tip of this type of the screwdriver is different from that of others. The tip is pointed and is made after cutting four flutes. Such screwdrivers are used where the screws have T cross-shaped groove on their heads. These screws are available in varying number of measurements.

These are made with cruciform tips that are unlikely to slip from the matching slots.

The sizes of Philips screwdrivers are specified by point size 1, 2, 3 and 4.

Watch makes screwdriver (Fig. 9.60): These screwdrivers are available in sets. In a set there are six screwdrivers, having number markings from 0 to 5. The smaller, the number of the screwdriver, the smaller and thinner is its bit. In place of handle, a cap is fixed to the shank in such a way that both the cap and the shank can be moved independently

For tightening or loosening a screw, tip is placed at the head of the screw and the cap is pressed with one finger, while with the other finger the shank is rotated. These can be used in all kinds of instruments and watches.

They are also called jeweller's screwdrivers.

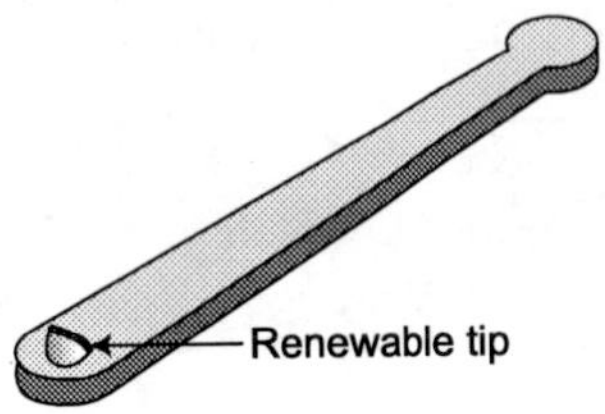

Fig. 9.58 Renewable tip ratchet off set screwdriver

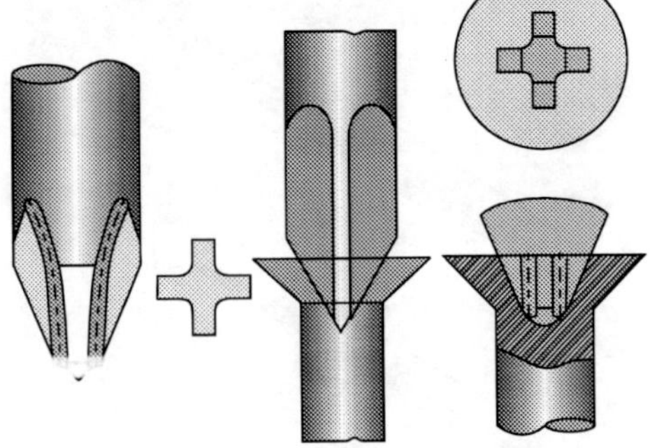

Fig. 9.59 Phillips screwsdriver

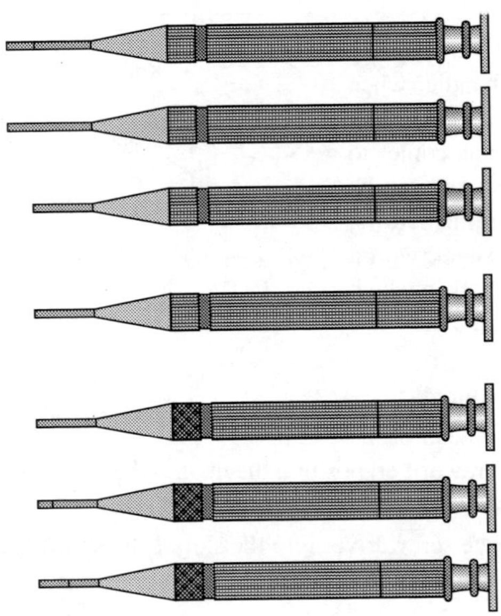

Fig. 9.60 Watch maker's screwdrivers

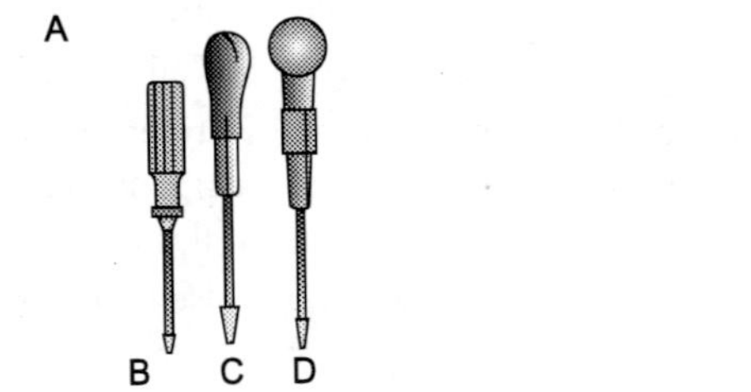

Fig. 9.61 Carpenter's screwdrivers

Carpenter's screwdriver (Figs. 9.61 and 9.62): This is made of a round rod and flat tip. Its handle is made flat from two sides. This enables the worker to exert full force on the screwdriver.

Yankee spiral ratched screwdriver **(Fig. 9.61B),** plastic handle screwdriver **(Fig. 9.61C),** Yankee ratchet screwdriver **(Fig. 9.61D),** and cabinet handle screwdriver.

Both hand and power operated screwdrivers are used, depending on production requirements. It is essential that screwdriver's points fit the slots in order to avoid damage to fastening or marring of the product by tool slippage.

Fig. 9.62 Carpenter's screwdriver

Torque screwdrivers are often provided by a slip clutch. In addition to providing a ready means of assuring uniform tightening, slip clutches offer positive insurance against inadvertent twisting off of screws **(Fig. 9.63).**

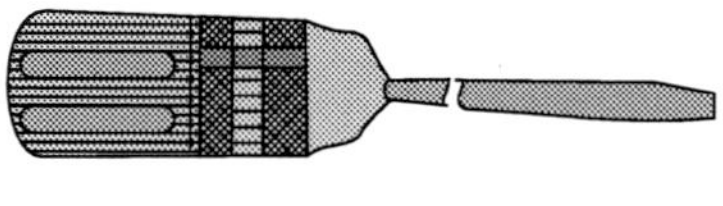

Fig. 9.63

Stumpy screwdrivers (Fig. 9.64): These are small but sturdy screwdrivers. They are used when other types of screwdrivers cannot be used due to the space limitation.

Specification of a screwdriver:

Screwdrivers are specified according to the

1. Length of the blade
2. Width of the tip.

Normal blade length: 45 to 300 mm. Width of blade: 3 to 10 mm.

The blades of screwdrivers are made of carbon steel or alloy steel, hardened and tempered.

Screwdrivers for special uses:

Small and sturdy screwdrivers are available for use where there is limited space **(Fig. 9.65).**

Screwdrivers with blades sheathed in insulation are available for the use of electricians **(Fig. 9.66).**

Stumpy screw driver

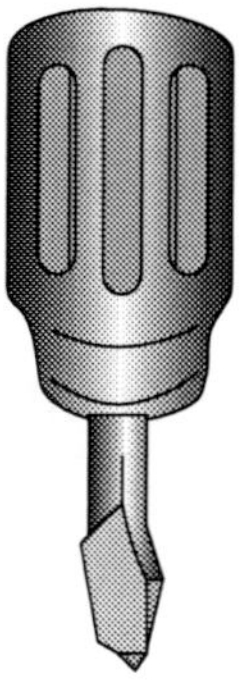

Fig. 9.64

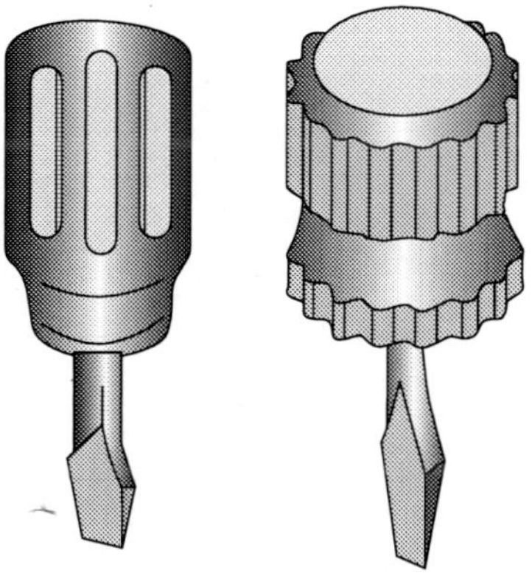

Fig. 9.65 Special screwdrivers used in limited space

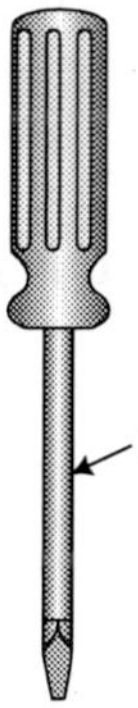

Fig. 9.66 Sheathed screwdriver

9.9 PRECAUTIONS

1. Use screwdrivers with tips correctly fitting into the screw slot **(Fig. 9.64).**
2. Make sure your hand and the handle are dry.
3. Hold the screwdriver with its axis in line with the axis of the screw.
4. While using a Philips screwdriver apply more downward pressure.
5. Keep your hand away to avoid injury due to slipping of screwdriver **(Fig. 9.65).**
6. Do not use screwdrivers with split or defective handles **(Fig. 9.69).**
7. In the case of damaged screwdrivers, the blades can be ground (the faces will be parallel with the sides of the screw slot) and used. While grinding ensure the end of the tip is as thick as the slot of the screw. **(Fig. 9.67)**
8. While using screwdrivers on small jobs brace the jobs on the bench or hold them in a vice.

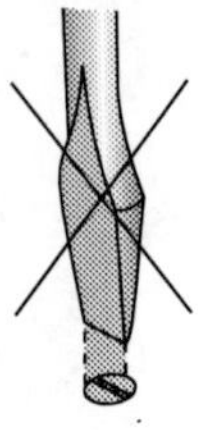

Fig. 9.67 Correct screwdriver to suit slot

Fig. 9.68 Proper holding of screwdriver

9.10 PLIERS

With the help of pliers small jobs and articles can be held. It is also used to cut or turn wires. Its jaws are hardened and tempered. Size of pliers varies from 4 to 10 inches. For different kinds of jobs, different types of pliers are used:

Pliers have a pair of legs joined by a pivot, hinge or fulcrum pin. Each leg consists of a long handle and a short jaw.

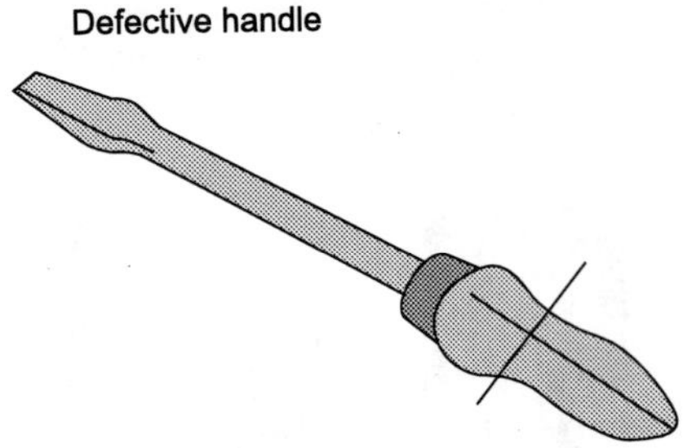

Fig. 9.69

9.11 ELEMENTS OF PLIERS WITH TWO-JOINT CUTTERS (COMBINATION PLIERS) (FIG. 9.70)

1. Flat jaw
2. Pipe grip
3. Side cutters
4. Joint cutters
5. Handles

Flat jaw tips are serrated for general gripping.

Pipe grip is serrated for gripping cylindrical objects **(Fig. 9.71).**

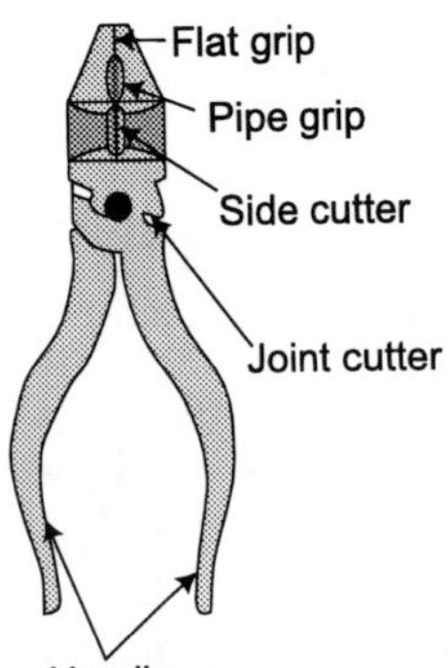

Fig. 9.70

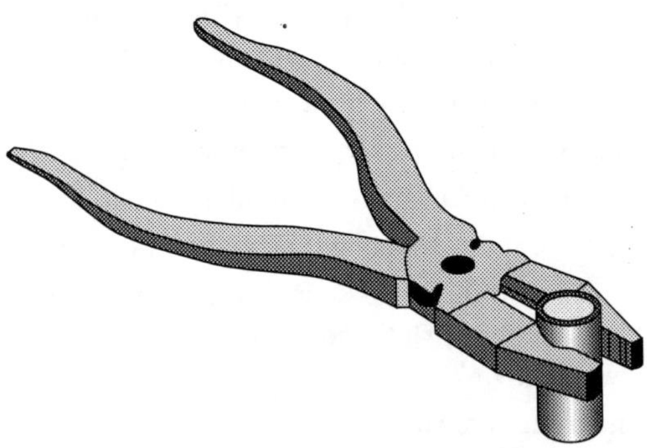

Fig. 9.71 Serration to grip pipe

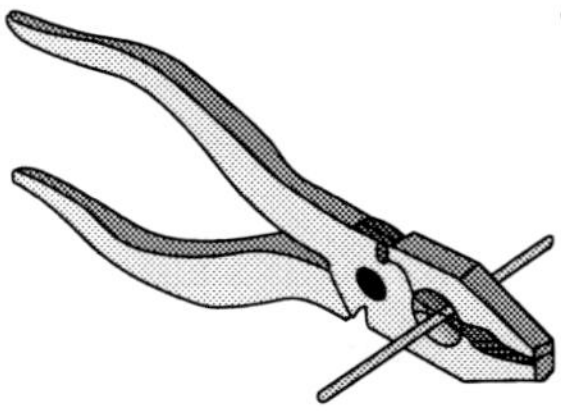

Fig. 9.72(A) Side cutter

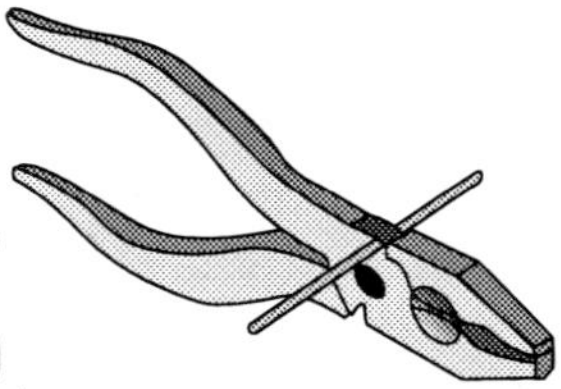

Fig. 9.72(B) Two joint cutter

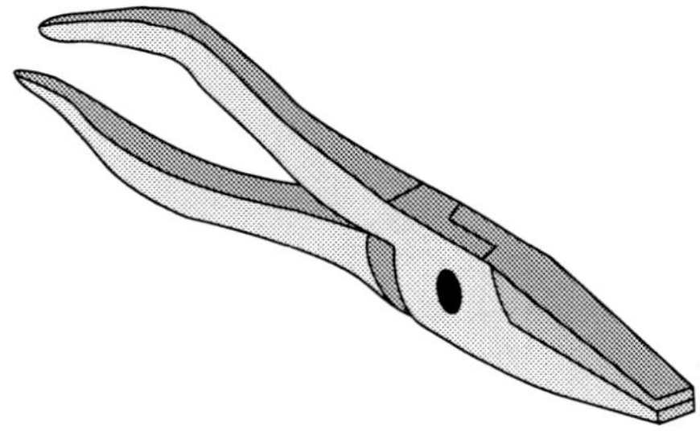

Fig. 9.73 Flat nosé cutter

Side cutters are provided for cutting soft wires **(Fig. 9.72A).**

Two joint cutters are provided for cutting or shearing steel wires **(Fig. 9.73B).**

Handles are used for applying pressure by hand.

Pliers are available in sizes from 150 to 230 mm.

9.12 TYPES OF PLIERS

Flat nose plier (Fig. 9.74): It has tapered wedge jaws with flat gripping surfaces which may be either smooth or serrated.

This plier has flat jaws. One side of the jaw is sharp for cutting purpose. These pliers are useful in a large number of jobs like cutting wires, turning nails and to hold small jobs.

It is used for bending and folding narrow strips of thin sheet metal **(Fig. 9.74).**

Pliers used in electrical jobs have insulated handles.

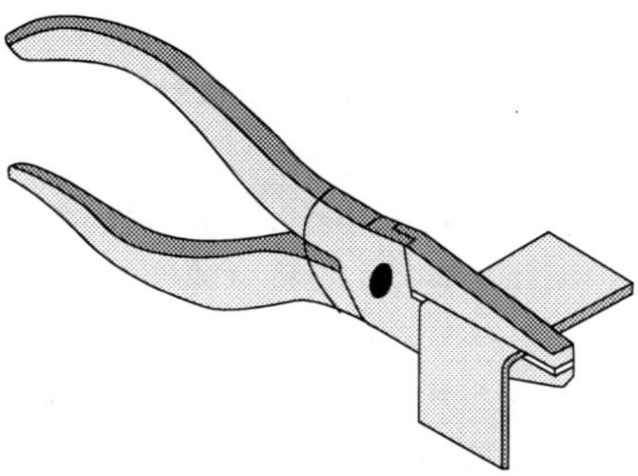

Fig. 9.74 Bending and folding

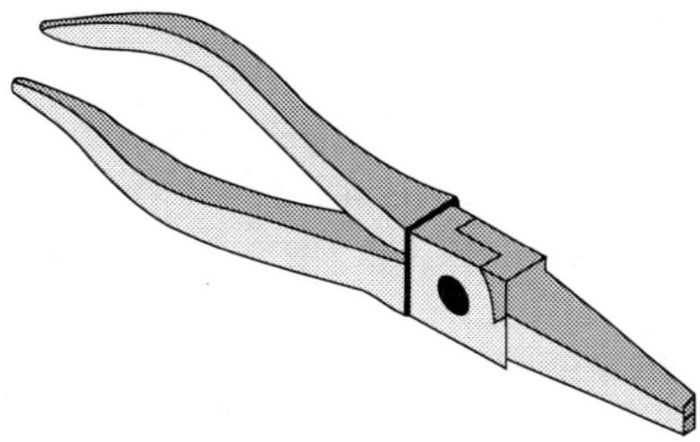

Fig. 9.75 Round nose pliers

Round nose pliers (Fig. 9.75): This type of plier is made with tapered round shaped jaws. They are used to shape loops in wires and to form curves in light metal strips **(Fig. 9.76).**

Slip-joint pliers (Fig. 9.78): These pliers are available in various ranges of positions with different shapes of pivot pins so that they have various ranges of jaw openings. These are mainly used for gripping.

End cutting pliers (Fig. 9.78): These pliers have the same uses as the side cutting pliers. It is also called pincer plier. It is used to pull out nails. The jaws are round shaped and tapered.

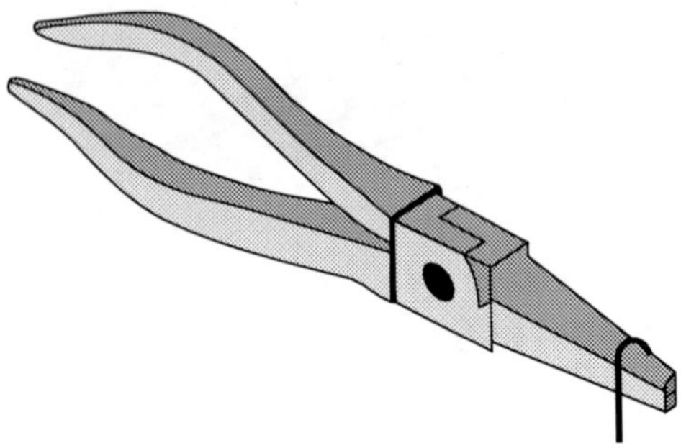

Fig. 9.76 Round nose pliers used to shape loops

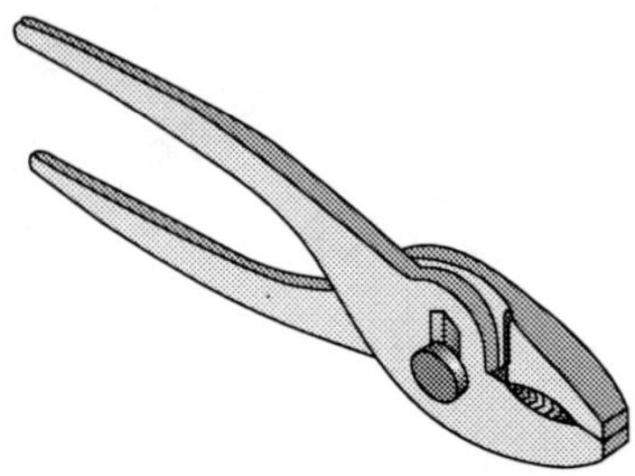

Fig. 9.77 Slip joint pliers

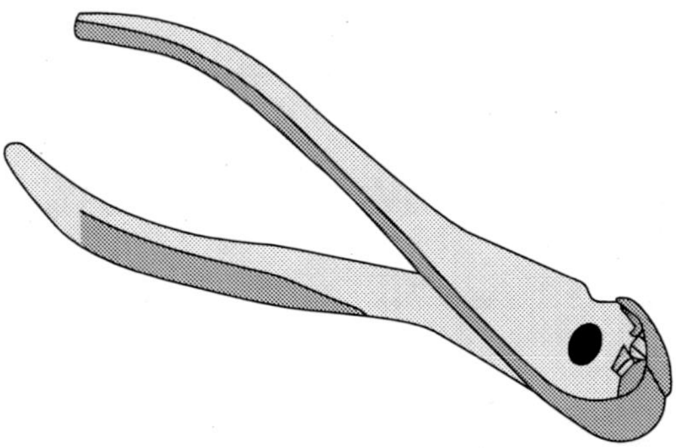

Fig. 9.78 End cutting pliers

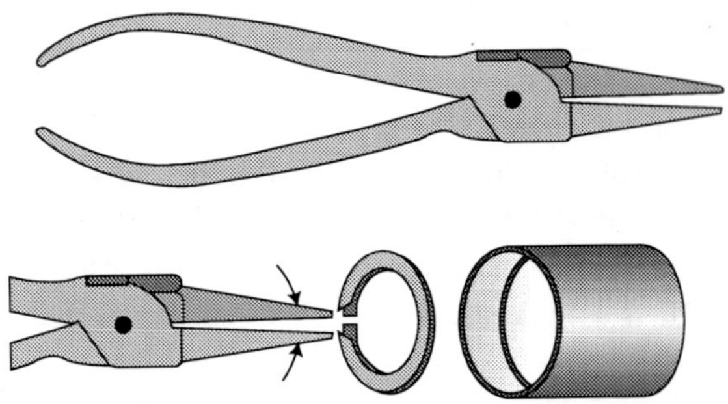

Fig. 9.79 Internal circlip pliers

Circlip pliers (Fig. 9.79): Circlip pliers are used for fitting and removing circlips in assembly works.

Internal circlip pliers (Fig. 9.79): It is used to fit and remove the internal circlip in the groove of the bore **(Fig. 9.79).**

Slip-joint, multi-grip pliers (Fig. 9.80): It is similar to the grip pliers but has more openings in the legs. It gives a range of jaw openings. It allows parallel gripping by the jaws in a number of positions.

The shape and length of the leg are different from those of the slip-joint pliers **(Fig. 9.80).**

Other of slip joint pliers

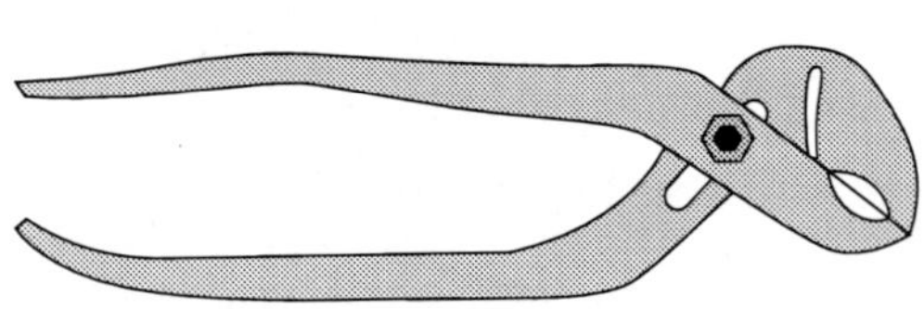

Fig. 9.80

Side cutting pliers (Fig. 9.82): It is made with jaws set at an angle. They are used for shearing off wires in confined spaces and cutting wires close to the surface level **(Fig. 8.83).** They are also used for spreading the cotter pin **(Fig. 9.84).**

External circlip pliers: External circlip pliers are used to fit and *remove* the external circlip in the grooves of the shafts.

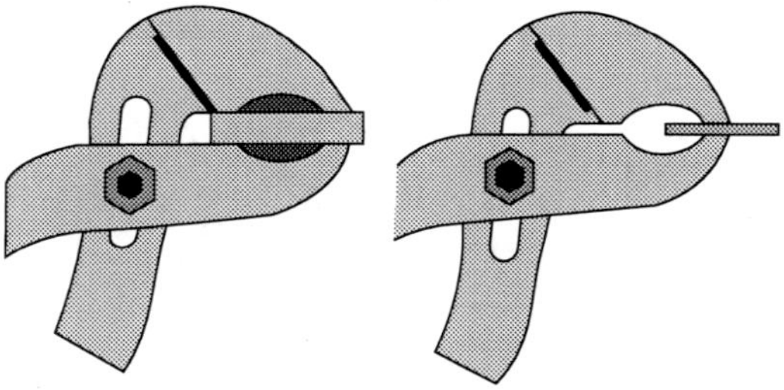

Fig. 9.81 Use of slip joint pliers

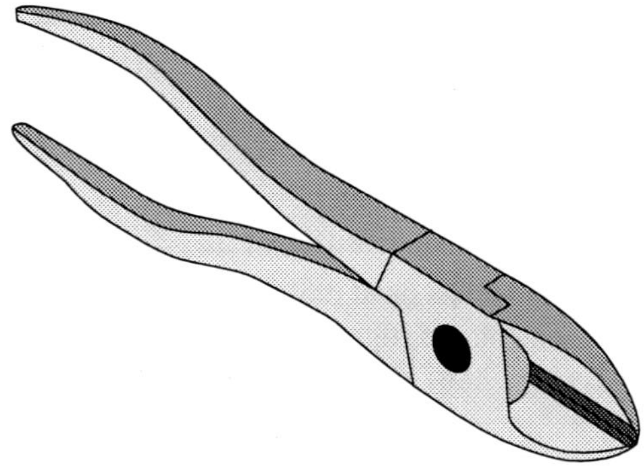

Fig. 9.82 Side cutting plier

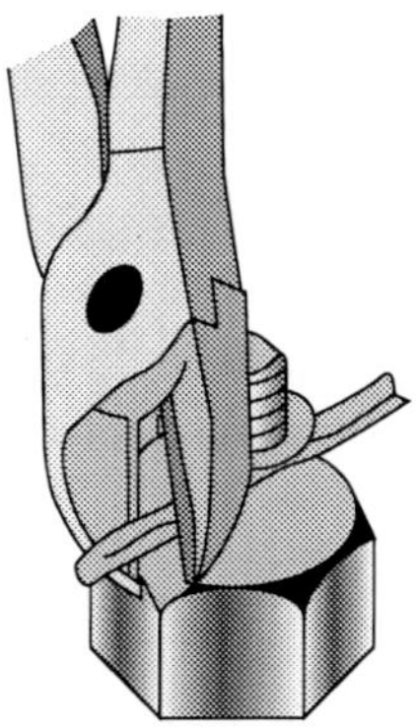

Fig. 9.83 Side cutting plier used for spreading cotter pin

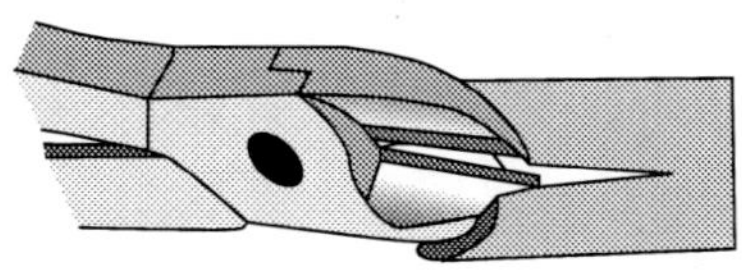

Fig. 9.84 Used for shearing in confined spaces

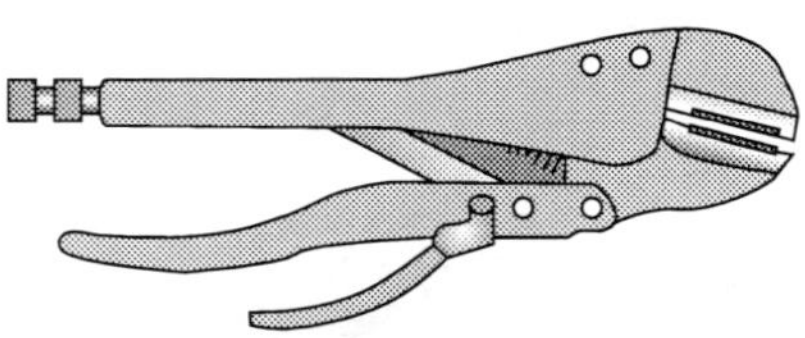

Fig. 9.85 Locking plier

Locking pliers (Fig. 9.86): The locking lever of the these pliers is attached with a movable handle which clamps the jaws on to an object of any shape. It has high gripping power.

The screw in the handle enables adjustment of the lever action to the work size.

The opening of its jaws can be enlarged or reduced. It is also used as a wrench or a vice.

9.13 PRECAUTIONS

1. Pliers should not be used in place of hammer, spanner or wrench.
2. For different types of jobs, different pliers should be used.

9.14 SPANNERS

Spanners are used more than any other tool for tightening or loosening of bolts and nuts of different jobs. These are generally made of drop forge steel or carbon steel of high tensile strength, or alloy steel and heat-treated for strength.

These are made in different shapes and sizes.

A spanner is a hand tool with jaws or opening or a ring at one end or at both ends for tightening or slackening nuts and bolts and screw heads **(Fig. 9.86).**

Fig. 9.86 Spanner

9.15 TYPES OF SPANNERS

1. Open end spanners
   - Single-ended spanners
   - Double-ended spanners
2. Ring spanners
3. Socket spanner
4. Box spanner
5. Combination spanner
6. Hook spanner
7. Adjustable spanner
8. "T" socket spanner
9. Magneto spanner
10. Allen key
11. Pin face adjustable spanner
12. Ring spanners

Open end spanners: They can be single ended or double ended.

Single-ended spanners (Fig. 9.87): These are general purpose spanners. Single-ended spanners are mostly supplied with machine tools for a specific purpose.

Only one end is open, in which a nut or bolt head is gripped to loosen or tighten it. The size is taken from the distance between their jaws. The selection of the spanner is made according to the sizes of the nut. (The size of the nut is the distance across the flats.)

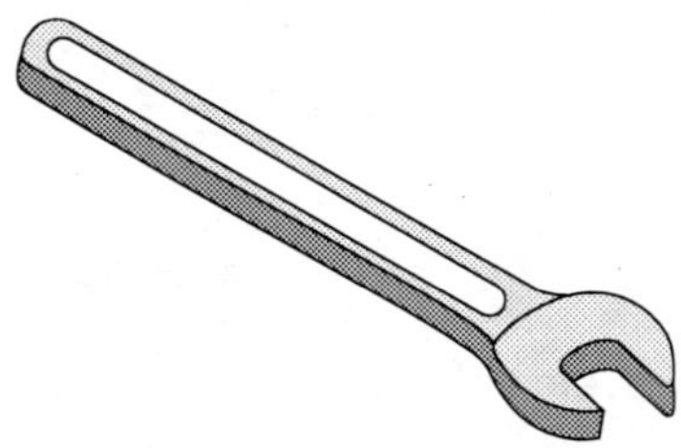

Fig. 9.87 Single ended spanner

In this type of spanner open ends are made at both the ends to hold nuts or bolt heads in them. These two ends are made at angles of 15° to 30° from the *in-between* body. These types of spanners are most widely used.

Double-ended spanners (Fig. 9.88): Double-ended spanners are standard spanners having two different size openings. Some spanners are made of chrome vanadium steel.

Open ends are made at both the ends to hold nuts or bolt heads in them. These two ends are made at angles of 15° to 30° from the *in-between* body.

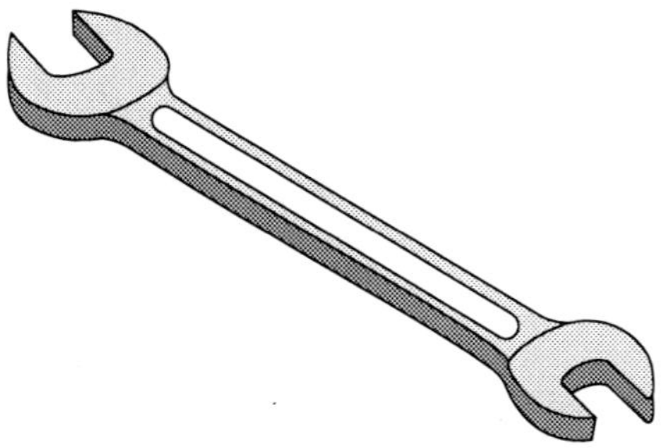

Fig. 9.88 Double ended spanner

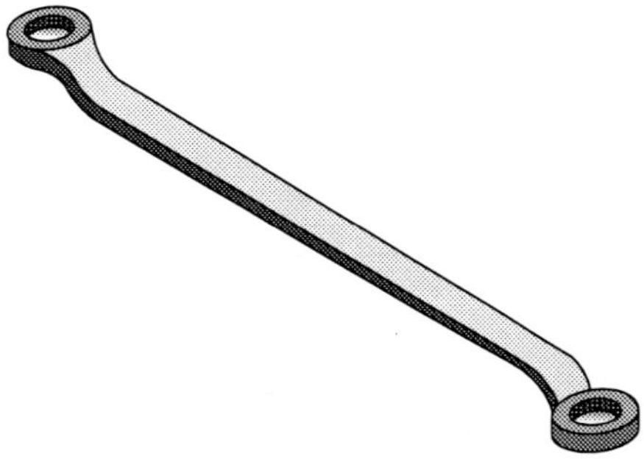

Fig. 9.89 Ring spanner

They are available in a set of 8, No. 8 to 27 mm. 8 × 10, 9 × 11, 12 × 13 14 × 15 16 × 17 18 × 19 20 × 22 and 24 × 27 mm.

Open ended spanners bigger than 27 mm size are also available.

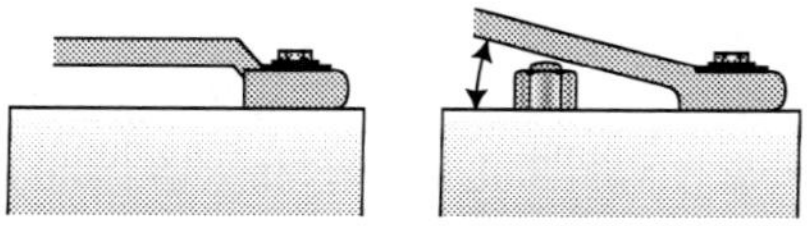

Fig. 9.90 Use of ring spanner

Ring spanners (Fig. 9.89): These types of spanners are used where obstruction close to the side of a nut prevails and application of open-ended spanners is not possible.

These are available in a set of 8 Nos. (8 to 27 mm) 8 × 9, 10 × 11, 12 × 13 14 × 15 16 × 17 18 × 19 20 × 22 and 24 × 27 mm.

On one or both the ends round rings are made to grip nuts and bolts in order to tighten or looseen them. These rings are bent a little in the opposite direction from each other. Inside these rings projected teeth are built with which six sided nut bolts can be easily gripped. As it grips all the sides, the possibility to slip is negligible..

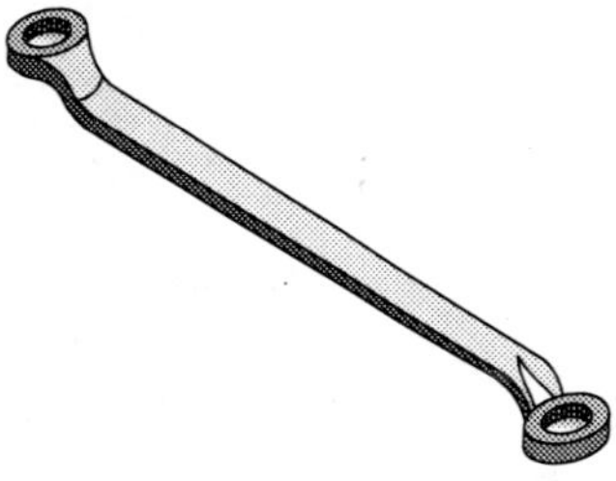

Fig. 9.91 Ring spanner

Socket spanner (Fig. 9.92): These spanners are of the form of round sockets. On one end of this spanner a square hole is provided. On the other end, grooves are cut to outside sizes of nut or bolt, as in a ring spanner. In the square hole, a Tommy bar is fixed whose shape is also square. These spanners are used where enough space is not available to use any other spanner.

Beside, these socket spanners are used where nuts and bolts are to be tightened by torque wrenches. These are available in small and big size in form of sets. Tommy bar universal tommy and ratchet Tommy along with these spanners.

The nuts and bolts slanting and in congested space can be tightened or loosend.

Box spanner (Fig. 9.94): These are made of pipes. One end or both the ends of this spanner have six sides, in which nuts or bolts are easily gripped. Holes are made across the rod on one or both the ends in which a rod is inserted and used as a handle. It is used in deeper spaces.

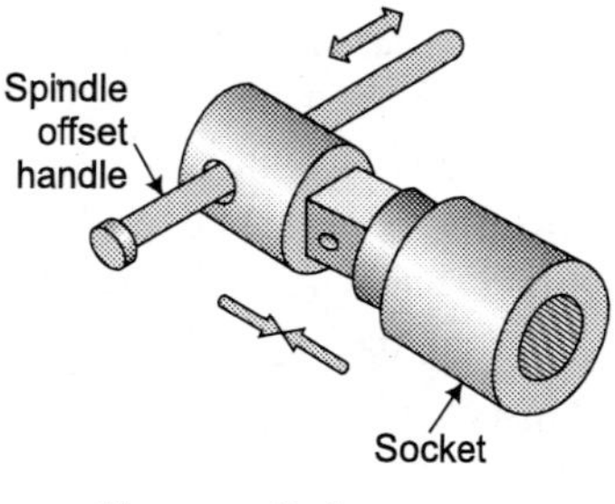

Fig. 9.92 Socket spanner

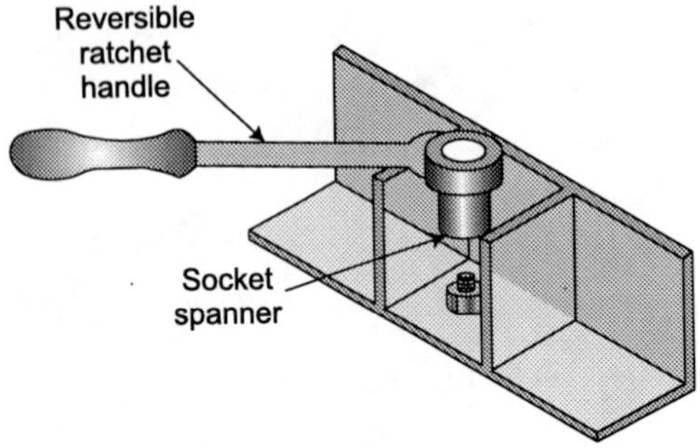

Fig. 9.93

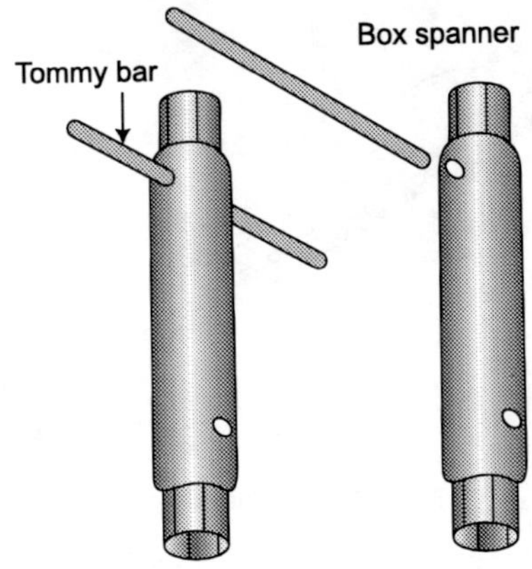

Fig. 9.94

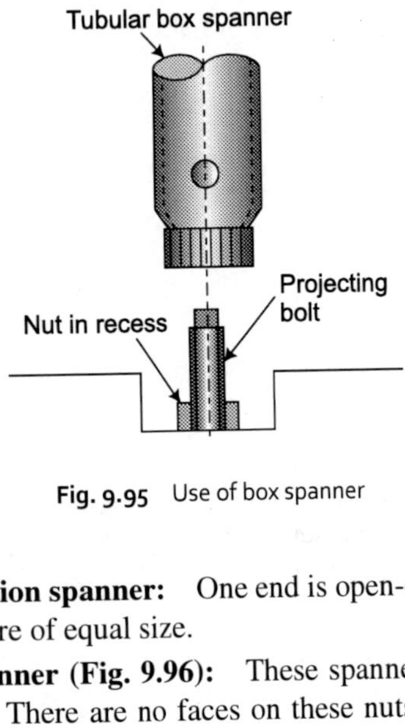

Fig. 9.95 Use of box spanner

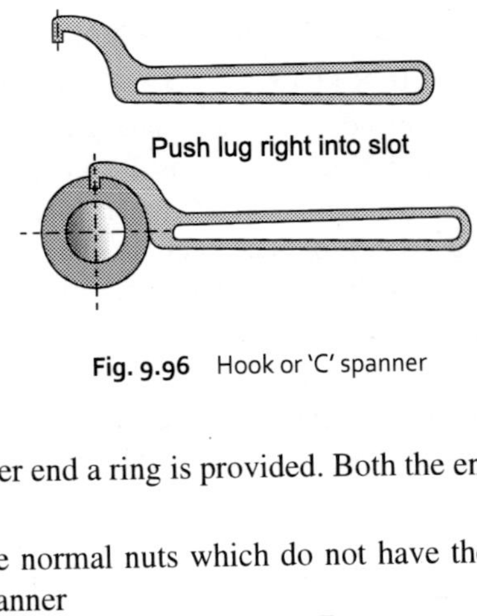

Fig. 9.96 Hook or 'C' spanner

Combination spanner: One end is open-ended and on the other end a ring is provided. Both the ends, the spanners are of equal size.

Hook spanner (Fig. 9.96): These spanners are used for those normal nuts which do not have the usual periphery. There are no faces on these nuts. The pin of this spanner gets fixed in the hole and performs the function of revolving it for opening or tightening.

Adjustable spanner (Fig. 9.98): This is a special type of spanner. It is also known as screw wrench. Its jaws can be expanded or narrowed down to suit the nuts or bolts.

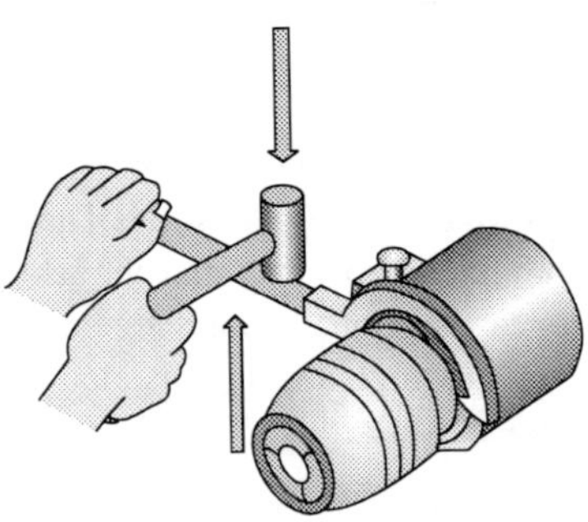

Fig. 9.97 Use of 'C' spanner

Most common types of adjustable spanners are similar to open end spanners, but they have one movable jaw. The opening between the jaws of a typical 250 mm spanner can be adjusted from zero to 28.5 mm. Adjustable spanners may range in length from 100 to 760 mm. This spanner **(Fig. (9.97)** has its jaws set at an angle of 22½° to the handle. Adjustable spanners are convenient for use where a full kit of spanners cannot be carried. They are not intended to replace fixed spanners which are more suitable for heavy service. If the nut or bolt does't match the size of spanner, adjustable spanner is more handy, the gap between the jaws of the spanner to match the nut or bolt nut.

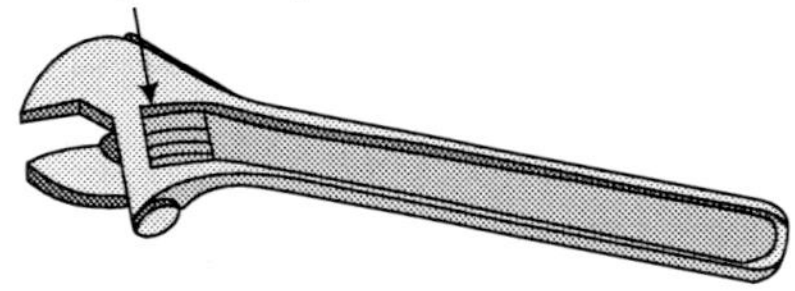

Fig. 9.98

"T" Socket Spanner: This is similar to the box spanner but it has a fixed hand on the top and in the bottom a socket. These sockets are of different sizes. They are used in deep spaces.

Magneto Spanner: These are bunch of small sized single-ended spanners of different sizes and tied together.

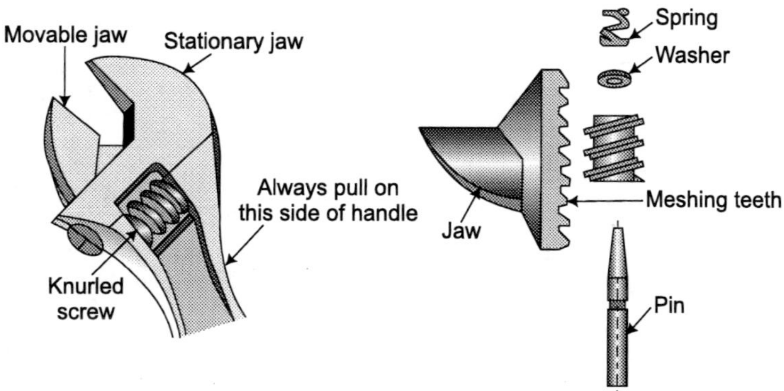

Fig. 9.99

Allen key (hexagonal socket wrenches) (Fig. 9.100): These are thin hexagonal steel rods; one end of which is bent to "L" shape at 90° angle. These are used to open such bolts whose heads are made round and have a hole of six faces on the upper side. Such bolts perform the function of tightening while keeping the jobs surface equal. These are available in numbers and sets of different sizes.

Figure 9.101 illustrates the types of screws used mostly in precision work (tool room work). The head of these screws will be in flush with the surface of the workpiece; they may not be projecting outside causing obstruction.

These are used to rotate set screws and socket headed screws having internal hexagon sockets. They are available in sets ranging in sizes from 1.5 to 20 mm, across the flats.

Make sure that the socket and wrench are cleaned before use. Use the correct size. Insert to the full depth. Discard if the ends become worn out or rounded as they are liable to slip under load.

Pin face adjustable spanner: This type of spanner is made by joining two legs with a rivet. At the end of its legs a pin is fitted on each leg. It is used on round nuts having holes in the periphery to receive the spanner pin. The distance between its legs can be adjusted to use it on all type of nuts.

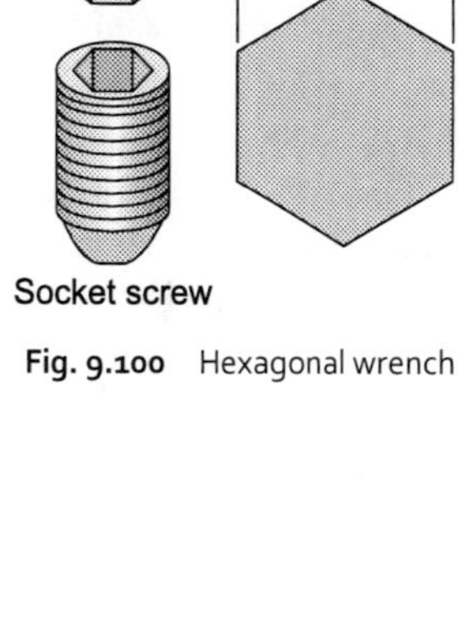

Fig. 9.100 Hexagonal wrench

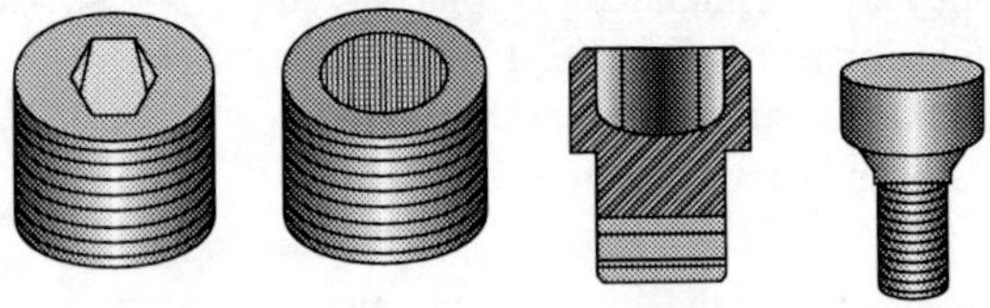

Fig. 9.101 Socket headed screws

9.16 SIZE OF THE SPANNER

All spanners are of a specific size. They are in units of mm, inches. Millimetre spanners are available in variations of 7 mm. Similarly, inch spanners are available in 1/16 inch and double ended and ring spanners are mostly in sets of 6, 8, 12 and 36.

In a set of 6 spanners in inch spanners, the double ended and ring spanners are available in the following sizes: 1 inch × 15/16 inch, 7/16 × 13/16, ¾ × 11/16, 5/8 × 19/32, ½ × 9/16 and 3/8 × 7/16 inch.

Similarly, millimetre spanners are in numbers. In double-ended and ring spanners on the one side one number is marked and on the other, another number one more than the previous, is marked. Socket spanners are available in different sizes.

9.17 WRENCHES

The word wrench means violent twist, turn, or pull. The wrench is a tool of various forms for gripping and turning bolt head, nut and round pieces like a pipe.

Apart from the spanners different types of wrenches are used for bolt head, nut and round pieces. Most of these wrenches are adjustable, with different opening ranges. Mainly the following types of wrenches are used:

1. Stillson pipe wrench
2. Foot print pipe wrench
3. Tension wrench
4. Hexagon socket wrench

Stillson pipe wrenches (Fig. 9.102): A jaw is fixed to the handle with onward facing teeth. Attached to the handle by a pivot pin is a spring-loaded casing that carries a knurled adjusting nut. This engages with a thread on the adjustable arm of a jaw with inward facing teeth.

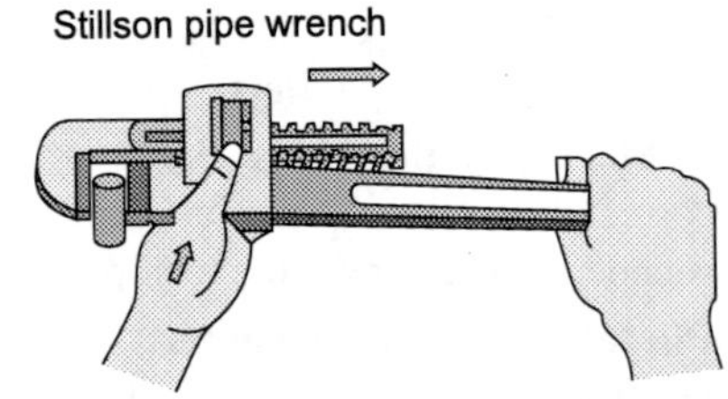

Fig. 9.102

Once the jaws are adjusted, the spring loading keeps them in contact with the work, and the toggle action causes the hardened serrations to bite into the work.

The jaws will mark the work. File off any burrs. Never use them on polished or plated surfaces. Never grip hardened materials with this type of wrench as this will damage the serrations.

Footprint pipe wrenches (Fig. 9.103): These are used for gripping and turning pipes and round stock, particularly in confined spaces.

Adjust the size by fitting the removable pin in the hole that allows the pipe to be gripped, with the handles a comfortable distance apart. Thrust the jaws fully on to the pipe. Squeeze the handles firmly. Pull on the folded steel handle to turn the pipe. Stop squeezing and slide the jaws back round the pipe, squeeze and pull again.

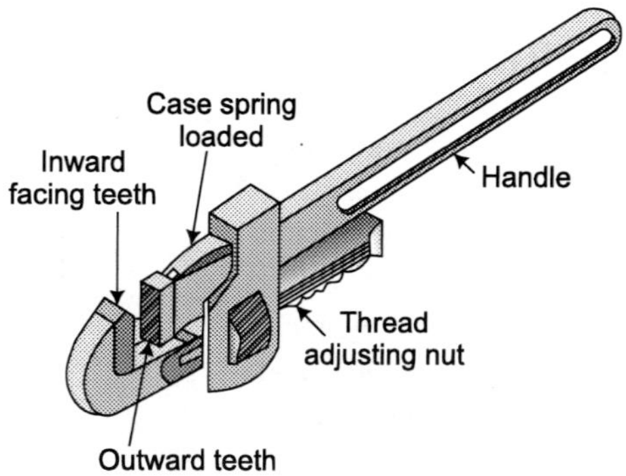

Fig. 9.103 Parts of pipe wrench

File off any burrs raised by the jaws on the pipe.

These are used for gripping and turning pipes and round stocks in confined places.

The required size is adjusted by placing the pivot pin in the different holes of the solid handle.

The grip is obtained by squeezing both the solid handles together.

The selection of hole should be such that the handles are not too far as this may result in uncomfortable holding of the handles.

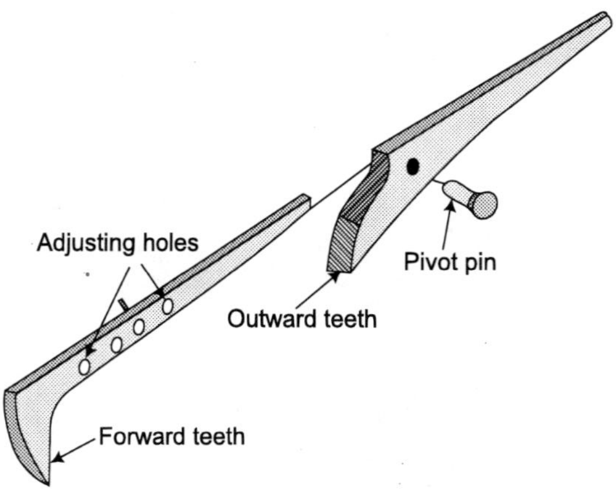

Fig. 9.104 Footprint wrench

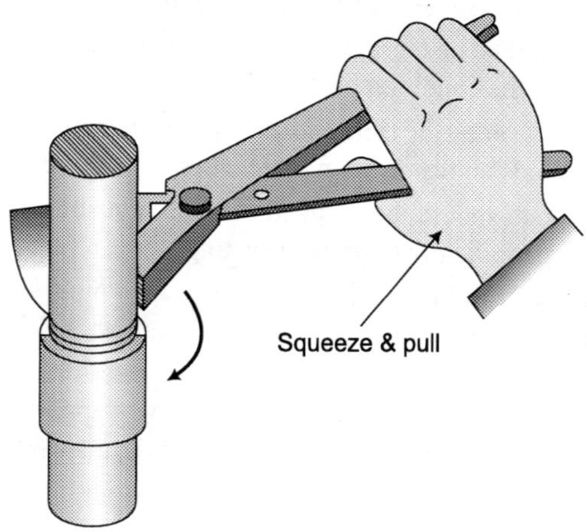

Fig. 9.105 Uses of footprint spanner

Tension wrenches (Fig. 9.106): A tension wrench acts as a torque limiting device for turning (rotating) nuts to a predetermined degree of tightness. This avoids breaking the fasteners. It is also essential to avoid warping or springing components held by multiple fasteners that could be unevenly or excessively tightened, for example, ylinder heads of engines.

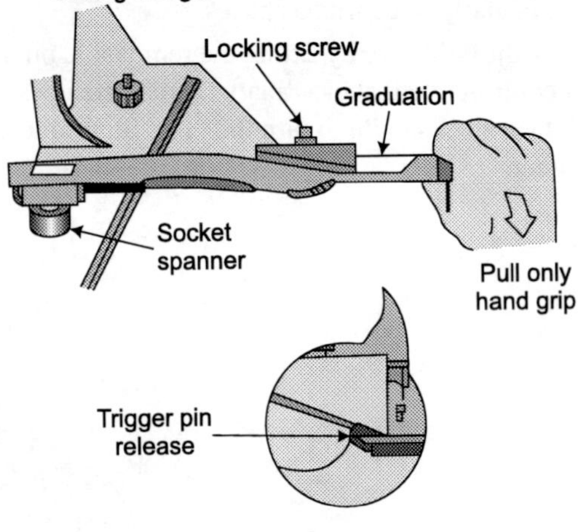

Fig. 9.106 Tension wrench

Some tension wrenches have direct reading indicators that you must watch as you pull the handle to the desired extent. With others, you preset to the desired graduation and pull until you detect a signal which may be an audible click, the release of a trigger pin or an automatic release within the wrench mechanism.

To apply the correct torque with a tension wrench, check that the threads of the nut and the bolt are clean and well formed, pull slowly with evenly increasing effort on the hand grip of the handle.

These wrenches have sharp serrated teeth and may damage polished or plated surfaces.

Chain pipe wrench (Fig. 9.107): The chain wrench is used for holding or gripping large diameter pipes where the Stillson wrench cannot be used.

This wrench is made of high tensile steel with hardened and tempered double-ended reversible jaws. A long tapered handle is bolted with the jaws.

The capacity of the chain wrench is approximately from 50 to 150 mm.

This can also be used for gripping irregular surfaces.

The ability of the pipe wrench to grip the pipe is directly related to the condition of the teeth.

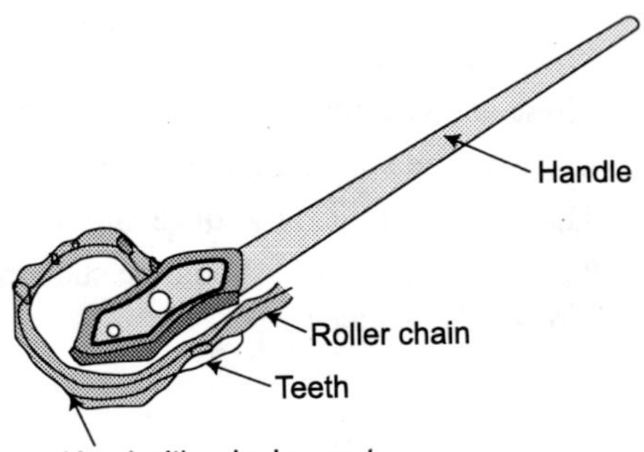

Fig. 9.107 Chain pipe wrench

- Cleaning the teeth and sharpening them with a triangular file can restore some wrenches to useful condition.
- Oil should be applied to the adjustment nut periodically to prevent rusting.

When we hold pipes and round jobs with pipe wrench it leaves some on the jobs. Therefore strap wrenches are used.

The chain pipe wrench is a heavy gripping tool and should not be used for pipes of less than 50 mm diameter.

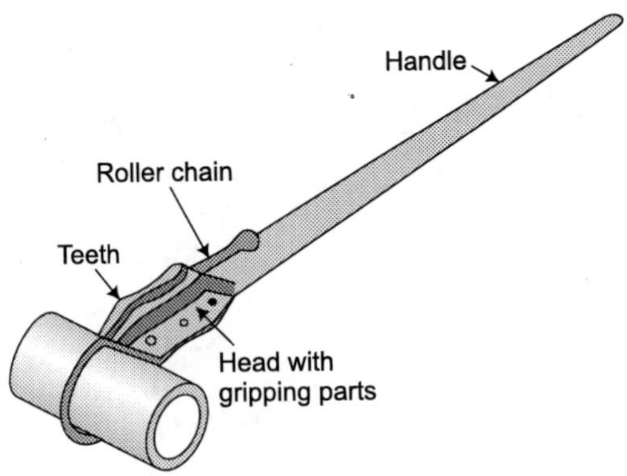

Fig. 9.108

Strap wrench (Fig. 9.109): Strap wrenches are used on finished tubular surfaces to avoid marking or damaging. These wrenches have metallic straps by which the surfaces can be tightly gripped.

A leather or canvas belt is provided which is rolled around the job and tightened with the help of its handle the pipe is revolved.

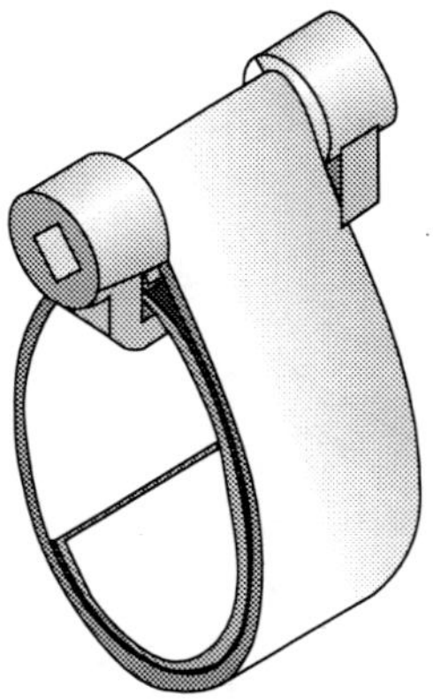

Fig. 9.109 Strap wrench

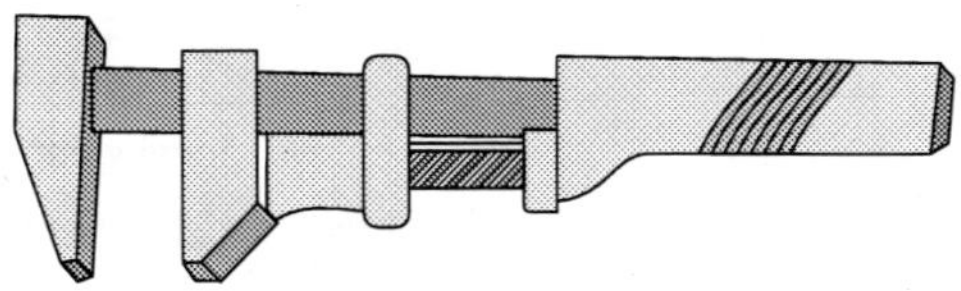

Fig. 9.110 Monkey wrench

Monkey Wrench (Fig. 9.110): It is also like an adjustable spanner. Its bigger jaw is built with the handle itself. The other jaw is movable on which threads are cut as in the pipe wrench. Its jaws are of plain surface like that of open-end spanner. These can be adjusted with a round nut. It is also used like a screw wrench.

9.18 SPANNERS: USE AND MAINTENANCE

Spanners are used for operating threaded fasteners, bolts and nuts. They are made with jaws or openings that fit square on hexagonal nuts and bolts and screw heads. They are made of high tensile or alloy steel. They are drop-forged and heat-treated for strength. Finally, they are given a smooth surface finish for ease of gripping.

Spanners vary considerably in shape to provide ease of operation under different conditions.

The correct spanner to use is the one that fits exactly and allows room for use. They should also permit the job to be done in a short time.

The following are the points to be noted for using spanners safely

Use open end and ring spanners by pulling on the shank. It is safest to pull as there is less chance of hitting your knuckles if the spanner or nut slips suddenly. If you are forced to push the spanner, use the base of your hand and keep your hand open.

Use both of your hands for large spanners.

Keep yourself balanced and firm to avoid slipping yourself, if the spanner slips suddenly. Hold on to some support, if there is any chance of falling.

Use both the hands, when using tubular box spanners.

Use two spanners to stop the head of the bolt rotating as the nut is operated.

Socket spanners may be turned by accessories which have square driving ends.

9.19 SIZE AND IDENTIFICATION OF SPANNERS

The size of a spanner is determined by the nut or bolt it fits. The distance across the flats of a nut or bolt varies both with the size and the thread system.

In the British system the nominal size of the bolt is used to identify the spanner.

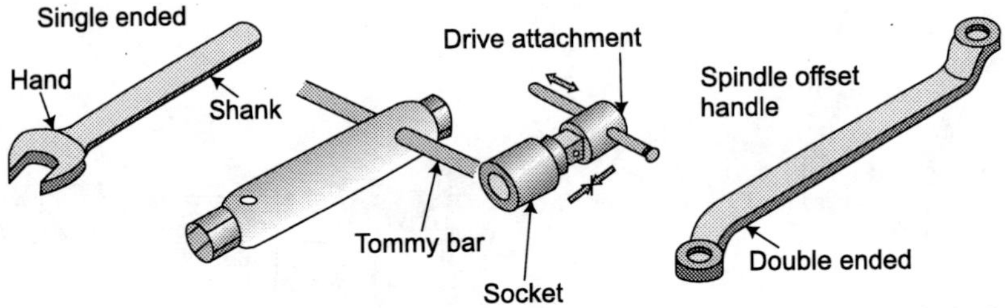

Fig. 9.111

In the unified standard system, the spanners are marked with a number based on the decimal equivalent of the nominal fractional size across the flats of the hexagon, following the sign A/F or with the fractional size across the flats following the sign A/F. In the metric system, spanners are marked with the size across the jaw opening followed by the abbreviation 'mm'.

To fit exactly, a spanner must be:

- Of the correct size
- Placed correctly on the nut
- In good condition.

Spanners have their jaws slightly wider than the width of the nuts so that they can be placed into position easily. Any excess more than a few hundredths of a millimeter clearance could cause the spanner to slip under pressure.

Place the spanner so that its jaws bear fully on the flats of the nut **(Fig. 9.112).**

Incorrect use damages the spanners.

Discard any defective spanners. The spanners illustrated here are dangerous to use **(Fig. 9.113).**

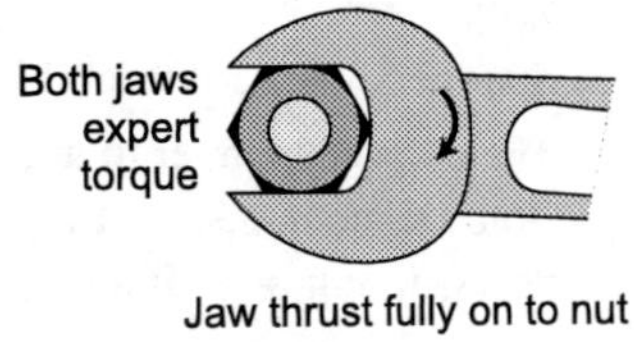

Fig. 9.112

Choose spanners that allow room for use **(Fig. 9.114).**

Nuts in inaccessible positions may be reached with socket spanners, with special drawing accessories **(Fig. 9.115).**

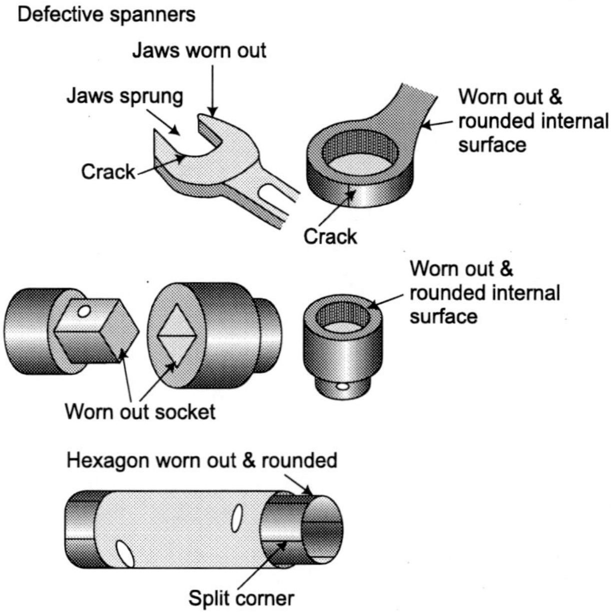

Fig. 9.113

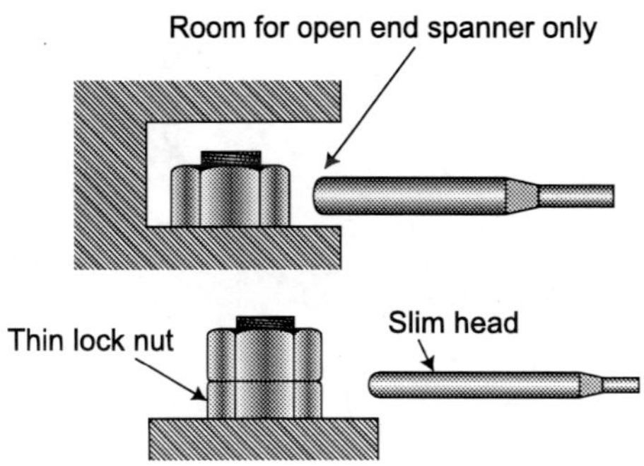

Fig. 9.114

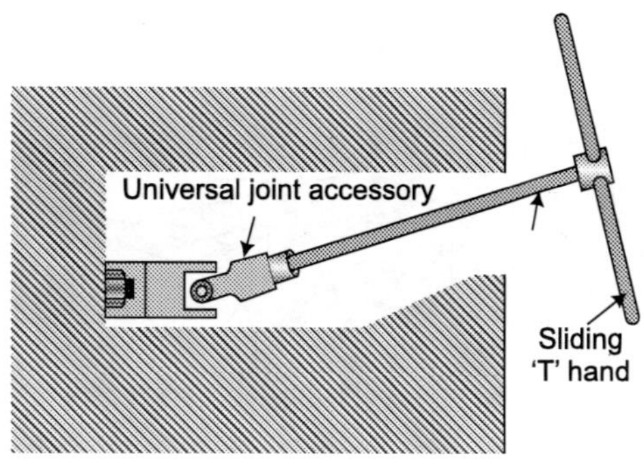

Fig. 9.115

9.20 LENGTH OF SPANNERS

Normally spanners have a length that is about ten times the width of the jaw opening.

Never exert excessive pull on a spanner, particularly by using a pipe to extend the length of the spanner.

Excess turning effect of the spanner could result in:

- Striping the thread
- Shearing the bolt
- Straining the jaws of the spanner
- Making the spanner slip and cause an accident

9.21 PIPE CUTTER (FIG. 9.116)

The wheel pipe cutter is used to make a square cut on the pipe. It consists of: (1) a cutter wheel, (2) two guide rollers and (3) an adjusting screw.

The cutter wheel tends to crush rather than cut the pipe. If it is blunt, it needs replacement.

This type of pipe cutter does not remove any materials but the cutter squeezes the metal and forces it ahead of the cutter until the pipe is cut through the wall thickness **(Fig. 9.116).**

This type of cutting leaves a large ridge on the inside of the pipe which would obstruct the flow **(Fig. 9.118).** The pipe must be deburred or reamed by a pipe reamer.

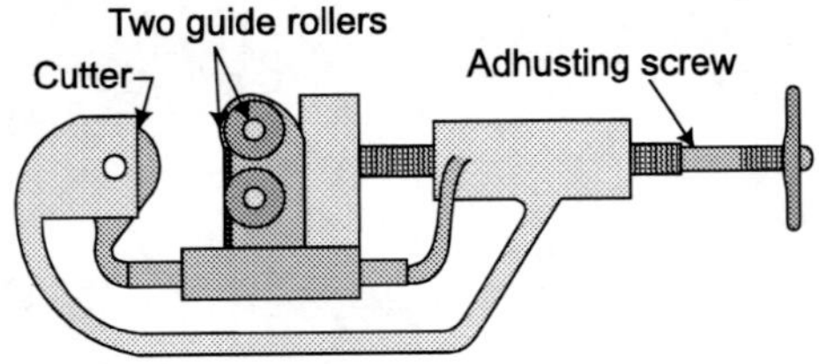

Fig. 9.116 Pipe cutter

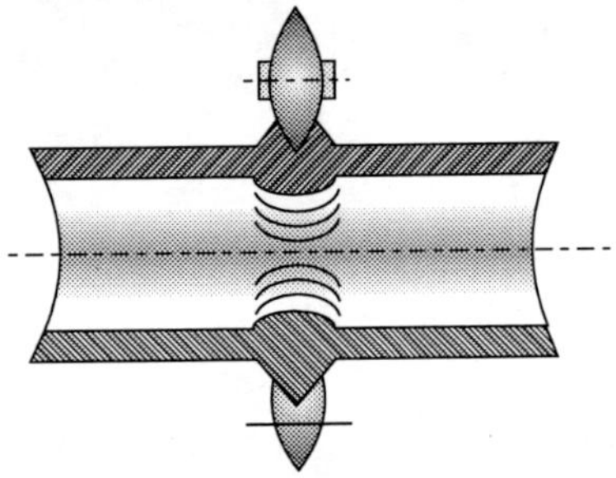

Fig. 9.117 Cutting action of pipe cutter

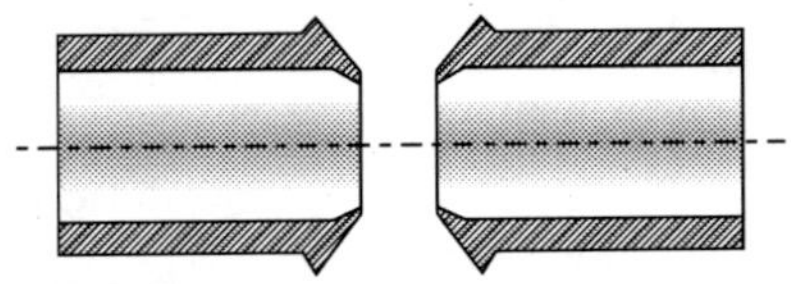

Fig. 9.118 Pipe after cutting

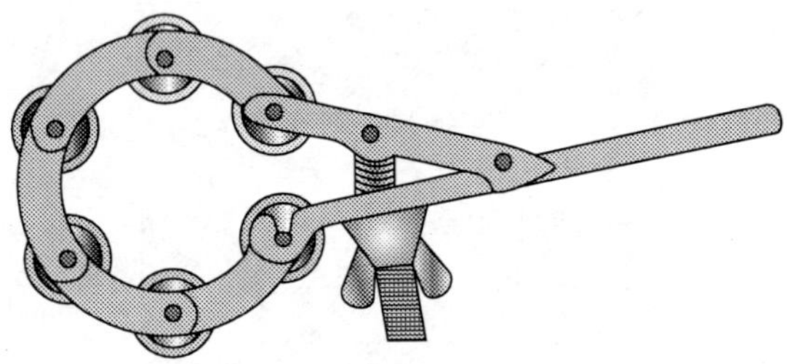

Fig. 9.119 Multi wheel chain pipe cutter

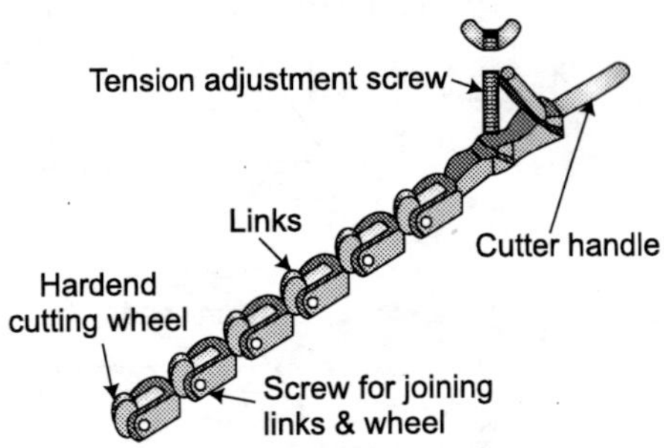

Fig. 9.120 Multi wheel chain pipe cutter

Care and maintenance of pipe cutters:

Before using the cutter check the wheels, pins and links for any damage.

Replace the wheels, pins and links if damaged.

As the wheel revolves around the pin, any wear on the pin will cause the wheel to wobble and the cut will not run square to the pipe. This may result in a:

- Chipped wheel (**Fig. 9.121**)
- Worn out pin (**Fig. 9.122**)

During pipe cutting, small flakes of metal break away and clog up the links and cutting wheels. Clean the links and wheels using a wire brush and soak the cutter in paraffin or kerosene to wash out the small particles of dirt and flakes.

After cleaning, apply light oil on all moving parts, links and wheels for easy cutting operation and to prevent rust forming on the tool.

Store the cutter and protect the wheels from possible damage when not in use.

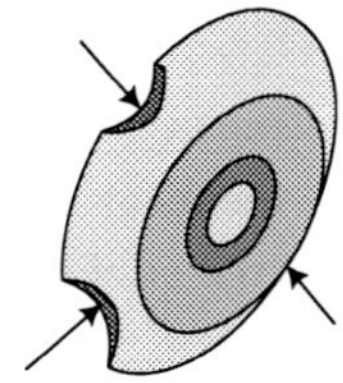

Fig. 9.121 Chipped wheel

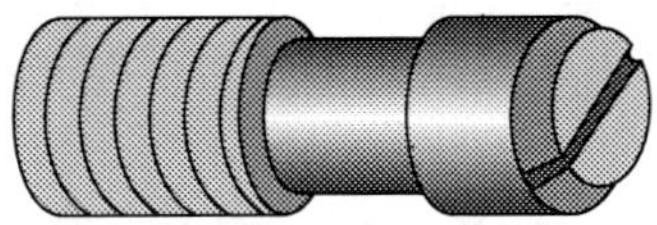

Fig. 9.122 Worn out pin

EXERCISE

1. Name the types of tools used in dismantling and assembling jobs under the fitter shop trade.
2. Draw a sketch of screwdriver and explain its parts.
3. Comment on offset screwdriver and Philips screwdriver.
4. What is the function of pliers and what are their different types?
5. Write brief comments on slip joint pliers and pincer pliers.
6. What is the use of a spanner and what are its different types?
7. How is a ring spanner different from socket spanner?
8. Where Hook spanner and Allen key are used?
9. What is meant by measurement of the spanner? Explain these measurements.
10. Write a brief note on the types of wrenchs.
11. What are the different kinds of hammers and explain their parts.
12. Draw a sketch of ball pane hammer and explain its parts.
13. Write a brief note on power hammer.
14. What precautions are needed while using a hammer?
15. What is the function of a vice and what are its different types?
16. Draw a sketch of parallel jaw vice and explain its parts.
17. What is special about a quick release vice?
18. Comment on hand vice, pin vice and tool maker's clamp.
19. What are the different types of vice packing clamps?
20. What is the use of "C" clamp?
21. What are different kinds of hammers?
22. Draw a sketch of ball pein hammer and explain its parts.
23. Write a brief note on power hammer.
24. What precautions are needed while using a hammer?
25. Draw a sketch of a screwdriver and explain its parts.
26. Comment on offset-screwdriver and Philips screwdriver.
27. Name the types of tools used in dismantling and assembling jobs under the fitter shop trade.

28. Draw a sketch of a screwdriver and explain its parts.
29. Comment on offset screwdriver and Philips screwdriver.
30. What is the function of pliers and what are its different types?
31. Write brief comments on slip joint pliers and pincer pliers.
32. What is the use of a spanner and what are its different types?
33. How is a ring spanner different from socket spanner?
34. Where hook spanner and Allen key are used?
35. What is meant by measurement of the spanner? Explain these measurements.
36. Write a brief note on the types of wrench.
37. What are the different kinds of hammers?
38. Draw a sketch of ball pein hammer and explain its parts.
39. Write a brief note on power hammer.
40. What precautions are needed while using a hammer?
41. What is the function of a vice and what are its different types?
42. Draw a sketch of a parallel jaw vice and explain its parts.
43. What is the speciality of quick release vice?
44. Comment on hand vice, pin vice and tool maker's clamp.
45. What are the different types of vice packing clamps?
46. What is the use of "C" clamp?

Bolts, Nuts, Studs and Machine Screws

Chapter 10

Assembly in the manufacturing process consists of putting together all the component parts and, assemblies of a given product, fastening, performing inspections and functional tests, installing, separating good assemblies from bad, and packaging and/or preparing them for final usage.

Assembly is unique in comparison with the methods of manufacturing such as machining; grinding and welding in that most of these processes involve only a few disciplines and possibly only one. Most of these operations cannot be performed without the aid of equipment, and components like bolts, studs, nuts, washers, etc.

There are several parts in any machine big or small. These parts are joined in different ways according to their role and position in a machine. Such holding devices as bolts, screws, keys are called fasteners. The act of holding together the different parts of a machine is called fastening.

10.1 TYPES OF FASTENING

1. Temporary fastening
2. Semipermanent fastening
3. Permanent fastening

10.1.1 Temporary Fastening

When parts of a machine are joined in such a way that these can be disjointed or joined again easily, it is called temporary fastening. There is no harm to the machine or its parts in temporary fastening. In temporary fastening nut, bolt, screw, lock pin and washers are used. These are called fasteners.

Bolts: This is widely used fastener. Bolt is made from a thin rod of metal.

On its upper side, different types of heads are made to fix them. On the remaining part, or a part of it, there are threads. From this side, it can be fixed in a job or a nut can be fitted.

When two flanges attached to two shafts, these flanges are joined together with the help of bolts and nuts **(Fig. 10.1)**.

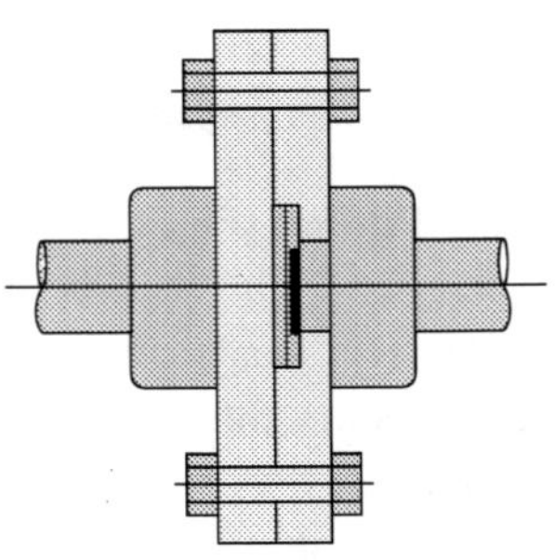

Fig. 10.1 Flanges of shaft bolted

According to the type of head, there are following kinds of bolts:

1. Hexagonal head bolt
2. Square head bolt

3. Cheese head bolt
4. "T" head bolt S. Hook bolt
5. Eye bolt
6. Cup head bolt
7. Counter sunk head bolt
8. Headless taper bolt.

Hexagonal bolt: This is most widely used. Six equal sides are made on its head. Size of head is standardized, i.e., it is made according to the diametric thickness of the bolt in different sizes. Its head fits with an ordinary double end or ring spanner. For convenience its head is chamfered to some extent. One advantage of making head of the size of bolt is that standard spanner fits well in it.

For example, when the diameter D-normal diameter of bolt then the difference between two sides =1 ½ D+ 1/2 and its chamfer will be at 30 0 angle and chamfer radius will be about 1 ¾ D. The thickness of the bolt would be ¾ D.

Depending on the application, different types of bolts are used.

Bolts with Clearance Hole (Fig. 10.2): This is the most commonly used type of fastening arrangement. The size of the bolt is slightly larger than the bolt (clearance hole).

Slight misalignment in the matching hole will not affect the assembly.

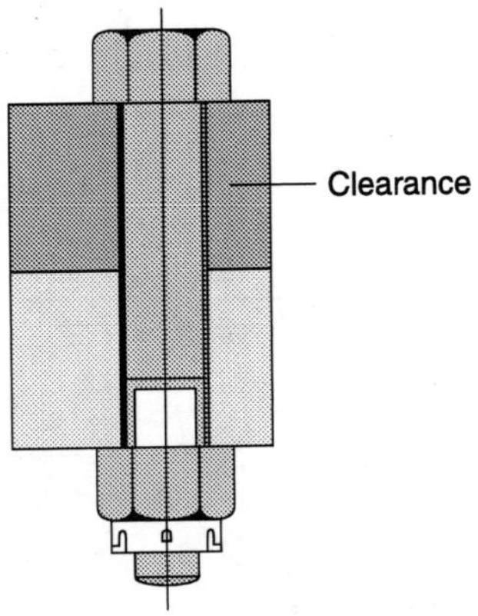

Fig. 10.2 Body with clearance hole

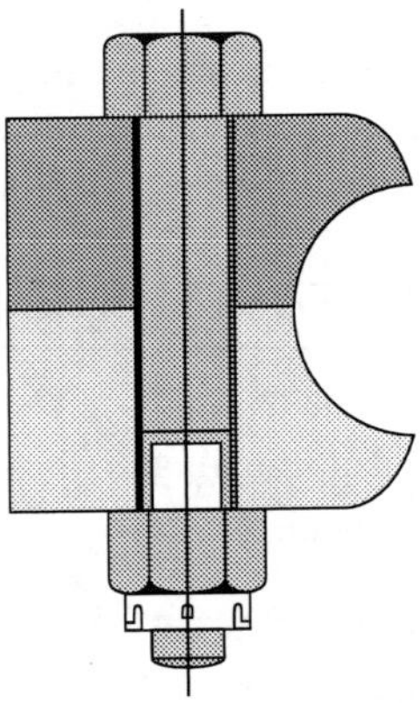

Fig. 10.3 Body fit bolt

Body Fit Bolt (Fig. 10.3): When the relative movement between the workpieces has to be prevented, this type is used. The diameter of the threaded portion is slightly smaller than the shank diameter of the bolt.

The bolt shank and the hole are accurately machined for achieving perfect matching.

Anti-fatigue Bolt (Fig. 10.4): This type is used when the assembly is subjected to alternating load conditions continuously. Connecting rod big ends in engine assembly are examples of this application.

The shank diameter is in contact with the hole in a few places and other portions are relieved to give clearances.

Square Head Bolt (Fig. 10.5): Head of this type of bolt is square. In common jobs it is used at such places where its head is gripped and it neither moves itself nor allows this part to move. Normally, these bolts are fitted in a slot or such a place where a square hole is there.

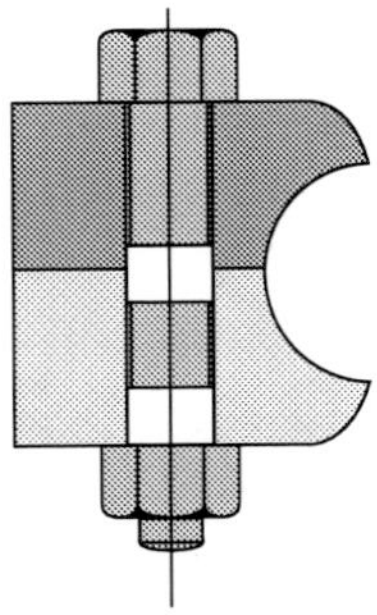

Fig. 10.4 Anti fatigue bit

Fig. 10.5 Square head bolts

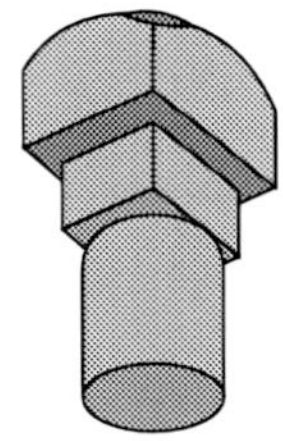

Fig. 10.5a Square head bolt

There are some bolts whose lower part below the square head is also made square. This part fits in the slot or square hole of the job and prevents the job from revolving. The standard of this bolt is **(Fig. 10.5 A).**

Diameter of bolt = D

Thickness of the head -= ¾ D

Distance between two sides = 3/2 D + 3/8

Chamfer radius = 7/4 D

Cheese Head Bolt (Fig. 10.6): At those places where there is inadequate space and there is difficulty in fitting the spanner, these types of bolts are used. Their heads are plain or round. But there is a pin below the head which saves it from rotating at the time of tightening or loosening the nut.

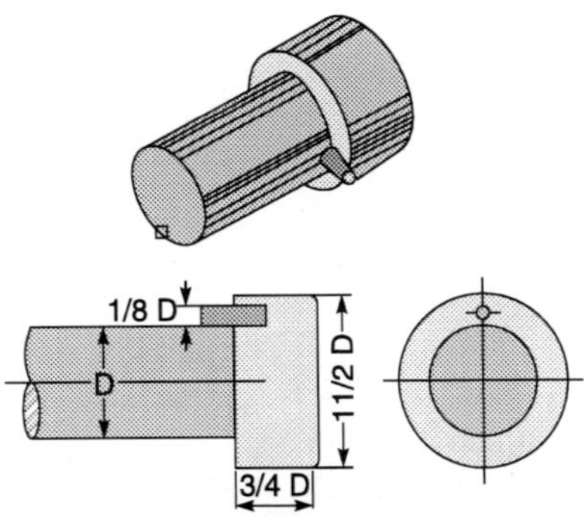

Fig. 10.6 Cheese head bolt

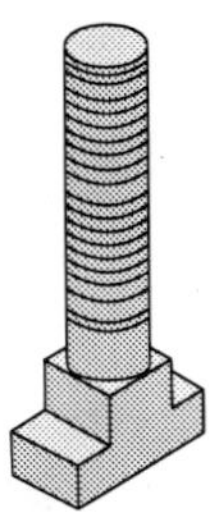

Fig. 10.7 "T" bolt

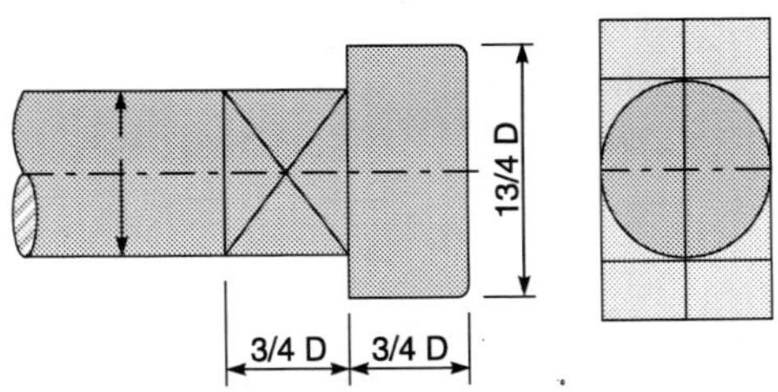

Fig. 10.7a "T" bolt

"T" Bead Bolt (Fig. 10.7): This type of bolts are used with machines. With the help of this bolt, the job can be fastened straight or with clamping devices to a machine table. Its head is square and below the head square shape is provided.

Hook Bolt (Fig. 10.8): This is a special type of bolt. Instead of square or hexagon head the blank is bent and a "U" shape is provided at right angles to the body of the bolt. It is used at places where there is a hole on one part and the other is used to hold the job.

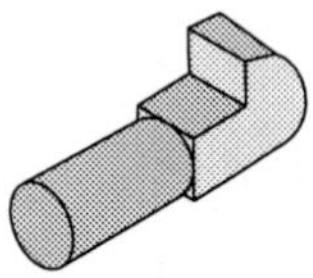

Fig. 10.8 Hook bolt

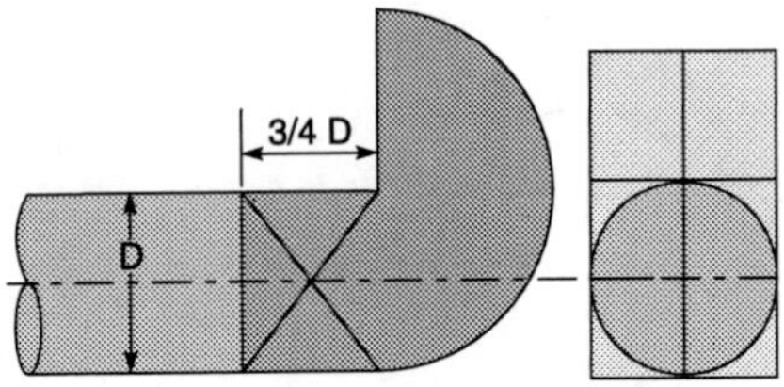

Fig. 10.8a Hook bolt

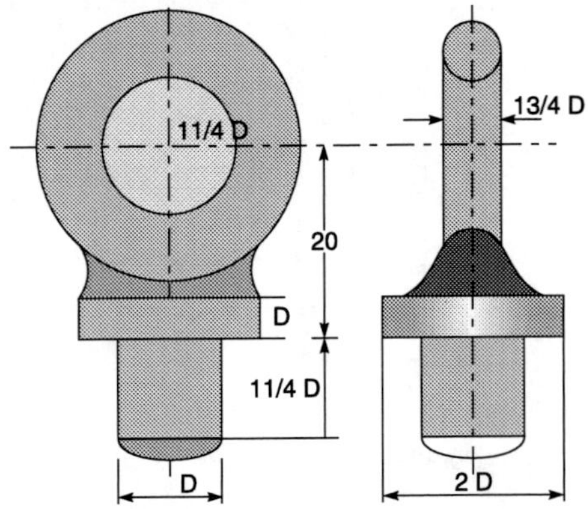

Fig. 10.9 Eye bolt

Eye Bolt (Fig. 10.9): These bolts are specially fitted with the body of machines like electrical motors. This eye is being provided to handle the machines lift or lower and shift them. It is a bolt having the head with an eye or hook.

Nut: It is a small block of metal or other material commonly square or hexagonal in shape. It has internal threads to receive a bolt. Bolts are mostly used with the help of nuts. According to their working position these are of the following types:

1. Hexagonal nut
2. Square nut
3. Flanged nut
4. Knurled nut
5. Wing nut
6. Cap nut
7. Dome nut
8. Capstan nut
9. Ring nut
10. Thumb nut

Hexagonal Nut (Fig. 10.10): This is an ordinary six sided form of nut and is most widely used. Because of its hexagonal shape (having six sides) it fits in the spanner easily and there is less chance of slipping. Its sides are at 60° angle. These are chamfered at 300 to 45°. Because of this chamfer, on the upper part of the nut a circle shape is made. Its standard is:

Thickness = T = D (diameter of its opposite side)

Distance of one side of nut to its opposite side = W + 1″ D + 1/8

Distance between two opposite diagonals = 2 D

Chamfer radius = 1 3/8 D or 1 ½ D

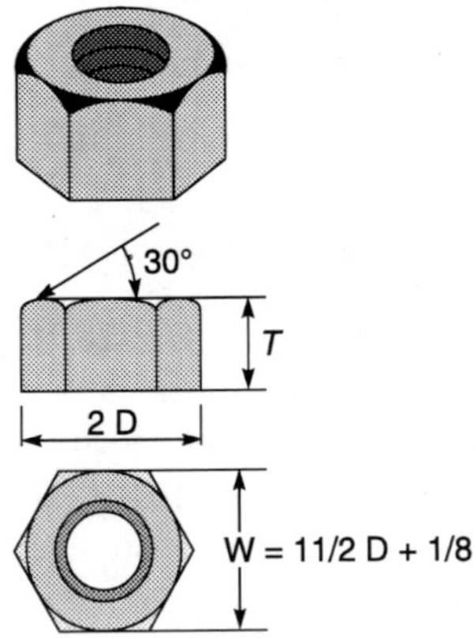

Fig. 10.10 Hexagonal nut

Square Nut (Fig. 10.11): It is used like square head bolt in normal jobs. Its four sides are made at 90° angle; therefore there is no possibility of slip of the spanner. This nut is also chamfered. Its standard is:

Thickness of nut T = D (diameter of nut)

Distance between opposite sides W = 1 ½ + 1/8 D

Chamfer angle = 30°

Chamfer radius = R = 2D

Flanged nut (hexagonal nut with collar Fig. 10.12)

This nut is similar to hexagonal nut but on its lower side a flange of greater diameter than that of the nut is built like a washer. It is used at places where holes have been enlarged or there is a need of washer.

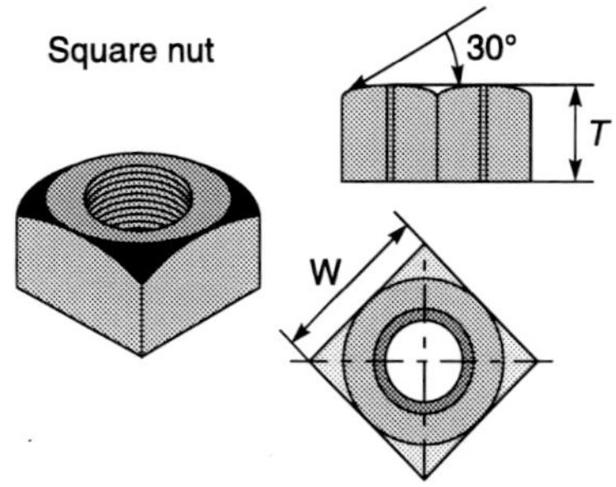

Fig. 10.11

Knurled Nut (Fig. 10.13 A, B): It is a nut of circular shape in which threads are cut inside it and knurling is done outside. These nuts can be taken out quickly by opening them with hands. They are used in measuring tools like micrometer etc.

Wing Nut (Fig. 10.13): This type of nut, on both the ends, a sort of wing is attached. With the help of these wings, the nuts can be easily tightened or loosened. These nuts are also known as fly nuts. These nuts are used at places where there is need to take out these nuts again and again such as in Hacksaw frame.

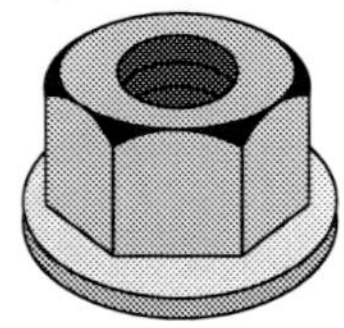

Fig. 10.12 Flanged nut

Cap Nut (Fig. 10.14): These types of nuts are similar to the ordinary hexagonal nuts but on the top side a cap is made with a bit small diameter. These nuts are used at places where chances of bolt end getting rusted. With the help of this type of nuts even the oil leakage through the threads is also checked.

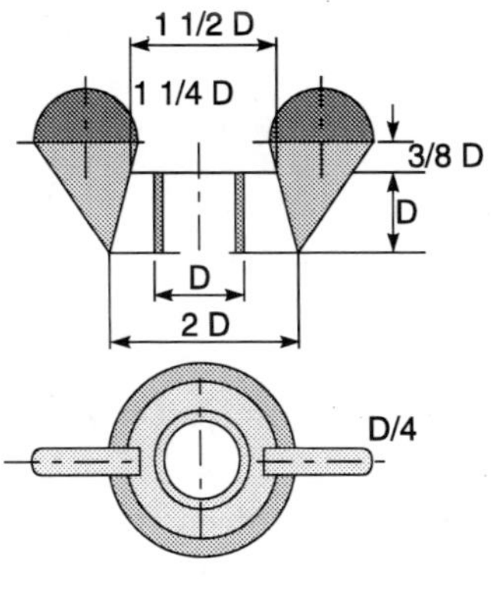

Fig. 10.13 Wing nut

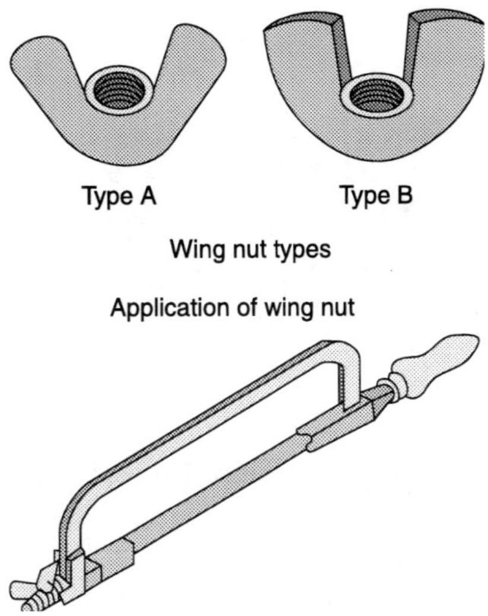

Fig. 10.13a

Dome Nut (Fig. 10.15): These nuts are similar to the cap nut but on their upper side semicircular dome is provided.

Slotted and Castle Nuts (Fig. 10.16): These nuts have special provision in the form of slots for fixing split pins for locking the nut.

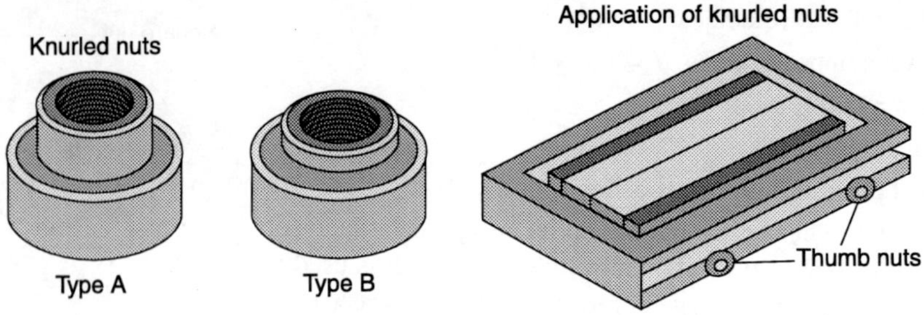

Fig. 10.13b

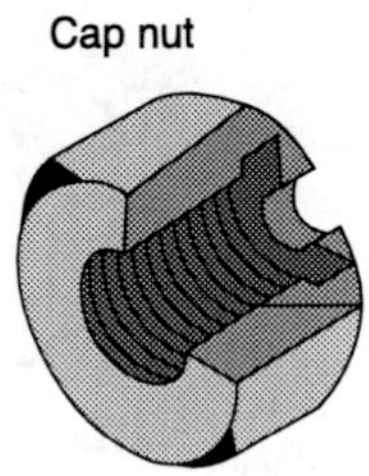

Fig. 10.14

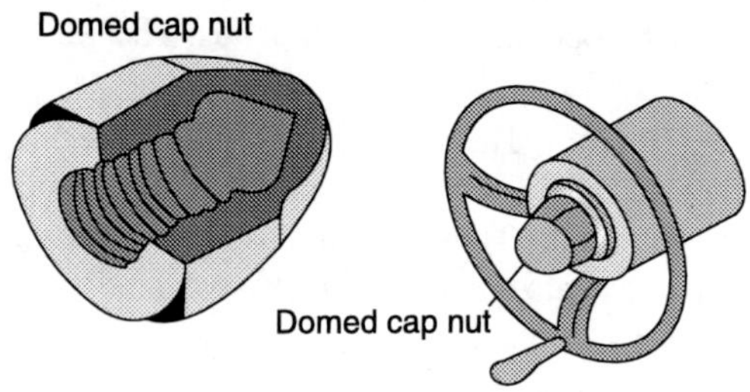

Fig. 10.15

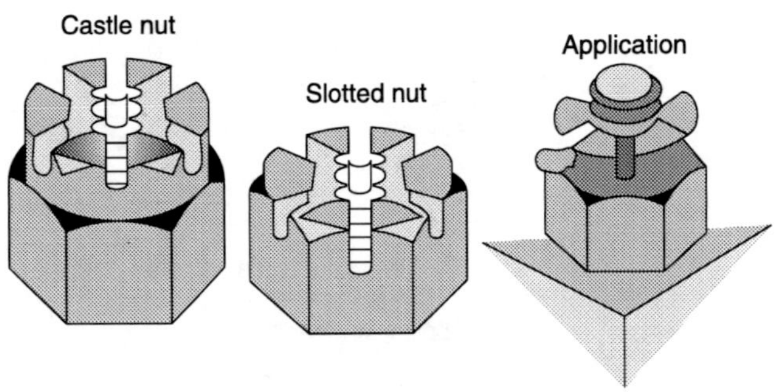

Fig. 10.16

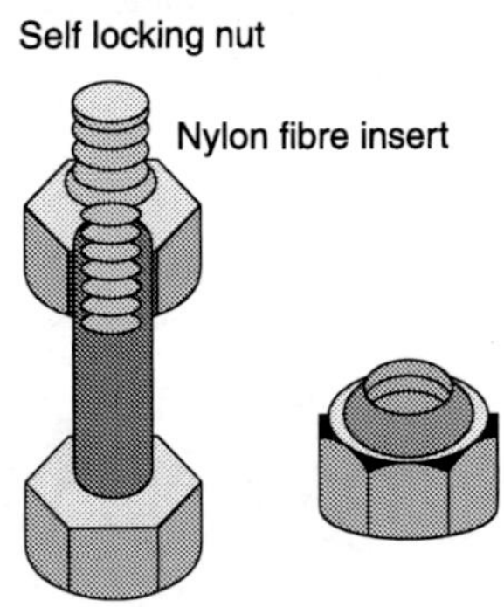

Fig. 10.17

Slotted nuts are hexagonal shaped throughout. In the case of castle nuts, the top part of the nut is cylindrical in shape.

There is a thin collar on the top side, which has six slots cut opposite to each other. The nut after tightening, a split pin is inserted in the hole and the other side of the split pin is opened, thus the locking is done.

Self-locking Nut (Simmonds Lock-nut (Fig. 10.17): This nut has an internal groove cut in which a fibre or nylon ring is inserted. This ring holds the nut tightly on the bolt and serves as a locking device.

Hexagonal Weld Nuts (Fig. 10.18): These are nuts used for welding on plate work. These nuts have:

- A spigot ring which fits in the hole of the plate.
- Three projections to provide a uniform contact on the surface that is to be welded.
- A hole countersunk on one end to protect the thread during welding.

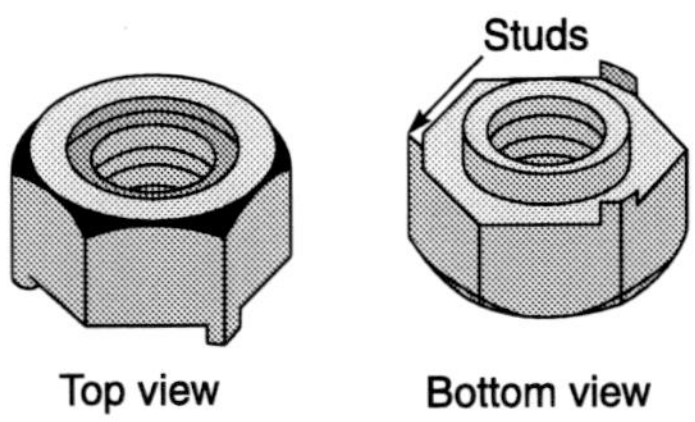

Fig. 10.18 Hexagonal weld nut

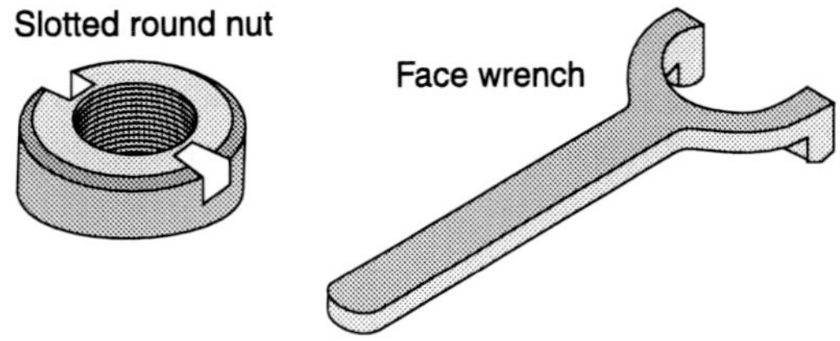

Fig. 10.19

Round Nuts: Round nuts of different types are available for special applications.These are:

Slotted round nut **(Fig. 10.19)**

Slotted round nut for hook wrench **(Fig. 10.20)**

Round nut with set pin holes on sides **(Fig. 10.21)**

Round nut with holes in face **(Fig. 10.22)**

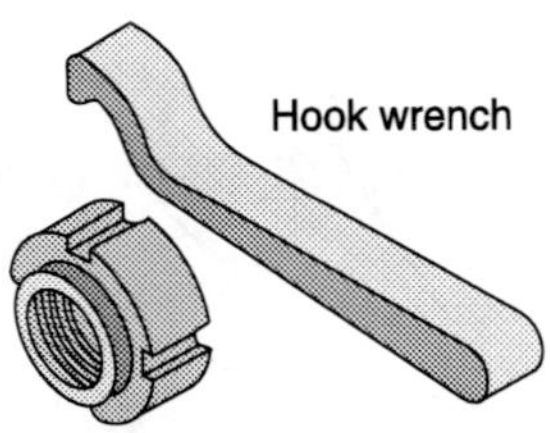

Fig. 10.20 Slotted round nut with hook wrench

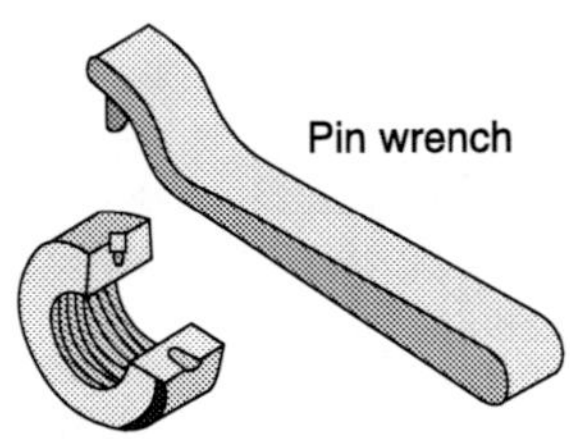

Fig. 10.21 Round nut with pin holes

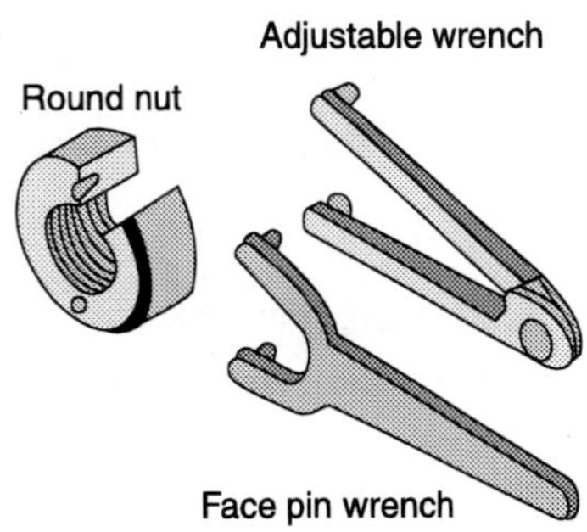

Fig. 10.22 Round nut with face pin wrench

10.2 LOCKING DEVICES FOR NUTS

A locking device is a device used to lock the threaded fasteners to prevent them from loosening. Due to vibration in the moving part, there is a tendency for the threaded fastener to get slack and to screw off. Then the assembled part will get loose and cause damages. Some examples are given below to illustrate the importance of the locking device.

In the case of a micrometer, the lock nut avoids the movement of the spindle after taking the reading. In the case of boilers and gas cylinders, locking of the nut avoids the leakage of steam or gas.

In automobiles the lock nut avoids the loosening of the assembled part.

10.3 CLASSIFICATION OF LOCK NUTS

Lock nuts are classified into two categories.

- Positive locking device
- Frictional locking device **(Fig. 10.24)**

10.3.1 Positive Locking Device

In the positive locking device, the locking action is positive. This locking device is difficult to fit and may take more time. But it is very essential to use this type of locking device in critical joints where failure could cause serious accidents.

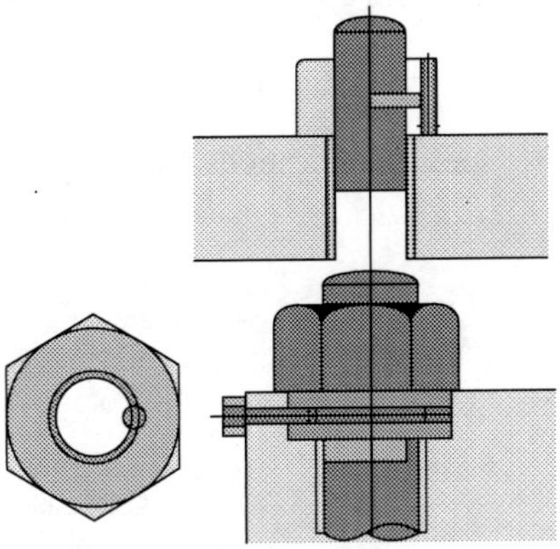

Fig. 10.23 Positive locking

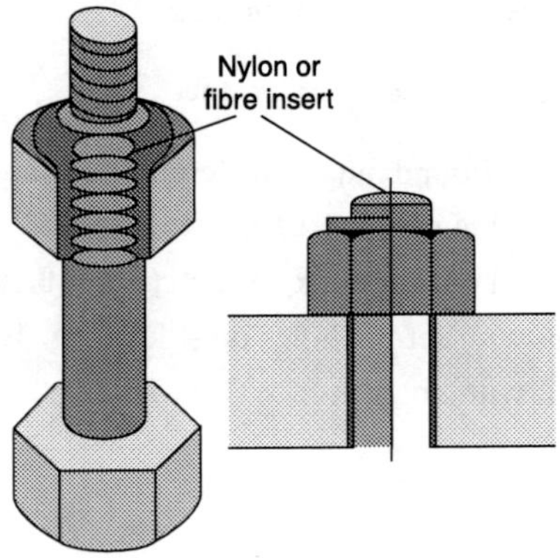

Fig. 10.24 Frictional lock

Example

Clutches, brakes, controls, etc.

The positive locking devices are:

- Standard hexagonal nut, cross-drilled and pinned
- Standard slotted nut
- Standard castle nut
- Hexagonal nut and locking plate
- Wiring bolt heads **(Fig. 10.25)**

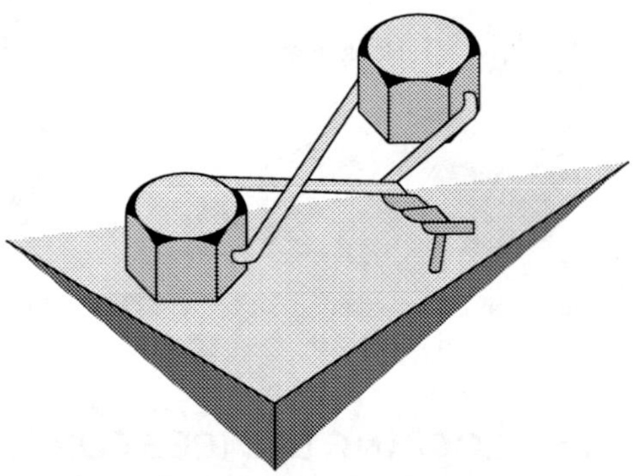

Fig. 10.25 Wiring bolt head

10.3.2 Frictional Locking Devices

These lock nuts are easy to fit and less time consuming.

The frictional locking devices are: Lock nut (chuck nut) **(Fig. 10.26)**

- Spring washer
- Wedge lock bolt
- Simmonds lock nut.

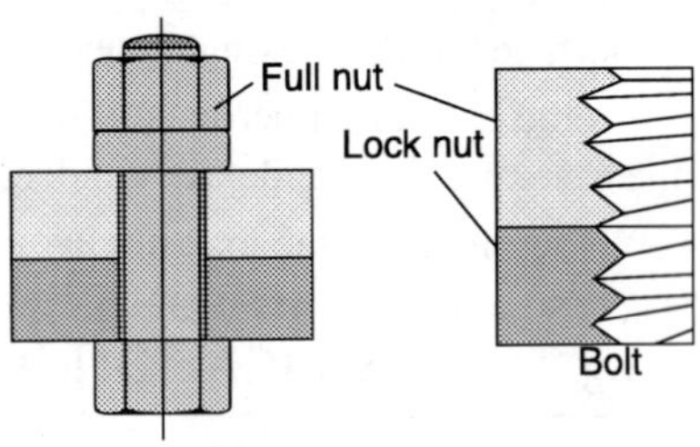

Fig. 10.26 Lock nut

10.3.3 Commonly used Locking Devices

Wing Nut: A wing nut is used where frequent adjustment or removal is necessary. It can be loosened or tightened rapidly without a wrench. These nuts are manufactured with the same material as is used for the bolts.

Thumb Nut: A thumb nut is used where the movement of the spindle is to be locked, such as in a micrometer. Stopping the movement of the spindle is necessary for taking a correct reading.

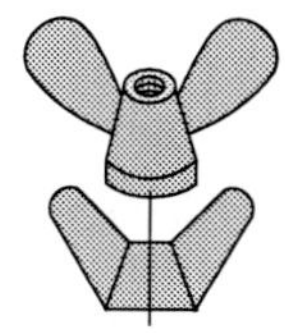

Fig. 10.27

Locking Ring: A locking ring is used in taper nose spindles of lathes to lock the chuck.

Castle Nut (Fig. 10.28): Slots are cut in a cylindrical collar provided on the top of the nut, thus overcoming the disadvantage of the slotted nut. Castle nuts are widely used in automobiles and locomotive engines to avoid sudden shock and vibration.

Circlips (Fig. 10.29): These are widely used to retain the component on a shaft or in a bore. Seating of these circlips in a slot by using a special type of pliers facilitates rapid assembly and disassembly.

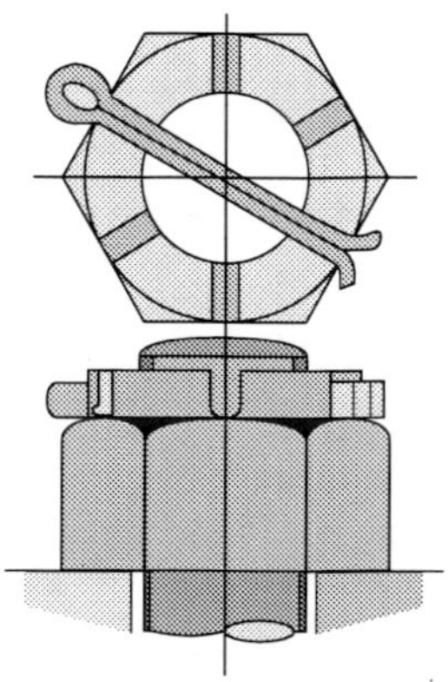

Fig. 10.28 Castle nut

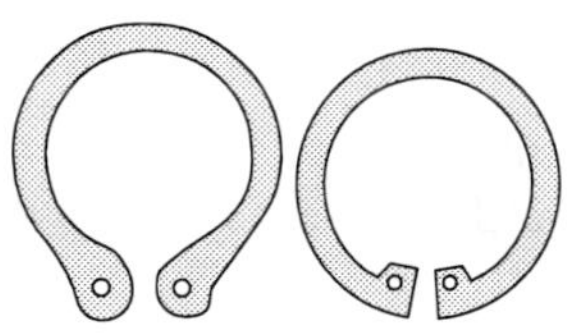

Fig. 10.29 Circlips

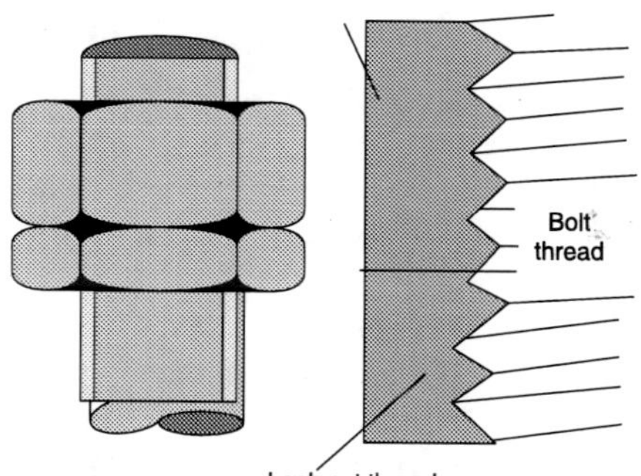

Fig. 10.30 Check nut and nut thread

Check Nut (Fig. 10.30): This nut is used along with one ordinary nut as shown in Fig. 10.30.

A check nut is also called a lock nut. The two nuts are thus locked or wedged tightly against each other and against the bolt. This will prevent slackening.

Self-locking Nut (Fig. 10.31): The self-locking nut will have a nylon insert to prevent the loosening of the nut from shock, vibration and temperature.

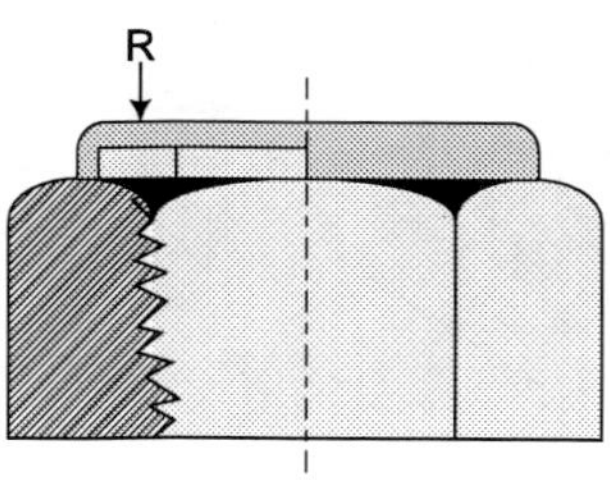

Fig. 10.31 Self locking nut

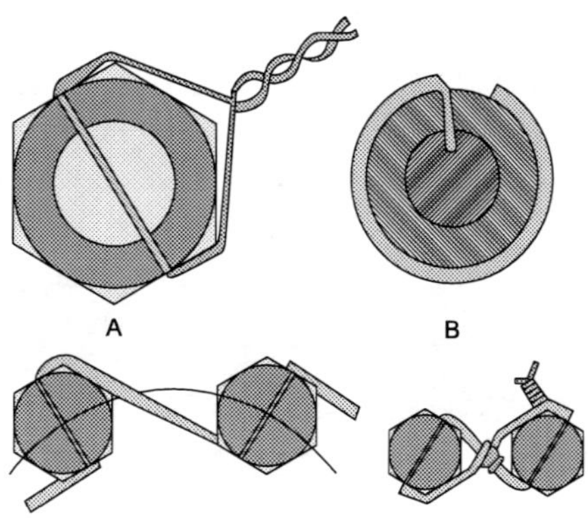

Fig. 10.32 Wire lock

Wire Lock (Fig. 10.32): Wire locks are used for light engineering works. The wire is passed through the groove.

Nut Applied with Sealant: These locking devices are for permanent locking in light works.

Split Pin (Fig. 10.33): A split pin is made from a steel wire of semicircular cross-section, bent as shown in the figure. It is inserted in a hole drilled in the bolt so that it exerts pressure on the top face of the nut to prevent it from turning.

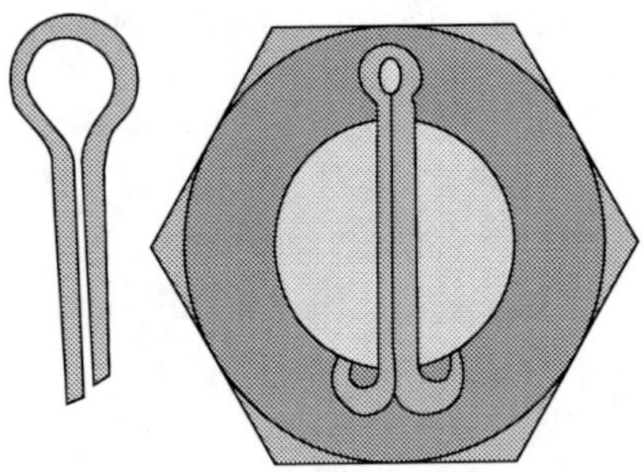

Fig. 10.33 Split pin

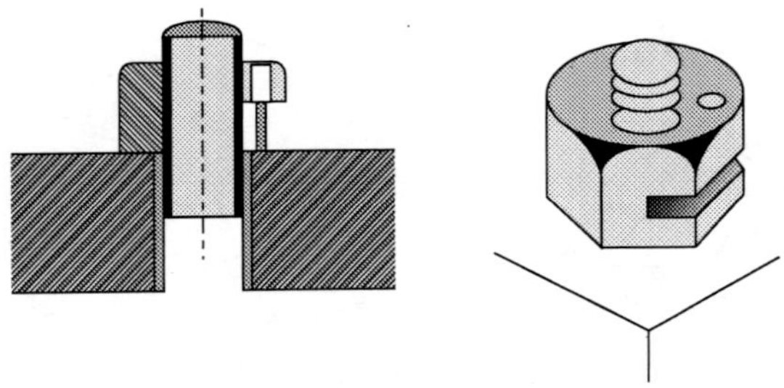

Fig. 10.34 Swan nut

Sawn Nut: (Wiles Nut) (Fig. 10.34): In this locking device, a slot is cut halfway across the nut. A screw is fitted with a clearance hole on the top part and a matching thread on the lower part of the nut. Tightening of the nut provides positive locking for the nut.

Slotted and Castle Nut with a Split Pin: The position of the nut can be locked using the split pin.

Split pins are designated by the nominal size, nominal length, the number of the Indian Standard and the material.

The nominal size is the diameter of the hole for receiving the split pins.

The nominal length is the distance from the underside of the eye to the end of the short leg **(Fig. 10.35).**

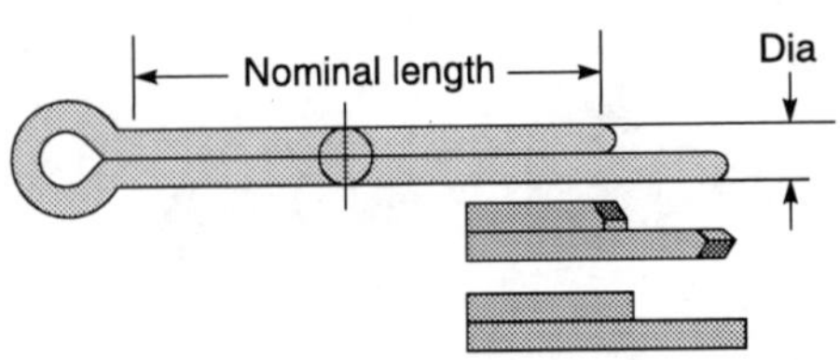

Fig. 10.35 Split pins

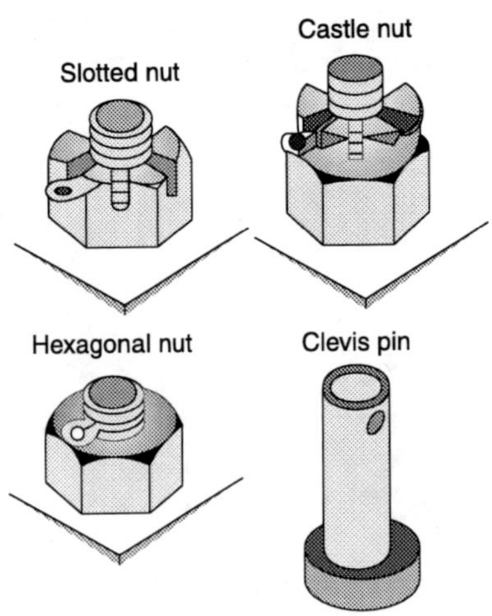

Fig. 10.36 Uses of split pins

Split pins are used for locking slotted nuts, castle nuts, hexagonal nuts, clevis pins, etc., and are used in different ways **(Fig. 10.36).**

Grooved Nut (Penning Nut) (Fig. 10.37): This is a hexagonal nut with the lower part made cylindrical. On the cylindrical surface there is a recessed groove in which a set screw is used to lock the nut.

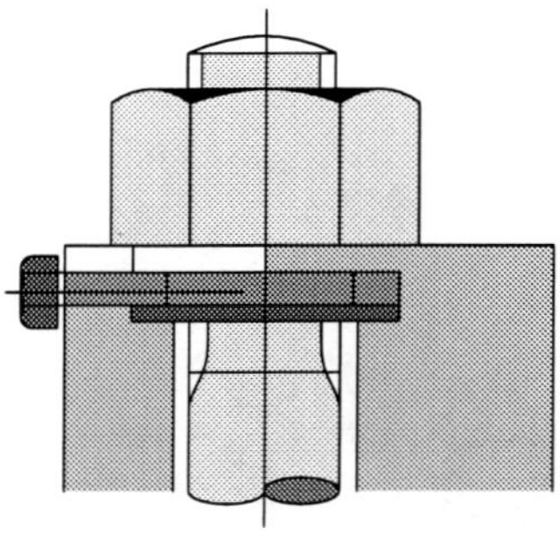

Fig. 10.37 Grooved nut

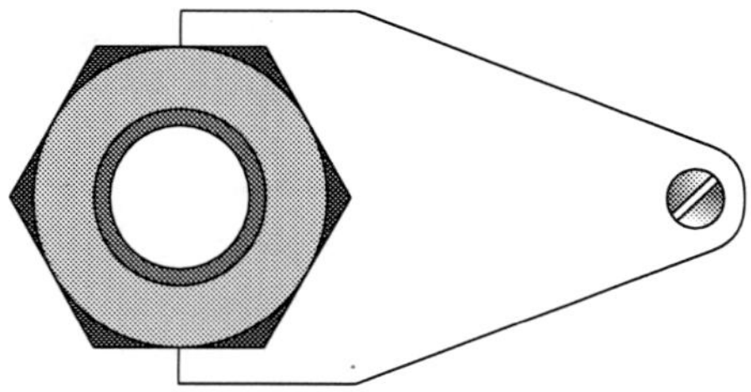

Fig. 10.38 Locking plate

Locking Plate (Fig. 10.38): For preventing the nut from loosening, locking plates are fixed on the outside of the hexagonal nut **(Fig. 10.36).**

Lock Washers with Lug (Fig. 10.39): In this arrangement of locking, a hole is drilled for accommodating the lug.

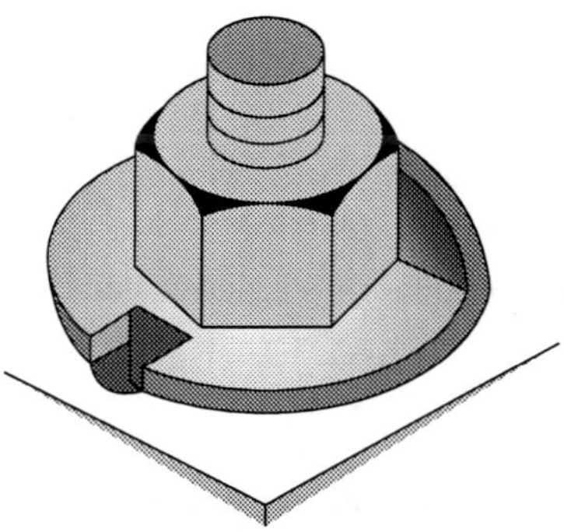

Fig. 10.39 Lock washer with lug

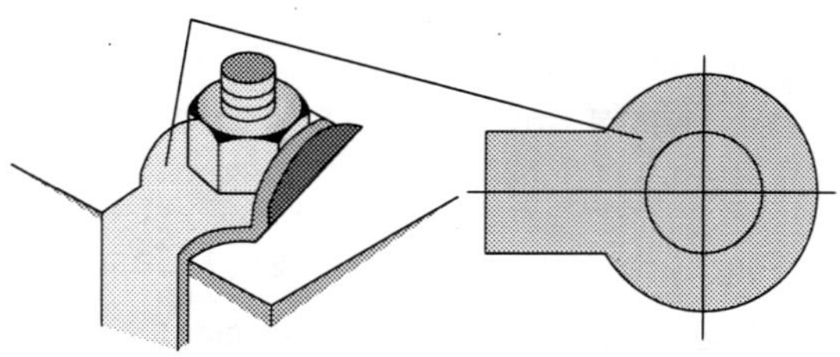

Fig. 10.40 Tap washer

The movement of the nut is prevented by folding the washer against the nut.

Tab Washers (Fig. 10.40): Tab washers can be used for locking the nuts which are located near an edge or corner.

Spring Washers (Fig. 10.41): Spring washers are available with single or double coils. These are placed under a nut in the assembly as washers. The stiff resistance offered by the washer against the surface of the nuts prevents loosening.

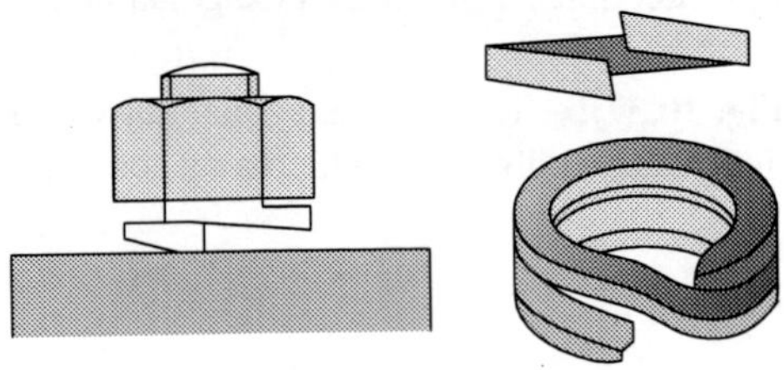

Fig. 10.41 Spring washer

10.4 STUDS

Studs are used in assemblies which are to be separated frequently.

Studs have two threaded ends. The threaded end which is fitted on the casting has coarse threads while the other end which takes the nut has fine threads **(Fig. 10.42).**

This variation in the thread pitch allows the fine thread or the nut end to strip, when excessively tightened, thus preventing damage to the casting. The threaded ends of studs may have difference in pitches, hand of thread and length of the threaded portion.

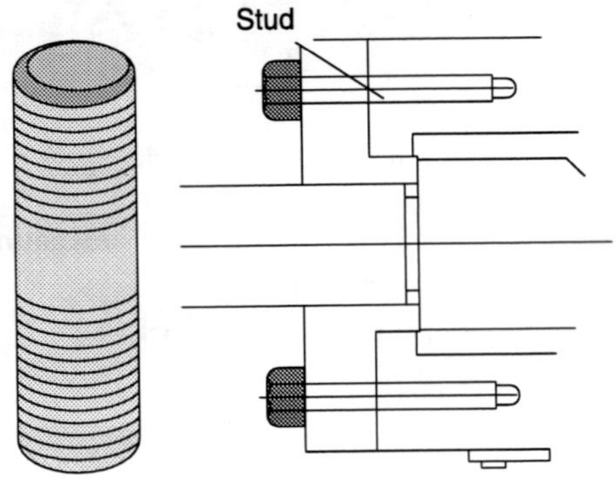

Fig. 10.42 Stud

10.4.1 Designation of Bolts as per BIS Specifications

Hexagon head bolts are designated by name, thread size, nominal length, property class and number of the Indian Standard.

Example

A hexagon head bolt of size M1 0, nominal length 60 mm and property class 4.8 is designated as:

Hexagon head bolt M1 0 × 40 - 4.8 - IS: 1363.

10.4.2 Explanation about Property Class

The part of the specification 4.8 above indicates the property class (mechanical properties). In this case it is made of steel with a minimum tensile strength = 40 kgf/ mm and having a ratio of minimum yield stress to minimum tensile strength of 0.8.

Note: Indian Standard bolts and screws are made of three product grades. A, B, & C, 'A' being precision bolts, and the others of lesser grades of accuracy and finish.

While there are many parameters given in the B.I.S. specification, the designation need not cover all the aspects and it actually depends on the functional requirement of the bolt or other threaded fasteners.

For more details on designation system Reference to IS: 1367 (Part XVI) 1979.

10.5 MACHINE SCREWS

Machine screws are used when a nut cannot be used in the assembly, and the component in the assembly has a threaded hole to receive the screw **(Fig. 10.43)**.

10.5.1 Types of Machine Screws

- Hexagon head screws
- Hexagon socket head cap screws
- Square head end screws

These are heavy duty screws.

Hexagon head screws are used when the projection of the screw head will not be an obstruction in the assembly.

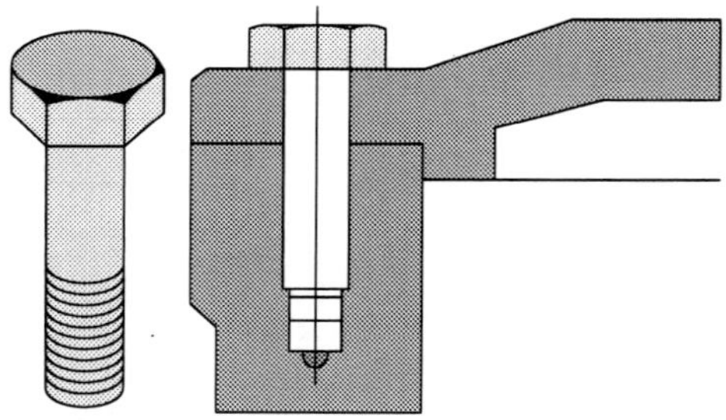

Fig. 10.43 Machine screw hexagonal head

Hexagon socket head cap screws are used when the projection of the screw head above the surface is to be avoided **(Fig. 10.44)**.

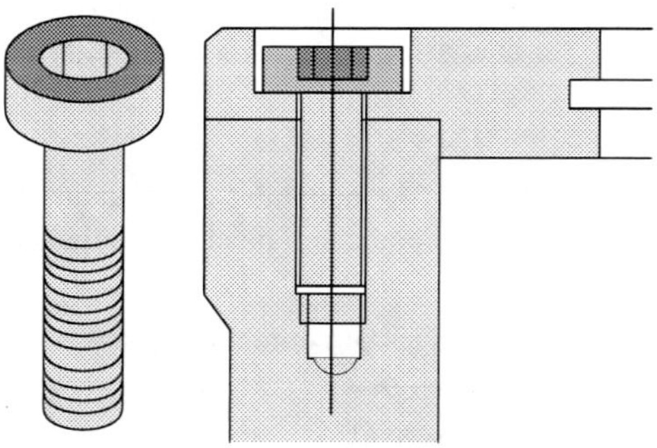

Fig. 10.44 Hexagonal socket head cap screw

The Indian Standard specification of socket head cap screws covers the range from 1.6 mm to 36 mm.

Hexagon head screws and hexagon socket head screws are made of steel. Hexagon head screws used in electrical work are made of brass.

Countersink Head Screws: There are four types of countersink head screws in common use.

They are:

- Slotted countersink head screws **(Fig. 10.45)**.
- Cross recessed countersink head screws **(Fig. 10.46)**.
- Slotted raised countersink head screws **(Fig. 10.47)**.
- Cross recessed raised countersink head screws **(Fig. 10.48)**.

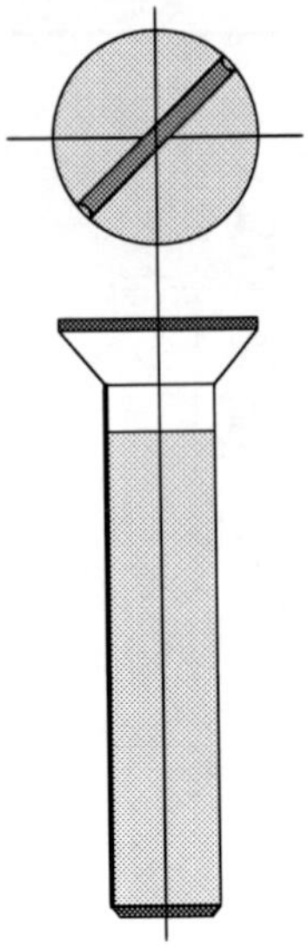
Fig. 10.45 Slotted counter sink head

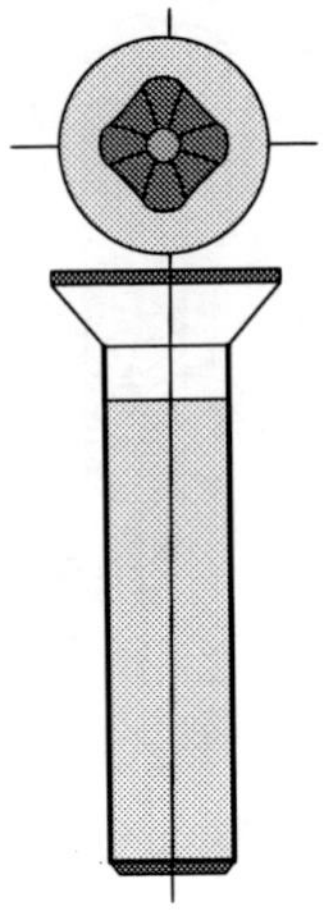
Fig. 10.46 Cross recessed countersink

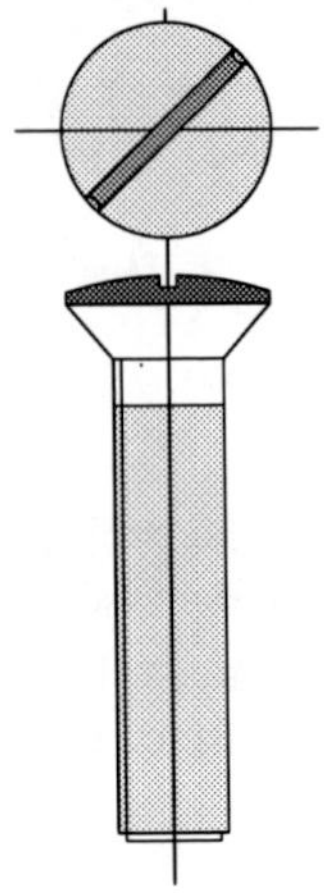
Fig. 10.47 Slotted raised countersink

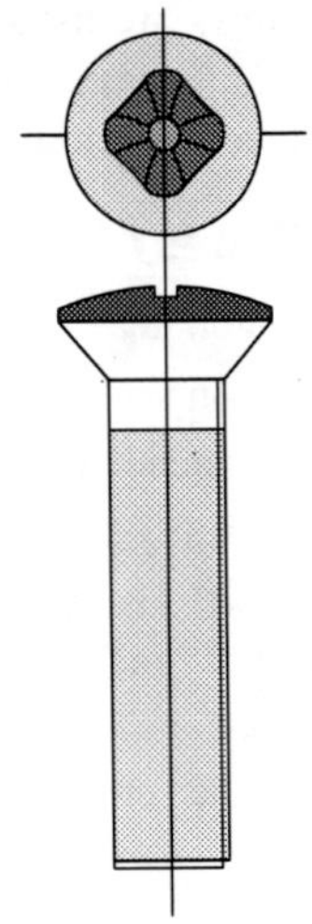
Fig. 10.48 Cross recessed raised counter-sink

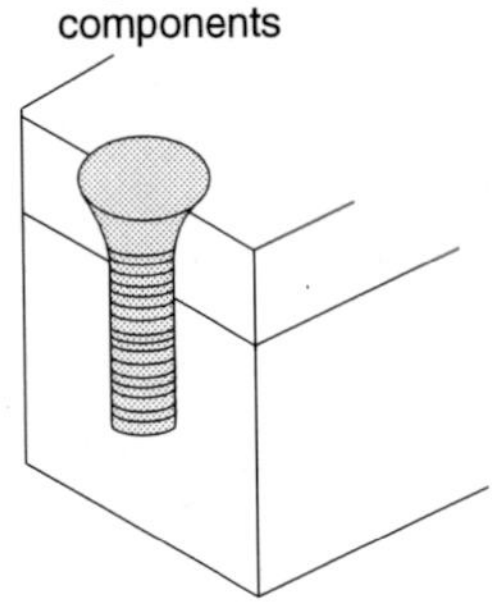

Fig. 10.49

Countersink screws are capable of aligning the matching component correctly with the threaded hole **(Fig. 10.49).** The projection screw head above the assembly is also avoided.

S.I.S. specification covers the following ranges of countersink head screw sizes in different types.

- Slotted countersink head screws - MI to M 20
- Cross recessed countersink head screws - M 1.6 to M 10
- Slotted raised countersink head screws - M I.6 to M 20

Square Head Screws: Square head screws are used in places where there is frequent removal and refitting of the assembly. These screws are tightened to a higher torque using a wrench **(Fig. 10.50).**

Square head screws are also available with collars. In these screws, there is a collar at the base which is an integral part of the head. The purpose of this collar is to protect the work surface from damages due to a constant use of wrenches.

Other types of machine screws used in light assembly work are:

- Pan head **(Fig. 10.51)**
- Cheese head **(Fig. 10.52)**
- Raised cheese head **(Fig. 10.53)**
- Round head **(Fig. 10.54)**

These screws are also available with slotted head or cross-recessed head.

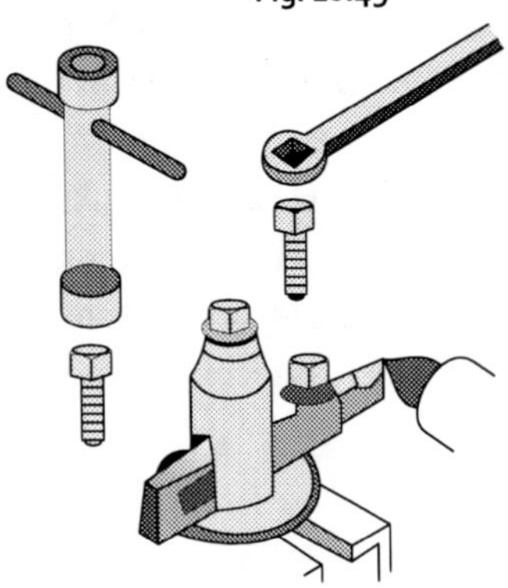
Fig. 10.50 Square head screw

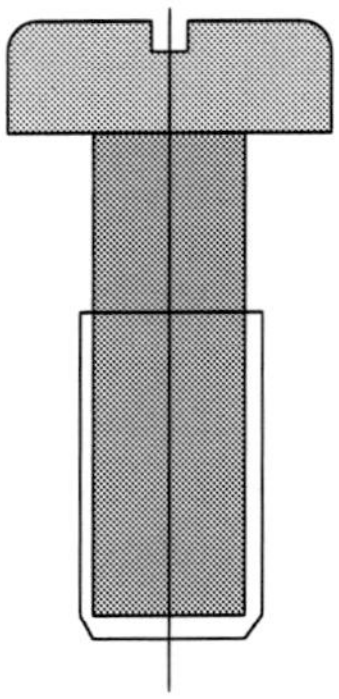

Fig. 10.51 Pan head screw

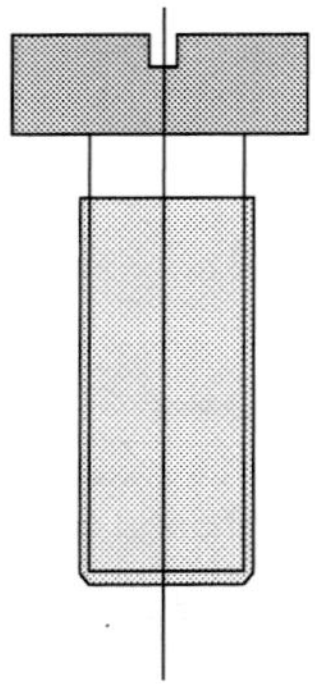

Fig. 10.52 Cheese head

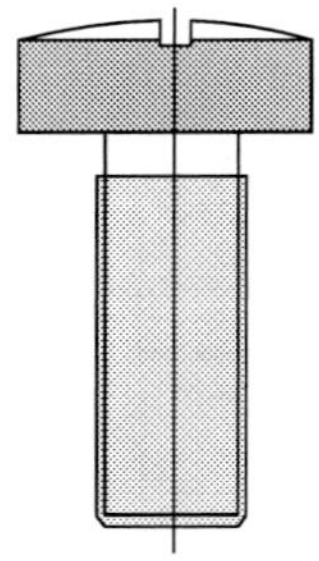

Fig. 10.53 Raised cheese head

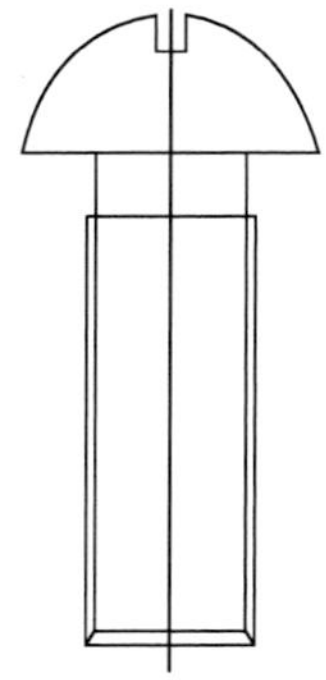

Fig. 10.54 Round head

Screws used for light duty are normally available up to 10 mm thread diameter.

These screws are made of steel, stainless steel or brass.

10.5.2 Set Screws

Hexagon Socket Set Screws: These are headless socket screws and are available with different points for various functional requirements **(Fig. 10.55).**

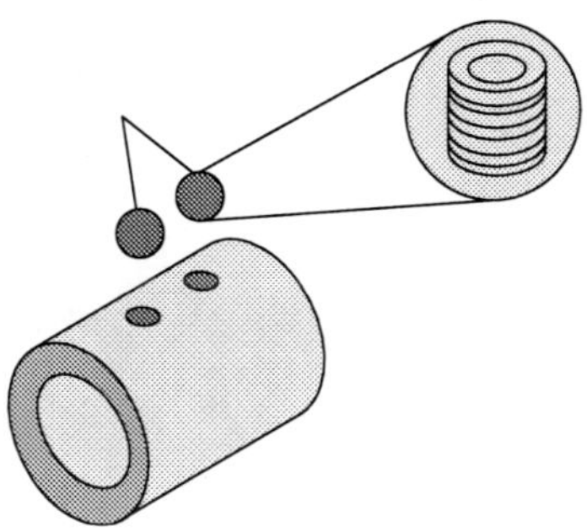

Fig. 10.55 Set screw

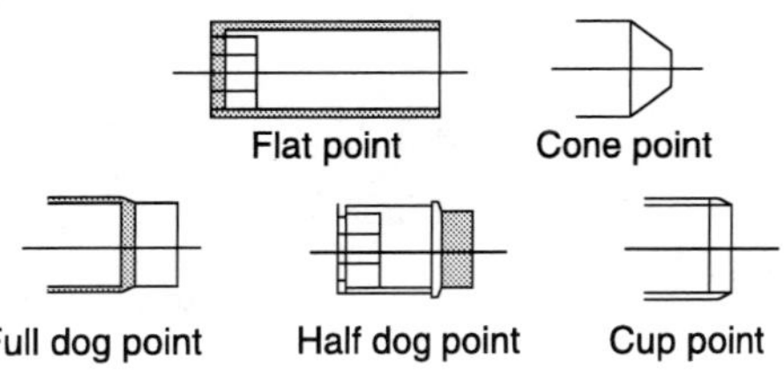

Fig. 10.56 Set screw points

The points either allow biting into the metal or tightening without damage to the work surface. They are used to fasten pulleys, collars to shafts, etc., for higher strength applications where space is limited.

Square Set Screws (Fig. 10.57): These set screws have similar applications as hexagon socket set screws but have square heads projecting above the work surface.

These are useful when the assembly needs frequent disassembly and setting.

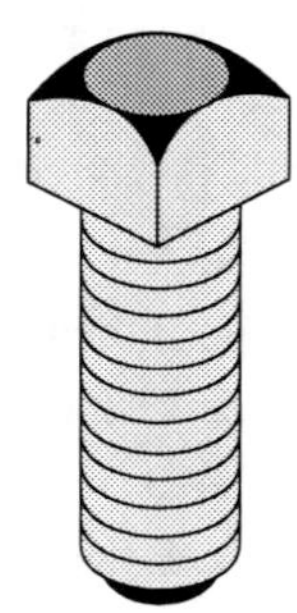

Fig. 10.57 Square head set screw

10.5.3 Grub Screws (Fig. 10.58)

Grub screws have similar application as hexagon socket set screws but are used for light holding. Grub screws are also available with different types of points.

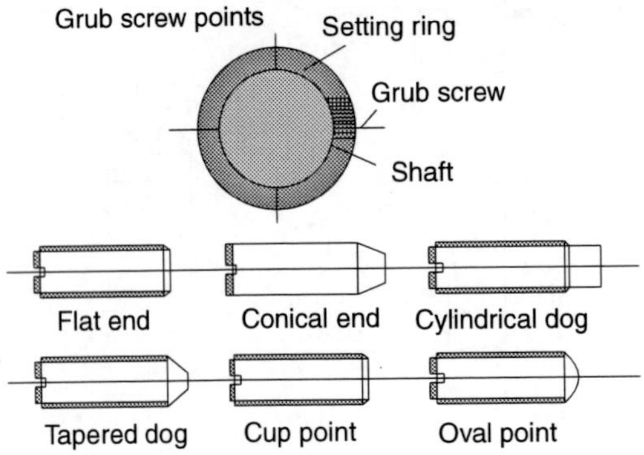

Fig. 10.58

10.5.4 Thumb Screws

Thumb screws are used in places where fixing and removal of components are frequent.

Tightening and loosening of the assembly is done by finger tightening only.

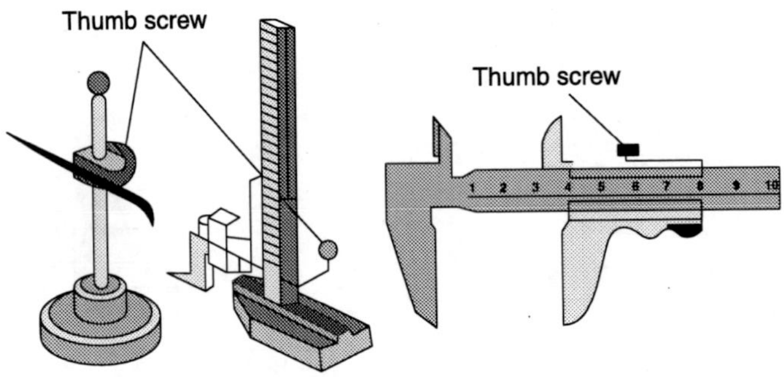

Fig. 10.59 Application of thumb screws

Examples

- Scribing block
- Vernier height gauge
- Vernier caliper, etc.

The Indian Standard specification IS: 3726-1972 specifies five types of thumb screws.

Type-A thumb screws, partially threaded

Type-B thumb screws, fully threaded.

Type-C slotted thumb screw, partially threaded.

Type-D slotted thumb screw, fully threaded.

Type-E flat thumb screws.

The type of thumb screw selected depends on the actual requirement in the assembly

Sizes

Thumb screws are available in the following sizes as per B.I.S.

M 1.6, M 2 M 2.S, M 3 M 4 MS, M 6 M 8 and M 10

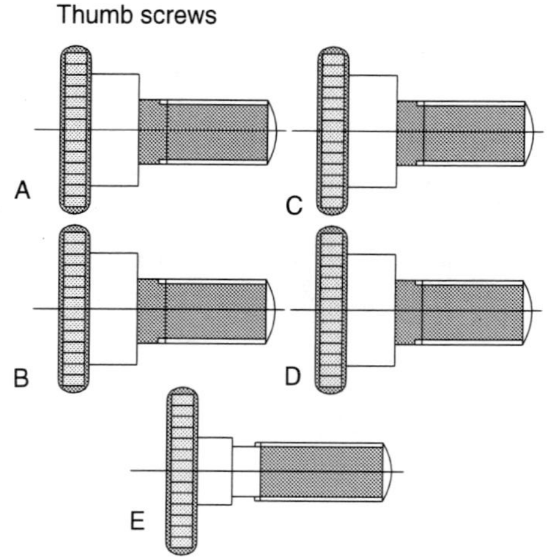

Fig. 10.60

10.5.5 Designation of Thumb Screws

The thumb screws are designated by the nomenclature type, thread size, nominal length, the number of Indian Standard and symbol for mechanical properties.

Example

A thumb screw of Type 'A', size M 6 nominal length 12 mm and of property class 4. 6 shall be designated.

Thumb screw A M 6 x 12 IS: 3726-46.

When brass or any other nonferrous metal is used for the manufacture of thumb screws the word 'brass' or the name of the nonferrous metal used will replace the property class number in the designation.

10.5.6 Self-Tapping Screws

Self-tapping screws are used in assembly where thin section metal sheets are used. Joints made using these screws are vibration-resistant and can be assembled and dismantled many times. The three types of self-tapping screws are:

- Thread forming
- Thread cutting
- Self piercing

The Thread Forming Type (Fig. 10.61): This type of metal screws produces the mating thread by displacing the material. These are useful for softer and thinner materials.

The Thread Cutting Type (Fig. 10.62): This type cuts the mating thread in the hole in the same way as a thread cutting tap. These screws will have projected ridges in the shape of thread for the cutting action. These are quite useful for self-tapping on hard or brittle materials with thin wall sections.

Self-piercing and Tapping (Fig. 10.63): These screws have special piercing points and twin start threads. These screws are used along with a special gun. In a single operation the hole is pierced and the screw also driven home.

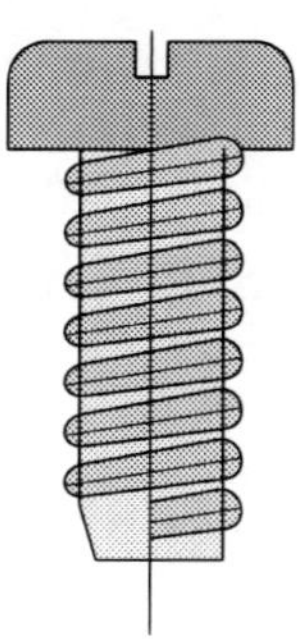

Fig. 10.61 Self tapping screws threads forming screws

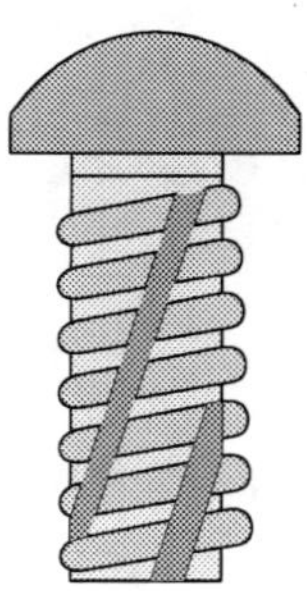

Fig. 10.62 Self tapping thread cutting

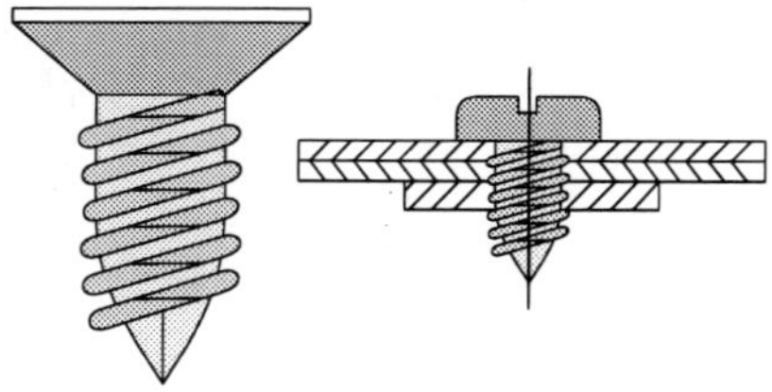

Fig. 10.63 Self piercing & tapping type

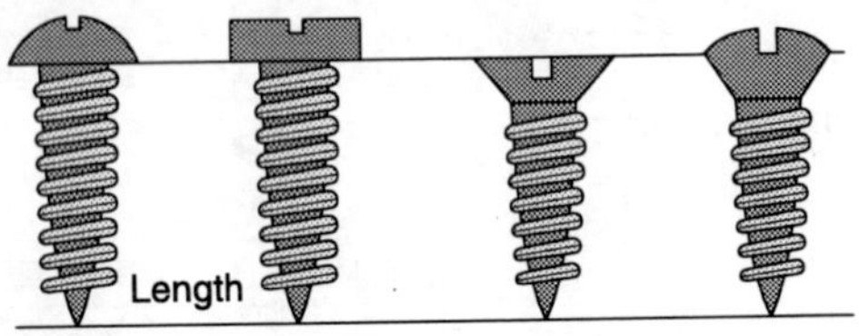

Fig. 10.64 Self tapping screws: different heads

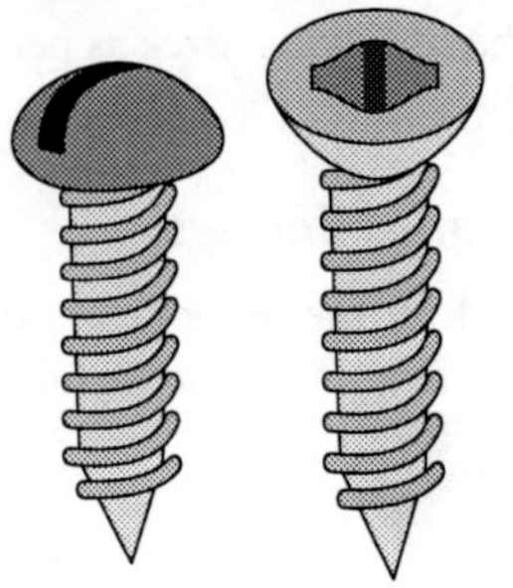

Fig. 10.65 Self tapping screws: cross recesed or slotted heads

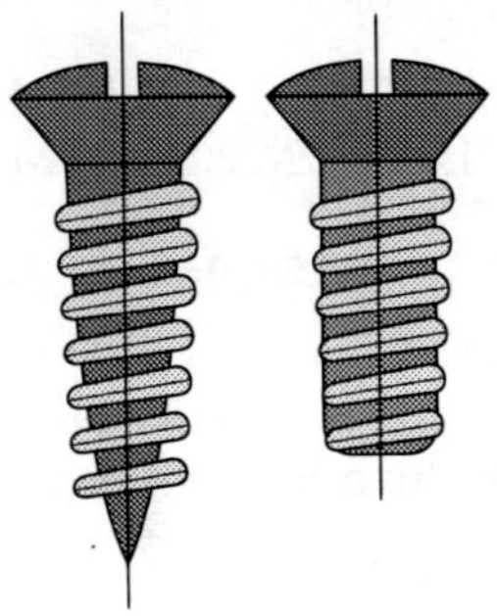

Fig. 10.66 Self tapping screws: types of end

Self-tapping screws are available in a variety of head shapes **(Fig. 10.64).**

They are either with cross-recessed or slotted heads **(Fig. 10.65).**

The ends of self-tapping screws are of two types as per Indian Standard. Type 'AB' and Type 'B' **(Fig. 10.66).**

Type 'AB' is a self-piercing type and Type 'B' is a thread forming type.

10.5.7 Designation of Self-Tapping Screws

Self-tapping screws are designated by the nomenclature, type, screw size number, nominal length and the number of the Indian Standard.

Example

A slotted raised countersunk (oval) self-tapping screw of type AB, screw size number 10, nominal length 25 mm shall be designated as:

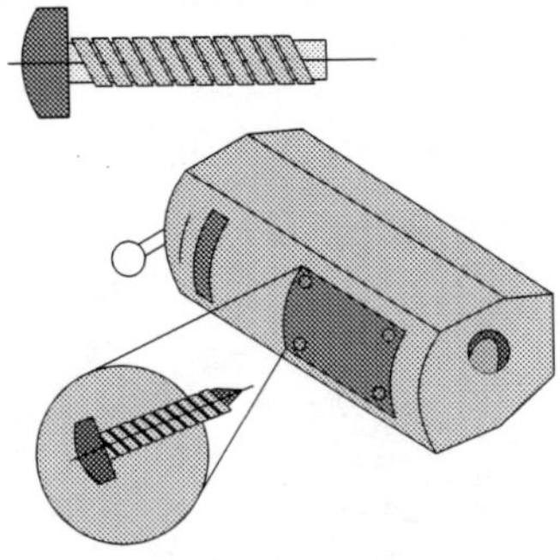

Fig. 10.67

Tapping screws: Oval head: as AB 10 × 25.

10.5.8 Hammer Driven Screws

These types of screws are used for fixing thin plates like nameplates, etc., on cast iron machine body. These screws have multi-start threads. For fixing this screw, a hole is drilled on the machine casting to the size of the pilot.

The screw pilot is then placed in the hole and driven home by a hammer.

These screws are available in size numbers from 00 to 14.

Screw size	0	1	2	(3)	4	(5)	6	(7)	8	10
Max.	1.52	1.90	2.24	2.57	2.90	3.30	3.53	3.91	4.22	4.80
Min.	1.45	1.83	2.13	2.46	2.79	3.20	3.43	0.078	4.09	4.65

Screw size	(12)	14	16
Max.	5.46	6.25	8.00
Min.	5.31	6.10	7.82

Designation

Example

A hammer driven screw with screw number 12 having length of 9.5 mm can be designated as: Hammer driven screw No. 4 x 9.5.

Note

Indian B.I.S. specification number may be included, if necessary.

The size of the screw (diameter) is indicated by a number which ranges from 0 to 16.

The B.I.S. specification number and the type of coating can be included at the end.

Screw size	Dia. D_1 Max.	Dia. D_1 Min.	Dia. of pilot d_2 Max.	Dia. of pilot d_2 Min.
00	1.52	1.45	1.25	1.17
0	1.90	1.83	1.60	1.52
2	2.54	2.46	2.11	2.03
4	2.95	2.85	2.44	2.34
6	3.56	3.45	2.95	2.85
8	4.24	4.12	3.45	3.35
10	4.62	4.50	3.84	3.71
12	5.38	5.23	4.50	4.39
14	6.15	5.99	5.13	5.03

10.6 WASHERS

Its shape is like a round ring. Before tightening the nut and bolt, it is fitted on to the stud or bolt. It is used to secure the tightness of a joint, screw, etc., and it becomes easy to open the nut/bolts and their threads are protected. It is a common practice to provide washers under the nuts in bolted joints.

Washers help to:

- Increase the frictional grip
- Prevent loosening of nuts due to vibration
- Prevent damage to the workpiece
- Distribute force over a larger area.

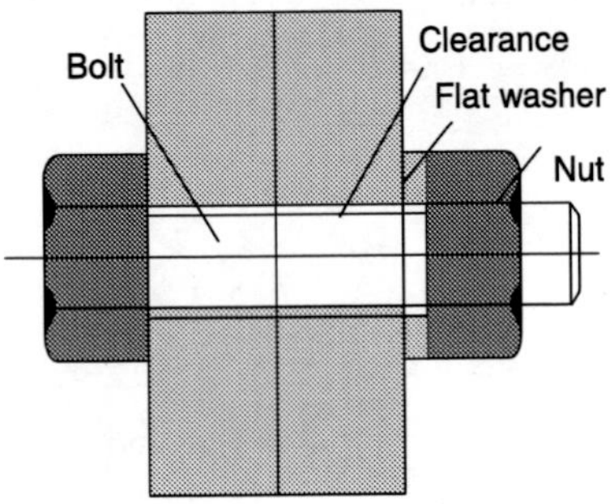

Fig. 10.68 Bolted connection

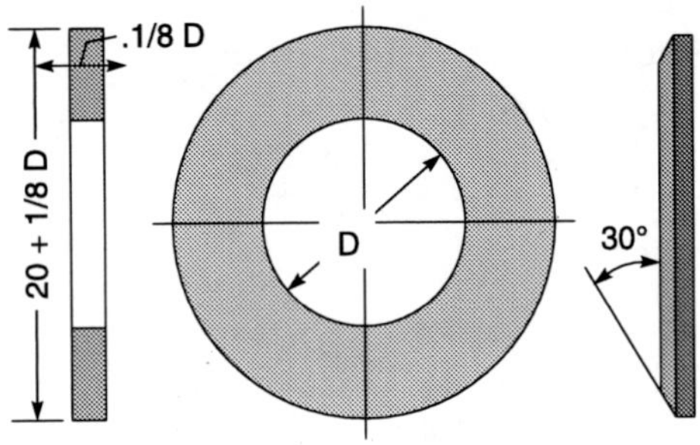

Fig. 10.69 Plain washer

Even if the diameter of the hole in which the bolt is to be fitted is a bit larger than the diameter of the bolt, with the washer's help the nut or bolt can be tightened easily.

10.6.1 Types of Washers

1. Plain washer
2. Taper washer
3. Spring washer
4. Toothed lock washer
5. Tab washer

Plain Washer (Fig. 10.69): A plain washer is a flat disc. It has a hole in its centre through which a bolt, stud or setscrew can pass. By using this, the surface below the bolts or nuts becomes plain. The pressure of the nut is uniform on the entire surface. This washer is used for bolting assemblies with flat surfaces. The diameter, thickness and the bore diameter are proportional to the diameter of the bolt. (IS 2016)

Diameter of the washer = 2D + 1/8 D

Thickness of washer = 1/8 D

Angle of chamfer = 30°

Diameter of inner hole = Diameter of the bolt + clearance.

Plain washers are available as machined or punched washers.

Machined Washers: These are used for assemblies using machined components. These are available with chamfer on one side or on both sides. They are heat treated and ground.

Punched Washers: These do not have chamfers and are commonly used in structural fabrication work.

Tapered Washers (Fig. 10.70): These washers are used in structural assemblies with tapered surfaces like inside ''I' beams, channels, etc., these washers help the bolt or nut to seat square to the hole.

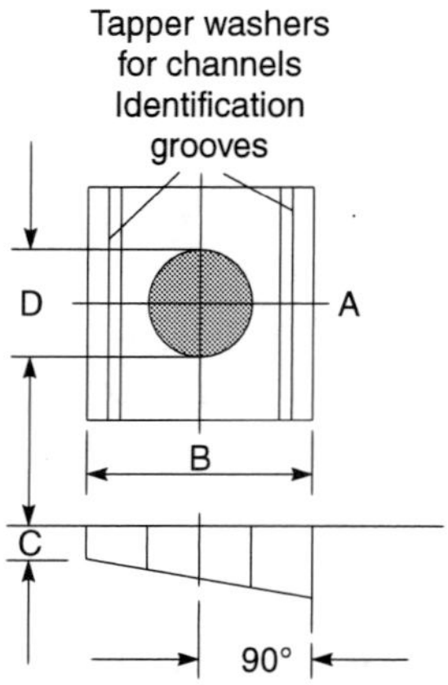

Fig. 10.70

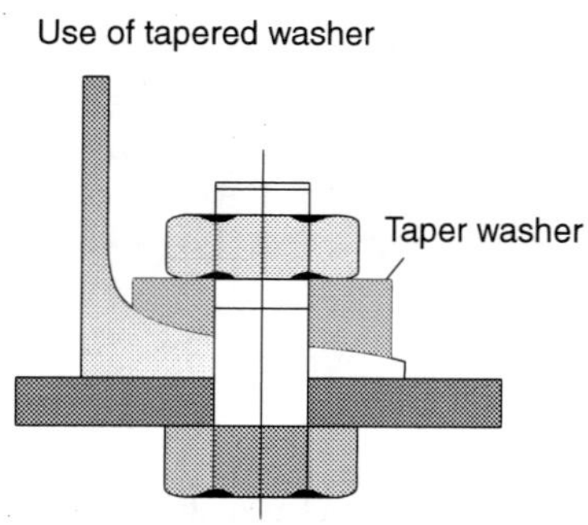

Fig. 10.71

Spring Washers (Fig. 10.72): These washers are in the shape of a ring cut by spring steel. Its ends are slightly raised up and down to give spring tension. These washers are placed above or below the nut or bolt and tighten so much that both their ends face each other and become levelled.

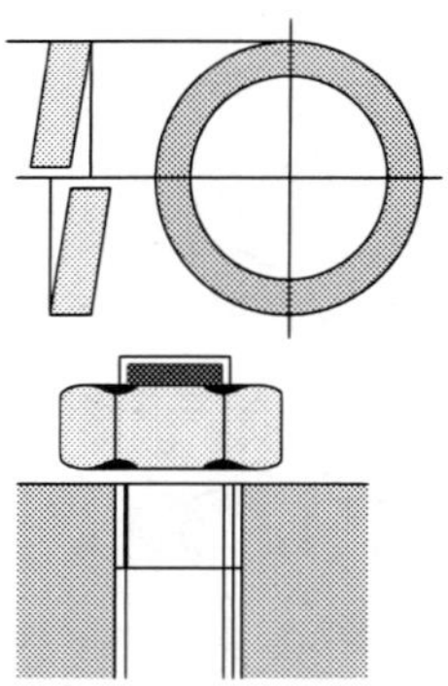
Fig. 10.72 Spring lock washer

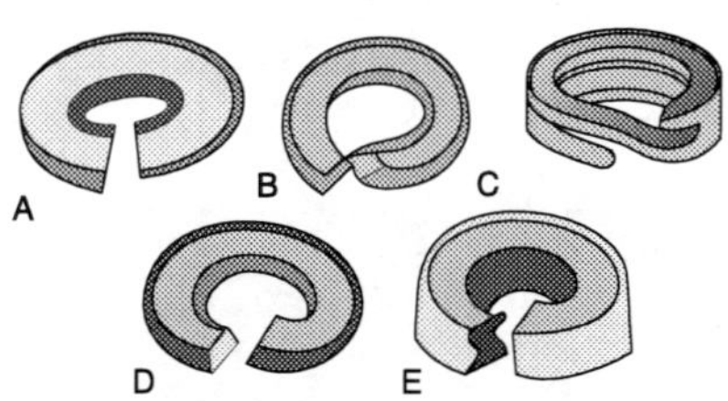

Fig. 10.72a Spring washers

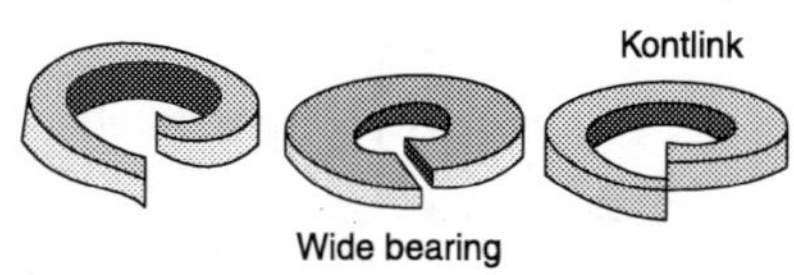

Fig. 10.72b Positive type 'D'

These washers are used to prevent a nut from loosening. These are used where there is vibration, or jolts occur. It exerts pressure against the nut and prevents the later (nut) from turning and working out loose. Each has its own advantages. According to their structures these are of the following forms:

- Single pattern spring washer
- Standard pattern spring washer

- Double pattern spring washer
- Grip pattern spring washer
- Girder pattern spring washer

Spring washers are used under nuts to prevent slackening of the nuts due to vibrations. They are made of spring steel, and when compressed they create tension between the bolt and the nut (**Fig. 10.73).**

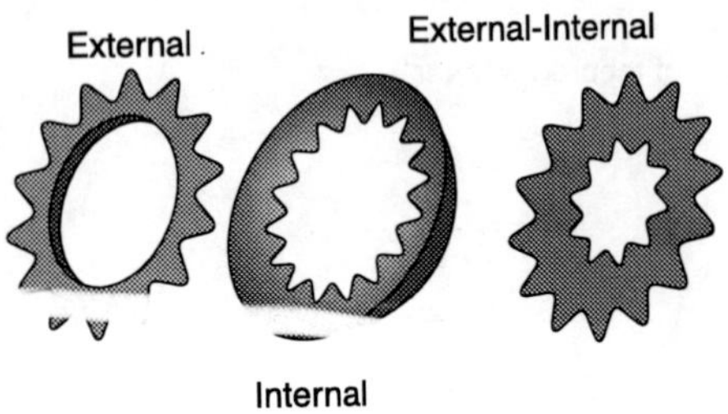

Fig. 10.73

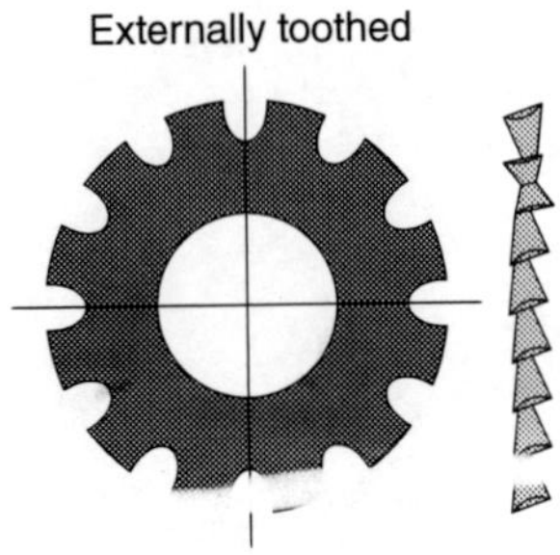

Fig. 10.74

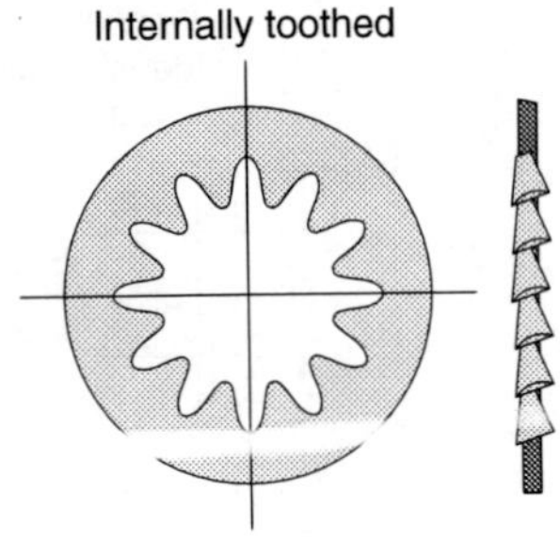

Fig. 10.74a

Lock Washers (Toothed Lock Washers): These washers are used to lock the nuts and bolts. These are of many forms. Generally, six or more leaves are extended beyond the main diameter of the lock washer. After fixing this washer its leaves are bent on the sides of nut or bolt. Thus, this washer prevents the nut bolts from getting loose. These washers are generally used in cars, trucks, etc.

These washers have serrations, cut and twisted. When placed between the nut and the assembly, this washer exerts friction on both the contacting surfaces. This prevents the nuts from slackening.

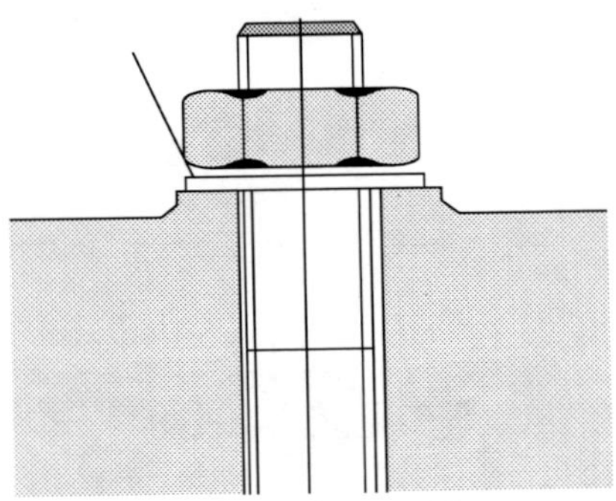
Fig. 10.75 Toothed lock washer

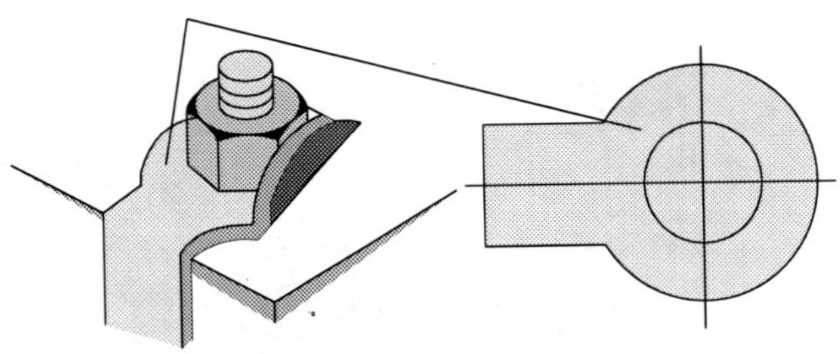
Fig. 10.76 Tap washer

Tab Washers (Fig. 10.76): These are also used for locking the nuts.

Specifications: The Indian Standard IS: 2016 -1967 designates a washer by name, type, size and number of the standard and material.

A machined washer of size 10.5 mm made of brass shall be designated as machined washer 1 0.5 IS: 2016 – Brass

Note: For detailed specification of different types of washers, refer to the following IS specifications

Taper washer - IS: 5374 and IS: 5372

Tab washer - IS: 8068

Toothed lock washer IS: 5371

Plain washer IS: 2016

10.7 THREAD INSERTS

A thread insert is used to repair a damaged internal thread or to provide a string thread in soft material such as aluminium.

A typical insert consists of a hardened steel cylinder, threaded inside and outside. The outside thread is tapered at one end in the same manner as a tap, while slots cut across the same end provide cutting edges as on a fluted tap. This type of insert is also known as thread bushes **(Fig. 10.77).**

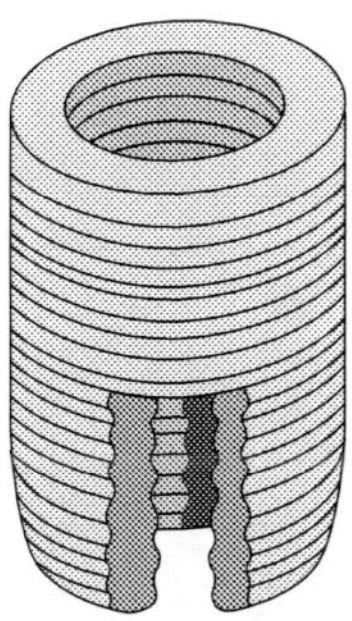

Fig. 10.77

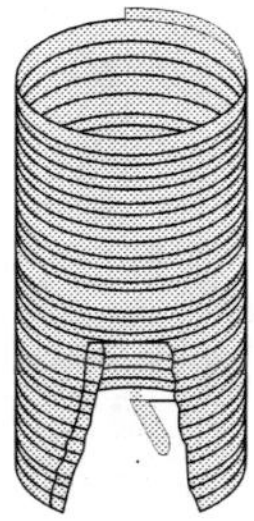

Fig. 10.78

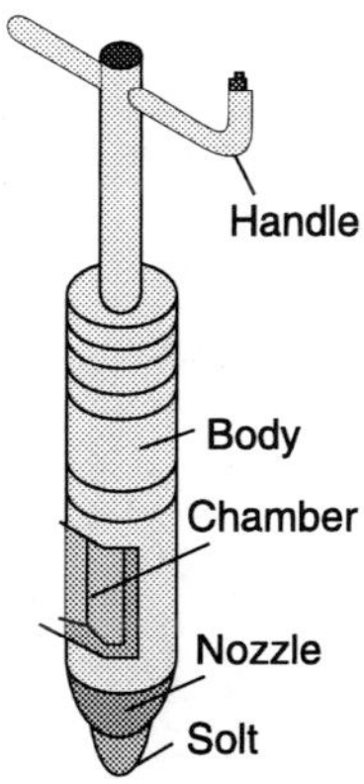

Fig. 10.79

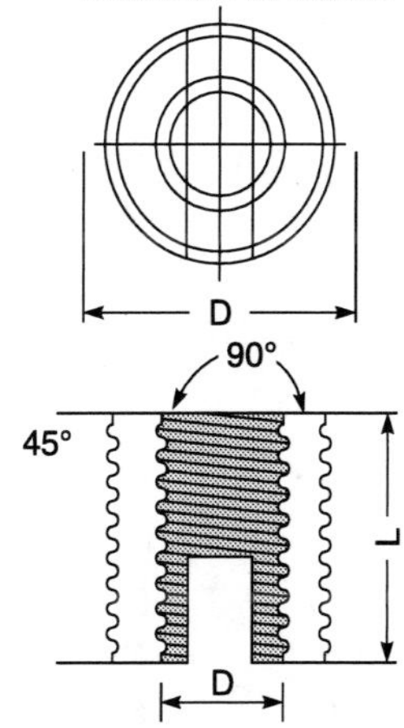

Fig. 10.80

An alternative form of insert consists of a coil wound from high tensile stainless steel wire of diamond cross section. The shape of the inside surface corresponds with the screw to be inserted (Metric, B.S.W., B.S.F., etc.) While the outer surface has similar threads but of greater diameter **(Fig. 10.78).** At the bottom of the insert, the wire is bent in across the centre line to form a tang.

The tang is gripped in a special tool **(Fig. 10.79)** to enable the insert to be screwed into its tapped hole.

To remove an insert use a special spade-like extracting tool **(Fig. 10.80).** The blade tapers to fit various sizes of inserts and is sharpened so that it cuts into the insert when turned anticlockwise.

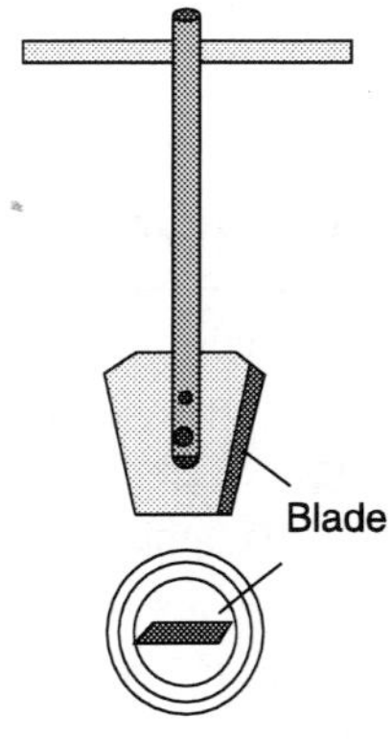

Fig. 10.81 Spade like tool to extract insert

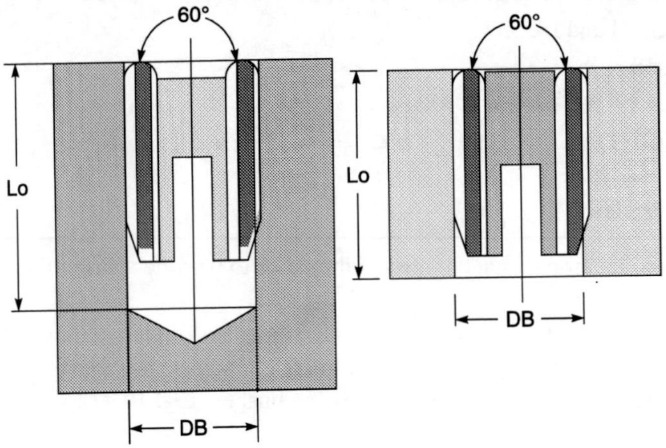

Fig. 10.81a Dimensions for inserts

Dimensions for the inserts:

Type	Inner	Thread	Outer	Thread	Length	Min. hole depth	Min. material thickness
	d	h	D	H	L	L	L
M 2.6	2.6	0.45	4.58	0.5	6	8	6
M 3	3	0.5	5	0.5	6	8	6
M 3.5	305	0.6	6	0.75	8	10	8
M 4	4	0.7	6.6	0.75	8	10	8
M 5	5	0.8	8	1	10	13	10
M 6	6	1	9	1	12	15	12
M 6	6	1	10	1.5	14	17	14
M 8	8	1.25	12	1.5	15	18	15
M 10	10	1.5	14	1.5	18	22	18
M 12	12	1.75	16	1.5	22	26	22
M 14	14	2	18	1.5	24	28	24

10.8 REMOVING BROKEN STUDS

When studs or screws are broken the following methods are used for removing them.

Screw Extractor (Fig. 10.82): Screw extractors are available in different sizes. Depending on the size of the broken stud a hole is first drilled. A screw extractor is then inserted into the hole and turned anticlockwise until it is tight. Turning further will loosen the stud.

Tapered Square Drift (Fig. 10.83): First a hole approximately to half the diameter and to about half the length of the broken stud may be drilled. A tapered drift with a square head is then driven into the hole.

Use a tap wrench or spanner and rotate the drift for unscrewing the stud.

Using a Screwdriver (Fig. 10.84): If there is sufficient projection of the broken stud, cut a slot with a saw and unscrew it with a screwdriver. This method is suitable only for small diameter studs.

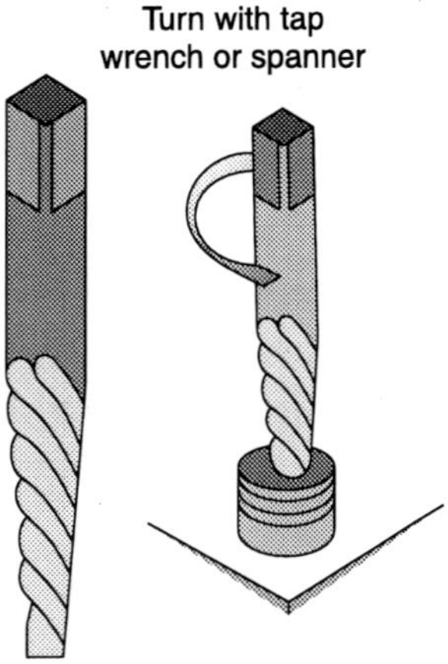

Fig. 10.82 Screw extaractor

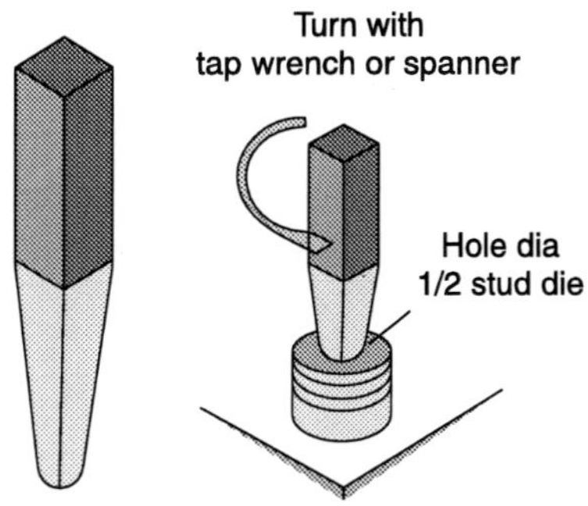

Fig. 10.83 Tapered square drift

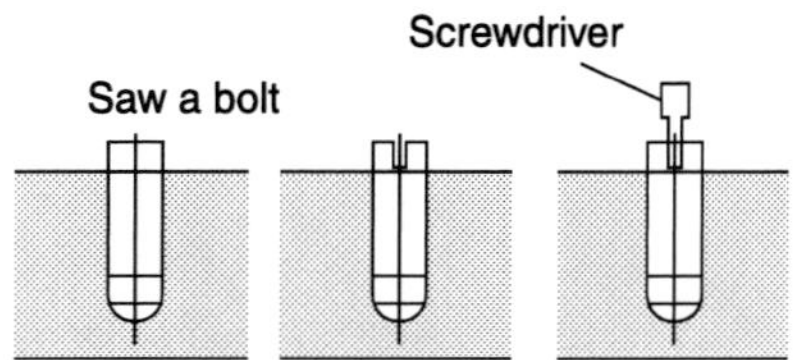

Fig. 10.84 Using a screwdriver

Punch and Hammer (Fig. 10.85): If the breakage of the stud is near the surface, sometimes it can be removed by using a punch. The punch is used to direct blows at different points to loosen the stud. Be sure the punch is used in the direction for unscrewing.

Using Spanner (Fig. 10.86): Large diameter studs which are broken above the surface can be removed by shaping a square head and then removing by a spanner.

By Drilling (Fig. 10.87): Broken studs which are very stubborn can be drilled through.

The remaining metal can be removed using a tap or a scriber point.

Sometimes it may be necessary to remove the stud completely by drilling and re-threading for oversize stud.

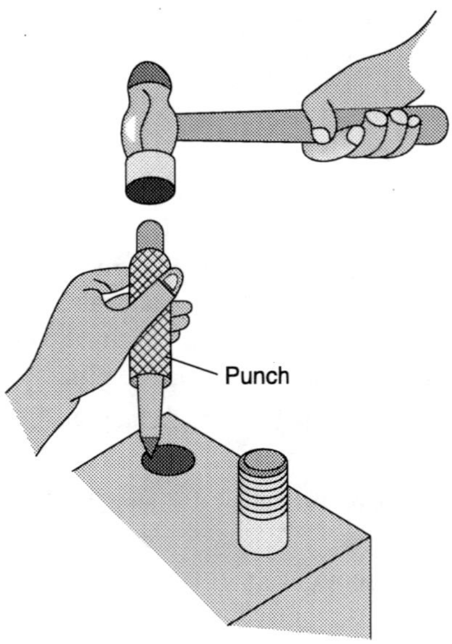

Fig. 10.85 Using punch & hammer

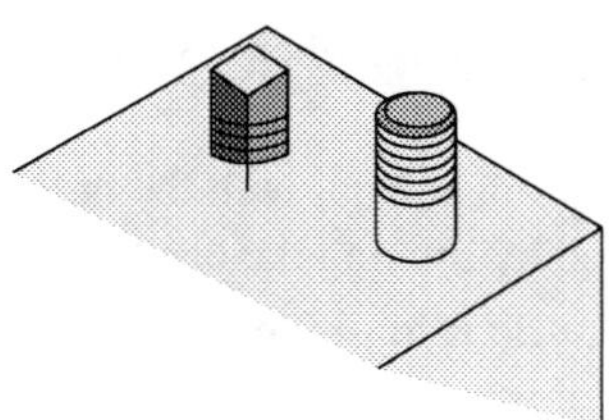

Fig. 10.86 File square to tit spanner

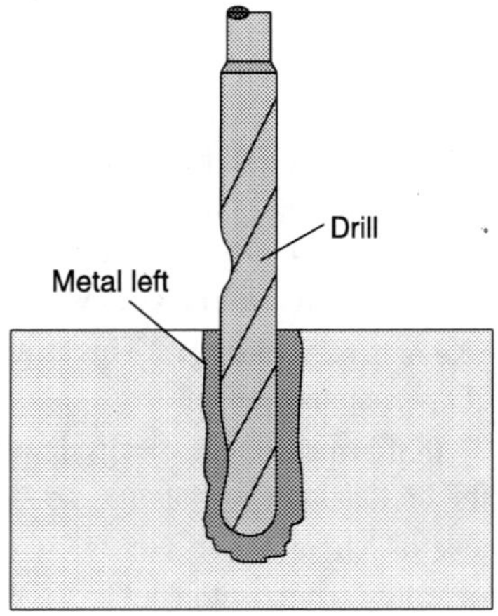

Fig. 10.87 Drilling

10.9 REMOVAL OF BROKEN TAPS

A tap broken above the surface of the workpiece can be removed using gripping tools like pliers.

Taps broken below the surface pose a problem for removing. Any one of the several methods given below can be used.

Use of Tap Extractor (Fig. 10.88): This is a delicate tool and needs careful handling.

This extractor has fingers which can be inserted on the flutes of the broken tap. The sliding collar is then brought to the surface of the work and the extractor turned anticlockwise to take out the broken tap.

A light blow on the broken tap with a punch will help to relieve the tap if it is jammed inside the hole.

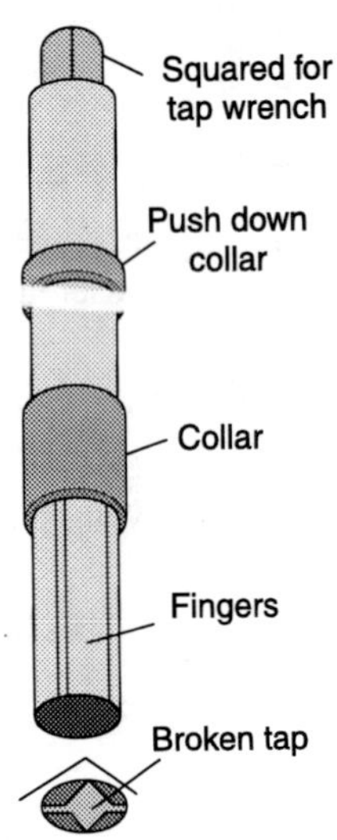

Fig. 10.88 Using tap extractor

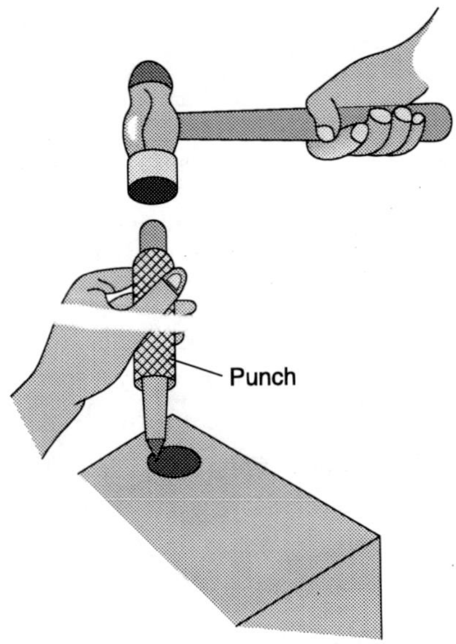

Fig. 10.89 Using punch & hammer

Use of Punch (Fig. 10.89): In this method the point of the punch is placed in the flute of the broken tap in an inclination and struck with a hammer. The positioning of the punch should be such that the broken tap is rotated anticlockwise when struck.

Annealing and Drilling the Tap: This is a method adopted when other methods fail. In this process the broken tap is heated by flame or by other methods for annealing. A hole is then drilled on the annealed tap. The remaining piece can be removed either by using a drift or using **EZY -OUT** (extractor). This method is not suitable for workpieces with low melting temperatures such as aluminium, copper etc.(**Fig. 10.90).**

Use of Arc Welding: This is a suitable method when a small tap is broken at the bottom of materials like copper, aluminum, etc.

In this method the electrode is brought in contact with the broken tap and stuck so that it is attached with the broken tap. The tap may be removed by rotating the electrode.

Use of Nitric Acid: In this method nitric acid diluted in a proportion of about one part acid to five parts of water is injected inside. The action of the acid loosens the tap and then it is removed with an extractor or with a nose plier.

The workpiece should be thoroughly cleaned for preventing further action of the acid. While diluting acid mix acid to water.

Use of Spark Erosion: For salvaging certain precision components damaged due to breakage of taps **spark erosion** can be used.

In this process, the metal (broken tap) is removed by means of repetitive spark discharges. The electrical discharge occurs between an electrode and the electro conductive workpiece (tap) and the minute particles are eroded both from the electrode and the workpiece. In many cases it may not be necessary to remove the broken tap completely. (After a small portion has been eroded, a screwdriver or punch can be used to remove the remaining portion of the tap.) The shape of the electrode also need not be round. It can be square or in the form of a slot on the workpiece for assisting the tools for rotating the broken tap.

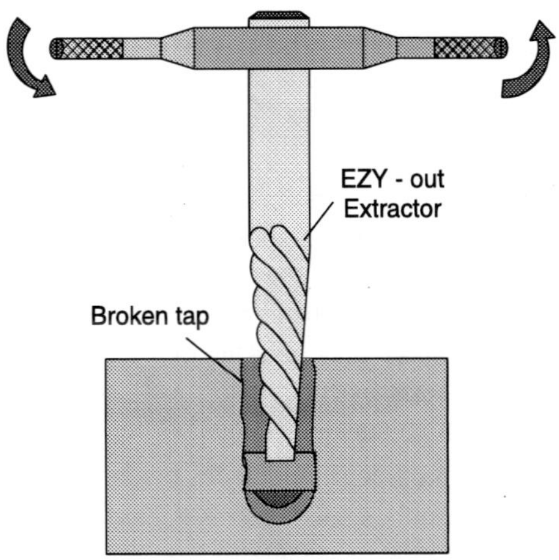

Fig. 10.90 Annealling & drilling the tap

EXERCISE

1. What are screw threads and state their uses.
2. What is meant by: (a) single and multi-start threads, and (b) right-hand and left-hand threads? Explain them with neat sketch.
3. Sketch a screw thread and level the designation of their different parts, (b) Explain clearly (i) pitch and (ii) lead.
4. State the various forms of screw threads used in general engineering.
5. With suitable sketches and give their specific uses.
6. Name the various methods of producing screw threads.
7. State where bolts and nuts are used, and specify each of them.
8. Describe with neat sketches the types of bolts and state their uses.
9. Describe the various forms of commonly used nuts and state their uses.
10. What are locking nuts? State their uses.
11. Distinguish between: (a) a stud and a bolt (b) a screw and a bolt. State the different types of cap screws commonly used.
12. Explain how riveting is made? Show some of the rivet joints.

Circlips

Chapter 11

Circlips are fastening devices used to provide shoulders for positioning or limiting the movement of parts in an assembly (**Fig. 11.1**).

Circlips are also called retaining rings.

The rings are generally made of materials having good spring properties so that the fastener may be deformed elastically to a considerable degree and still spring back to its original shape. This permits the circlips to spring back into a groove or other recess in a part or they may be seated on a part in a deformed condition so that they grip the part by functional means. Circlips are manufactured from spring steel with high tensile and yield strength.

Circlips are fastening devices used to provide shoulders for positioning or limiting the movement of parts in an assembly.

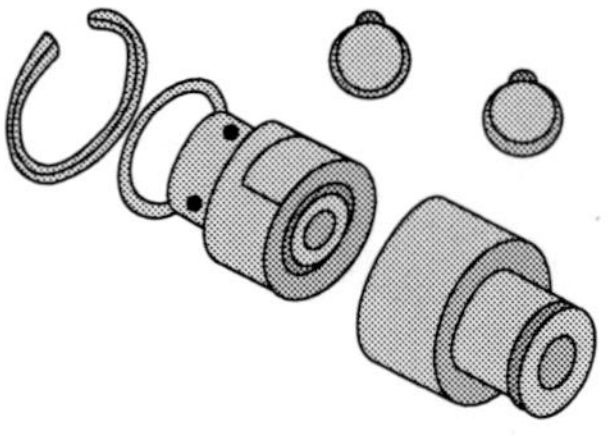

Fig. 11.1

11.1 TYPES OF CIRCLIPS

1. **Internal circlip (Fig. 11.2):** These types of rings are assembled in holes, bores or housing.
2. **External circlip (Fig. 11.3):** These types of rings are installed on shafts, pins, studs and similar parts.

Both types offer a number of advantages over other fastening methods.

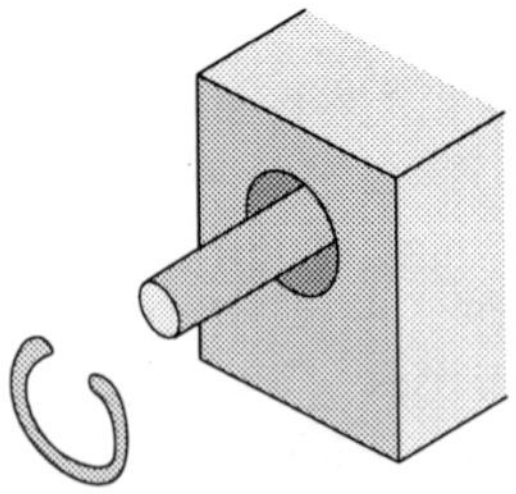

Fig. 11.2 Internal circlip

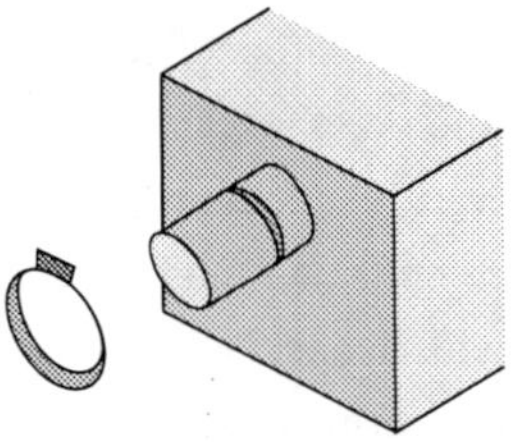

Fig. 11.3 External circlip

- Their cost is relatively low when compared with other types of fasteners.
- Their use often results in savings in raw material and simplified machining operations for other parts in the assembly.
- One circlip often can replace two or more parts.

Assembly tooling developed for circlips usually permits very rapid assembly of the fasteners, even by unskilled workers.

Material: Because retaining rings depend for their function largely on their ability to be deformed elastically during assembly and disassembly; the materials must have good spring properties. Circlips are manufactured from spring steel with high tensile and yield strength.

11.2 AXIALLY ASSEMBLED CIRCLIPS

Circlips are classified according to the loading methods and may be classified as:

1. Axially assembled rings
2. Radial assembled rings.

Axially Assembled Rings (Fig. 11.4): These are installed in a direction parallel to the axis of the shaft or bore in which the rings will be used. Most axially assembled rings have small gaps. They contact grooves in which they are seated and transmit the thrust loads to the groove wall around most of the groove circumference. As a result, most axially assembled rings have relatively high thrust - load capacity. There are different types of axially assembled rings as discussed below.

Basic Internal Rings (Fig. 11.5): They have a relatively small gap. The outside diameter of the ring is concentric with the bore or groove diameter to give maximum ring contact over most of the groove circumference and provide for high thrust capacity of the assembly.

Basic External Rings (Fig. 11.5): They have only a minute gap. The tapered section (external) is almost completely around the ring, and as is the case with the internal type, it permits maximum flexibility without excessive permanent set. The inside diameter is concentric with the diameters of the shaft and groove.

Inverted Internal Rings (Fig. 11.6): They have a relatively small gap for contraction in a bore or housing. The inside diameter is concentric to the bore and

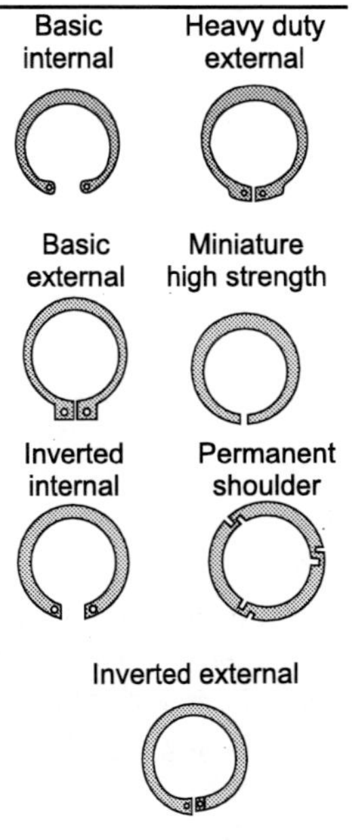

Fig. 11.4 Axially assembled circlips

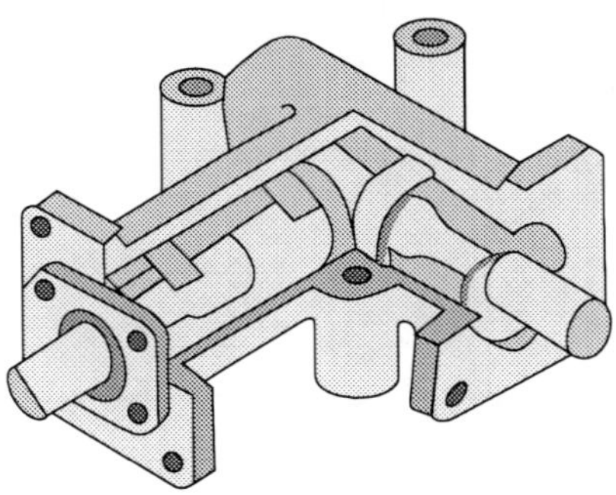

Fig. 11.5 Basic external circlips

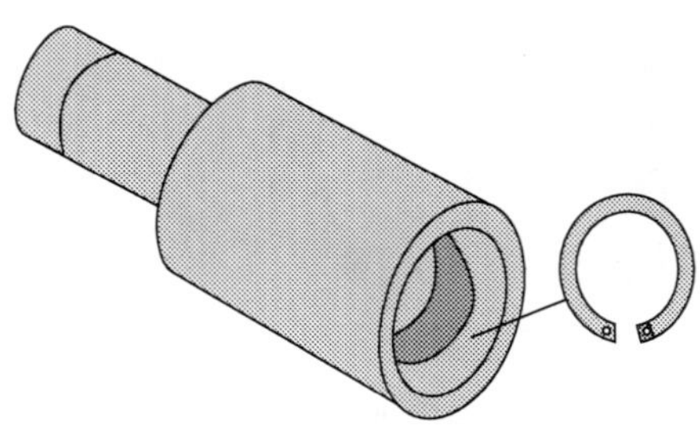

Fig. 11.6 Inverted international rings

groove diameters. The tapered section extends around a major portion of the circumference for maximum flexibility and seating in grooves of adequate depth. The ring provides a uniform shoulder against the retained part and greater assembly clearance.

Inverted External Rings (Fig. 11.7): They have the same design like inverted internal rings except that the lugs are located in the inner circumference. When installed in a groove on a shaft or similar part, they provide a uniform shoulder against the parts to be retained.

Heavy Duty External Rings (Fig. 11.8): In this type, the average radial section height and thickness are greater and the inside is somewhat smaller. The rings take a permanent set on installation and grip the bottom of the groove very lightly. Because of its heavy construction, it secures against heavy thrust load. The rings are useful where repeated loading and fatigue are likely to occur; they eliminate the need for support washers.

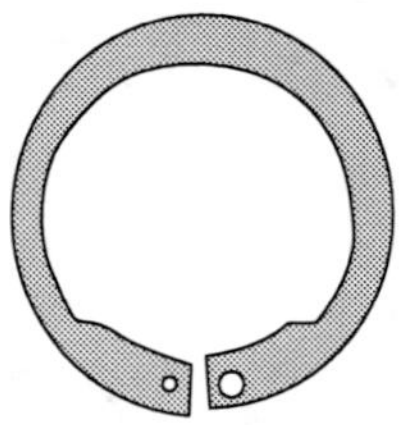

Fig. 11.7 Inverted external ring

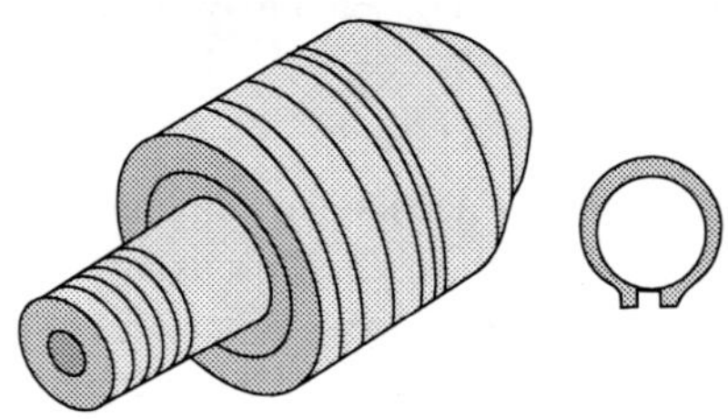

Fig. 11.8 Heavy duty external ring

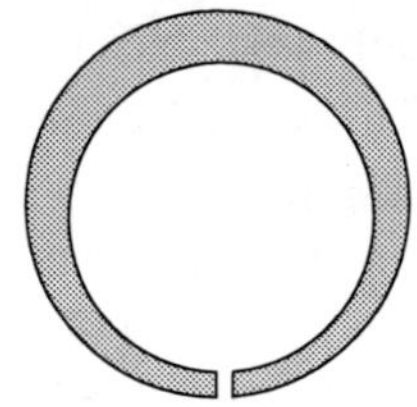

Fig. 11.9 Miniature high strength rings

Miniature High-Strength Rings: They are thicker than standard types and have higher radial sections. They do not have lugs. The rings provide high thrust capacity in relatively shallow grooves and small shafts.

Permanent Shoulder Rings (Fig. 11.10): Permanent shoulder rings have no gap and cannot be expanded or contracted. They have notches around their circumference and are crimped into a groove by means of special crimping pliers. The notches deform during the crimping process so that the ring diameter is reduced and the ring engages with the groove. Once installed, the ring forms a permanent shoulder on the shaft and must be destroyed to remove. The rings are useful where a tamper-proof construction is desirable and relatively high thrust capacity is applied. Permanent shoulder rings are made of soft materials such as low carbon steel and aluminium.

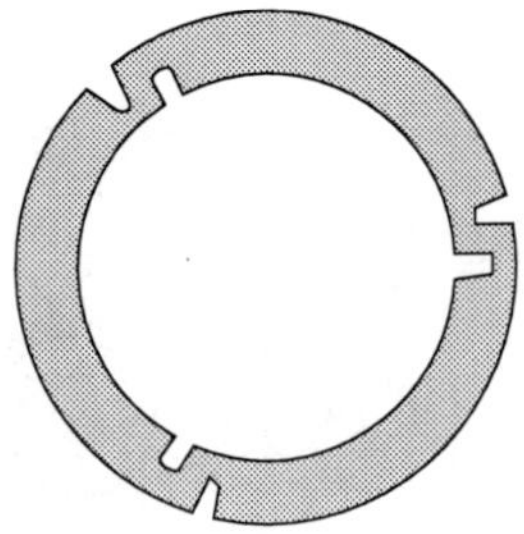

Fig. 11.10 Permanent shoulder ring

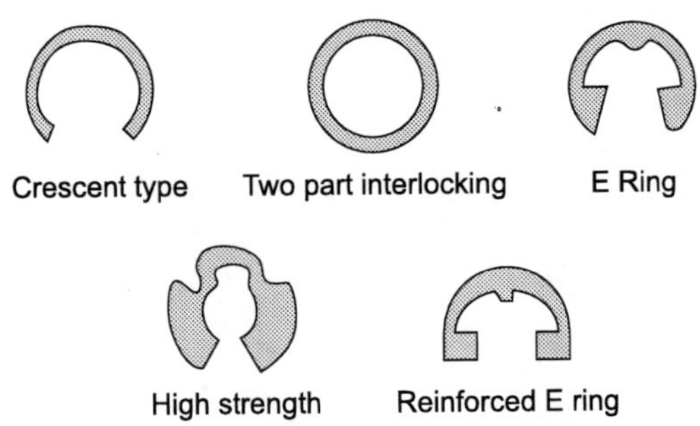

Fig. 11.11 Radially assembled rings

Radially Assembled Rings (Fig. 11.11): They are installed in a radial direction to the axis of the part on which they will be used. Radially assembled rings have wide gaps for installation, since the gap portion cannot be used for transmitting thrust loads to the groove wall. The radial types generally have lower thrust capacity than the axial rings.There are different types of radially assembled rings as discussed below.

These rings have a wide gap for radial assembly into a groove on a shaft and provide a low shoulder against a retained part.

Crescent-Shaped Rings: These rings have a wide gap for radial assembly into a groove on a shaft and provide a low shoulder against a retained part.

'E' Rings (Fig. 11.12): They have a large gap and a large outside diameter. Three lugs inside the ring circumference seat the ring in its groove and provide contacting surfaces for the groove wall. 'E' rings are used to provide relatively large diametre shoulders on small shafts.

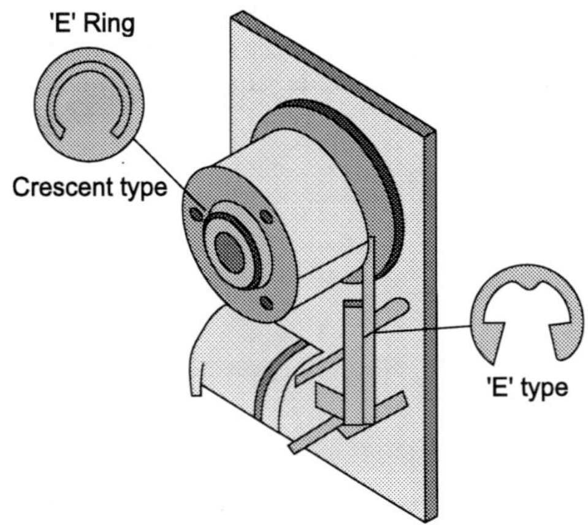

Fig. 11.12

Reinforced 'E' Rings: They differ from conventional 'E' rings in that the bending portions between the lugs are tapered, and thus can provide a greater radial bending section than the standard type. These rings grip their grooves very tightly and provide high resistance against radial dislodgement from the groove.

Two-part Interlocking Rings (Fig. 11.13): They are external radial assembled rings which form a uniform shoulder completely around the shafts on which they are used. The ring halves lock together by means of inter-engaging lugs at each end. Each half has a reduced section at the centre for necessary flexibility in bending during installation or removal. When the rings are assembled in grooves, they resist high rotational speeds.

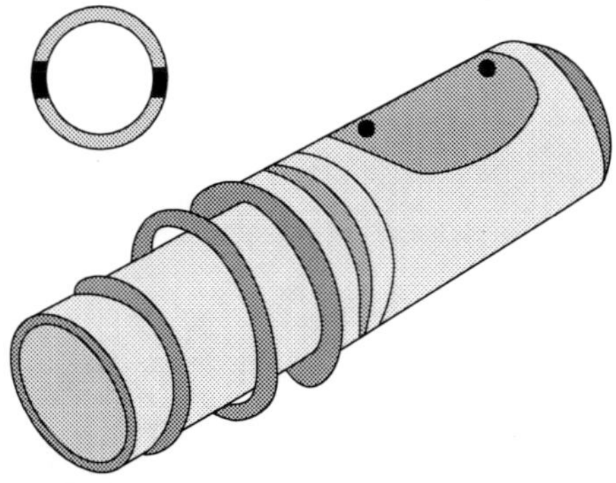
Fig. 11.13 Two-part interlocking rings reinforced 'E' ring

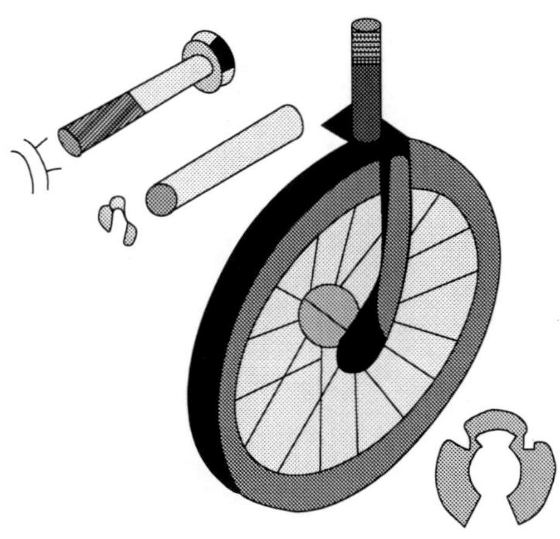

Fig. 11.14

High Strength Types (Fig. 11.14): They are radially assembled rings and have two large lugs which contact a major portion of the groove around its circumference. This provides good resistance to thrust loads. The rings are thicker than the comparable size 'E' or crescent type rings. The rings are seated in deep grooves and provide a large shoulder on small diameter shafts.

Thinner Gauge High Strength Rings: They have the same configuration like the high strength rings but with reduced thickness. They may be used in the same width grooves as 'E' rings for similar shaft sizes where a higher thrust capacity is required and when the rings must accommodate large groove diameter tolerances.

Rivets and Riveted Joints

Chapter 12

12.1 RIVETS

Rivets are cylindrical rods having heads of various shapes. They are used for assembling the parts of a workpiece together.

Rivets are fasteners used for joining metal sheets and plates in fabrication work such as bridges, ships, cranes, structural steel work, boilers, aircraft, etc.

The rivet is then inserted and closed by force so that it completely fills the hole and forms a rigid joint (**Fig. 12.1**).

Riveting is a method of making permanent joints. For riveting the plates to be joined are drilled or punched.

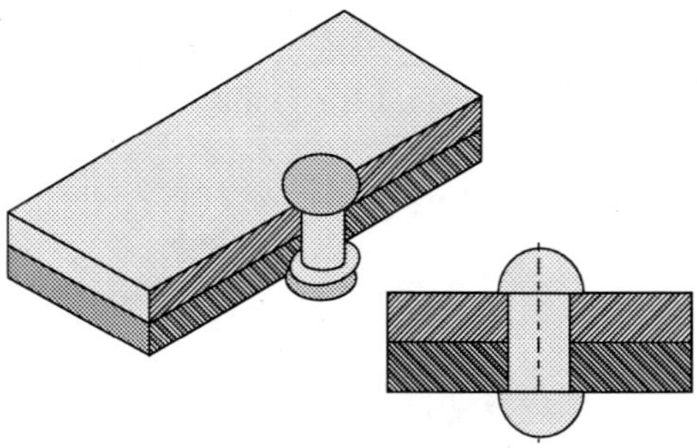

Fig. 12.1 Riveted joint

12.2 SPECIFICATIONS OF A RIVET

The rivet is specified by the shape of the head diameter, and the length.

Shape of Head: The shape of the rivet head is to be selected according to the intended use of the workpiece to be riveted.

Diameter: The diameter is to be selected depending on the required strength.

Length: Length is to be selected depending upon the thickness of the components to be riveted.

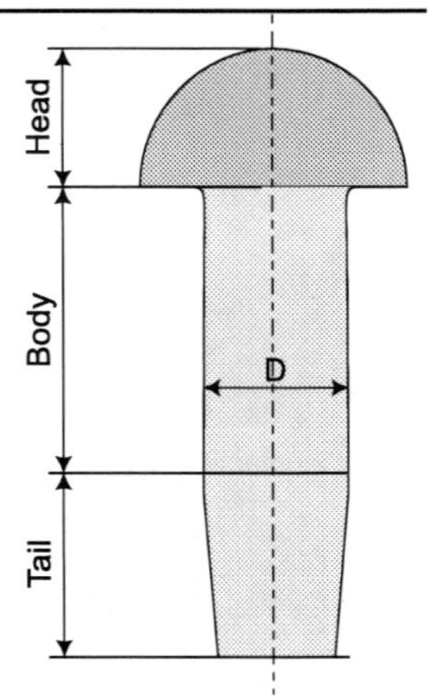

Fig. 12.2 Specification or rivet

12.3 TYPES AND USES

Snap Head (Fig. 12.3): It is the most commonly used form, and it gives a very strong joint and used for structural work. The opposite end of the rivet is shaped similar to the head.

Pan Head (Figs. 12.4-12.5): It is a very strong rivet. The opposite end is usually shaped to snap the head shape. These rivets are used in heavy constructions. It is used where strength is the only consideration such as heavy structural work.

Conical Head (Fig. 12.6): It is generally used in light assembly where riveting is done by hand hammering.

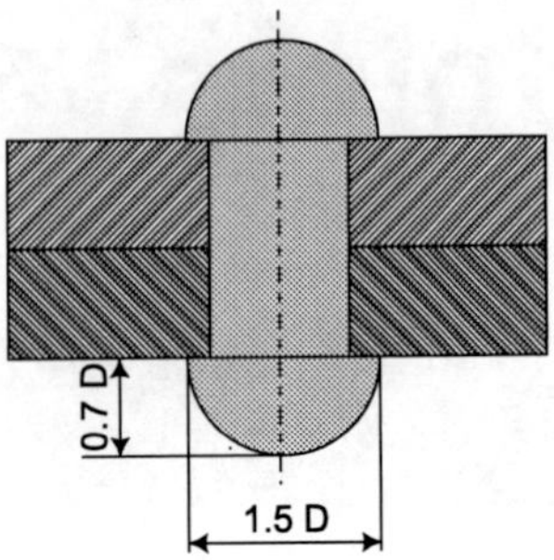

Fig. 12.3 Snap head rivet

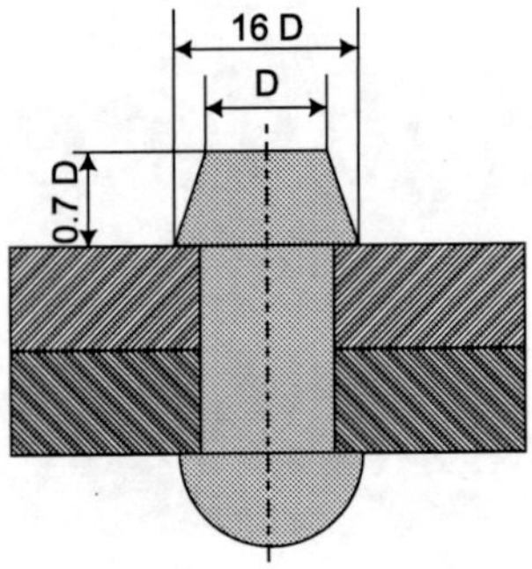

Fig. 12.4 Pan head rivet

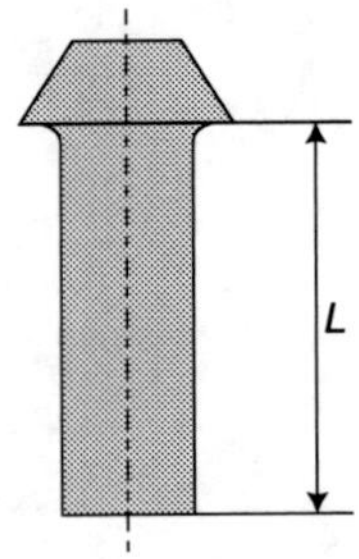

Fig. 12.5 Pan head

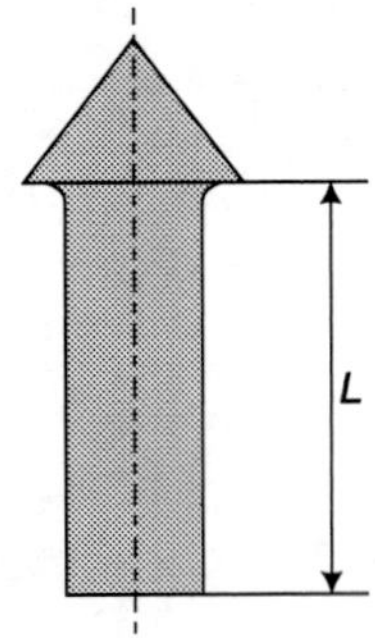

Fig. 12.6 Conical head

Countersink Head (Fig. 12.7 & 7A): This rivet is mainly used in fabrication where projection of the rivet head has to be avoided. The opposite end is finished either to snap head or countersunk head. Countersunk heads are available with different angles. The standard countersunk head has an angle of 90°. Rivets with 120° angle are used for joining thin plates. A method of joining thin sheets to thick plates is shown in **Fig. 12.5A.**

It is used where projection of the rivet head is to be avoided.

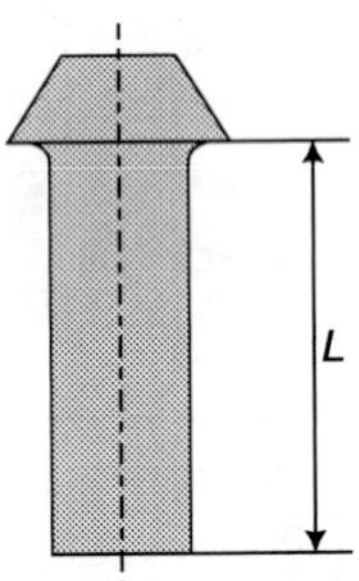

Fig. 12.7 Countersink head

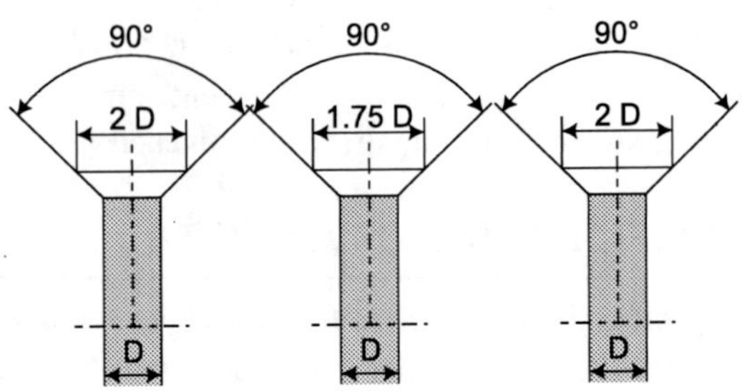

Fig. 12.7a Countersink head

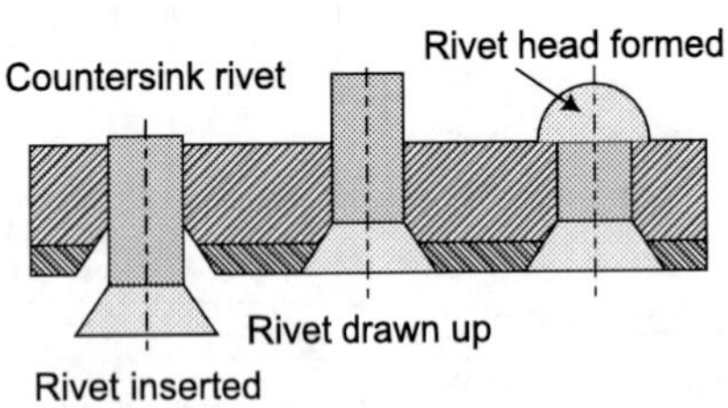

Fig. 12.7b Method of joining thin plates

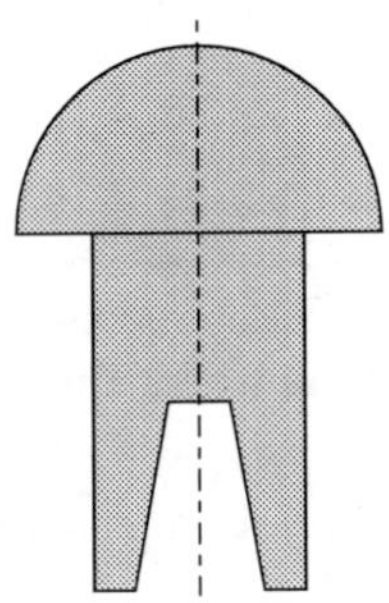

Fig. 12.8 Bifurcated rivet

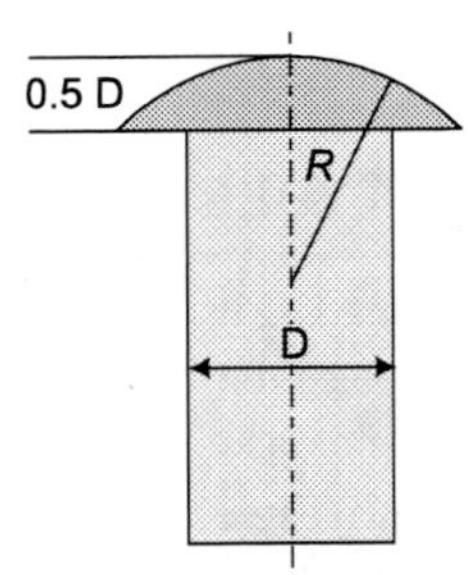

Fig. 12.9 Mushroom head

Bifurcated Rivet (Fig. 12.8): The shape of the head is shown in the figure and the bifurcated portion is used for fastening light parts i.e., tin plates, leather, plastics, etc.

Mushroom Head: These rivets are used to reduce the height of the rivet head above the metal surface **(Fig. 12.9).**

Flat Head (Fig. 12.10): Flat head rivets are most commonly used in sheet metal fabrication, where the metal is very thin and a projecting head of the rivet is objectionable.

Flat head rivets are also available for heavy fabrication work.

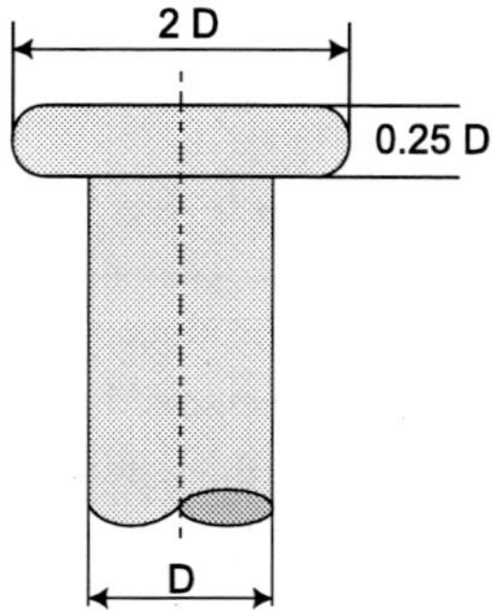

Fig. 12.10 Flat head

12.4 DESIGNATION OF STEEL RIVETS

A mild steel snap-head rivet of 16 mm diameter having a length of 70 mm shall be designated as snap-head rivet 16 × 70 IS.

A high tensile steel snap-head rivet of 16 mm diameter having length of 70 mm shall be designated as snap head rivet 16 × 70 HT IS

Other types of rivets are also designated in a similar manner.

Note: The length of rivets is indicated by the shank length **(Fig. 12.11).**

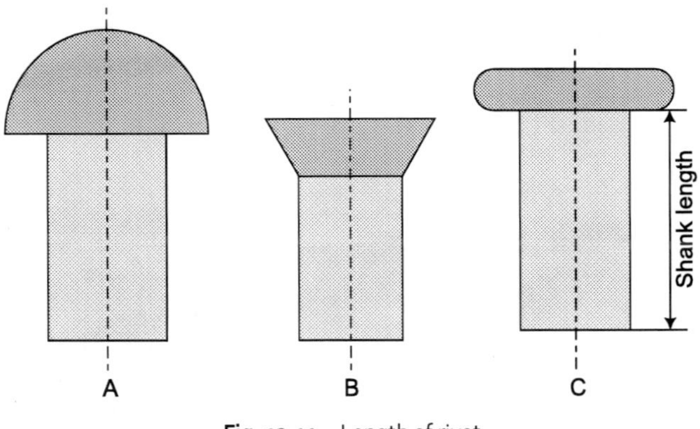

Fig. 12.11 Length of rivet

12.5 RIVETING (HOT AND COLD WORKING)

Riveted joints can be formed either hot or cold.

Smaller diameter rivets used in light fabrication are worked cold.

Larger diameter rivets are usually worked hot.

For hot working, the holes in which rivets are inserted are made larger for accommodating thermal expansion.

The clearance provided in the holes for cold working is less than that provided for hot working.

Materials: In riveting, the rivets are secured by deforming the shank to form the head, and hence these are made of ductile materials.

Examples

Low carbon steel, brass, copper and aluminium.

12.6 RIVET PROPORTIONS

In order to produce efficient and good quality riveted joints the following aspects are important:

- The size of hole drilled for inserting the rivets.
- The diameter of the rivet in proportion to the thickness of the plates/sheets to be joined.
- The length of the rivet according to the type of rivet and the thickness of plate/sheets.

The Size of the Rivet and the Hole: The size of the hole to be drilled is according to the diameter of the rivet used.

A formula generally used for determining the diameter of the rivet is:

$$D = 1.2\ t.$$

D is the diameter of the rivet and t is the thickness of the plates. The size of the hole has to be slightly larger than the nominal diameter of the rivet **(Table 12.1).**

For hot-working rivets more clearance is needed than for cold-working rivets.

Table 12.1 Length of rivets

Rivet Nominal Dia. mm	**2**	**3**	**4**	**5**	**6**	**8**	**10**	**12**	**15**	**15-40**
Hole diameter	2.2	3.2	4.2	5.3	6.3	8.5	11	13	16.5	Holes larger than nominal Dia. by 1.5 to 2.0 mm

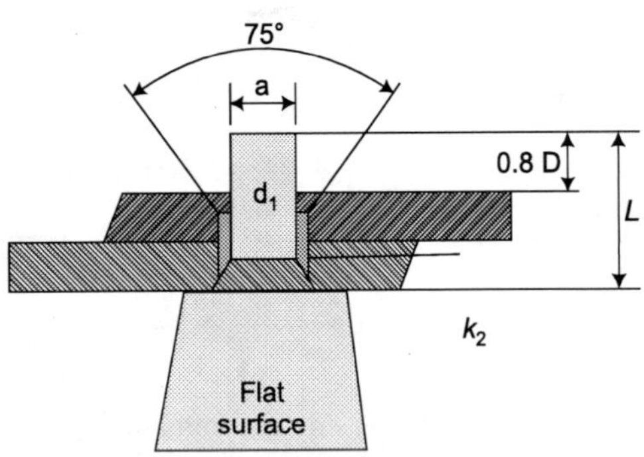

Fig. 12.12 Length of countersink rivet

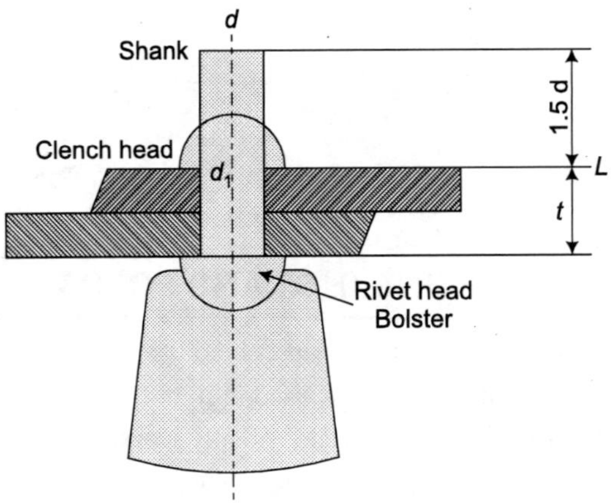

Fig. 12.12a Length of rivet

The length of a rivet is the shank length. This will vary according to the thickness of the plates to be riveted and the type of the rivet head. A formula generally used in the shop floor is: **(Fig. 12.1a)**

$$L = t + 1.5\ d$$

Length of countersink head rivets **(Fig. 12.12)**

$$L = t + 1.5\,d$$

L = shank length

t = total thickness of plates to be joined

d = rivet diameter

d_1 = hole diameter

12.7 RIVETED JOINTS

In construction and fabrication work different types of riveted joints are made.

The commonly used joints are:

- Single riveted lap joint
- Double riveted lap joint
- Single strap butt joint
- Double strap butt joint.

Single Riveted Lap Joint (Fig. 12.13): This type of joint will have two rows of rivets. The overlap is large enough to accommodate two rows of rivets.

Double Riveted Lap Joint (Fig. 12.14): This type of joint will have two rows of rivets. The overlap is large enough to accommodate two rows of rivets.

Double Riveted (Zigzag) Lap Joint (Fig. 12.15): This provides a stronger joint than the single lap joint. The rivets are placed either in a square formation or in a triangular formation. The square formation of rivet placement is called chain riveting. The triangular formation of rivet placement is called zigzag riveting.

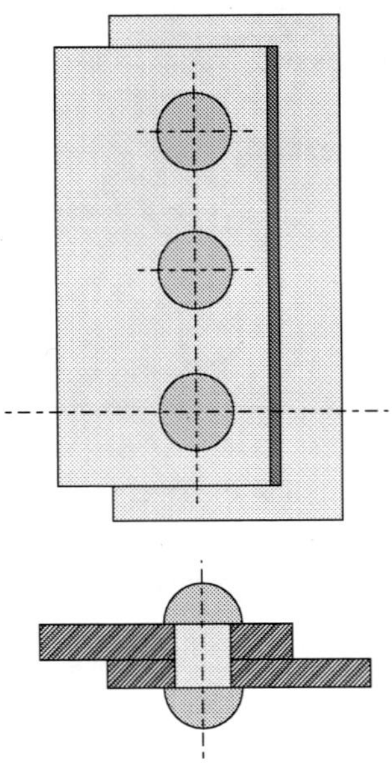

Fig. 12.13 Single rivet lap joint

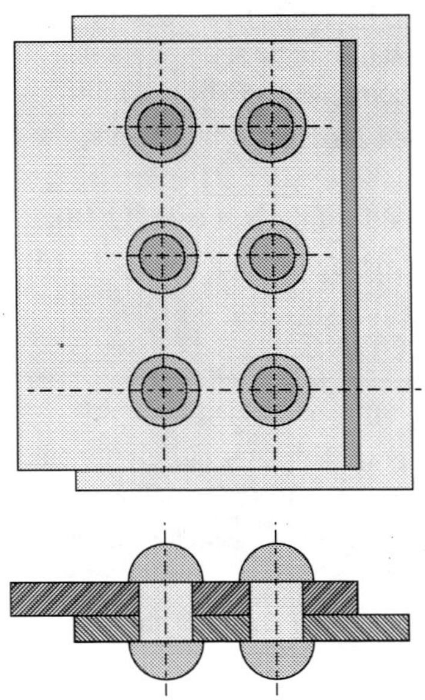

Fig. 12.14 Double rivet lap joint

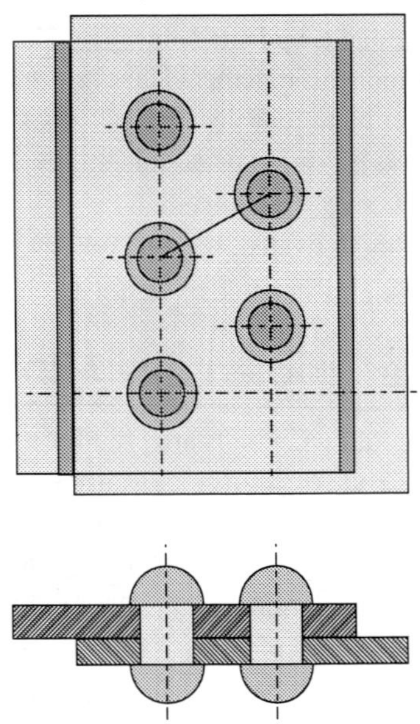

Fig. 12.15 Double riveted lap zig zag

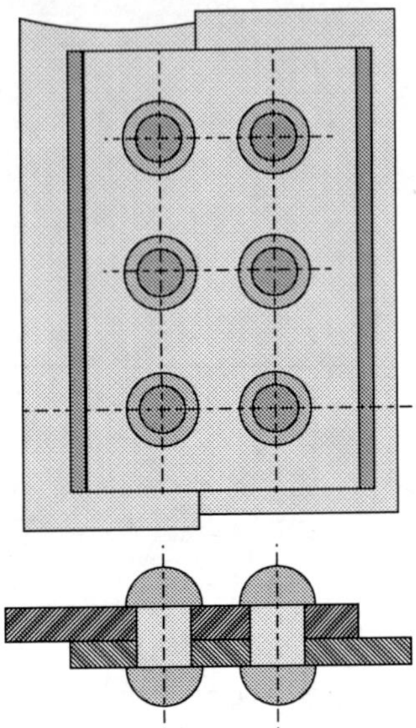

Fig. 12.16 Single strap butt joint

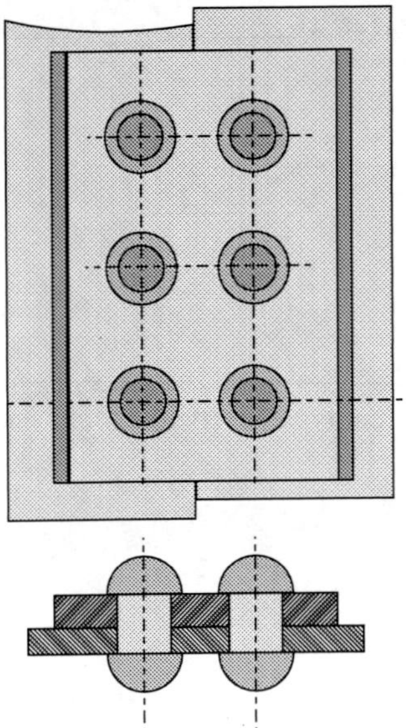

Fig. 12.17 Single strap butt joint

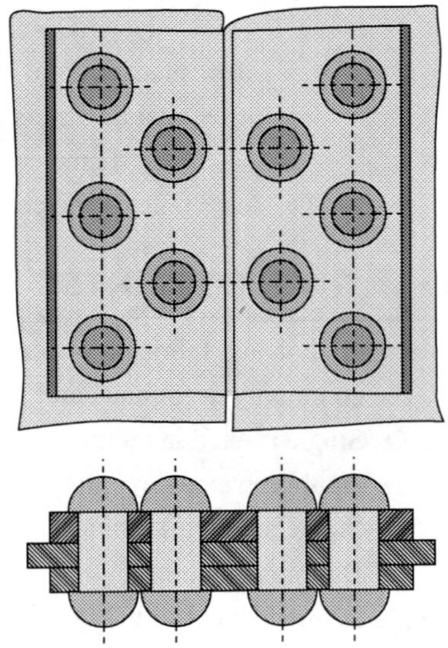

Fig. 12.18 Double for triple rivets chain or zig zag single or double strap

Single Strap Butt Joint (Fig. 12.16): This method is used in situations where the edges of the components are to be joined by riveting.

A separate piece of metal called strap is used to hold the edges of the components together.

Double Strap Butt Joint (Fig. 12.17): This joint is also used for joining the edges of components together. This is stronger than the single strap butt joint. This joint has two cover plates placed on either side of the components to be assembled.

When single or double straps are used for riveted butt joints, the arrangement of rivets may be:

Single riveted, i.e., one row on either side of the butt **(Fig. 12.18).**

Double or triple riveted with chain or zigzag formation **(Fig. 12.19).**

Rivet set Dolly

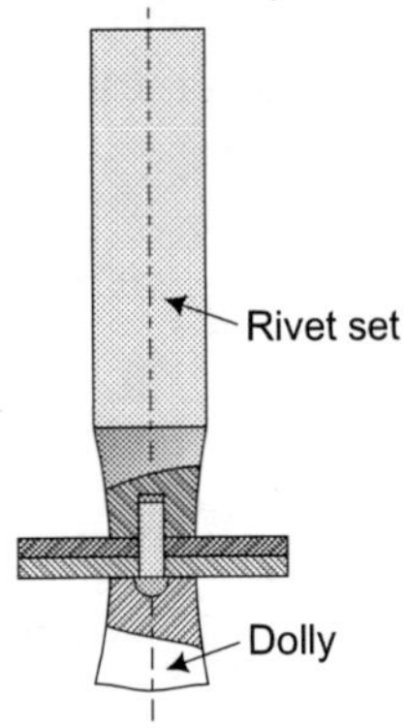

Fig. 12.19

12.8 TOOLS FOR HAND RIVETING

The following tools are used for making efficient riveted joints.

Rivet Set: A rivet set is used for bringing the plates closely together after inserting the rivet in the hole. This is required while riveting thin plates or sheets with small rivets (**Fig. 12.19).**

Dolly: This is used to support the head of the rivet which is already formed and also to prevent damage to the shape of the rivet head (**Fig. 12.20).**

Snap: The rivet snap is used to form the final shape of the rivet during riveting. Snaps are available to match the different shapes of rivet heads (**Fig. 12.21).**

Combined Rivet Set: This is a tool which can be used for setting and forming the head **(Fig. 12.21).**

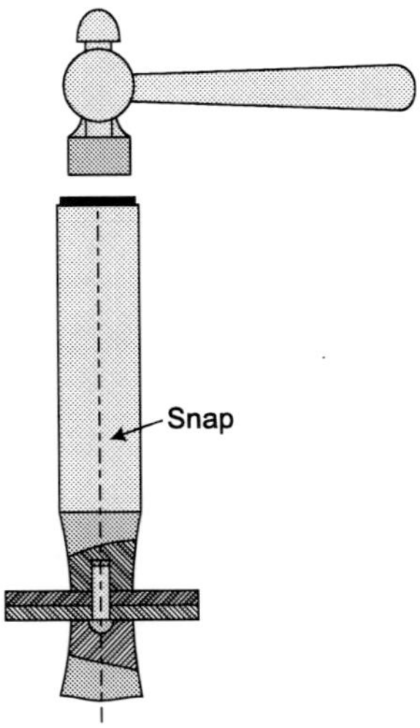

Fig. 12.20 Snap joint

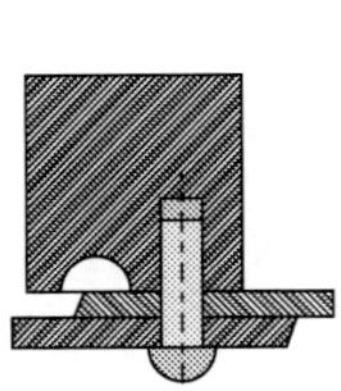

Fig. 12.21 Combination rivet set

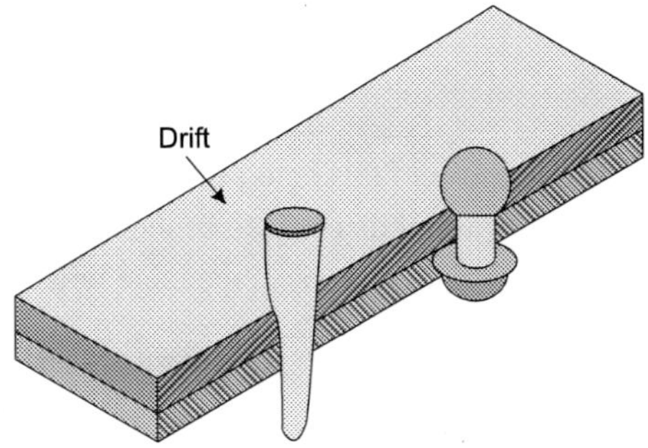

Fig. 12.22

Drift: This is used to align the holes to be riveted (**Fig. 12.22**).

Hand Riveter (Fig. 12.23): This has a lever mechanism which exerts pressure between the jaws when the handle is pressed. This is useful for riveting copper or aluminium rivets. Interchangeable anvils can be provided.

Pop Riveter (Fig. 12.24): This is used for riveting pop rivets by hand. The trigger mechanism squeezes the rivet and separates the mandrel of the rivet. In this method, as the mandrel is separated from the rivet, the head is formed on the other end.

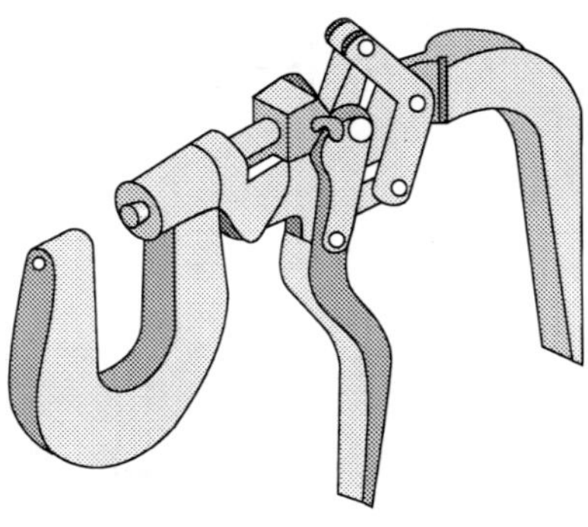

Fig. 12.23 Hand riveter

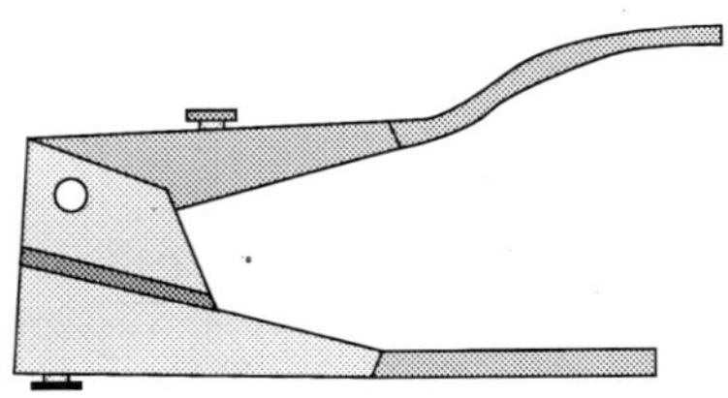

Fig. 12.24 Pop riveter

12.9 SPACING OF RIVETS IN JOINTS

The spacing of the rivet holes depends upon the job. Given below is a general approach in determining this.

Distance from the Edge to the Centre of the Rivet (Fig. 12.25): The space or distance from the edge of the metal to the centre of any rivet should be at least twice the diameter of the rivet.

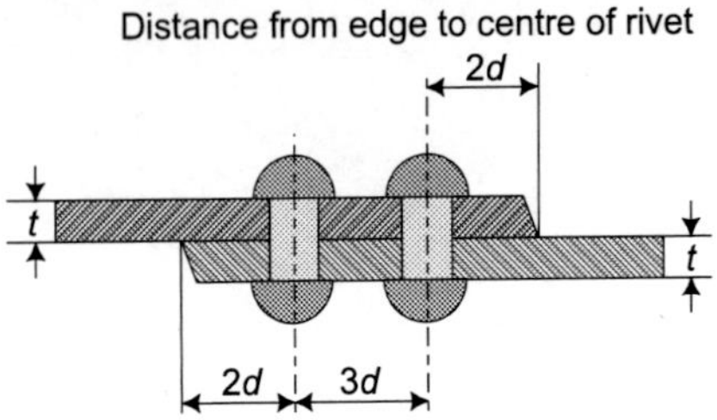

Fig. 12.25

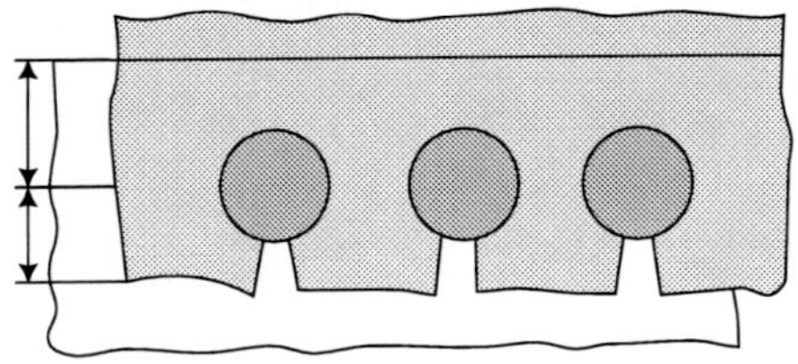

Fig. 12.26 Spliting effect

The purpose of this is to prevent the splitting of the edges (**Fig. 12.26**).

The maximum distance from the edge should not be more than ten times the thickness of the plate.

Too much distance from the edge will lead to **gaping (Fig. 12.27).**

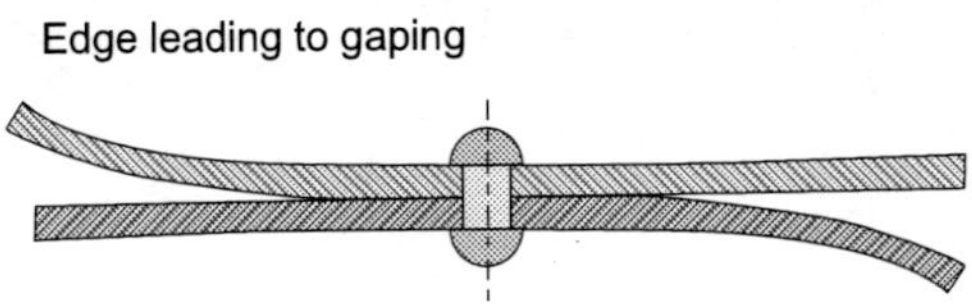

Fig. 12.27

Pitch of Rivet (Fig. 12.28): The minimum distance between rivets should be three times the diameter of the rivet **(3D).**

This distance will help to drive the rivets without interference.

Too closely spaced rivets will tear the metal along the centre line of the rivets (**Fig. 12.29).**

The maximum distance between the rivets should not exceed twenty-four times the thickness of the metal.

Too far a pitch will allow the sheet/plate to buckle between the rivets (**Fig. 12.29).**

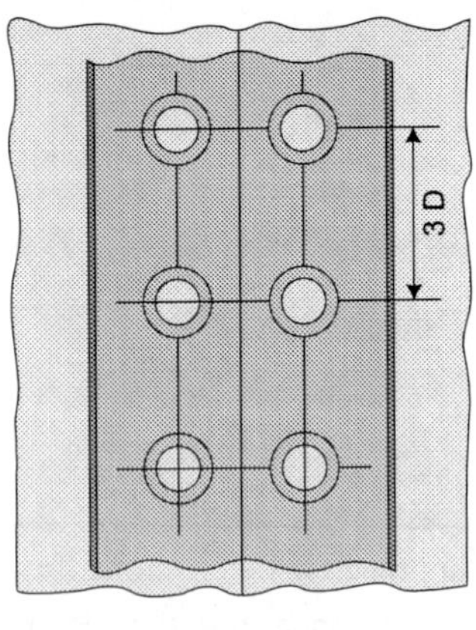

Fig. 12.28

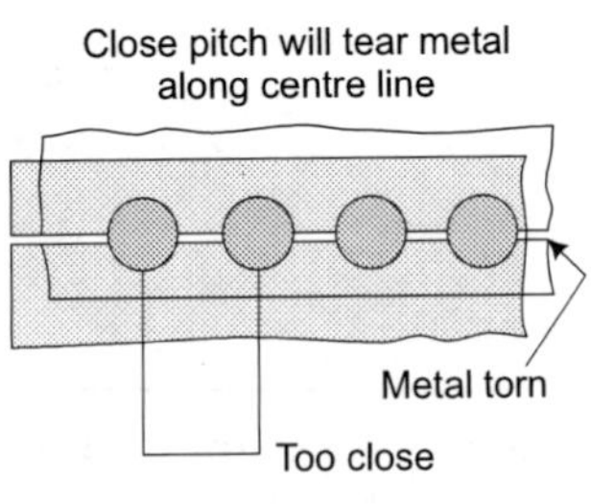

Fig. 12.29

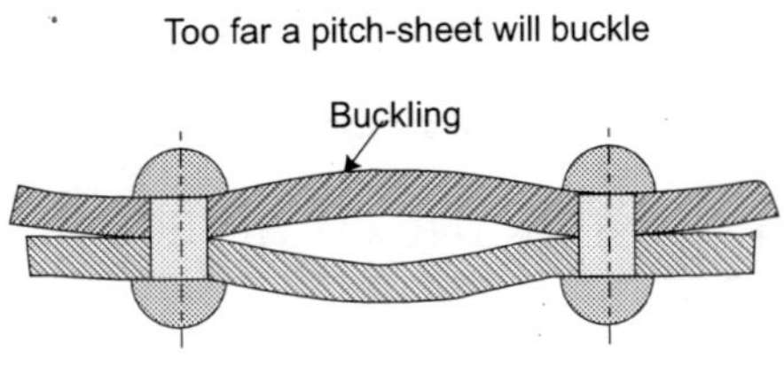

Fig. 12.30

12.10 DEFECTS IN RIVETED JOINTS

While making riveted joints certain precautions are to be exercised to avoid defects in the joints.

A few common causes and defects and the resultant effects in riveting are given below.

Causes of riveting defects	Resulting effect
Holes wrongly aligned **Fig. 12.31**	Rivet deformed & does not completely fill the hole 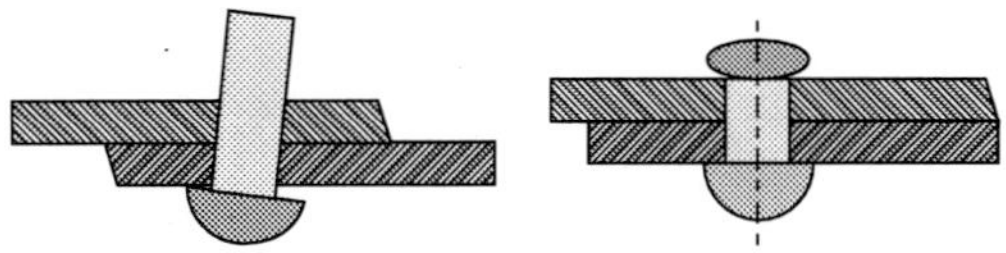
Rivet too short **Fig. 12.32**	Rivet too short 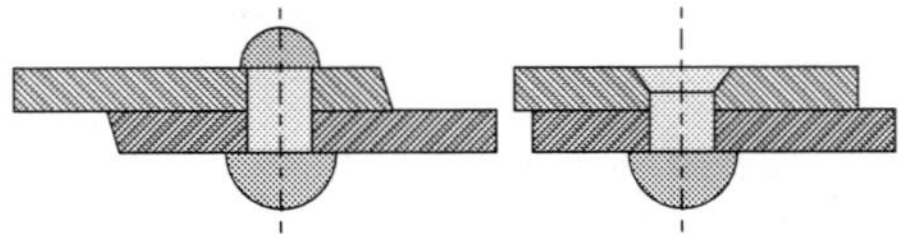 Plate surface damaged, counter sinking not completely filled
Hole too large **Fig. 12.33**	Hole too large 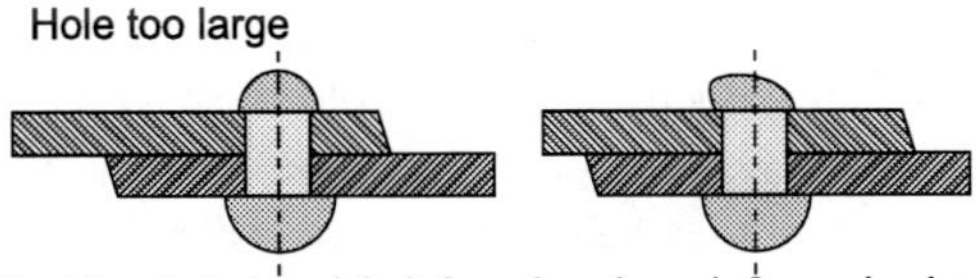 Rivet tends to bend & deform head weak & poorly shape(
Burrs in drilling **Fig. 12.34**	Hole too large 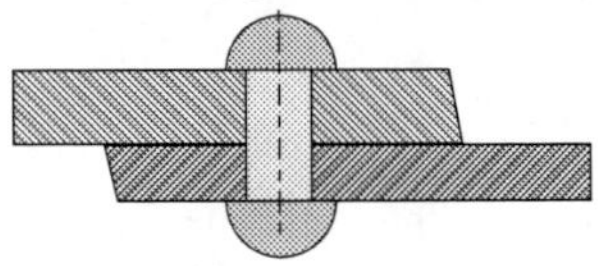 Plates/sheets not closed together unequal heads
Burrs between plates **Fig. 12.35**	Burr between plates 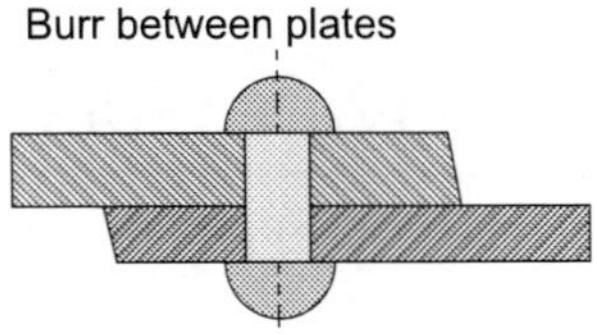 Gap between plates weak joint
Rivet not set correctly **Fig. 12.36**	Rivet not set correctly 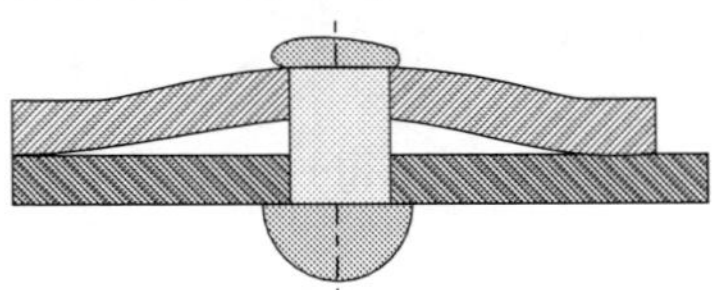 Rivet shank swells between plays weak joint

Rivet length too long
Fig. 12.37

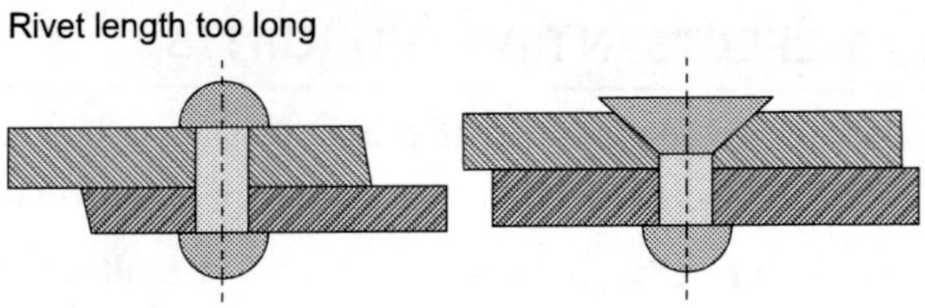

Head formed out of centre
Fig. 12.38

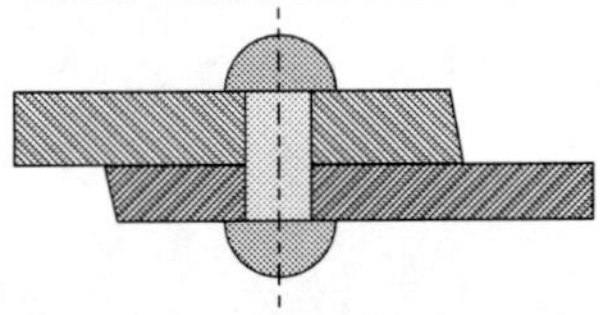

Caulking and Fullering: In order to provide a leak-proof joint in the construction of fluid containers, caulking and fullering are carried out after riveting.

Caulking: Caulking is an operation of closing down the edges of the plates and heads of the rivets to form a metal-to-metal joint.

The edge of the rivet head is tightly pressed and expanded on the plate by a caulking tool which looks like flattened cold chisel **(Fig. 12.39).**

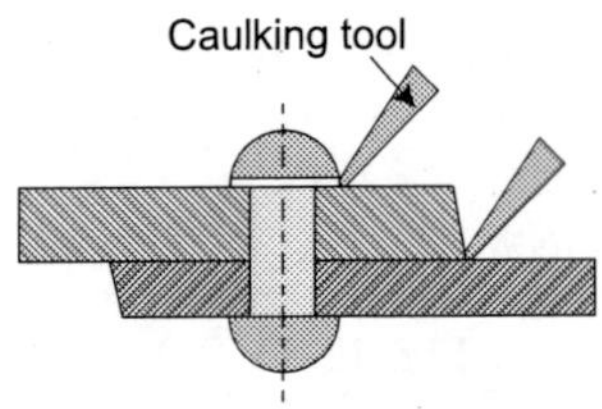

Fig. 12.39

Fullering: Fullering is an operation of pressing the whole surface of the edge of the plate. It is done by a fullering tool **(Fig. 12.39).**

When the caulking tool is about as thick as the plate, it is called a fullering tool.

The whole surface of the edge of the first plate is tightly pressed on the second plate.

A better fluid-tight joint is achieved by fullering.

Caulking is done on the edges of the plates as well as on the edges of the rivet heads. But fullering is done on the edges of the plate only. To facilitate caulking and fullering on the plates, the edges of the plates are beveled about 80° to 85° **(Fig. 1.40).**

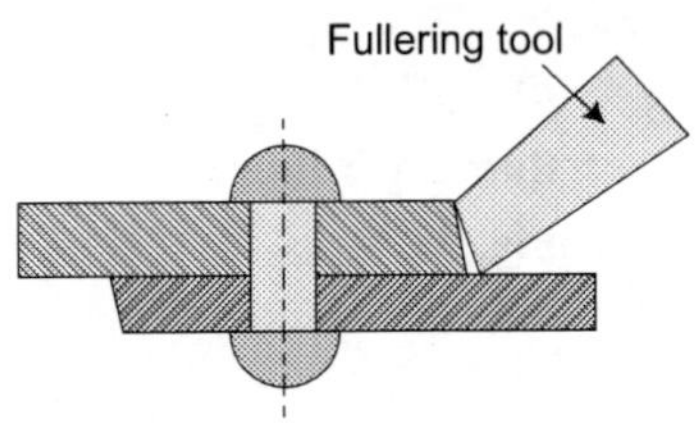

Fig. 12.40

12.11 THE STRENGTH OF RIVETED JOINTS

A riveted joint is only as strong as its weakest part and it must be borne in mind that it may fail in one of the following four ways.

- Shearing of the rivet
- Crushing of the metal
- Splitting of the metal
- Rupture or tearing of the plate

These four undesirable effects are illustrated in the table below:

Riveted joint	Effects	Cause	Prevention
Shearing of rivets 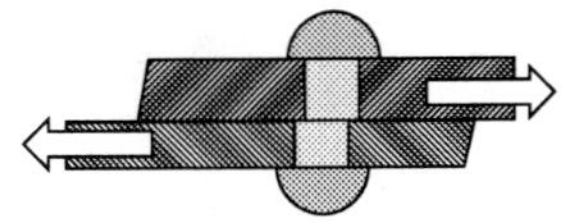	Shearing of the rivet	Diameter of the rivet too small compared with the thickness of the plate.	Select the correct diameter rivet to suit the thickness of the plate.
Crushing of metal 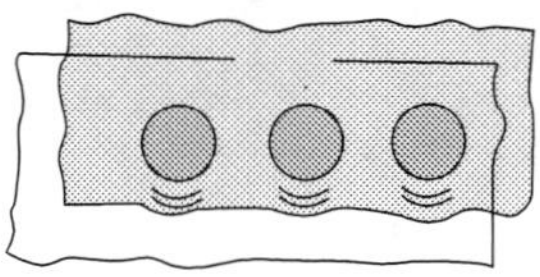	Crushing of the metal		Select the correct diameter rivet for the thickness of the metal plate.
Splitting of metal 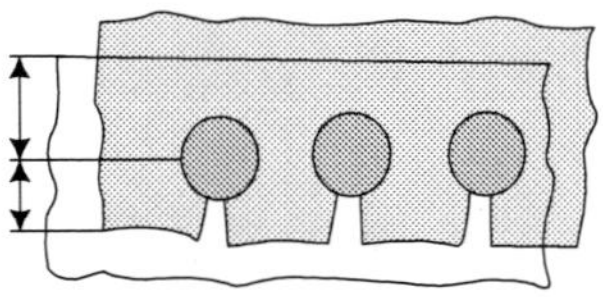	Splitting of the metal		Drill or punch the rivet at the correct distance from the edge and use the correct lap allowance for the diameter of the rivet.
Tearing of sheet 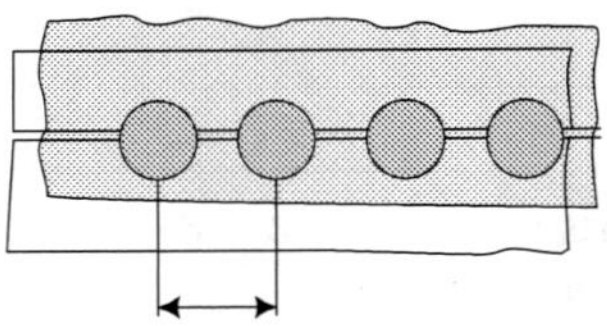 Tool close	Tearing of the plate		Punch or drill rivet holes at the correct spacing or 'pitch'. In addition remove all burrs from the holes before final assembly

EXERCISE

1. How many types of rivet joints are there?
2. Draw figures of main rivet heads. Write about the main tools used for riveting.
3. What are the causes of defects in riveted joints? Explain.
4. Explain how riveting is made. Show some rivet joints.

Cylindrical and Taper Pins

Chapter 13

In machine shop assembly various methods are used for securing components together. A few of the common methods are:

- Dowelling
- Pinning
- Staking
- Brazing/hard soldering
- Using of adhesives

13.1 DOWELLING

This is used for accurate positioning of two or more parts. This allows the parts to be separated and relocated in position. Different types of dowels are used depending on the type of assembly.

The components dowelled are always fixed with retaining screws in the assembly.

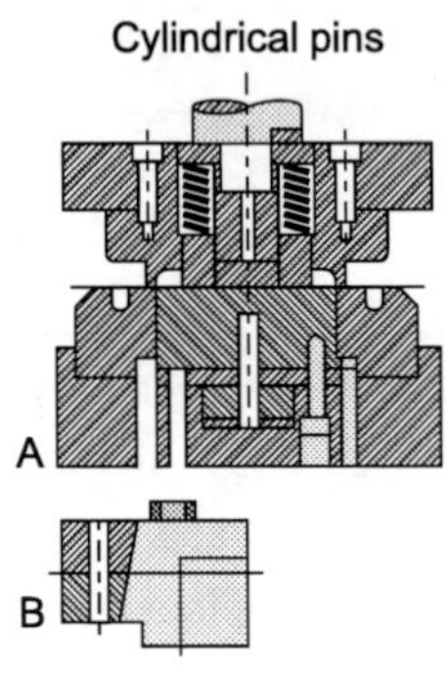

Fig. 13.1

13.1.1 Cylindrical and Taper Pins

These are used for:

- Locating hole positions for assemblies whenever they are dismantled and assembled. (**Examples:** jigs and fixtures, cover plates, machine tool assembly etc.) **(Fig. 13.1 A &B).**
- Assembling components (**Examples:** wheels, gears, levers, cranks, etc., to shafts) **(Fig. 13.2 A and B).**

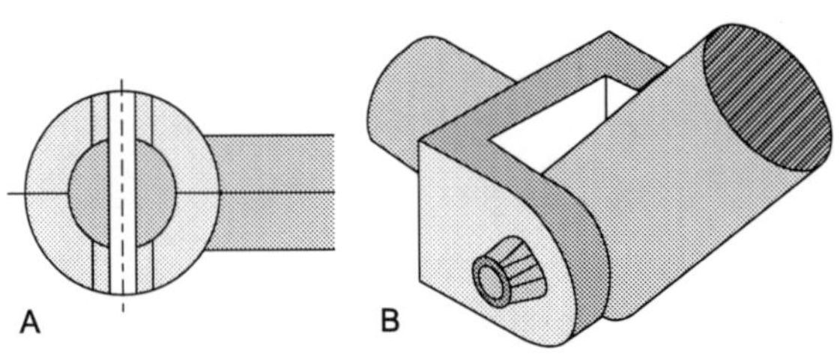

Fig. 13.2 Cylindrical pins

Cylindrical pins are available with different types of:

- Ends
- Tolerances
- Surface quality
- Material.

Cylindrical pins are also available in unhardened and hardened conditions.

Unhardened cylindrical pins are of three types **(Fig. 13.3).**

- Chamfered and rounded end
- Chamfered end
- Square end

They are useful in general assembly work.

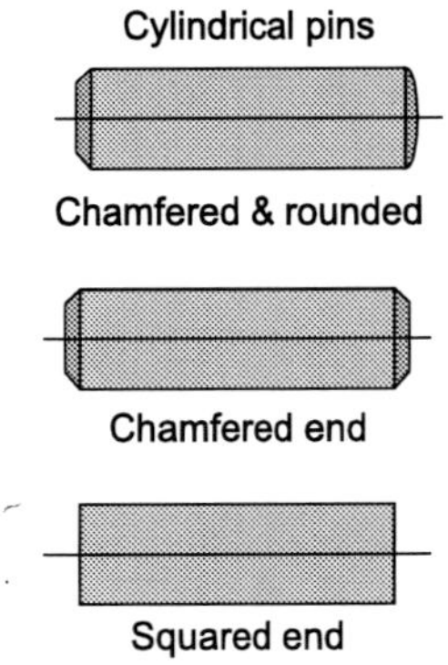

Fig. 13.3

Hardened cylindrical pins are made of high grade steel and are finished by grinding **(Fig. 13.4).** These pins can withstand higher shearing force. These pins are used in precision assemblies like jigs and fixtures and other tool making works.

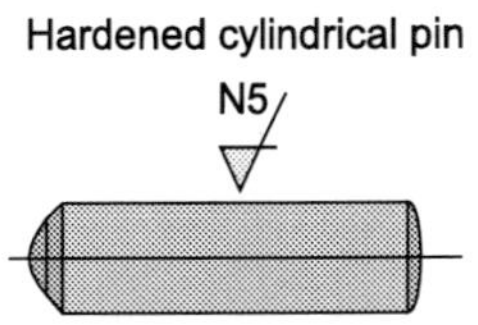

Fig. 13.4

In tool assemblies the parts will be fixed by screws or bolts **(Fig. 13.5)** and are located by using cylindrical pins.

Hardened cylindrical pins are available with dimensional tolerance m 6.

Unhardened and hardened cylindrical pins are made to fit in the holes finished by standard reamers.

Cylindrical pins are designated by the name, nominal diameter, tolerance on diameter, nominal length and the number of B.I.S. Standard.

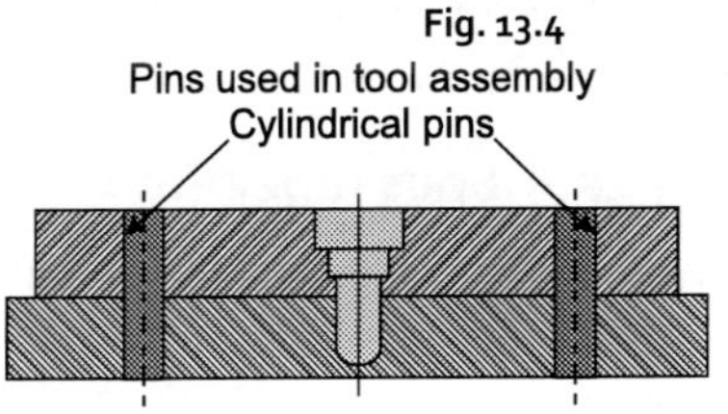

Fig. 13.5

Example:

A cylindrical pin of nominal diameter 10 mm, tolerance h 8 and nominal length 20 mm shall be designated as cylindrical pin 10 h 8 × 20 15:2393.

Note

The I.S. number refers to unhardened cylindrical pins. Cylindrical pins are also referred to as dowel pins.

13.1.2 Taper Pins

Taper pins of different types are used in assembly work.

Taper pins allow for frequent dismantling and assembling of components without disturbing the precise nature of location. They are used to transmit small torques **(Fig. 13.6).**

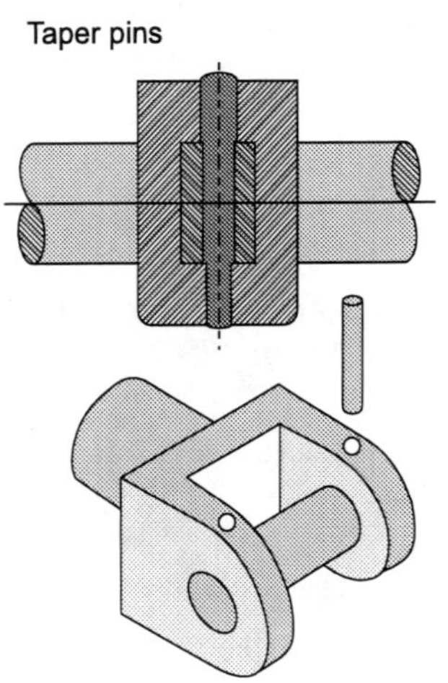

Fig. 13.6

Taper pins are of three types **(Fig. 13.7).**

- Type A - taper pins with a surface finish of N 6.
- Type B - taper pins with a surface finish of N 7.
- Type C - split taper pins with a surface finish of N 7.

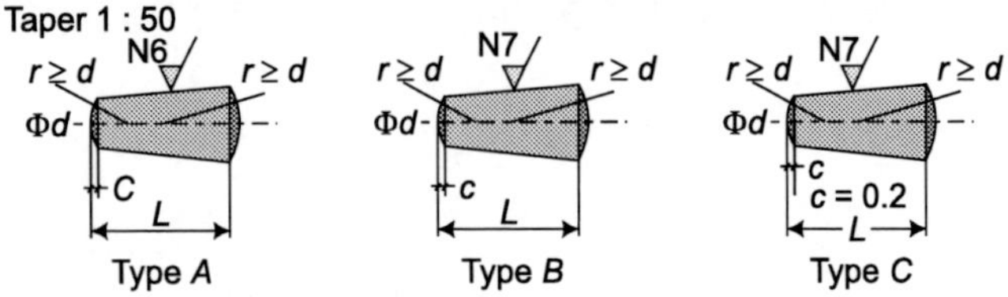

Fig. 13.7

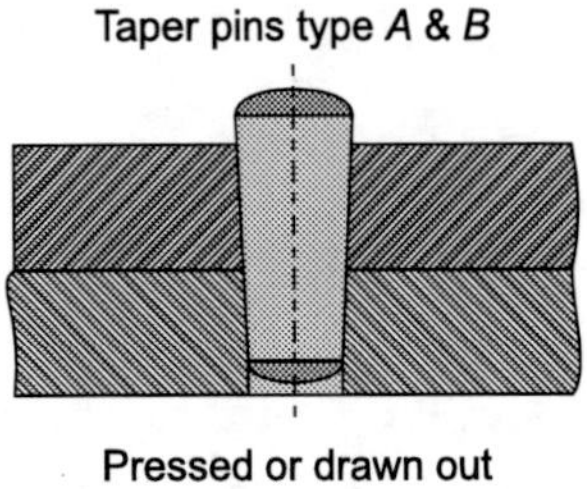

Fig. 13.8

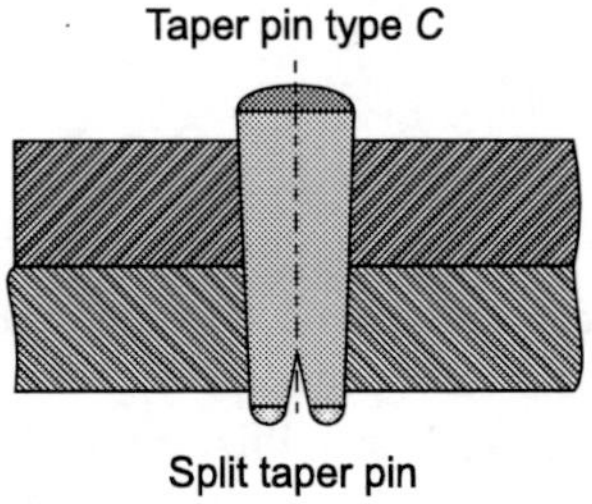

Fig. 13.9

All taper pins have a taper of 1:50 and are finished within a dimensional tolerance of h 10.

Taper pin types A & B assembly (**Fig. 13.8**) and type C (**Fig. 13.9**).

13.1.3 Split Taper Pin

In the case of split taper pins the split end can be slightly opened to ensure a more positive locking.

Taper pins are designated by name, type (A, B or C) nominal diameter, nominal length and number of the standard.

Examples

A taper pin of Type A of nominal diameter 10 mm and nominal length 50 mm shall be designated as Taper pin A 10 × 50 IS: 6688.

A split taper pin of nominal diameter 10 mm and nominal length of 60 mm shall be designated as split taper pin C 10 × 60 IS: 6688.

The nominal diameter in the case of taper pins is the diameter at the small end of the taper.

Threaded taper pins are available for:

Locking the pins and preventing loosening due to vibration (**Fig. 13.10**).

Assisting in drawing the pins out of the blind holes (**Fig. 13.11**).

Threaded taper pins with internal threads are also available (**Fig. 13.12**).

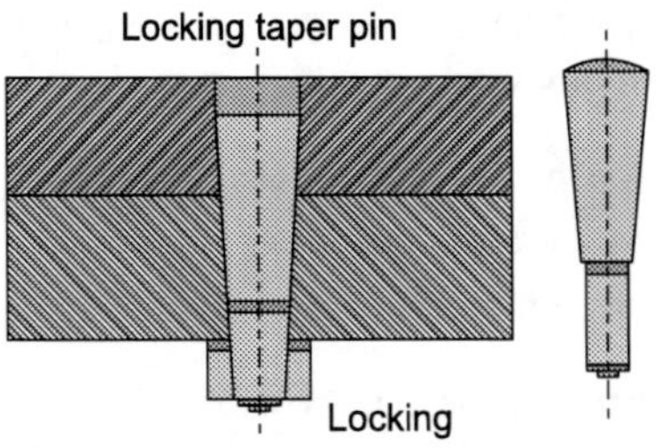

Fig. 13.10

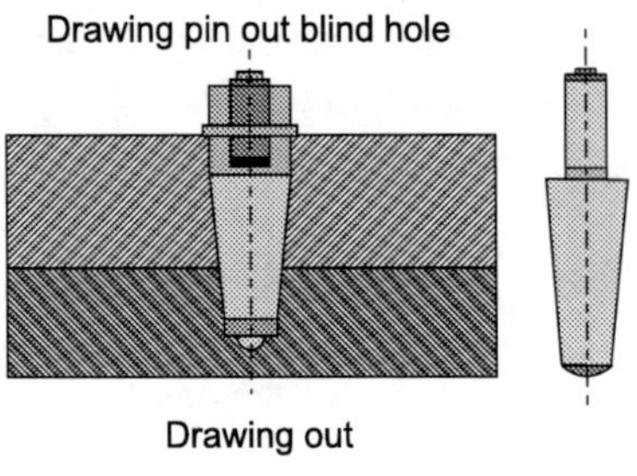

Fig. 13.11

13.1.4 Grooved Pins

These pins have three slots rolled on the outer surface. The sides of the grooves/slots bulge out. The holes in which slotted pins are used

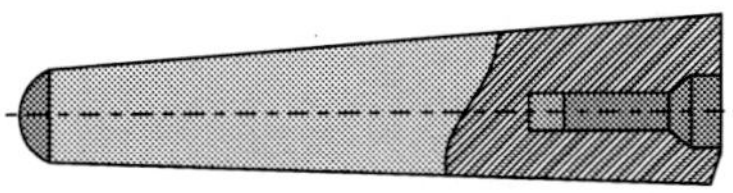

Fig. 13.12 Taper pin with internal threads

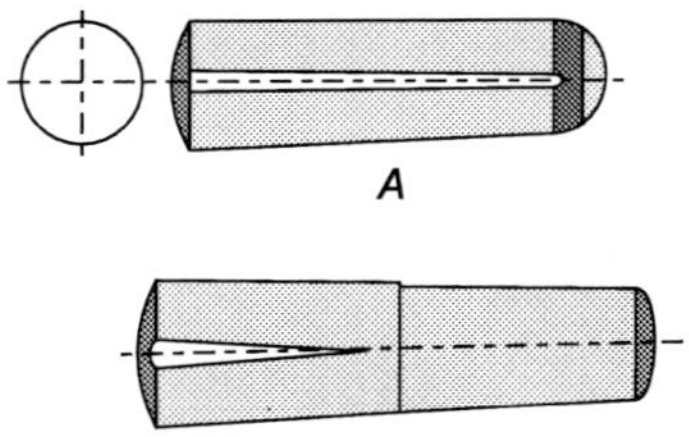

Fig. 13.13 Grooved pins

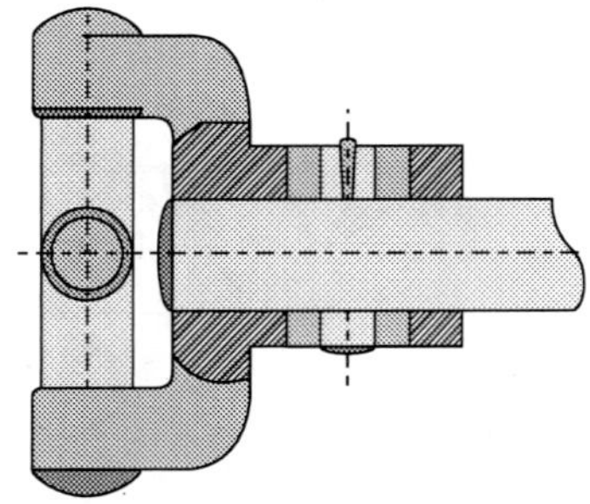

Fig. 13.14 Grooved pins

are not finished by reaming. Grooved pins are available as straight pins (**Fig. 13.13 A**), and tapered pins (**Fig. 13.13 B**). These are used in assemblies which are not dismantled frequently and where high accuracy is not required **Fig. 13.14).**

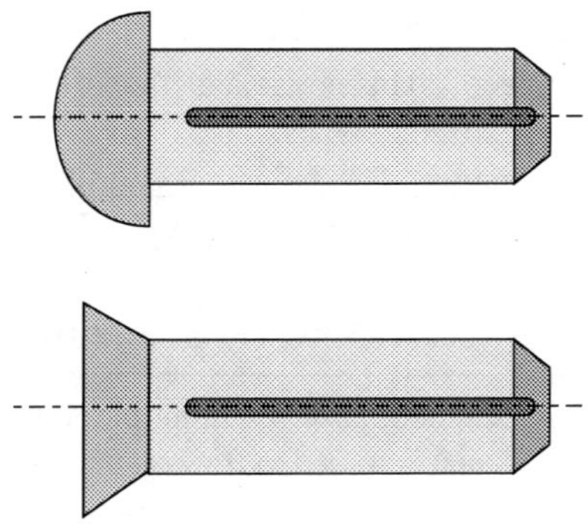

Fig. 13.15

13.1.5 Spring Pins (Fig. 13.16)

Spring pins are used for locating assemblies with wide tolerance in the corresponding holes. These pins are manufactured from flat steel bands and rolled to form a cylindrical shape. These springs will stay tight in the fitting hole because of the spring action.

This eliminates the need for drilling and reaming of the assembly together. The spring pin adjusts itself in the case of slight misalignment.

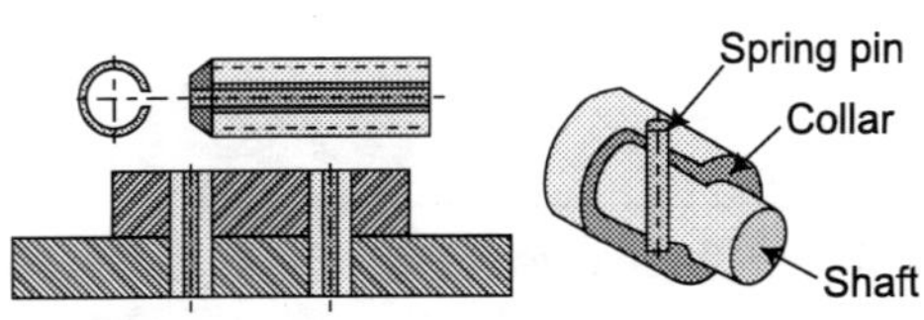

Fig. 13.16 Spring pins

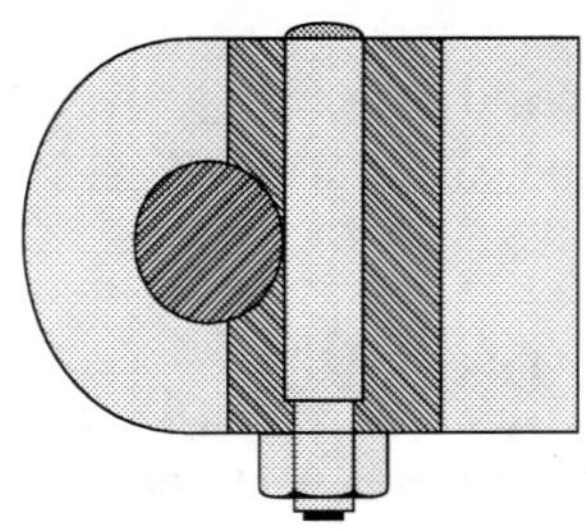

Fig. 13.17 Cotter pin

13.1.6 Taper Pins

Taper pins will position parts accurately. The component can be dismantled easily and assembled without any change in location.

The holes for fitting taper pins are finished using taper pin reamers.

Peening (Fig. 13.19): When parts are to be assembled together this is one method of assembly. Basically this is similar to riveting

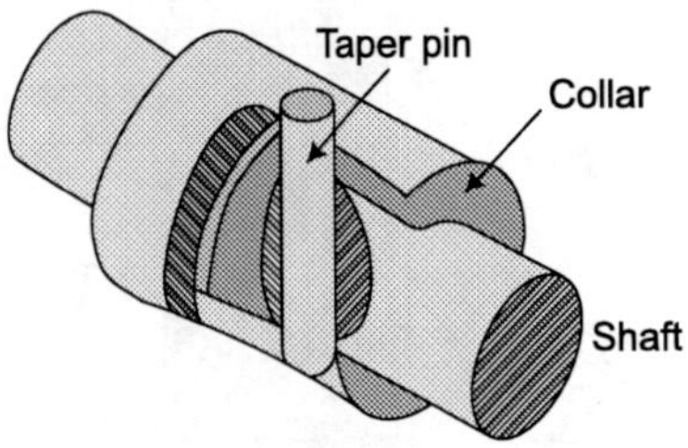

Fig. 13.18 Taper pins

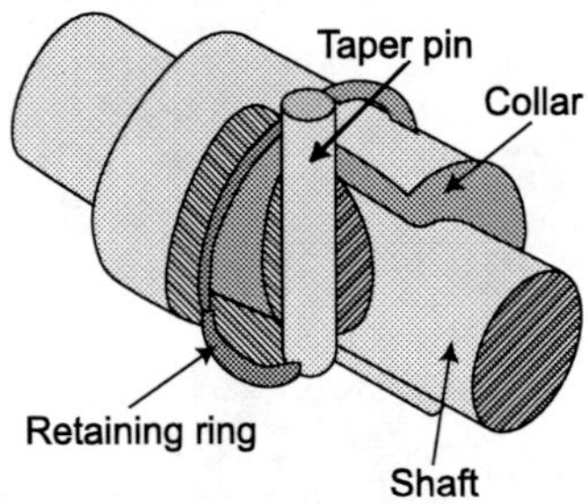

Fig. 13.18a Taper pins

Grooved pins with head are also used in assembly involving small components.

Cotter Pins (Fig. 13.17): These are fitted like dowels in reamed holes and held in position by a retaining ring.

Pinning: This is also a method of locating and securing components together. Pins are of different types.

Parallel Pins: These are fitted like dowels in reamed holes and held in position by retaining rings (**Fig. 13.18 A).**

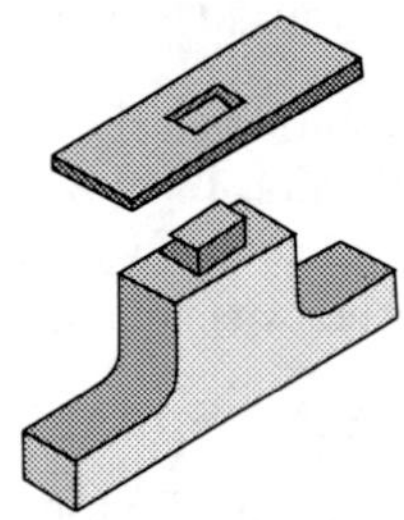

Fig. 13.19 Peening

13.2 STAKING (FIG. 13.20)

This is a method of retaining parts in an assembly in which a portion or all of a component is forced to flow on the other component. This increases the efficiency of the fit.

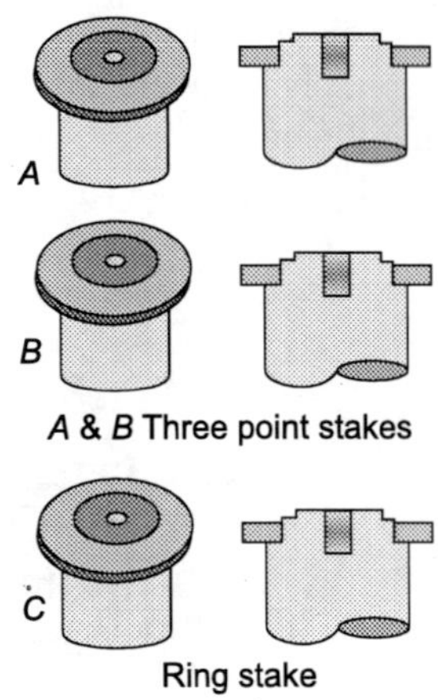

Fig. 13.20 Staking

13.3 BRAZING AND HARD SOLDERING

This is a process of joining metals by fusing a layer of nonferrous metal between the surfaces to be joined.

The alloy used for brazing is known as spelter (combination of copper and zinc).

13.4 ADHESIVES

The adhesives commonly used are epoxy adhesives. This adhesive gives a strong bond between materials to be assembled. This is not affected by moderate moisture or heat. It is usually supplied in two containers/tubes. One is resin and the other is the hardener.

Adhesives in Press Tool Making: Since the age old days the most common technology used for giving a desired shape and size to a particular component from the raw material is the chip-forming method like turning, drilling and milling.

The essential prerequisites for the most of the manufacturing processes are high quality tools, jigs, fixtures and gauges. These aids play very important and vital role in ensuring the accuracy in tool making.

Limits, Fits and Tolerances and Allowances

Chapter 14

It is a well known fact that no two things in nature can be identical, they may be found to be closely similar. This is true of production of component parts in engineering also. We know that every process is a combination of three elements, viz., man, machine and material. A change in any one of these will constitute a change in the process. All these three elements are subjected to inherent and characteristic variations. These variables result in the variation of size of components. If the process be under control, i.e., all the assignable causes of variation and controllable causes of variation have been removed or controlled, the size variation will be within reasonable limits. Thus when process is in control, most of the items will be on or about the mean value and other values will be distributed in a more or less symmetrical way about the mean value within expressible limits. It is therefore obvious that it is impossible to produce a part to an exact size and some allowance known as tolerance has to be allowed. The tolerance allowed also depends on the functional requirements which can't be sacrificed. No difficulty arises when the tolerance allowed is sufficiently greater than the process variation (i.e., capability of machine). The designer should ensure that the tolerance on any component is neither restrictive nor permissive, i.e., it should be as wide as the process demands.

Thus, in brief it can be said that due to inevitable inaccuracy of manufacturing methods it is not possible to make any part precisely to a given dimension and it can only be made to lie between two limits, maximum and minimum. The difference between these two limits is called the permissive tolerance. When tolerance distribution is on only one side of basic size, it is referred to as unilateral tolerance. If permissible variation is both plus and minus, it is referred to as bilateral tolerance.

Generally in engineering, any component manufactured is required to fit or match with some other component. The correct and prolonged functioning of the two components in match depends upon the correct size relationships between the two, i.e., the parts must fit with each other in a desired way, e.g., if a shaft is to rotate in a hole, there must be enough clearance between the shaft and hole to allow the oil film to be maintained for lubrication. Similarly, if the shaft is to be held tightly in the hole then there must be enough of interference between the shaft and the hole.

The desired condition of match between the shaft and the hole could be obtained by specifying a definite size to one and varying the other. But actually, this is not possible to make a perfect size and if at all, it is made there are no means to measure it accurately and economically during machining.

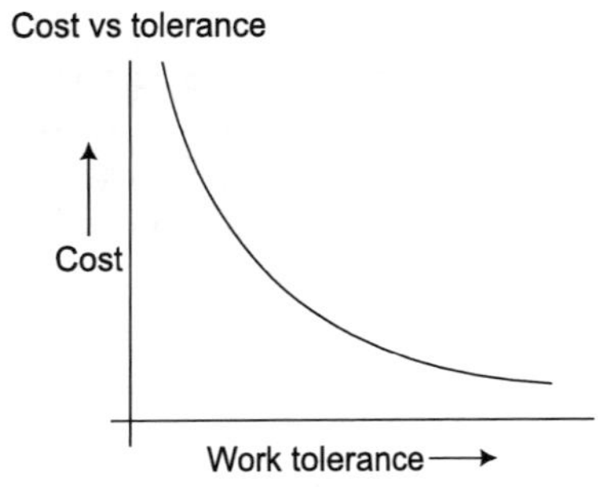

Fig. 14.1

Also it is important to note that the cost of production keeps on increasing tremendously for very precise tolerance (**Fig. 14.1**). As the tolerance approaches zero, the task of achieving it becomes enormous and finally impossible. At the other end, fabrication cost decreases with the relaxing of tolerance, but a value is soon reached beyond which little manufacturing relief is gained. It is difficult to assign any typical numerical value to tolerance and fabrication cost because of the interdependence of tolerance upon the dimension. From above it would be clear that magnitude of tolerance must be set so as to permit as low manufacturing cost as possible and still achieve the necessary functional performance.

In the early days the major of the components were actually mated together, their dimensions being adjusted until the required type of fit was obtained. But the interchangeable production and continuous assembly require some standard procedure to be followed. In order to obtain various fits, it is possible to vary the hole sizes and shaft sizes. Thus, by variation of hole and shaft sizes, innumerable types of fits are possible. All this has therefore been standardized and the primary aim of any general system of standard limits and fits should be to provide guidance to the users in selecting basic functional clearances and interferences for a given application or type of fit, and in providing tolerances which provide a reasonable and economical balance between fits, consistency and cost.

14.1 TOLERANCE

As already indicated tolerance is the magnitude of permissible variation of a dimension or other measured or control criterion from the specified value. Tolerances have to be allowed because of the inevitable human failings and machine limitations which prevent ideal achievements during fabrication. Due to its inevitability, tolerances constitute an engineering legality for deviation from the ideal value, and like any other legal matter, formulation of tolerances must also be given due consideration, and much thought and planning should go into it. The various factors affecting the choice of tolerances should be given due consideration, as the setting of tolerances is not an arbitrary matter. Though functional requirement is the primary consideration, there are other factors like standardization, methodisation, manufacturing needs, etc., which also influence bearing on the choice of tolerances and the primary consideration of functional requirement sometimes requires compromise with these factors.

The primary purpose of tolerances is to permit variation in dimensions without degradation of the performance beyond the limits established by the specification of the design. Where high performance is the criterion, the functional requirements will be the dominating factor in setting tolerances. However, where the functional performance provides some latitude, then tolerance choice may be influenced and determined by factors like standardization, methods of tooling, and available manufacturing equipment.

The numerical values of tolerances may range across the entire spectrum of measurements and if in all the cases, the functional requirement is taken as the only criterion to decide the value of tolerance. Then there may be serious disadvantages like too many different tolerances which means excessive amounts of calculations and studies to establish tolerance values, excessive amount of special tooling, complications in inspection, etc. If a limited number of standard tolerances are established, and the tolerances are chosen from these so that these are slightly closer than the function dictates, then we get the advantages of fewer varieties of tooling, few calculations and increased unit quantities because of repeated uses of the same designs.

14.2 INTERCHANGEABILITY

In earlier times production used to be confined to small number of units and the same operator could adjust the mating components to obtain the desired fit. With time the concept of manufacturing techniques kept on changing and today the same operator is no more responsible for manufacture and assembly too. With economy oriented approach mass production techniques were inevitable, that led to breaking up of a complete process into several smaller activities and this led to specialization. As a result various mating components will come from several shops; even a small component would undergo production on several machines. Under such conditions, it becomes absolutely essential to have strict control over the dimensions of portions which have to match with other parts. Any one component selected at random should assemble correctly with any other mating component that too selected at random. When a system of this kind is ensured it is known as interchangeable system. Interchangeability ensures increased output with reduced production cost.

In interchangeable system, as an operator is concerned only with a limited portion of overall work, he can easily specialize himself in that work and give best results leading to superior quality. He need not waste his skill in fitting the components by hit and trial and assembly time is reduced considerably. In the case of big assemblies, several units to manufacture individual parts can be located in different parts depending on the availability of specialized labour, raw material, power, water and other facilities and fine assembly of all individual components manufactured in several units can be done at one place. The replacement of worn or defective parts and repairs is rendered very easy and the cost of maintenance is very much reduced and shut down time also reduced to minimum.

Interchangeability is possible only when certain standards are strictly followed. Universal interchangeability (i.e., parts drawn from any two altogether different manufacturing sources for mating purposes) is desirable and for this it is essential that common standards be followed by all, and all standards used by various manufacturing units should be traceable to a single source, i.e., international standards. When all parts to be assembled are made in the same manufacturing unit, then local standards may be followed (condition being known as local interchangeability) but for reasons of obtaining spares from any other source it is again desirable that these local standards be also traceable to international standards.

The required fit in an assembly can be obtained in three ways:

1. Universal or full interchangeability
2. Selective assembly.
3. Standard

14.2.1 Universal or Full Interchangeability

Full interchangeability means that any component will match with any other mating component without classifying manufactured components in subgroup or without carrying out any minor alterations for mating purposes. This type of interchangeability is not must for interchangeable production and many times not feasible also as it requires machines capable of maintaining high process capability and very high accuracy, and very close supervision on production from time to time (process capability of a machine is defined as its $\pm 3\sigma$ spread of dimensions of components produced by it. If a plot is drawn of the actual dimensions of the similar components produced by a machine, it is found to follow natural law of distribution, i.e., having meant of all the components at central value with a spread of $\pm 3\sigma$ value, being known as standard deviation, and $\pm 3\sigma$ as the process capability of machine). For full assembly it is essential that only such machines be selected for manufacturing whose process capability is equal to or less than the manufacturing tolerance allowed for that part. Only then every component will be within desired tolerance and capable of mating with any other mating component.

14.2.2 Selective Assembly

Today the consumer not only wants quality and precision trouble-free products but also wants them at attractive prices. This has become possible only by adopting automatic gauging for selective assembly whereby parts: manufactured to rather wider tolerances fit and function as though they were slowly manufactured in a precision laboratory in very close tolerances. This is a concept which does away with old idea of inspection in which part is identified as "good" or "bad"; good part being used for assembly and bad used to be scraped.

In selective assembly the components produced by a machine are classified into several groups according to size. This is done both for hole and shaft and then the corresponding groups will match properly.

If some parts (shafts and holes) to be assembled are manufactured to normal tolerances of 0.01 mm (and, both are within the curve of normal distribution) an automatic gauge can segregate them into ten different groups with a 0.001 mm limit for selective assembly of the individual parts.

Thus parts with tolerances of 0.001 mm are obtained (due to segregation) and both the conditions of high quality and low cost can be served by selective assembly technique. However, it is very important that the two component parts to be fitted together must he kept within the normal distribution, i.e., the central or mean value should be at desired calculated value and the process capability of two machines producing shafts and holes must be identical, otherwise for some components the mating components will not be available. If not so, then other techniques need to be adopted which are described below.

Figure 14.2 shows a case in which the process capability of both shaft and hole producing machines is same but tolerances on parts are desired as one-tenth of process capability of machines. In such a case the parts are segregated by automatic inspection into ten groups and parts in shaft region S_1 are matched with parts in hole region H1, S2 with H2 and so on. This results in matching of parts having tolerances $1/10^{th}$ of machine capability. In this case as the process capability of both machines is same, equal number of parts are available in each segregated zone and no wastage will be there.

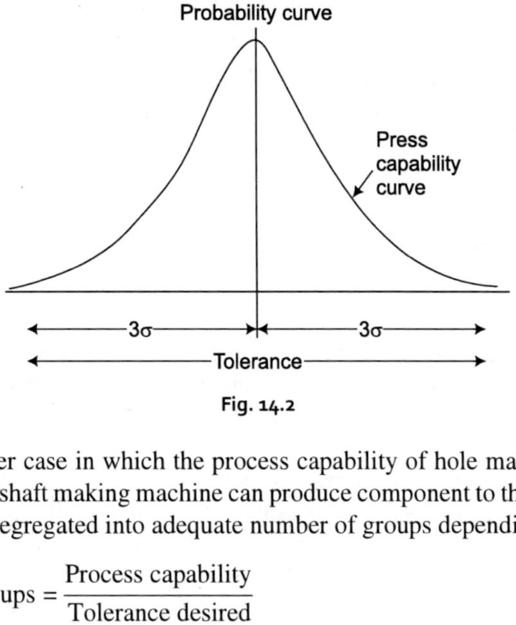

Fig. 14.2

Figure 14.3 shows another case in which the process capability of hole making machine is much wider than the tolerance of part but shaft making machine can produce component to the desired tolerance. In such a case the parts with hole are segregated into adequate number of groups depending on the desired tolerance.

$$\text{No. of groups} = \frac{\text{Process capability}}{\text{Tolerance desired}}$$

This curve when broken into several groups enables to determine the number of components likely to be produced in each group.

Although modern machine tools are capable of machining to high levels of accuracy, their ability to duplicate a specific dimension is still limited. Temperature changes, tool wear, deflections and vibrations of the machine and the work, in addition to human errors, are inherent in the machining of any part. The influence of these factors precludes the exact duplication of size. Most mechanisms will function satisfactorily when the tolerances of the individual components are defined.

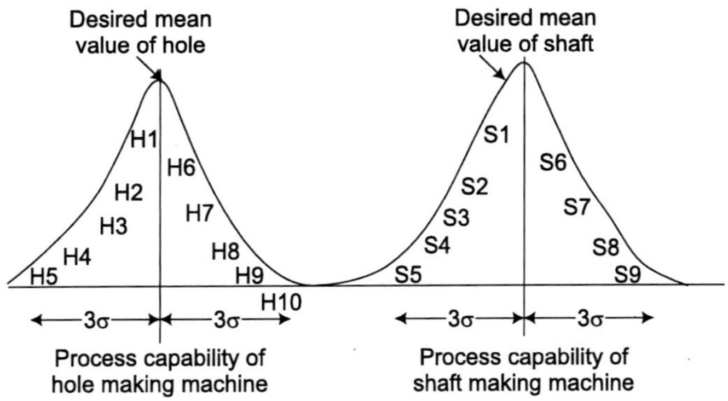

Fig. 14.3 Process capability

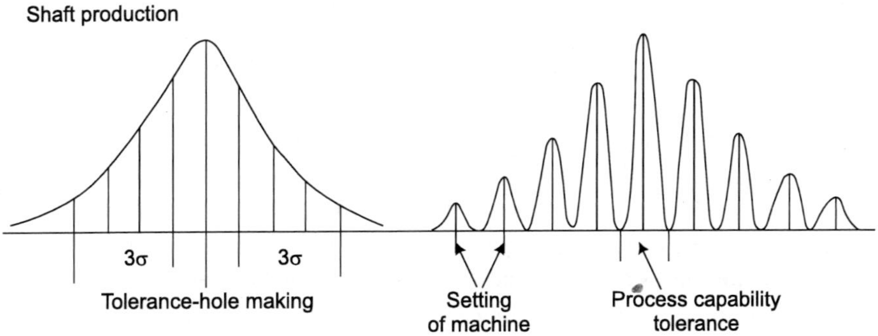

Fig. 14.4

Preferred limits and fits for cylindrical parts consists of a series of preferred basic sizes for tolerances and allowances, a system of 13 tolerance grades, tables of recommended fits, and definitions. This standard is in accord with recommendations of B.S.I. In general, and so far as applicable, it is in substantial agreement with the system developed under the auspices of the International Organization for Standardization (ISO).

Essential data are reproduced in this section for helping to improve parts interchangeability, and to simplify products, practices, and facilities. Many factors such as diameter, length of engagement, bearing load, speed, lubrication, temperature, humidity, materials, and surface roughness must be considered in selecting tolerances and allowances for a particular application.

14.3 INDIAN STANDARD LIMITS AND FITS (IS: 919-1961)

Definitions: In general, the definitions of the terms listed apply to fits between plain (non-threaded) cylindrical parts.

Nominal size is the designation which is used for the purpose of general identification. Dimension is a geometrical characteristic, e.g., diameter, length, angle, or centre distance. Size is a designation of magnitude. When a value is assigned to a dimension, it is referred to hereinafter as the size of that dimension.

Note: It is recognized that the words "dimension" and "size" are both used at times to convey the meaning of magnitude.

Allowance is an intentional difference between the maximum material limits of mating parts (see definition of fit). It is a minimum clearance (positive allowance) or maximum interference (negative allowance) between the mating parts.

Tolerance is the total permissible variation of a size, i.e., the difference between the limits of size.

Basic size is that size from which the limits of size are derived by the application of allowances and tolerances.

Design size is that size from which the limits of size are derived by the application of tolerances. When there is no allowance, the design size is the same as the basic size.

Actual size is a measured size.

Limits of size are the applicable maximum and minimum sizes.

Maximum material limit is the maximum limit of size of an external dimension or the minimum limit of size of an internal dimension.

Minimum material limit is the minimum limit of size of an external dimension or the maximum limit of size of an internal dimension.

Tolerance limit is the variation, positive or negative, by which a size is permitted to depart from the design size.

Unilateral tolerance is a tolerance in which variation is permitted only in one direction from the design size.

Bilateral tolerance is a tolerance in which variation is permitted in both directions from the design size.

Fit is the general term used to signify the range of tightness which may result from the application of a specific combination of allowances and tolerances in the design of mating parts.

Actual fit between two mating parts is the relationship existing between them with respect to the amount of clearance or interference which is present when being assembled.

Note: Fits are of three general types: clearance, transition, and interference.

Clearance fit is one having limits of size so prescribed that a clearance always results when mating parts are assembled.

Interference fit is one having limits of size so prescribed that an interference always results when mating parts are assembled.

Transition fit is one having limits of size so prescribed that either a clearance or interference may result when mating parts are assembled.

Basic-hole system is the system of fits in which the design size of the hole is the basic size and the allowance is applied to the shaft.

Basic-shaft system is the system of fits in which the design size of the shaft is the basic size and the allowance is applied to the hole.

Tolerance systems

Unilateral Tolerance: In the unilateral system, variation on each part from the design size is in only one direction and may be either plus or minus, depending on whether the tolerance applies to a hole or a shaft. In this system, the minimum (low) limit of the hole and the maximum (high) limit of the shaft are considered the design sizes of the respective parts, from which sizes the tolerances are applied.

The unilateral system permits changing tolerances on holes or shafts without impairing the essential character of the respective fits. This is an advantage in engineering a new product. It may be desirable on initial tooling to adopt close tolerances H1 and S1 for the mating parts.

However, after manufacturing technique has been developed and the product proved in the field, it may be found possible to increase the tolerances of the mating parts to H2 and S2 if the increased maximum looseness is not objectionable to good performance. The adoption of larger manufacturing tolerances will realize, in most cases, lower manufacturing costs.

Bilateral Tolerance System: The tolerance is applied in both directions from design size on holes H3 and H 4 together with shafts S3 and S4 It is not possible to change the tolerances on holes and shafts of this set of examples and retain the same class of fit

Application of the unilateral system of tolerances can be found in most standard systems of fits.

When the conception of tolerances was originally developed, the use of bilateral tolerances was the result of the designer's aiming at a definite size and specifying a maximum permissible deviation in both directions from that size. The study of a tolerance system, as applied to the manufacture of interchangeable parts and mass production, working between two limits, found the unilateral system to be preferable.

All national standard systems developed during the past 40 years and the ISO systems are based exclusively on unilateral tolerances.

Basic-hole and Basic-shaft Systems: The same fits can be obtained in either the basic hole or the basic shaft system and the designer must decide on the adoption of either system to secure general interchangeability. A shaft dimensioned to give a certain class fit with a hole in the basic hole system differs in size from the shaft giving the same fit in the basic shaft system. In The shaft for a loose fit in diagram 'A' is smaller than the nominal diameter (size of basic hole), while in diagram 'B' the size of the corresponding shaft is equal to the nominal diameter **(Fig. 14.5).**

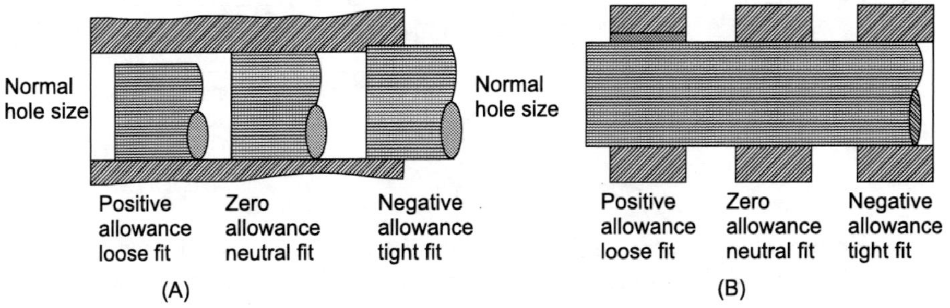

Fig. 14.5 (A) Basic hole system (B) Basic shaft system

Application of either system depends upon many conditions, including the product to be manufactured, the method of manufacturing or processing, and the state or condition of raw stock prior to processing. In the basic hole system, shaft members made to the specifications of one class of fit may be mated with hole members made to the specifications of another class without destroying the essential character of the fit. Similarly, in the basic-shaft system, hole members made to specifications of one class of fit may be mated with shaft members made to the specifications of another class without destroying the essential character of the fit.

In manufacturing practice, the basic hole system is most commonly used. The main reason for its preference is that the hole cannot be varied conveniently because of fixed sizes of drills, reamers, broaches, plug gages, etc., whereas the shaft size is generally controlled with adjustable cutting tools or grinding wheels and can be measured with adjustable comparators, limit snap gages, etc. The standardization of non-adjustable tools provides economies in production. Also, a change in a shaft diameter, after production has started, causes little or no trouble since non-adjustable tools are not affected.

The basic shaft system is preferred in some branches of industry that use a large volume of cold-finished shafting. This permits the use of stock without machining it and the mounting of the same shaft, without stepping it, of several parts with different fits.

14.4 GEOMETRICAL TOLERANCES

The possibility of the free assembly of two components as shown in (**Fig. 14.6**) depends upon the combined effect of the actual sizes, and the error of geometric features. These features are straightness, flatness, parallelism, squareness, angular displacement, symmetry, concentricity, roundness, and eccentricity. They are defined as geometric tolerances or tolerances of form and state how far actual surfaces are permitted to vary from the perfect geometry implied by drawings. For example, on parts longer than 25 mm, specifications relative to the geometry of an object can be based on the permissible variation per mm for flatness, straightness, parallelism, or squareness.

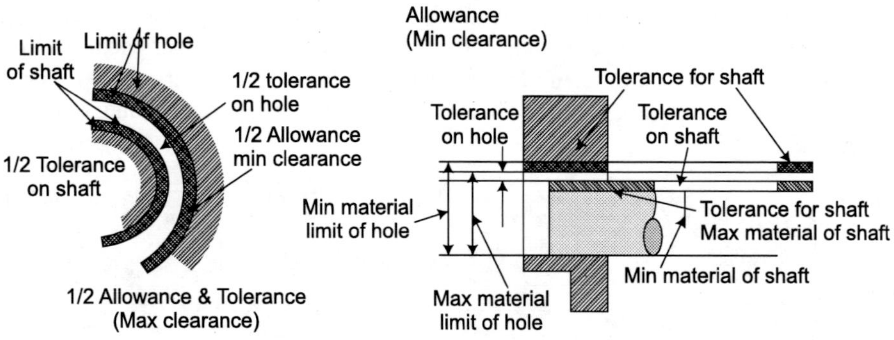

Fig. 14.6

Shapes or forms into which material is fabricated may be defined by use of geometric terms, such as plane (surface), a cylinder, a cone, a square, or a hexagon. The geometric definition assumes a perfect form, but because a perfect form cannot be produced, variations must be restricted if a specific quality is to be maintained. When a feature has errors of form or position, its size is virtually altered; the size of a hole is reduced and the size of a shaft is increased. Therefore, care should be exercised in selecting fits so that consideration is given to geometric tolerances. These tolerances should be specified where appropriate, especially where close fits are involved, for all requirements critical to functioning and interchangeability.

Methods for specifying geometrical tolerances are based upon the variation of conditions found within the feature size of a part, regardless of part size, or the maximum or minimum material conditions. Each company must determine their basis for interpretation of the tolerances relative to the need, application, and operating conditions with the company.

Acceptability: A part shall be acceptable if its actual size does not exceed the limits of size specified in numerical values on the drawing or in writing. It does not meet dimensional specification if its actual size exceeds those limits.

Reference Temperature: The actual size of a part shall be considered to be that size which the part has at the international standard reference temperature of 68° F (20°C).

14.5 CLASSES OF STANDARD FITS

Standard fits are designated by means of symbols to facilitate reference to classes of fit for educational purposes. These symbols are not intended to be shown on manufacturing drawings; instead, sizes should be specified on drawings.

The letter symbols used are as follows:

- RC running or sliding clearance fit
- LC location clearance fit
- LT transition clearance or interference fit
- LN location interference fit
- FN force or shrink fit

These letter symbols are used in conjunction with numbers representing the class of fit; thus "FN 4" represents a class 4 force fit.

Each of these symbols (two letters and a number) represents a complete fit, for which the minimum and maximum clearance or interference, and the limits of size for the mating parts, are given directly in the tables of standard fits.

Standard Fits: Standard fits have been developed to give a series of standard types and classes of fits on a unilateral-hole basis; such that the fit produced by mating parts in any one class will produce approximately similar performance throughout the range of sizes. These prescribe the fit for any given size, or type of fit; they also prescribe the standard limits for the mating parts which will produce the fit.

In developing standard fits it has been recognized that any fit will usually be required to perform one of the three functions, as indicated by the three general types of fits: running fits, location fits, and force fits.

14.6 SELECTION OF FITS

In selecting limits of size for any application, the type of fit must be determined first, based upon the use or service required from the equipment being designed; then the limits of size of the mating parts can be established, to ensure that the desired fit will be produced.

It does not necessarily embrace other factors which affect the fit of mating parts such as straightness, ovalness, lobbing, and bow. Some applications are best served by considering the dimensional tolerance as an envelope within which all deviations are to be maintained. Other applications require closer control of some characteristics such as straightness or roundness. In such instances, all factors must be carefully weighed and adequate provision to guard against them must be made when establishing production processes and tooling. Proper inspection techniques should be utilized to detect their presence and to permit corrective action to be taken promptly.

Running and Sliding Fits: Running and sliding fits, for which limits of clearance are intended to provide a similar running performance, with suitable lubrication allowance, throughout the range of sizes. The clearances for the first two classes, used chiefly as slide fits, increase more slowly with diameter than the other classes, so that accurate location is maintained even at the expense of free relative motion.

These fits may be described briefly as follows:

Close sliding fits are intended for the accurate location of parts which must assemble without perceptible play.

Sliding fits are intended for accurate location, but with greater maximum clearance than close sliding fits. Parts made to this fit move and turn easily but are not intended to run freely, and in the larger sizes may seize with small temperature changes.

Precision running fits are about the closest fits which can be expected to run freely; they are intended for precision work at slow speeds and light journal pressures but are suitable where appreciable temperature differences are likely to be encountered.

Close running fits are intended chiefly for running on accurate machinery with moderate surface speeds and journal pressure, where accurate location and minimum play are desired.

Medium running fit are intended for higher running speeds, or heavy journal pressures, or both.

Free running fits are intended for use, where accuracy is not essential, or where large temperature variations are likely to be encountered, or under both these conditions.

Location fits are intended for use where wide commercial tolerances may be necessary, together with an allowance, on the external member.

Location fits are intended to determine only the location of the mating parts; they may provide rigid or accurate location, with interferences fit or provide some freedom of location, as with clearance fits. Accordingly, location fits are divided into three groups: clearance fits, transition fits, and interference fits.

These types of fits are described as follows:

Location clearance fits are intended for parts which are normally stationary which can be freely assembled or disassembled. They run from snug fits for parts requiring accuracy of location, through the medium clearance fits for parts such as ball races, and housing, to the loose fastener fits where freedom of assembly is of prime importance.

Location transition fits are compromise between clearance and interference fits, for application where accuracy of location is important but either a small amount of clearance or interference is permissible.

Location interferences fits are used where accuracy of location is of prime importance and for parts requiring rigidity and alignment with no special requirements for bore pressure. Such fits are not intended for parts designed to transmit frictional loads for one part to another by virtue of tightness of fit, as these conditions are covered by force fits.

Force fits or **shrink fits** constitute a special type of interference fit, abormally characterized by interference of constant bore pressure throughout the range sizes. The interference varies almost directly with diameter, and the difference between its minimum and maximum value is small, to maintain the resulting pressures within resonable limits.

These fits are described briefly as follows:

Light drive fits require light assembly pressures, and produce more or less permanent assemblies. They are suitable for thin sections or long fits, or in cast iron external members.

Medium drive fits are suitable for ordinary steel parts, or for shrink fits on light sections. They are about the tightest fits that can be used with high-grade cast iron external members.

Heavy drive fits are suitable for heavier steel parts or for shrink fits in medium sections.

Force fits are suitable for parts which can be highly stresses, or for shrink fits where heavy pressing forces required are impractical.

14.7 MODIFIED STANDARD FITS

Fits having the same limits of clearance or interference may sometimes have to be produced by using holes or shafts other than those recommended This may be accomplished by using one of the following:

1. **Bilateral holes** will result in nonstandard holes and shafts.
2. **A basic-shaft system** will result in nonstandard holes and shafts.

Bilateral-hole Fits: The common case is where holes are produced with fixed tools, such as drills or reamers; to provide a longer wear life for such tools a bilateral tolerance is desired.

Basic-shaft Fits: For these fits the maximum size of the shaft is basic. The limits of clearance or interference are identical with those hole systems, for the corresponding fits. The symbols used for these fits are identical with those used for standard fits produced on a basic-shaft basis.

14.8 GRADES OF FITS

The grade of a fit is determined by the permissible variation in its relative looseness or tightness. This variation depends on the tolerances given on the mating parts, i.e., on the grade of workmanship or, briefly, the grade of the individual parts.

The concepts "class" and "grade" are illustrated by **(Fig. 14.7)**. The four tolerances shown at the left **(Fig. 14.7A)** apply to basic holes of the same basic diameter and vary from 0.0007 to 0.0002 inch. Since all the holes are basic, they are of the same class. However, their grades are different, improving from left to right. A similar series of tolerances is shown for basic shafts of the same class but different grades.

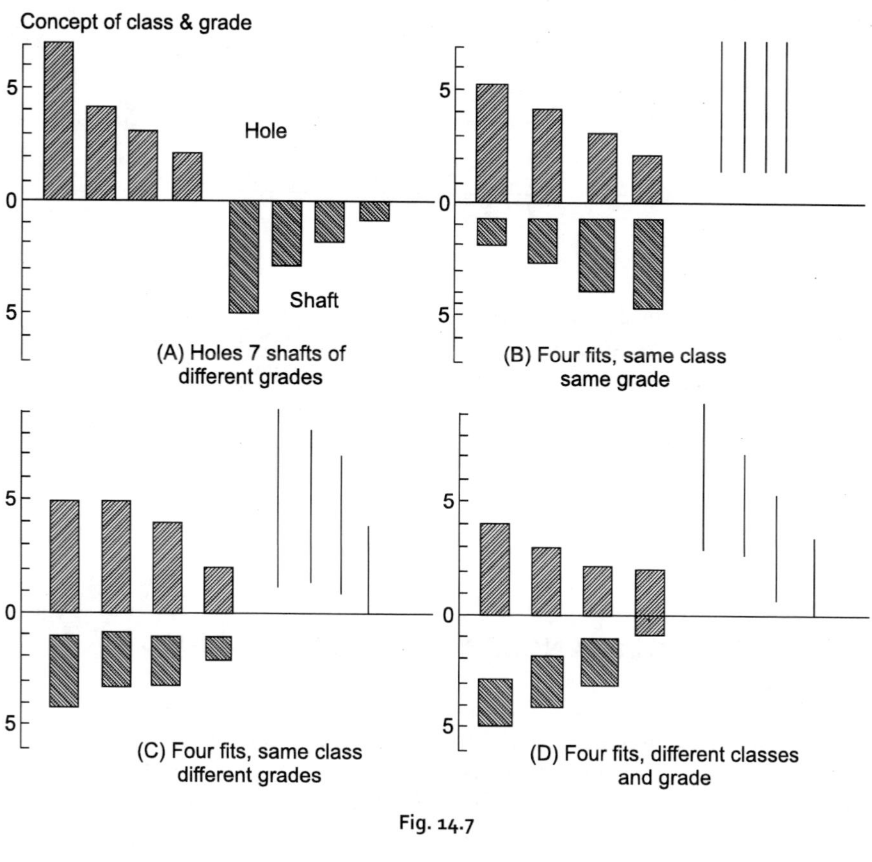

Fig. 14.7

Figure 14.7B shows four fits having the same allowance of 0.0001 inch. Therefore, these fits are of the same class. They have the same maximum clearance (0.0007 in) and, hence, the same range of variation in clearance (maximum clearance minus minimum clearance). This range is represented for each fit on the right side by the length of a heavy line. Since the four ranges are equal, the four fits are of the same grade.

The four fits **(Fig. 14.7 C)** have the same allowance of 0.003 mm and, therefore, are of the same class. The maximum clearances are different (0.023, 0.0203, 0.018, and 0.01016 mm, respectively), and also the ranges of variations in clearance, as indicated again by the four heavy lines on the right. These show that the grade of the fits improves from left to right.

Figure 14.7 D represents four fits which are different in regard to class, as well as to grade.

From left to right, the allowance decreases from 0.0076 to zero, and the grade of the fit Improves.

The definition may be repressnted as in **Fig. 14.8.**

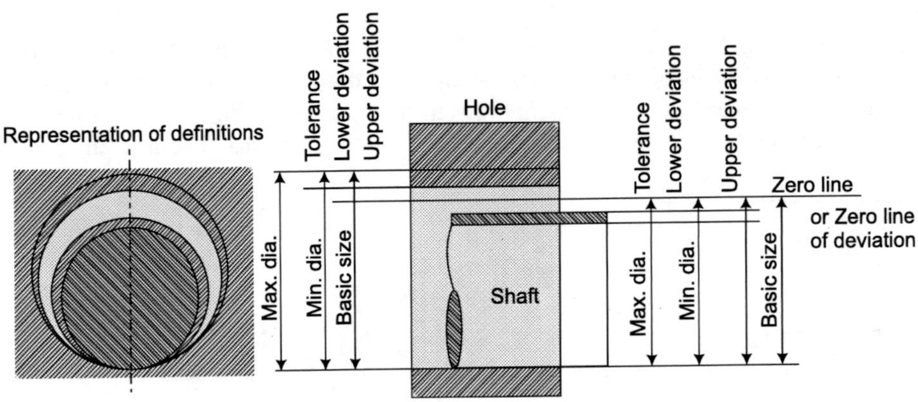

Fig. 14.8

Figure 19.8 which illustrates the definitions, is replaced by a schematic diagram similar to **Fig. 19.9** for the sake of simplicity. In this simplified schematic, the axis of the part, which is not represented, always lies, by convention, below the diagram (in the example illustrated, the two deviations of the shaft are negative and those of the hole positive).

Two of the most commonly used methods of applying fits are the hole basis system and the shaft system **(Fig. 19.9).**

The schematic diagram of tolerance zone in various cases is shown in **Fig. 14.10.**

Two commonly used methods

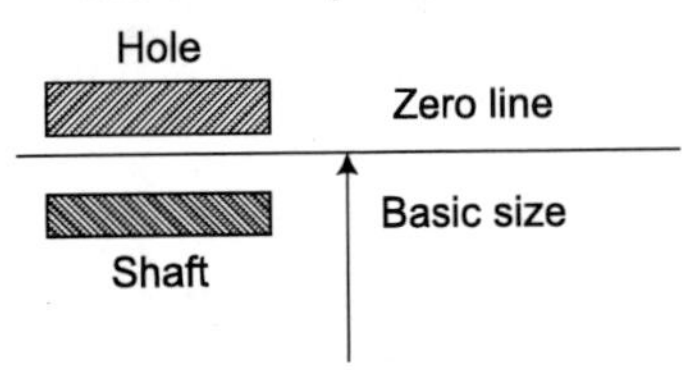

Fig. 14.9

14.9 SYMBOLS FOR TOLERANCES AND DEVIATIONS AND SYMBOLS FOR FITS

In order to satisfy the usual requirements both of individual parts and of fits, the system provides, for any given basic size, a whole range of tolerances together with a whole range of deviations defining the position of these tolerances with respect to the line of zero deviation, called the zero line

The tolerance, the value of which is function of the basic size, is designated by a number symbols, called the grade.

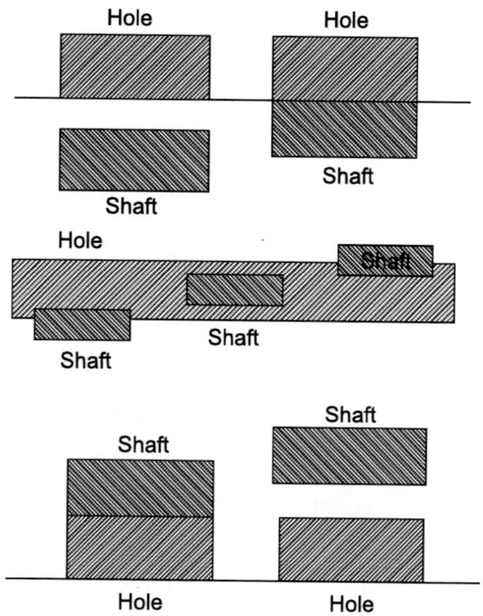

Fig. 14.10 Schematic representation of transition fit

The position of the tolerance zone with respect to the zero line, which is a function of the basic size, is indicated by a letter symbol (in some cases, two letters), a capital letter for holes, a small letter for shafts **(Fig. 14.10)**.

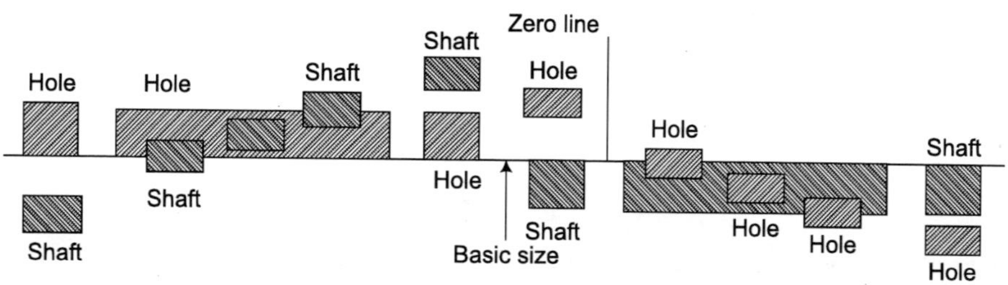

Fig. 14.11 Example from hole basis system

Fit System

Two of the most commonly used methods of applying are the hole-basis system and shaft-basis system **(Fig. 14.12).**

Examples: 4 S g7

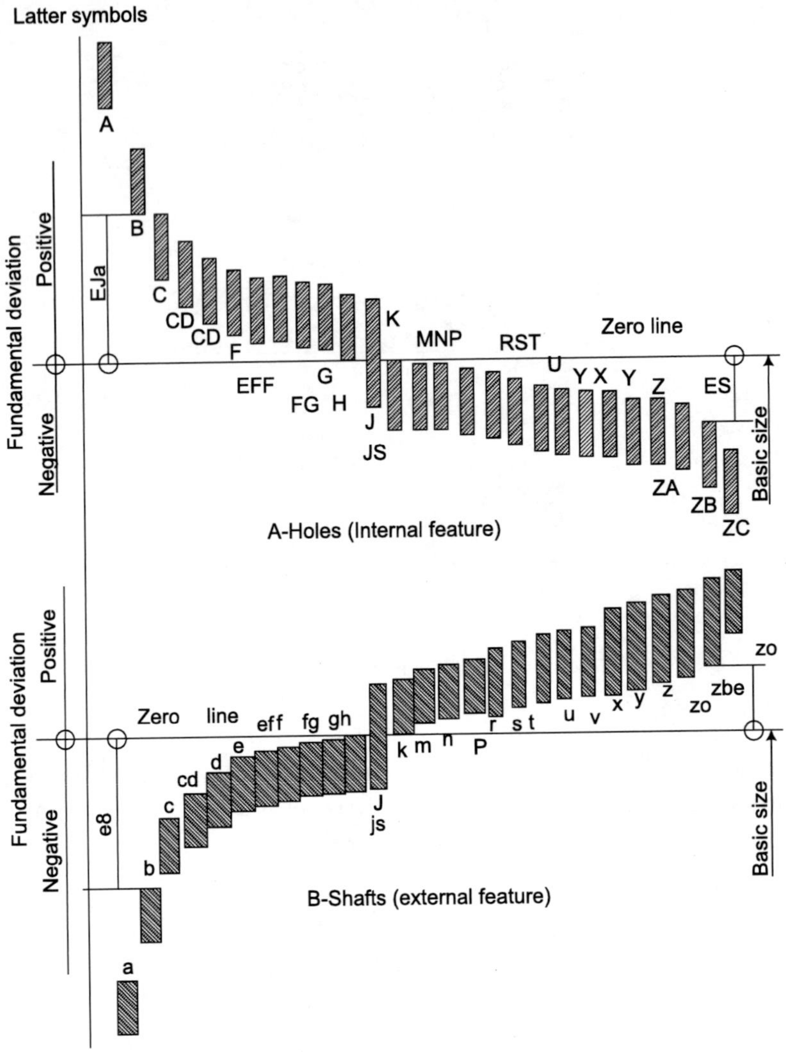

Fig. 14.12

A fit is indicated by the basic size common to both components, followed by symbol corresponding to each component the hole being quoted first.

Examples: 45 H8/g7 (possibly 45 H8-g7 or 45 H8/g7).

14.10 GUIDE FOR SELECTION OF FITS

This is the recommendation for the selection of fits in engineering problems concerned with the mating of a shaft and a hole. These recommendations are also applicable to non-cylindrical fits.

This gives the representative usage of various classes and grades to fit. These examples are only of an illustrative character and they do not specify any design details.

14.11 SYSTEM OF FITS

The hole-basis system is the system of fits in which design size of the hole is the basic size and the allowance is applied to the shaft **(Fig. 14.13).** In the shaft-basis system the design size of the shaft is the basic size and the allowance is applied to the hole.

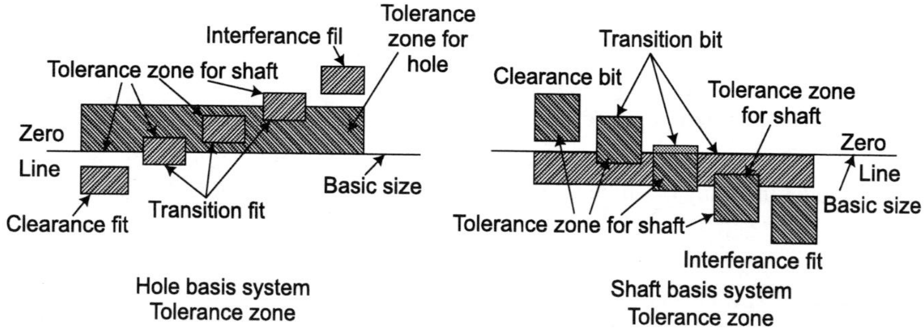

Fig. 14.13 System of fits

The application of either system depends on many conditions, such as, the nature of product, the manufacturing methods, the condition of the raw material, etc.

The hole-basis system is the extensively used system. This is because a hole is more difficult to produce than a shaft due to the fixed character of hole producing tools. The shaft basis system should only be used where it will convey unquestionable economic advantages, that is, where it is necessary to be able to mount several parts with holes having different deviations on a single shaft of drawn steel bar without machining the latter.

The designers should decide on the adoption of either system to secure general interchangeability. A shaft dimension to a certain class of fit with a hole in the hole basis system differs from the shaft giving the same fit in the shaft-basis system.

14.12 CLASSIFICATION OF FITS

The system of fits may be broadly classified as clearance fit, transition fit and interference fit.

Clearance fit results in a positive clearance over the whole range of the tolerance. Shafts a to g produce a clearance fit with the basic hole H.

Transition fit may result in either a clearance fit or interference fit depending on actual value of the individual tolerances of the mating components. Shaft, is to n produce transition fit with the basic hole H.

Interference fit results in a positive interference over the whole range of tolerance. Shafts p to produce interference fit with the basic hole H.

Tolerances and deviations for both holes and shafts offer a wide range of fits of which many of the possible combinations may not be of practical use. Majority of common engineering requirements may be satisfied on the basis of a restricted selection of tolerance grades resulting in economy and ease of standardization, yet leading to universally applicable and recommended fits. The commonly used fits are given in reference books.

14.13 CHOICE OF FITS AND TOLERANCES

Fits: Fits may be selected on the hole-basis system or the shaft-basis system. The choice of fits considerably depends on the material of mating parts, workmanship, length of engagement, bearing load, speed, type of lubrication, temperature, humidity, surface finish, etc.

Tolerances: In order to obtain the most economic manufacture consistent with satisfactory quality, it is recommended that as wide a tolerance shall be provided as is possible. In the allocation of tolerances to the hole and the shaft members of a fit, it is generally advisable to give a larger tolerance to the hole than to the shaft due to the fact that hole is more difficult to produce than a shaft. The exception is in the case of very large sizes where the effects of temperature play a large role.

The tolerances chosen shall be the largest compatible with the conditions of use the hole being the more difficult member to machine, may often be allocated a tolerance one grade coarser than that of the shaft (for example, H8-f7).

14.14 RECOMMENDATIONS

Recommendations for selection of tolerance zones for general purposes are given in Table available in reference books.

Wherever possible the tolerance zones shall be chosen from the corresponding symbols for shafts and holes as indicated. The first choice shall preferably be made from the symbols enclosed in the frames.

Dimensions for the fits chosen from the tolerance zones enclosed in the frames in Tables.

14.15 THE INDIAN STANDARD SYSTEM

The system of limits and fits recommended in **IS: 919-1963** comprises 19 grades of fundamental tolerances or, in other words, grades of accuracy of manufacture, with designations IT 01, IT 0 and IT 1 to IT 16; and 25 types of fundamental deviations indicated by letter symbols for both holes and shafts (capital letters A to Z C for holes and small letters a to z C for shafts) in diameter steps ranging from 1 to 500 mm. The 25 deviation are represented by A, B, C, D, E, F, G, H, I, J, K, M, N, 0, P, R, S, T, U, V, X, Y, Z, ZA, ZB and Zc.

Innumerable fits ranging from extreme interference to those of extreme clearance can be obtained by suitable combination of fundamental tolerances and fundamental deviations mentioned above. All but very exceptional engineering requirements are covered, from very coarse work to fine gauge manufacture. A unilateral hole basis system is recommended, but a full range of holes suitable for use on shaft basis system, unilateral or bilateral, is also included in **IS: 919-1963.** Two Indian Standards have already been published, namely, IS: 2101-1962 for sizes above 500 mm and up to 3150 mm and IS: 2709-1964

For the sake of convenience, a *basis size* is ascribed to the part and each of the two limits is defined by its deviation from this size (the magnitude and sign of the deviation is obtained by subtracting the basic size from the limit in question).

- Upper deviation of a hole ES — high
- Lower deviation of a hole EI— low
- Upper deviation of shaft es —high
- Lower deviation of shaft ei — low.

Further explanation with sketches **Fig. (14.4), (14.15) (14.16) (14.17) (14.18)**.

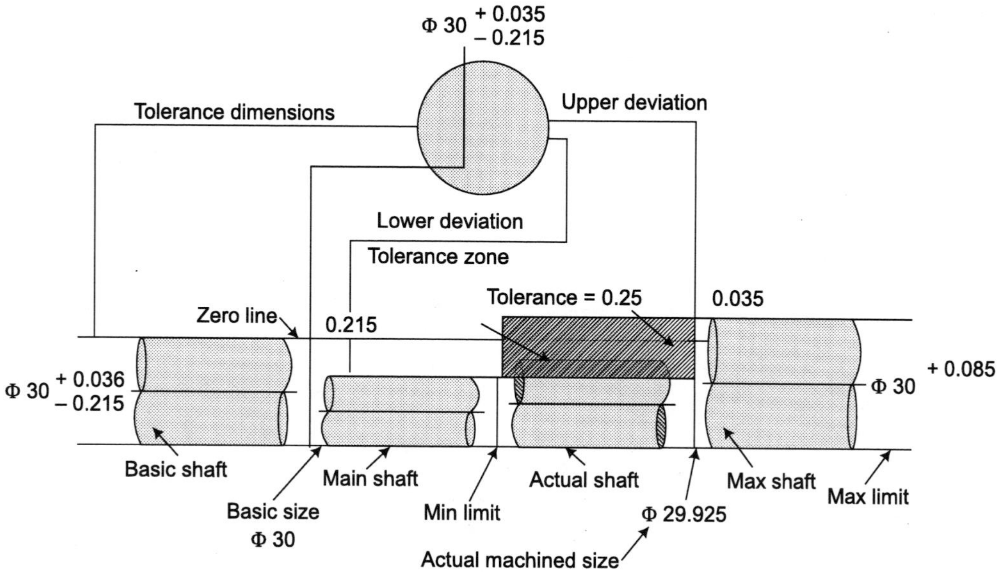

Fig. 14.14 Graphical representation of limits & tolerances

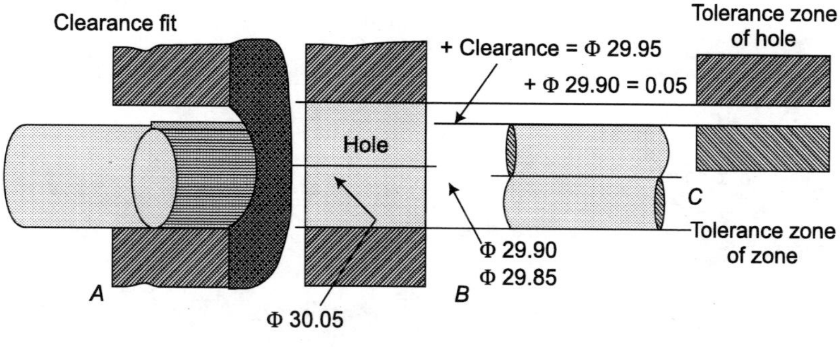

Fig. 14.15

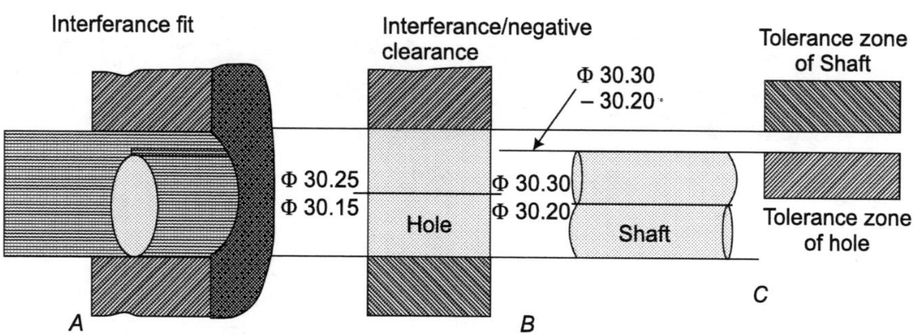

Fig. 14.16

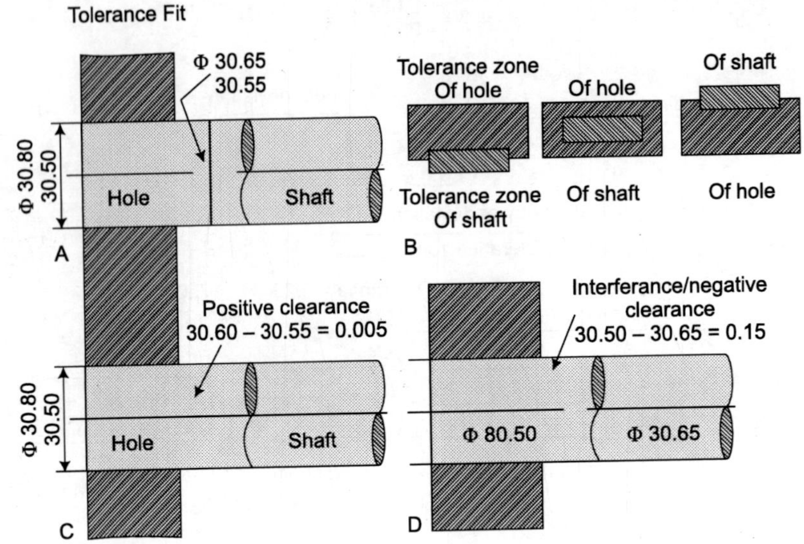

Fig. 14.17

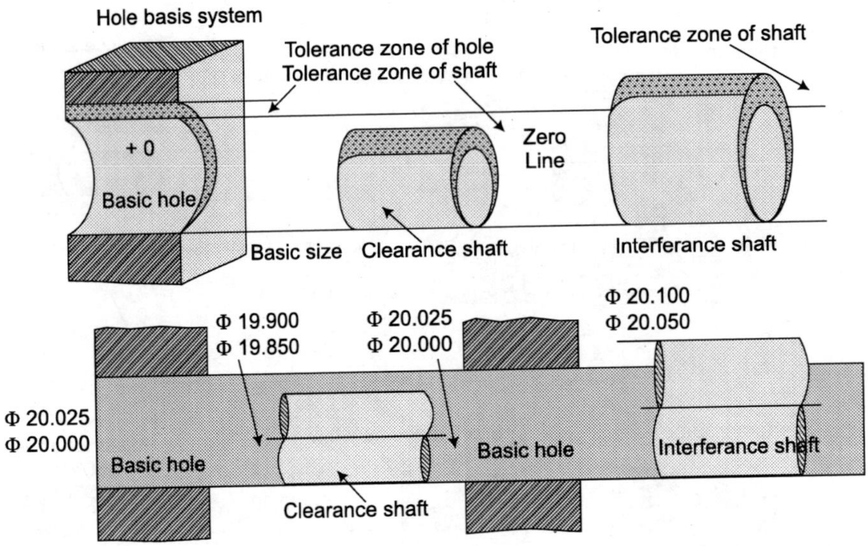

Fig. 14.18

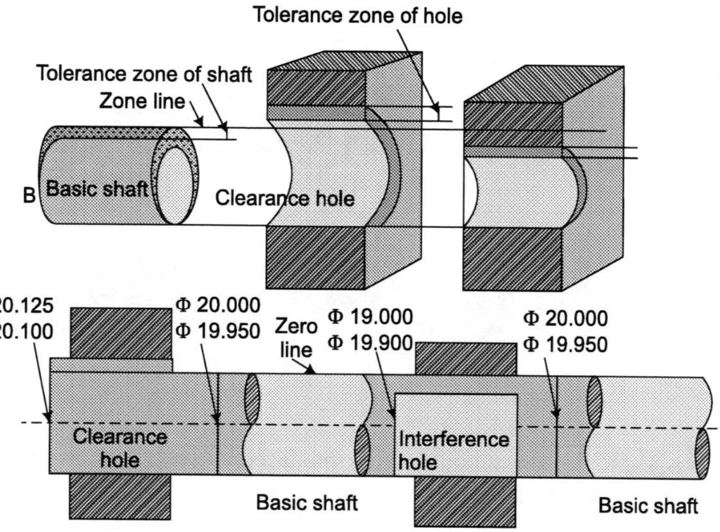

Fig. 14.19

EXERCISE

1. What do you understand by interchangeability? Why is it necessary?
2. Write definitions of the following:
   (a) High limit and low limit.
   (b) Tolerance
   (c) Allowance.
3. What is the difference between unilateral tolerance and bilateral tolerance?
4. Define a "fit".
5. What are the methods used for a "Fit"?
6. Write comments on the following:
   (a) Clearance fit
   (b) Transition fit
   (c) Interference fit.

Gauges

Chapter 15

With the help of precision measuring tools, parts produced / manufactured are inspected or tested. Measurements with these tools are no doubt accurate but some probability difference in the sizes pose difficulty in large quantity.

The technological developments have led to various techniques of manufacture. Manufacturing of parts is done anywhere nationally or internationally and used anywhere irrespective of the place of manufacture. Standards are maintained at all levels.

The standards have volunteered limits, fits and tolerances and the concept of interchangeability has come a long way in the manufacturing field. But there is a difficulty securing this quality or a lot of time is being consumed in this regard because of exact measurement by precision instruments. Therefore, in large scale production gauges are used. These gauges are made in different rates of tolerance, in different shapes and designs.

There are numerous advantages in using the gauges.

1. Saves time as a number of jobs can be checked in less time.
2. Semiskilled technicians can also help in production by using these gauges.
3. Jobs done with these gauges have the property of interchangeability.
4. It consumes less time in assembling these jobs.
5. Jobs are produced with proper tolerance.
6. There is no need of employees for inspection and examination at production stage,
7. There is little possibility of defects in the jobs.
8. Rate of production is increased.
9. As the cost on the job is reduced, its price can be reduced.
10. Those who produce such parts earn goodwill in the market.

15.1 CONSTRUCTION OF GAUGES

In shape and design gauges are made in such a way that they can measure and check length thickness, depth and taper of a job easily and accurately. These are generally made of high carbon steel and they are made hard and accurately ground. In addition, some gauges which are widely used are made of vanadium steel or chromium steel. After manufacturing these gauges, they are hardened, ground, lapped and finished, and then seasoned by keeping in hot water for 10 or 12 hours. Strict checking is done to ensure that the size of the gauge is accurate. Electroplating is done on gauges so that their sizes do not change because of long use. If there is a reduction in their size, it is brought to the desired size through electroplating or through lapping.

15.2 CLASSIFICATION OF GAUGES

Machine parts, are manufactured with different accuracies. Gauges can be classified into the following three categories, according to their use and accuracy.

1. **Worker Gauge:** This type of gauge is used by the technicians at the time of production of jobs. By using these gauges we need not set the tools repeatedly. These gauges are made generally to check the measurement with accuracy up to 0.0005" or 0.025 mm.
2. **Inspection Gauge:** Various jobs/parts manufactured by the technicians, some short- coming are possible. Therefore, it is necessary to get the jobs checked by inspectors. This gauge can check the jobs with accuracy to the extent of 0.0003" or 0.0076 mm.
3. **Master Inspection Gauge:** These types of gauges are also known as reference gauges. The veracity of inspection gauges is verified with it from time to time because the inspection gauges also wear out and there is possibility of their giving in accurate results. Master inspection gauges can check the inspection gauges with accuracy to the extent of 0'0001"/.0.0025 mm.

15.3 GAUGES AND GAUGING

Gauging is the process of investigating or determining that the product conforms dimensionally to drawings, specifications, or other prescribed requirements. Gages are the means by which products are checked with the prescribed dimensions.

There are two general methods of approach to the problem of gauging product. According to the method of approach taken, the purposes are:

- By attributes to assure that the product is within the prescribed limits of size.
- By variables, to measure the extent of deviation of product from a normal or basic size.

Gauging by attributes is usually accomplished by use of limit gages. Gauging by variables is accomplished by the use of indicating gauges. A sharp line of demarcation cannot always be made between the two methods, as both fixed limit gauges and indicating gauges are used under some conditions for gauging by attributes, while gauging by variables is accomplished only by the use of indicating gauges.

15.3.1 Fixed-limit Gauges

1. Maximum-material-limit gauges, known as go gauges, which control the minimum allowable looseness or maximum allowable tightness in the fit of mating parts and, accordingly, control interchangeability.
2. Minimum-material-limit gauges, known as not-go gauges, or, in the cases of thread gauges, high or low gauges, which control the maximum looseness or minimum interference between mating parts. Fixed-limit gauges thus control in large measure the tightness or looseness of the assembled components.

Gauging by attributes through the use of fixed gauges is a rapid method of determining whether or not the product is within the prescribed limits. Gauging by variables provides a means of evaluating the size of each part establishing its relationship to limits of size, and determining to what extent limits of size are exceeded by rejected parts. This gauging process is particularly useful where classification or statistical quality control of the limits of size is desired.

15.4 ACCURACY ATTAINABLE IN MEASURING

Refinement in gauge making and gauging practice has kept pace with the increasing tendency toward closer tolerances. Plain limit plug, ring, and snap gauges are utilized where product dimensions are 115 mm or smaller, to gauge the product to not less than 0.0025 mm tolerance. Considerable care must be used with snap or other open type gauges in gauging the product to close tolerance, because it is possible to spring them by hand gauging pressure. Where extremely small tolerances must be gauged and where straightness and roundness are of major importance, other types of measuring or indicating instruments offer greater accuracy.

Extensive practice in gauging a product is to use the 10 per cent rule. Under this practice, the product should be checked with gauges made to tolerances which do not in total embrace more than one-tenth of the product tolerance. The gauges themselves should be measured with equipment with a rated accuracy not exceeding one-tenth of the gauge tolerance. Thus, if the product tolerance is 0.076 mm, the total of gauge tolerances and wear allowance should not exceed 0.0076 mm and the gauge, in turn, should be measured on equipment having a rated accuracy not exceeding 0.00076 mm.

Gauges suitable for gauging by either attributes or variables usually require master gauges or instruments for setting or measuring them to the maximum and minimum limits of size. Master gauges comprise those which are used.

- As a reference for accuracy
- As a means of setting fixed gauges for gauging by attributes and indicating gauges for gauging variables.

Because the accuracy of gauges or comparators depends upon the accuracy of the masters, the masters must be made definitely within their limits of size. This entails measurement in controlled-temperature laboratories at 68° F. Measuring pressures must be in accordance with prescribed standards. Compensation must be made for any residual errors in measuring wires, measuring rolls, and end standards when used. The accuracy of comparators and measuring machines should be checked as near to the desired measurement as possible, using masters with all errors known. Finally, care must be exercised to keep the observational error to a minimum

15.5 GAUGING TERMS

Unit of Length: A nominal distance in space fixed by definition and independent of temperature, pressure, or any other physical condition.

Standard of Length: The physical embodiment of a defined unit of length. The established standard is the meter, which by international agreement is defined as: I M= 1,650763.73 wavelengths (in a vacuum) of the orange-red radiation corresponding to the transition between certain energy levels of the krypton-86 atom — the most uniform known to date.

Dimension: A dimension is a geometrical characteristic such as diameter, length, angle, circumference, or centre distance, of which the size is specified.

Gauge Blocks: The basic industrial standards used as reference and setting means for comparative measurement in precision manufacturing.

Measuring Wires and Rolls: Hard cylinders of extreme accuracy used as a means of measuring angle, taper, pitch diameter of screw threads, etc.

Plug Gauge: A gauge the outside measuring surfaces of which are designed to verify the specified uniformity of holes. A plug gauge may be straight or tapered, plain or threaded, and of any cross-sectional shape.

Taper-lock Plug Gauge: A plug gauge in which the gauging member has a taper shank, which is forced into a taper hole in the handle. American Gauge Design (AGD) for this style covers the range above 1.50 to 38.354 mm.

Wire-type Plug Gauge: A plug gauge comprising a gauging member of straight cylindrical section throughout its length and which is held in a collet-type handle. This design is an AGD standard in the range above 0.076 to and including 19.304 mm.

Trilock Plug Gauge: A plug in which three wedge-shaped locking prongs on the end of the handle are engaged with corresponding locking grooves in the end of the gauging member by means of a single screw, thus providing a self-entering support with a positive non-rotating lock. ACD for this style covers the range above 19 to 200 mm.

Annular Plug Gauge: A shell-type plug gauge, where the gauging member is in the form of a ring, the external surface of which is the gauging section. The central portion of the web is machined away to reduce weight. Handles are provided for convenience when using these gauges. This construction is employed for plain and thread plug gauges in the ranges above 200 mm

Flat Plug Gauge: A plug gauge made in the form of a central axial segment of a plain cylindrical plug gauge.

Plain Cylindrical Plug Gauge: A complete internal gauge of single- or double-ended type. It consists of a handle and gauging member or members, with suitable locking means.

Progressive Cylindrical Plug Gauge: A complete internal gauge consisting of handle and gauging member in which the go and not-go gauging sections are combined in a single unit secured to one end of the handle.

Taper Plug Gauge: An internal gauge in the form of a frustum of a cone having diameter, taper, and length suitable for the internal gauging of taper dimensions in accordance with the specifications of the product. An extended portion may be provided for checking a tang slot when specified.

Thread Plug Gauge: A complete internal thread gauge of either single- or double-ended type, comprising handle and threaded gauging member or members, with suitable locking means.

Ring Gauge: A gauge the inside measuring surfaces of which are circular in form. They may be cylindrical or conical, plain or threaded.

Plain Ring Gauge: An external gauge of circular form employed for the size control of external diameters. In the smaller sizes it may consist of a gauge body into which is pressed a bushing, the latter being accurately finished to size for gauging purposes.

Taper Ring Gauge: An external gauge the internal contour of which conforms to the frustum of a cone having diameter, taper, and length suitable for the gauging of external taper dimensions in accordance with the specifications of the product. An extended portion may be provided for checking a driving tang when specified.

Thread Ring Gauge: An external thread gauge employed for the size control of threaded work. It may be of solid design or provided with means for adjustment. The thread-ring-gauge locking device provides an effectual means of expanding, contracting, and locking the thread ring gauge during the manufacturing or resizing processes. It comprises an adjusting screw, a locking screw, and a sleeve.

Snap Gauge: A fixed gauge arranged with inside measuring surfaces for callipering diameters, length, thickness, or width.

Plain Adjustable Snap Gauge: A complete external-calliper gauge employed for size control of plain external dimensions, comprising an open frame, in both jaws of which gauging members are provided. One or more pairs of gauging members can be set and locked to any predetermined size within the range of adjustment.

Plain Solid Snap Gauge: A complete external-calliper gauge employed for size control of plain external dimensions, comprising an open frame and jaws, the latter carrying gauging members in the form of fixed, parallel, nonadjustable anvils.

Master Disk: A cylinder provided with insulating grips, used for setting comparators, snap gauges, and other gauge types.

Thread Snap Gauges: These consist of conical points or serrated or threaded blades, plates, or cylinders which are adjustable mounted in a snap-gauge frame.

Spline Plug Gauge: A plug gauge having a series of projecting keys equally spaced about the periphery, which fit into the spline ways to be gauged.

Spline Ring Gauge: A ring gauge having keys which are complementary to the splined shaft to be gauged.

Indicating Gauge: A dial indicator is a mechanism for amplifying and measuring the displacement of a movable contact point, thereby measuring a dimension or variations from a standard dimension. These are comprised essentially of a case with means for mounting the indicator, a spindle carrying the contact point, an amplifying mechanism, a pointer, and a graduated dial.

Flush-pin Gauge: A gauge for checking the distance between two surfaces, comprising a body having a through hole, and a pin in the hole which projects from a face of the body a distance equal to the dimension to be gauged when the opposite or indicating end of the pin is flush with the opposite face of the body. The indicating end of the pin or the adjacent face of the body has a step of a depth equal to the tolerance on the dimension gauged.

Combinations of these types of gauges are incorporated into devices for measuring concentricity, for location, cutter setting, fixture gages, etc. Gauge blanks are standardized in construction, design, and range in accordance with commercial standards.

15.6 GAUGE TOLERANCES

A gauge is a manufactured article and must be made to manufacturing tolerance. While the gauge tolerances are small, they may consume a part of the product tolerance and should be taken into consideration.

15.6.1 Gauge Maker's Tolerances

In general commercial practice, gauge tolerance for work gauges is held to 10 per cent of the component total tolerance and to 0.5 per cent for inspection gauges when no gauge tolerance is specified, unless the tolerance so determined materially exceeds the accepted standard for precision-gauge manufacture.

Gauge makers tolerances for plain cylindrical plugs and rings have been standardized in four classes, XX, X, Y, and Z according to nominal size and to the degree of accuracy necessary in the gauging operation.

It is commercial practice to apply unilateral tolerance to go and not-go gage members.

The tolerances on go plain plug gauges are applied plus. Not-go plain-plug-gauge tolerances are applied minus. The go plain ring tolerances are applied minus; not-go plain ring tolerances are applied plus. There are exceptions to this, one being master rings and plugs where bilateral tolerances are commonly used; a second is where encroachment on product limits is minimized by applying tolerance in not-go gauges bilaterally.

In most gauging systems it is common practice to relate the gauge makers' tolerance to the component total tolerance. The magnitude of the tolerance affects the cost of the gauge as well as its useful function. If the tolerance of the gauge is unnecessarily small, the cost may rise sharply for both the gauge and its application; on the other hand, if the gauge tolerance is too large, it may not fulfil its intended function and may cause needless rejection of good parts. A gauging system is incomplete unless definite values are established for permissible gauge tolerances and wear limitations.

15.7 GAUGING SYSTEMS

The principal gauging systems are based on either the bilateral or the unilateral system, or on combinations of the two.

Bilateral (Split-Tolerance) System: The maximum or minimum work limits, or both, bisect the gauge tolerance; i.e., gauge tolerance is equally split and applied one-half plus and one-half minus.

Unilateral System: A unilateral gauge tolerance is encompassed by the maximum and minimum work limits. ANSI has specified in ATS B4a-1925 that the extreme size for all plain limit gauges shall not exceed the extreme limits of the parts to be gauged. All variations in the gauges, whatever their cause or purpose, shall bring these gauges within these extreme limits. Thus a gauge which represents a minimum limit may be larger, but never smaller, than the minimum size specified for the part to be gauged; likewise, the gauge which represents a maximum limit may be smaller, but never larger, than the maximum size specified for the part to be gauged.

Some exceptions to the rigid interpretation of this system are noted in "Handbook H28 :Screw Thread Standards for Federal Services".

Wear allowances may be applied to go plain plug and ring gauges at the user's discretion. Any application of wear allowances uses a part of the product tolerance. Plain gauges made to MIL specifications provide a wear allowance equal to approximately 5 per cent of the product tolerance in applications where the product limits are 0.025 mm.

The National Bureau of Standards and ANSI publish standards, for gauges to be used for the production and inspection of screw threads.

15.8 ALLOCATION OF GAUGE TOLERANCES

The limits of size for screw threads specified in "Handbook H28" and in "American Standard ANSI Bl.I-1960" establish the extreme limitation of an acceptable product thread and the tolerance permissible in manufacture. Strict preservation of the integrity of both the product limits and the manufacturing tolerance depends on the accuracy of the gauges, and the amount and direction of the gauge tolerance and wear allowance. If gauge variation is inside the product limits, it has the effect of reducing the tolerance available to the manufacturer of the product; if the variation is outside the product limits, the product itself may exceed the limits. The choice of such direction is affected by differing requirements at the maximum-material and minimum-material product limits.

In general, the maximum-material limits should not be exceeded because these limits constitute definite lines of demarcation between a satisfactory product and one which will not satisfactorily interchange or assemble according to predetermined standards of manufacture. When there is no allowance in the product limits of size, the maximum-metal limits define the point where interference of mating parts begins, that is, the point where parts may cease to be interchangeable. Even when an allowance is provided, its entire amount is usually required to be available for its purpose. As the maximum-metal limits are controlled by go-gauges, it follows that both the gauge tolerance and the wear allowance for go-gauges should be within the product limits, despite the fact that the tolerance available to the product manufacturer is thereby reduced. When so provided, the wear will be towards the maximum-material limit, and the gauges should not be used when they have worn sufficiently to exceed the maximum-metal limit.

To assure the production of all parts within the minimum-material limits and the acceptance of all satisfactory parts, preferred practice is that tolerance for manufacturer's Hi-Low gauges be specified as within the product limit. As minimum-material limit acceptance gauges wear within the product limits, they should be discarded or reconditioned when worn beyond the applicable gauge tolerance.

15.9 THREAD PLUG AND RING GAUGES

The class W and X thread-gauge tolerances are normally applied unilaterally and are encompassed within the extreme product limits. The selection of suitable gauges for use in the inspection of threaded products depends entirely upon the specification of the product.

Taper Pipe-Thread Gauges: There are three specifications for standard taper pipe threads in common use.

The standard-type plug gauge incorporates a single gauging notch; the matching ring is fitted to come flush to this notch. The limit-type plug gauge retains the basic notch, and in addition two limit notches or steps are incorporated indicating maximum and minimum size. The ring usually uses a standoff from the basic step. ANPT and NPTF gauging also involves the use of an L-3 thread plug and an L-2 thread ring gauge for checking product threads beyond those checked by the L-1 gage. Coordination of the results also checks excessive taper errors.

ANPT gauges also include plain taper plugs and rings for gauging crest truncation. Although there are no official standards for six-step plain taper plugs and rings for gauging crest truncation of dry seal NPTF threads, such gauges are used unofficially.

Straight Pipe Threads: Many types of straight pipe threads are currently in use. These include the following:

- NPS. American National Standard straight pipe thread of special diameter. NPSC. American National Standard straight pipe thread for pipe couplings.
- NPSH and NH. American National Standard internal straight pipe thread for hose couplings and nipples.

Specifications for these straight pipe threads can be found in Handbook H28,5 and ANSI B2.1.

NPSF and NPSI. American National Standard intermediate internal straight pipe thread (dry seal).

NPSL. American National Standard straight pipe thread for lock nuts and lock nut pipe thread.

NPSM. American National Standard straight pipe thread for mechanical joints.

NPSC, NPSI, and NPSF threaded products are usually gauged with taper thread plug gauges.

15.10 FURTHER DISCUSSIONS ON GAUGES

1. Plug gauge
2. Ring gauge
3. Snap gauge
4. Radius gauge

5. Profile gauge
6. Centre gauge
7. Receiving gauge
8. Caliper gauge
9. Drill gauge
10. Drill point gauge II. Feeler gauge
11. Screw pitch gauge
12. Angle gauge
13. Taper gauge
14. Wire gauge
15. Gauge block
15. Indicator gauge
17. Sin bar

Plug Gauge: This type of gauge is used to verify measurement of different kinds of holes. These gauges are manufactured according to different job requirements and are known by their shape and use such as:

1. Cylindrical plug gauge
2. Taper plug gauge
3. Thread plug gauge
4. Square plug gauge
5. Hexagonal plug gauge, etc.

All these plug gauges are one-ended. The inspection plugs are made as per the actual measurement of the job and this measurement is indicated, on the plug itself. These gauges are called nominal size plug gauge. These gauges are used as reference gauges.

In construction jobs, the operators generally make use of double-ended plugs. On one end of this type of plug "Go" is written while on the other end "Not Go" is indicated.

Both its ends "Go" and "Not Go" are made according to the limit of the hole of the job which is to be tested. The size of "Go" end is kept slightly more than the low limit and the measurement of the "Not Go" end is kept slightly more than high limit. For example, if we have to measure a hole of size 2" and its approved

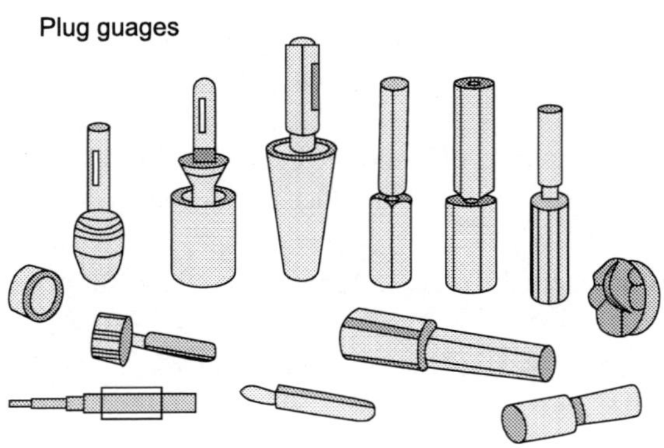

Fig. 15.1

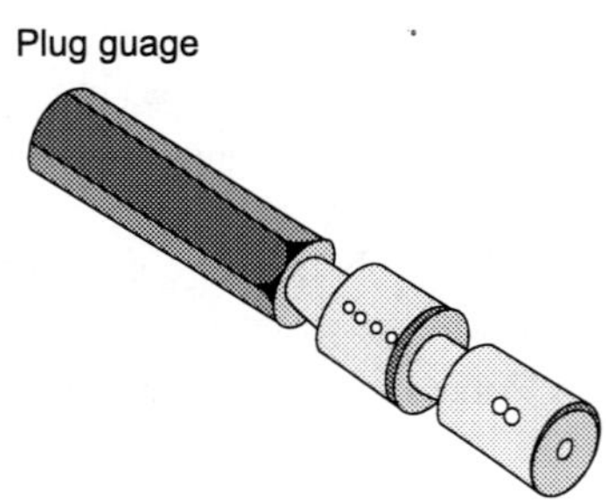

Fig. 15.1a

tolerance is + O.003 or-0·002 the "Go" end of the checking plug gauge should be (2"-0'002 = 1.998"). In the same way, the "Not Go end should be (2" +0·003" = 2·003").

While checking the bore (hole), the "Go" end of the plug gauge should go inside the hole and the "Not Go" end should not go inside the hole. Therefore, these types of gauges are called limit plug gauge. For the purpose of convenience, the plug with "Go" indication is made smaller or short. On the contrast the plug with "Not Go" indication is made bigger or longer in comparison with the other end. As such it takes little time in recognizing the "Go" and "Not Go" ends. As the "Go" end has to be inserted in the hole, it is made in length, as per the job. On the other hand "Not Go" end is not be inserted in the hole, therefore there is no difficulty in making it bigger in size. In this type of gauges, a knurled or square or six-phased handle is fitted between the plugs. This contains the name, special comment, if any, and the measurement, etc. There are some limit plug gauges which have both "Go" and "Not go" plugs at one end itself.

These types of limit plug gauges are used to test (or verify) in short time the jobs having a hole across them. But we cannot check the closed holes with these limit plug gauges.

Plug gauges having diameter up to 2" are made in solid form along with the handle but in plug gauges of measurement exceeding 2", handles and plugs are built separately and these are fitted together through threads. In order to reduce the weight of bigger plug gauges, holes are drilled. The most important thing in these gauges is that since the "Go" end is inserted in the hole.This end wears out more than any other part. Therefore, by replacing or repairing only one part, the plug gauge becomes fit for accurate checking.

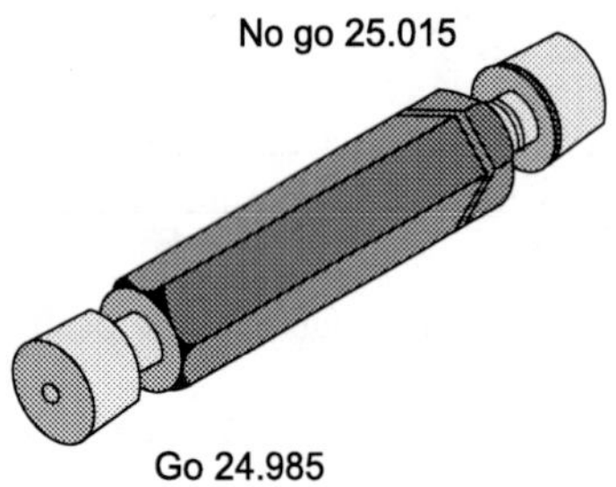

Fig. 15.2 Plug guage

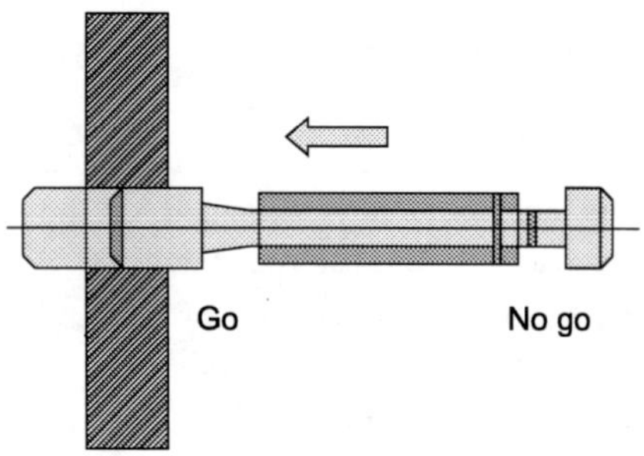

Fig. 15.2a Plug guage

15.11 USE AND PRECAUTIONS

For using plug gauges these should be rotated inside the job in revolving motion. The hole of the job should be properly cleaned before using the plug gauge. Before and after using the plug gauge they should be properly cleaned.

Ring Gauge: This gauge is used for checking external diameters of round and cylindrical jobs. The diameter of the hole is made according to the actual measurement of the job. Their outer part is knurled so that these can be handled easily

In any round job, there are generally four types of external surfaces. According to these surfaces, these gauges are also of four types.

1. Plain Ring Gauge: This gauge is used for checking the external measurement of a round or cylindrical job.

Fig. 15.3

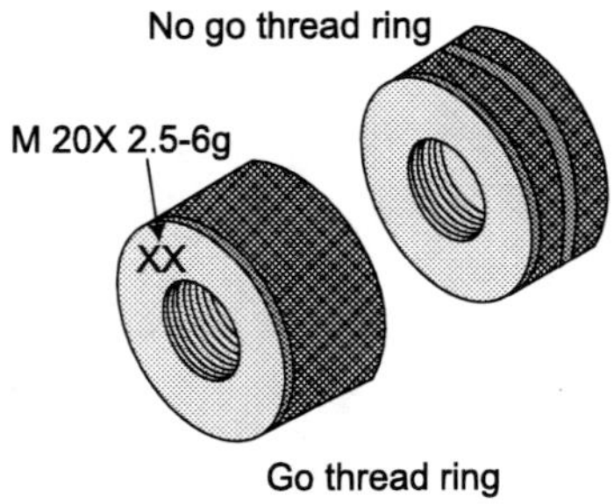

Fig. 15.4

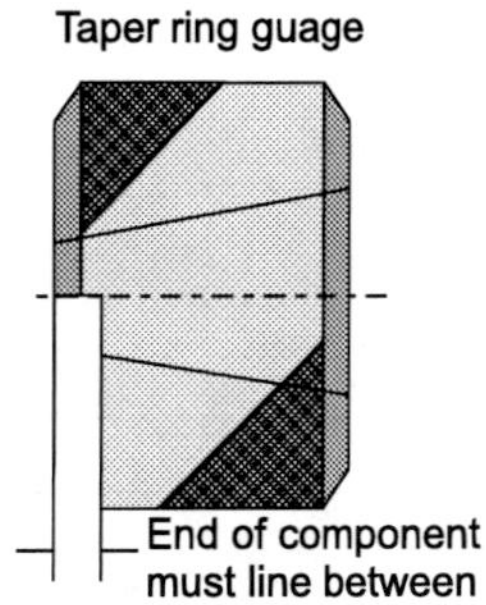

Fig. 15.5

2. Thread Ring Gauge (Fig. 15.4): This type of gauge is used to check the external threads cut on cylindrical jobs.

2. Taper ring gauge (Fig. 15.5): These gauges are used for checking the external taper angles of round or cylindrical job.

1. Form Ring Gauge

Form ring gauge is used especially for external measurement of those jobs which have abnormal shapes.

Some ring gauges are made according to the gauge limits also. Such gauges are made of rectangular steel by making holes on both the sides. On one end, "Go" is indicated, while on the other "Not Go" is written.

The holes on both these ends are made according to the limits of the jobs.

2. Snap Gauge This is non adjustable and used as a quick means for checking length, breadth, thickness or its external diameter. These gauges ate broadly of two types:

(i) Solid Snap Gauge: This type of gauge is almost similar in shape to the English alphabet 'c'. This is available in the form of one-end and two-ends limit snap gauge. The checking sides of this gauge have very plain surface which is made very smooth and accurate. **(Fig. 15.6) & (Fig. 15.6a)**

One-end gauge is meant for checking the actual size and two-end gauge is known as limit snap gauge. With this gauge, checking can be done as per limit of the job. "Go" is indicated on its one end and "Not Go" is written on the other end, their size is also indicated. While checking the size of a job, the job should go into the "Go" end. If it does not go into "Go" end, it means the size of job is larger. The job should not go into "not go" end, if it goes, it means that the size of the job has become less than the actual size. **(Fig. 15.6b)**

Some snap gauges have one end and are in "C" shape but these are made according to the limit of the job.

In this type of snap gauge both "Go" and "Not Go" sizes tire at one end. The main advantage of these gauges is that it saves time.

(ii)Adjustable Snap Gauge: The shape of this gauge is also like "C" shape like one-ended solid snap gauge. On their inspecting surface three or four anvils are fitted. These anvils are adjusted with the help of screws in proper size. Therefore, we can check different jobs according to their limit with this snap gauge. In order to control the size, gauge block is used. This type of gauge is with four anvils. At the time of checking a job, the job should go into the external pair of anvils but it should not go into the internal pair of anvils. This gauge can be adjusted up to 3/4". **(Fig. 15.7)**

Radius Gauge (Fig. 15.8, 15.8A, 15.8B, 15.8C): Shape of this gauge is like a small pocket knife having a number of leaves fixed on both its ends with a screw or a rivet. These leaves have concave or convex shape

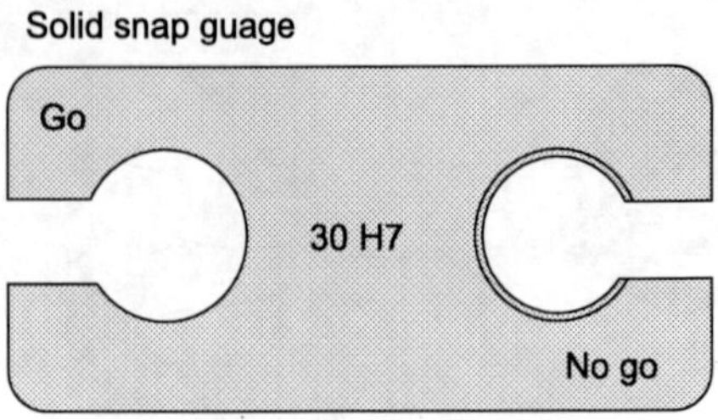

Fig. 15.6

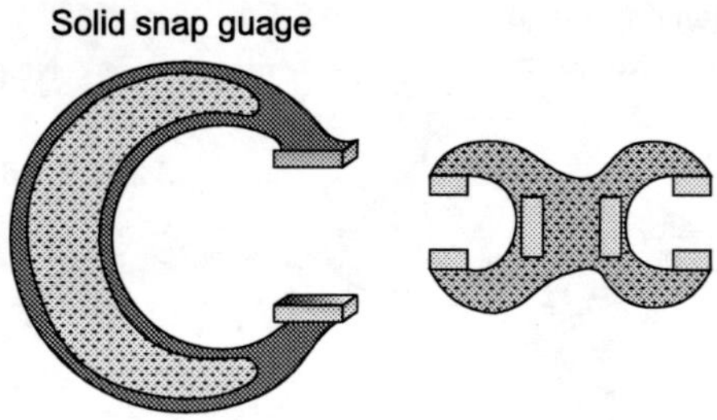

Fig. 15.6a

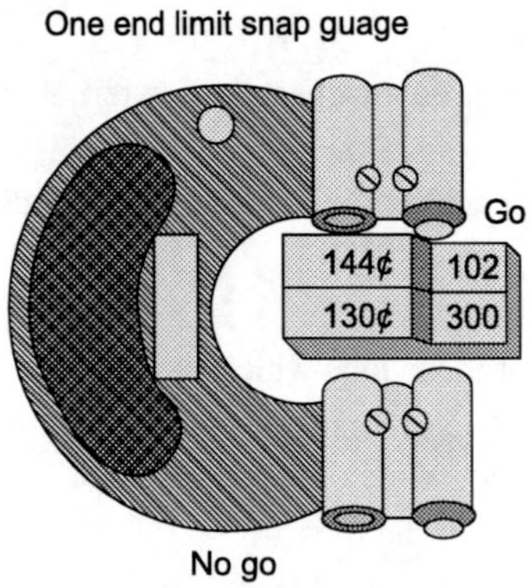

Fig. 15.6b

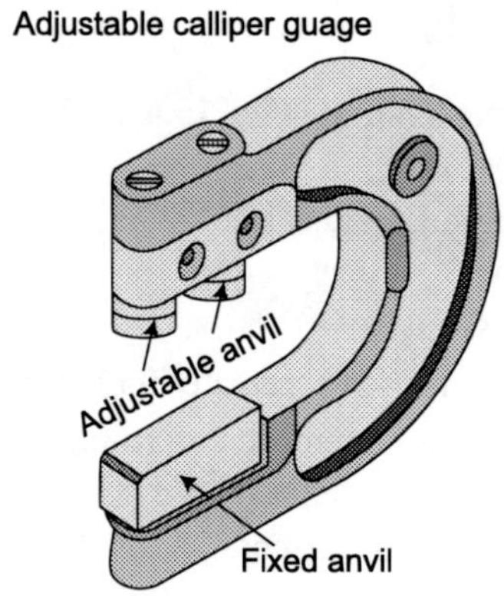

Fig. 15.7

at their ends. Size of their radius is indicated on the leaves. With this gauge the radii of fillets and rounded corners can be measured. It is also known as **fillet gauge.** One radius gauge has leaves of one range. Different ranges of radius are available.

Radius gauges are classified in several ways according to their ranges, These are available both in English and Metric ranges, English radius gauges are in-range of 1/64, 1/8, 9/64, 5/32, 11/64, etc. Similarly, metric radius gauges are available in 0 25 mm, range such as 1.5, 1.75, 2.25, 2.75, and 3 mm etc. These are called "A" and "B" sizes.

Some special types of radius gauges are also available. On both the ends of each leaf, there is a radius of concave and convex range.

The size of the radius is marked on the individual blades of the gauge. These are available in different combinations with internal and external radius.

These are made of high quality steel sheets, and are finished to accurate radius. Radii of parts are checked by comparing the radius of the gauge sticking it, on the job like an Engineer Square. While checking, it should be ensured that even light cannot pass between the job and the gauge.

Profile Gauge (Fig. 15.9): These gauges are specially made for checking the size of external shapes of special jobs. The main advantage of this gauge is that various sizes of job are checked simultaneously.

A, B, C are three different types of profile gauge. In A its measuring surface is like knife edge.

Centre Gauge (Fig. 15.10): It is a fiat gauge used for setting a tool for the cutting of "V" threads. This is made of spring steel. "V" is made of angles of 60°, 55° and 29° (**Fig. 15.10**).

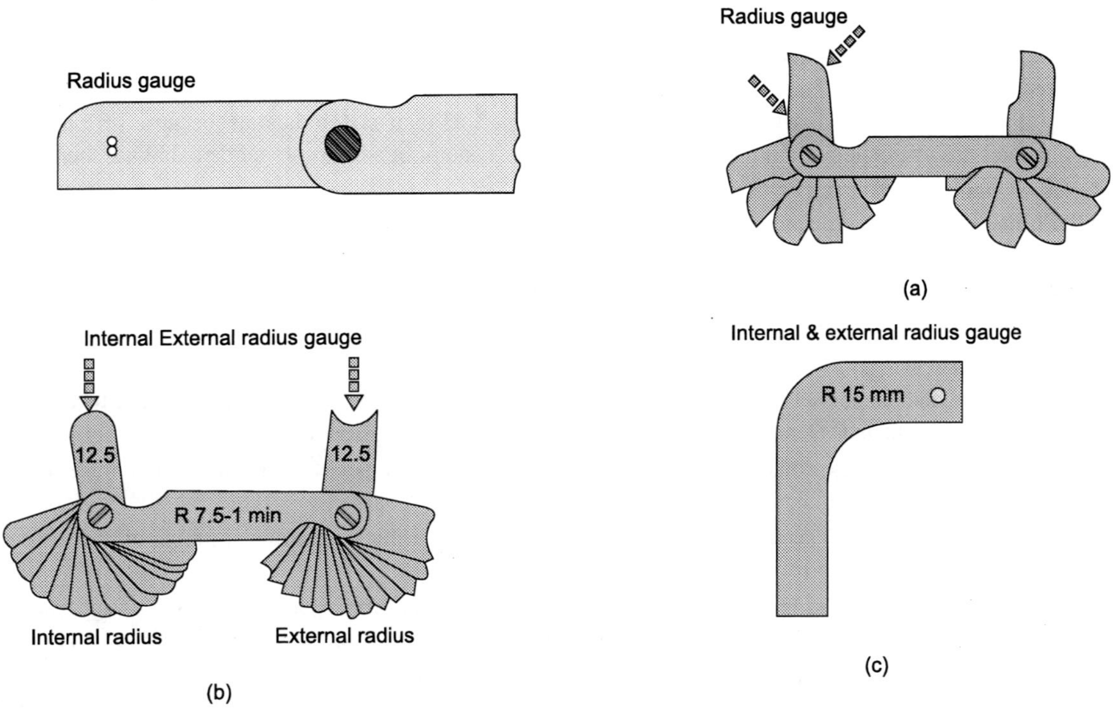

Fig. 15.8

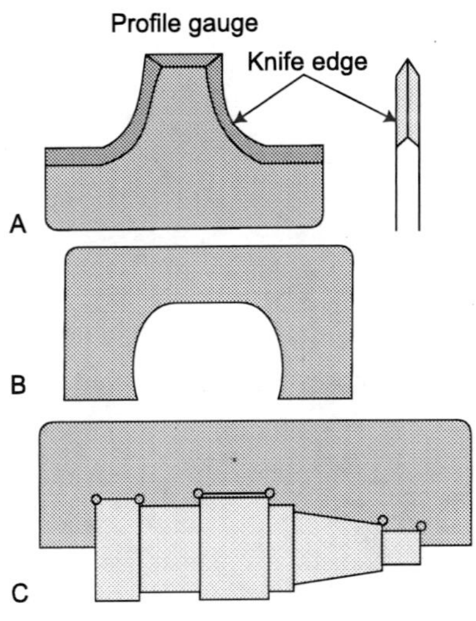

Fig. 15.9

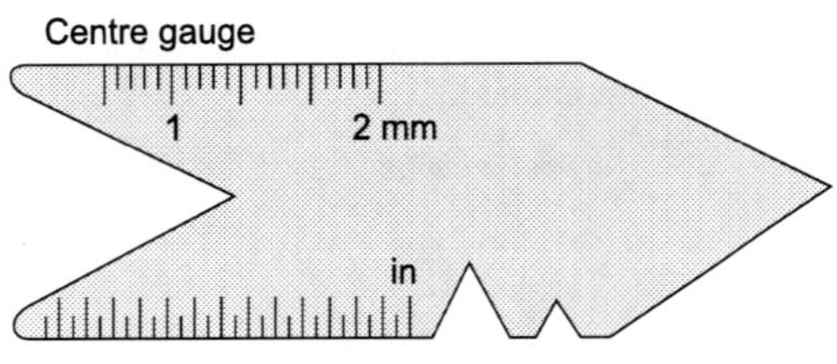

Fig. 15.10

Calliper Gauge: The main feature of this gauge is that we can check the internal or external size of a job with one gauge. Its shape is almost similar to the snap gauge but on one of its end there are jaws for checking internal size and on the other end, there is an anvil for checking the external size.

Drill Gauge: It is a flat steel plate drilled with holes of different sizes and properly marked so that the size of a drill may be easily determined by fitting it to the plate. Its size is 61/4" X 2 5/ 16" X 5/32". Its length is measured by its radius from both the sides. The size of the holes is indicated according to the drill. We can measure the size of those drills with this gauge which have lost their size indication because of long use. These gauges are available both in metric as well as British system. In addition letter drill and number drills are also available.

Drill Point Gauge: In order to drill properly and make holes of proper size, it is essential that drill is ground at proper angle and both its lip centres are equal from the centre line. After grinding the drill, drill point gauge is used to check the drill point.

It has two parts:

(i) Steel rule

(ii) Gauge, which can be set anywhere with a thumb screw. This gauge can be used like other measuring tools as well such as Depth Gauge, Engineers Square and Hook Rule, etc.

Feeler Gauge: This gauge consists of a number of hardened and tempered blades of various thicknesses mounted in a steel case. The thickness of individual leaves is marked on it.

The sizes of the feeler gauges in a set are carefully chosen in order that a maximum number of dimensions can be formed by building up from a minimum number of leaves.

The dimension being tested is judged to be equal to the thickness of the leaves used, when a slight pull is felt while withdrawing them. Accuracy in using these gauges requires a good sense of feel. **(Fig. 15.11)**

B.I.S. Set: The Indian Standard establishes four sets of feeler gauges Nos. 1, 2, 3 and 4 which differ by the number of blades in each and by the range of thickness (minimum is 0.03 mm to 1 mm in steps of 0.01 mm). The length of the blade is usually 100 mm.

Example

Set No. A of Indian Standard consist of 13 blades of different thicknesses.

0.03,0.04,0.05,0.06,0.07,0.08,0.09,0.10,0.15,0.20, 0.30, 0040, 0.50.

Uses:

Feeler gauges are used:

- To check the gap between the mating parts
- To check and set the spark plug gaps and tappet clearance in an engine, etc.
- To set the clearance between the fixture (setting block) and the cutter/tool for machining the jobs.
- To check and measure the bearing clearance, and for many other purposes where a specified clearance must be maintained.

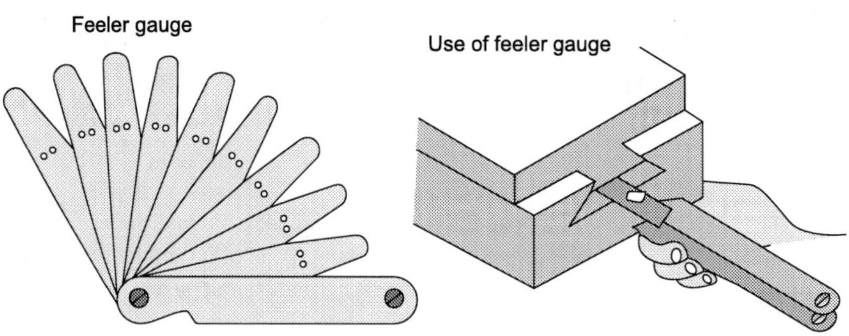

Fig. 15.11

Screw Pitch Gauge: This gauge is mainly used to check the pitch of external and internal threads.

The surface of a screw has a complex shape. The following elements of a screw thread are to be considered in thread measurement.

- Major diameter
- Minor diameter
- Pitch
- Effective diameter
- Thread angle
- Form of root and crest

The above elements contribute to the strength and interchangeability of the threads.

The most important elements to be checked are:

- The pitch of the thread
- The angle
- The effective diameter.

Screw Pitch Gauge: This gauge is mainly used to check the pitch of external and internal threads.

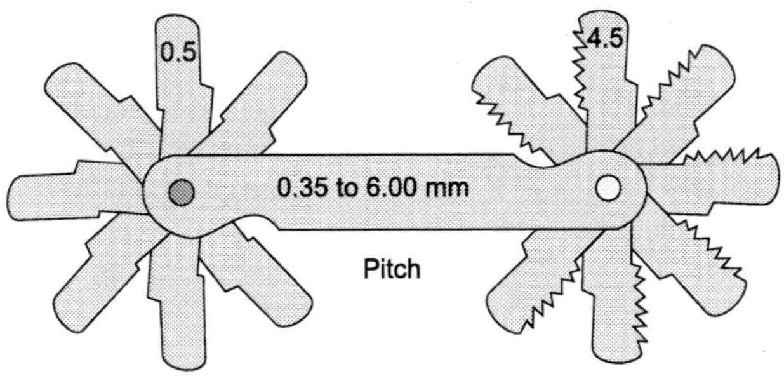

Fig. 15.12 Screw pitch gauge

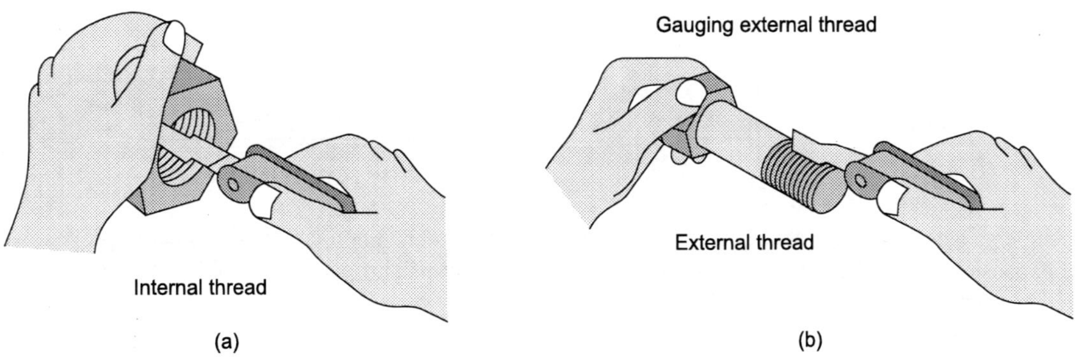

Fig. 15.12a Gauging internal thread

1. British Standard or Whitworth Pitch Gauge:

   This type of thread has 55° angle. In one set there are 22, 24, 26, 28 or 30 leaves. In a screw pitch gauge with 22 leaves, leaves are from 9 T.P.1 to 30 T.P.1. Leaves in the screw pitch gauge with 24 leaves, are of 4 T.P.I to 30 T.P.I. Screw pitch gauge with 20 leaves are of two kinds. In one set there are leaves of 4 T.P.I. to 60 T.P.I. and in the other set 32 T .P.I. to 80 T.P.I. In screw pith gauge with 28 leaves, leaves are from 4 T.P.I. to 62 T.P.I. Screw pitch gauges with 30 leaves, have leaves of two sets. In one type the leaves are of 3t to 60 T.P.I. and in the other set from 4 to 42 T.P.I.

2. American National Standard Screw Pitch Gauge

   This type of thread has an angle of 60°. This gauge has 27 leaves. These leaves are of 21 T.P.1 to 28 T.P.I.

3. Bicycle Screw Pitch Gauge

   This type of threads are made for electrical goods and cycle parts. There are 22. leaves. These leaves are of 32 T.P.I. to 74 T.P.I.

4. Metric Screw Pitch Gauge

   These are of two kinds. First set has 28 leaves, which have leaves of pitch ranging from 25 to 250 mm. In the second set there are 22 leaves with pitch ranging from 1 mm to 11.5 mm.

5. International Standard Screw Pitch Gauge

   With this gauge, threads of international standard are measured. There are 17 leaves in this gauge. Its leaves are of 5 mm to 7 mm pitch.

Angle Gauge (Fig. 15.13): This gauge is made of steel leaves in a case. it contains leaves on both the ends. Their ends are made of angles and the degree of these angles is marked. These leaves are made of steel and are duly tapered. Their size is 9/32, X 1 1/16 X 43/16'. One set contains 18 leaves, 9 leaves on one end and 9 on the other. We can check angles of

1°, 2°, 3°, 4°, 5°,7°,8°,9°,10°, 12°, 14°, 14!0 15, 20°, 25°, 30°, 35° and 45° with this angle gauge.

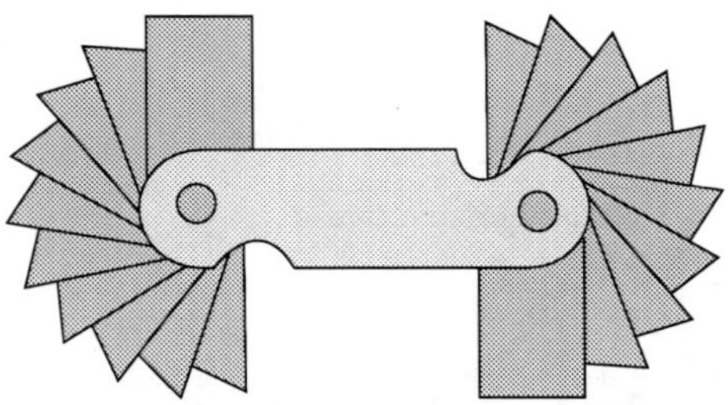

Fig. 15.13 Angle gauge

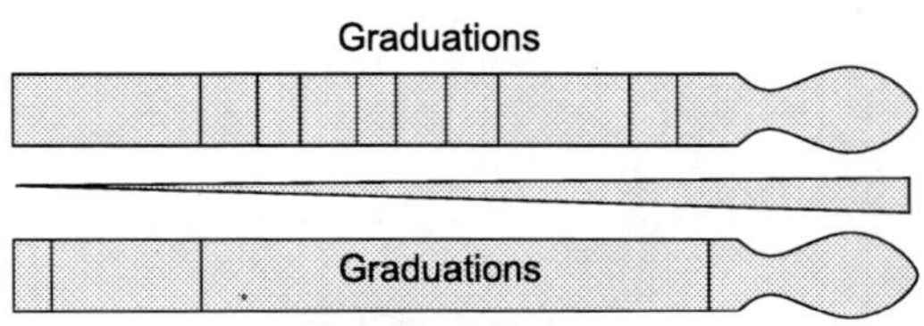

Fig. 15.14 Taper gauge

Angle gauges are wedge-shaped steel blocks made from hardened non-creep alloy steel to prevent wear and expansion errors. Its faces are lapped in the same manner as slip gauges, enabling them to be wrung together in combinations.

Applications

- To measure the angle of a workpiece.
- To set up a component upon a machine table.
- To calibrate adjustable angular measuring instruments.

Advantages

- They can be used either horizontally or vertically.
- They can be used to check jobs on which it is not easy to use other precision instruments.
- They are not restricted to angles smaller than 45.

Angle gauges are also available in different sets.

A typical set comprises twelve pieces divided into three series i.e., degrees, minutes and seconds as follows.

There are two sets of angle gauges available in the above typical set, designated as 'A' and 'B'. The standard 'B' contains all the 12 gauges, while standard 'A' contains one additional gauge of 3 seconds which enables combinations to be built up to 3 seconds

The angle of the gauge and the arrowheads are etched on each gauge to indicate the angle direction and size.

Note: The angle of the gauge and the arrowheads are etched on each gauge to indicate the angle direction and size.

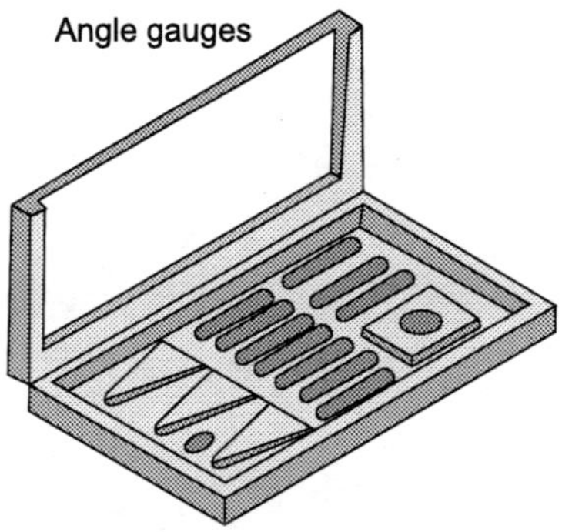

Fig. 15.15

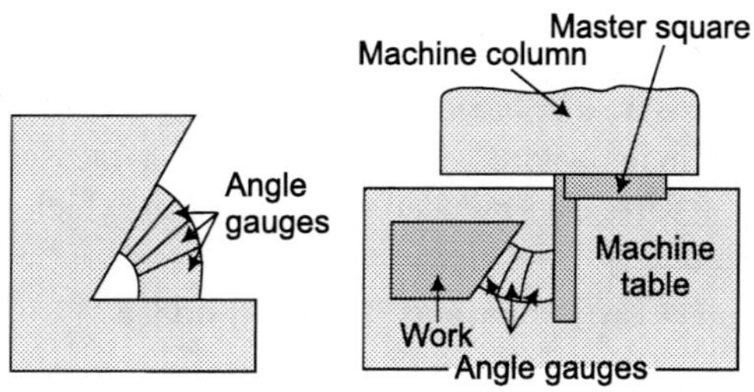

(a)

Fig. 15.15a

Degrees	1,3,9,27&41	5 pieces
Minutes	1,3,9 &27	4 pieces
Seconds	8,18 and 30	3 piece

Taper Gauge (Fig. 15.14): This a gauge used for checking the accuracy of tapers, inside or outside as well as to measure ina slot or internal or external diameter of a pipe. It is 1/16 inch broad and 6 feet in length. It is made of tool steel. Its thickness is in taper. On one side the taper is from 0.10 inch to 150 divisions is indicated, each part is bigger to the other by 0.001 inch on the other side, difference between each part is of 1/20 mm and parts are indicated from 0.3 mm to 4 mm. Besides this, another type of taper gauge is also available. In this type of set there are four or eight leaves. These leaves are made of 0.012 thick spring steel and their shape is tapered towards the end.

Uses of angle gauge

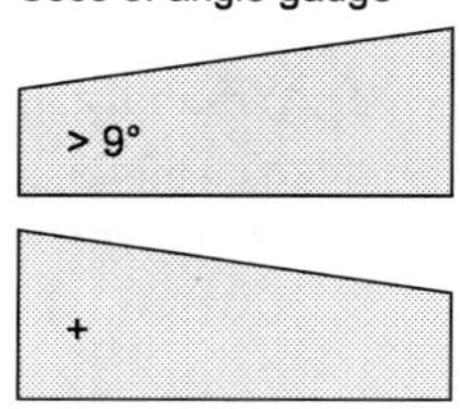

Fig. 15.16

Taper Gauge: With eight leaves **(Fig. 15.17) is** a set 8 leaves, set A are 2 feet in length. The accuracy of measurement of taper from 1/10 inch (0.100 inch) to ½ inch (0.5 inch) up to 0.00 1 inch. The leaves of set B are 2 ¾ in length. We can measure taper from ½ inch (0.500 inch) to 1 inch. For measuring taper, leaf of appropriate size is put inside the holes. At this moment, both sides of the gauge leaf should be stuck to the

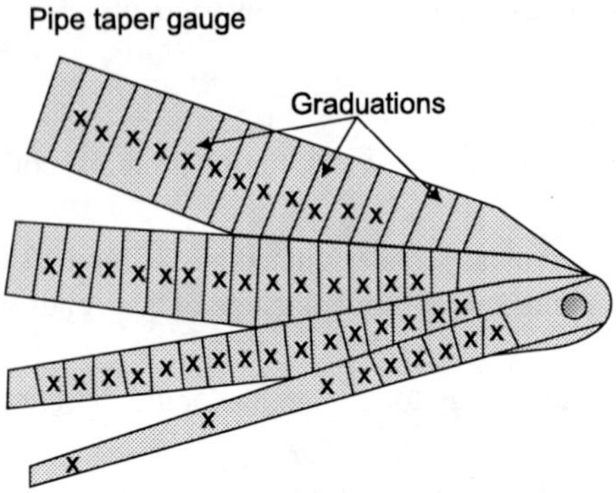

Fig. 15.17

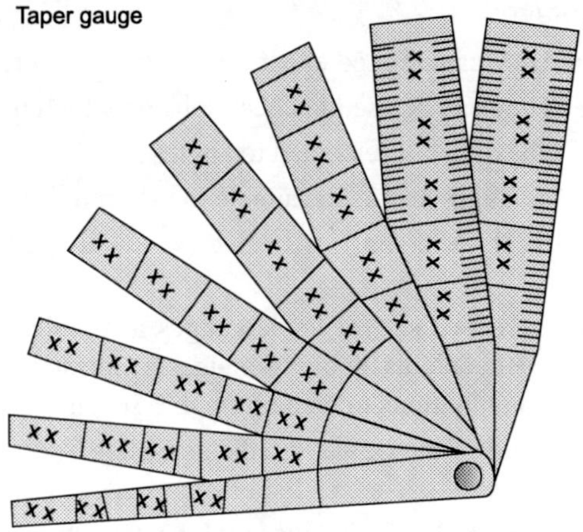

Fig. 15.18

job. We can determine or read accurately the measurement of outside caliper with taper gauge.

Wire Gauge (Fig. 15.19): It is a notched plate having a series of slots. These are numbered according to the sizes of wire and sheet metal manufactured. These are used for measuring the diameter of wire. The number of the slots in which wire or sheet fits properly, is its gauge.

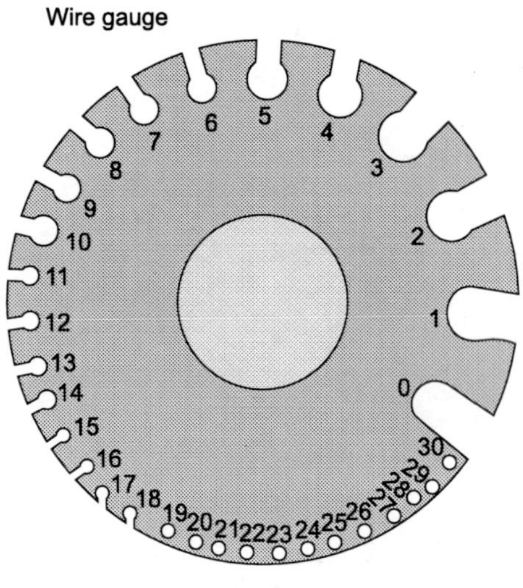

Fig. 15.19

1. **American (Brown and Sharp) Standard Wire Gauge:**
   These gauges are used for measuring copper, brass, aluminum and all non-ferrous metal wires, sheets and plates. The gauge contains slots of numbers from 0000 to 40.
2. **The Birmingham or Stubs Iron Wire Gauge:**
   It is used for measuring the hot and cold rolled steel and sheets. It contains slots of numbers 0000 to 36.
3. **Washburn and Moyen Standard Wire Gauge:**
   With this gauge we can measure drill rods, in addition to wires and sheets of 0000 to 36 numbers.
4. **Washburn and Menoy Steel Music Wire Gauge:**
   With this guage we can measure special wire which are used in musical instruments, Slots of numbers ranging from 00000000 to 30 are cut in this gauge.
5. **S.W. American Music Wire Gauge:**
   This guage is also used for measuring wires used in musical instruments but it contains slots of numbers ranging from 000000 to 33.
6. **Imperial Standard Wire Gauge:**
   This contains slots of numbers ranging from 000000 to 40 for measuring wires or sheets.
7. **Stubs Steel Wire Gauge:**
   Slots from number 1 to 40 are cut in this guage.

8. **United States Standard Wire Gauge:**

   Iron and steel plates are measured with this gauge. Slots of numbers from 000000 to 36 exist in this gauge.

In all the wire gauges, slot numbers have their own standard sizes. Excepts in the case of music wire gauge, in all other wire gauges, as the number of the slot increases, the-stze-decreases. Only in W. and M. Steel Wire Gauge and American S.W. Co. Musical gauge, as the number of the slot increases so does the size of the slot.

Slip Gauge or Gauge Block

It is necessary to obtain accuracy in sizes in production of different machine parts. This accuracy depends upon the approved limit of a job to be manufactured. In ordinary jobs where limit is up to (+) or (–), it can be measured with ordinary steel foot rule. But in the production of various tools and machine parts where limit of tolerance is '0001 cannot be measured with foot rule. In certain jobs the limit of tolerance is approved to a very small extent. For measuring or checking the size of such jobs slip gauges are used. These are generally made of tool steel. Their shape is square, rectangular, or triangular. Their gauging surface is quite plain. Such is the extent of their accuracy that when two pieces of gauges are kept properly one over the other through proper ringing action, they stick to each other becuase of vacuum. In such a state, not even a ray of light can pass through them.

Their manufacturing process is also different from that of ordinary gauges. These are cut by machines with appropriate allowance, and then machining is done. They are manufactured in almost correct size and then seasoning is done in natural or artificial way. After seasoning, tempering, grinding and lapping is done.

These slip gauges are always in sets. In one set there are 5 to 103 gauge blocks. In each set-range, size is chosen in such a way that gauge can be prepared to take any measurement within that range, by adding the gauges through the gauge ringing method. These gauges have been divided into three different sections on the basis of their utility and accurancy. These are "B", "A" and "AA"

1. **Worker Slip Gauge Set.**

   These gauges belong to the "B" class. It is used by the technician in workshops. This gague is for accurancy of '000008'.

2. **Inspection Slip Gauge Set.**

   This gauge belongs to the "A" class. This is used by the inspectors to test the accuracy of the jobs made by the workers. These are made for accuracy of 000004"

3. **Master Slip Gauge Set.**

   This guage belongs to "AA" class. It has the highest accuracy. These are made for '000002" accuracy. The accuracy of other gauges and tools is tested with these sets.

Shapes of gauges: According to the jobs and utility of testing, the shapes of gauges are of three types:-

1. Rectangular gauge block
2. Square gauge block
3. Angle gauge block

For using the above gauge blocks, different types of attachments have also to be used such as foot block, straight edge, scriber, adjustable holder, etc.

Uses of Slip Gauge. This gauge is generally used for the following functions:

- To inspect (or verify) micrometer and vernier calipers.
- To set adjustable gauges such as snap gauges
- To use sin bar
- For exact marking of a job
- To verify measurement of a job, with the help of comparator.
- To have comparative measurement of any job, with the help of an indicator.

For using slip gauge in any kind of job, more than one gauge blocks are joined. The act of joining gauge blocks in known as ringing.

Winging Action

1. Two or more gauge blocks are joined in the following manner:-
2. The gauging surface of the gauge blocks should be cleanded properly with tissue paper or chamois leather. At times, carbon tetrachloride is also used to clean these blocks.
3. The gauging surface of both the gauge blocks should be kept in a crossed manner.
4. Now the gauge blocks should be turned in such a way, that their gauging surfaces come in a line.
5. Again, the gauge blocks should be withdrawn and brought out to about fifty per cent.
6. Now again, the upper gauge block should be pushed back and brought back to parallel position.

Metric Slip Gauge

Apart from inch measurement gauges, metric slip gauges are generally used in most of the countries throughout the world. This type of block of gauges are also available in sets, One set consists of 113, 111, 88, 86 or 45 slip gauges, with these gauges, measurements can be taken.

15.12 GRADES

Grade "OO" accuracy: It is a calibration grade used as a standard for reference to test all the other grades.

Grade "O" accuracy: It is an inspection grade meant for inspection purposes.

Grade "i" accuracy: Is the workshop grade for precision tool room applications.

Grade "ii" accuracy: Is for general workshop applications.

Three grades of slip gauges are recommended as per IS - 2984. They are:

- Grade '0'
- Grade i
- Grade ii.

Care and maintenance points to be remembered while using slip gauges:

- Use a minimum number of blocks as far as possible while building up a particular dimension.
- While building the slip gauges, start wringing with the largest slip gauges and finish with the smallest.

While holding the slip gauges do not touch the lapped surfaces.

If available use protector slips on exposed faces.

After use, clean the slips with carbon tetrachloride and apply petroleum jelly for protection against rust.

Before use, remove petroleum jelly with carbon tetrachloride. Use chamois leather to wipe the surfaces.

Table 1 Set of 112 pieces

Different sets slip gauges

Range (mm)	Steps (mm)	No. of pieces
Special piece	1.0005	1
1st series 1.001 to 1.009	0.001	9
2nd series 1.01 to 1.49	0.01	49
3rd. series 0.5 to 24.5	0.5	49
4th series 25.0 to 100.0	25.0	4
Total pieces		112

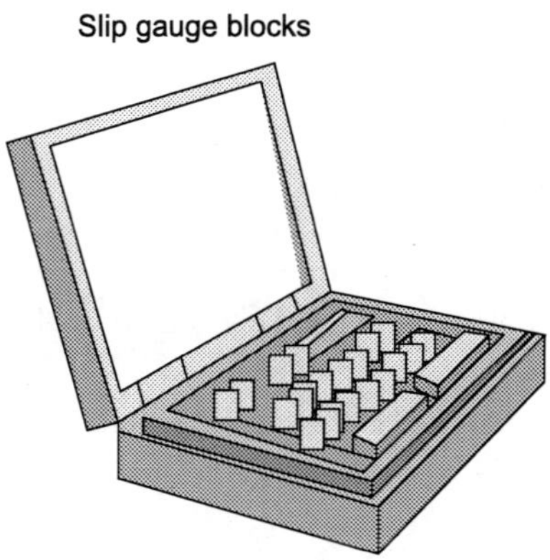

Fig. 15 20

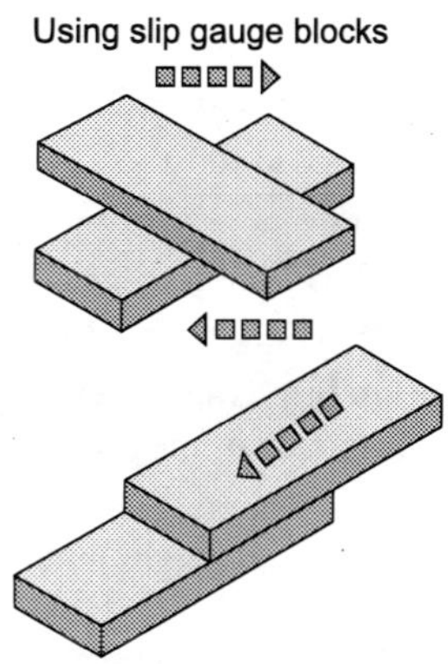

Fig. 15 21

Set of 103 pieces:

Different sets slip gauges

Range (mm)	Steps (mm)	No. of pieces
1st series 1.005		1
2nd series 1.01 to 1.49	0.01	49
3rd. series 0.5 to 24.5	0.5	49
4th series 25.0 to 100.0	25.0	4
Total pieces		103

Set of 78 pieces:

Range (mm)	Steps (mm)	No. of pieces
1.0025	—	1
1005	—	1
1.0075	—	1
1.01 to 1.49	0.01	49
0.5 to 9.5	0.5	19
10.0 to 50.0	10.0	5
75.0 & 100.0	—	2
Total pieces		78

Set of 47 pieces:

Range (mm)	Steps (mm)	No. of pieces
1st series 1.005	—	1
2nd series 1.01 to 0.09	0.01	9
3rd. series 1.0 to 1.9	0.1	9
4th series 1.0 to 24.0	1.0	24
5th series 25.0 to 100..0	25.0	4
Total pieces	—	47

1. **Worker Slip Gauge Set:** These gauges belong to the "B" class. It used by the technicians in workshops. This gauge is for lesser accuracy.
2. **Inspection Slip Gauge Set:** This gauge belongs to the "A" class. This is used by the inspectors to test the accuracy of the jobs made by the workers.
3. **Master Slip Gauge Set.** This gauge belongs to "AA" class. It has the highest accuracy.. The accuracy of other gauges and tools is tested with these gauge sets.

15.13 SHAPES OF GAUGES

According to the jobs and utility of testing, the shapes of gauges are of three types:

- Rectangular gauge block
- Square gauge block
- Angle gauge block

For using the above gauge blocks, different types of attachments have also to be used such as foot block, straight edge, scriber, adjustable holder, etc.

15.14 USES OF SLIP GAUGE

This gauge is generally used for the following functions:

1. To inspect (or verify) micrometer and vernier callipers.
2. To set adjustable gauges such as snap gauges.

3. To use sin bar.
4. For exact marking of a job.
5. To make solid snap gauge and caliper gauges.
6. To verify measurement of a job with the help of comparator.
7. To have comparative measurement of any job, with the help of an indicator.

For using slip gauge in any kind of job, more than one gauge blocks are joined. The act of joining gauge blocks is known as ringing.

15.15 RINGING ACTION

Two or more gauge blocks are joined in the following manner:

1. The gauging surface of the gauge blocks should be cleaned properly with tissue paper or chamois leather. At times, carbon tetrachloride is also used to clean these blocks.
2. The gauging surface of both the gauge blocks should be kept in a crossed manner.
3. Now the gauge blocks should be turned in such a way that their gauging surfaces comes in a line.
4. Again, the gauge blocks should be withdrawn and brought out to fifty per cent.
5. Now again, the upper gauge block should be pushed back and brought back to parallel position.

After this action both the gauge blocks will stick with each other and even air cannot pass through these blocks. If we want to separate them by pulling them directly, we cannot do so. These gauge blocks can only be separated by pushing them back. In the same way, other gauge blocks can also be joined. When all the blocks stick together they are in a position to stick together when they are hanged.

Fig. 15.22 Ring of slip gauges

15.16 SINE BAR

A sine bar is a precision measuring instrument for checking and setting of angles.

The principle of a sine bar:

The principle of the sine bar is based on the trigonometrical function.

In a right angled triangle the function known as sine of the angles is the relation ship existing between the opposite side to the angle the hypotenuse.

It may be noted that for setting the sine bar to different angles slip gauges are used.

A surface plate or marking table provides the datum surface for the setup.

The sine bar, the slip gauges and the datum surface upon which they are set form a right angled triangle.

Sine of the angle ϕ = opposite side/hypotenuse.

Features

This is a rectangular bar made of stabilized chromium steel.

The surfaces are accurately finished by grinding and lapping.

Two precision rollers of the same diameter are mounted on either end of the bar. The centre line of the rollers is parallel to the top face of the sine bar.

There are holes drilled across the bar. This helps in reducing the weight, and also facilitates fixing on angle plate.

The length of the sine bar is the distance between the centres of the rollers. The commonly available sizes are 100 mm, 200 mm, 250 mm and 500 mm. The size of a sine bar is specified by its length.

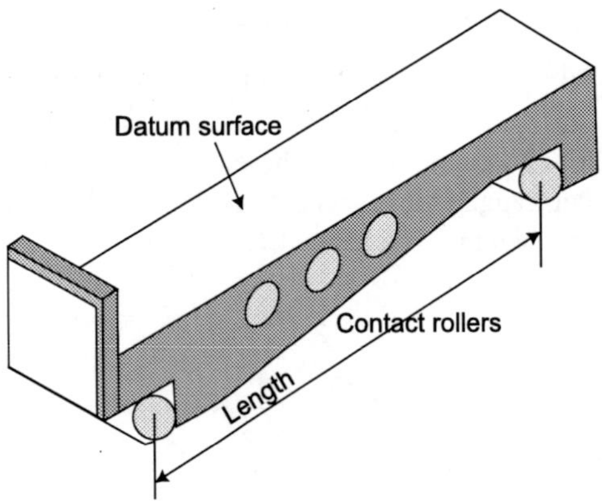

Fig. 15.23 Sine bar

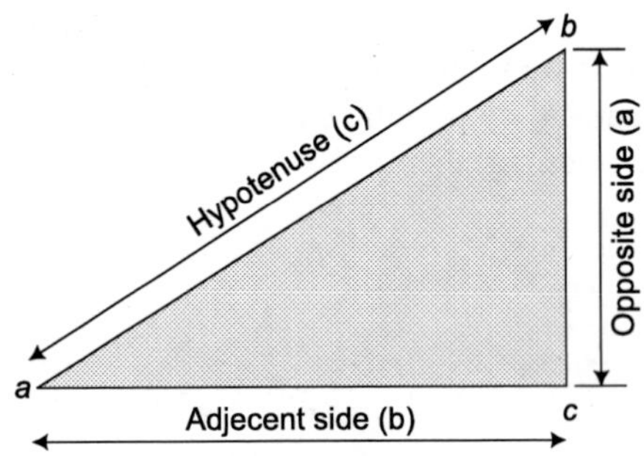

Fig. 15.23a Principle of sine bar

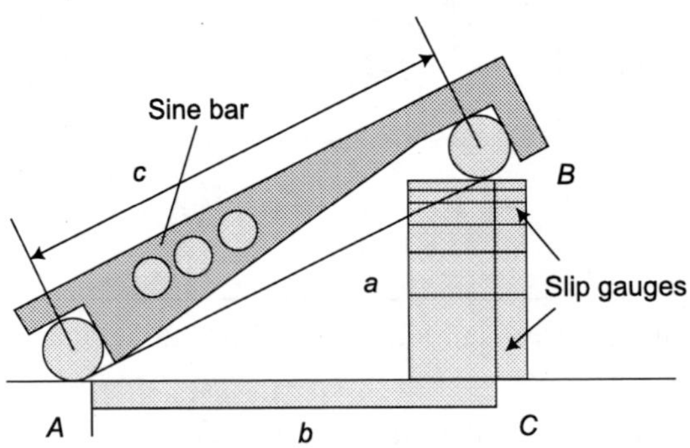

Fig. 15.23b Use of sine bar

Determining Taper Using Sine Bar and Slip Gauges: Sine bars provide a simple means of checking angles to a high degree of accuracy.

Determining taper using sine bar

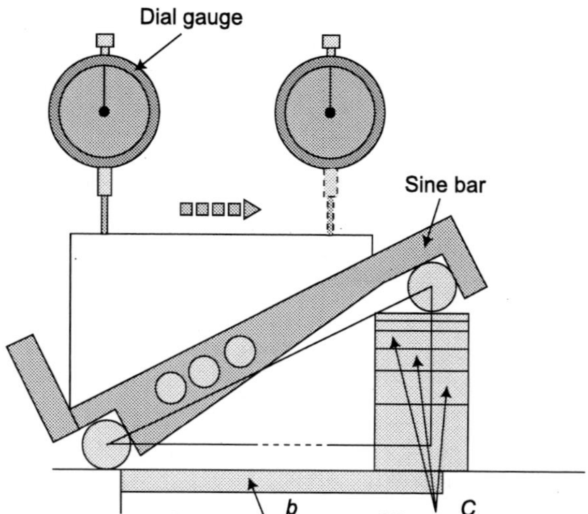

Fig. 15.24

The use of a sine bar is based on trigonometric function. The sine bar forms the hypotenuse of the triangle and the slip gauges the opposite side.

Checking the Correctness of a Known Angle: For this purpose first choose the correct slip gauge combination for the angle to be checked.

The component to be checked should be mounted on the sine bar after placing the selected slip gauges under the roller.

A dial test indicator is mounted on a suitable stand or vernier height gauge. The dial test indicator is then set in the first position and the dial is set to zero.

Move the dial to the other end of the component (second position). If there is any difference then the angle is incorrect. The height of the slip gauge pack can be adjusted until the dial test indicator reads zero on both ends. The actual angle can then be calculated and the deviation, if any, will be the error.

Method of Calculating the Slip Gauge Height: To determine the height of slip gauges for an angle of 25° using a sine bar of 200 mm long.

$$\text{sine } \Phi = a/c = 25°;\ a = c \text{ X sine } \Phi = 200 \times 0.4226;\ a = 84.52 \text{ mm.}$$

Therefore the height of the slip gauge required is 84.52 mm.

Note: The value of sine Φ can be got from mathematical tables (natural trigonometrical function).

Tables are also available with readily worked out sine bar constants for standard length sine bars.

15.17 DIAL TEST INDICATORS

Dial test indicators are instruments of high precision, used for comparing and determining the variation in the sizes of a component. These instruments cannot give the direct reading of the sizes like micrometers and vernier calipers. A dial test indicator magnifies small variations in sizes by means of a pointer on a graduated dial. This direct reading of the deviations gives an accurate picture of the conditions of the parts being tested.

The magnification of the small movement of the plunger or stylus is converted into a rotary motion of the pointer on a circular scale.

Two types are in use according to the method of magnification

1. Plunger type (Fig. 15.25)
2. Lever type. (Figs. 15.26 & 15.27)

The plunger type dial test indicator:

1. Pointer
2. Rotatable bezel
3. Bezel clamp
4. Back lug
5. Transparent dial cover
6. Stem
7. Plunger
8. Anvil
9. Revolution counter

For converting the linear motion of the plunger, a rack and pinion mechanism is used.

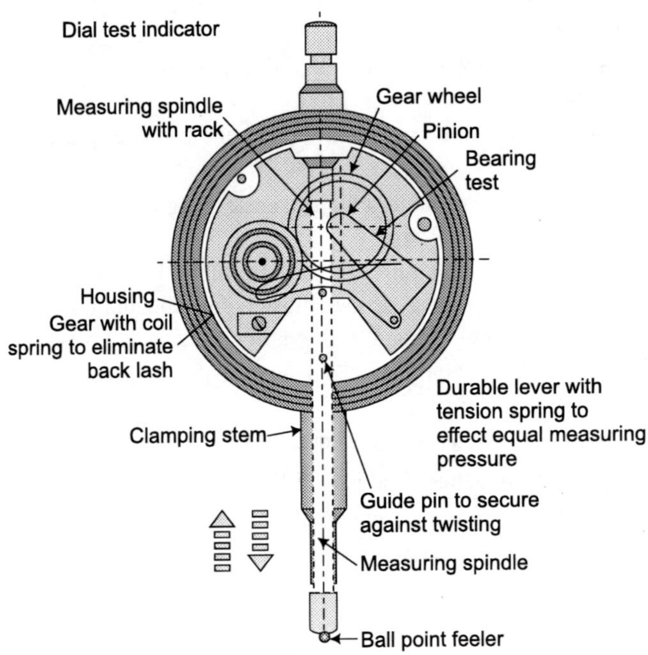

Fig. 15.25

The Lever Type Dial Test Indicator (Fig. 15.27 and 15.28): In this type of dial test indicators, the magnification of the movement is obtained by the mechanism of the lever and scroll.

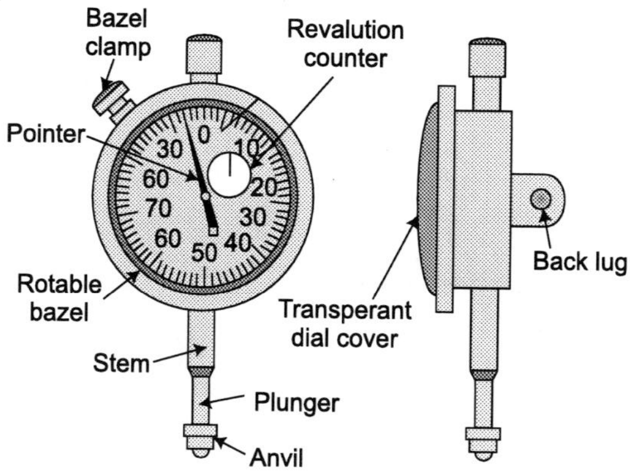

Fig. 15.26

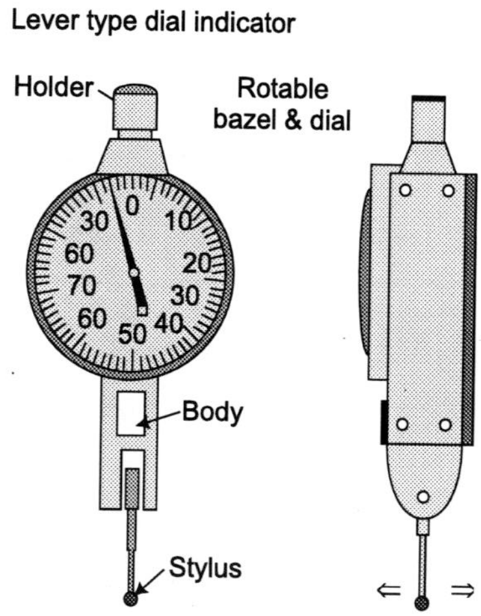

Fig. 15.27

It has a stylus with a ball- type contact, operating in the horizontal plane.

This can be conveniently mounted on a surface gauge stand, and can be used in places where the plunger type dial test indicator application is difficult.

Features of Dial Test Indicators: An important feature of the dial test indicator is that the scale can be rotated by a ring bezel, enabling it to be set readily to zero.

Many dial test indicators read plus in a clockwise direction from zero, and minus in the anticlockwise directions so as to give plus and minus indications.

Uses:

- To compare the dimensions of a workpiece against a known standard, e.g., slip gauges.
- To check plane surfaces for parallelism and flatness.
- To check parallelism of shafts and bars.
- To check concentricity of holes and shafts.

Indicator Stands

Dial test indicators are used in conjunction with stands for holding them so that the stand itself may be placed on a datum surface or machine tools.

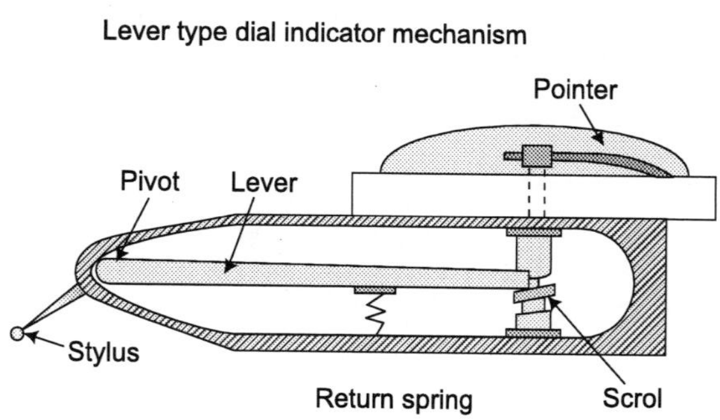

Fig. 15.28

Types of stands for dial indicators

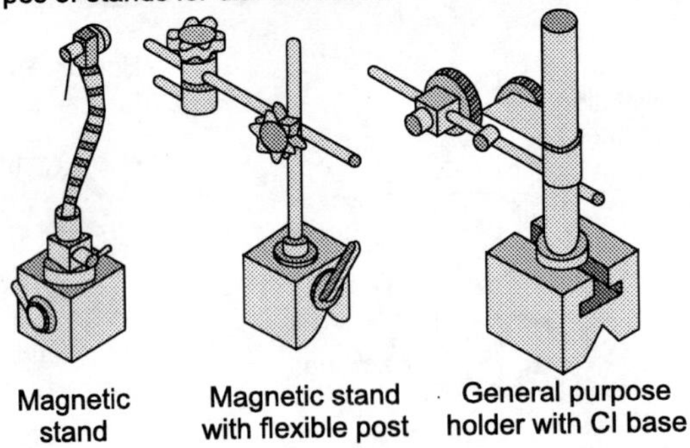

Fig. 15.29

Different types of stands are:

- Magnetic stand with universal clamp
- Magnetic stand with flexible post
- General purpose holder with cast iron base.

EXERCISE

1. What are the uses of gauges in workshops?
2. Which are the main gauges used by a fitter?
3. How do we classify gauge?
4. Why do we use master inspection gauge?
5. What is the difference between gauge and other measuring tools?
6. Briefly describe plug gauge, ring gauge and span gauge.
7. What are the different types of radius gauge?
8. What is the difference between drill gauge and point gauge?
9. Draw a sketch of filler gauge and describe.
10. What is the use of screw pitch gauge?
11. What is the working principle of a dial test indicator. Explain with the help of a sketch?
12. Explain the use of sine bar with the help of a sketch.
13. Of how many types wire gauges are there and what are their uses?
14. Draw a sketch of try square and explain its uses.
15. What do you mean by "Go" and "NOT GO" gauges?
16. Comment on the following:
    (a) Centre gauge
    (b) Receiving gauge
    (c) Caliper gauge
    (d) Angle gauge
    (e) Taper gauge

Part Two

Drilling Machines

Chapter 1

For boring holes in metal or wood, drill and drilling machine is used. Drill is a tool which is fitted in a drill machine and then moved. When the drill rotates on the metal it bores a hole. As different types of drills are used for boring holes in different metals likewise there are different types of drilling machines. These machines can be divided into two parts.

1.1 PORTABLE DRILL MACHINE

For boring holes in small jobs, this type of drill machine is used. Besides, it is used for those jobs which cannot be brought to the other fixed drilling machines. The portable drill machines are generally of five types. These can be easily moved from one place to another.

1.1.1 Hand Drill Machine (Fig. 1.1)

It is driven manually. It has a handle on the upper side, which we hold with our left hand. Pressure is given with this handle. A chuck is fixed in the spindle on the lower side to hold the drill. In between a gear is fixed in which a handle is attached for driving the drill. When we move this handle, the drill moves through the medium of a small gear. Holes can be bored with this up to ¼ inch.

They are used in structural fabrication, sheet metal and carpentry, particularly where electricity or pneumatic supply is not available.

The bevel gear type drilling machine is used for drilling small diameter holes up to 6 mm.

1.1.2 Breast Drill Machine (Fig. 1.2)

This drill is bigger in size and capacity in comparison to hand drill machine. Its shape is like a hand drill machine. But instead of a handle on the upper side there is a coveted iron breast plate. This breast plate is pressed with breast and pressure is exerted on the drill. Two additional bevel gears are also provided which are fitted in the spindle in order to increase the speed. It is generally used in repair works. Holes up to 1 inch diameter can be bored with this machine.

The breast drilling machine is used for drilling holes of larger diameter as more pressure can be exerted. Drills between 6 mm and 12 mm can be used on this machine.

1.1.3 Ratchet Drilling Machine (Fig. 1.3)

Its capacity is more than hand drill machine and breast drill machine. It is used for boring holes in those heavy jobs, which cannot be fitted on the fixed drill machines. In this type of drill machines ratchet shank twist is used which is square and also tapered. In this machine, in place of chuck, tapered square hole is there in its spindle in which ratchet shank drill fits easily. A straight drill post is fitted in its base plate. There is an arm on its drill post with which it can be set at the desired height. Ratchet and drill is also attached to the arm. It is driven (or operated) with a handle of 2 to 4 feet in length. Two dowel are there in its handle, which fit in the ratchet and move the spindles and the drill. When the drill is moving with the dowel, if we move the handle in the opposite direction the dowel ratchet slips and for that period the spindle remains fixed. In this way, slowly holes are made in heavy jobs. In order to give feed to the machine, a sleeve having threads is fitted above the spindle which is gradually moved with hands.

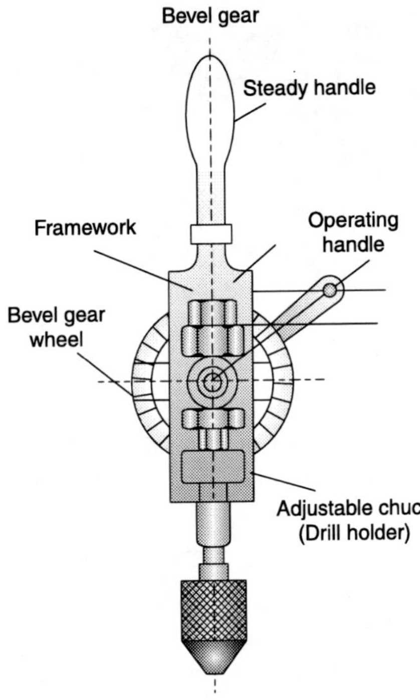

Fig. 1.1 Hand drilling machine

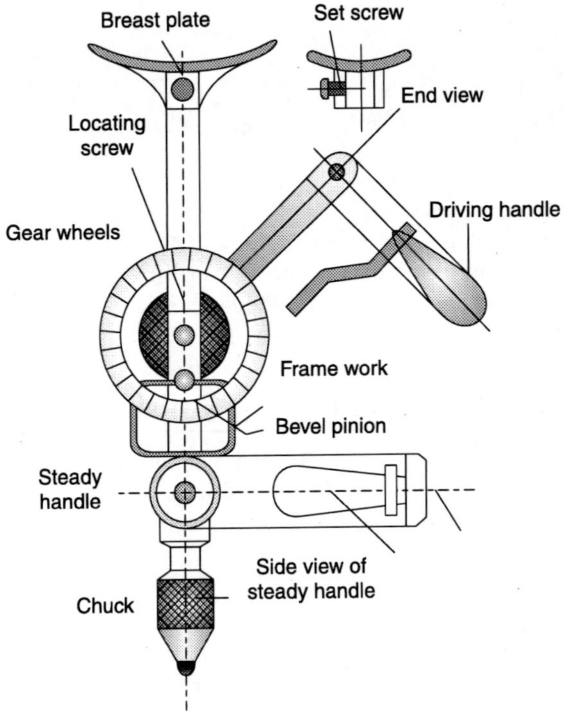

Fig. 1.2 Breast drill

1.1.4 Electrical Hand Drill Machine (light duty) [Fig. 1.4]

This type of machine is driven by electric power instead of hand power. An electric motor is fitted with this machine. A spindle is fitted on its motor shaft with the help of a few gears. On this spindle the drill chuck is mounted. On one side a handle is built according to convenience and a switch to put on and off the machine is fitted within the reach of fingers. These machines can be driven both with A/C and D/C current. According to their working capacity they are of two types: Light duty electrical hand drill machines, in which drills of 1/4 inch diameter are used.

Electric hand drilling is a light duty machine and has, usually, a single speed.

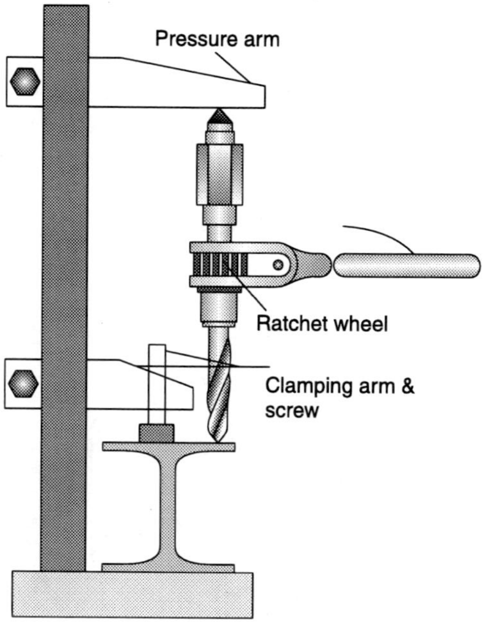

Fig. 1.3 Ratchet drilling machine

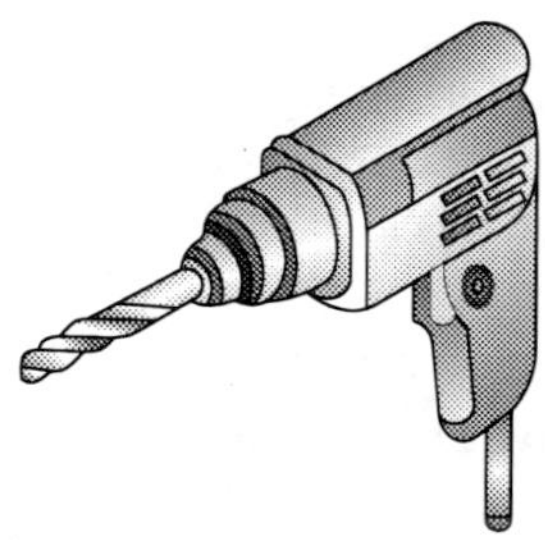

Fig. 1.4 Electrical hand drill

1.1.5 Heavy Duty Electrical Hand Drill Machine (Fig. 1.5-1.6)

This drill has an additional feature; the drill speed can be varied through a system of gears. This is particularly useful for drilling larger diameter holes, i.e., 1 inch diameter drills are used. Neat and clean holes are bored with these machines without using more power. These can also be used for other works such as filing or grinding; attaching rotary files or grinding wheels.

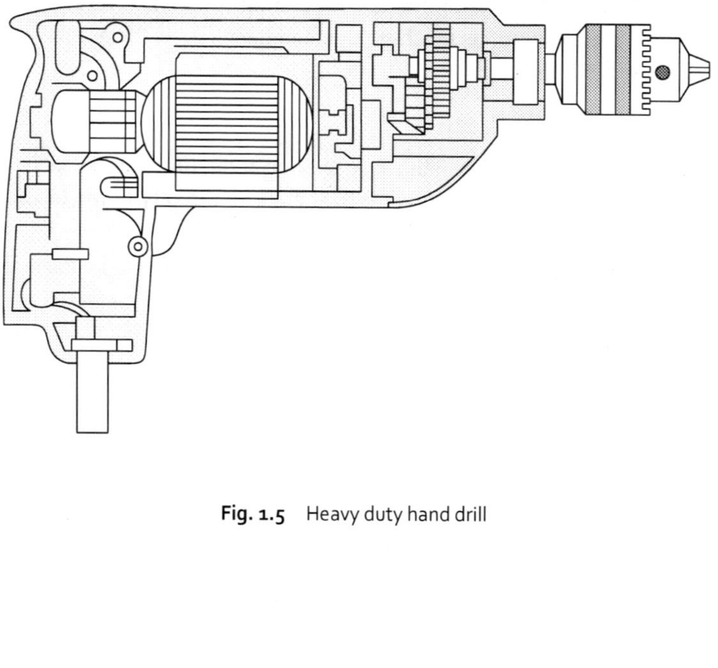

Fig. 1.5 Heavy duty hand drill

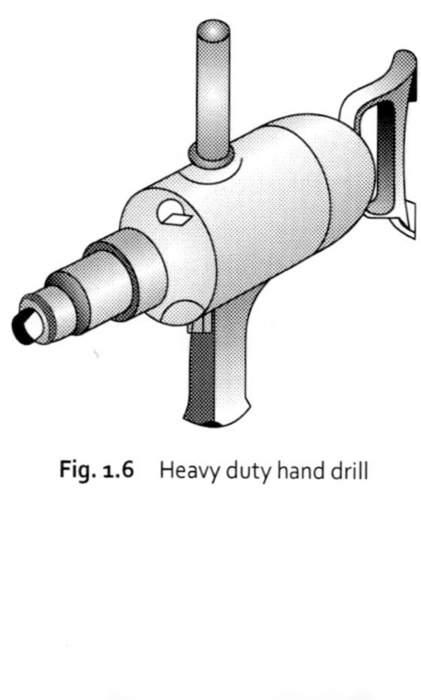

Fig. 1.6 Heavy duty hand drill

1.1.6 Pneumatic Drill Machine (Fig. 1.7)

Its construction and way of working is almost similar to the electrical hand drill machine. The only difference is that it does not have electric motor and wires etc. Drill is rotated by the compressed air power. To provide air, air valve and plunger is fitted. Air is supplied by an air compressor which is connected with the pressure line. A flow control valve provided on the handle itself to start or stop the supply of air and varies the speed of rotation. Where use of electrical hand drill is risky, pneumatic drill machine is used where electrically operated drills are prohibited; e.g. in the sea or other places of water. In modern days use of pneumatic power is very common and handy.

1.1.7 Stationary Drill Machine

Those machines which can be fixed at one place on the foundation are called stationary or fixed drill machines. These are used for drilling, reaming and tapping etc. of heavy jobs. These are of varied capacity. For using these machines, it is necessary to bring the jobs to the machine table.

These drill machines are of the following types:

1. Sensitive Drill Machines.
2. Pillar Type Drill Machines.
3. Column Type Drill Machines.
4. Radial Drill Machines.
5. Gang Drill Machines.
6. Multi Spindle Drill Machines.

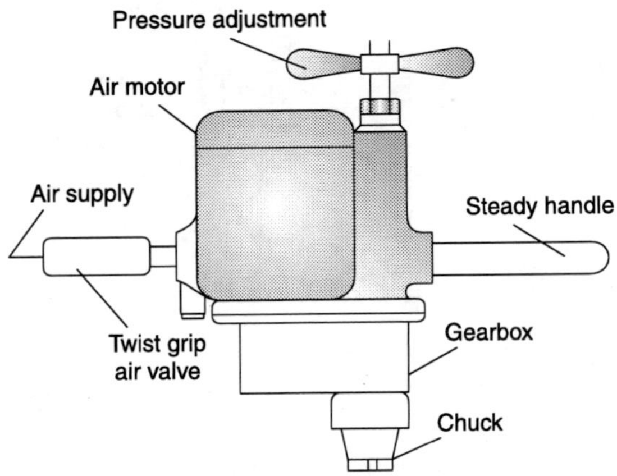

Fig. 1.7 Pneumatic hand drill

1.1.8 Sensitive Drill Machine (Fig. 1.8)

This is a medium class drill machine. Its speed is comparatively very fast. These are of hand-operated as well as electrically operated types. These are generally fixed on work bench by bolt-nut. These machines are considered best for drilling and tapping. There is no provision for self-regulating feed. Therefore, feed is given by hand with the help of rack and pinion. Because in this machine the feed depends on the alertness of the worker that is why it is also called the sensitive drill machine.

1.1.9 Bench Type Sensitive Drill Machine (Fig. 1.9)

The simplest type of drilling machine is the sensitive type. This is used for light duty work. This is capable of drilling holes up to 12.5 mm diameter. The drills are fitted in the chuck or directly in the tapered hole of the machine spindle.

For normal drilling, the work – surface is kept horizontal. If the holes are to be drilled at an angle, the table can be tilted.

Different spindle speeds are achieved by changing the belt in the stepped pulley.

1.1.10 Pillar Type Drill Machine (Fig. 1.10)

This type of machines is highly used in the workshops. Holes up to diameter of 25 mm. can be done

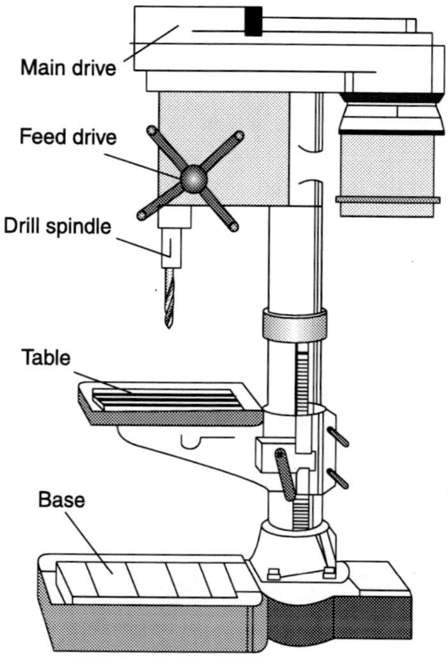

Fig. 1.8 Sensitive bench drill

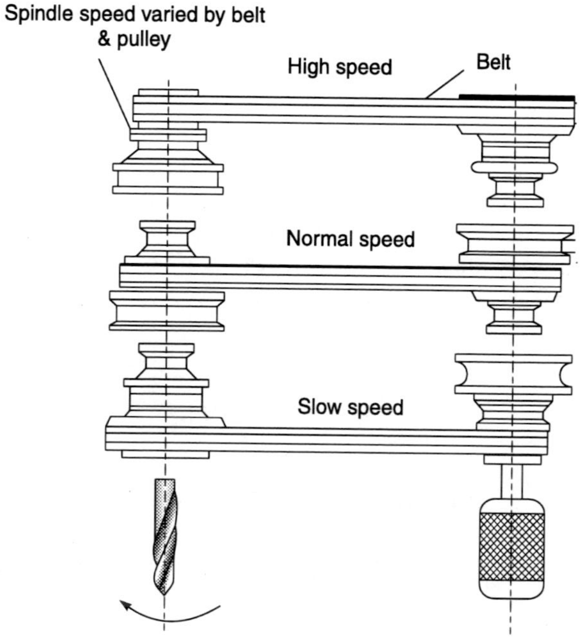

Fig. 1.9 Bench type sensitive drill machine

This is an enlarged version of the sensitive bench drilling machine. These drilling machines are mounted on the floor and driven by more powerful electric motors. They are used for heavy duty work. Pillar drilling machines are available in different sizes

Large machines are provided with a rack and pinion mechanism for moving the table for setting the work.

Main motion and feed are imparted to the drill by the drilling machine. The variety of workpiece shapes, size, quality of holes to be made has led to the designing of several types of drilling machines. On most of the drilling machines, besides normal drilling work, countersinking, reaming and threading can also be performed.

The column holds the spindle, the main and the feed drive and machine table.

The drill spindle

It takes up the drilling tool. The drill spindle is guided in the sleeve. At the end of the spindle, towards the machine table, a taper hole is provided for fixing the drilling tools.

The main drive

It transmits the rotary motion from an electric motor or from a transmission to the drill spindle. In order to have different speeds, it is designed with step pulleys or change gear drive. There are also machines with infinitely variable transmission.

The feed drive

The feed drive gives linear feed motion to the drill spindle. The drill sleeve carries the gear rack **Fig. (1.11).** This is meshing with a gear which is turned by a hand lever The drill sleeve can be moved up and down in a bearing. In order that the drill spindle can follow this motion, it is held at the upper end by two ring nuts and at the lower end by the collar of the drill-spindle head. To reduce the friction an anti friction bearing is situated between the collar and sleeve. The upper end of the drill spindle is slideable in the V-belt pulley or in the gear which drives the drill spindle by a key, sliding in a longitudinal slot of the drill spindle. For vertical movement of the drill spindle in larger machines, a worm in connection with a worm wheel, is used. Very often, the automatic feed is affected by a sliding gear or dive key drive which is actuated by the main drive. By lever shifting, various feeds can be set.

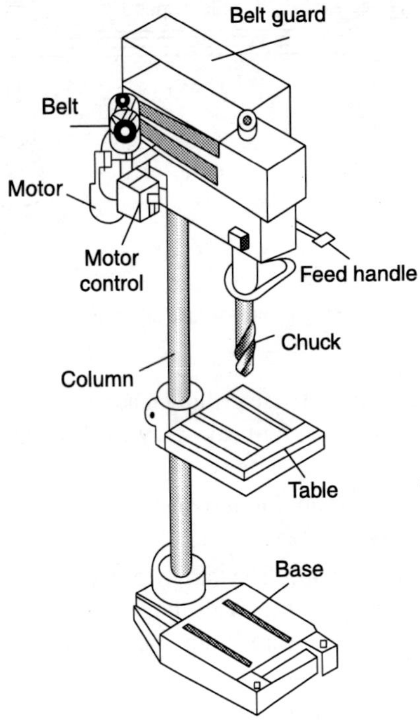

Fig. 1.10 Pillar type drill

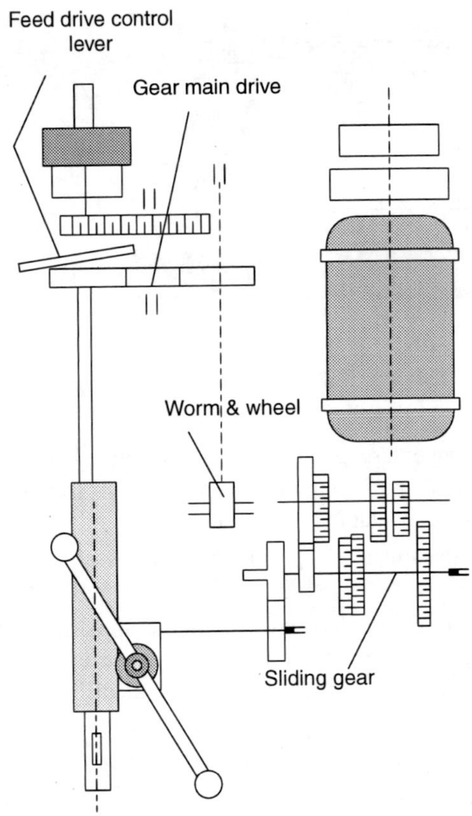

Fig. 1.11

In order to drill a hole of a certain depth, a stop **Fig. (1.12)** is used. Frequently, an automatic tripping device is provided which disengages the feed at reaching the desired depth of hole.

The working table

It takes up the workpiece. "T" slots serve for clamping. Grooves around the table drain off the coolant. With a crank handle, the table can be moved up and down by a gear and gear rack, and can be tightened with a lever.

Use of a column drilling machine: As a rule, for drilled holes, up to a diameter of 25 mm, it is disadvantageous that while drilling deep holes, the drill spindle emerges far from its bearings; the drill can then easily run off the centre.

The vertical drilling machine belongs to the category of stationary drilling machines, because it is fixed at one place in the workshop. Besides, there are also electric portable drilling machines and hand drilling machines of various designs.

1.1.11 Multi-Spindle Drilling Machine (Fig. 1.13)

The drilling head is equipped with several spindles which are driven by the main spindle. Several holes can be drilled in one operation. The machine is used in mass production.

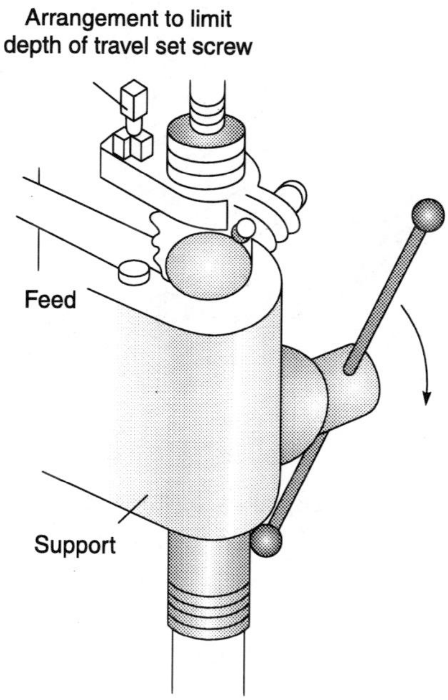

Fig. 1.12

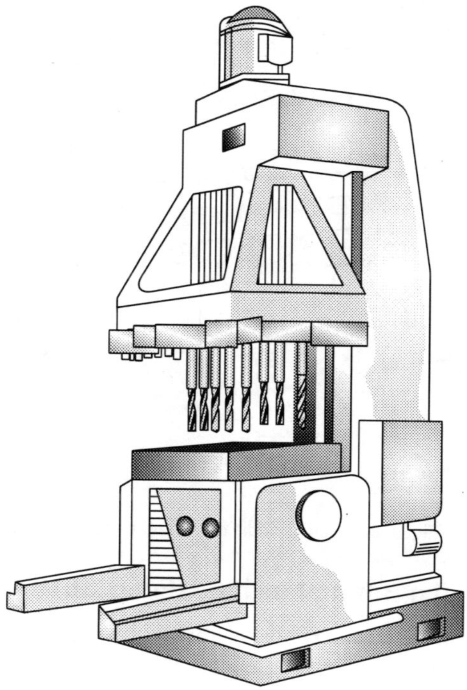

Fig. 1.13 Multi-spindle drilling machine

1.1.12 Gang-Spindle Drilling Machine (Fig. 1.14)

On one workpiece several working operations can be performed successively, e. g., drilling, countersinking and reaming. The machine is used in mass production.

1.1.13 Radial Drilling Machine (Fig. 1.15)

The drill spindle head is carried by the arm and can be shifted in radial direction. The arm can be swung around the column and adjusted in height. In modern machines, the drill spindle is driven by a flange motor which is seated on the drill spindle head. A large speed range enables the drilling of big holes and small holes. The workpiece is clamped on the working table which is provided with T-slots. Because of the many adjusting possibilities drilling can be done in different positions without reclamping being necessary.

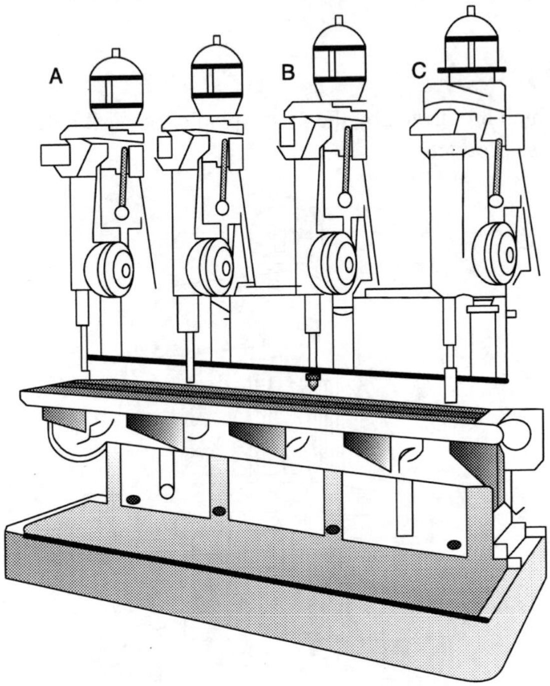

Fig. 1.14 Gang spindle drilling machine

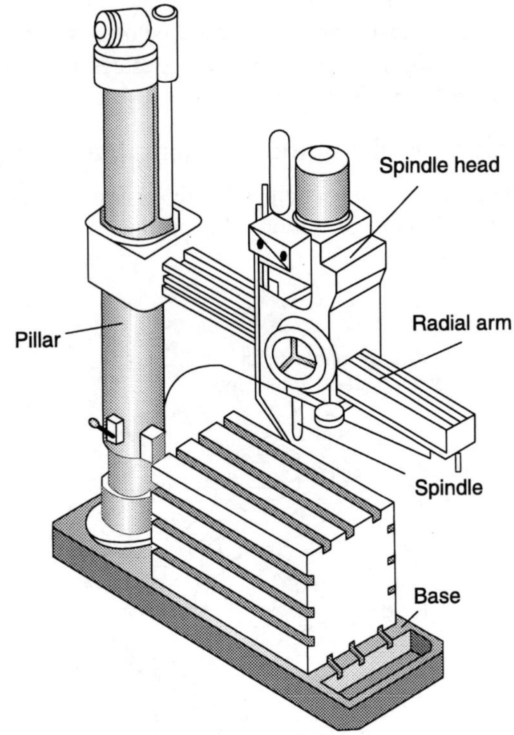

Fig. 1.15 Radial drilling machine

1.1.14 Jig Boring Machine (Fig. 1.16)

On a jig boring machine, holes with highly accurate centre-to-centre distances are bored. The work spindle is held in accurate bearings. The workpiece is clamped on the machine table which is designed as a compound table. This is adjustable in length and crosswise by means of spindles. With the help of measuring devices, centre distances with tolerances as small as 0.001 mm can be set.

1.1.15 Horizontal Boring Machine (Fig. 1.17)

It is used for boring, milling and turning work on complicated workpieces. The horizontal main spindle serves for holding boring and milling tools. It is rotated by a flange motor and is adjustable in longitudinal direction. By means of a gear, which is mounted in the headstock various number of revolutions and feeds can be obtained headstock can be moved up- and downwards on a vertical column.

An auxiliary column serves to support long boring bars. The workpieces are clamped on the machine table. It can be rotated and, moreover, can be shifted in longitudinal and transversal directions, thus the workpiece can be machined at various places in one setup.

There are also boring machines with fixed machine tables. In this case, the vertical column can be shifted in transversal direction. For the convenience of the operator and for the fast operation of the machine, all the levers are attached to the headstock.

Horizontal boring machines belong to the most versatile machine tools.

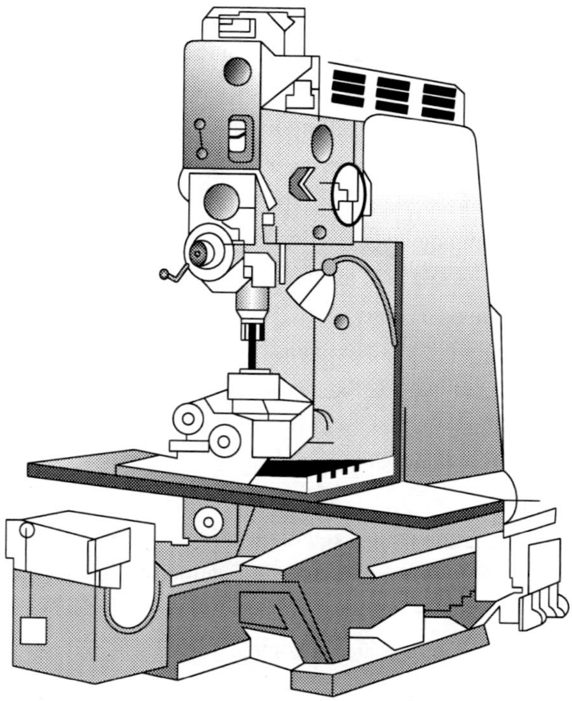

Fig. 1.16 Jig boring machine

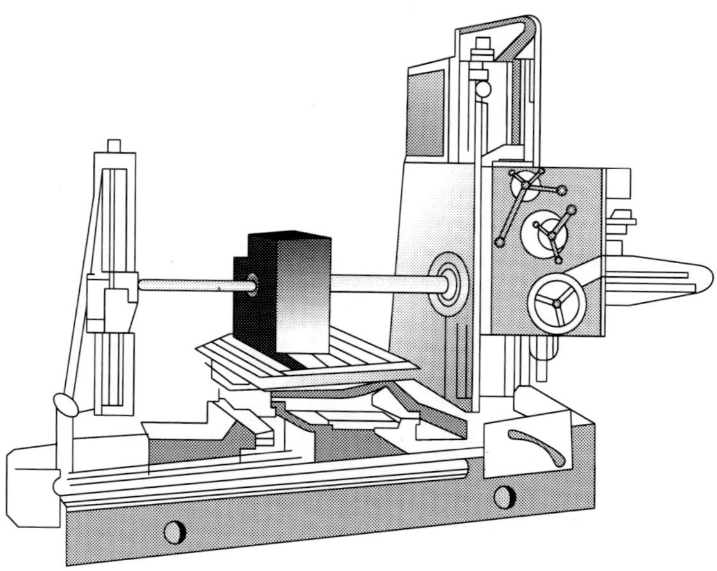

Fig. 1.17 Horizontal boring machine

EXERCISE

1. Name different types of drilling machines.
2. Sketch and describe in brief of a radial drilling machine.
3. How the size of a drilling machine is specified? Discuss.
4. Name various work holding devices of drilling machine. Describe one with sketch.
5. Explain the construction of the following parts of a drilling machine:
   (a) Base, (b) Drill head, (c) Spindle drive and feed mechanism.
6. What are different tool holding devices in a drilling machine?
7. Describe a tapping attachment in a drill machine.
8. List various drilling machine operations.
9. How the drill size for tapping is fixed? Explain.
10. What is a spade drill? When is it used? Sketch one.
11. What is trepanning? Describe.
12. Explain counter-boring and counter-sinking operations.
13. List various types of drills.
14. Describe twist drill nomenclature using sketches.
15. How is the drill size specified? Discuss of the codification of drill sizes and types.
16. Draw a sketch of a simple twist drill with tapered shank and show its various elements.
17. Name the drill bit materials. Which material is mostly used?

Screw Thread

Chapter 2

An introduction is given in the chapter "7 and 8 of Part in Thread cutting by hand."

Screw is a cylinder or pipe with spiral ridge thread running around it outside (male) or inside (female).

In other words, spiral grooves of equal pitch and equal depth made on parabolic round cylindrical surface or conical round surface.

A screw thread is a ridge of uniform section formed helically on the surface of a cylindrical body.

Threads play a very important role in engineering field for manufacturing various articles of daily use. In this engineering and scientific age, machines are used every moment at every stage. Without threads most of the machines cannot exist.

2.1 SCREW-THREAD NOMENCLATURE

Knowledge of accepted terminology and symbols is necessary in order to identify or specify screw threads correctly.

The terms commonly applied to screw threads may be classified in four general groups, namely, those relating to

(1) Types of threads,

(2) Sizes and fit of mechanical parts in general,

(3) Geometrical elements of both straight and taper screw threads, and

(4) Dimensions of screw threads.

2.2 TERMS RELATING TO TYPES OF SCREW THREADS

Screw Thread

A screw thread (hereafter referred to as a thread), is a ridge, usually of uniform section and produced by forming a helical groove on the external or internal surface of a cylinder, or by forming a conical spiral on the external or internal surface of a cone or frustrum of a cone. A screw thread formed on a cylinder is known as a straight or parallel thread, to distinguish it from a taper screw thread, which is formed on a cone or frustrum.

Thread

A thread is a portion of a screw thread encompassed by one pitch. On a single start thread it is equal to one turn.

Single-start Thread

A single-start thread is one having the lead equal to the pitch.

Multiple-start Thread

A multiple-start thread is one in which the lead is an integral multiple of the pitch.

External Thread

An external thread is one on a cylindrical or conical external surface.

Internal Thread

An internal thread is one on a cylindrical or conical internal surface.

Right-hand Thread

A thread is a right-hand thread if, when viewed axially, it winds in a clockwise and receding direction. A thread is considered to be right-hand unless specifically indicated otherwise.

Left-hand Thread

A thread is a left-hand thread if, when viewed axially, it winds in a counterclockwise and receding direction. All left-hand threads are designated LH.

Complete Thread

The complete or full-form thread is that cross section of a threaded length having full form at crest and root. Note: Formerly in pipe-thread terminology this was referred to as "the perfect thread," but that is no longer considered desirable.

Incomplete Thread

An incomplete thread is a threaded profile having either crests or roots, or both crests and roots not fully formed, resulting from their intersection with the cylindrical or end surface of the work or the vanish cone. It may occur at either end of the thread. Note: formerly in pipe-thread terminology this was referred to as "the imperfect thread," but that is no longer considered desirable.

Lead Thread

The lead thread is that portion of the incomplete thread that is fully formed at root but not fully formed at crest which occurs at the entering end of either external or internal threads.

Vanish thread (partial thread, washout thread; or thread run out)

A vanish thread is that portion of the incomplete thread which is not fully formed at root or at crest and root. It is produced by the chamfer at the starting end of the thread-forming tool. Note: Threads produced employing a cam-actuated single-tool process (Cridan process) or similar type of equipment may have fully formed roots which run out on a vanish cone which is formed by the tool withdrawal pattern.

Effective Thread

The effective (or useful) thread includes the complete thread, and those portions of the incomplete thread which are fully formed at the root but not at the crest (in taper pipe threads this includes the so-called "black-crest threads"), thus excluding the vanish thread.

Total Thread

The total thread includes the complete and all the incomplete threads, including the vanish thread.

Classes of Threads

Classes of threads are distinguished from each other by the amounts of tolerance or tolerance and allowance specified.

Thread Series

Thread series are groups of diameter/pitch combinations distinguished from each other by the number of threads per inch applied to specific diameters.

2.3 TERMS RELATING TO SIZE AND FIT OF MECHANICAL PARTS

Dimension

A dimension is a geometrical characteristic such as diameter, length, angle, or centre distance.

Size

Size is a designation of magnitude. When a value is assigned to a dimension, it is referred to hereafter as the size of that dimension. Note: It is recognized that the words "dimension" and "size" are both used at times to convey the meaning of magnitude.

Nominal Size

The nominal size is the designation which is used for the purpose of general identification.

Basic Size

The basic size is that size from which the limits of size are derived by the application of allowances and tolerances.

Reference Size

A reference size is a size without tolerance used for only information purposes and does not govern manufacturing or inspection operations.

Design Size

The design size is the basic size with allowance applied, from which the limits of size are derived by the application of tolerances. If there is no allowance, the design size is same as the basic size.

Actual Size

An actual size is a measured size.

Limits of Size

The limits of size are the applicable maximum and minimum sizes.

2.4 MAXIMUM MATERIAL LIMIT

A maximum material limit is that limit of size that provides the maximum amount of material for the part. Normally, it is the maximum limit of size of an external dimension or the minimum limit of size of an internal dimension.

2.5 MINIMUM MATERIAL LIMIT

A minimum material limit is that limit of size that provides the minimum amount of material for the part. Normally, it is the minimum limit of size of an external dimension or the maximum limit of size of an internal dimension. Note: An example of exceptions is an exterior corner radius where the maximum radius is the minimum material limit and the minimum radius is the maximum material limit.

Allowance

An allowance is a prescribed difference between the maximum material limits of mating parts. It is the minimum clearance (positive allowance) or maximum interference (negative allowance) between such parts.

Tolerance

A tolerance is the total permissible variation of a size. The tolerance is the difference between the limits of size.

Tolerance limit

A tolerance limit is the variation positive or negative, by which a size is permitted to depart from the design size.

Fit

Fit is the general term used to signify the range of tightness or looseness which may result from the application of a specific combination of allowances and tolerances in the design of mating parts.

2.6 TERMS RELATING TO GEOMETRICAL ELEMENTS OF SCREW THREADS

Axis of Thread

The axis of a thread is the axis of its pitch cylinder or cone.

Major Cylinder

The major cylinder is one that would bind the crests of an external straight thread or the roots of an internal straight thread.

Sharp Major Cylinder

The sharp major cylinder is one that would bind the sharp crests of an external straight thread or the sharp roots of an internal straight thread.

Major Cone

The major cone is one that would bind the crests of an external lapel thread or the roots of an internal taper thread.

Sharp Major Cone

The sharp major cone is one having an apex angle equal to that of the pitch cone, the surface of which would bind the sharp crests of an external taper thread or the sharp roots of an internal taper thread.

Pitch Cylinder

The pitch cylinder is one of such diameter and location of its axis that its surface would pass through a straight thread in such a manner as to make the widths of the thread ridge and the thread groove equal. It is therefore located equidistantly between the sharp major and minor cylinders of a given thread form. On a theoretically perfect thread these widths are equal to one-half of the basic pitch.

Pitch Cone

The pitch cone is one of such apex angle and location of its vertex and axis that its surface would pass through a taper thread in such a manner as to make the widths of the thread ridge and the thread groove equal. It is therefore located equidistantly between the sharp major and minor cones of a given thread form. On a theoretically perfect taper thread these widths are equal to one-half of the basic pitch.

Minor Cylinder

The minor cylinder is one that would bind the roots of an external straight thread or the crests of an internal straight thread.

Sharp Minor Cylinder

The sharp minor cylinder is one that would bind the sharp roots of an external straight thread or the sharp crests of drill's internal straight thread.

Minor Cone

The minor cone is one that would bind the roots of an external-taper thread or the crests of an internal-taper thread.

Sharp Minor Cone

The sharp minor cone is one having an apex angle equal to that of the pitch cone, the surface of which would bind the sharp roots of an external-taper thread or the sharp crests of an internal-taper thread.

Pitch Line

The pitch line is a generator of the cylinder or cone specified in the definitions of pitch cylinder and pitch cone.

Form of Thread

The form of a thread is its profile in an axial plane for a length of one pitch of the complete thread.

Basic form of Thread

The basic form of a thread is the theoretical profile of the thread for a length of one pitch in an axial plane, from which the design forms of the threads for both the external and internal threads are developed.

Design forms of Thread

The design forms for a thread are the maximum material forms permitted for the external and internal threads. In practice, however the forms of roots are indeterminate contours not encroaching on the maximum material form of the mating thread when assembled.

Fundamental Triangle

The fundamental triangle is the triangle whose corners coincide with three consecutive intersections of the extended flanks of the basic form.

Flank

The flank (or side) of a thread is either surface connecting the crest with the root.

The flank-surface intersection with an axial plane is theoretically a straight line.

Leading Flank

The leading flank of a thread is the one which, when the thread is about to be assembled with a mating thread, faces the mating thread.

Following Flank

The following flank of a thread is the one that is opposite to the leading flank.

Load Flank

The load flank is that which takes the externally applied axial load in an assembly. The term is used particularly in relation to buttress and other similar threads.

Clearance Flank

The clearance flank is that which does not take the externally applied axial load in an assembly.

Crest

The crest is that surface of the thread which joins the flanks of the thread and is farthest from the cylinder or cone from which the thread projects.

Root

The root is that surface of the thread which joins the flanks of adjacent thread forms and is identical with or immediately adjacent to the cylinder or cone from which the thread projects.

Sharp crest (crest apex)

The sharp crest is the apex formed by the intersection of the flanks of a thread when extended, if necessary, beyond the crest.

Sharp root (root apex)

The sharp root is the apex formed by the intersection of the adjacent flanks of adjacent threads when extended, if necessary, beyond the root.

Base

The base of a thread section is that part which coincides with the cylindrical or conical surface from which the thread projects.

Vanish Cone

The vanish cone is a cone the surface of which would bind the roots of the vanish thread formed by the lead or chamfer of the threading tool, or in the case employing a cam-actuated single-tool process (Cridan process) or similar type of equipment, which is formed by the tool withdrawal pattern **Fig. (2.1)**.

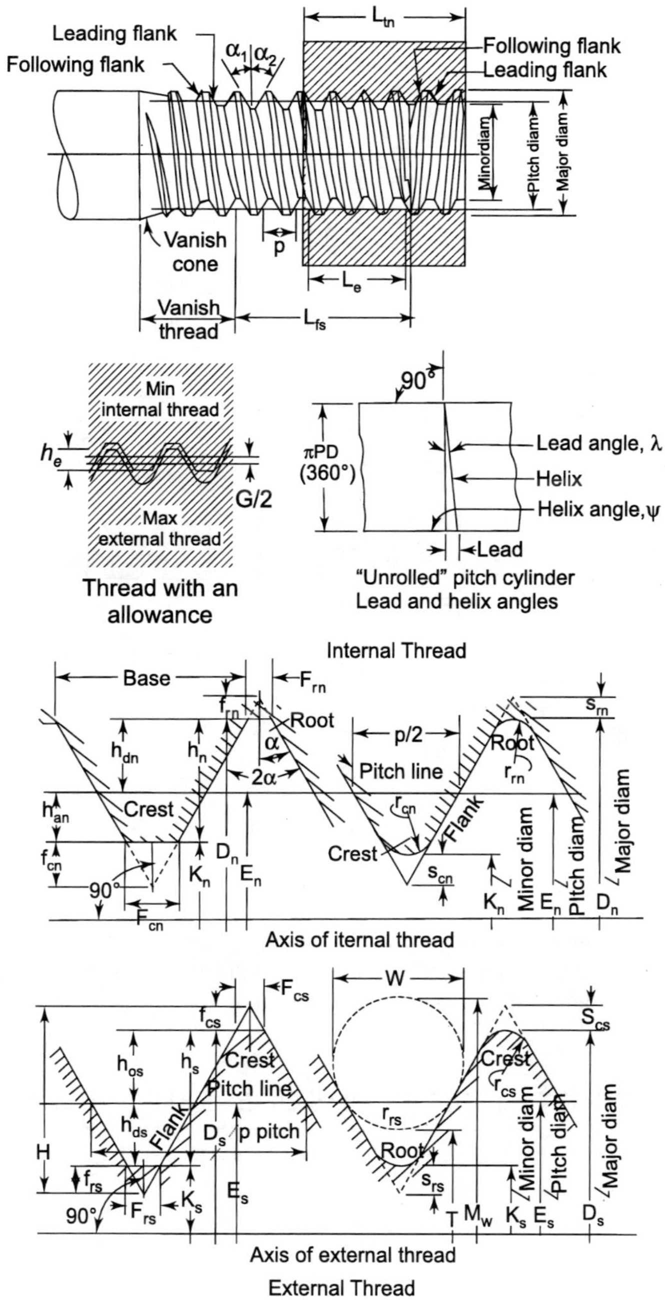

Fig. 2.1

Plane of Vanish Point

The plane of vanish point of an external thread is the intersection of generators of the vanish cone with generators of the largest major diameter of the thread, **Fig. (2.2)**.

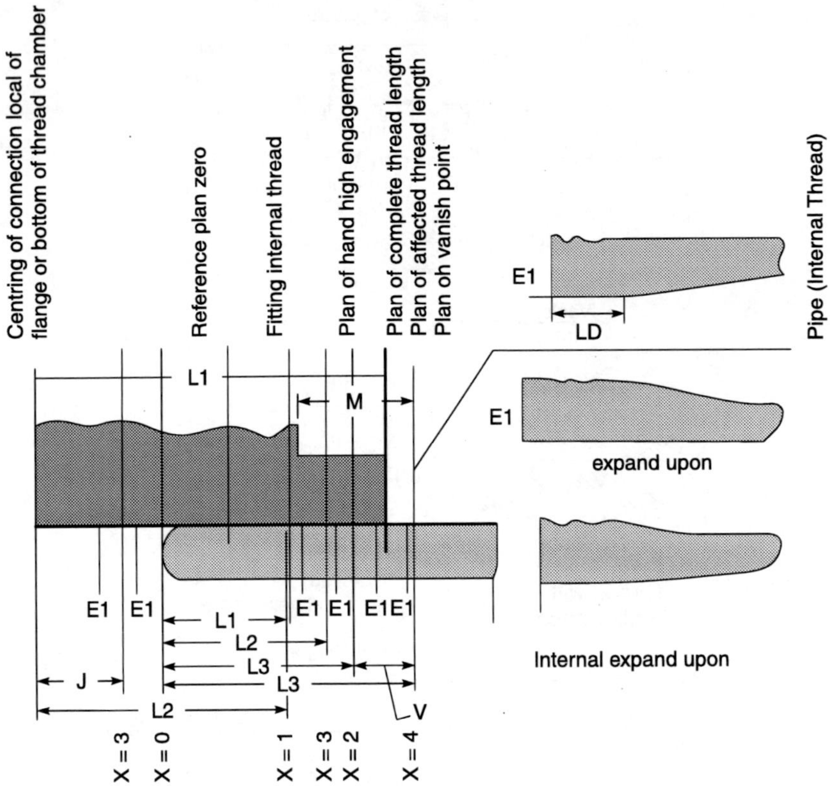

Fig. 2.2 Plane of vanish point

Blunt-start or Blunt-end Thread

"Blunt start" or "blunt end" designates the removal of the incomplete thread at the end of the thread. This is a feature of threaded parts that are repeatedly assembled by hand, such as hose couplings and thread plug gages, to prevent cutting of hands and crossing of threads, and which was formerly known as a Higbee cut **Fig. (2.3)**.

2.7 TERMS RELATING TO DIMENSIONS OF SCREW THREADS

Pitch

The pitch of a thread having uniform spacing is the distance, measured parallel to its axis, between corresponding points on adjacent thread forms in the same axial plane and on the same side of the axis. The basic pitch is equal to the lead divided by the number of thread starts.

Lead

When a threaded part is rotated about its axis with respect to a fixed mating thread, the lead is the axial distance moved by the part in relation to the amount of angular rotation. The basic lead is commonly specified as the distance to be moved in one complete rotation. It is necessary to distinguish measurement of lead from measurement of pitch, as uniformity of pitch measurements does not assure uniformity of lead. Variations in either lead or pitch cause the functional diameter of thread to differ from the pitch diameter.

Helix Variation

Helix variation of a thread is a wavy deviation from true helical advancement. The "helical path" includes the helix with its superimposed variation and is measured either as the maximum deviation from the true helix or as the "cumulative pitch." The cumulative pitch is the distance measured parallel to the axis of the thread between corresponding points on any two thread forms whether or not they are in the same axial plane.

Threads Per Inch

The number of threads per length is the reciprocal of the pitch.

Turns Per Length

The number of turns per length is the reciprocal of the lead.

Included Angle

The included angle of a thread (or angle of thrcad) is the angle between the flanks of the thread measured in an axial plane.

Flank Angle

The flank angles are the angles between the individual flanks and the perpendicular to the axis of the thread, measured in an axial plane. A flank angle of a symmetrical thread is commonly termed the half angle of thread.

Lead Angle

On a straight thread, the lead angle is the angle made by the helix of the thread at the pitch line with a plane perpendicular to the axis. On a taper thread, the lead angle at a given axial position is the angle made by the conical spiral of the thread, with the plane perpendicular to the axis, at the pitch line **Fig. (2.1)**

Helix Angle

On a straight thread, the helix angle is the angle made by the helix of the thread at the pitch line with the axis. On a taper thread, the helix angle at a given axial position is the angle made by the conical spiral of the thread with the axis at the pitch line. The helix angle is the complement of the lead angle **Fig. (2.1)** Note: The helix angle was formerly defined in accordance with the present definition of lead angle.

Thickness of Thread Ridge

The thickness of thread ridge is the distance between the flanks of the thread ridge, normally measured parallel to the axis at the specified pitch radius. The thickness of thread ridge may be specified and measured parallel to the axis at any other specified radius. Note: The pitch radius is equal to one-half of the pitch diameter.

Width of Thread Groove

The width of the thread groove is the distance between the flanks of adjacent thread ridges normally measured parallel to the axis at the specified pitch radius. The width of a thread groove may be specified and measured parallel to the axis at any other specified radius.

Height of Fundamental Triangle

The height of the fundamental triangle of a thread, that is, the height of a sharp V thread, is the distance, measured radially, between the sharp major and minor cylinders or cones, respectively.

Height of Thread

The height (or depth) of thread is the distance, measured radially between the major and minor cylinders or cones, respectively. Note: In American practice the height of thread is often expressed as a percentage of three-fourths of the height of the fundamental triangle.

Addendum

The addendum of an external thread is the radial distance between the major and pitch cylinders or cones, respectively. The addendum of an internal thread is the radial distance between the major and pitch cylinders or cones, respectively.

Dedendum

The dedendum of an external thread is the radial distance between the pitch and minor cylinders or cones, respectively. The dedendum of an internal thread is the radial distance between the major and pitch cylinders or cones, respectively.

Crest Truncation

The crest truncation of a thread is the radial distance between the sharp crest (crest apex) and the cylinder or cone that would bound the crest.

Root Truncation

The root truncation of a thread is the radial distance between the sharp root (root apex) and the cylinder or cone that would bound the root.

Major Diameter

On a straight thread the major diameter is that of the major cylinder. On a taper thread the major diameter at a given position on the thread axis is that of the major cone at that position.

Minor Diameter

On a straight thread the minor diameter is that of the minor cylinder. On a taper thread the minor diameter at a given position on the thread axis is that of the minor cone at that position.

Pitch Diameter

On a straight thread the pitch diameter is the diameter of the pitch cylinder. On a taper thread, the pitch diameter at a given position on the thread axis is the diameter of the pitch cone at that position. On a single-start thread of perfect form and lead, it is also the length between intercepts of a line which is perpendicular to the thread axis and intersects thread flanks on opposite sides of the thread axis. Note: When the crest of a thread is truncated beyond the pitch line, the pitch diameter, pitch cylinder, or pitch cone would be based on a theoretical extension of the thread flanks."

Pitch diameter on the buttress casing thread is defined by the American Petroleum Institute, in API Standard 5B, as being midway between the major and minor diameters.

2.8 THREAD-GROOVE DIAMETER (SIMPLE EFFECTIVE DIAMETER)

On a straight thread the thread-groove diameter is the diameter of the coaxial cylinder, the surface of which would pass through the thread profiles at such points as to make the width of the thread groove equal to one-half of the basic pitch. It is the diameter yielded by measuring over or under cylinders (wires) or spheres (balls) inserted in the thread groove on opposite sides of the axis and computing the thread groove diameter as thus defined.

On a taper thread the thread-groove diameter is the diameter at a given position on the thread axis of the coaxial cone, the surface of which would pass through the thread profiles at such points as to make the width of the thread groove (measured parallel to the axis) equal to one-half of the basic pitch. It is the diameter yielded by measuring over or under cylinders (wires) or spheres (balls) inserted in the thread groove on opposite sides of the axis and computing the thread-groove diameter as thus defined.

Thread-ridge Diameter

On a straight thread the thread-ridge diameter is the diameter of the coaxial cylinder whose surface would pass through the thread profiles at such points as to make the thickness of the thread ridge equal to one-half of the basic pitch. On a taper thread the thread-ridge diameter is the diameter at a given position on the thread axis of the coaxial cone whose surface would pass through the thread profiles at such points as to make the thickness of the thread ridge (measured parallel to the axis) equal to one-half of the basic pitch.

Functional (virtual) Diameter

The functional diameter of an external or internal thread is the pitch diameter of the enveloping thread of perfect pitch, lead, and flank angles, having full depth of engagement but clear at crests and roots, and of a specified length of engagement. It may be derived by adding to the pitch diameter in the case of an external thread, or subtracting from the pitch diameter in the case of an internal thread, the cumulative effects of deviations from specified profile, including variations in lead and flank angle over a specified length of engagement. The effects of taper, out of roundness, and surface defects may be positive or negative on either external or internal threads. (A perfect go thread plug or ring gage, having a pitch diameter equal to that specified for the maximum material limit and having clearance at crest and root, is the enveloping thread corresponding to that limit. Note: Also called the virtual diameter, effective size, or virtual effective diameter.

Form Diameter

The form diameter is the diameter at the point nearest the root from which the flank is required to he straight.

Length of Complete Thread

The length of complete thread is the axial length of a part where the thread section has full form at both crest and root; that is, the vanish threads are not included. However, on commercial fasteners where there are unfilled crests at the start of rolled threads or a chamfer at the start of a thread, not exceeding two pitches in length, this is traditionally included in the specified thread length.

Note: When designing threaded products, it is necessary to take cognizance of (1) such permissible length of chamfer and (2) the first threads which by virtue of gauging practice may exceed the product limits and which may be included within the length of complete thread. However, when the application is such as to require a minimum or maximum number, or length, of complete threads, the specification should so state. Similar specification is required for a definite length of engagement.

Length of Thread Enagement The length of thread engagement of two mating threads is the axial distance over which two mating threads are designed to contact.

Depth of Thread Engagement

The depth (or height) of thread engagement between two coaxially assembled mating threads is the radial distance by which their thread forms overlap each other.

Standoff

The standoff is the axial distance between specified reference points on external- and internal-taper threaded members or gages, when assembled with a specified torque or under other specified conditions.

Letter symbols and Designations

Symbols associated with screw threads are of two kinds: (1) letter symbols for designating dimensions of screw threads and threaded products and (2) abbreviations used as designations for various standard thread forms and thread series.

Dimensional Symbols

Standard letter symbols to designate the dimensions of screw threads in text and formulas are available in the standards.

Thread-element designations are capital-letter abbreviations based on names of various thread dimensions in thread designations.

2.9 UNIFIED AND AMERICAN NATIONAL SCREW THREADS

In use today in American industry are two screw-thread systems-the older American National, a standard originally developed for the United States, and the Unified, an international standard between the United States, United Kingdom, and Canada. In 1948 the United State: reached an accord with Britain and Canada on Unified thread specifications. Included were the form of thread, the pitches, and the tolerances for sizes X in and larger in the coarse, fine and special threads to be designated by UNC, UNF, and UNS, respectively (the equivalent American National thread designations were NC, NF, and NS). The form adopted for Unified thread is substantially the 600 American National form, and threads of both standards having the same diameter and pitch are mechanically interchangeable. The principal differences between these standards relate to the application of allowances and differences in tolerances, as well as the differences in designations. The Unified standard threads generally supersede the previously published American National Standard threads.

Standard Series

The various diameter-pitch combinations for the eleven series in the Unified standard are available in the standards.

Thread Classes

Thread classes are distinguished from each other by the amounts of tolerance and allowance. Classes: 1 A, 2 A, and 3 A apply to external threads only, and classes 1 B, 1 B, and 3 B apply to internal threads only. Classes

2 A and 2 B are the most commonly used thread standards for general applications, including production of bolts, screws, nuts, and similar threaded fasteners. Classes 3 A and 3 B may be used for applications requiring tolerances closer than 2 A and 2 B Classes 1 A and 1 B are used on threaded components where quick and easy assembly is necessary and where a liberal tolerance is required to permit ready assembly, even with bruised or dirty threads. Any combination of external thread class and internal-thread class may be based on assembly requirements. For example, a class 2 A external thread may be used with a class 1 B, 2 B, or 3 B internal thread.

2.10 BASIC PROFILES OR FORMS OF SCREW THREADS

The profile of a screw thread is based on whether it functions as a fastening device or as power transmission aid. The profile is triangular, known as V thread in the former, and square or its modifications in the latter. Thus the two basic profiles or forms of screw threads are:

(i) V thread and

(ii) Square thread.

Transmission: They should have greater load carrying capacities and offer less frictional resistance to motion.

Why 'V' threads are used for fastening and square threads for power transmission?

The main consideration in the threads used for fastening is that they should offer high frictional resistance to motion so as to prevent the slackening of the nut while in power transmission; they should have greater load carrying capacities and offer less frictional resistance to motion. The following comparative force analysis acting on the V and the square threads will decide the suitability of a particular type of thread for the specific use.

Figure (2.4) shows the profiles of V and square threads of same pitch.

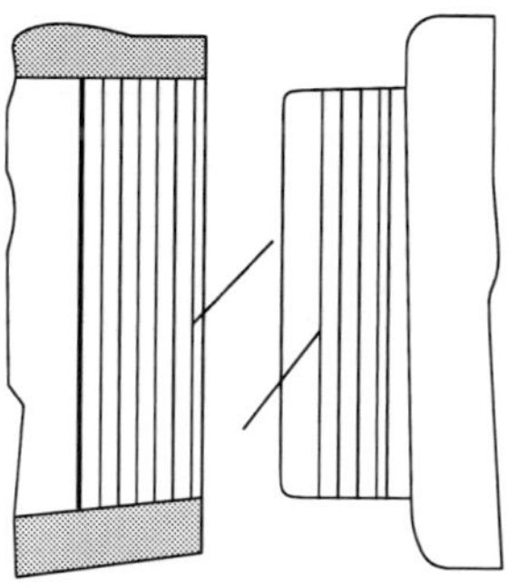

FIG. 2.3

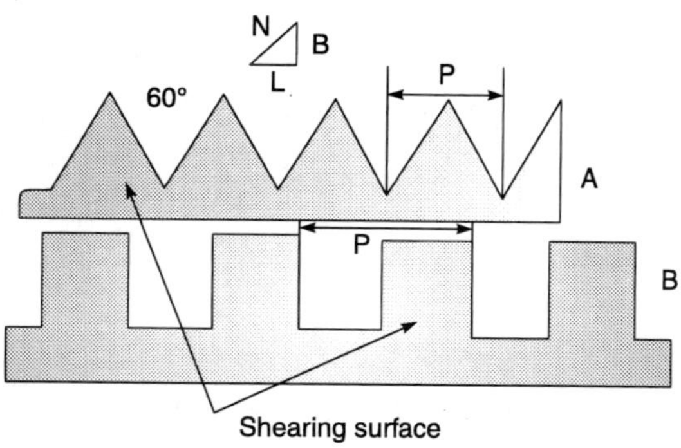

FIG. 2.4 Thread profile

The principal force or the load acting on the screw rod is axial. The axial load 'L' acts on the 'V' and square threads. Obviously this load should be supported by the resultant pressure acting between the surfaces of contact of the mating threads. Since this resultant pressure acts normal to the surfaces of contact of the mating threads, it is called normal pressure N. Since the contact surface of 'V' threads are inclined, the normal pressure which acts perpendicular to them may be resolved into axial component, and an another component perpendicular to the axial component. Since the load on the thread is axial, the axial component carries the load, hence called load carrying component, L. Since the perpendicular component acts perpendicular to the axis and away from it, it base got the tendency to break the nut, hence called bursting component, B. The friction between the contact surfaces depends on the normal pressure. For the given axial load if the angle of the V thread increases, both the normal pressure and the bursting force increases. The increase of normal pressure causes an increase in friction which requires increased effort. Although the bursting force reduces as the angle of the V thread decreases, it also decreases the normal pressure which in turn decreases the friction between the contact surfaces, thus renders it unsuitable for fastening. Since the sides of the square threads are parallel, there is no bursting force and therefore the entire axial load will be equal to the normal pressure, hence the square threads have greater load carrying capacity than V threads. This renders the square threads ideal for power transmission.

As shown in **Fig. (2.4)**, for the given pitch, the V thread has twice the amount of material than the square thread at the core, therefore has greater resistance to bending and shear, so more suitable for fastening where strength is the main consideration. The inclined faces of the V thread render greater area of surface contact than the square thread, this offers high frictional resistance to motion. This is an advantage in fastening as it prevents the nut from slackening back. These reasons render V threads ideal for fastening.

2.11 DIFFERENT TYPES OF V THREADS

British Standard Whitworth or BSW Thread

This thread was introduced by Sir Joseph Whitworth and was standardized as British Standard Thread, abbreviated as **BSW.** The profile of this thread with standard proportions is shown in **Fig. (2.5).** It has a thread angle of 55° and is rounded off at the crest and root as to render it less liable to damage than a sharp V thread.

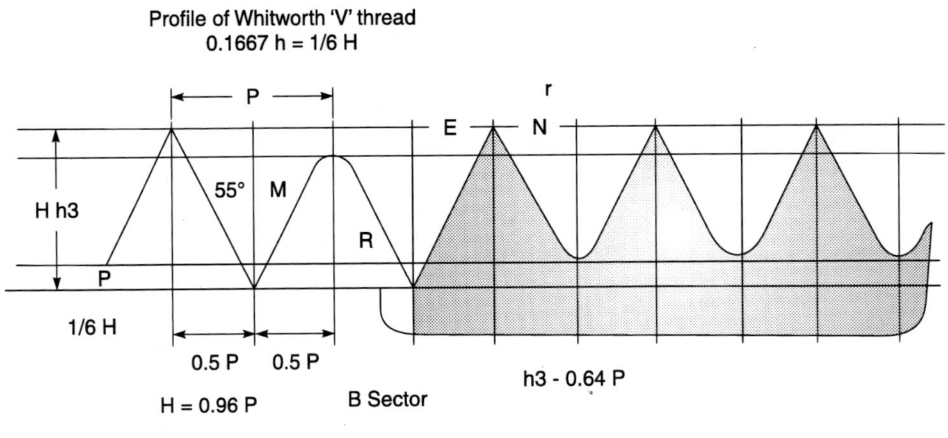

Fig. 2.5

British Standard Fine or BSF Thread

The profile of this thread is similar to that of BSW, but pitches are closer and hence the depth of the thread will be small. This has a better finish comparatively. These are commonly used in aircraft and automobiles.

British Standard Pipe or BSP Thread

This thread is also similar to **BSW** profile with finer pitches so as to suit thickness and the material of the pipe.

British Association or BA Thread

This thread was standardized by British Association and used extensively for screws used for electrical and other instrument work. The profile of these thread standard proportions is shown in **Fig. (2.6)**. It has a thread angle of 47.5°.

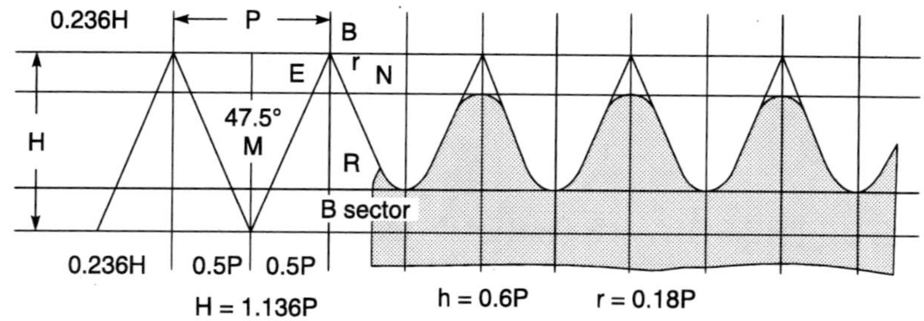

Fig. 2.6 Profile of British Standard Association thread

American National or Sellers Thread

This thread was adopted as the American Standard Thread by the Standards Institution. The profile of this thread with standard proportion is shown in **Fig. (2.7).** It has 60° thread angle and crests and root are kept flat.

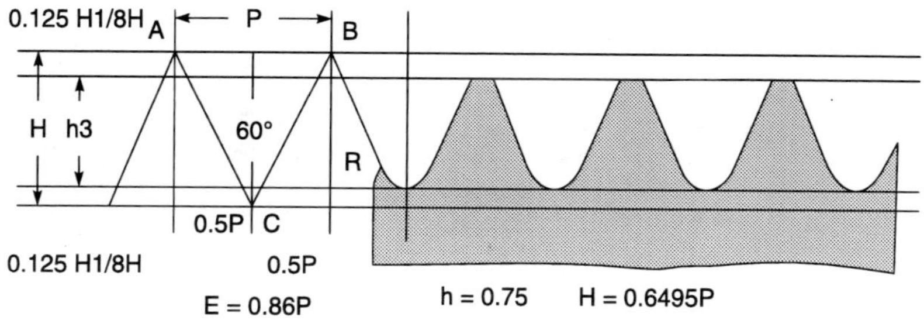

Fig. 2.7 Profile of Sellers thread

ISO Metric Thread or Unified Thread

The Indian Standard Institution has adopted V-thread profile recommended by International Organization for Standards, ISO, and metric screw thread for use in this country. This thread is also known as Unified Thread. It has 60° thread angle. The profile of this thread with standard proportions is shown in **Fig. (2.8)**

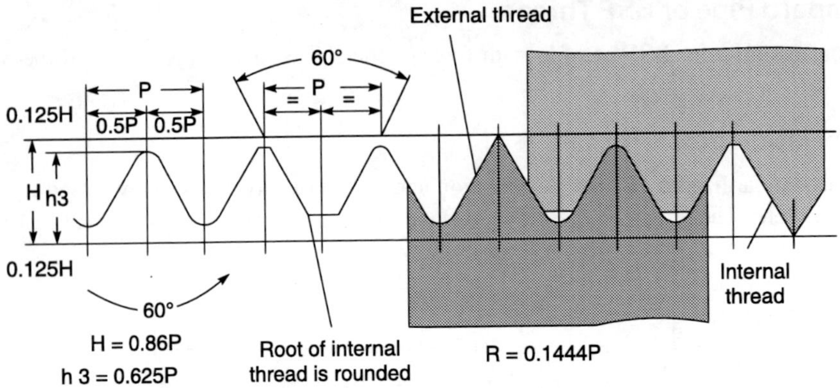

Fig. 2.8 ISO Metric thread

2.12 PROFILE OF THE SQUARE THREAD AND ITS MODIFICATIONS

Square Thread

The profile of this thread with standard proportions is shown in **Fig. (2.9).** The depth of the external threads is equal to half the pitch. The sharp corners at the roots of both external and internal threads are rounded off to radius R. The sharp corners at the crest of the internal thread are chamfered to b X 45°.

A square thread is designated by the letters SQ followed by the nominal diameter and pitch, the two being separated by X sign. For example, a square thread of nominal diameter 40 mm and pitch 10 mm is designated as SQ 40 X 10.

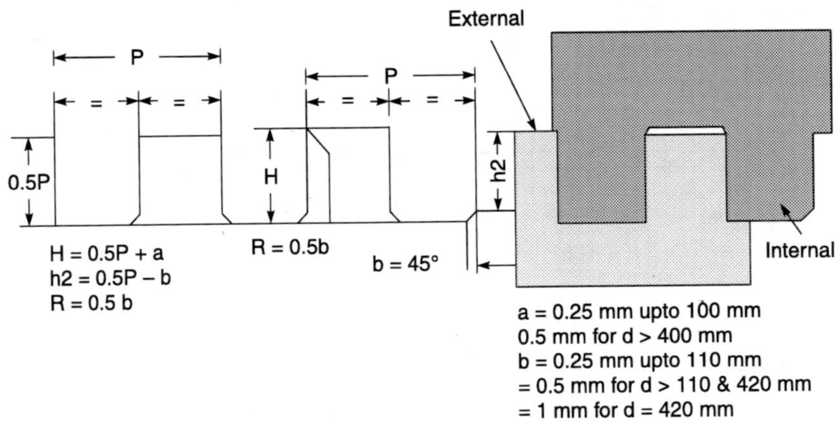

Fig. 2.9 Profile of square thread

Acme Thread

This thread is a modified form of a square thread. Unlike the square thread it is easier to cut and is stronger at the root. The angle of the thread is 29°. The inclined sides of the thread facilitate quick and easy engagement and disengagement, for example, the split nut of the lead screws of a lathe. It is satisfactorily used for power screw, like brake screws, jack screws and on the valve operating screws for axial power transmission. The section of this thread with standard proportion is shown in **Fig. (2.10).**

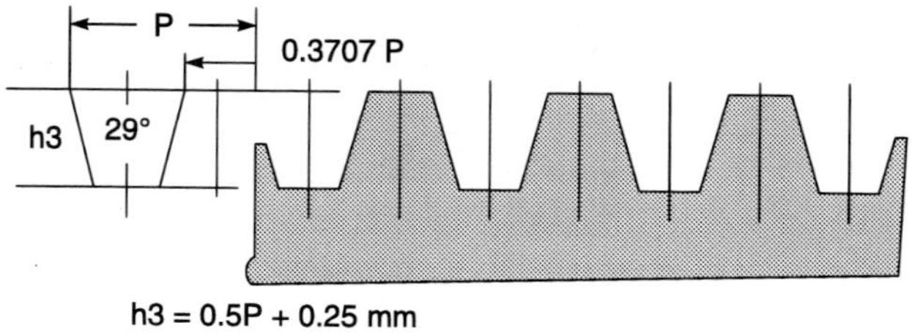

Fig. 2.10 Profile of ACME thread

2.13 ISO METRIC TRAPEZOIDAL THREAD (SIMILAR TO ACME THREAD)

The profile of the trapezoidal thread shown in **Fig. (2.11)** is the ISO Metric thread profile recommended by the Indian Standards Institution for use on parts in preference to Acme thread. It has a thread angle of 30°.

A trapezoidal thread is designated by the letters **Tr** followed by the nominal diameter and pitch, the two being separated by X sign. For example, a trapezoidal thread of nominal diameter 40 mm and pitch 7 mm is designated as:

T_r 40 X 7.

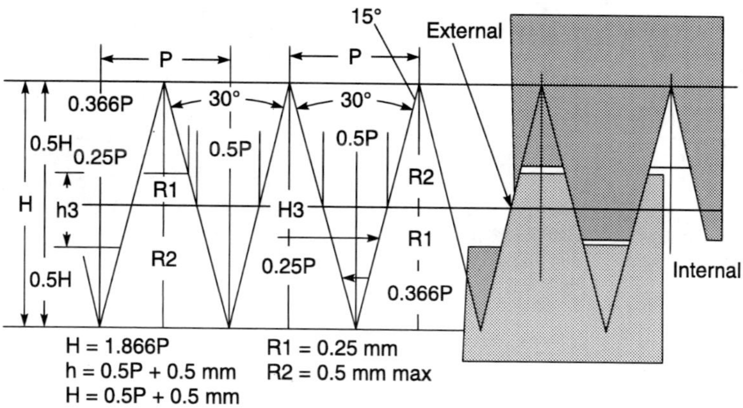

Fig. 2.11 Profile of trapezodial thread

Buttress Thread

The profile of this thread is a combination of square and V threads, combining the advantages of the square thread like ability to transmit power and low frictional resistance with the strength of the V thread. It is used for transmitting unidirectional load as in the case of screw press, vices and breech locks of guns to resist the recoil. The section of this thread with standard proportions is shown in **Fig. (2.12).**

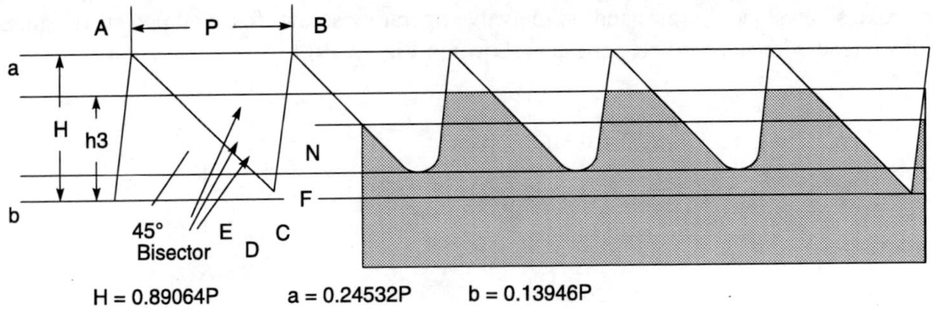

Fig. 2.12 Profile of butress thread

2.14 SAW TOOTH THREADS (SIMILAR TO BUTTRESS THREAD)

The profile of the saw tooth thread shown in **Fig. (2.13)** is the ISO Metric thread profile recommended by the Indian Standards Institution for use in parts in preference to buttress thread. It has a thread angle of 33°.

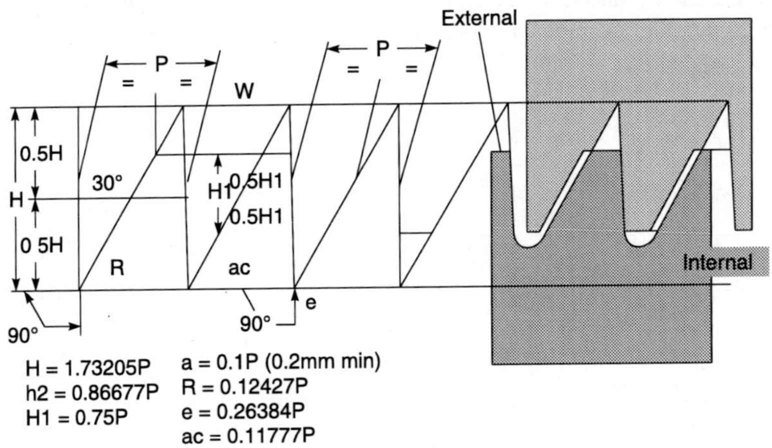

Fig. 2.13 Profile of saw-tooth thread

Knuckle Thread

The profile of this thread is yet another modification of the square thread. The sharp corners of the square thread which are liable for damage are rounded off as shown in **Fig. (2.14)**. This thread is used where heavy wear and rough use is expected as in railway carriage coupling screws. The advantage of this thread is that it can be rolled in sheet metal or cast so as to have shallow depths makes its use on electric bulbs and sockets, bottle tops, etc. The basic profile of the knuckle thread is shown in **Fig. (2.14).** The modified profile shown in **Fig. (2.15)** is with standard proportions recommended by the Indian Standards Institution for use in this country.

A knuckle thread is designated by the letter **K** followed by the nominal diameter d, pitch, the two being separated by the X sign. For example a knuckle thread of nominal diameter 10 mm and pitch 2.54 mm is designated as K10X2.54.

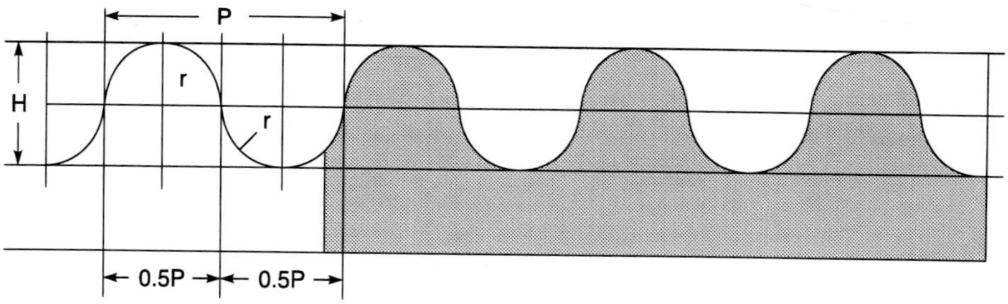

Fig. 2.14 Profile of knuckle thread

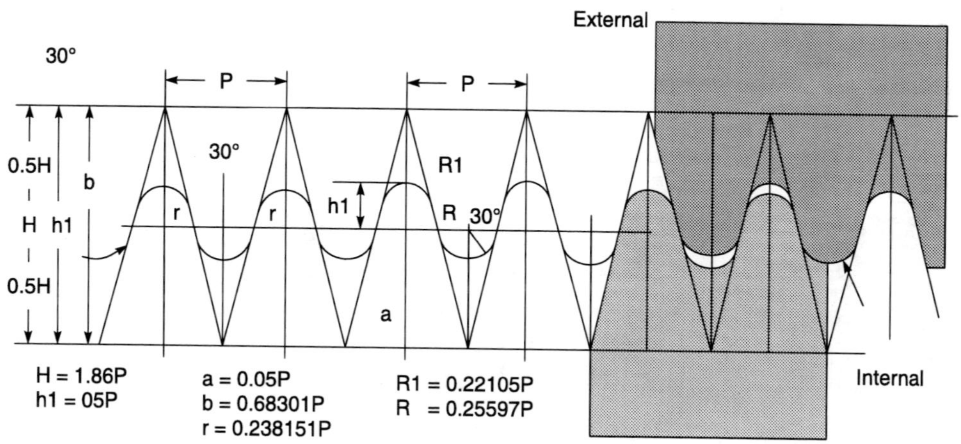

Fig. 2.15 Profile of ISI knuckle thread

2.15 RIGHT-HAND AND LEFT-HAND THREADS

A thread is right handed if, when viewed axially from the nut end; it winds away in the clockwise direction. Thus on a screw rod having right hand threads, a nut advances axially when turned clockwise as shown in **Fig. (2.16A)**. A thread is left handed if, when viewed from the nut end, it winds away in the anticlockwise direction. Thus on a screw rod having left hand thread, a nut advances axially when turned anticlockwise as shown in Fig. **(2.16B)**. A thread is always considered to be right handed unless otherwise specified. A left handed thread is abbreviated as LH on drawing.

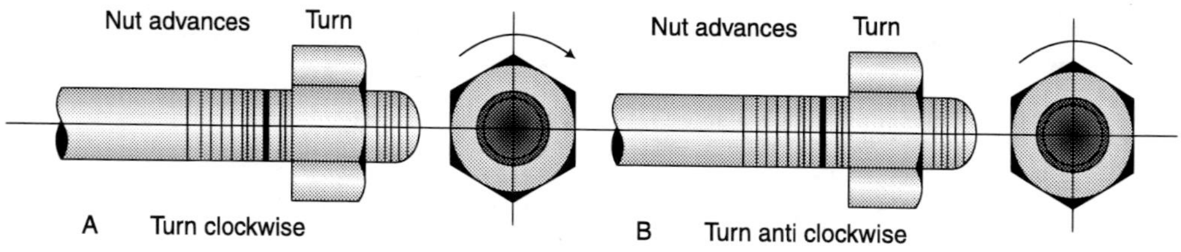

Fig. 2.16 Right hand & Left hand threads

2.16 SINGLE AND MULTI START THREADS

When a nut is turned on a bolt having a single continuous thread cut on it by one full turn, it advances axially through a distance equal to the pitch. As explained, lead is defined as the axial distance advanced by the nut in one full turn. Thus in a single continuous thread called single start thread, the lead is equal to pitch. Single start square thread is shown in **Fig. (2.17)** and single start V thread is shown in **Fig. (2.18)**. In certain mechanisms, for the quick transmission of motion, a large axial motion per revolution of either the screw or the nut will be required. Therefore in a single start thread, a large pitch is necessary for the increased axial advance per revolution of the screw or nut. But the increase in pitch for the given diameter of the screw rod increases the depth of the thread of both V and square types. In square threads, the depth of the thread increases with the increase in pitch because it is equal to half the pitch as shown in **Fig. (2.17B)**. Similarly, in V threads the depth of the thread increases with the increase in pitch because the angle of the thread is same as shown in **Fig. (2.18 B)**. As seen in these figures, the increase in pitch of the single start thread increases the depth of the thread and consequently reduces the core diameter of the screw rod as shown in

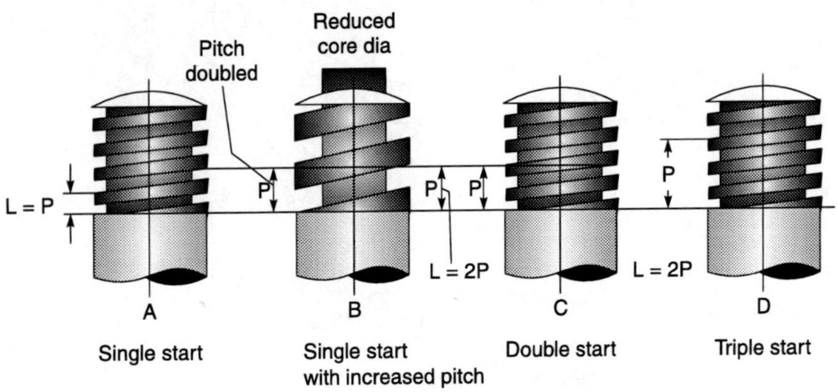

Fig. 2.17 Multi-start square thread

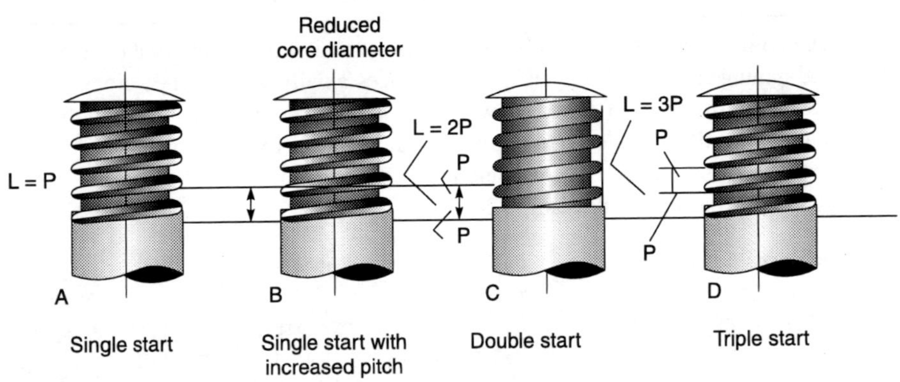

Fig. 2.18 Multi-start 'V' threads

Fig. (2.17 B) and **Fig. (2.18 B)** for square and V threads respectively. The reduction in the core diameter of the screw rod reduces the strength of the screw rod. In such cases, two or more separate threads having the same pitch as in single-start threads, but with the increased lead, are cut side by side so as to run parallel to one another. Since lead is equal to the product of the pitch and number of starts, in multi-start threads lead may be increased by increasing only the number of starts without increasing the pitch and thus the depth and core diameter are maintained same as in single start thread. For a double start thread, the lead is equal to twice the pitch as shown in **Fig. (2.17 C)** and **Fig. (2.18 C)** and for a triple start thread the lead is equal to thrice the pitch as shown in **Fig. (2.17 D)** and **Fig. (2.18 D)** for square and V threads respectively.

For double start threads, two threads are cut separately with their starting points and 2, diametrically opposite as shown in the top views in **Fig. (2.17 C)** and **Fig. (2.18 C).** Smilarly, in triple start threads, the threads are cut separately with their starting points 2, and 3 at 120° apart on the circumference of the screw as shown in the top views in **Fig. (2.17 D)** and **Fig. (2.18 D)**.

Multistart threads are used wherever quick motion with less power is desired to be transmitted as on fountain pens, cameras to adjust the focus, automobile starters, valve stems, tooth paste tube caps, bottle caps, etc.

EXERCISE

1. Define threads. What are the advantages of threads?
2. What is the difference between *left-hand* threads and *right-hand* threads?
3. Comment on the following:
   (a) Pitch of threads (b) Pitch diameter
   (c) Minor diameter (d) Major diameter
   (e) Depth of thread.
4. Show the above by drawing their figures.
5. What do you understand by pitch of thread and lead of thread?
   What is the function of worm threads ? Comment on the following:
   (a) Crest (b) Root
   (c) Thread angle (d) Helix angle
   (e) Number of threads.
   Explain in brief the main sources of thread making.
6. What are the different standards of thread?
7. Draw a figure of Acme thread and mention its uses.

Epoxy Resins in Press Tool Making

Chapter 3

The use of plastics in tooling applications was introduced during the Second World War to compensate a critical shortage of tooling metals. The epoxy, resins were introduced in 1950 s and till date a large varieties have been developed for various purposes.

Generally, the epoxy resins have not introduced new concepts in tooling but rather have been used to save time and money over metal tooling or to provide performance superior to that with older tooling methods.

Tooling is a term which covers a variety of different types of equipment. The tool may be a working model of the part to be produced; it may be a model from which the part is produced either directly or indirectly it may be a mould in or against which the part is produced; it may be an equipment used to form under heat and or pressure a contour, sheet metal or plastic; or it may be an accessory item used for guiding and checking other production operations.

The wide variety of tools, of all sizes and for many different applications can be made from epoxy resins with a few basic production techniques. Although there are many possible variations of the basic methods and a wide range of resins to choose from, the construction of a specific tool for a particular purpose is basically simple as it involves the choice and only one production, method and of one resin system for each stage of production.

The resins have all been formulated for their particular applications and are supplied mainly in pre-weighed packs. The resins and hardener in the correct mixing ratios designated by the manufactures are to be used strictly adhering to the instructions of the manufacturer. Also a range of auxiliary products such as mould sealers, release agents, fillers, glass cloth and colouring agents are being produced.

3.1 BASIC CONSTRUCTION METHODS OF PRESS TOOLS

Epoxy resin tools (Araldite) for press-forming sheet metal are particularly favoured in prototype, short-run and medium run production for which quality steel tools are uneconomic. To machine a complex-contour die from tool steel is costly, this may involve sophisticated machinery, right type of materials, and right type of skilled labour. To cast the same die from resins will result in a sharp reduction in labour costs and construction time. The machining and finishing processes are totally eliminated.

Press tools and drop hammers can also be designed for long production runs (10,000-50,000) or more depending on malleability and form. The concentration of high compressive shear and friction forces at severe draw radii are borne by metal inserts (metal inserts are normally used where the draw radius is less than five times the workpiece thickness). Cast in position when the die is poured. Trim dies using metal inserts to carry out actual shearing have proved highly successful in long production runs. This combination of resins with metals, gives advantages unobtainable from either tooling material when used alone.

Metal-forming tools normally comprise a low cost rigid core faced with the appropriate resin. This resin is rigid on both tool when work-piece thickness is allowed for in tool construction; where this is not allowed for (Elimination of allowance for workpiece thickness by use of a resilient Araldite facing resin depends on metal gauge and draw conditions.

Construction of Permanently Located Guide Bushes for Pillar Guide Frames

One can understand the expensive conventional process in which the bush is pressed in the top plate.

The location of the centres of the holes for housing the hardened and ground bushes in regard to the pillars pressed in the bottom frame require special type of machine (jig boring machine) and highly skilled labour. For drop hammer work on light alloys no allowance is required for (1) 20 swg (0.9 mm) and thinner material on non-vertical draws (2) 22 swg (0.7 mm) and thinner on small vertical draws) the facing resin is rigid on one member but resilient on the other.

Zinc-alloy cores are used for drop hammer dies araldite cores are used for punches and other types of dies.

The production of apertures for profile punches in blanking and piercing dies is difficult and time consuming even for skilled workers. Much time can be saved by making the openings larger than required and filling up the space with araldite epoxy resin. This technique saves time and thus reduces production cost; it also releases skilled personnel for other tasks, since highly developed manual skills are no longer required. In certain cases blanking and piercing dies of higher quality and longer service life can be made by the new method. Repairs can also be carried out quickly simply and securely by casting. The parts of blanking and piercing dies which can be successfully located using cast resins include:

- Punches with sliding clearance in guide plates.
- Interchangeable punches in punch support plates.
- Punches permanently fixed in punch holder plates.
- Guide pillars lath sliding clearance in guide framer.

With all this the guide bushing is likely to seize

By locating the guide bushes permanently with cast resin in the top frame, a great advantage can be experienced over the pressed in bushes.

The pillars and the bushes are turned, hardened and ground. The pillars are pressed into the bottom frame after the holes are drilled and bored. The holes drilled in the top frame should be 3 - 6 mm larger than the external diameter of the guide bushings. Surfaces which will be bonded by the casting resin should be abraded and degreased.

The top and bottom frames are assembled and aligned on parallel supports. The bushings are then fitted over the pillars. Their position is fixed by a supporting ring of plasticine (moulding clay; underneath, which also seals off the casting space.

The top and bottom frames are assembled and aligned on parallel supports. The bushings are then fitted over the pillars. Their position is fixed by a supporting ring of plasticine (moulding clay; underneath, which also seals off the casting space.

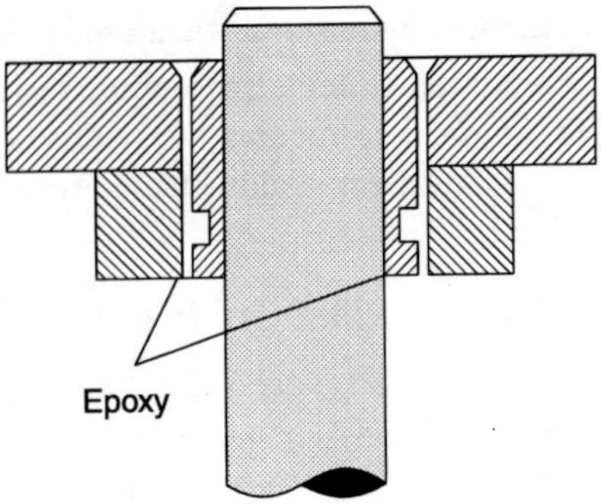

Fig. 3.1 Anchoring of punch

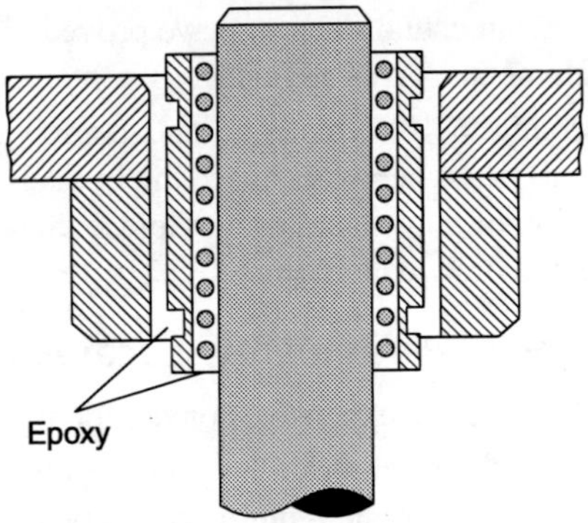

Fig. 3.2 Bush with ball bearing

Punches with Sliding Clearance in Guide Plates

The shape of the opening in the guide plate is marked out, starting from the die without clearance. The aperture is cut with an extra of 1-3 mm. The cut surface of the aperture is decreased and the mating surfaces of the plates are treated with release agents.

The punch, coated with release agent is pushed into the opening centred and held vertically. A plasticine collar around the opening serves as a pouring aid and prevents overflow of the casting compound.

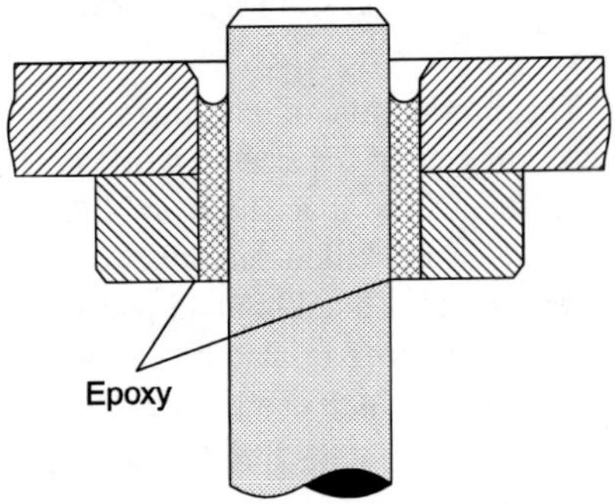

Fig. 3.3 Steel bush replaced by epoxi

The shaped aperture in the guide plate can for simplicity be replaced by a drilled hole.

Where high stripping pressures are encountered increased wear may be prevented by placing a perforated metal sheet on the underside of the guide plate. This procedure may also be adopted for thin punches which require deep guides to provide support.

Interchangeable Punches in Punch Support Plates

In this instance no sliding clearance is required because the punch need only be removed from its plate for maintenance or repair. A close fit is achieved by applying a thin coat of release agent to the surface of the punch.

The position of the punch is determined, depending on the type of tool, by the die guide, stripping or ejector plate. In certain cases the punch can also be fixed by at least two screws.

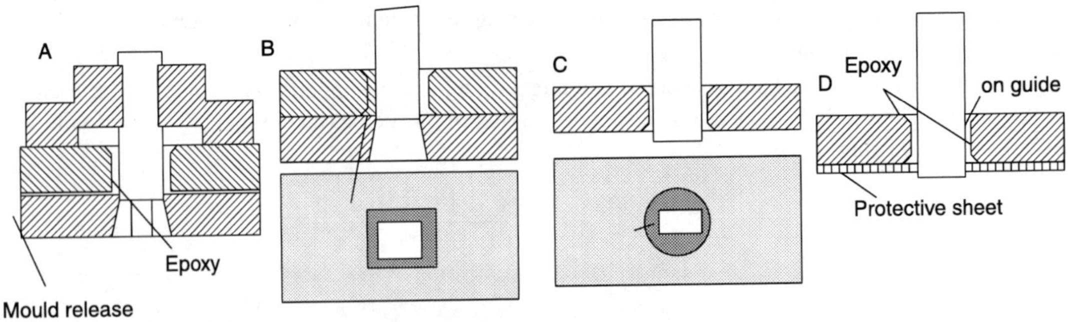

Fig. 3.4 *Guide plate with resins*

Punches Permanently Fixed in Support Plates

If the punch is to be fixed in the punch holder plate, the punch head must be roughened, degreased and if necessary provided with some form of mechanical keying device. The plate and the punch are then aligned and assembled.

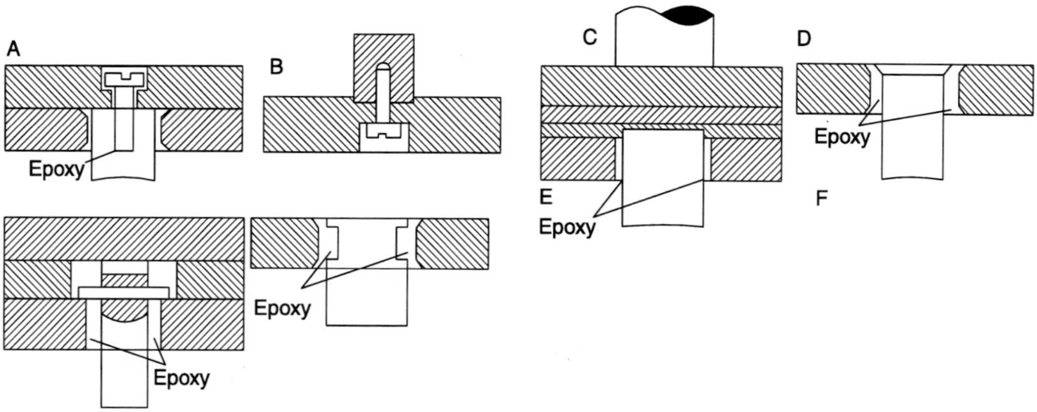

Fig. 3.5 Anchoring of punches

Location of Punch and or Die when they are in a Number of Pieces (Split Dies)

In case of punches and or dies when they are manufactured in segments and assembled, the normal practice is to fix the same with at least two dowel pins and two screws. The intricacies involved in providing the pins are completely eliminated by using resins for locating the punch or die in segments.

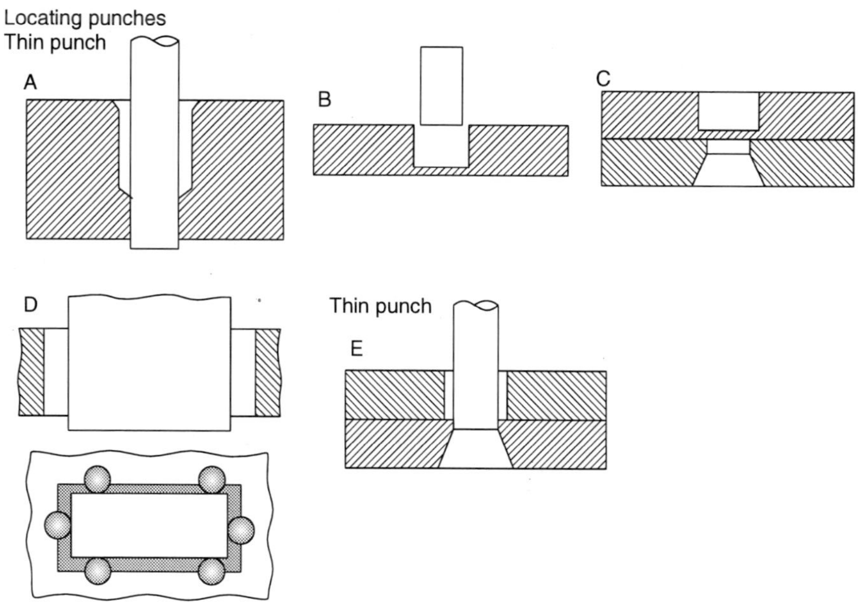

Fig. 3.6

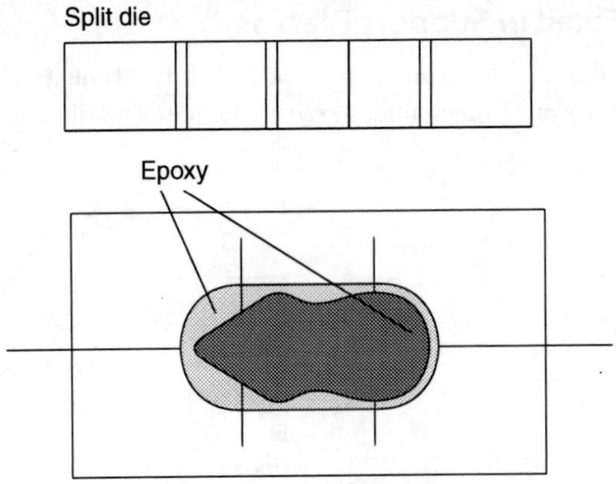

Fig. 3.7

Mechanical Power Transmission

Chapter 4

Transfer of mechanical power from one place to another is necessary for operating a machine. This process is known as "mechanical power transmission". All machines are operated with one 'or the other power. This power is made available through electric motor or automotive engines like steam engines, diesel engines, and petrol engine. The power generated by these engines is transferred through one or the other medium. Generally, the following mediums are applied for this purpose:

- Lever mechanism
- Pulley and belts
- Chain and sprocket
- Gears
- Clutch
- Shaft and coupling
- Cam
- Rope

4.1 LEVER MECHANISM

Evolution of power transmission started with the principle of the lever mechanism. A lever is a bar used to raise heavy or fixed object. Straight bar or other rigid structure of which one point (fulcrum) is fixed, another is connected with the force (weight) to be resisted or acted upon, and a third is connected with the force (power) applied.

This simple phenomenon led to many lever mechanisms.

1. First order lever
2. Second order lever
3. Third order lever

The Mechanical Advantages of the Lever

$F \times a = Q \times b$ (law of moments)

Force $F \times$ lever arm $a =$ load $Q \times$ load arm b.

Moment of force = moment of load

Sum of clockwise moment = sum of anticlockwise moment

Lever arm is the perpendicular distance of the fulcrum from the line of action of the force acting on the lever.

Based on the first order lever principle crowbars are used to lift or move / shift heavy loads.

The lever most commonly used is a crowbar.

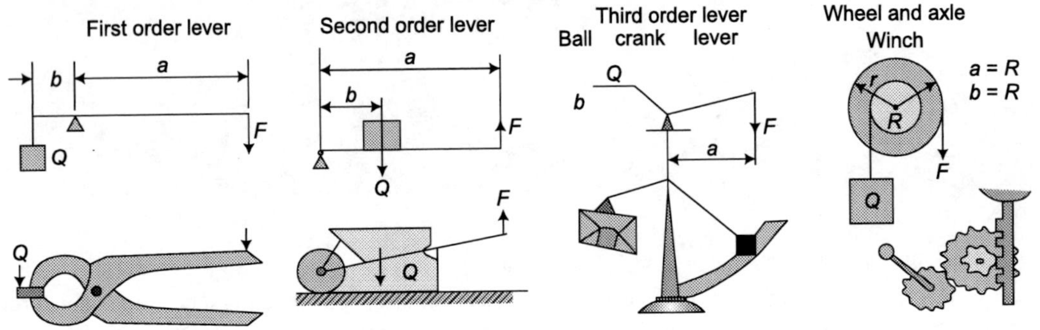

Fig. 4.1 Lever mechanism

Crowbars

Crowbars give leverage, so that heavy loads can be lifted or moved. They are made in different lengths with hexagonal or octagonal steel bars. Short crowbars are easier to handle and the point will fit into a narrow gap, but requires more force. Long crowbars provide a greater leverage.

Types of crowbars (Fig. 4.2)

There are two types of crowbars, single ended and double ended. A single ended crowbar is safer to use as the handle has a rounded end. The double ended crowbar normally has a curved end used for lifting, and a straight end used for pushing.

Lifting Equipment by Crowbars

If the gap under the machine is not good enough to accept the tip of the crowbar, tap a small steel wedge under the machine to increase the gap and place the toe of the crowbar under the machine and press the other end down to lift the machine (**Figs. 4.2 & 4.4**).

Position the handle so that no one will be endangered if the crowbar slips. When pushing or lifting, never push the crow bar close to the load or to the ground, as your fingers might be caught if the bar slips.

Always use both hands and hold close to the end of the crowbar to get maximum leverage **(Fig. 4.5)**

Stand with the legs apart so that the balance is not lost if the crowbar slips **(Fig. 4.6)**

Equipment are provided usually with a lifting pocket. Place the toe of the crowbar in it for lifting the machine and moving it.

The fulcrum point must be firm enough to take the force. If the point of the crowbar is used as the fulcrum, it must be dug in firmly to prevent slipping. (**Fig. 4.7**)

Check the condition of the crowbar, and if found bent or cracked, it must not be used. The burrs or sharp edges on the crowbar must be removed before using it.

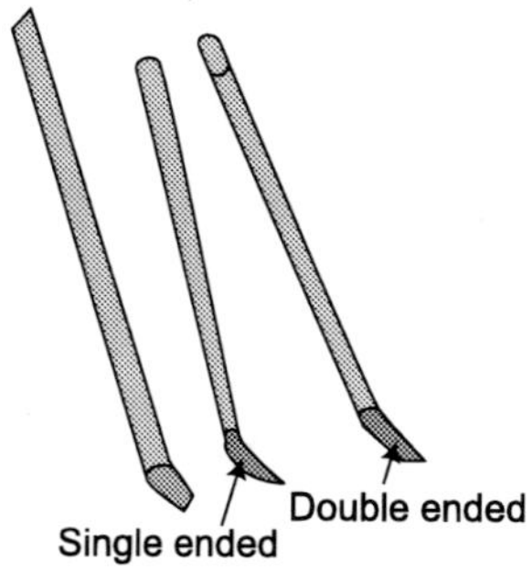

Fig. 4.2 Types of crow bars

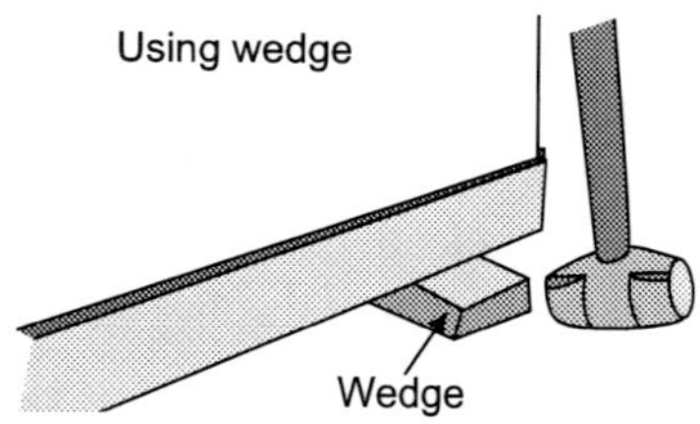

Fig. 4.3

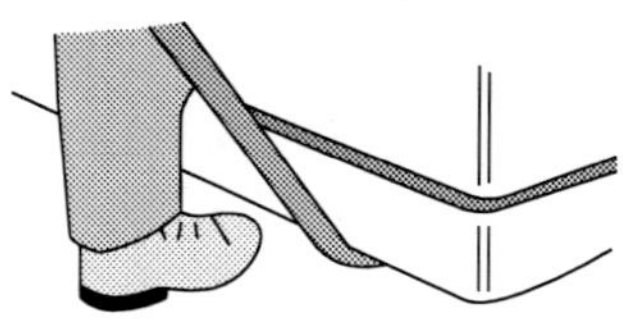

Fig. 4.4

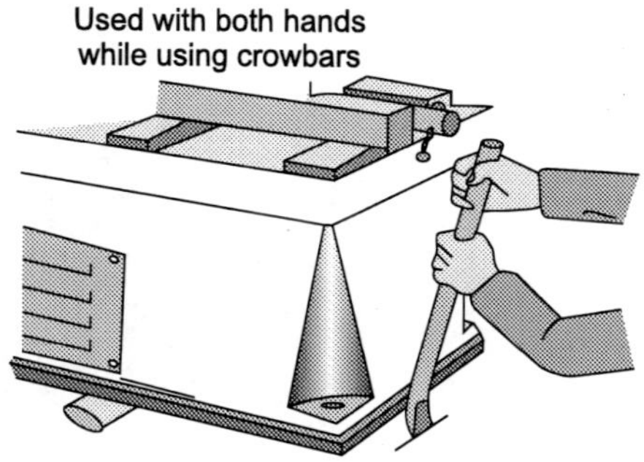

Fig. 4.5

Fig. 4.6

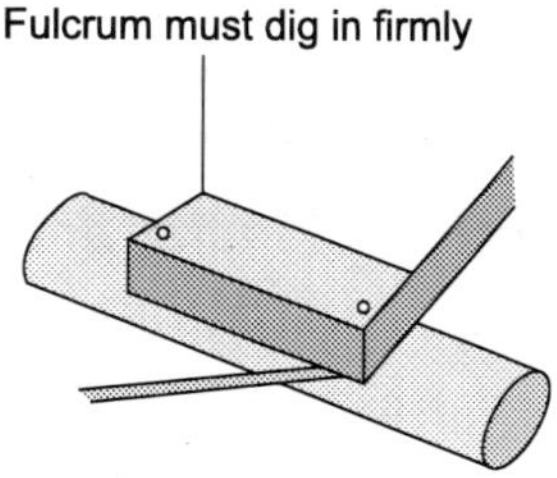

Fig. 4.7

Equipment are provided usually with a lifting pocket. Place the toe of the crowbar in it for lifting the machine and moving it.

The fulcrum point must be firm enough to take the force. If the point of the crowbar is used as the fulcrum, it must be dug in firmly to prevent slipping **Fig. (4.7).**

Rollers

Rollers are placed under the equipment so that they can be moved easily. Mild steel or G.I pipes of sufficient wall thickness can be used as rollers. The rollers should be long enough to project from both sides of the load so that they can be positioned easily. The diameter must be large enough to roll over any unevenness along the route but small enough so that they can be lifted easily **Fig. (4.8).**

Moving Equipment Using Rollers

Before starting to move a load, check the route and remove any obstructions. The route should be flat and firm enough to take the weight of the moving equipment.

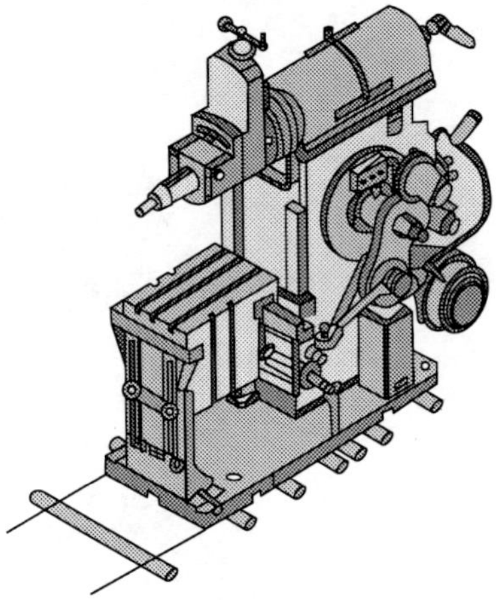

Fig. 4.8 Roller

4.2 PULLEY AND BELTS

The most easy and common medium of power transmission is "pulley and belts". Pulleys are available in small and big sizes. These are operated with the help of belts. Pulley fixed on the shaft or pulley which moves other pulleys are called driver pulleys. The pulleys moved by the driver pulley are called follower pulleys. Pulley is like a round wheel, which is fitted on the shaft. We can increase or reduce the speed of power transmission as in the case of gears and shafts by using big and small pulleys.

Lever Principle of Pulley and Belt Drive

Lever principle as applied to the pulley and belt drive:

$d_1 \times n_1 = d_2 \times n_2$ and the same in the compound drive is

$d_1 \times n_1 = d_2 \times n_2 = d_3 \times n_3 = d_4\, n_4\ (n_2 = n_3)$

Types of Belt Drive (Flat Belts)

The power from one pulley to another may be transmitted by any of the four types of belt drives.

Open Belt Drive (Fig. 4.9) and (Fig. 4.11)

An open belt drive is used between parallel shafts in which the direction of rotation of driving and driven shafts is the same. When the centre distance between the two shafts is large, then the tight side of the belt should be lower one.

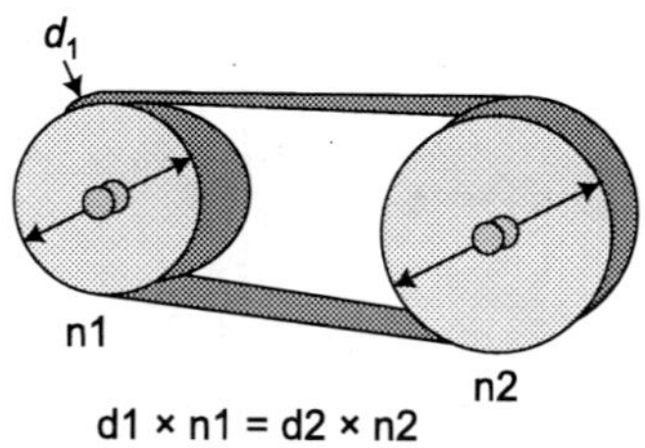

Fig. 4.9 Pulley & belt

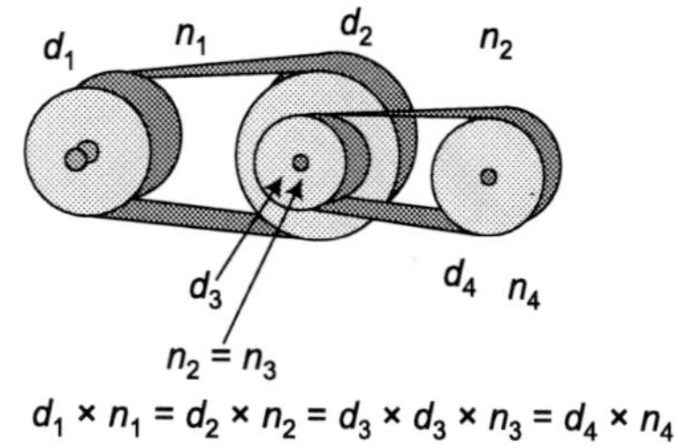

Fig. 4.10 Compound drive

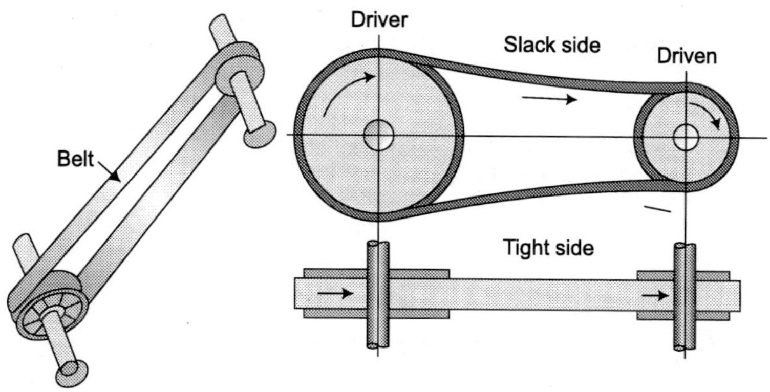

Fig. 4.11 Open belt drive

Crossed or Twist Belt Drive

This type of arrangement is used with shafts parallel and rotating in the opposite directions. A little consideration will show that at the point where the belt crosses, it rubs against each other and there will be excessive wear and tear. In order to avoid the shafts should be placed at maximum distance of 20 b where b is the width of belt and the speed of the belt should be less than 15 m/s.

Quarter Turn Belt Drive (Fig. 4.13)

The quarter turn belt drive **(Fig. 4.13 A)**, is used with shafts arranged at right angles and rotating in one definite direction. In order to prevent the belt from leaving the pulley, the width of the face of the pulley should be greater or equal to 1.4 *b*, where *b* is the width of belt.

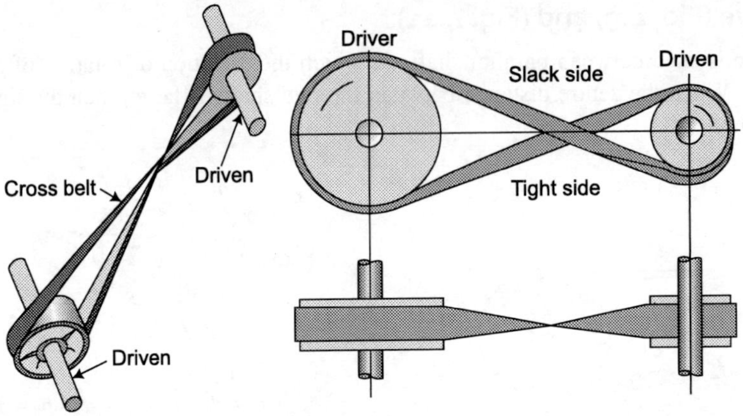

Fig. 4.12 Cross belt drive

In case the pulleys cannot be arranged as shown in **Fig. (4.13 B)** or when the reversible motion is desired, then a quarter turn belt drive with a guide, pulley, as shown in **Fig. (4.13 B)**, may be used.

A cross-belt drive in which the direction of rotation of driving and driven shafts is opposite.

Right Angled Drive (Fig. 4.13 B)

This drive is employed between shafts at right angles using guide pulleys. In this the horizontal drive is converted into vertical drive with the help of the guide pulleys.

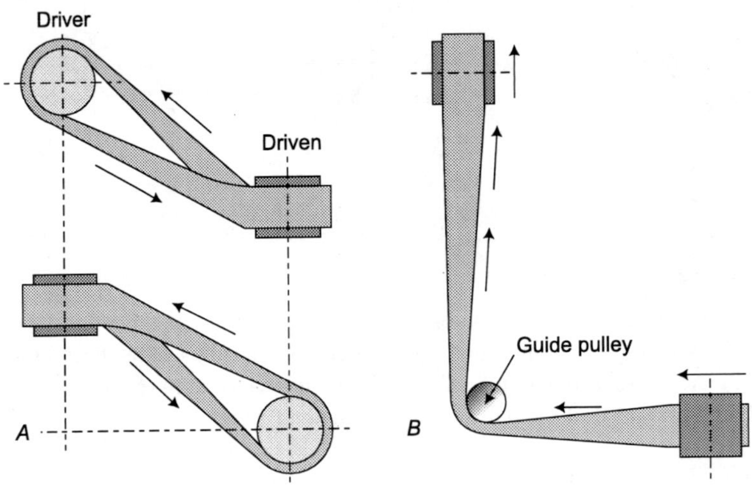

Fig. 4.13 Quarter turn belt drive

Compound Bell Drive (Fig. 4.14)

A compound belt drive is used when power is transmitted from one shaft to another through a number of pulleys.

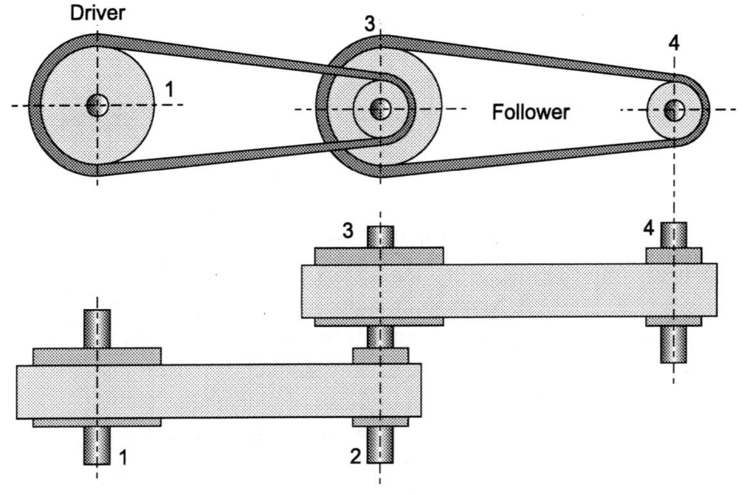

Fig. 4.14 Compound belt drive

Jockey or Idler Pulley (Fig. 4.15)

This type of drive is used with shafts arranged in parallel and when an open belt drive cannot be used due to small angle of contact on the smaller pulley. This is provided to obtain high velocity ratio and when required belt tension cannot be obtained by other means.

If the centre distance between the pulleys exceeds 3 times the larger pulley diameter and the ratio of the pulley diameters exceeds 6 to 1, a jockey or idler pulley is fitted near the driver pulley on the slack side of the belt to increase the arc of contact. This increases the wrapping angle and tension on the belt, necessary for the transmission of the torque.

When it is desired to transmit motion from one shaft to several shafts, all arranged in parallel, a belt drive with many idler pulleys may be employed **Fig. (4.16).**

Stepped Drives or Compound Drive (Fig. 4.17)

Stepped drives are used to obtain different speed ratios. Pulleys of different sizes are employed.

Three different speeds can be obtained by changing the belt position from one step to another.

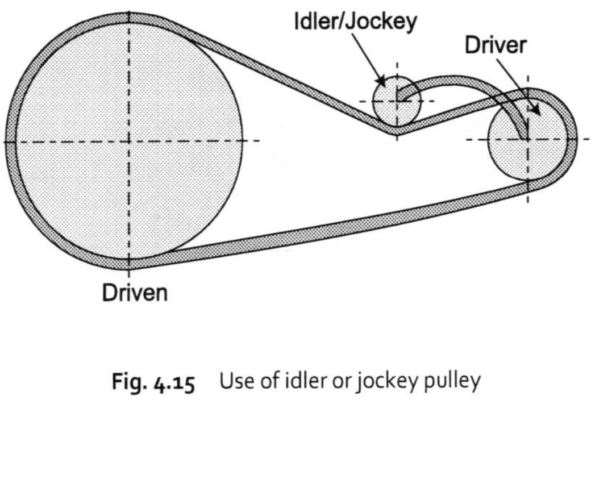

Fig. 4.15 Use of idler or jockey pulley

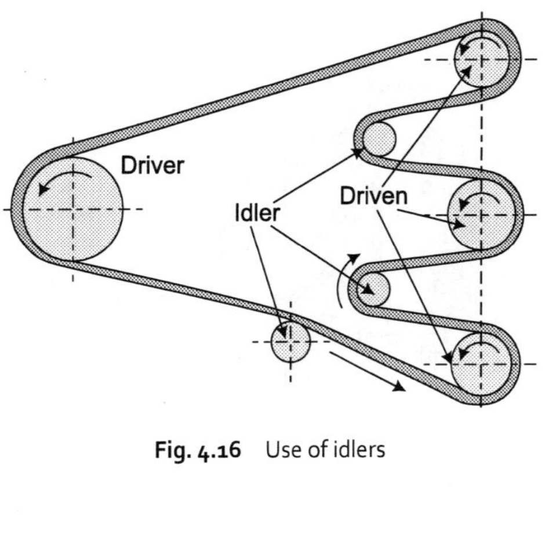

Fig. 4.16 Use of idlers

This can be termed as a group of several pulleys. Pulley faces are made in the step pulley with pulleys of different diameters fixed in order. As it is made like steps it is known as step pulley. It is used at such places where there is need to increase or decrease the speed of driven pulley for various jobs, e.g., drill machine, lathe machine etc. These are often used in pairs.

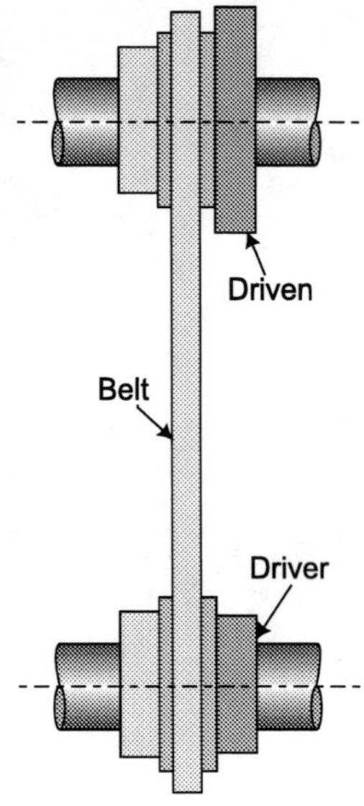

Fig. 4.17 Stepped pulley

Pulley (Fig. 4.18)

Pulleys are used to transmit power from one shaft to the other at a moderate distance away by means of a belt or strap running over them. They may be made of cast iron, wrought iron, pressed steel or wood. Main shaft pulleys are generally made of wrought iron, pressed steel or wood which gives them suitable strength combined with lightness, while counter shaft pulleys which are usually smaller than main shaft pulleys are made of cast iron. These pulleys have a thin rim of rectangular section over which the belt runs. Usually, pulleys are provided with arms which may be straight or curved and the cross section is usually described as "oval", i.e., roughly elliptical. The central part of the pulley is called the "nave", "eye", or "boss". To add strength and stiffness large pulleys are provided with ribs between the rim and the boss. Sometimes these ribs are reinforced with arms for greater stiffness and durability. The rims of all cast iron pulleys are generally "crowned", i.e., slightly greater in diameter at the centre than at the edges. As the belt seeks the highest point on the pulley, the effect of crowning is to keep the belt in a central position. The crowning of pulleys, in most cases, should not exceed 25 mm on the diameter per metre of width, and the width of the pulley should be one-fourth greater then the width of the belt used. The minimum diameter of pulleys should be at least 24 times the thickness of the belt used to run on the pulley.

A pulley contains

1. Face
2. Hub
3. Arm
4. Bore

Face

The part containing periphery of the pulley is called 'face'. It is slightly bulges from the centre and is descends from the sides. The protruding middle part is called crown. Face is either plain or a "V" groove is built on it.

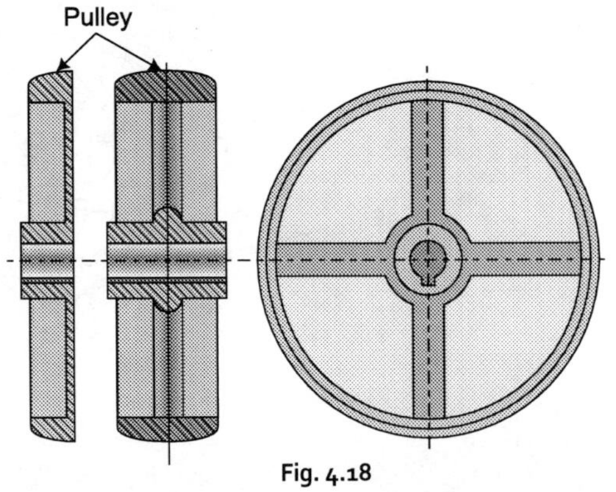

Fig. 4.18

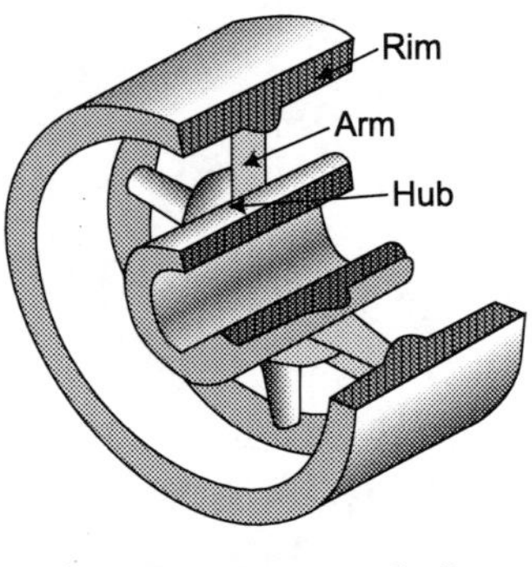

Fig. 4.18 A Overview of pulley

Hub

The central part of a pulley, which has a bore for the shaft, is called hub. These are of three types.

(i) Solid hub

In this type of hub only shaft of one particular size is fitted. For shaft of different sizes, it is used with different bores.

(ii) Slotted hub

This type of hub is slotted from one side which is tightened with a nut and bolt after fitting the shaft. It is also made in one piece. It can be used with shaft of different diameter.

(iii) Split hub

This type of hub is made in two pieces (parts). For fitting these two parts on the shafts two nuts and bolts are used. Their faces are made of two pieces. These can be used for big and small shafts of the same type.

(iv) Arms

The joining part which join pulleys face and hub, is called arm. The number of arms is according to the diameter of the pulley and its use, it may be four, six or eight.

(v) Bore

Hole in the hub is called "bore". A key way is cut in the hole. With the "key" shaft and pulley is locked.

For different machines and jobs the following types of pulleys are used.

Types of Pulleys

1. Step Pulley. **Fig. (4.17)**
2. "V" Groove Pulley
3. Split Pulley
4. Cone pulley
5. Jockey Pulley
6. Round Pulley
7. Fast and Loose Pulley
8. Drum Pulley

Step Pulley (Fig. 4.17)

This can be termed as a group of several pulleys. Pulley faces are made in the step pulley with pulleys of different diameters fixed in order. As it is made like steps it is known as step pulley. It is used at such places where there is need to increase or decrease the speed of the driven pulley for various jobs as drill machine, lathe machine, etc. These are often used in pairs.

Grooved Pulleys: Grooved Pulleys for Rope, V-Belt, Etc.

These are generally made entirely of cast iron, but occasionally the arms are made of wrought iron. They are mainly used for the transmission of large powers over great distances and the effect of the groove is to increase the frictional grip of the rope on the pulley and thus reduce the tendency to slip. The grooves on the pulleys are V-shaped, the angle between the two faces being from 40° to 60°. The rope rests on the two sides, and not on the bottom of the groove **Fig. (4.20)**. In guide pulleys, the grooves are semicircular, and the rope rests on the bottom of the groove. The diameter of a rope pulley is generally not less than thirty times the diameter of the rope.

A groove like the English letter "V" is cut on the face of this pulley. It is used in electric motors and automobiles. A special "V" shape belt is used with these pulleys. In some pulleys more than one "V" grooves are cut. These types of pulleys are used for power transmission for heavy jobs. With the use of more than one "V" belts there are less chances of slip of power.

Split Pulley

These types of pulleys are made in two parts of face and hub. Its speciality is that it can be used on the pulleys of different sizes, big and small. It is used in saw machines.

Cone Pulley (Fig. 4.19)

It is used in a set of two pulleys (pairs). It is tapered. Similar taper is cut on two pulleys. These are fitted in such a way that large end of one being opposite the small end of the other, so that the shifting of the belt will make a change of speed. By this means, we can get appropriate speed for different jobs.

Jockey Pulleys or Rider Pulleys (Fig. 4.15) and (Fig. 4.16)

Jockey pulleys are used to increase the area of contact, and also to produce more tension in the belt. They are mounted near the smaller of the two pulleys in a belt drive and always ride on the slack side of the belt. The pulley presses down the slack side due to having loaded bearing in which it runs.

One problem which we usually come across while using pulleys is that after some time their belts get loose and start slipping, which results in loss of power. In order to overcome this problem a third pulley is used with the pair of two pulleys. It is called jockey pulley. This pulley is a self controlled pulley which keeps the belt tight, when it becomes loose.

Round Pulley

A round groove is cut on the face of this type of pulley. Round belts or ropes are used to operate it. It can also be made like the step pulley.

Fast and Loose Pulleys

The methods of drive that in workshops and factories, power is in certain cases transmitted to the machines from a main line-shaft called main shaft, through an intermediate shaft, called counter shaft, and two pulleys known as fast pulley and loose pulley that are mounted on the counter shaft. A fast and loose pulley arrangement enables a machine to start or stop at will, without stopping the belt. Loose pulleys should revolve freely on the shaft, but the fast pulley or fixed pulley is firmly keyed on the shaft. To stop the machine shaft, the belt is moved from the fast pulley to the loose pulley by means of a belt shifier. The diameter of the loose pulley is sometimes made slightly smaller than that of the fast pulley so that when the belt is on the loose pulley there is very little tension in the belt. A loose pulley is usually provided with a brass or gun metal bush and needs efficient lubrication for smooth running.

These pulleys are used for the machines which are quickly started or put off. These are used in pairs of two small pulleys, which are fitted on a counter shaft. One pulley is tightened on the counter shaft with key or some other medium and other pulley is fitted through a provision for free rotation. When the machine is to be operated, the belt is mounted on the tight pulley. This pulley is known as fast pulley. When the machine is to be put off, the belt is slipped onto the independent pulley, which is known as loose pulley. Its major advantage is that we can put off the machine yet the main shaft being run with the electric motor or oil engine keeps on rotating. A lever is provided for slipping the belt.

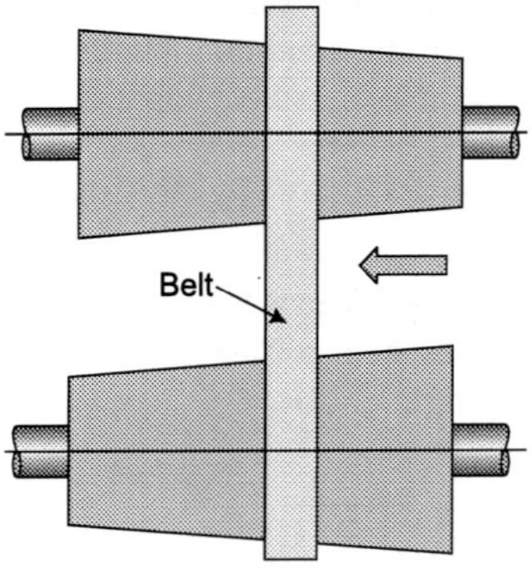

Fig. 4.19 Cone pulley

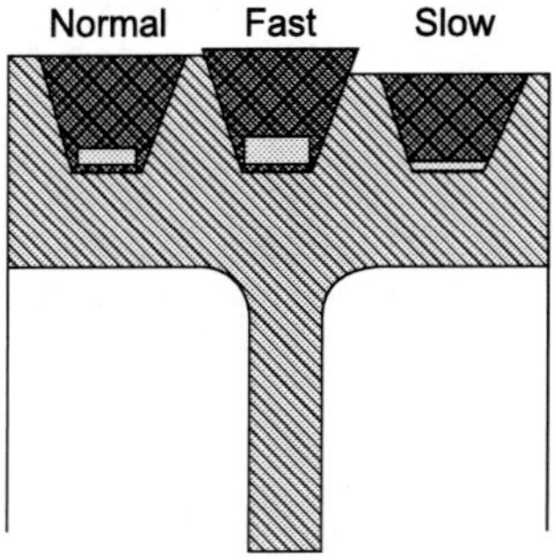

Fig. 420 Pulley with 'V' grooves

Speed Cones or Stepped Pulleys

Speed cones are cast iron pulleys having several steps of different diameters on which a belt may run. Speed cones are used for varying the velocity ratio between a pair of parallel shafts by simply shifting the belt from one step of the pulley to the other. They work in pairs, one on the counter shaft and the other on the machine spindle. The diameters of the corresponding steps must be such that the same belt can operate, no matter how many steps are employed.

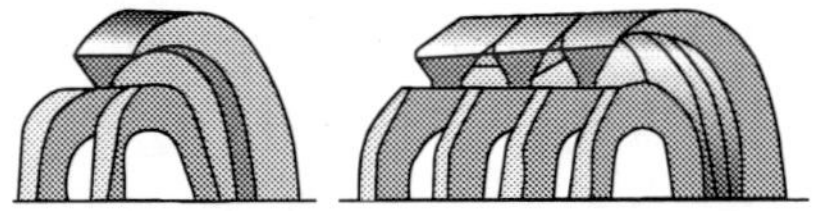

Fig. 4.20 A 'V' groove & 'V' Belt

Guide Pulleys

Guide pulleys are used to connect non-parallel shafts those which intersect and those which do not intersect to guide the belt into the proper plane. Guide pulleys are also used when the two shafts to be connected are close together. Each portion of the belt, as it passes from one pulley to the other, being taken round part of the periphery of a guide pulley, which is suitably placed **Fig. (4.13).**

Wrought-Iron Pulleys

They are now used in great numbers for several important advantages. Wrought iron pulleys are light, strong, and durable, and are entirely free from the initial strains which exist in cast iron pulleys due to unequal contraction in cooling. In wrought iron pulley the rim is made of sheet wrought iron or pressed steel sheet and riveted to the arms which are made from bar-iron or mild steel. The boss is sometimes also made of wrought iron, but generally it is made of cast iron. To facilitate the erection of pulleys on the main shaft they are usually made in halves, and the parts securely bolted together. Steel often takes the place of wrought iron in the construction of pulleys.

Wooden Pulleys

Wooden pulleys are made from 150 to 3000 mm in diameter, the rim built up in thickness, each composed of a number of segments arranged to break joint, the whole properly fitted and nailed and glued together. The arms are put in when the rim is being built up, and extend entirely through to the outside. The smaller sizes are made in block form. The pulleys are made in halves, and are fitted with interchangeable bushings so that they may be clamped tightly upon the shaft, dispensing with the use of a key.

As the coefficient of friction of leather on wood is higher than on the metal, it is claimed that wooden pulley will transmit more power than an iron pulley under the same condition, but they are not so durable as iron. Wooden pulleys can only be used in situations free from excessive dampness or moisture, and not subjected to high or varying temperatures.

Drum Pulley

This pulley is used as a driver pulley on the counter shaft of those machines which are operated by one single electric motor or oil engines and by one counter shaft. Their faces are made appropriately broad and flat belts are used.

Types of belts

Basically five types of belts are used for the transmission of power.

- Flat belt **(Fig. 4.21 A)**
- V-belt and multiple V-belts **(Fig. 4.21 B & C)**
- Ribbed belt **(Fig. 4.21 D)**
- Toothed or timing belt **(Fig. 4.21 E)**
- Link belt **(Fig. 4.21 F)**

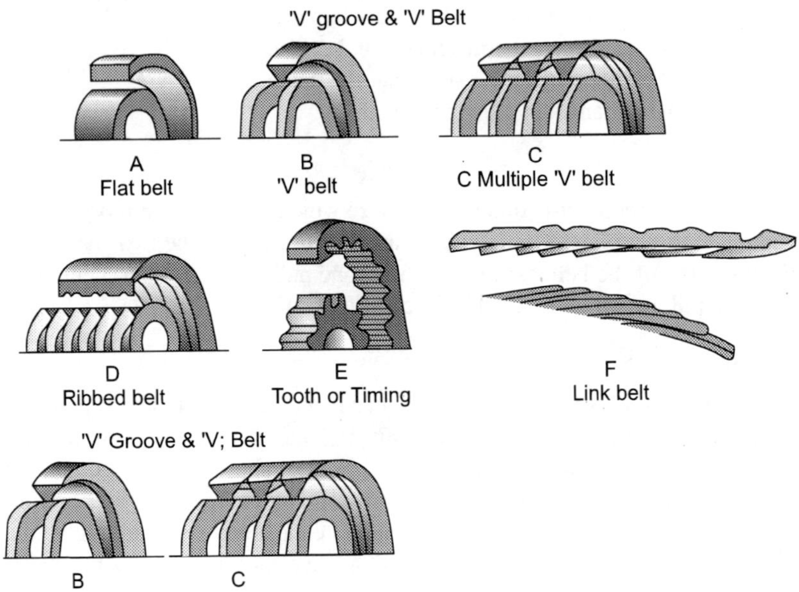

Fig. 4.21 Types of belts.

The choice of a particular belt depends upon speed ratio, centre distance, flexibility, strength, and economy and maintenance consideration of the driving system.

Belting is made of different materials and of varied cross sections either flat or "V" shaped. The materials of belting in common use for power transmission are: (1) leather, (2) cotton and canvas, (3) India rubber, and (4) steel.

Leather belts are made from the "prime" or "butt" portion of the hide. The method of cutting up the butt for the production of belting of the highest quality is very important. The butt is only about 1.5 m long, but belts of any length can be made by joints about every 1.5 m. Belts are also made of a single and double thickness.

Single belting, i.e., with the thickness composed of one piece only, is now made in four standard thicknesses denoted by the numbers 1 to 4, and of 4, 5, 6, and 7 mm in thickness respectively. Some manufacturers grade their belting by the weight in gm per sq cm, varying between certain fixed limits according to width.

Double belts formed by cementing, sewing, or riveting together two thicknesses of leather are sometimes employed for heavy duty, but should be avoided as far as possible, and should never be used on pulleys less than 900 mm in diameter. Where circumstances will allow, single-belts are preferred, as they are more flexible, absorb less power in bending round pulleys, and are in consequence more durable.

Leather belts may preferably be used both in dry and wet places at ordinary temperatures.

Canvas or woven belts are manufactured from cotton or camel hair. They are made in two distinct varieties, known commercially as canvas and solid woven respectively. Canvas belting is made from stout canvas or cotton "duck" 'folded to the required breadth and thickness, the latter dimension being denoted by the number of folds or plies. This is made in thickness from 3 to 10 ply.

Solid woven belting is produced in the loom in one piece of the required breadth and thickness from yarns spun from long stapled cotton. Hair belting is growing in favour, and may be woven entirely from camel hair or hair may be used in combination with weft and binders of cotton.

The canvas and woven belting are stronger and they are preferable to leather in warm climates, in damp atmosphere, and in exposed positions.

India-rubber belts are made by cementing together the canvas plies with a composition of vulcanized India-rubber. This kind of belting is considered the best in damp situations, but is expensive, and must be kept free from oil or grease, which are ruinous to rubber.

Halata belting has also wide applications in industry. This is primarily cotton canvas belting prepared in a similar manner with halata in place of rubber. Halata is a gummy substance resembling gutta-percha, obtained from the milky juice of a tree which grows in British Guinea. Halata belting is often preferred in the heavily saturated steam-laden atmosphere of a dye house, or when subjected to chemical fumes, or again, for out-of-door use.

Flat Belt (Fig. 4.21 A)

These appear rectangular in cross section and run on flat pulleys. The ends of the belt are joined by pins or stitched to make the belt endless.

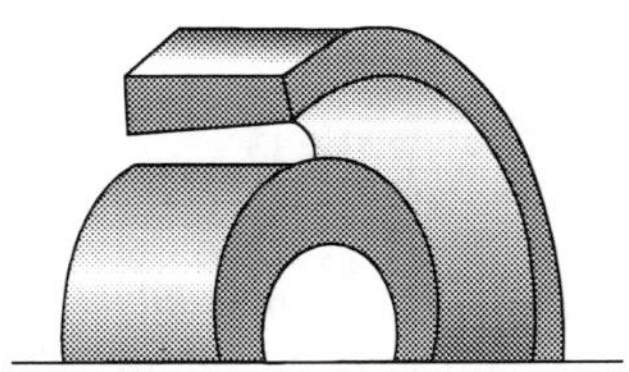

Fig. 4.21 A Flat belt

V-Belts (Figs. 4.22 & 23)

'V' belt drives are generally used when the distance between the shafts is too short for flat belt drives: Owing to the wedge action between the belt and the sides of the groove in the pulley, the V belt is less likely to slip hence more power can be transmitted.

V-Belt: When a belt is trapezoidal in section, designed to run in a 'V' shaped groove, it is known as a V-belt. The modern V-belts are made of fabric and vulcanized rubber with a cotton-cord tension element. The belt runs in 40° V-grooves turned in the pulleys **Fig. (4.20)**. The cross sectional area varies, depending on the dimensions, the smallest dimensions is 10 and 6 mm, and the largest 50 and 30 mm, respectively.

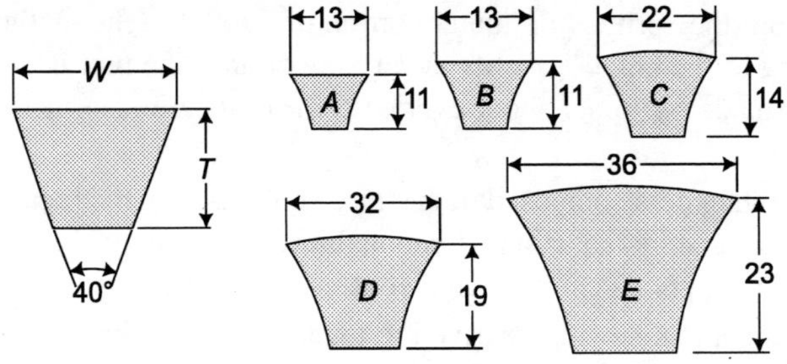

Fig. 4.22 Dimensions of 'V' belt

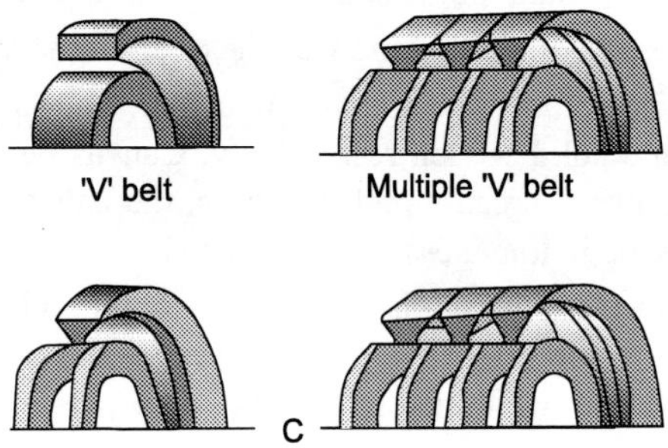

Fig. 4.23 'V' Grooves & 'V' belts

V-belt drives are used when centre distance between shafts is short and transmission numbers are large. The distance, in general, should not be less than the larger pulley diameter and not more than the sum of the two pulley diameters. V -belt transmits a larger amount of power from a pulley of a given width of face, and being almost positive and slip less in action, it is displacing

Dimensions of 'v' Groove of Sheave

Belt cross section	Power kW	Width	Thickness
A	075 - 5.0	13	8
B2 – 15	17	11	
C	705 – 75	22	14
D	20 – 100	32	19
E	> 100	38	23

Individual manufacturer's belts may deviate slightly from the dimensions for various constructional reasons. Crowning, if any in belts should be regarded for the measurement of thickness.

Matching Tolerance for Multiple 'V' Belt Sets

A, B, C, D, E		3 V, 5 V, 8 V	
Std. length	**Matching tolerance**	**Std. length**	**Matching tolerance**
26 – 75	2.54	250 – 1000	3.18
80 – 144	5.08	1060 – 2000	6.35

168 – 270	7.62	2120 – 2800	9.53
300 – 390	10.16	3000 – 5000	12.7
420 – 480	12.7		
540 – and above	15.24		

Care and Maintenance of Belt

The life of a belt will be prolonged and its driving powers kept at capacity by giving it proper attention.

Driving surface of the belt should be kept clean and free from any dirt and other foreign matter. Such dirt, if allowed to get into the surface, forms into lumps, tearing or distorting the driving face of the belt and preventing it from forming proper contact with the face of the pulley.

Care should be taken to prevent oil or grease from getting at the belt. It will cause loss of power by slip, and is particularly harmful to blatta or cotton belts, tending to separate the plies. It will also rot a leather belt and rapidly lessen the efficiency. An application of ground chalk will absorb the oil on a leather belt and make it workable for a time.

To keep a leather belt from becoming dry and lifeless, it should be given an occasional washing and brushing in warm water, and afterwards a good greasing with tallow or with one of the patent dressings sold for the purpose.

If belt slip becomes troublesome there are many compositions on the market to increase the friction between the belt and the pulley. A cheaper and better known remedy is powdered resin.

The face of the belt is also very important. The canvas side of a blatta belt is the driving face, whether the drive is crossed or open. As regards leather belts, the grain side is the correct driving side. It should transmit nearly twice the power conveyed by the flesh side. The flesh side, which has the greatest tensile strength, will stand the stretching strain necessary in the outside bend around the pulleys.

The endless V belt is shaped roughly like a trapezium in cross section, and is made of cord and fabric, and is treated with rubber and moulded together in a uniform manner and shape.

Advantages of Flat-Belt Drive

- Simplicity, smooth operation, low cost, low maintenance and long life.
- Flexibility, shock absorption and efficiency at high loads.

Disadvantages of V-Belt Drive

- They are not positive drives.
- Less efficiency due to slip and creep.
- Not suitable for short centre distance.
- Belt joints reduce the life of the belts.

Advantages of V-Belt Drive

- It is compact, so installation is possible in limited space.
- It is used when the centre distance between the driver and the driven pulleys is short.
- Less vibration and noise.
- Cushions the motor and bearing against load fluctuation.
- Easy replacement and maintenance.
- Transmit higher torque than flat belts.

- Suitable for short centre drives.
- Available in standard sizes.
- Run vibration free and are durable.

Designation of V. belt as per IS: 2494

The V belt confirming to the standard shall be designated by the cross section symbol, normal inside length and number of IS standard

Example

C 3048 IS: 2494

C = V-belt cross-section

3048 = Nominal inside length in mm in untensioned state.

1. Belt Length Calculation: (Flat Belts)

The best way to determine the correct length of a belt is by wrapping a steel tape around the pulley and recording the measurement. If the belts are to spliced and cemented, an additional amount of belt length must be provided for making the lap. The formula to determine the belt length is given below:

Calculation

Open belting

Length of open belting:

D, C, and R = diameter, circumference and radius respectively, of large pulley; d, c, and r = diameter, circumference and radius respectively of smaller pulley x = centre to centre distance.

$$L = \frac{D + d}{2} \times 3\frac{1}{7} + 2x, \; L_1 = \frac{(D + d)}{2} \times 3\frac{1}{7} + 2x$$

Length of cross belting:

$$L\,c = C/2 + c/2 + 2\sqrt{x^2 + (R + r^2)}$$

The performance and life of the belt depends considerably on the selection of the drive system.

2. Correct Pulleys are Very Important

They should preferebly be obtained from standard menufactureres and of good quality material, groove dimensions and finish, the distance between centres, the pitch diameter, and the tolerance on machines finishing and balancing are very important and must be inspected to ensure that they meet IS 3142 before acceptance.

Special care should be taken to check that the bore centre is concentric to the pitch diameter/outside diameter and when fitted to the shaft, the wobble should not exceed standards laid.

How to Check Your Drives

After the pulleys are fitted to the drives, check:

- The pulley shafts should be parallel and the pulleys are in line.
- The sheave of grooves are correctly aligned.
- The sheaves are properly locked and do not wander.
- The sliding rails of the motor take up should be sufficient to handle 41/2 % variation in belt length and normal take up due to wear and belt stretch.

The motor should be so positioned during installation that 1/3 of this movement is towards the driving unit to allow for ease of belt fitment while 2/3 is available for take up purpose.

3. Storing Belts for Future Use

New belts should be kept in shade, away from heat, oil or direct sunlight. Belts should not be stored on floor or near windows, heaters, etc.

During storage, avoid excess weight on and distortion of belts. When hung on pegs, the longer belts should be coiled so that loops are not greater than approximately 2500 mm in circumference. Pegs should be crescent shaped and large enough to avoid compression set from corners or from acute bends of the belt.

Unfavourable storage conditions can lead to reduced belt life and to variation in belt lengths. Some fibres used in strength members of belts are subject to shrinkage in storage, the amount depends on the treatment technology, relative humidity and storage time.

4. The Art of Building "V" Belts

Belts with polyester strength members are far superior to belts with rayon strength members but this is not obtained by just substitution of the members. The two belts although they look alike are generations different resulting from changes in rubber compounding required for different areas of the belt, the treatment of the polyester cord and the type of envelope fabric. Proven and trusted technology usage and proper manufacturing technique using sophisticated equipment is the key to a successful polyester belt product, resulting in customer satisfaction. Polyester belts not only carry more loads but also in a proper designed drive have very much longer life. Initial higher costs are more than compensated by higher rating and longer life. It therefore pays to buy the best.

5. Installing for a Satisfying Performance

For proper appreciation of belt performance and satisfaction, the installation procedures are very important.

(a) Clean both pulleys, especially the grooves and check for burrs, wear down, etc.

(b) Fit a belt in each groove, one at a time, press it down and see how the belt fits in the groove. If belts fit wrongly as shown in diagram below, check pulley grooves and belt for correct dimensions and change as required.

(c) Slacken motor base and slide inwards to maximum position.

(d) Fit belts into grooves freely. Do not roll on or force over the grooves with a tool. In case of multi-belt drive, please ensure that belt slack of all belts is either on top or at bottom. See sketch for details.

(e) Slide the motor back, thus tensioning the belts. Make sure that the pulley shafts are parallel and in line. This is very important. Also pulley should be keyed firmly to the shaft.

(f) Check position of belts in groove. Variation in pitch lines will result in belts operating at different speeds. Some of the belts will slip and belt life considerably reduced. **Do not mix new and old belts. Belts fitted on a drive should be from one manufacturer only.**

Work belts around in the groove by hand, so that the slack of all belts is on the top like **Fig. (4.25)** or the slack of all belts is on the bottom **Fig. (4.26).**

With the slack of some belts on bottom (see solid line) and the slack of others on top (see dotted line). Since V belts will not slide in the groove, belts thus applied will be injured when tightened for operation.

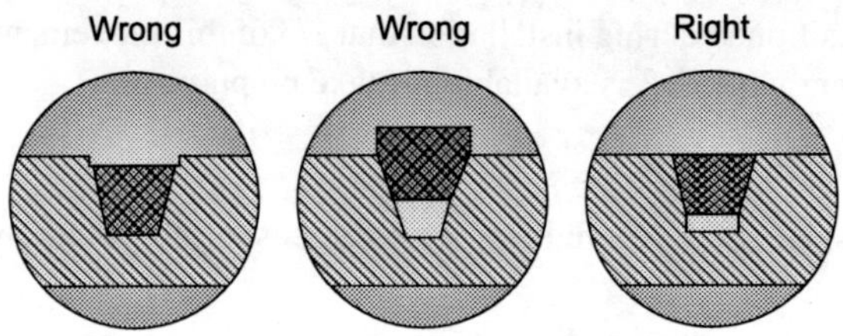

Fig. 4.24 Correct way of fixing belts

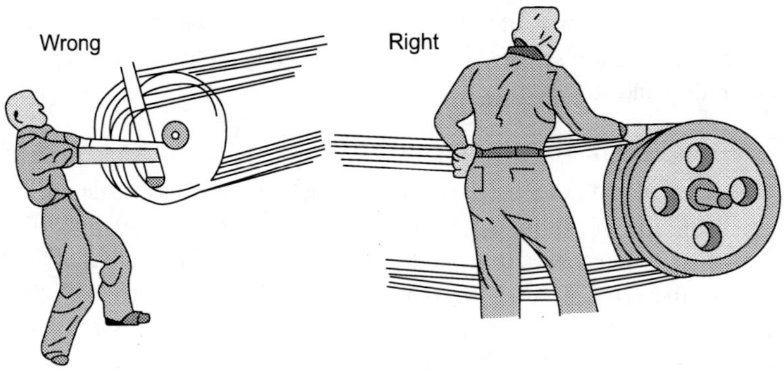

Fig. 4.25

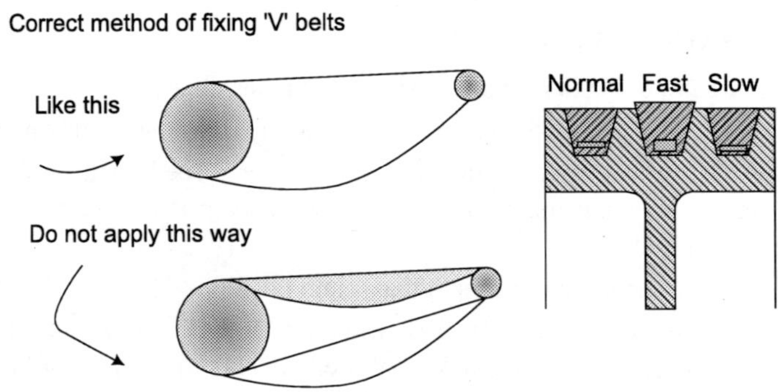

Fig. 4.26 Correct method of fixing 'V' belts

6. Precautions for Running in

After all checking and corrections as required have been done, the belt is tensioned by sliding back the motor on the rails. The motor is started and tension is applied until slip ceases to exist. This can be checked from speed ratio data. When the drive is in motion slight sag will be noticed on the loose side.

Following initial tension adjustment, additional tension should be applied to minimize slippage during the 48-hour "running in" period. During this period two things happen

(a) The initial stretch is being taken out of the belt.

(b) The soft rubber surface of the outer envelope is being abraded and the belt is seating itself at a slightly lower position in the sheave, allowing the belt to run slack.

Unless new belts are taken-up, there is apt to be considerable slipping, burning or other irreparable damage.

Polyester belts made to proper technology and under correct methods have very small shrinkage in storage and less stretch on fitment. Consequently they take less time to get adjusted to working conditions and thereafter require very little attention on maintenance.

Check the condition: All belts are operating on different pitch circles. All are running at different speeds **Fig. (4.26).**

7. The Final Step to Full Satisfaction

Because of the high power ratings it is necessary to be able to measure belt tensions with sufficient accuracy to avoid belt slip or overloaded bearing or to meet particularly arduous conditions. The following procedure is recommended for drives coming within the normal range for each belt section. Measure the length of the span in millimetres. At the centre of the span apply a force with a spring scale in a direction perpendicular to the span, until the belt is deflected from the normal by an amount equal to 0.015 mm for every millimetre of span length. For example, the deflection for a span of 1 m would be 1000 mm × 0.015 mm or 15 mm. Note the force and compare it with the value of P given in table. In all cases it is essential that the pulley centres be fixed and that the larger pulley be then rotated at least four times before making the measurement. On a multiple belt drive it is essential that a matched set of belts be used and the above procedure be carried out on each belt, the average value of these forces being compared with the specified values of P in table.

The belt tension should be such that the measured force falls within the values given in table. A measured force below the lower value indicates under-tensioning whilst a measured force above the higher value indicates over-tensioning. However, when starting up a drive with new belts, the drive should be tensioned to the higher value since the tension falls rapidly in the early stages of running in. Some difficult drives may need a tension toward the higher value to be maintained, such drives usually have one or more of the following properties:

- High belt speed;
- Low belt speed;
- Small arc of contact;
- High overload on start-up;
- Frequent high torque start-up.

All belt drives should be watched carefully during the running-in period after initial start-up. Re-tensioning should be carried out by the above procedure when necessary.

Deflection force required for measuring installation tension in V-Belt drives in Newton.

Cross section symbol	Smaller pulley diameter mm	Required deflection Force P at centre of span for speed range of		
		0 m/s-10 m/s	1.0m/s-10 m/s	20 m/s – 30 m/s
'V' belt		N	N	N
A	80 – 140	12- 18	10-15	8-12
	Above 140	18-27	15-22	12-18

B	125-200	24-36	20-30	16-24
	Above 200	36-54	30-45	24-36
C	200-400	48-72	40-60	32-48
	Above 400	72-110	60-90	48-72
D	355-600	96-140	80-120	65-96
	Above 600	140-210	120-180	96-140

N = Newton = 0.102 kg.

Belt Tension

Belts must be tensioned correctly to transfer the torque from the driving pulley to the driven pulley to prevent unnecessary wear.

Too much of belt tension curtails belt and bearing life. As the belts stretch in use, it is necessary to check and adjust the belt drive tension.

When a drive is transmitting power the belt pulls or the belt tensions. There is the tight side tension (Tt) and a slack side tension (Ts) **Fig. (4.28).**

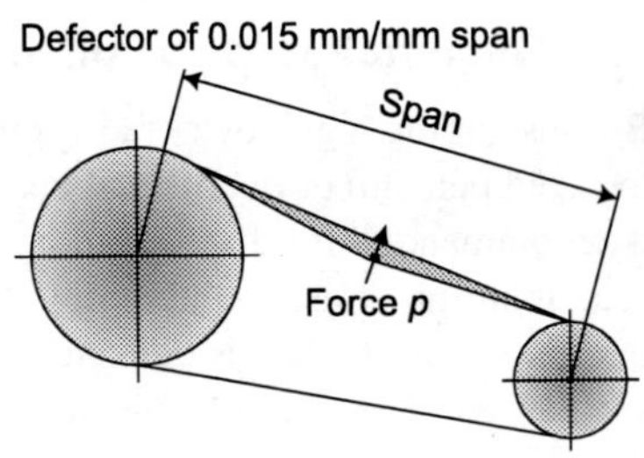

Fig. 4.27 Belt defection

Tension Ratio

The ratio of the tight side to the slack side tension is commonly referred to as the tension ratio. A higher ratio between the tight side and slack side tension makes the belt loose and slip.

This causes lack of effective pull for transmitting the required power.

Adjustment of Tension

When the distance between two pulleys is fixed, the tension of a belt is adjusted by an idler. **Fig. (4.29)**

When the distance between two pulleys is not fixed, the tension of the belt is adjusted by the adjustment screw **Fig. (4.30).**

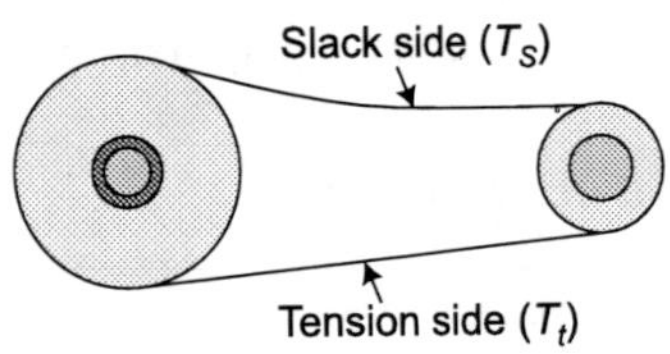

Fig. 4.28 Belt tension

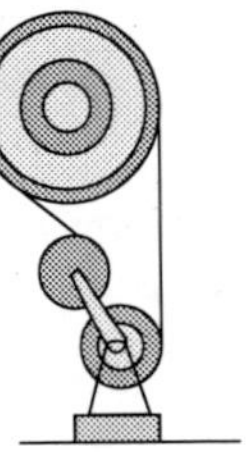

Fig. 4.29 Distance betwen pulleys fixed tension is adjust by an idler

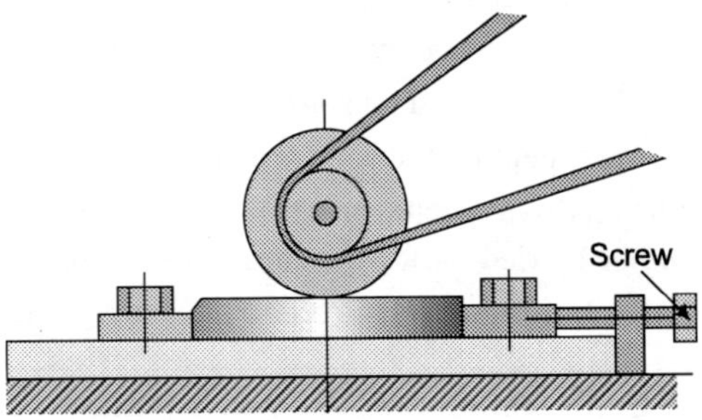

Fig. 4.30 Distance between pulleys adjusted by screw

Arc of Contact

Tension is necessary to create friction between the pulleys and the belt. Torque transmission depends on the contact area of the belt over the pulleys **Fig. (4.31).**

If the wrapping angle is big, the pulley can transmit high torque **Fig. (4.32).**

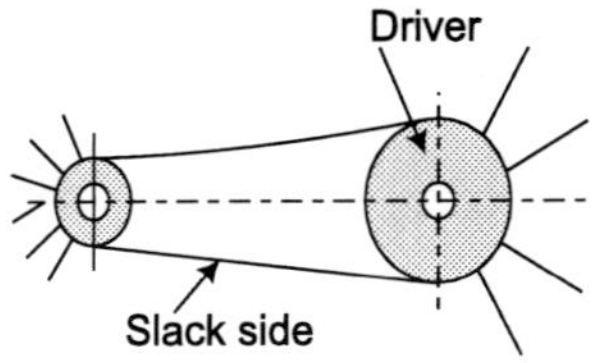

Fig. 4.31 Transmission depends on contact area

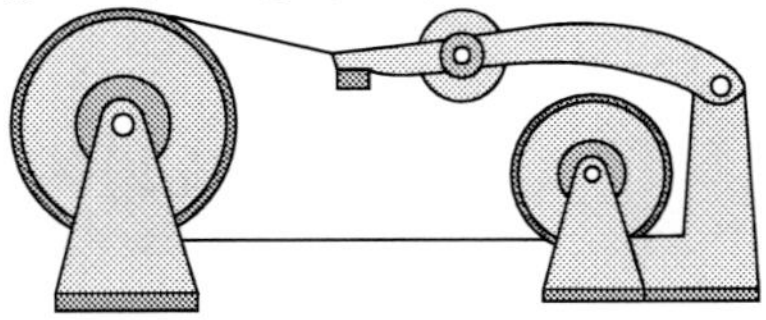

Fig. 4.32 Jockey pulley used to increase area of contact

Belt Efficiency

To provide maximum arc of contact the following points should be considered.

- Heavy belts of multi ply construction should not be used on small diameter pulleys.
- If the arc of contact is insufficient because of the short centre distance between the pulleys, a jockey pulley should be introduced as near to the small pulley as possible **Fig. (4.32).**

Excessive tension in the belt reduces the arc of contact, and introduces additional stresses which drastically reduce the life of the belt and bearings **Fig. (4.33).**

Vertical drives should definitely be avoided because the belt tension necessary to withstand gravitational pull **Fig. (4.34)** and accompanying slippage would result in adverse effects.

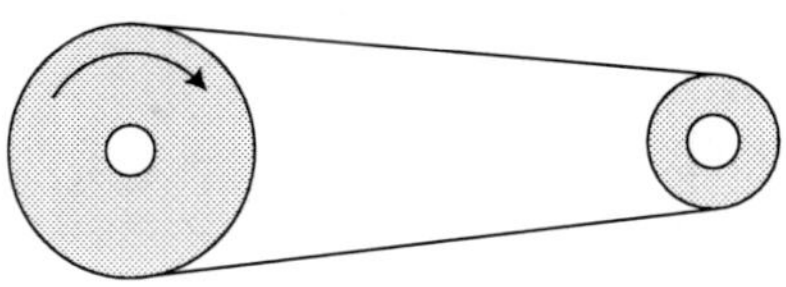

Fig. 4.33 Reduced arc of contact because of excessive tesion

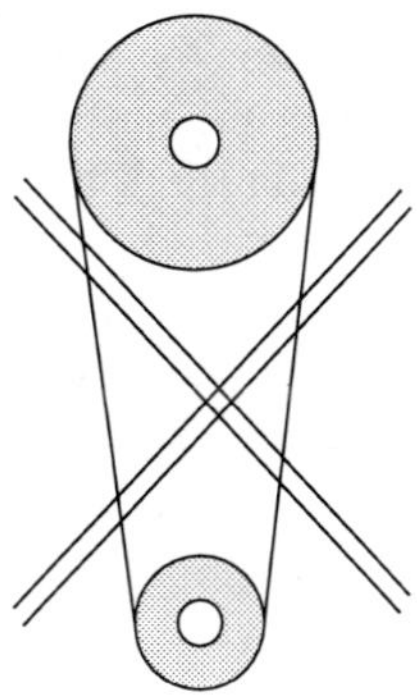

Fig. 4.34 Avoide vertical drive

On the open belt drive, the slack side **Fig. (4.35)** must be at the top and the centre distance between the pulleys should be the maximum.

To Measure Tension of V-Belt Drives

To determine the force required to deflect one belt per 25 mm span length, apply a force perpendicular to the span at the centre of the belt large enough to deflect one belt to 0.5 mm span from its normal position **Fig. (4.36).**

- Compare this deflection force with the range of forces.
- If it is less than the minimum recommended deflection force, the belts should be tightened.

If it is more than the maximum recommended deflection force, the drive is tighter than it needs to be.

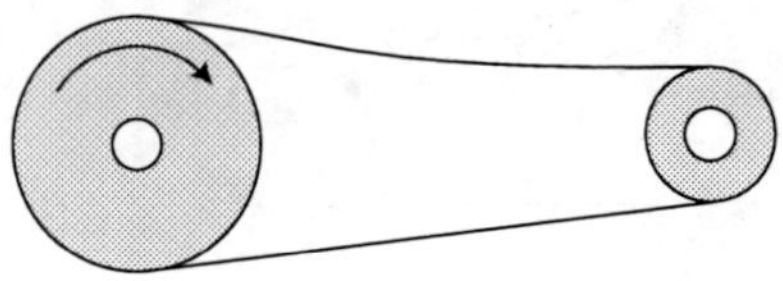

Fig. 4.35 Open belt drive

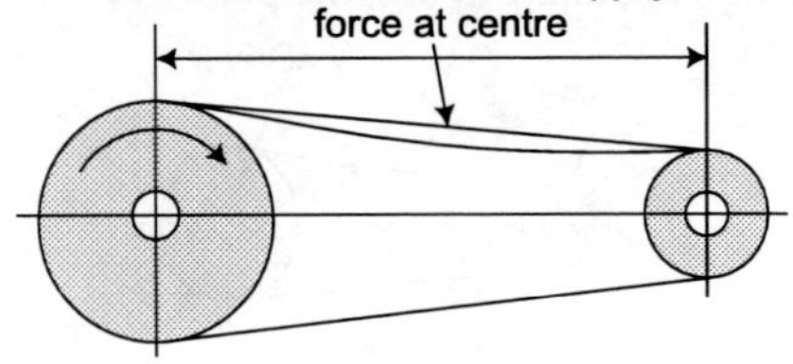

Fig. 4.36

Care and Maintenance

- Keep the pulley faces and belts free from foreign material which may cause slips.
- When the 'V' belts begin to show signs of wear they should be replaced. Replace all the belts in a multiple 'V' belt drive rather than a single one.
- Check and adjust drive tension periodically.
- Store belts in a cool, dark and dry place.

The belt tension should be adjusted in such a way that the deflection force is in between the maximum and minimum.

Creep and Slip of Belt (Fig. 4.37)

As the belt turns on a pulley it tends to stretch on the contact area of the driving pulley and shorten on the driven pulley. This localized movement of the belt is a direct result of the elastic stretch and is known as creep. Greater the load more will be the creep. The figure shows the condition of belt as a result of creep.

Slip is the actual difference caused between the surface speed of the belt and pulley. The effect of slip may be reduced by decreasing the pulley ratio and maintaining proper alignment. Creep, being the physical characteristic of the belt, cannot itself be controlled. Slip and creep jointly cause power loss.

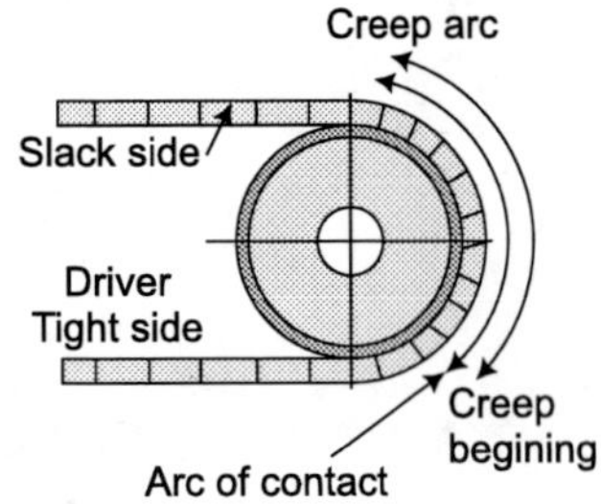

Fig. 4.37 Creep of belt

Belt Dressing

Due to continuous rubbing of the belt on the surface of the pulley the belt gets dried up due to friction, and heat is generated. This causes the belt to slip.

To keep the belt supple and free from cracks, belt dressing is applied. Tallow or powdered resins are good dressing materials which are applied on the inner face of the belt. This improves the gripping property of the belt.

Types of Fasteners

Joints in Belting

Belting is generally supplied by the manufacturers in one continuous length ready for use, so that when placed upon the pulleys it only remains to make the closing joint by some form of fastener to produce an endless belt. Three methods are generally used for such fastening, namely, lacing, metal fastener of various kinds and endless, cemented or by applying solution joints.

Leather belts are often joined by raw hides. For lacing the ends of the belts are cut square and butted together, and the lace is threaded in round or oval holes which are made with a hand-tool known as belt-punch.

Leather and balata belts may be made endless by splicing and cementing exactly in the same way as the permanent joints employed in the manufacture of the belt itself. The belt is opened up by heating with a hot iron to enable the plies to be separated. They are then cut to stepped form, so that when the ends are placed together, the thickness of the joint in any part is the same as in the body of the belt. The plies are successively covered with solution, and the joint after thorough heating, is fixed in a clamp or press for ten to thirty minutes until it is set. There are two kinds of belt cements, one for dry and the other for damp conditions. The first is available in cakes or lumps, which dissolve in a glue pot. The other type is sold ready to apply.

The belt fasteners commonly used in addition to the alligator type are as follows.

Wire Type Belt Fastener

Fig. (4.38) shows the wire type fastener generally used on light duty machines

'Lagrelle' Type Belt Fastener

Fig. (4.39) shows a Lagrelle type fastener used on heavy duty machines.

Jackson-Type Belt Fastener

The Jackson-type fastener is used on medium duty machines **Fig. (4.40).**

Crescent Plate Belt Fastener (Fig. 4.41)

This is a mechanical type belt fastener which is used on medium duty machines.

Belt Fasteners (Alligator type) Fig. (4.42)

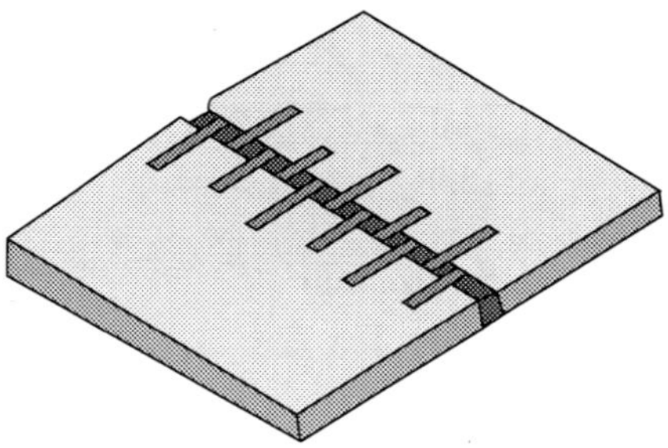

Fig. 4.38 Wire type belt fastener

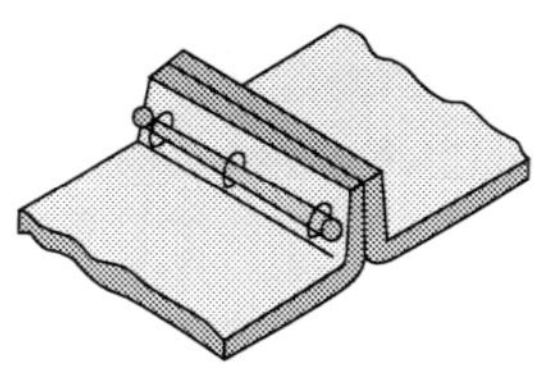

Fig. 4.39 Lagrelle type belt fastener

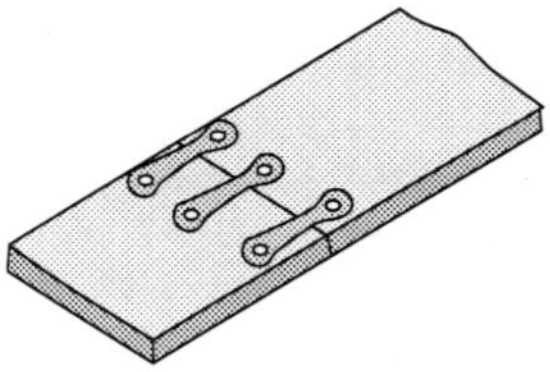

Fig. 4.40 Jackson type belt fastener

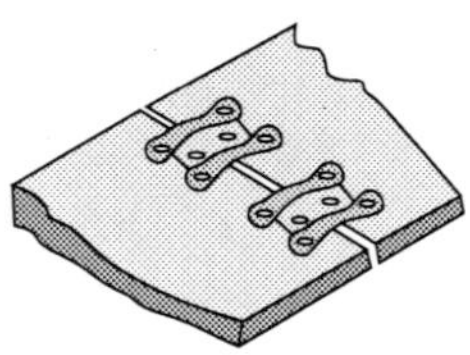

Fig. 4.41 Crescent type belt fastener

Alligator type fasteners are used in joining belting for industrial purposes. The belt fastener is made of steel sheets conforming to IS: 513-1973. The pins shall be made from mild steel wire conforming to IS: 280-1972. The position of the pin in a joint is illustrated in **Fig. (4.42).**

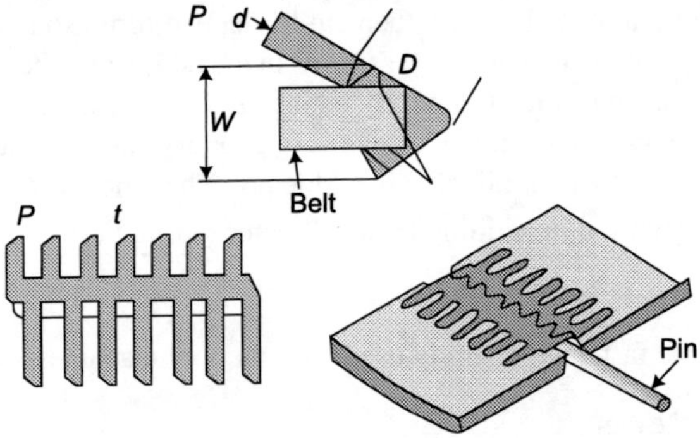

Fig. 4.42 Specification for aligator type belt fastener

Designation	Thickness of belt	Metal thickness (Sheet)	Point depth	Approx Overall width	Approx Overall Deptht, Min
15	3 to 4	1.0	5.0	18	13
20	4 to 5	1.1	6.5	22	17
25	5 to 5.5	1.2	7.0	25	21
27	5.5 to 7	1.2	8.0	29	24
35	7 to 8	1.8	9.5	32	30
45	8 to 9.5	1.8	11.0	38	31
55	9.5 to 11	2.0	14.0	48	40
65	11 to 13	2.0	16.0	54	41

Width of Bar prong	Pitch of prong	Fastener Designation	Pin size mm
2.5	-	15,20,25	2.64
3	-	27,35	3.25
3	8		
3	8	45, 55, 65.	4.06
4	10		
5	12		
6.5	16		
6.5	16		

Problem	Probable Cause	Solution
Short belt life	Worn sheaves, leading to abnormal envelope wear oil or grease on belts, leading to opening of envelope seams or belt softening or swelling. High ambient temperature gouges or extreme cover wear caused by belts rubbing on drive guards or other objects. Spin burns from belt, slipping on driver sheave under starting or stalled load.	Replace sheaves or recut to correct dimensions. Check drive for excess oil, leaky bearings. Clean belts and sheaves with cloth soaked in gasoline. Shield belts, provide ventilation, and check lubrication. Remove obstruction or realign drive to provide clearance. Tighten belt.
Belts turn over in groove	Misalignment of pulleys and shafts. Worn sheaves. Flat idler pulley misaligned. Excessive vibration. Damaged tensile section in belts; frayed or gouged belts.	Realign drive. Replace sheaves or recut to correct dimensions. Realign idler. Adjust tension. Replace frayed belt or belts with tensile member broken. Replace belts.
Belt squeak	High starting load or excessive overload	Tighten belts.
Belt squeak or "chirp"	Movement of belts on flat pulley. Dust or rust on side of belt.	Realign idler. Squeak may be annoying, but will not harm belts. Forget it.
Belt break prematurely	Shock or extreme overload. Belts damaged during installation. Foreign material falling into drive causing abnormal abrasion. Small sheaves causing transverse bottom breaks. Small pulleys causing separation. Excessive tension, causing separation	Operate machine properly. See dealer for proper belt. Follow recommended installation procedure. Provide drive guard. Check sheave dimensions. Check pulley diameters and ratios Adjust tension
Belts stretch beyond take up	Drive operating under excessive tension. Poor adjustment under belted drive. Belt stored in damp area. Worm sheaves.	Slack off drive to proper tension Adjust take up Check dealer for proper belt. Dry in oven at 2500 F for two hours Replace sheaves or recut to correct dimensions.
Excessive vibration	Belts too lose. Damaged tensile member in belt.	Tighten belts. Add kiss idlers. Improper installation. Replace with new belts properly installed

Belts too long at installation	Wrong size belts. Drive set up improperly. Insufficient take up.	See dealer for correct belt. Re-check machinery. See dealer for correct belt.
Belts too short at installation	Wrong size belts. Drive set up improper. Insufficient take up.	See dealer for correct belt. Recheck machinery.
	Belts stored in damp area.	Dry in oven 250°F for two hours.
Belts mismatched at installation	Poorly matched at factory. Old belts and new belts mixed on same drive. Mixed set of belts from more than one manufacturer.	Replace with another set of belts. Never install new belts with old belts. Replace all belts. Replace with new set from manufacturer.
	Pulley shaft not parallel, causing unequal stretch. Worn or improperly machined sheaves.	Realign drive. Replace sheaves or recut to correct dimensions
Belts mismatched after	Old belts and new belts mixed on same drive. Mixed set of belts from more than one manufacturer.	Never install new belts with old belts. Replace with new set from one manufacturer.
	Pulley shafts not parallel. Worn or improperly machined sheaves	Realign drive. Replace sheaves or recut to correct dimension.
	Tensile member damaged during installation.	Replace with new belts, properly installed.
Belts mismatched after service	Old belts and new belts mixed on same drive. Mixed set of belts from more than one manufacturer.	Never install new belts with old belts. Replace with new set. Replace with new set from one manufacturer. Realign drive.
	Pulley shafts not parallel. Worn or improperly machined sheaves	Replace sheaves or recut to correct dimension. Replace with
	Tensile member damaged during installation.	new belts, properly installed.
Loss in driven speed	Pulley ratios not correct. Excessive slip. lose belt.	Check ratio of sheaves diameter against ratio of RPM-change to correct pulleys. Check-shut down drive-sheave temperature by feel. A badly slipping belt will heat sheave so that it will nearly burn operator's bare hand. Check for proper tensions.
Localized wear	Incorrect sheave diameters	Check cross section dimension. If narrow pulley spin. If full internal breakdown and resultant swell.
Belt envelope hardening and cracking	Heat and chemical fumes	Check working conditions, provide ventilation

4.3 CHAIN DRIVE

Chains are used for high transmission numbers (up to 15) and can impart as much as 5000 hp. They are mostly used when the distance between centres is short. But they are also employed when the centre distance is as much as 8 m. They are now in general use for the transmission of power in cycles, motor vehicles, agricultural machinery, road rollers, etc. and for gearing in workshops and factories, and are continually being installed to displace belt or rope drive and wheel gearing.

Well-conceived chain drives are highly effective, are positive in motion, i.e., permit no slip, and may be used where an exact average velocity ratio is essential. A chain drive takes up less space than a belt drive and has no initial tension. For belt or rope drives a certain minimum distance between the shafts is necessary unless jockey pulleys are to be used. This prevents the use of belts or ropes for connecting directly to shafts which are closed together.

A chain may be regarded as a belt built up of rigid links, which are hinged together in order to provide the necessary flexibility for the wrapping action around the driving and driven wheels. Wheels having teeth especially designed for chains are known as chain sprockets and bear a superficial resemblance to spur gears.

Chain drives are used for transmission of motion at constant velocity ratio without creep and slippage. Chains are used in conjunction with sprocket pinions and sprocket wheels. Chains and sprockets are available in both British and metric standards. The sprockets are generally keyed to the shafts.

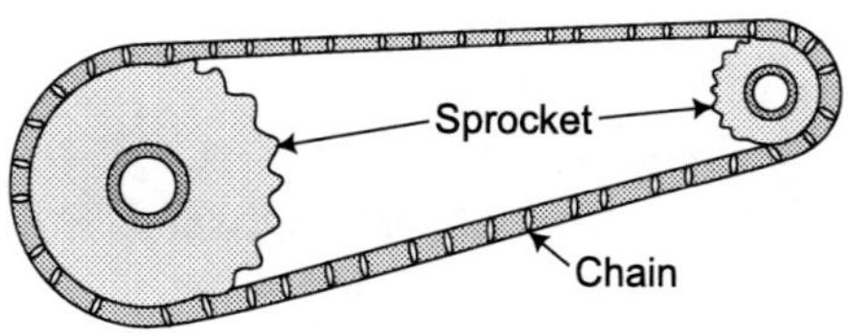

Fig. 4.43 Chain and sprocket drive

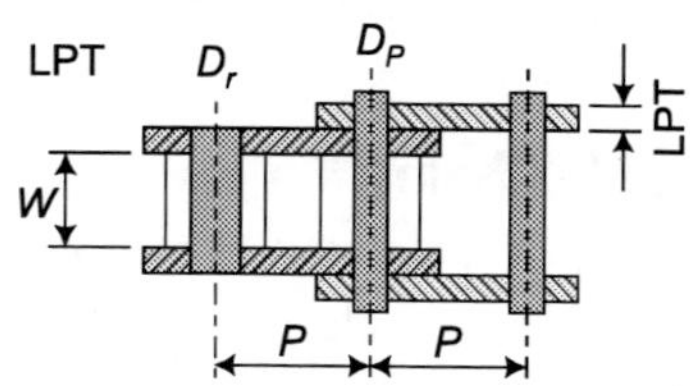

Fig. 4.43a Specification of chain links

Chain drives consist of an endless series of chain links which mesh with toothed wheels called sprockets **Fig. (4.44).**

Roller Link D: An inside link consisting of two inside plates, two bushings, and two rollers.

Pin Link G and E: An outside link consisting of two pin-link plates assembled with two pins.

Inside Plate A: One of the plates forming the tension members of a roller link.

Pin Link Plate E: One of the plates forming the tension members of a pin link.

Pin F: A stud articulating within a bushing of an inside link and secured at its ends by the pin-link plates.

Bushing B: A cylindrical bearing in which the pin turns.

Roller C: A ring or thimble which turns over a bushing.

Assembled Pins: G.-Two pins assembled with one pin-link plate.

Connecting-Link G and I: A pin link having one side plate detachable.

Connecting-Link, Plate l: The detachable pin-link plate belonging to a connecting link. It is retained by cotter pins or by a one-piece spring clip (not shown).

Offset Link L.: A link consisting of two offset plates assembled with a bushing and roller at one end and an offset link pin at the other.

Offset Plate J: One of the plates forming the tension members of the offset link.

Offset Link Pin K A pin used in offset links.

Roller Diameter D are approximately *% P.*

Width, W is defined as the distance between the link plates. It is approximately the chain pitch.

Pin diameter are approximately 5/16 P or 1/2 of the roller diameter.

Thickness of link plates for heavy series of any pitch is approximately that of the next larger pitch standard series chain.

Maximum height of roller link plate = 0.95 pitch.

Maximum height of pin link plates = 0.82 pitch.

Maximum pin diameter = nominal pin diameter + 0.128 mm.

Maximum hole in bushing = nominal pin diameter + 0.04 mm.

Maximum width of roller link= nominal width of chain + 2.12 X nominal link plate thickness.

Minimum distance between pin link plates = maximum width of roller link + 0.128 mm.

In addition to the standard roller and inverted tooth types, a wide variety of drive chains of different construction are available. Such chains are manufactured to various degrees of precision ranging from unfinished castings or forging to chains having certain machined parts. Practically, all these chains as well as standard roller chains can be equipped with attachments to fit them for conveyor use. A few such types are briefly described in the following paragraphs. Detailed information about them can be obtained from the manufacturers.

Detachable Chains

The links of this type of chain, which are identical, are detachable. Each has a hook-shaped end in which the bar of the adjacent articulates. They are available in malleable iron or pressed steel. The advantage is the ease with which any link can be removed.

Cast Roller Chains

Cast roller chains are constructed, wholly or partly of cast metal parts and are available in various styles. In general, the rollers and side bars are accurately made castings without machine finish. The links are usually connected by means of forged pins secured by nuts or cotters. Such chains are used for slow speeds and moderate loads, or where the precision of standard roller chains is not required.

Pintle Chains

Unlike the roller chain, the Pintle chain is composed of hollow cored cylinders cast or forged integrally with two offset side bars and each identical. The links are joined by pins inserted in holes in the ends of the bars and through the cored holes in the adjacent links. Lugs prevent turning the pins in the side bars and thus ensure articulation of the chain between the pin and the cored cylinder.

Standard Roller Transmission Chains

A roller chain is made up of two kinds of links: roller links and pin links alternately spaced throughout the length of the chain **Fig. (4.44)**.

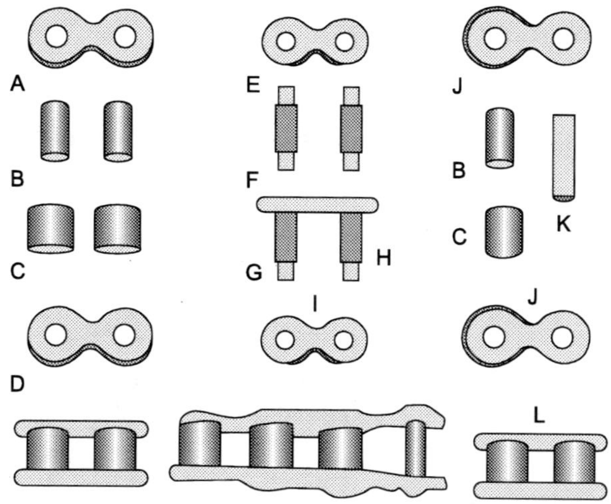

Fig. 4.44 Chain links

Roller chains are manufactured in several types, each designed for the particular service required. All roller chains are so constructed that the rollers are evenly spaced throughout the chain. The outstanding advantage of this type of chain is the ability of the rollers to rotate when contacting the teeth of the sprocket. Two arrangements of roller chains are in common use: the single-strand type and the multiple-strand type. In the latter type, two or more chains are joined side by side by means of common pins which maintain the alignment of the rollers in the different strands.

Standard Roller Chains

These are manufactured to the specifications in the American National Standard for precision power transmission roller chains, attachments, and sprockets (ANSI B29.I-I975) and, where indicated, the data in the subsequent tables, have been taken from this standard. These roller chains and sprockets are commonly used for the transmission of power in industrial machinery, machine tools, motor trucks, motorcycles and tractors and similar applications.

Non-Standard-Roller Chains

These developed individually by various manufacturers prior to the adoption of the ANSI standard are similar in form and construction to standard roller chains but do not conform dimensionally to standard chains. Some sizes are still available from the originating manufacturers for replacement of existing equipment. They are not recommended for new installations, since the manufacture is being discontinued as rapidly as possible.

Standard Double-Pitch Roller Chains

These are like standard roller chains, except that their link plates have twice the pitch of the corresponding standard-pitch chain. Their design conforms to specifications in the ANSI Standard for double-pitch, power transmission roller chains and sprockets (ANSI B29)-1954, RI974). They are specially useful for low speeds, moderate loads or long center distances.

Standard Roller Chain Nomenclature and Load

Chain Pitch

Distance between centres of adjacent joint members Other dimensions are proportional to the pitch.

Tolerance for Chain Length

New chains under measuring load, must not be under-length, over-length tolerance is 0.0025 per pitch in mm. length measurements are to be taken over a length of at least 300 mm.

Measuring Load

This is the load in kilograms under which a chain should be measured for length. It is equal to one percent of the ultimate tensile strength with a minimum of 10 kg and a maximum of 500 kg for both single and multiple strand chains.

Maximum Ultimate Tensile Strength

For single strand chain this is equal to or greater than 6000 kg. The minimum tensile strength or breaking strength of a multiple strand chain is equal to that of a single strand chain multiplied by the number of strands.

Standard Roller Chain Numbers

The right-hand figure in the chain number is zero for roller chains of the usual proportions, 1 for a lightweight chain and 5 for a roller less bushing chain. The numbers to the left of the right-hand figure denote the number of ¼ inches in the pitch. The letter H following the chain number denotes the heavy series; thus the number 80 H denotes a I-inch pitch heavy chain. The hyphenated number 2 suffixed to the chain number denotes a multiple strand, 3 a triple strand, 4 a quadruple strand chain and so on.

Heavy Series

These chains, made in 3/4-inch and larger pitches, have thicker link plates than those of the regular standard. Their value is only in the acceptance of higher loads at lower speeds.

Light weight Machinery Chain

This chain is designated as No. 4 I. It is ½ inch pitch; ½ inch wide; has 0.306-inch diameter rollers and a 0.I4 I-inch pin diameter. The minimum ultimate tensile strength is 1500 pounds.

Multiple-strand Chain

This is essentially an assembly of two or more single strand chains placed side by side with pins that extend through the entire width to maintain alignment of the different strands.

Types of Sprockets

Four different designs or types of roller-chain sprockets are shown by the sectional views, **Fig. (4.45)**

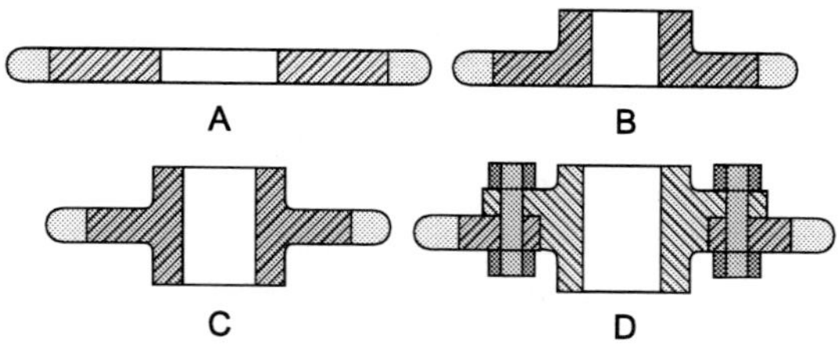

Fig. 4.45 Types of sprockets

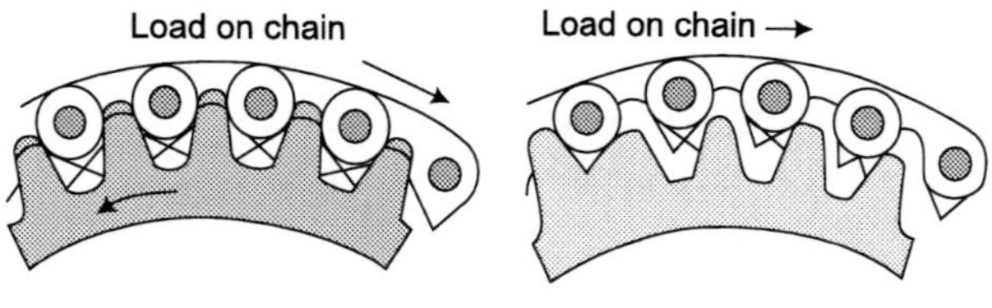

Fig. 4.46 Silent chain & sprocket load on chain

Type A is a plain plate; type B has a hub on one side only; type C, a hub on both sides; and type D, a detachable hub. Also used are shear pin and slip clutch sprockets designed to prevent damage to the drive or to other equipment caused by overloads or stalling.

Silent or Inverted Tooth Chain

Silent or inverted tooth chain consists of a series of toothed links alternately assembled either with pins or with a combination of joint components in such a way that the joints articulate between adjoining pitches. Side guide chain has guide links which engage the sides of the sprocket. Center guide chain has guide links which engage a groove or grooves in the sprocket.

Characteristics of Silent Chain Drives

The silent or "inverted-tooth" driving chain has the following characteristics: The chain passes over the face of the wheel like a belt and the wheel teeth do not project through it; the chain engages the wheel by means of teeth extending across the full width of the under side, with the exception of those chains having a central guide link; the chain teeth and wheel teeth are of such a shape that as the chain pitch increases through wear At the joints, the chain shifts outward upon the teeth, thus engaging the wheel on A pitch circle of increasing diameter; the result of this action is that the pitch of the wheel teeth increases at the same rate as the chain pitch. Another distinguishing feature of the silent chain is that the power is transmitted by and to all the teeth in the arc of contact, irrespective of the increasing pitch due to elongation. The links have no tiding action either on or off the teeth, which results in a smooth and practically noiseless action, the chain being originally designed for the transmission of power at higher speeds than are suitable for roller chains. The efficiency of the silent chain itself may be as high as 99 per cent, and for the complete drive, from 96 to 97 percent, under favourable conditions; from 94 to 96 per cent can be secured with well-designed drives under average conditions.

The life and upkeep of silent chains depend largely upon the design of the entire drive, including the provision for adjustment. If there is much slack, the whipping of the chain will greatly increase the wear, and means of adjustment may double the life of the chain. A slight amount of play is necessary for satisfactory operation. The minimum amount of sag should be about 1/8 inch. Although the silent chain shifts outward from the teeth and adjusts itself for an increase of pitch, it cannot take up the increased pitch in that portion of the chain between the wheels; therefore, the wheel must lag to the extent of the increased pitch in the straight portion of the chain.

Standard Silent Chain Designation

The standard chain number or designation consists of: (1) a two-letter symbol SC; (2) one or two numerical digits indicating the pitch in eighths of an inch; and (3) two or three numerical digits indicating the chain width in quarter-inches. Thus, SC302 designates a silent chain of 3/8-inch pitch and 1/2-inch width, while SCIOI2 designates a silent chain of 1 ¼-inch pitch and 3-inch width.

Advantages of a Chain Drive

Positive contact between the chain and the drive sprockets eliminates the possibility of slips.

Has a wide range of driving power.

- Can be used where there is a large distance between the driving and driven shafts. Useful for low speed and high torque transmissions. Can absorb shocks.
- Chain drives are compact.
- Chain drives withstand heat, dirt and weather exposure when properly lubricated.

Jockey Sprocket (Fig. 4.47)

A spring-loaded jockey sprocket can be used to tension a chain which transmits the drive between the sprockets with fixed centres.

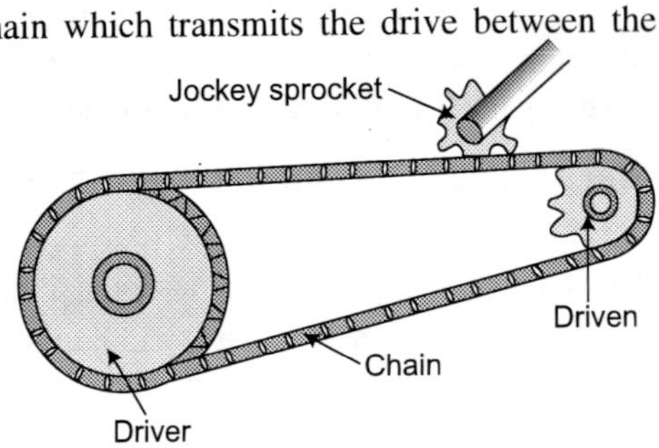

Fig. 4.47

4.4 GEAR DRIVES

In a belt drive, slipping may occur if the power to be transmitted is too great for the frictional grip between the surfaces, and friction wheels do not give a positive drive.

Gear drives are drives which transmit motion and thus power from one shaft to another with the help of gears, when the distance between them is relatively small. The axes of the shaft may be parallel, non-parallel or intersecting. These are positive drives. They transmit power at constant speed ratio.

It is possible to drive shafts that are parallel, intersecting, or neither parallel not intersecting by the use of toothed gears. Toothed gearing is often used in preference to belting, friction drives, or chain drives, where moderate or large amounts of power must be transmitted at a constant velocity ratio.

A gear is a toothed member. It is designed to transmit motion and thus power to another toothed member by successively engaging its teeth with those of the other gear. The teeth may be cut on cylindrical or conical or rectangular or elliptical blanks/blocks. The tooth shape and sizes of gears are standardized.

Gears are used to transmit torque/motion from the driving shaft to the driven/follower shaft:

- to change the velocity ratio
- to change the direction of rotation.
- to get a positive drive.

Gears are made from cast iron, steel, nonferrous, plastic or fibber material.

Types (Classification) of Gears

Gear-tooth forms are conjugate cams, the one acting against its mate to produce the action required, which is generally uniform rotary motion. Many different forms are needed and used in a variety of applications. A few of these forms have been standardized, economy and potential accuracy of production being major considerations in their adoption as standards. These forms may be produced either by form cutting or by generation. Form-cut gears have fixed tooth proportions while generated gears may vary to meet specific needs and still employ the standard generating tool.

Gears are used for many purposes, ranging from transmitting motion at high or low speeds under heavy or light loads to holding some element in a fixed position. Those used to transmit uniform rotary motion have definite pitch surfaces, the forms of which depend upon the types of the tooth elements and the position relation of the axes of the two gears.

Axis Positions as Related to Gear Forms

Parallel axes are associated with pitch surfaces that consist of two tangent pitch cylinders with spur or helical gear-tooth forms. Intersecting axes are associated with pitch surfaces that are cones with straight or spiral bevel gear teeth. Axes that are not parallel and do not intersect each other may be associated with anyone of several types of gears, as follows:

1. **Crossed helical gears:** with pitch surfaces consisting of two cylinders and two planes, both with helical teeth.
2. **Worm gears:** with pitch surfaces consisting of a plane for the worm and a cylinder for the worm gear, with helical threads or teeth on the worm and enveloping teeth on the gear.
3. **Hypoid gears:** with pitch surfaces in the form of hyper boloids of revolution with conjugate teeth which generally are of "formate" form (one form selected arbitrarily and the other made conjugate with it).
4. **Hourglass worms:** of many types, usually with indeterminate pitch surfaces; the hourglass worm being a barrel cam with a multiple-tooth follower.

Many different forms of gears are used and the types most commonly used in industry are:

1. Spur gear.
2. Spiral gear.
3. Helical gear.
4. Bevel gear.
5. Worm and worm wheel.
6. Rack and pinion.

Classification of toothed gearing

Name	**Kind**	**Relation of axes**	**Pitch surface**
Spur gears	Straight	Parallel	Cylinders
Bevel gears		Intersecting	Cones

	Spiral skew	Intersecting	Cones
	Hypoid	Not in one plane	Hyper boloids
Helical gears	Parallel	Not in one plane	Cones
	Crossed	Parallel	Cylinders
Worm and worm wheel	Not in one plane	Cylinders	
		Not in one plane	Cylinders

Spur Gear Drive

Gears on parallel axes

Gears on parallel axes are of either spur or helical tooth forms.

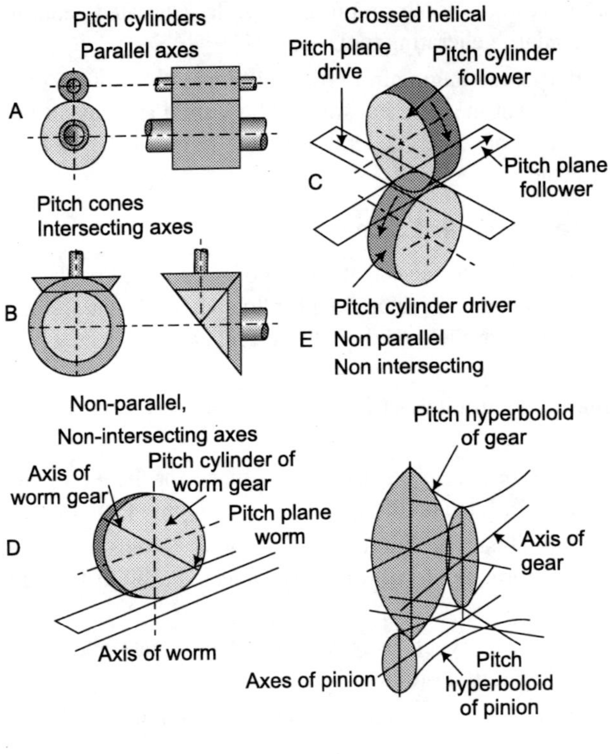

Fig. 4.48

Spur Gears (Fig. 4.49)

The teeth are cut parallel to the axis of rotation. The spur gears are used to transmit power between two parallel shafts.

For any spur-gear tooth-form system there is a basic rack form. The rack is a gear of infinite pitch diameter with a plane pitch surface. Its basic form is that of a rack that will mesh with the given tooth form. Conversely, it is the form of a rack-shaped cutter that will generate the given tooth form. The simplest and most effective way to specify the tooth forms for any given gear-tooth system is in terms of the basic rack form.

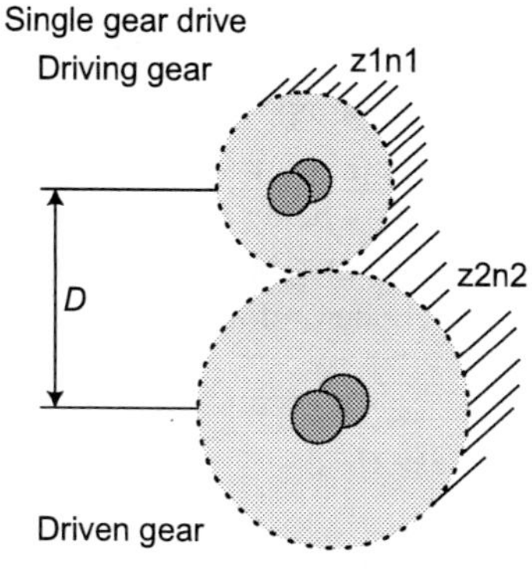

Fig. 4.49

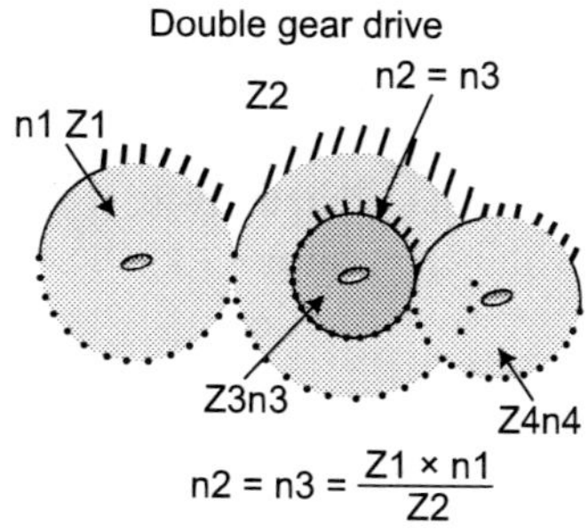

Fig. 4.49 A

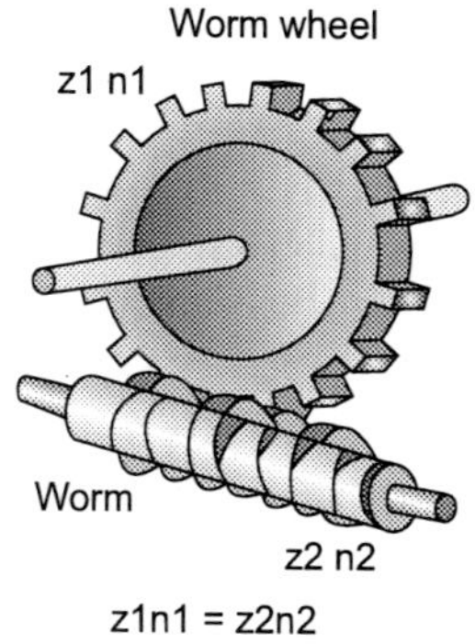

Fig. 4.49 B

Standard Forms for Spur Gears

There are four standard forms for spur gears:

- 14.5° composite form
- 14.5° full-depth involutes system
- 20° or 25° full-depth involutes system
- 20° stub-tooth involutes system

14.5° Composite Tooth Form

This form was developed to be cut with formed milling cutters, to operate on standard centres with mating gears having equal addendum, and so that a 12-tooth pinion could operate with anything from an equal 12-tooth pinion to a rack. The gears are conjugate to the basic rack.

14.5° Full-Depth Involutes System

This form is used for gears produced by gear hobs and pinion-type shaping cutters. The tooth proportions of this basic rack form are the same as those for the 14.5° composite system. The conventional design of gears of this system is also the same as for the 14.5° composite system.

20° and 25° Full Depth Involutes System

The basic rack is used for specifying tooth proportions in designing spur gearing with 20 and 25° pressure angle, full-depth tooth forms so that cutting tools can be designed to closely approximate this form. The number of teeth is equal to or exceeds 18 in the pinion or 36 in the pair. For 25° pressure angle, equal-addendum teeth are used where the number of teeth is equal to or exceeds 12 in the pinion or 24 in the pair. Long- and short-addendum teeth are used with pinions having numbers of teeth as specified in AGMA 201.02 and ANSI B6. 1-1968. However, the use of long-addendum pinions and short-addendum gears as specified in these standards should be limited to speed-reducing drives only.

20° Stub-Tooth Involutes System

This form was developed to obtain a stronger tooth for heavy loading, mostly at low speeds; also to avoid undercutting on the smaller gears. There are several different designs of this stub-tooth form with slightly different tooth proportions. The basic rack form will mesh with all of them; the differences make only a slight change in the amount of the clearance at the roots of the teeth. The tooth proportions of these gears are standardized:

Helical Gears (Fig. 4.51)

These are widely used in automobiles. The application of helical gears in an oil pump.

The teeth of helical gears are cut along a helix on the pitch cylinder; otherwise they are similar to spur gears and the action on any given plane of rotation as for spur gears. All other factors being equal, helical gears run more smoothly and quickly. The helical teeth introduce an end thrust, which must be met in the design of mounting; or it may be neutralized within the gear itself by the use of herringbone or double-helical teeth.

Fig. 4.50 Spur gears

Fig. 4.51 Helical gears

The same basic rack forms are used for helical gears as for spur gears. When these forms are milled or hobbed, this form is the normal basic rack form. In these cases the diameters of the gears are larger than the equivalent spur gears. When helical gears are shaped on the Fellows gear shaper, the forms and pitches are standard in the plane of rotation of the gears and the sizes are identical to those of equivalent spur gears. Hobs can be swivelled to cut any helix angle; Fellows and Sykes cutters are limited to the helix angles of the cutters and guides.

The end thrust is exerted by the driving and driven gears in the case of helical gears and the thrust may be eliminated by using double helical gears. These gears are called herringbone gears **(Fig. 4.55).**

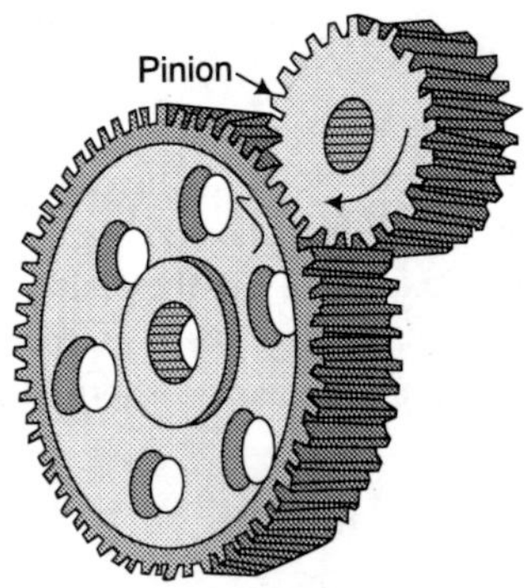

Fig. 4.52 Spur gear

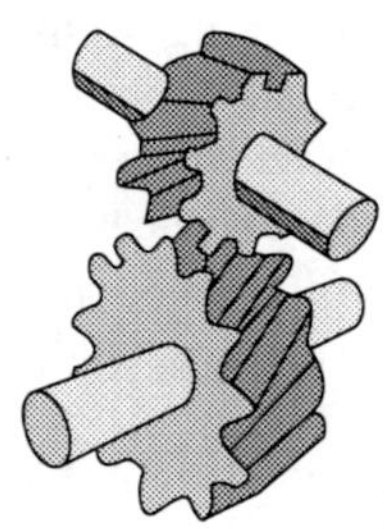

Fig. 4.53 Helical gear

Gears On Intersecting Axes

Gears on intersecting axes may have straight bevel, spiral bevel, or zero bevel teeth. The pitch surfaces of these gears are cones, and the point where the apexes of the two pitch cones meet (cone apex) is the intersection point of the two axes.

Bevel Gears (Fig. 4.54)

The crown gear for bevel gears corresponds to the basic rack for spur gears. Its pitch surface is a plane. The elements of its tooth form converge, and all meet at the cone apex. The form of the teeth on the crown gear determines the form of the teeth of all bevel gears of the system. Almost universally, bevel gears should make contact externally, and internal bevel-gear drives should be avoided.

Bevel-Gear Tooth Proportions

Tooth proportions are based on the tooth depth along the back cone. The teeth are tapered to the cone apex. The length of the back cone element from its apex to the pitch circle is used as the pitch and radius of the equivalent spur gear. The tooth proportions are translated into their angular values in relation to the pitch cone of the bevel gear.

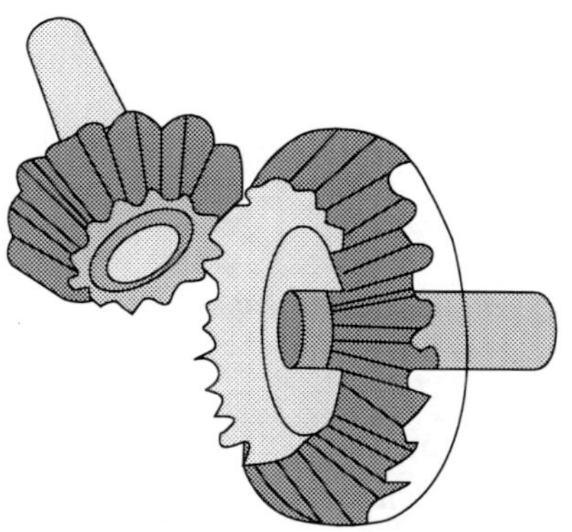

Fig. 4.54 Bevel gears

Fig. 4.55 Double helical gears or Hearring bone gears

Straight bevel gears cut on modern bevel-gear generators have a localized tooth contact pattern and are known as "coniflex" gears. The superiority of the Coniflex gears over straight bevels cut with full-length tooth contact lies in the control of the tooth contact length. This localization of contact permits some adjustment of the gears in assembly and compensates for small displacement due to deflection under operating loads without concentrating the load on the ends of the teeth.

The bevel gears are used to transmit motion between shafts at various angles to each other. The teeth profile may be straight or spiral.

In a hand driller, the bevel gears transmit motion when the shafts are at right angles to each other.

Miter Gears

If two bevel gears are symmetrical to each other and transmit motion at right angles, such gears may be called mitre gears.

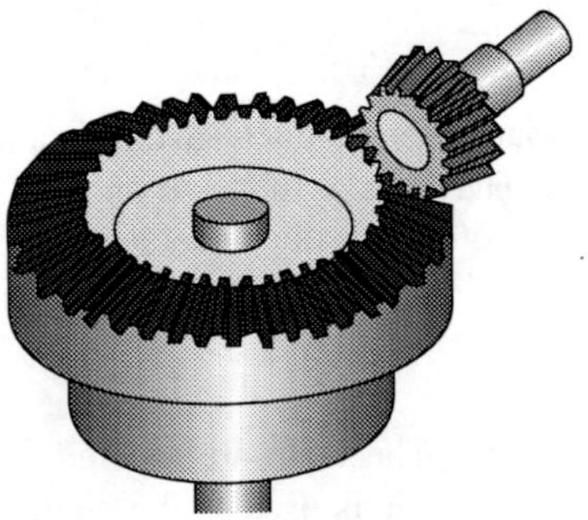

Fig. 4.56 Bevel gear shafts at various angles to each other

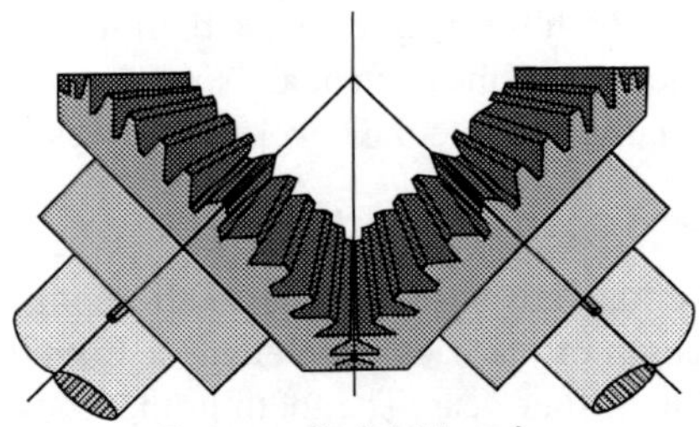

Fig. 4.57 Mitre gears

Worm Shaft and Worm Gear (Fig. 4.58)

The worm shaft has spiral teeth cut on the shaft and the worm wheel is a special form of gear teeth cut to mesh with the worm shaft.

These are widely used for speed reduction purpose.

The application of worm and worm gear is in the index-head gear mechanism **Fig. (4.59).**

Worm Forms

In many cases the shape of the worm is limited by the hobs available to cut the worm gear.

The common practice is to use axial pitches of even fractions of an inch because of change-gear problems associated with some machines used to finish worms.

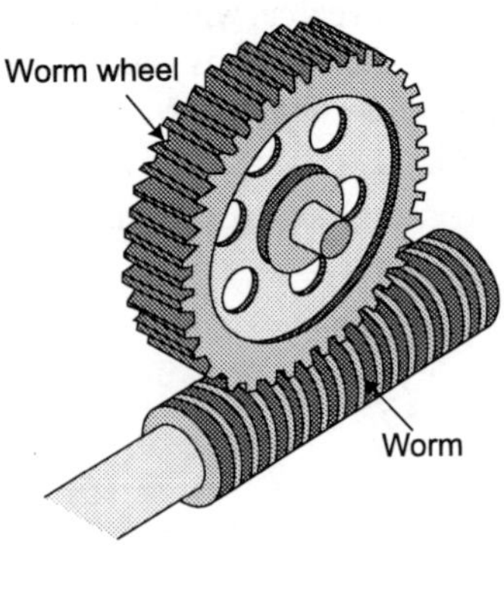

Fig. 4.58 Worm shaft & worm gear

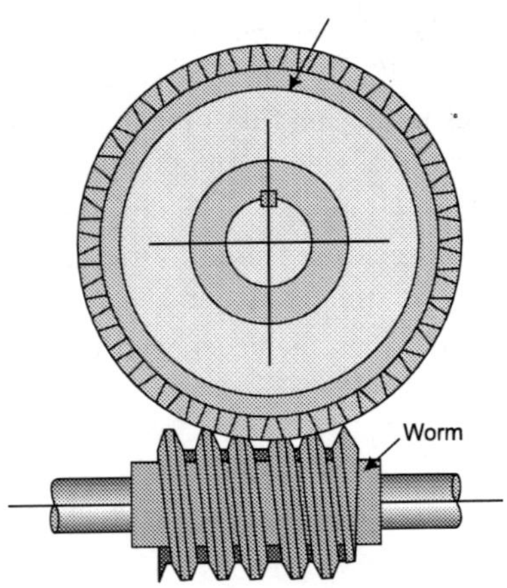

Fig. 4.59 Worm shaft and worm gear

Single-thread worms have been used frequently because of the greater cost of multiple threaded worms and hobs. As a result, the pitches employed have averaged from two to three times as coarse as required by the application, although the general trend in all other types of gearing has been toward the finer pitches, which are smooth-running and more efficient than the coarser pitches.

Fig. 4.60 Application of Worm gears

Thread Angles and Lead Angles

The conjugate action between the teeth is that of a rack of changing form across the face mated with a spur gear. The change in the form of this rack tooth depends upon the thread angle, its lead, the diameter of the worm, and the distance of any specific section from the axial section. The average pressure angle is reduced on one side of the axial section, while it is increased on the opposite side. This change in pressure angle becomes more pronounced as the leads are increased, particularly on the side with the less pressure angles. This makes it necessary to use larger thread angles for the larger lead angles than for the smaller ones and to use narrower faces on the worm gears for the higher lead angles.

Thread Forms on Worms

The form of the thread on a worm depends upon the form and the type of the cutting tool, its setting in relation to the axis of the worm, and the lead and diameter of the worm. Thus, the axial form of the thread produced by a straight-sided threading tool with its cutting edges in the axial plane of the worm is a straight line. That produced by the same tool tipped to the lead angle of the worm is concave. The form produced by a straight-sided, cone-shaped milling cutter with its axis tipped to the lead angle of the worm is generally convex. When its diameter is small and the lead angle is large, the form may be concave near the root and convex at the tip of the thread. Any of these forms may be and have been used as the thread form for worms.

Design Considerations

The starting point for the design of a worm gear drive is the worm itself and the equivalent hob, which should always match the worm. In general, the best form depends upon the type of equipment available to produce the worm, and the tool engineer must know the method that is to be used before he is in a position to design a hob to match the worm. The exact type of helicoids used for the worm is of small importance except in the cases of small-diameter, multiple-thread worms where the thread form and worm gear face come close to or overlap the limits of conjugate action. For the very small diameter worms, the forms produced by the straight-sided chasing tool permit conjugate action closer to the worm axis than does any other form. The type of helicoids is determined by the method used to take the finishing cut (or grind) on the worm threads. If a screw helicoid is required, the grinding wheel must be suitably formed to produce it.

The worm gear face should be inside the field of conjugate action. The total number of teeth in the worm and worm gear should never be less than 40 unless a very narrow face is used on the worm; more teeth will be better. It is apparent from only a casual study of helicoidal section that undercut starts on the leaving side of the worm gear face very much sooner than on the axial section, owing to the changing profile of the basic rack form of the worm thread on the off centre section.

This system transmits motion at right angles to the axis of motion at different planes.

Rack and Pinion

The rack and pinion can change rotary movement into linear movement and vice versa **Fig. (4.61).**

This mechanism is used in drilling machines for pushing the drill bit to cut material **Fig. (4.62).**

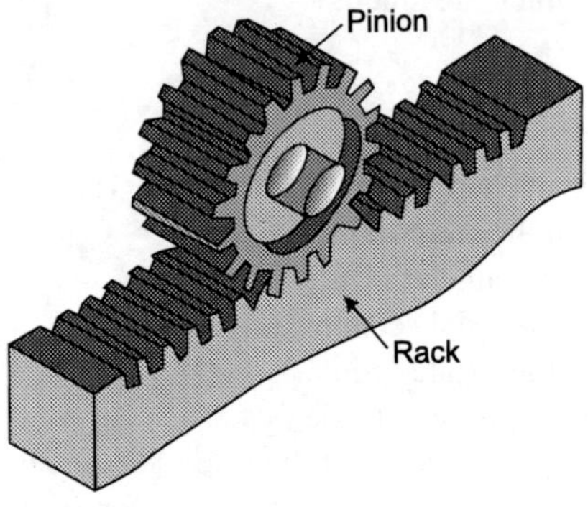

Fig. 4.61 Rack and pinion

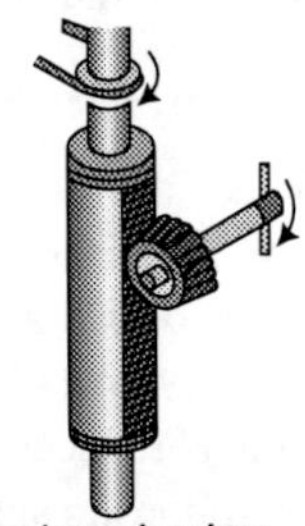

Fig. 4.62 Rack and pinion

Hypoid Gears (Fig. 4.63)

The hypoid gears are used in automotive differential gearboxes. A pair of hypoid gears is similar to the spiral bevel gear but with the shafts offset. The tooth action between each gear is a combination of rolling and sliding action along a straight line. The pitch surfaces are hyperboloids of revolution; as such the gears are called hypoid gears.

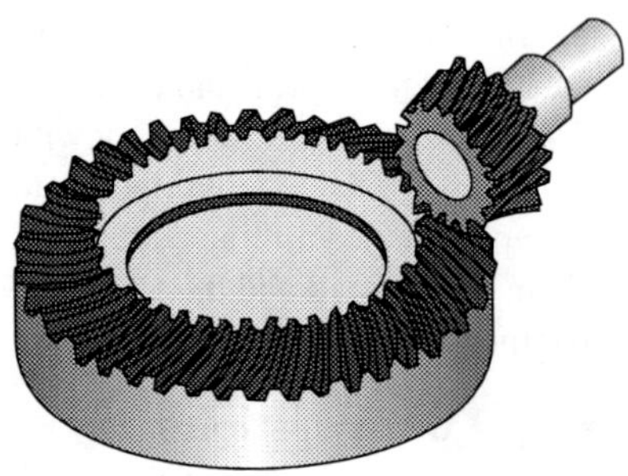

Fig. 4.63 Hypoid gear

Velocity Ratio of Gear Train

The gear train transmits motion without slip.

Formula for velocity ratio of gear train

N 1 T 1 = N 2 T 2 where

N 1 = RPM of driver gear

T 1 = number of teeth in the driver gear N 2 = rpm of the follower/driven gear

T 2 = number of teeth in the driven gear.

4.5 SHAFTS, KEY AND KEY–WAY

Turning Moments

Suppose, a force P (kg) acts at right angles to a handle or lever, R meter long, capable of rotating about an axis C. Then the product P × *R* is called the turning moment or torque of P. Here, then, the turning moment is P × R meter kilogram.

When the lever rotates once the force passes through a distance 2π R meter, and thus the work done by P in one rotation is 2π R × P meter kilogram. This may be written $\pi \times 2$ RP. But as RP is the turning moment or torque, we may call it T. Then work done is expressed as 2π T or $\pi \times 2$ T meter kilograms. In N revolutions work done will be $\pi \times 2$ T × N meter kilograms.

Alternatively we may say that work done = turning moment × angle turned through in radians.

Horsepower

Horsepower is a measure of the rate of doing work, i.e., it gives the amount of work done in a given time. James Watt introduced the standard of horsepower, viz., 33, 000 foot-pounds of work per minute.

To find the horsepower of any machine or other agent we divide the number of foot-pounds of work done per minute by 33, 000.

$$\text{H. P.} = \frac{\text{Work done per minute (foot-pounds)}}{33{,}000}$$

(In electrical calculations 1 H.P. = 746 watts. Watts = volts × amperes.)

Again consider the crank or wheel in **Fig. (4.64).**

$$\text{H. P.} = \frac{\pi \times 2\,\text{T} \times \text{N}}{12 \times 33000}$$

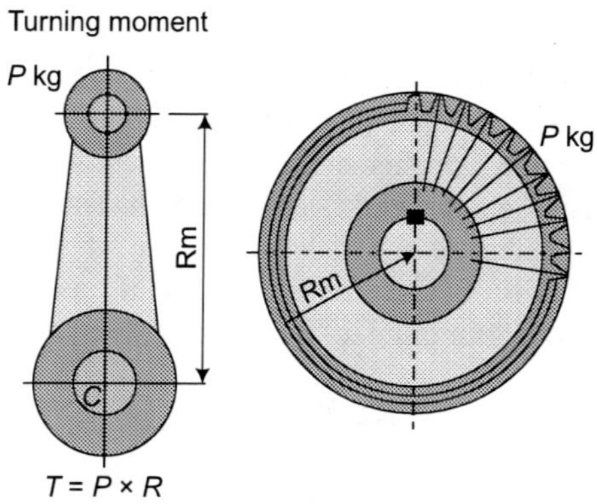

Fig. 4.64

Example

The average turning moment on a crankshaft is 2500 lb-ft when the shaft turns at 125 r.p.m. What is the H.P. transmitted by the shaft?

Solution:

Write down the formula

$$\text{H. P.} = \frac{\pi \times 2\,\text{T} \times \text{N}}{12 \times 33000}$$

In this formula T is expressed in pound-inches.

The Resistance of Solid Round Shafts to Torque varies as the cubes of their diameters. When a shaft is subjected to a pure twist the stress produced is a shear stress. This stress is greatest at the outside of the shaft, its intensity growing progressively and uniformly less towards the centre (where it disappears).

Let D = diameter of solid shaft:

F_S = shearing stress at its circumference, i.e., its skin stress.

T = twisting moment or torque, in lb-in.

Then moment of resistance of the shaft to twisting:

$$= \frac{\pi\, D^2 f_S}{16} = \text{twisting moment (T)}$$

Learn: Equate the twisting moment to the moment of resistance to twisting. Thus,

$$\frac{T = \pi D^2 f_S}{16} \quad \text{Calculate the value of D}$$

Moment of Resistance of Twisting (T)

$$T = \frac{\pi (D^2 - d^2)}{16 (D)} F_s$$

Replacement of solid shaft by hollow one of same material.

Let d_1 = diameter of solid shaft,

$$\frac{\pi d_1^3 f_S}{T} = \frac{\pi (D^4 - d^4)}{16 D} F_s$$

Angle of Twist

A straight line drawn on a shaft not transmitting power becomes a spiral when power is being transmitted. The term torsional stiffness means the resistance to twisting when a torque is applied. Torsional stiffness of a cylindrical shaft is proportional to the fourth power of its diameter and inversely proportional to the length subjected to torque.

Consider the piece of shafting shown in **Fig. (4.49 B)**. Suppose it is a part of the back axle of a car and that the propeller shaft, being in gear with the end A, applies a torque which twists that end relative to end *B*. Thus, a point a takes up position *b*. The angle *acb* is called the angle of torsion, and is extremely small. Now consider the end view of the axle where the arc of twist is shown as *de*. The angle *eod* is called the angle of twist. Practice has shown that if this twist does not exceed 1 degree in 20 diameters the axle will be stiff enough for ordinary purposes.

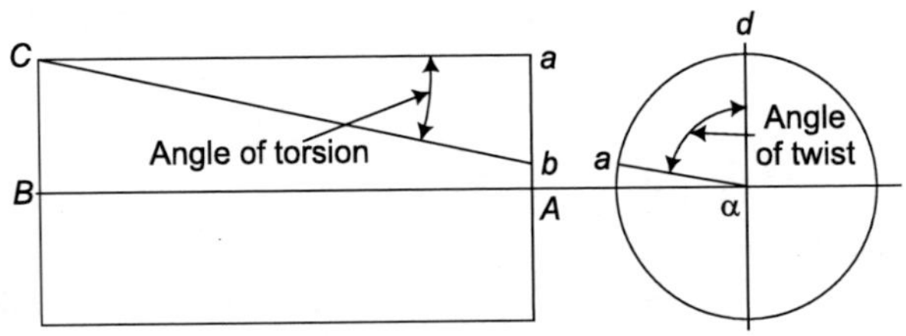

Fig. 4.65 Angle of twist

Ultimate Torsional Strength of Various Materials

(Round bars 1 meter diameter; load applied at 1 meter radius.)

Cast steel	Average 475 kg
Mild steel	Average 540 kg
Wrought iron.	Average 360 kg
Cast iron	Average 345 kg.
Copper	Average 180 kg.

These, although practical test loads, are somewhat higher than are usually adopted in practical calculations.

Terms in Common Use

The term **shaft** is applied to a bar (usually of cylindrical form) capable of turning about its axis and of transmitting work. Those parts of the shaft resting within **bearings** are called journals. The term **axle** is properly applied to a shaft subjected to bending only, but in motorcar work the term is more often than not indiscriminately used. The abutting ends of lengths of shafting are joined by **couplings** or **clutches.** Cylindrical shafts are solid or hollow.

We generally use a gear or pulley for moving another shaft through a shaft already in motion. It is used to locate and secure an object in position such as gear or pulley. With the help of key, gear or pulley cannot move on their shaft but the shaft and gear or pulley move together as one unit. A key is inserted between a shaft and a wheel in the key way. This makes the shaft and gear or pulley (or wheel") one unit. Keys are generally made of good quality steel in order to withstand the stresses.

A key is an insert which is housed in the key way to fit together a hub or a pulley to transmit torque. A key way is provided on the shaft and also on the hub or on a pulley to connect together the conjugate parts by inserting the key in between. The key can be withdrawn at will to disengage the mating components.

The following definitions are given in the standard:

Key and Key Way

A key is defined as **a piece inserted between the joint of two pieces, to prevent relative movement, or as a piece inserted in an axial direction between a shaft and a hub, to prevent relative rotation.** Keys are temporary fastening and are always made of steel, for they are subjected to considerable crushing and shearing stresses. A **key way** is a recess in a shaft or hub to accommodate a key.

A demountable machinery part which, when assembled into key seats, provides a positive means for transmitting torque between the shaft and hub.

Key Seat

An axially located rectangular groove in a shaft or hub.

The standard recognizes that there are two classes of stock for parallel keys used in industry. One is a close, plus toleranced key stock and the other is a broad, negative toleranced bar stock. Based on the use of two types of stock, two classes are:

1. A clearance of metal-to-metal side fit obtained by using bar stock keys and key seat. This is a relatively free fit and applies only to parallel keys.
2. A side fit, with possible interference or clearance, obtained by using key stock and key seat. This is a relatively tight fit.
3. This is an interference side and not standardized, however it is suggested that the top and bottom fit range given under Class 2 for parallel keys be used.

Special considerations dictate the use of a key seat in the hub shallower than the preferred nominal depth shown, it is recommended that the tabulated preferred nominal standard key seat be used in the shaft in all cases.

Key Seat Alignment Tolerance

A tolerance of 0.010 inch, max is provided for offset due to parallel displacement of key seat centre line from centre line of shaft or bore of key seats in shaft and bore. The following tolerances for maximum lead (due to angular displacement of key seat centre line from centre line of shaft or bore and measured at right angles to the shaft or bore centre line of key seats in shaft and bore are specified: 0.002 inch for key seat length up to and including 4 inches; 0.0005 inch per inch of length for key seat lengths above 4 inches to and including 10 inches.

Common Types

Different types are used depending on the requirement of transmission.

Saddle Key

- Hollow saddle key,
- Flat saddle key

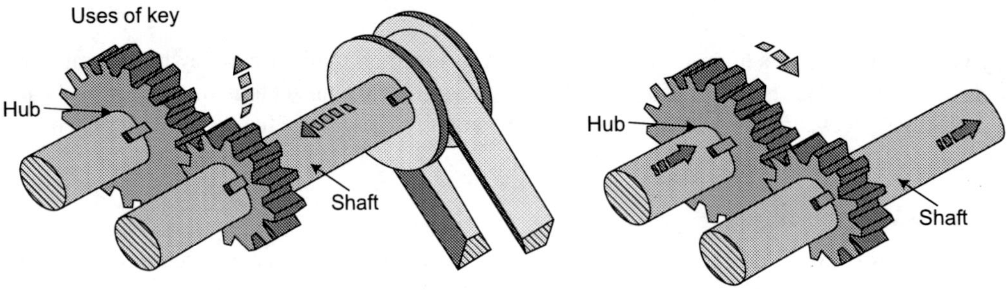

Fig. 4.66

Hollow Saddle Key

One face of this key has a curvature to match with that of the shaft surface. It has a taper of 1 in 100 and is driven in through the key way, of the hub **Fig. (4.67)**.

The hub is held on the shaft due to friction. This is useful only for light duty transmission.

Flat Saddle Key

This key has a rectangular cross-section assembly. A flat surface is machined on the shaft **Fig. (4.68).** The key is placed between the flat surface of the shaft and the key way on the hub. This is considered to be stronger than the hollow saddle key. This is not suitable for heavy duty transmission.

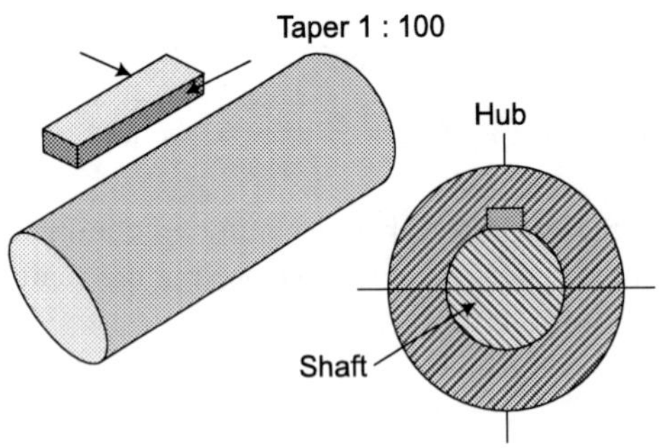

Fig. 4.67 Hollow saddle key

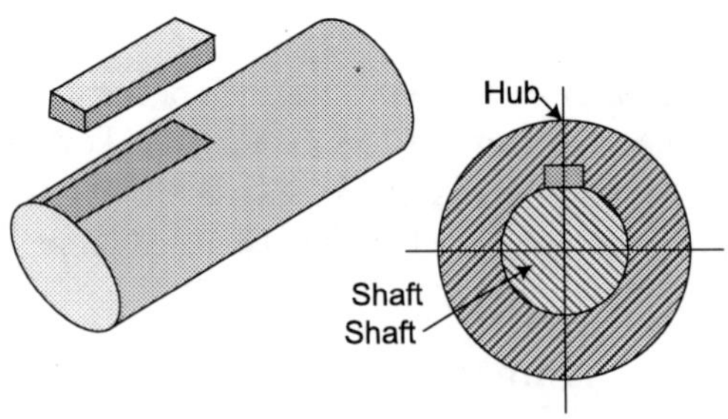

Fig. 4.68 Flat saddle key

Circular Taper Key

In this case both the shaft and the hub have semicircular key ways cut on them **Fig. (4.69).** The taper key is driven in while assembling. This key is also suitable only for light transmission.

Standard taper is 1:100, nominal thickness is measured at the large end. A key of rectangular or square section is uniform in width; tapered in thickness. The ends may be square or rounded.

Sunk Key

Five kinds of sunk keys are available. (i) rectangular taper key, (ii) square taper key, (iii) rectangular parallel key, (iv) square parallel key, (v) woodruff key included in the parallel keys are feather keys.

This key has a rectangular cross section and it fits into the key way cut on both the shaft and the hub. Sunken keys are either parallel or tapered **Figs. (4.70 & 4.71).**

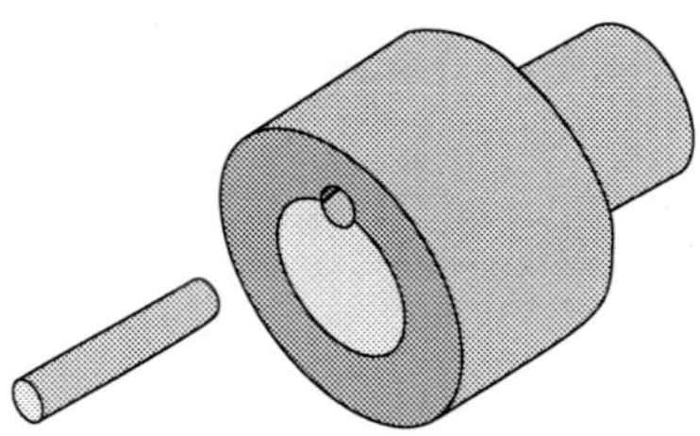

Fig. 4.69 Circular taper key

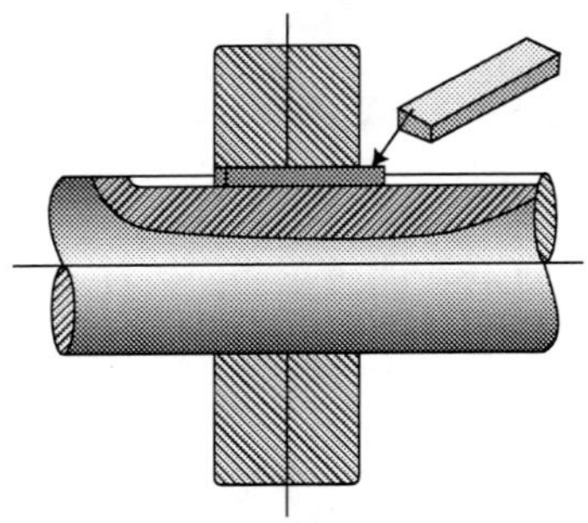

Fig. 4.70 Parrallel sunk key

Gib Head Key

This is another type of sunk key. This has a gib-head to assist in fixing and removing the keys **Fig. (4.72).**

The gib keys are sometimes provided to facilitate withdrawal.

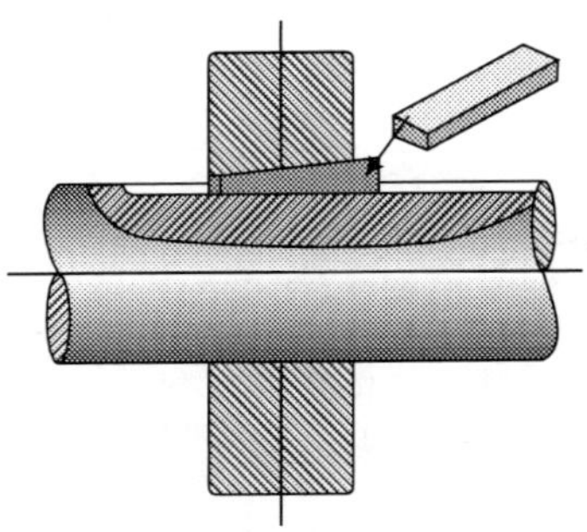

Fig. 4.71 Taper sunk key

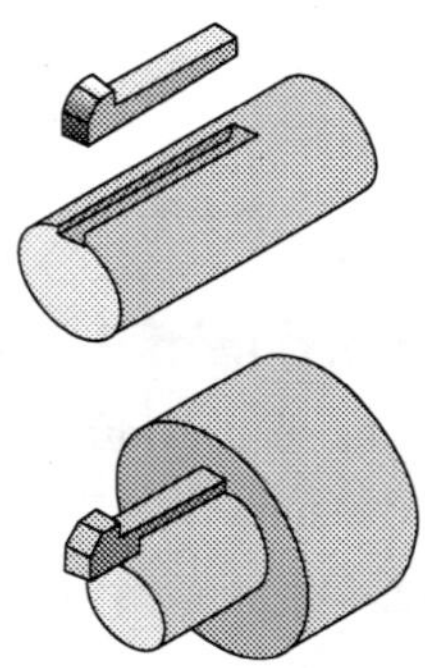

Fig. 4.72 Gib headed key

Parallel Key or Feather Key

This is the most commonly used key, used for transmitting unidirectional torque from a rotating shaft to a hub/wheel or from hub/wheel to shaft. A hub or a pulley or a wheel or a gear is engaged to the shaft by a key which prevents relative motion.

In many cases the key is screwed to the shaft key way **Fig. (4.74).**

Where axial movement of the hub is required, a clearance fit is provided between the hub and the shaft and the hub and the key. Three types of fits are shown for feather key in **Fig. (4.75).**

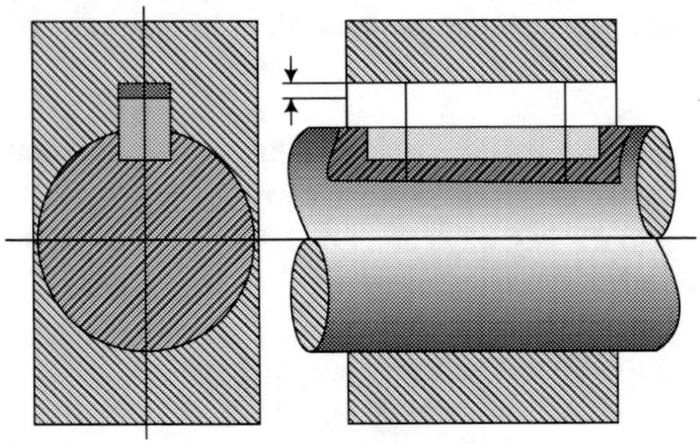

Fig. 4.73 Function of feather key

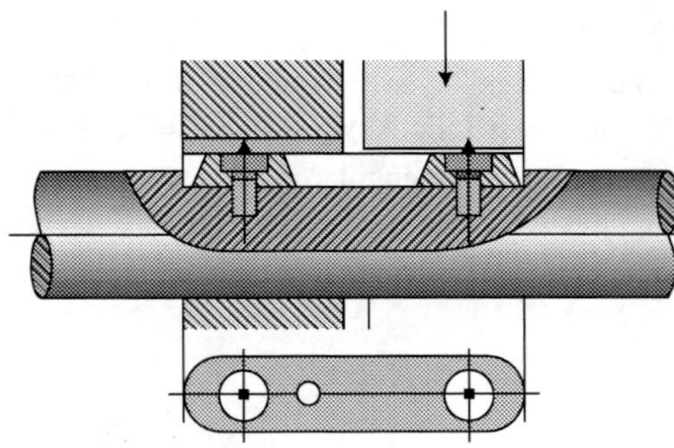

Fig. 4.74 Parallel or feather key

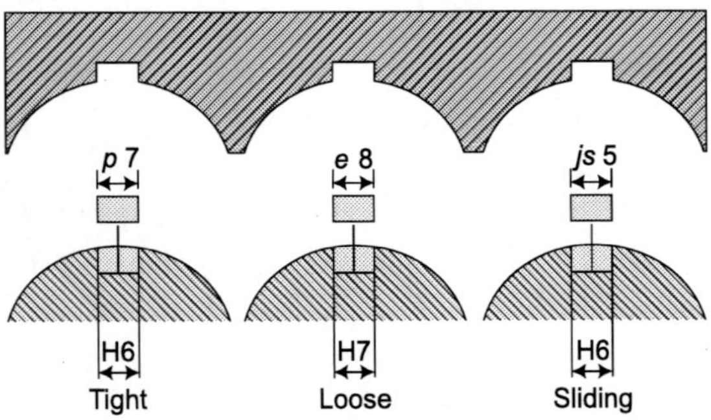

Fig. 4.75 Clearance fit for feather key

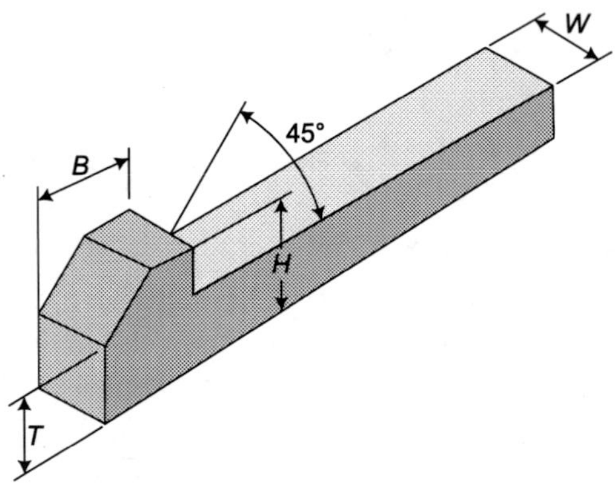

Fig. 4.76 Taper gib headed key

Approximate Proportion of Parallel or Taper Keys

If D is the diameter of the shaft, width of the key W = 1 /4D + 2 mm. Nominal thickness T = 2/3 w.

Diameter of shaft = 40 mm

Width = ¼ × 40 + 2 = 12 mm.

Thickness = 2/3 × 12 = 8 mm.

Thickness at the large end is the nominal thickness of the taper key.

Taper is 1 in 100 on the top face only.

Taper and Jib-Headed Key (Fig. 4.77)

The key has a jib-head with a taper (1 in 100) on the top face. It is driven on to the key way by hammering on the jib to have a tight fit. The taper rectangular key without a jib-head is also in use. A jib-headed key can be withdrawn easily and used for transmitting more torque. It is not good for high speed applications.

Approximate Proportion of Jib-Headed Key (Fig. 4.77)

H = 1.75 T

B = 1.5 T

W = ¼ D + 2

Nominal thickness T = 2/3 W

Angle of chamfer = 45 0

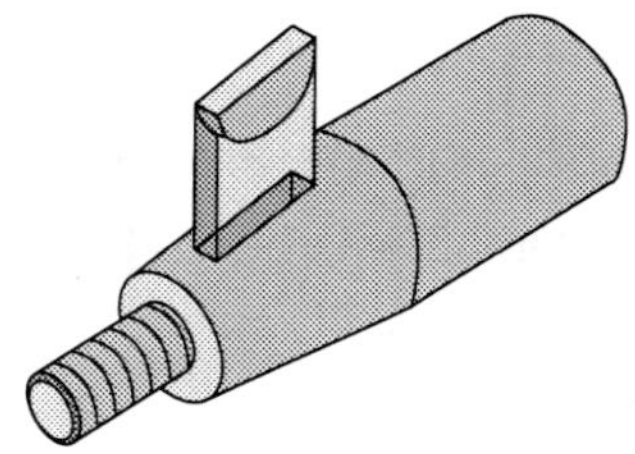

Fig. 4.77 Woodruff key

Example

Diameter of shaft = 46 mm,

Width (W) = ¼ × 46 + 2 11.5 + 2

= 13.5 rounded off to 14 mm.

Thickness (T) = 2/3 × 13.5 = 9 mm

H = 1.75 × 9 = 15.75 say 16 mm,

B = 1.5 × 9 = 13.5 mm.

Woodruff Key (Fig. 4.78)

It is a semicircular key used for transmitting light torque. Its key way is milled to the profile of the key on the shaft which tends to weaken the shaft. This type of key positions itself in the key way to accommodate the hub to have an easy assembly.

A demountable machinery part which, when assembled into key provides a positive means for transmitting torque between the shaft and hub.

This is an easily adjustable sunk key. It is segmental in front view, being part of cylindrical disk capable of tilting in a recess milled in the shaft by a cutter having the same curvature as the disk from which the key is made. This is largely used on automobiles and machine tool work.

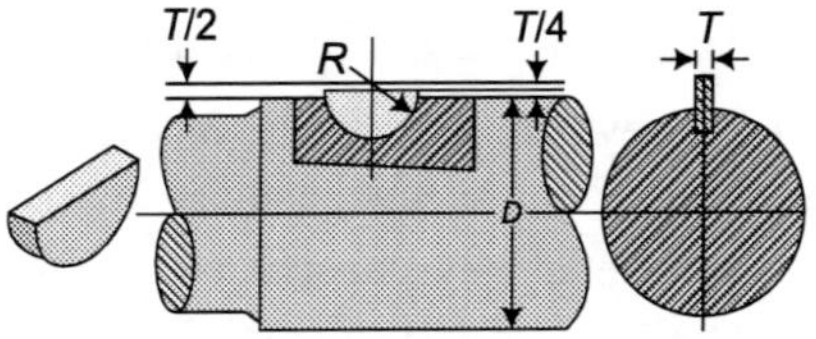

Fig. 4.78 Specification of Woodruff key

The main advantages are (1) it accommodates itself readily to any taper in the hub, nave or boss of the mating piece, (2) it is useful on coned ends of shafts, where its extra depth in the shaft prevents any tendency of the key to turn over in its key way (the usual form of sunk key is very apt to turn in its key way unless well fitted at its sides).

Disadvantages: (1) the relatively great depth of the key way weakens the shaft; (2) it cannot be used as a feather key.

Woodruff Key Number

The number is an identification number by which the size of key may be determined.

Woodruff Key Seat-Shaft

This is the circular pocket in which the key is retained.

Woodruff Key Seat-Hub

This is an axially located rectangular groove in a hub and is referred to as a key way.

Approximate Proportion of Woodruff key (Fig. 4.78)

Radius of the key (R) = D /3

Thickness (T) = D/6

Example

For shaft dia 30

Round Key

This is a key of circular section, fitted in a hole drilled partly in the shaft, and partly in the hub **Fig. (4.81).**

It is of cylindrical cross section and is used in assemblies to secure the mating components where the torque is light. The key is fitted parallel to the shaft into the drilled hole made partly onto the shaft and partly onto the mating part.

Approximate Proportion

If diameter of shaft = D

Diameter of key (d) = 1/6 D

Example

Diameter of shaft = 30 mm

Diameter of key = 1/6 × 30 = 5 mm.

Tangential Key (Fig. 4.82)

These keys are used when very high torque of impact type is to be transmitted in both the directions of rotation. Common applications are found in flywheels, rolling mills, etc.

Tangential key consists of two taper rectangular wedges, positioned one over the other in opposite directions. Two sets of keys are fixed at 120° angle and should be such that the broad side is directed along a tangent to the shaft circle while the narrow side sits along the radius of the shaft.

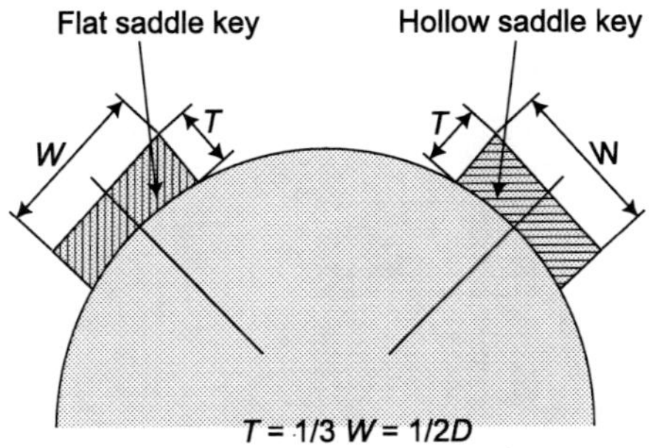

Fig. 4.79 Flat & hollow saddle key

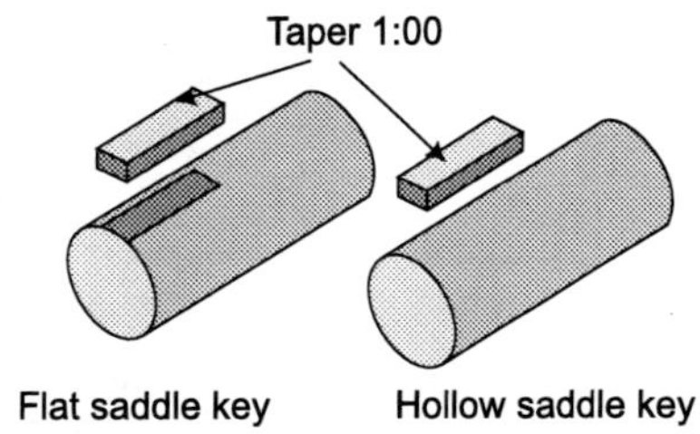

Fig. 4.80 Flat & hollow saddle key

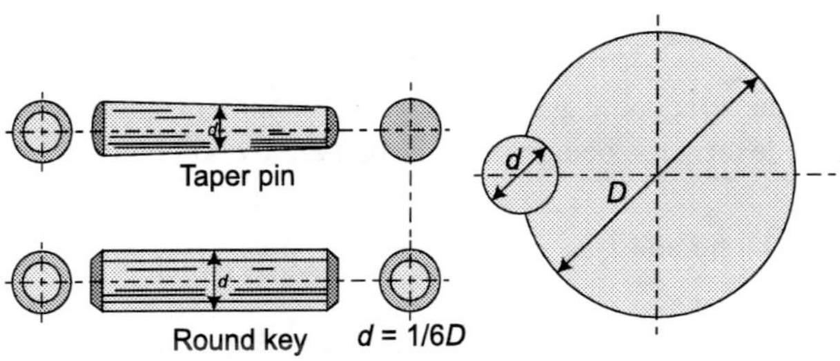

Fig. 4.81 Proportions of taper round key

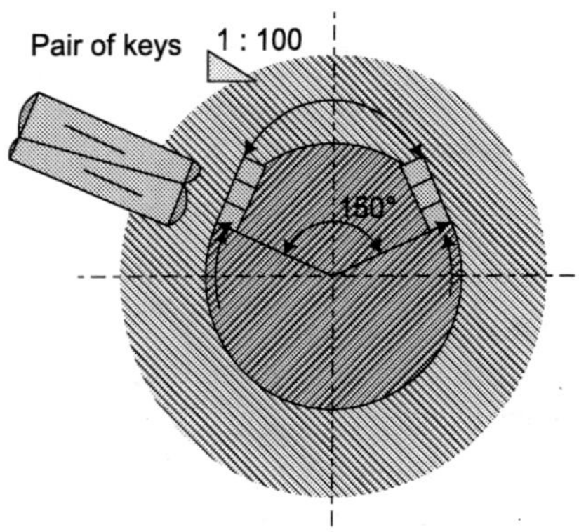

Fig. 4.82 Tangential keys

4.6 SPLINES AND SERRATIONS

A splined shaft is one having a series of parallel keys formed integrally with the shaft and mating with corresponding grooves cut in a hub or fitting; this is in contrast to a shaft having a series of keys or feathers fitted into slots cut into the shaft. This latter construction weakens the shaft to a considerable degree because of the slots cut into it and, as a consequence, reduces its torque-transmitting capacity.

Splined shafts are most generally used in three types of applications:

1. For coupling shafts when relatively heavy torques are to be transmitted without slippage.
2. For transmitting power to slideably-mounted or permanently-fixed gears, pulleys, and other rotating members.
3. For attaching parts that may require removal for indexing or change in angular position.

Splines having straight-sided teeth have been used in many applications; however, the use of splines with teeth of involute profile has steadily increased since

1. Involute spline couplings have greater torque transmitting capacity than any other type;
2. They can be produced by the same techniques and equipment as are used to cut gears; and
3. They have a self-centred action under load even when there is backlash between mating members.

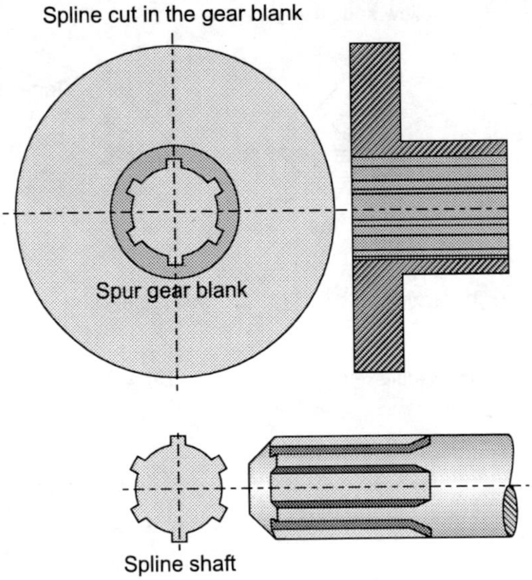

Fig. 4.83

Spline shafts have one or more splines of the same size, being designated by the number of splines, e.g., 6 spline shaft. The sliding piece (hub, nave, wheel, lever, etc., being similarly prepared we obtain free axial movement as well as a positive drive.

In motor cars gearbox work both these requirements must be met, and whilst square shafts have been employed it is now common to fit spline shafts.

In motor cars gearbox work it is common to fit cylindrical shafts having four or six splines or feather ways.

On splined shafts the splines are usually cut by hobs, but splined holes are usually cut by broaches.

A serrated shaft has a series of adjacent triangular projections or teeth formed around its periphery. The theoretical major diameter is that of a circle passing through the apices of the fundamental triangles formed by the flanks of the teeth. The theoretical minor diameter is that of a circle passing through the points where the bases of the fundamental triangles touch. The effective diameter of the serrated shaft is that of a circle midway between the theoretical major and minor diameters.

Similarly, a serrated hole has on its surface a series of adjacent triangular teeth from which fundamental triangles can be formed. A circle passing through the apices is the theoretical minor diameter and that passing through the points where the bases of the adjacent triangles touch is the theoretical major diameter. The effective diameter is midway between the two.

The apex of the tooth on the shaft is truncated and a radius is provided at the root of the serration (or space) on the shaft. The same applies to the crests of the splines and the roots of the serrations in the hole. The serrations in the hole are so proportioned that the nominal diameter is that of a circle which passes through the points of tangency of the tooth flanks with the bending radius, the tooth flanks of the nominal hole being formed by the sides of the fundamental triangles. The nominal hole is referred to as the low limit or minimum hole. To guard against interference the maximum major diameter of the shaft is made smaller than the nominal diameter of the serrated hole.

The motorcycle gear lever is an example of serrated shaft.

Strength of Keys

A key fastening a pulley to a shaft transmits a turning moment to or from the shaft and the greatest turning moment it would have to transmit would be equal to the moment of resistance of the shaft to twisting. This would only be the case in shaft couplings and main drives, for on a line-shaft only a small amount of the whole work transmitted by a shaft is taken off at each pulley.

If

w = width of key

l = length of key

t = thickness of key

Resistance of key to shearing = l × w × fs

Resistance to compression = l × ½ t × fc (approximately.

Examining these two statements we see that the strength of a key varies directly with its length.

1. Suppose we imagine the resistance of key to shearing to be equal to the moment of resistance of the shaft to twisting we have:
   i. l × w × fs × d/2 = π/16 d 2 fs (d/2 is the distance of the resisting section from the centre of the shaft axis.)
2. Suppose we imagine the resistance of key to compression equal to the moment of resistance of the shaft twisting, we have,
   i. l × t/2 × fc × d/2 = π/16 d 2 fc,
3. Suppose crushing resistance of key equals its shearing resistance, we have,

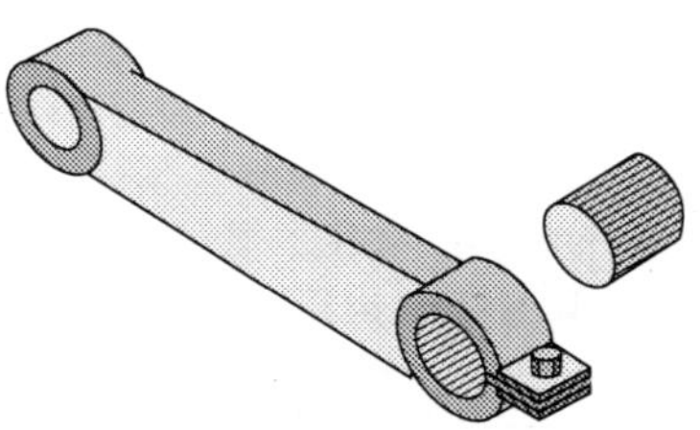

Fig. 4.84 Serrated shaft

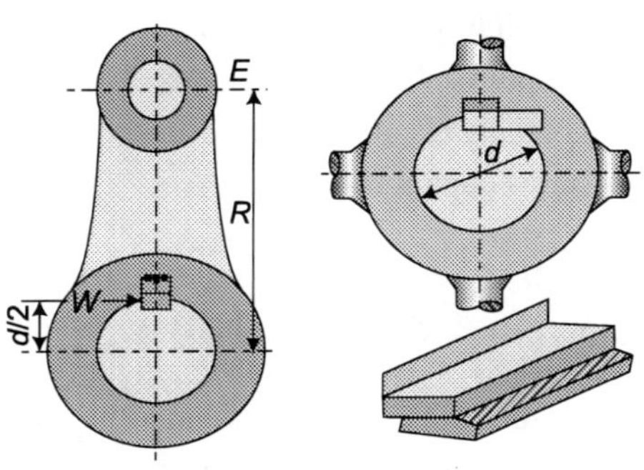

Fig. 4.85 Strength of key

The last statement only holds for feather keys not fitted tightly at the sides. Where a key is tightly wedged the crushing action on the sides is considerably less than in the case of sliding keys. If we take safe f sc = twice safe f s the statement w f s = ½ t f c gives w = t, in other words a square key. This is often the shape of feathers or sliding keys. In wedge-shaped keys tightly driven home t is considerably less.

4.7 SHAFT COUPLINGS

Mild steel shafting is available in standard lengths of 6 metres. When the length exceeds the standard, which is available in the market, two or more shafts will be coupled using the couplings. Over this length there is considerable expense and inconvenience in transport. Thus a long line-shaft is composed of a number of lengths coupled end to end by couplings. The two main classes of couplings are **rigid** and **flexible.**

Coupling is a device to connect or "couple" together two shafts to transmit power from one rotating shaft to the other. As a mechanical joint it is used for various purposes with wide applications. Prime movers such as engines, generators, motors, etc., are be coupled to the other machines such as pumps, blowers etc., which are to be driven by couplings. They also facilitate disengagement of the lines for repairs, etc., when necessary. This way it is flexible also. Some of the couplings are also used to reduce the transmission of shock loads from one shaft to the other. Some couplings facilitate the coupling of the two misaligned shafts. They are also used to alter the vibration characteristics of the rotating shafts and sometimes to protect the machines against overloads.

Couplings are classified as:

1. Rigid coupling,
2. Flexible couplings,
3. Disengaging couplings and
4. Nonaligned couplings.

Rigid Couplings

Couplings are used where a highest degree of axial alignment of the two shafts is required. They connect the two shafts rigidly and do not permit any relative rotation and axial motion between them. They are also further classified as:

1. Box, or muff, or sleeve coupling,
2. Half lap coupling,
3. Split muff coupling,
4. Flanged coupling of unprotected type,
5. Flanged coupling of protected type and
6. Compression coupling.

Flange Couplings - Unprotected (Fig. 4.86)

The flanges are iron castings keyed to shafts (sometimes shrunk on as well as keyed). Keys are driven from the inside faces of flanges. Bolts are made a tight fit in the holes to reduce the tendency to shear. Faces of flanges are turned true with the shaft axes after the flanges are keyed on.

A flanged coupling is the simplest type of rigid coupling most extensively used. Normally, it consists of two cast iron or steel "bosses" provided with projecting flange plates at one of their ends. The flange plates are drilled with a number of equidistant bolt holes on their flat faces with their centres lying on an imaginary circle, called "pitch circle". Each of the flanged bosses is securely keyed to the end at the each shaft using a tapered key driven from inside. Generally, the keys are mounted in the two shafts at 90° to each other. Then the flanges are bolted together by a number of bolts and nuts. Power is transmitted from one shaft to the other through the bolts. These bolts are in running fit in the holes which are drilled and reamed in the flanges in order that the load is taken smoothly without any impact which would take place if the bolts are fitted loose in the holes. Correct alignment of the two shafts is ensured irrespective of the bolts by allowing the end of one shaft to enter a small distance in the boss bore of the other flange. Protruding of the bolt heads and nuts beyond the flat faces of the flanges renders the risk of accidents. Hence, this type of flange coupling is classified as unprotected type.

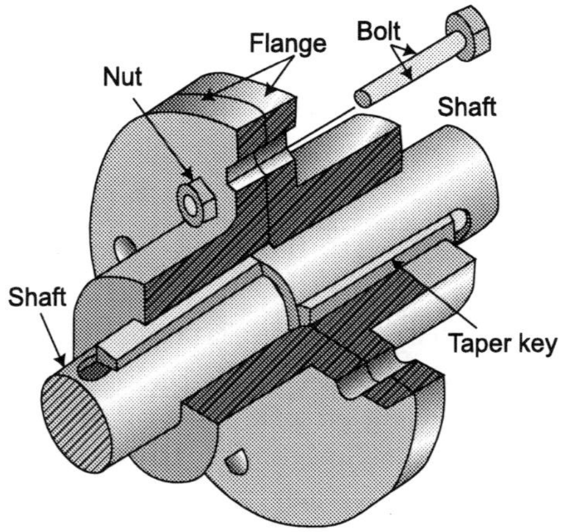

Fig. 4.86 Unprotected flanged coupling

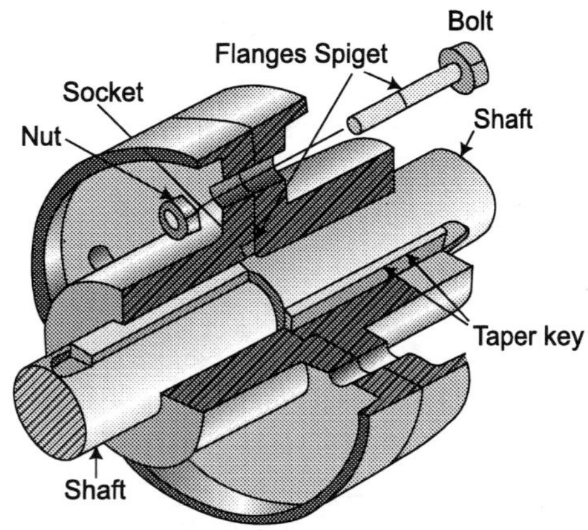

Fig. 4.86 A Protected flange coupling

Flanged Coupling-Protected Type (Fig. 4.86 A)

This type is same as the unprotected flanged couplings; as protective measure the bolt heads and nuts are shielded by a peripheral protrusion, called 'shroud', on each flange. Alignment of the two shafts is independent of the bolts by the provision of a turned projection called 'spigot', on one of the flanges which fits into a corresponding recess, called 'socket', in the other flange. The length of the spigot projection is kept slightly less than the depth of the socket. The socket and spigot are turned to give a push fit for accurate alignment of the two shafts. The bolt holes are drilled and reamed to give a running fit for the bolts in order that the load is taken smoothly without any impact.

Each of the flanged bosses is securely keyed to the end of the each shaft using a tapered key is driven from inside. Generally, the keys are mounted in two shafts at 90° to each other. The two flanges are bolted together by bolts and nuts. Power is transmitted from one shaft to the other through the bolts.

Flanged Couplings: Bolt Dimensions

We have seen that the bolts used in flanged shaft couplings are subjected to both shear and tensile stresses. The former is more important in connection with our present purpose. We understand, then, that the bolts transmit the torque from shaft to shaft and in doing so are subjected to shear stress.

Let

D = diameter of solid shaft in inches;

d = diameter of bolts in inches;

R = radius of bolt circle in inches;

N = number of bolts;

fs = maximum shear stress (i.e., skin stress) in shaft (kg per sq cm.);

fs' = shear stress in bolts (kg per sq cm.);

Then the resistance to shear of one bolt = area × stress = π/4 d 2 (kg)

This is a tangential force. The moment of which about the shaft axis

$= \pi d^2$ f.s × R (cm kg)

Then the moment of resistance of N bolts

$= \pi d^2$ f.s × R × N (cm kg)

Moment of Resistance of shaft to size = moment of Resistance n bolts

$\pi/16\ D^3\ f\,s = \pi/4\ d^2\ f\,s \times R \times N$

$1/4\ D^3\ f\,s = d^2\ f\,s \times R \times N$

Hints

1. When shafts run out of line bending stress is set up in the bolts. Furthermore, in small couplings with small bolts there is no doubt that tensile stress is more important than shear stress.
2. In order to use the formula given above it will be necessary to settle on some of the unknowns. D will generally be given and N will generally follow from tables to be found in the various engineering pocket books. The following are average values and will serve as a rough guide: Shafts up to 88 mm diameter, N = 4 d = 16 mm to 25 mm; shafts 88 to 150 mm N = 6, d = 25 mm to 10 mm the effective resistance of coupling bolts increases with their distance from the centre.

Comparative Shear Stress Intensity in Bolts and Shafts

Nowadays bolts and shafts are usually of the same material. But shear stress intensity allowed in the bolts varies from, say, 25 per cent, in small couplings; to 80 per cent, in large couplings, of that allowed for the shaft. Thus, the ratio of shear stress: 25 to 80: 100. To arrive at the stress in the bolts work outwards from the shaft.

The Equation Applied to Hollow Shafting

Let

D = inside diameter of hollow shaft

D 1 = outside diameter of hollow shaft

Then π/16 (D 14 – D 4) / D1 fs = π/4 d 2 fs × N × R

Horsepower Transmitted by a Shaft or Shaft Coupling

= twisting moment (kg meter) × radians turned per min. divided by equivalent unit.

Muff or Box Coupling (Fig. 4.88)

A muff coupling consists of a cylindrical sleeve called "muff" passed over the two shafts which almost butt each other. Single taper sunk key is used to join the shafts to the muff. Since a single key is to used to join the two shafts to the muff, key ways in them have to be cut in one setting, because the key ways cut in the two shafts in different settings render misalignment of the key ways. This may result in uneven tightness between the key and connected members in the assembly later. Removal of the muff entails a lot of wasteful time having to· remove or push the elements like, key, bearings, shafts themselves and rework the whole process if the shafts are to be coupled again. This difficulty is tided over by improving; design of the muff coupling. Split muff coupling, is one such type of improved version.

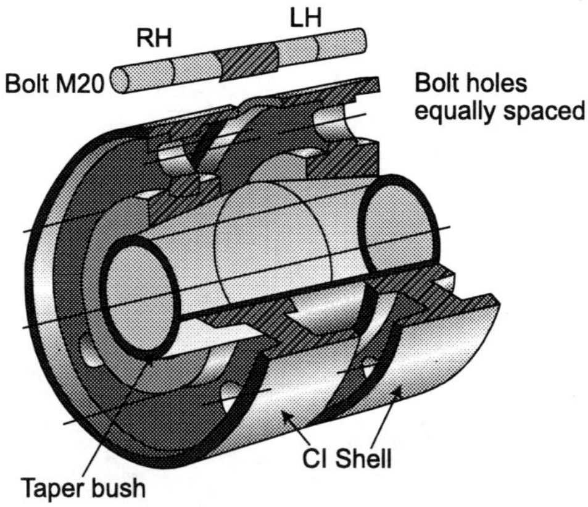

Fig. 4.87 Frictional grip coupling

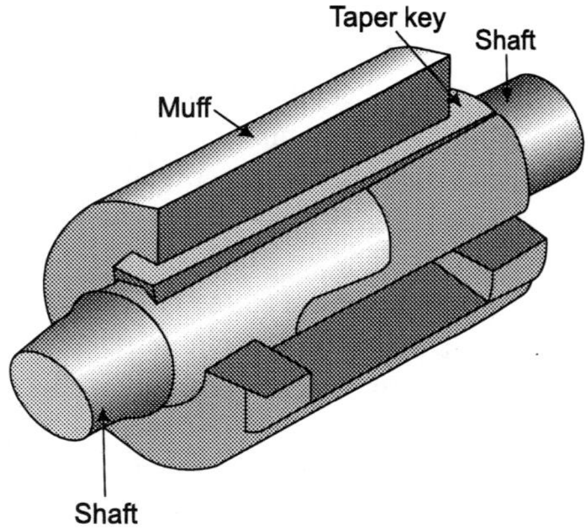

Fig. 4.88 Muff coupling

Half lap Muff Coupling (Fig. 4.89)

This type of coupling is improvised solid muff coupling. But the ends of the two shafts overlap. Nearly fifty percent of the material is removed at the ends of the two shafts to provide their overlapping, the two shafts become weak. So that portion of the shafts within the muff is enlarged. A hollow saddle key is used to connect the muff and the two shafts. It prevents the axial movement of the muff over the two shafts also. The torque is transmitted by the overlaps of the two shafts. The taper 1:12 provided for the overlapping surfaces prevents the separation of the two shafts due to unbalanced axial tensile forces, if any. The inherent disadvantage of the muff or sleeve coupling, while disassembling to remove the muff is true of this type also.

Split Muff Coupling (Fig. 4.90)

This type of coupling as the name itself implies, the muff is split longitudinally into two halves and the joint faces are perfectly machined. The two split muffs are recessed to accommodate the bolt heads and nuts. The two split halves are clamped together to obtain the sufficient grip between the split muffs and the two shafts by bolts and nuts. Initially, a parallel key is placed in the common key way in the shafts and then the split muffs are placed in position and clamped together. Dismantling is greatly simplified with this type of coupling. By simply removing the bolts, muffs can be easily removed from the shafts without disturbing the setup. The torque transmitted is greater than the ordinary muff coupling since the key and the friction grip transmit the torque. Hence, this type of coupling is used for heavy duty work.

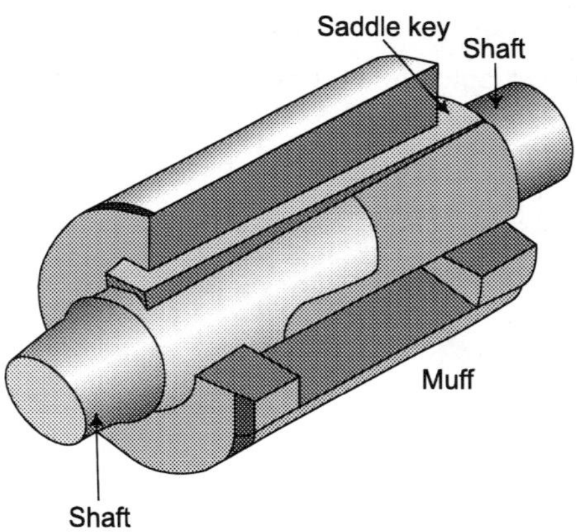

Fig. 4.89 Half lap muff coupling

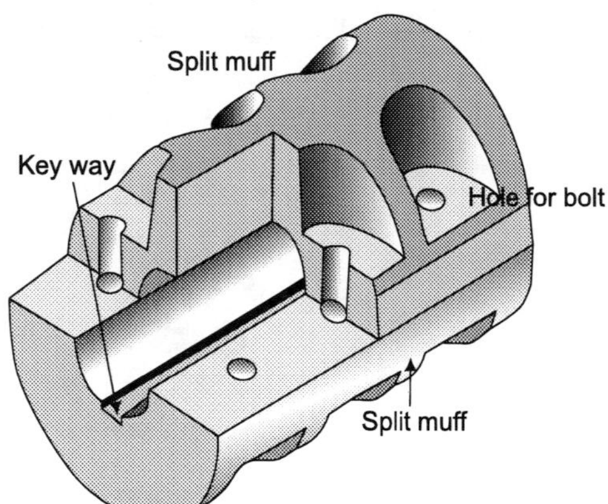

Fig. 4.90 Split muff coupling

Compression Coupling (Fig. 4.91)

This is a rigid type of coupling and is an alternative to the flanged coupling. This coupling requires no keys and can easily be fitted and removed. It consists of a compressible steel sleeve which fits on the ends of the two shafts which are joined together. The external surface of the sleeve is machined to a double conical form. The sleeve is cut through at six equidistant points on the circumference, one saw cut running from one end to the other end whileothers are cut alternatively from the opposite ends to about 85% of its length. The sleeve is encased by cast iron flanges with conical bores to suit the taper of the sleeve. When the nuts are tightened, the two flanges will be drawn together, the sleeve being readily flexible gets compressed on to the two shafts firmly. The friction grip between the compressible sleeve and the shafts helps to transmit the power from one shaft to the other and the bolts do not take any load. Two tapped holes are provided in the second flange, meant for removing the flange off the sleeve using the puller.

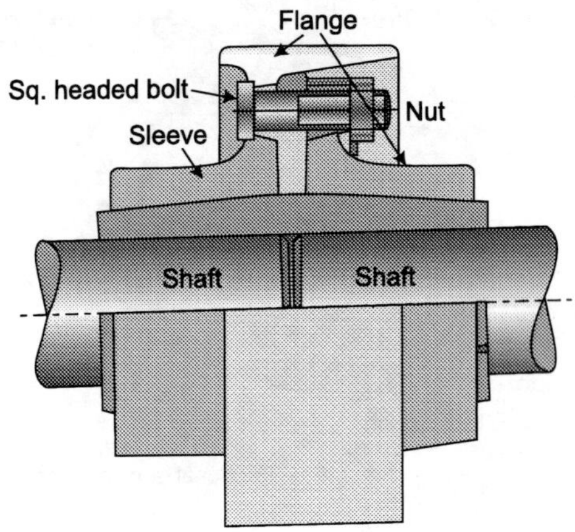

Fig. 4.91 Compression coupling

Flexible Couplings

In the direct drive between a prime mover and an electric motor or between an electric motor and a machine there may be four bearings to maintain in alignment. The power absorbed and wasted by badly aligned shafting often comprises quite a large amount of that generated. Flexible couplings are nowadays used in such cases and many forms are on the market. They should permit the two shafts to vary in axial alignment or to float slightly. Many are designed to act as electrical insulators between the shafts. They should absorb shocks and stresses as well as take up mal-alignment and wear of bearings.

Pin Type

Two flanges, one or both are provided with studs engaging in holes in the other disk. Studs have insulating and renewable driving surfaces built up of leather washers. Used for electrical drives, motors to reduction units, etc.

Grundy Type (Fig. 4.91A)

Two cast iron flanges and one centre piece of non-conducting material such as leather.

H.P. at 100 r.p.m. given by makers as ¼ to 35 according to size, adapted for connecting motors to machine tools, sewing machines, pumps, fans and blowers, etc.

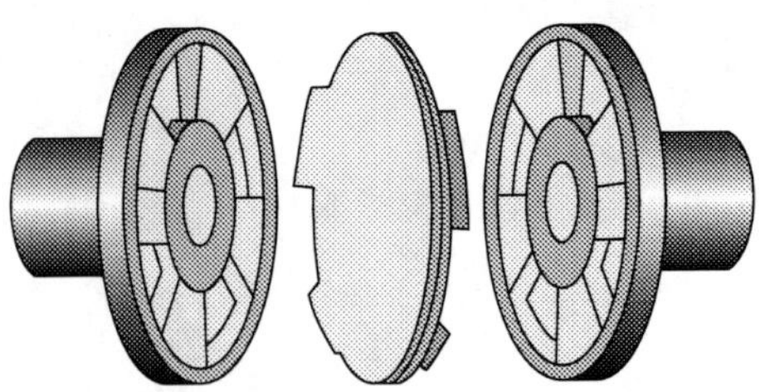

Fig. 4.91 A Groundy type flexible coupling

Croft's Multi-flex

These are driver and driven flanges in iron or steel. Driven half has two sets of lugs cast on the flange on two diameters, spaced to accommodate a corresponding set of lugs on the driving half. Connection is made by two flat belts threaded around the lugs of the driven half. For larger sizes the detachable flange renders the belt easily accessible.

Bibby Type (Fig. 4.91 B & C)

For high degree of resilience, for accommodating loss of alignment and for damping out shocks and vibrations in the heavier drives, the Bibby couplings should be mentioned.

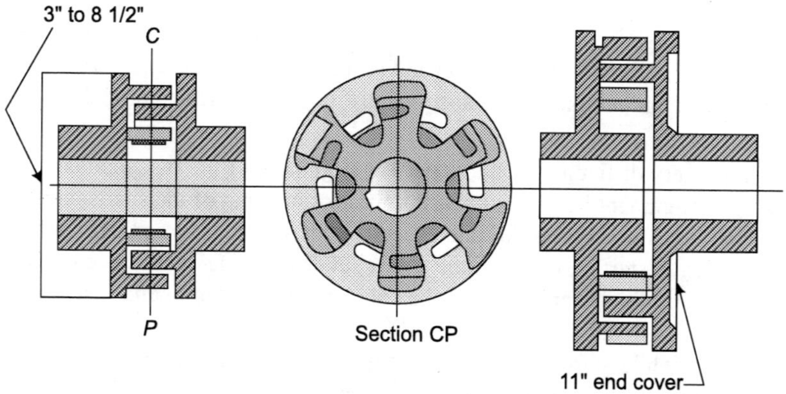

Fig. 4.91 B Croft's multiplex coupling

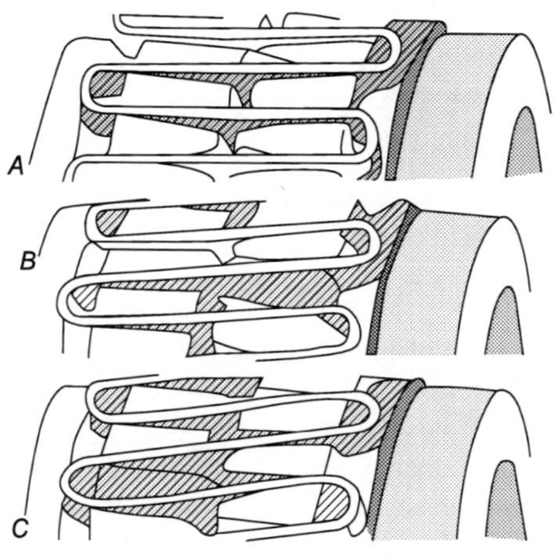

Fig. 4.91 C Bibby coupling

It consists of two serrated disks connected by a grid spring. On the grooves are flared so that the many spring members have long flexible spans at normal loads, but become supported by sides if the grooves when overloads occur. The springs are made of steel having an elastic limit of 32, 800 kg per sq centimetre. When hardened and tempered. They are enclosed in a grease-tight cover.

Applications

- Diesel engines where high speed engines need gearing up with low speed propeller shaft.
- Oil engines geared up with vertical pumps.
- Steam engines geared up with rotary pumps.
- Airship engines geared up with propellers.
- Turbine drives between steam turbine and generator.

Fabric Flexible Coupling

This is used at the end of a propeller shaft in automobiles.

It consists of two drop-forged spiders each consisting of a boss with three arms. The disks are of flexible material (leather or fabric) which acts as a cushion and transmits power to the rear-axle-without shocks. It is unaffected by mud or water and, unlike the metal-to-metal "universal" used on propeller shafts, it requires no lubrication.

In practice, it will be very difficult to achieve perfect alignment of the two shafts and will be still more difficult to maintain the alignment because of bearing wear or bending of the shafts, etc. The misalignment of the shaft causes continuous stress reversals and excessive bearing wear. When a propulsion unit is directly coupled to a machine, while starting or when there is a sudden change in the speed of the driver shaft or during a quick acceleration, the driven shaft will he subjected to sudden undue load variation resulting in a jerk or a shock. In order to avoid undue load variation on the shaft, it is often necessary to allow a slight slip when speed changes while starting, running, accelerating. This is achieved in flexible couplings by transmitting the power through a resilient medium, such as rubber reinforced with a fabric, which absorbs shocks and stresses as well as take up misalignment of the shafts. The flexible couplings are used to connect in cases like prime mover to an electric generator or to couple an electric motor to a machine tool.

There are several types of flexible couplings in use.

1. Bush type flexible couplings,
2. Fabric disc type flexible couplings,
3. Fabric flexible couplings,
4. Band type flexible couplings.

Bush Type Flexible Coupling (Fig. 4.92)

This coupling is the modified form of protected type of flange coupling. It consists of cast iron or steel flanges keyed to the shafts. One of the flanges is provided with larger bolt holes and the other flange is provided with smaller bolt holes. Instead of bolts, threaded pins are used. The pins pass through the flexible media such as rubber bushes or leather washers in the larger holes of the flange and rigidly fastened to the other flange by nuts. The rubber bushes act as cushions and absorb the shock, and permits all degrees of flexibility, rotary, transverse and longitudinal misalignment. This coupling is largely used for directly coupling electrical machines of small and medium power, to pumps, machine tools, etc.

Fabric Disc Type of Flexible Coupling (Fig. 4.93)

This type of coupling is similar to the protected type of flange coupling. The drive is effected by means of pins which bear on a fabric disc situated between the two flanged bosses which are keyed rigidly to the shafts. The two flanges are provided with large cylindrical holes and small conical holes with their centres

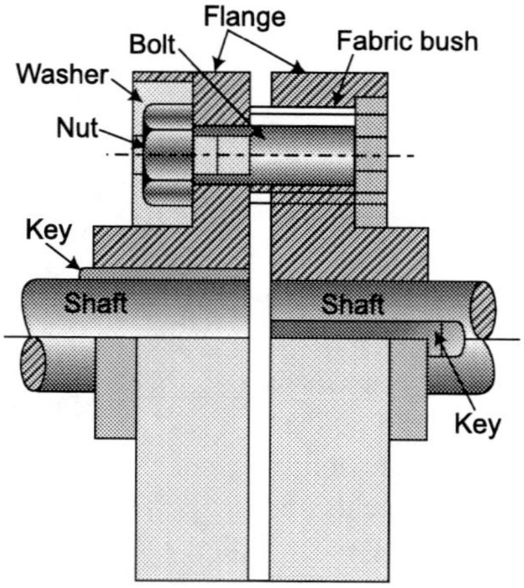

Fig. 4.92 Bush type flexible coupling

lying alternately and evenly spaced on the pitch circle. The pin is cylindrical at one end, threaded at the other and being conical between the two ends. The pins are inserted so that their conical portion fits in the conical holes. Tightening of the nuts on the pins will hold them rigidly to the flanges. The fabric disc is "provided with cylindrical bolt holes on the pitch circle at diameter equal to the diameter of the pitch circles on the two flanges. Steel washers are riveted to the fabric disc on its both the flat faces for each of its bolt holes. The fabric disc is placed in between the two flanges so that the pins are passed into the bolt holes provided in it with the cylindrical ends of the pins being projected into the large holes provided in the other flange. The resilient nature of the disc provides a fair degree of flexibility because the pins are secured to the two flanges alternately. The fabric disc will yield as power is transmitted from one shaft to the other. This type of coupling is largely used for directly coupling electrical prime movers to the other machines.

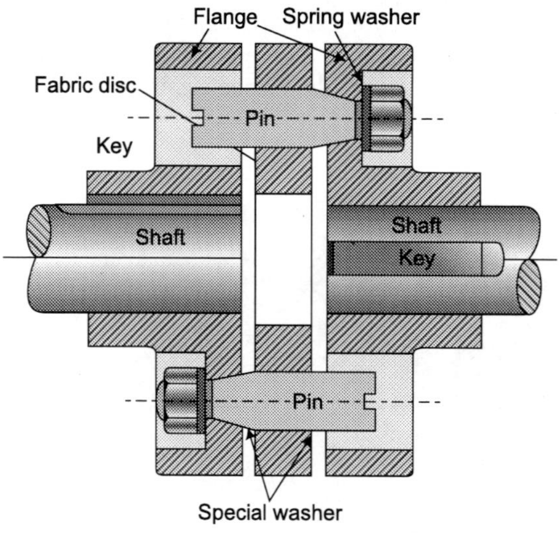

Fig. 4.93 Fabric disc flexible coupling

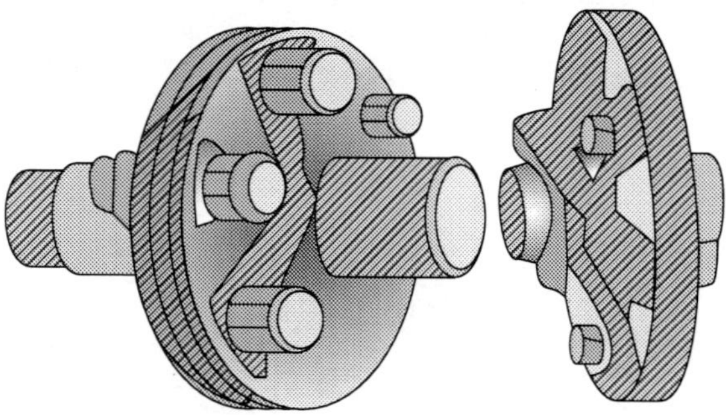

Fig. 4.93 A Fabric Flexible coupling

Fabric Flexible Coupling (Fig. 4.94)

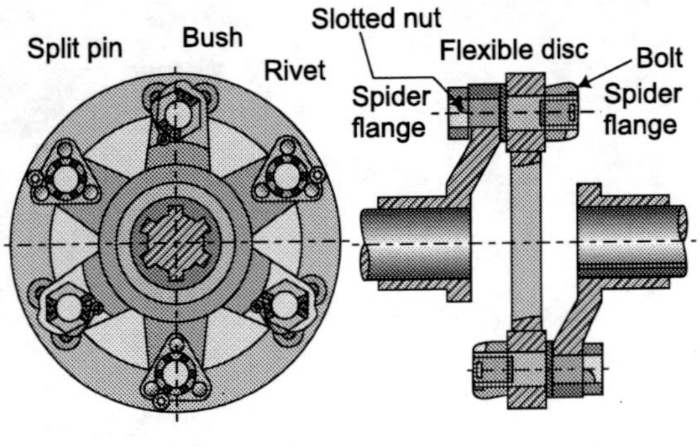

Fig. 4.94 Fabric type flexible coupling

This type of coupling is similar to the fabric disc type of coupling, but the solid circular flanges are replaced by the two drop forged spiders each consisting of a boss with three projecting radial arms in the ends of which the bolts are fitted. The bolts have collars to enable them to be secured in the arms. The bolts are engaged in a flexible disc of tough rubber or leather. The transmission bushes protect the flexible disc where the bolts engage. The disc is bolted to the spiders placed in diagonally opposite positions; flexibility in transmission is obtained as the disc 'yields' as the power is transmitted from one shaft to the other. Splined shafts are fitted into the splined bosses. This coupling also known as fabric universal joint is suitable for medium drives at all speeds and is largely used for the transmission of power in automobiles. It has advantages over the machined type of universal joint, because not only it is cheaper to manufacture, but also (1) it requires no lubrication, (2) it is silent, (3) there are no wearing parts and it is unaffected by mud and water.

Band Type Flexible Coupling (Fig. 4.95)

This type of coupling transmits the power through· an endless leather belt. It consists of an outer flange keyed to one of the shafts. A loose ring having eight projecting segments spaced uniformly around its circumference by eight bolts is bolted to the inner face of the outer flange. The centre ring keyed to the other shaft, and like the loose ring is also formed with eight projecting segments spaced uniformly around its circumference. When the two shafts are aligned, the centre ring and the loose ring are so designed that the two sets of segments lie in the same plane at right angles to the axes of the two shafts. An endless leather belt is passed around these two sets of segments as shown in the assembled view. The leather belt before placed in position has its two ends riveted together with copper rivets. To keep the belt in position and for general protection, a cover plate is fixed to the segments of the centre ring using set screws. When the two shafts are running, there will be a small amount of "play" between them. This coupling is more suitable for transmitting the power between the two shafts which are, or may become slightly out of line and when the driven shaft is liable to shock and vibration.

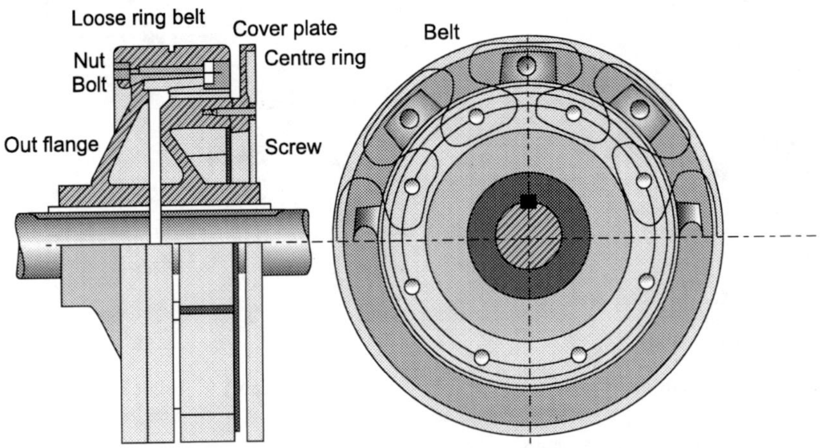

Fig. 4.95 Bend type flexible coupling

Disengaging Couplings

The rigid couplings and the flexible couplings are used for connecting shafts in order to rotate them as single unit. Cases however occur in which it is desirable that a drive from a shaft may be engaged or disengaged occasionally as and when required while the driver shaft is rotating. This can be effected in disengaging couplings. Essentially a disengaging coupling consists of one part secured to the driving shaft, and the other part connected to the driven shaft in such a way that it can move axially to engage or disengage from the rotating driving shaft. This type of coupling is used for slow running shafts.

The two types of disengaging couplings are:

1. Claw coupling and
2. Cone coupling or cone friction clutch.

Claw Coupling: or Love Joy Coupling (Fig. 4.96)

This coupling consists of two flanged bosses having number of claws or jaws projecting from their inner flat faces. The number of claws may be three or four and may be either squared or spiralled. The square claw coupling can be driven in both the directions, whereas the spiral jaw coupling is designed to be driven only in one direction. The advantages of the spiral type are, engaging is smooth and higher shaft speeds. The driven half of the coupling consists of a sliding flange having an extended boss to accommodate the fork of the operating lever. A parallel key is screwed to the driven shaft. It facilitates the axial movement of the sliding flange on the shaft. This sliding flange is moved axially along the shaft by operating the fork so as to engage with the jaws of the other half of the coupling. The other half of the coupling comprises a rigid flange secured rigidly to the driver shaft by a gib head key. This type of coupling is used for slow speed shafts in machine work and automobile practice.

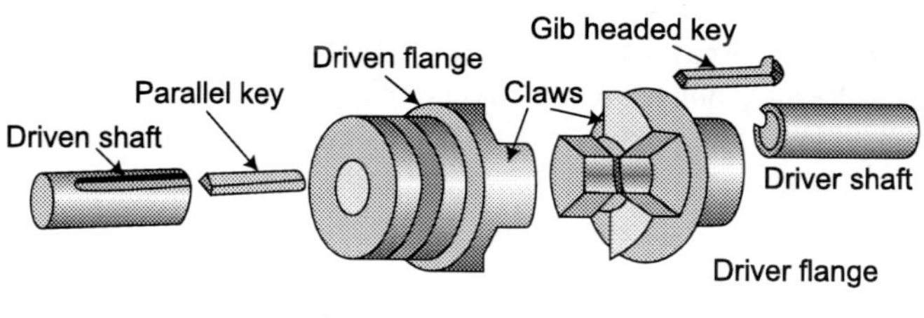

Fig. 4.96 Claw coupling

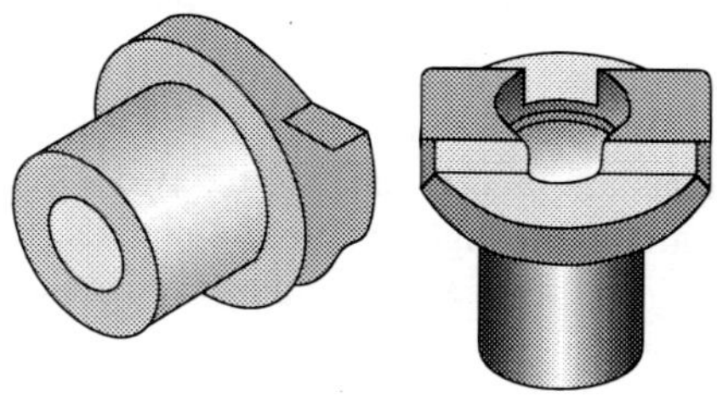

Fig. 4.96A Claw coupling

Non-Aligned Couplings

The rigid couplings and disengaging couplings have coaxial shafts. The flexible couplings permit slight misalignment of the axes of the two shafts. When the axes of the two shafts are not coaxial, "non-aligned couplings" are used. The two types of nonaligned couplings are:

1. Oldhams coupling and
2. Universal coupling.

When the axes of the two shafts are parallel but out of line considerably, Oldham's coupling is used. When the axes of the two shafts intersect universal coupling is used.

Oldhams Coupling (Fig. 4.97)

This consists of two flanges and a centre disc. Each of the flanges is securely keyed to its respective shaft. The flanges are made, either of cast iron or steel and are provided with grooves in order to accommodate the lugs which protrude from the centre disc. The centre disc is made of steel and the lugs which project from it are set at right angles to each other on its opposite flat faces, thus preventing them from working out of the flange grooves during their rotation.

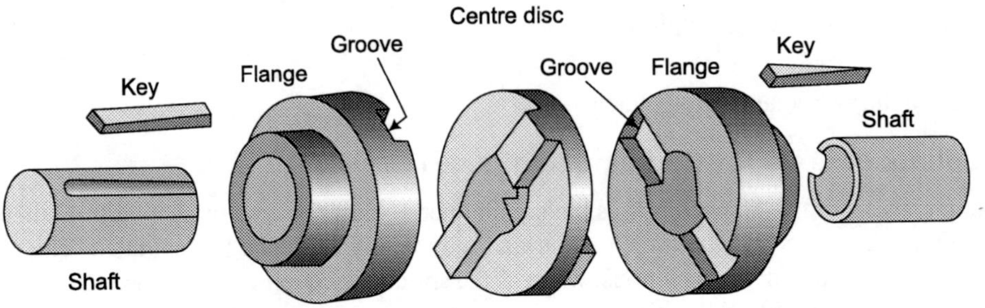

Fig. 4.97 Odhams coupling

The lugs are consequently sliding a fit in the grooves of the flanges which enable them to "float" or slide slightly. This therefore provides slight axial movement of the shafts, and compensates for very slight angular misalignment.

A coupling of this type is suitable for transmitting heavy loads running at slow speeds.

Universal Coupling or Hookes Joint (Fig. 4.98 A) and (Fig. 4.98 B)

This is used to connect two shafts whose axes intersect each other. The advantage of this coupling is that the angle between the two shafts may be varied while the shafts are in motion. Several types of universal joints are used in practice the two types are most commonly used. When a single joint connects the two shafts at an angle rotation is not transmitted uniformly, but by using two couplings, exact uniformity is achieved. Universal couplings or Hooke's joints are used in machine tools such as milling machines and in automobiles.

The universal coupling consists of two forks keyed to the ends of the two shafts, two trunnion plates and four bushes. The trunnion plates are placed in between the two forks such that their projecting pin ends are fitted in the two halves of the bushed portions of the trunnion plates. The two trunnion plates are joined by hexagonal bolts and nuts.

Universal Coupling (Fig. 4.99)

Another type of universal coupling consists of two forks keyed to the ends of the two shafts. A central block consisting of two cylindrical bushes cast or welded at right angles, is placed between the two forks and connected to them by two pins.

Knuckle Joint

It is a means of connecting round bars of wrought-iron so as to permit axial adjustment in one plane. The bars sustain a tensile load of 6 tons. The main feature of the design is an endeavour to obtain a rod, eye and double-eye or fork of equal strength. Let us examine this typical knuckle joint and calculate the stresses at various sections.

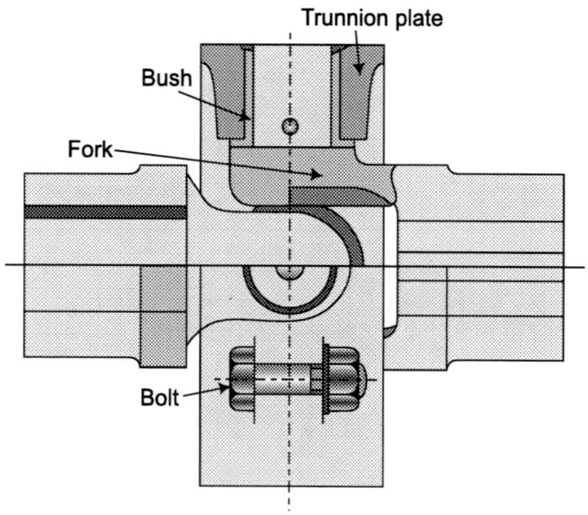

Fig. 4.98A Universal coupling

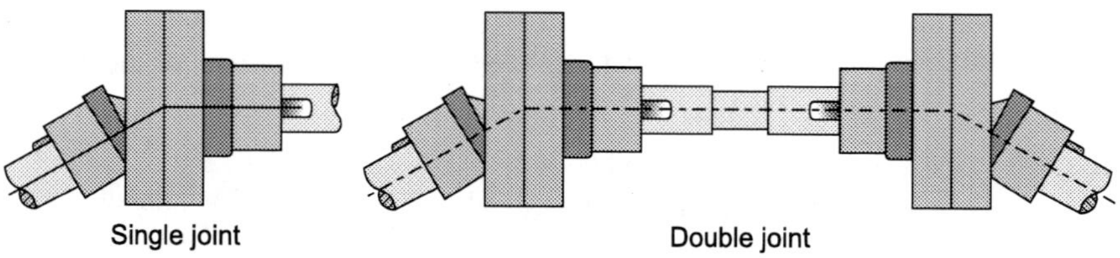

Fig. 4.98B Universal coupling

Fig. 4.99 Universal coupling/Knuckle joint

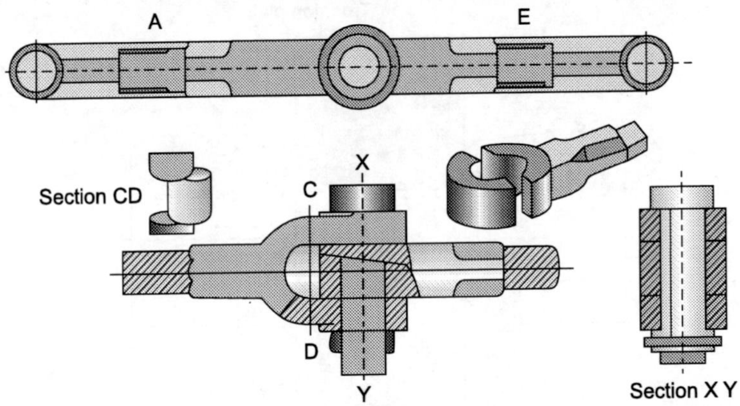

Fig. 4.100 Knuckle joint details

The joint may fail by breaking across the plain portion of 25 mm diameter.

(1) Therefore, tensile stress at A or B = 6/0.7857 = 7.64 tons per sq in.

(2) The section of the eye at the end of rod B produced by a plane X Y passing through its centre is shown as part of "section X Y" in the diagram. The area of the hatched portions of the eye = 2(1.2 × 0.5) = 1.2 sq inch.

Tensile stress in eye = 6/1.2 = 5 tons per sq inch.

(3) A similar section of the double-eye at the end of rod A shows four rectangles the total area being, 4(0.75 × 0.5) = 1.5 sq inch.

Tensile stress = 6/1.5 = 4 tons per sq inch.

(4) The section CD across the fork shows two rectangles 1.2 × 0.6 inch each.

Area = 2(1.2 × 0.6) = 1.44 sq inch.

Tensile stress = 6/1.44 = 4.17 tons per sq inch.

(5) The pin is in double shear and is likely to break across two sections 1.2 inch apart. Each section has an area at 0.7854 sq inch.

Thus the combined area resisting shear = 1.5708 sq inch.

Shear stress = 6/1.5708 = 3.82 tons per sq inch.

(6) The pin may crush. Consider the portion of the pin within the single eye at the end of rod B. Its bearing area = 1.2 × 1= 1.2 sq inch.

Compressive stress = 6/1.2 = 5 tons per sq inch.

Now consider the portion of the pin within the double eye at the end of rod A. Its bearing area = 2(0.75 × I) = 1'5 sq in.

Compressive stress = 1'5 = 4 tons per sq inch.

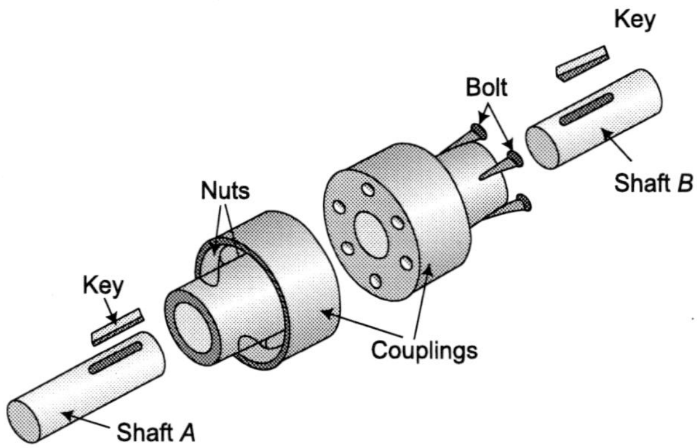

Fig. 4.101 Position of shafts to be coupled

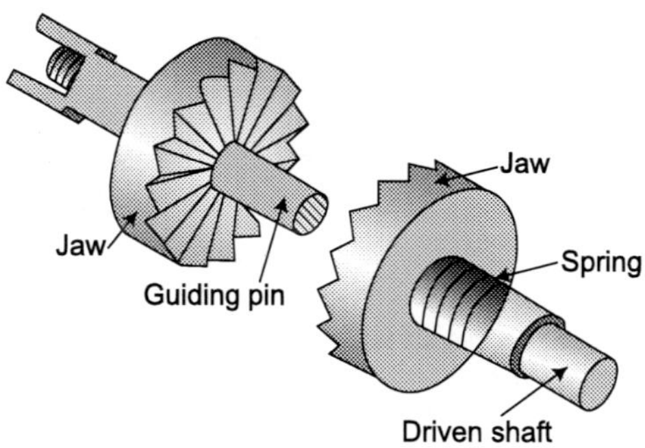

Fig. 4.102 Parts and position of safety coupling

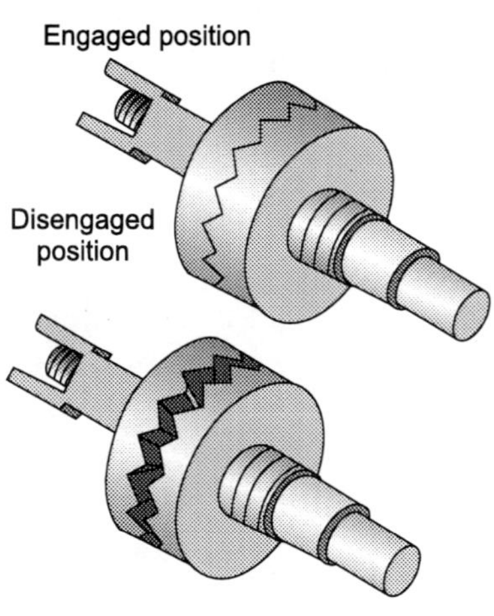

Fig. 4.103 Position of safety coupling

Safety or Slip-type Coupling

A safety coupling is self-disengaging when the torque is higher than the friction shaft couplings generated by the spring and jaw.

Plate Coupling (Fig. 4.104)

When shaft ends can be brought into perfect alignment, such shafts can be coupled together by means of a plate coupling.

The shaft cannot be shifted in the axial direction.

The shafts are secured with keys between the shafts and hubs.

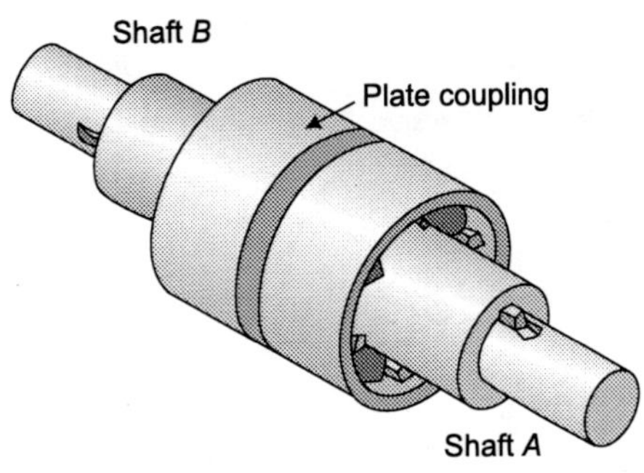

Fig. 4.104 Plate coupling

Clamp Coupling (Fig. 4.105)

This coupling is built up of two parts which are placed over the connecting ends of the shafts.

The clamp coupling joins two shafts when they are in alignment, and this coupling does not permit any axial movement of the shafts. This is used on line shafts in textile mills. Torque is generally transmitted by friction. Sometimes a key is used for more power requirements.

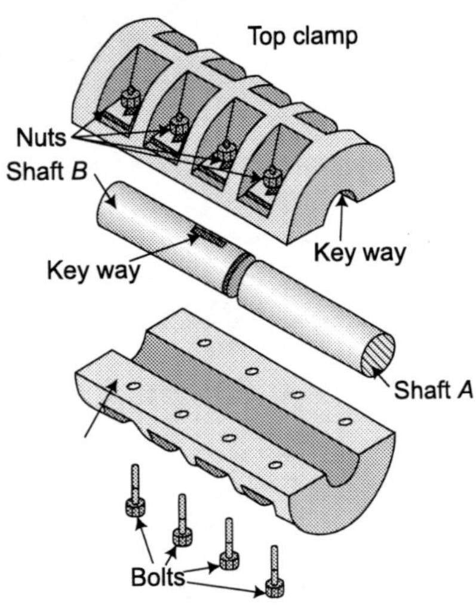

Fig. 4.105 Clamp coupling

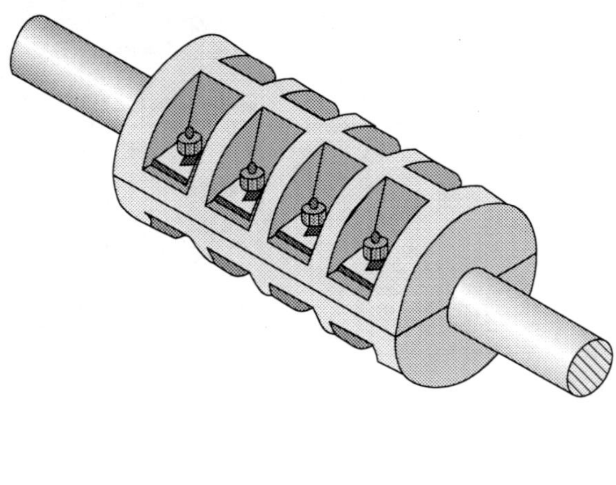

Fig. 4.106 Clamp coupling assembled

Couplings are not shiftable but clutches are shiftable during work and this is the difference between couplings and clutches.

In machinery and automobiles it is often necessary to couple a driven shaft at rest to a rotating driving shaft when transmission between the two is desired, or disengagement of the already coupled shafts is required. These situations arise in machinery when quick operational changes are required on the driven side without stopping the driving unit. In automobiles, for example, while starting and when changing gears according to the gradients, it will be necessary to run the engine "free", that is, the engine shaft and the propelling shaft, which in turn is connected to the wheel axle, have to be disconnected. When starting an automobile, the engagement should be gradual and smooth. It is here friction clutches are the most commonly employed means of performing effectively the engagement and the disengagement of the driving and the driven shafts while the driving shaft is still rotating. The basic principle of operation of all types of friction clutches is as follows. Both the driving and the driven units are provided with the friction surfaces, either flat or conical. To engage the driving and the driven units, the friction surface of the driven unit which is stationary is moved gradually towards the rotating friction surface of the driving unit and held pressed to it so that a firm friction grip is established between the two. To maintain the grip throughout the period of engagement, pressure is exerted on both the friction surfaces by springs. To disengage the driven unit from the driving unit, the friction surface of the driven unit is moved apart against the tension of the springs so as to separate it from the friction surface of the driving unit. The desirable properties of the friction materials are, less wear, withstand high temperature of operation and high coefficient of friction. The friction materials commonly employed are asbestos base materials and powder metal materials.

4.8 CLUTCHES

Types of Clutches

1. Claw coupling
2. Fluid coupling
3. Electromagnetic clutch

These serve the same purpose with a difference. The difference being that the friction clutches can be directly engaged to get gradual increment in speed while the other types need speed reduction of the driving unit before the engagement and then run at the desired speeds.

Friction clutches are classified as:

1. Single plate clutch,
2. Multiple disc clutch,
3. Cone friction clutch
4. Centrifugal clutch.

Single Plate Clutch (Fig. 4.107)

The single plate clutch also called "disc clutch" is most commonly employed in automobiles.

The driving unit of the clutch consists of the driving shaft, fastened to the driving flange rivets. The rim of the driving flange is cut with gearing teeth to engage with the pinion of the self starter. A self aligning ball bearing is socketed inside the driving flange. This bearing facilitates the driven shaft to align itself so as to be coaxial with the driving shaft. Friction discs are riveted to the support plate, which itself is riveted to the clutch plate. The support plate with twelve 3 mm width radial slots, prevents the possible distortion of the plate when the plate becomes hot due to slipping. Two oil throw rings are fastened to the support plate to prevent the lubricating oil from reaching the friction faces. The lubricating oil which might have been passed along the driven shaft is thrown into the circular grooves of diameter 252 mm and width 8 mm provided in the driving flange and the driven flange. From there it escapes through the inclined oil holes of diameter 3 mm drilled at 45° in the walls of the two flanges. The bore in the clutch plate is machined with six internal straight sided splines to permit the axial movement of the driven shaft. The driven unit of the clutch consists of the flange into which two bushes are fitted through which the driven shaft. The friction discs are held pressed to the driving flange by six equi-spaced and mounted in the driven flange itself. Three toggle levers at equal angular distances serve to withdraw the flanges and separate the friction surfaces in contact. These toggle levers pivot at the outer ends of the pins inserted in brackets, screw fixed to the driven flange. The lower ends of the toggle levers butt against the end collar of the sliding sleeve which itself is mounted over the projecting cylindrical end of the driven flange. Pressure applied on the foot pedal as indicated by the arrowhead, withdraws the sliding sleeve which moves the lower ends of the toggle levers outwards. This movement of the lower ends of the toggle levers outwards makes the driven flange move out against the spring pressure which relieves pressure on the friction discs and disengages the driving and the driven shafts. The driven shaft is given a running fit in the driven flange and the sliding sleeve is free to rotate over the projecting end of the driven flange. When the pressure on the foot pedal is released the sliding sleeve moves inwards. The plate springs withdraw the toggle levers back to their position. This relieves the pressure on the pins and so the friction discs, the driven flange and the driving flange are brought to the engaged position due to the action of helical springs.

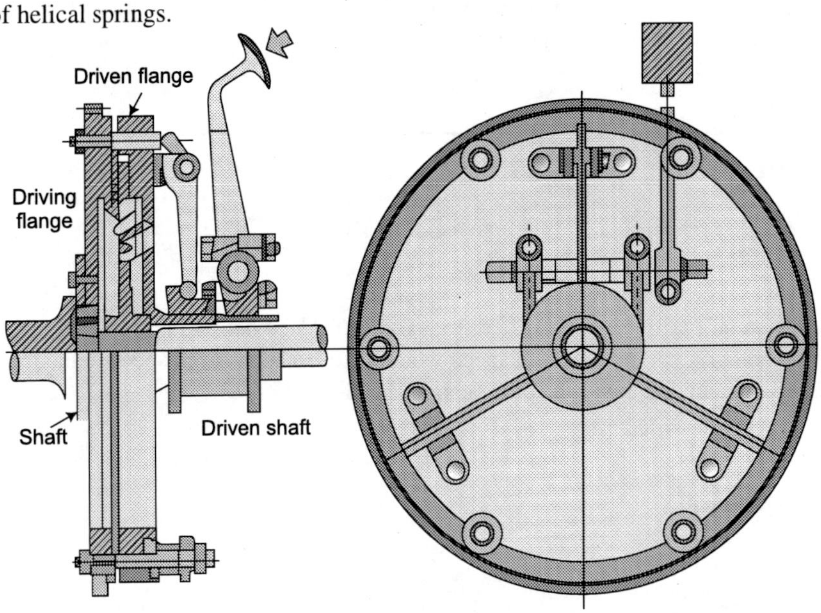

Fig. 4.107 Single plate cluth

This is used in automobile vehicles. The rubbing surface is covered with friction lining of asbestos/plastic/cotton, with steel wires. The contact force is produced by springs which effect the continuous closing of the clutch. The pedal force acts against the spring force and opens the clutch.

Multiple Disc Clutch (Fig. 4.108)

Multiple disc or plate clutch with a number of clutch plates is capable of heavier duty due to the increased area of contact, that is, it is capable of transmitting more power.

The driving unit consists of the driving shaft is keyed to the centre boss by a feather key and four inner friction discs. The centre boss is provided with six straight sided external serrations. The friction discs are also provided with the corresponding six internal serrations. The driven unit consists of a shell mounted freely over the driving shaft, three outer friction discs, a stud carrier disc, a cover and a sliding sleeve. The shell is provided with six internal serrations to receive the six external serrations provided in the outer friction discs. The inner and the outer friction discs are arranged alternately in the assembly. The stud carrier disc is also provided with six external serrations which fit into the six internal serrations provided in the shell. Three studs are screwed into the threaded holes in the stud carrier disc such that their outer projecting ends butt against the vertically projecting arms of the L levers. The shell, the cover and the sliding sleeve are mounted freely on the driving shaft. The sliding sleeve and the cover are connected through two sets of levers by pin joints. The linkages consist of L levers and the jaw levers. The cover is fixed to the shell by six hexagonal head screws. The operation of the clutch is as follows. The sliding sleeve is moved axially inwards on the driving shaft by the operating lever, not shown in the figure, which actuates the levers to exert pressure on the studs, which in turn press the stud carrier to provide the necessary friction grip between the inner and to outer discs which makes the shell to rotate along with the driving shaft.

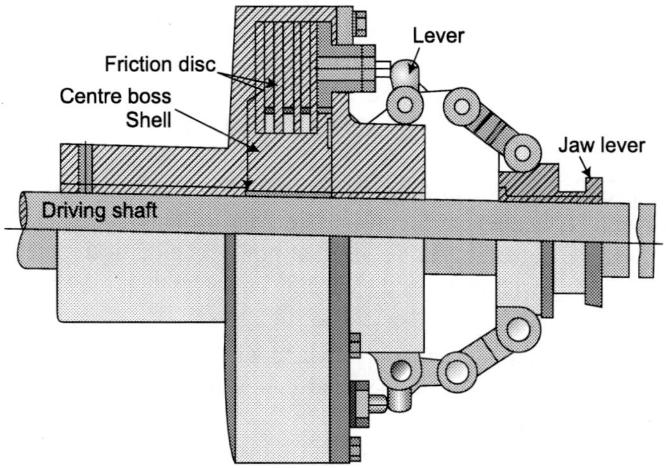

Fig. 4.108 Multiple disc clutch

Centrifugal Clutch (Fig. 4.109)

Centrifugal clutches unlike the plate clutches operate automatically by the action of the centrifugal force. They gradually engage and disengage the driving and the driven machines at the predetermined speeds, and slip at overloads. This kind of operation finds the application of this type of clutch in squirrel cage induction motors, drums of the washing machines, belt conveyors, mopeds, etc., which have their starting torques comparatively lower than the maximum torque developed during running.

This type of clutch transmits the power from the driving shaft to the pulley mounted freely over the driving shaft itself. The spider hub keyed to the driving shaft carries four radially projecting arms bolted to it. Four leaf springs are mounted as freely supported beams between the successive arms. Each of these leaf springs carries a shoe which bears against its ends and is loaded by the adjusting screw.

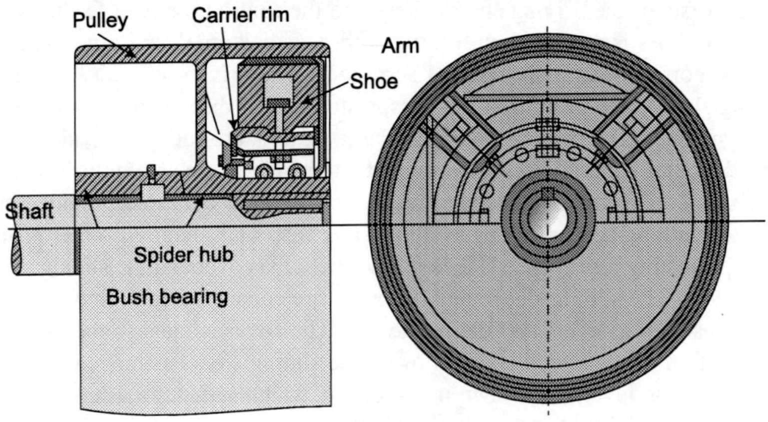

Fig. 4.109 Centrifugal clutch

When the inner piece has achieved a sufficiently high speed, the centrifugal weights swivel towards the outside, press the jaws on the outer piece with the friction lining and the clutch is closed. When the speed is reduced, the clutch opens by itself, e.g., mopeds.

Friction Clutches

In friction clutches we have an example of the transmission of power by means of friction. The force of friction is pressed into useful service when machinery is driven by belts, ropes, etc., as well as in traction. When we consider brakes as on automobiles, we see that power can be overcome by friction. In general, however, especially when dealing with surfaces which slide on each other, the engineer directs his energies to the reduction of friction.

Friction clutches are used to communicate, gradually and without shock, the motion of a revolving body to one at rest. Again, they are used where sudden and complete disconnection of two rotating shafts is necessary, where such shafts are in alignment.

The clutch on a motor car is a "flexible member" designed to connect, or disconnect, engine and back axle. On machine tools we find that ingenious friction clutch designs have enabled fast-and-loose pulley arrangements to be dispensed with.

Conical Friction Clutch

This simple clutch consists of two iron castings A and B. Of these B is secured to its shaft by means of a feather along which it can slide so as to make frictional contact with the inside of casting A (which is keyed to the driven shaft and bears against a solid collar).

Movement of B is obtained by means of a forked lever or other "shifting gear" bearing in the groove G. This brings the two inclined conical surfaces into close contact and a drive ensues. Disengagement is effected with equal speed.

Coefficient of Friction

- Cast iron on cast iron – 10°
- Leather on greasy cast iron – 12°
- Ferodo fabric on cast iron – 18°

Note that the angle is greater when the male cone is covered with leather, ferodo, or some friction-creating material.

Coefficients of Friction

- Cast iron on cast iron — 0.15 to 0.2
- Dry leather on metals – 0.5 to 0.6
- Ferodo fabric on metal – 0.3 to 0.4

Disadvantages

Compared with other types, considerable axial force is necessary to run the halves into working engagement. This force sets up an end thrust referring to **Fig. (4.110)**

$F = P(\mu \cos \theta + \sin \theta)$ where F = axial force in pounds

P = normal pressure between conical surfaces, in pounds

θ = angle between axis of shaft and conical surface

μ = coefficient of friction.

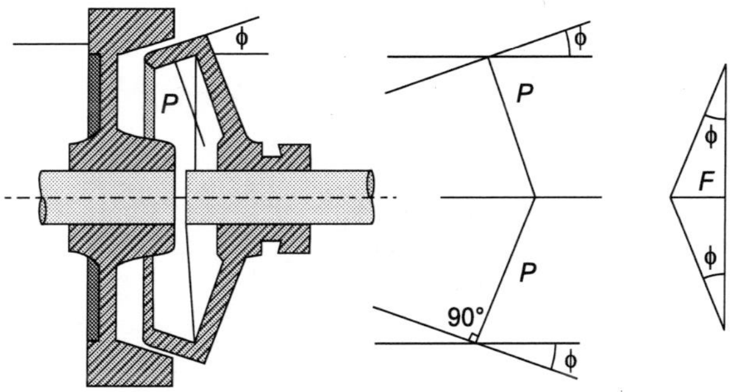

Fig. 4.110 Conical friction clutch

Other Types of Clutches

- Dog clutch D Cone clutch
- Multi-plate clutch
- Electromagnetic multiple disc clutch. D Overriding clutch
- Air clutch
- Centrifugal clutch
- Single plate clutch

Dog Clutch (Fig. 4.111)

The dog clutches provide a positive drive but can only be engaged when two elements of the clutch are stationary or are being gently moved by hand.

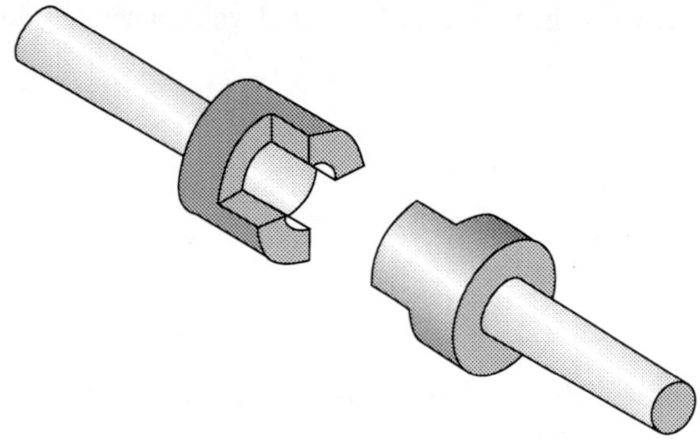

Fig. 4.111 Dog clutch

Cone Clutch (Fig. 4.112)

The cone clutch can be engaged progressively whilst one or both of the elements are rotating. It can transmit low power.

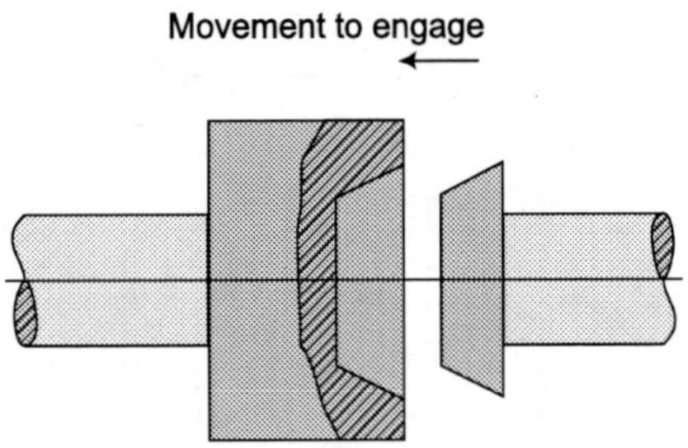

Fig. 4.112 Cone clutch

Multi-plate Clutch (Fig. 4.113)

The multi-plate clutch is widely used in machine tools to connect the transmission gearbox to the driving motor. It is compact, smooth in operation and very powerful. A brake is frequently built into the clutch so that the transmission gearbox is rapidly brought to rest when the clutch is disengaged.

Electromagnetic Multiple Disc Clutch (Fig. 4.114)

This clutch joins the shaft and the gear. It can be operated through a cable from a distance. If direct current is applied, it builds a magnetic field on a magnetic coil. It flows through the discs and firmly pulls and attracts the armature disc. The armature clamps the plates together so that they transmit the drive.

Air Clutch (Fig. 4.115)

An air clutch requires no mechanical adjustment since the moving parts automatically take up any wear on the friction surface. Air pressure must be maintained continuously while the clutch is engaged.

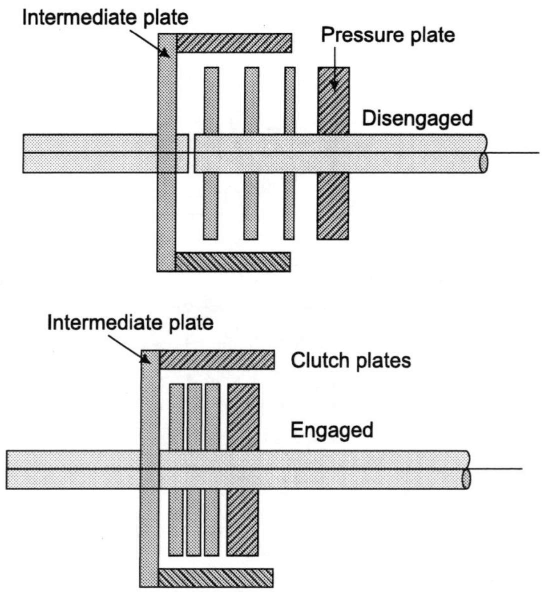

Fig. 4.113 Multiplate clutch

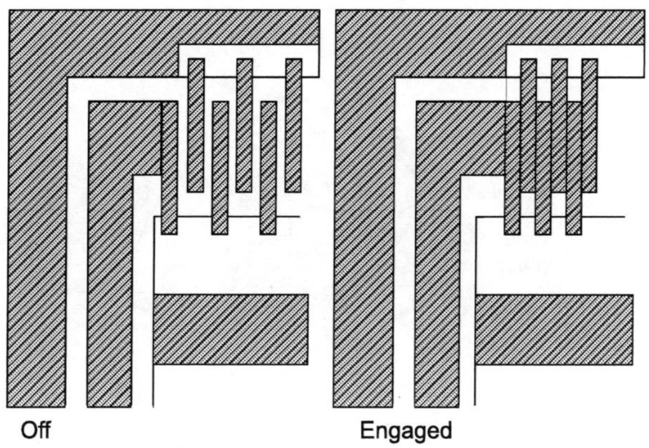

Fig. 4.114 Electromagnetic multiple disc

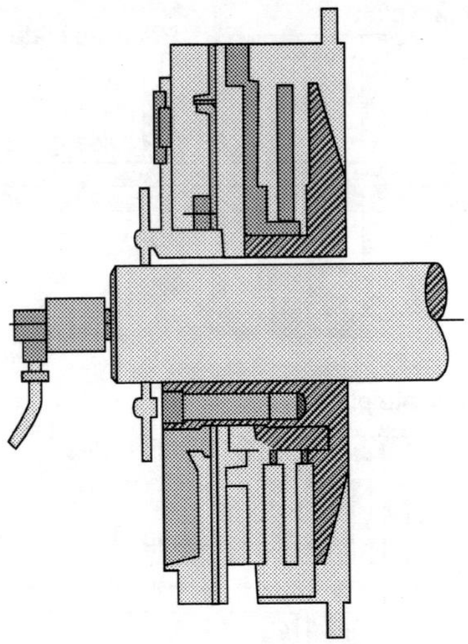

Fig. 4.115 Air clutch

Centrifugal Clutch (Fig. 4.116)

When the inner piece has achieved a sufficiently high speed, the centrifugal weights swivel towards the outside, press the jaws on the outer piece with the friction lining and the clutch is closed. When the speed is reduced, the clutch opens by itself, e.g., moped

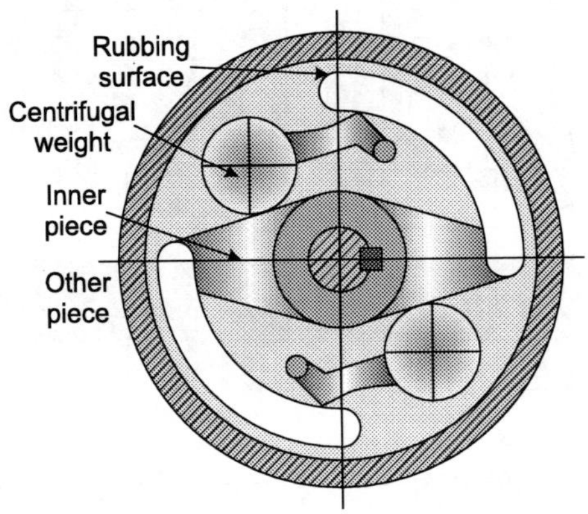

Fig. 4.116 Centrifugal clutch

Overriding Clutch (Fig. 4.117)

When the inner piece has to be faster the overriding clutch transfers the turning moment by the climbing of balls or the cylindrical rollers. It opens in the opposite case.

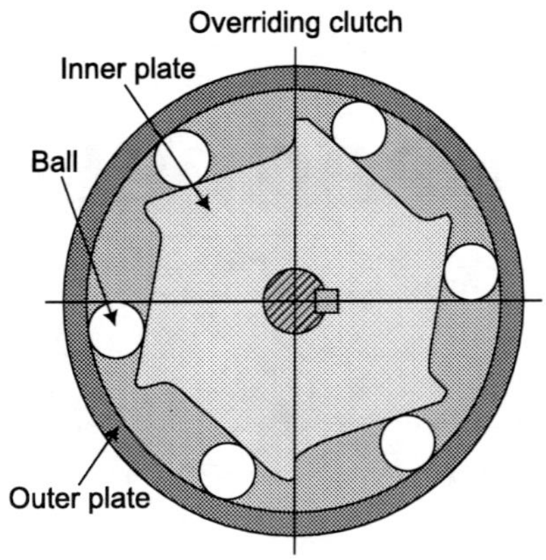

Fig. 4.117

EXERCISE

1. What is meant by power transmission? Why do we need it.
2. Describe the different mediums used for power transmission.
3. (i) How can speed be increased or decreased by using gears.
   (ii) In what way the direction of a driven gear can be changed.
   (iii) How the speed of a driven gear can be known at the time of power transmission.
4. Describe the types of pulleys.
5. What is meant by "Jockey Pulley".
6. What is meant by fast and loose pulley.
7. What are the advantages of chain and sprocket in power transmission.
8. Where do we use flexible coupling.
9. In what different ways belts are used on pulleys and when.
10. What do you understand by creep of belt.
11. What is meant by key and key way?
12. How many kinds of keys are used?
13. What are couplings?
14. Why couplings are necessary?
15. Classify the types of couplings.
16. List out all the various types of couplings in use.
17. Draw proportional sketches of the couplings.
18. What are clutches?
19. Why clutches are necessary?
20. Classify the types of clutches.
21. List out all the various types of clutches in use.
22. Draw proportional sketches of the clutches.

Bearings

Chapter 5

5.1 FRICTION

When one body slides over another, its motion is opposed by resistance along the surface of contact of the two bodies. The force which resists motion is known as friction. Within broad limits sliding friction depends on the degree of roughness of the surfaces and the materials of which they are made. Again it is proportional to the force pressing the bodies together. Sliding friction between two bodies can be reduced by arranging that the surfaces shall not actually make contact but are separated by a thin film of fluid or semi-fluid lubricant.

A very thin film of lubricant suffices to keep surfaces "slippery" and unless the pressure, is very great they will not come into contact, i.e., if the lubricant is sufficiently viscous. It has been shown that viscosity of a lubricant diminishes with increase of temperature the oil in its thinned condition being liable to be squeezed out, with consequent metal to metal contact and seizure.

Instances of Sliding Friction

Slide valves of steam engines, journals of shafts in bearings, piston rods in stuffing boxes, plain footstep bearings, ball and roller bearings, rolling wheels of any kind.

Coefficient of Friction

A sliding block is on a horizontal surface. The slider can be weighed and pulled along the board by loading the scale pan which is attached to the block by a string passing over a pulley. It is found that when the scale pan is sufficiently loaded and is given a slight downward jerk the slider will move and continue in motion. The weight of the slider plus the load it carries is the vertical pressure between the surfaces. We may call it W. The horizontal force required to keep the slider in motion equals the weight of the scale pan plus the weight it carries. Let us call horizontal force P. Experiment shows that for the same slider and horizontal board, P is proportional to W. The ratio is called the coefficient of sliding friction (μ) for the surfaces considered. As P is equal and opposite to the force of friction we could write frictional resistance.

P = frictional resistance divided by normal pressure = P/W or $P = \mu \times W$

The result is independent of the areas of the surfaces in contact.

Coefficients of friction: Fig. 5.1)

- Metals on metals (well lubricated) – 0.03 to 0.06
- Cast-iron on cast-iron or brass – 0.15 to 0.2
- Cast-iron on steel – 0.2 to 0.3
- Brass on steel – 0.15 to 0.17

- Steel on steel – 0.14
- Dry leather on metals – 0.5 to 0.6
- Oily leather on metals – 0.15 to 0.18
- Ferodo on metal – 0.3 to 0.4

Friction at Journals

The frictional resistance acts with a leverage *r* mm to prevent rotation of the journal. Thus, with a given value of *F*, the external moment required to rotate the shaft will vary with the length of *r*. Thus, a journal should be as small in diameter as is consistent with strength and stiffness.

$F = W \times \mu$ where μ = coefficient of friction.

Assume that frictional resistance in the case of a cylindrical shaft running in. Its bearing is not materially different from the case of a slider moving along a horizontal surface.

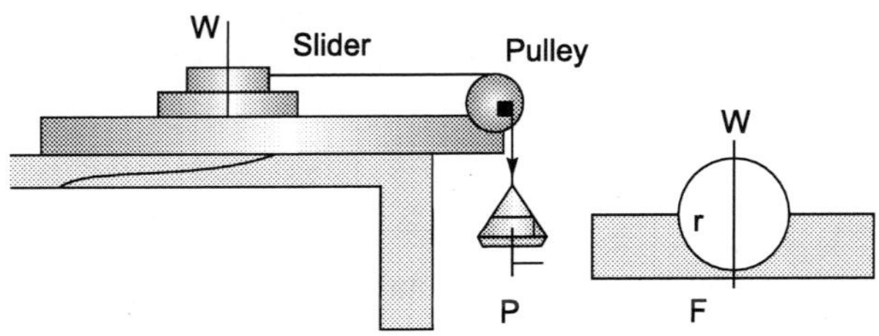

Fig. 5.1 Frictional force

Velocity of Sliding Surfaces

Its velocity of sliding in meters per minute = π DN/ 1000

A shaft of diameter *D* mm. makes *N* revolutions per minute.

Work lost at bearings by friction:

If

μ = coefficient of friction,
D = diameter of shaft in mm
N = speed in revolutions per minute,
W = total load on bearing in kg,
T = resisting moment due to friction,
H.P. = horsepower used in overcoming friction.

Then $T = \mu \times W \times D$ divided by 1000×2 meter kg

Chafing or rubbing or attrition of two bodies is friction. The resistance offered by one surface over another surface when the surfaces move with respect to one another. This action leads to wear and tear of the metal, the rubbing cause heat, both these actions increases the gap between the two mating parts and this widened gap makes the parts to rattle and make noise. Also the parts vibrate. All these actions create major maintenance problems.

In other words friction is the resistance to motion of two surfaces in contact. One distinguishes between gliding friction and rolling friction. The rolling friction is lower than the gliding friction.

Rotating shafts transmitting power need supports. The problems that arise at the supports are the wear of the rotating shaft and the support itself, generation of heat due to friction at the support, loss of power, maintenance, safety, etc. The inevitability of the support attended by the undesirable factors as mentioned above is to a large extent answered by "bearings", thwarting the undesirable factors. Inasmuch, good bearings offer suitable fits to the shafts, offer provisions for good lubrication and easy replacement of the worn-out surfaces. The necessity of the bearings are seen in supporting the spindles of machine tools, crankshaft of engines, axles of automobiles, shafting in workshops, etc. Bearings, to fulfil the above-mentioned requirements under different conditions take to different shapes and structure. However they are generally grouped under two categories, namely, the sliding contact bearings, and rolling contact bearings or anti-friction bearings.

In all types of machines power is transmitted through the many media. Power transmission. In these mediums or methods different types of shafts are used. Besides this, there are several such parts in a machine, in which shafts are used. Bearing is used to give support to these shafts while they are in motion. In addition to providing support, these bearings also provide base to long shafts.

5.2 BEARINGS

Bearings facilitate the relative motion of surfaces in contact with limited consumption of energy:

- The lower the friction the lower the energy required.

Broadly there are the following advantages of bearings:

- Shaft gets strong base.
- It provides stable speed.
- Lubrication becomes possible between the base and the shaft.
- Speed of the machine increases.
- Parts of the machine do not wear out early.
- It saves time and power.

Sliding Contact Bearings

In bearings of this type, the relative motion between the shaft and the bearing surface is of "pure sliding". Due to the surface contact, the friction between surfaces of the rotating shaft and till bearing is relatively high. This is minimized by lubricating the two surfaces. Depending upon the type of load that the bearing supports, sliding contact bearings are classified as below.

Types of Bearings

Bearings can be divided into three types

1. Journal bearing.
2. Foot-step bearing.
3. Thrust bearing.

Journal Bearing

It is supporting portion of a shaft i.e., the part which revolves in the bearing. It is of the following three kinds.

Solid Bearings (Fig. 5.2)

This is the easiest means of bearing. It is a one -piece rigid bearing. Its use is limited as it must be slipped over the end of the shaft which is to support. If it wears out it can not be set right by adjusting it.

These are the types of the sliding contact bearings that are used to support shafts subjected to loads perpendicular to the shaft axis. The portion of the shaft supported by bearing is called "journal". In the journal bearings, the inner surface of a hollow cylindrical bush usually in halves, forms the contacting bearing surface. The bushes, commonly termed "brass", should be made of metals which have sufficient strength to resist distortion under load a high melting point and high resistance to abrasion without scoring the journal. Since gun metal an alloy of copper and tin, has exceedingly good wearing qualities, bushes are generally made out of it. Other materials such white metal, an alloy of lead or tin, copper and zinc, Babbitt, Magnolia, etc., are used to 'line' the bushes when the bearings are especially of larger size.

Some commonly used types of journal bearings are:

1. Bushed bearing,
2. Plummer block,
3. Swivel bearing,
4. Axle bearing.

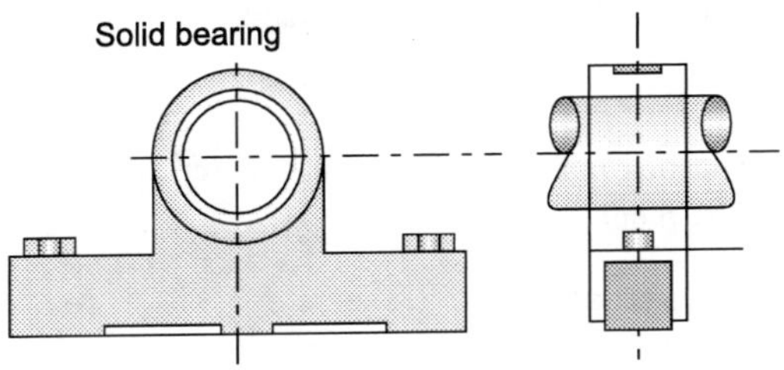

Fig. 5.2

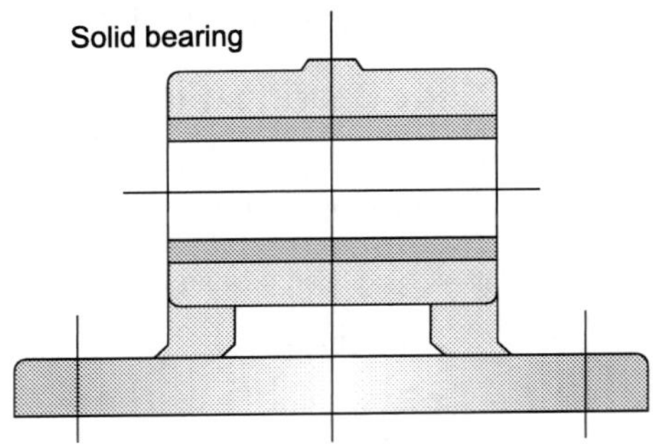

Fig. 5.2A

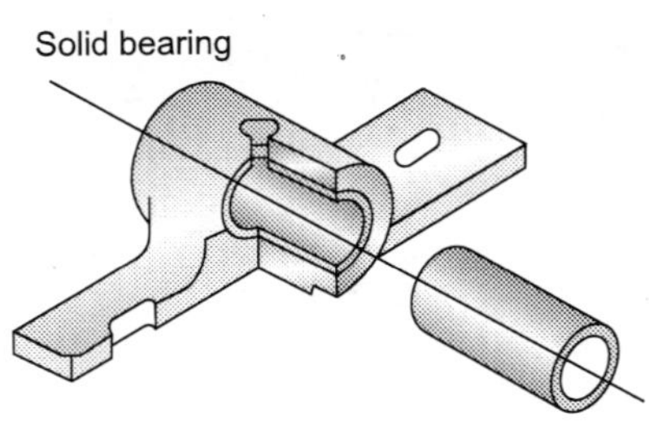

Fig. 5.2B

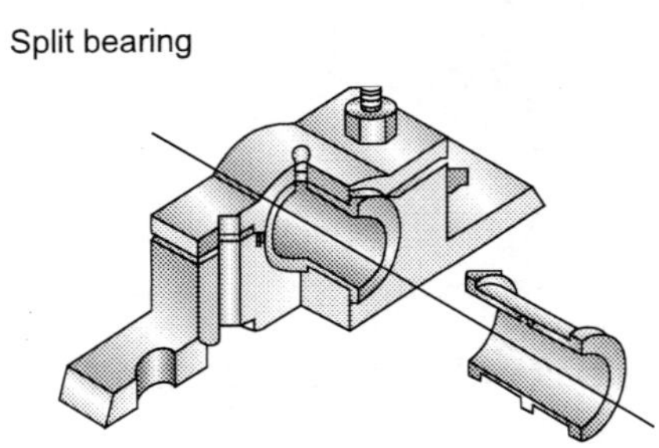

Fig. 5.2C

Bush Bearing

It consists of cast iron block, brass or gun metal solid bush. The block consists of a horizontal cylindrical part with an integral rectangular base plate. The base plate is provided with bolt holes for bolting down the bearing in position. The bolt holes are rectangular with semicircular ends. This type of large bolt holes facilitates proper alignment while bolting. The bottom of the base plate is relieved leaving a narrow machined strip all round. This bottom relief of the base plate requires machining of only narrow strip all round instead of the entire bottom surface and also reduces the extent of preparation and levelling of the bearing resting area relatively. The outer surface of the bush is finished to give a light press fit when fitted in the bearing block. The inner surface of the bush is finished to H7 tolerance. The shaft may be finished to give the required kind of clearance fit depending on the application. The main disadvantage of this type of bearing is the introduction and the removal of the bush pose maintenance difficulty.

This can be called the improved form of solid bearing. A bush made of brass or gun metal is used in this bearing. When the bush wears out, it can be easily changed.

These are made of bearing materials in the form of bush and are press-fitted in fabricated or cast iron housing.

Split Bearing

These bearings are made in halves and assembled in special plumber blocks.

Self-aligning Bush Bearings (Fig. 5.2 D)

In this type, the bearing bush is pressed into a special sleeve for self-aligning, in case slight angular misalignment or deflection due to the load between the bearing and the support points occurs.

Adjustable Slide Bearing (Fig. 5.2 E)

This type of bearing has provision for wear adjustment. The bearing is fitted in the tapered hole of the housing for adjustment of wear. The bearing is drawn inside by means of a nut.

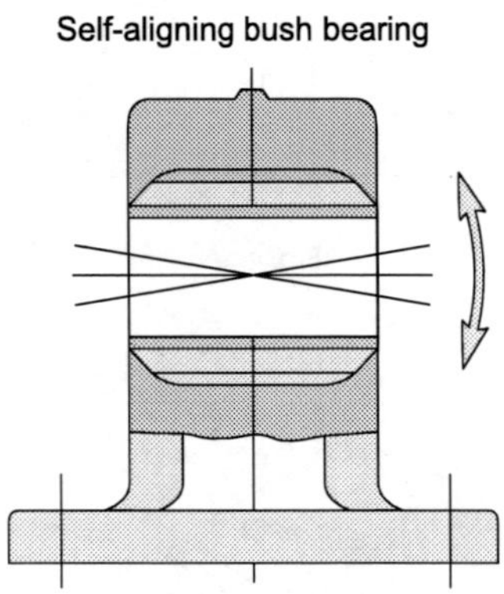

Fig. 5.2D

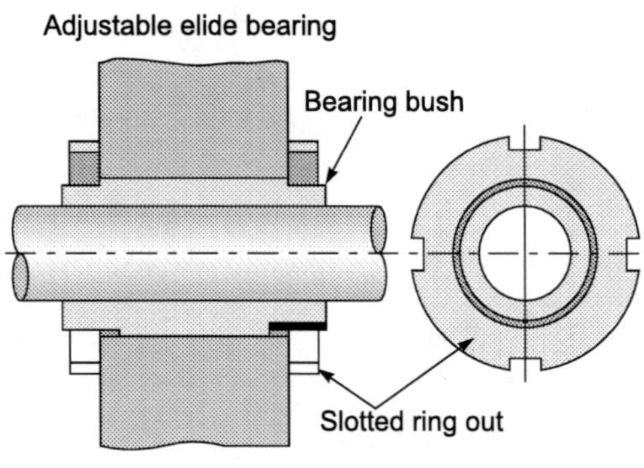

Fig. 5.2E

Plummer Block or Pedestal Bearing (Fig. 5.3)

When a long shaft is to be supported at intermediate points, bushed bearings are not used due to their limitations, in such cases Plummer Blocks are used. In the Plummer Block Bearing the body is split horizontally along or near the axis of the shaft. This will enable to overcome the disadvantage of the bushed bearing.

This consists of a pedestal called body, a cap, two split bushes, two square head bolts, and hexagonal nuts. The body, made of cast iron, is provided with two bolt holes in the base plate for bolting the bearing in position. These bolt holes will be rectangular with semicircular ends to enable alignment of the bearing to the shaft while bolting. The bottom of the body is relieved leaving a narrow machined strip all round. This relief at the bottom requires the machining of narrow strip all around instead of the entire bottom surface and also reduces the extent of ratio and levelling of the bearing area relatively. Another set of two long bolt holes with recess at their bottom ends are provided in the body. Two long square head bolts are introduced in these bolt holes such that their square heads are accommodated in the recesses. The split bush has a projecting snug. When this split bush is fitted in the body, the projecting snug of the split bush fits into the corresponding hole drilled in the body. This prevents the rotation of the bushes along with the shaft. In some of the Plummer blocks instead of providing a snug, the external surface of the ends of the split bushes are made semi-octagonal for a short length adjacent to the collars. Similar octagonal recesses are made in the body and the cap at the place where the octagonal faces of the split bushes take their seating. The axial movement of the split bushes is prevented by collars. The other split bush, provided with the oil hole is placed over the bottom bush. This bush is placed after the shaft takes its position. The cap mounted over the top is held down by bolts and nuts. After assembling, a small clearance will be left between the cap and the body. This clearance provides the necessary clamping grip.

This bearing is used for large size and fast moving mild steel shafts. Bushes of brass, bronze or gun metal or two pieces of bushes are used. These are fixed in stationary blocks of cast-iron and then fitted with nut and bolt.

These are generally called "brass". Provision is there for lubrication in these bearings. Provision is also made to see that bearing does not slip in left, right, ahead or backward directions.

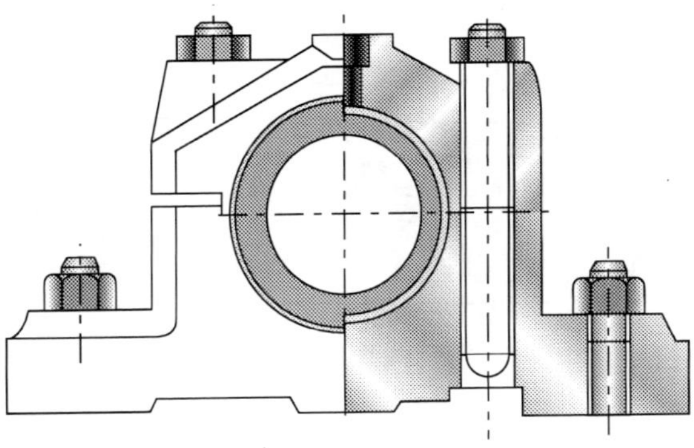

Fig. 5.3 Plumber blick

Foot-Step Bearing (Fig. 5.4)

It is also known as pivot bearing or two-bearing. These are used where there is need to provide support to the shaft in standing position. In short, it is used at the lower end of a vertical shaft or spindle to carry the end thrust. Bushes of brass, bronze or gun metal are fixed in the blocks of cast-iron to make use of this bearing. In order to protect the end of shaft, its face is slightly hollowed.

Thrust Bearing (Fig. 5.5)

Thrust bearings are used to support the shafts subjected to axial forces. The axial forces exert pressure, called end thrust on the supporting bearings, hence called "thrust bearings". The design of the thrust bearings depends on the location of the axis of the shaft. When the design permits a vertical rotating shaft subjected to axial forces can be supported with its lower end to rest on the bearing surface. This type of thrust bearing is called "footstep bearing". Suppose a vertical or horizontal rotating shaft subjected to axial forces acting in one direction, has to run beyond the bearing, is employed in which the collar formed on the shaft rests on the machined face of the supporting bearing. This type of the thrust bearing is called "collar bearing". When the thrust is more, or when the axial forces acts in both or either of the directions, number of collars are formed on the shaft, which run in the circular grooves formed in the bearing.

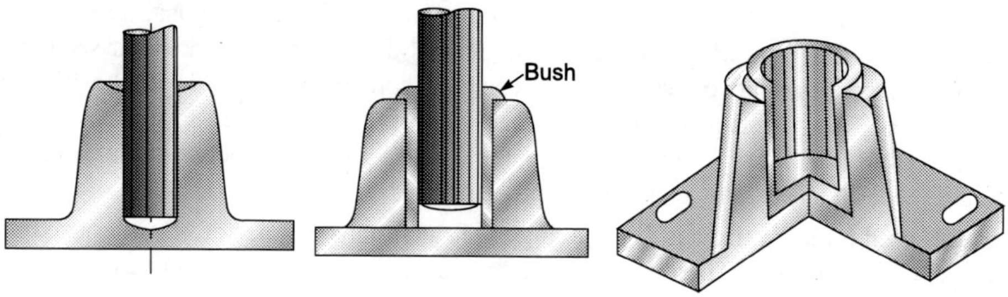

Fig. 5.4 Foot step bearing

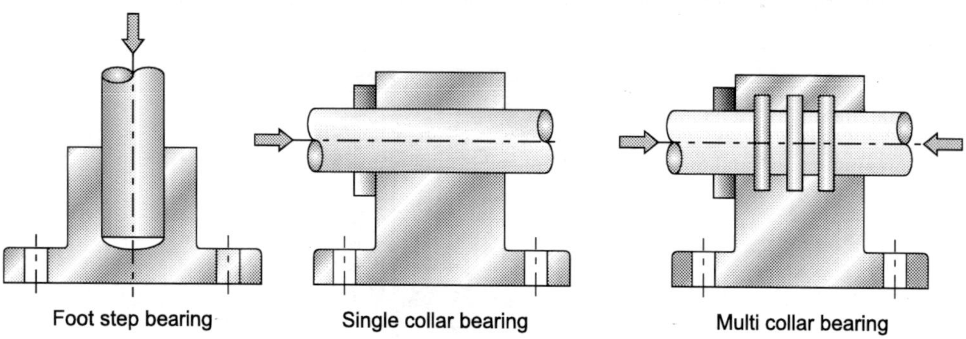

Fig. 5.5 Thrust bearing

This bearing is used to receive the longitudinal thrust or pressure of shaft. They absorb pressure through the collars made on the shaft. Its main advantage is that the shaft rotates at a fixed point.

5.3 BALL BEARING

Bearings facilitate the relative motion of surfaces in contact with limited consumption of energy:

The lower the friction, the lower the energy required.

Friction is the resistance to motion of two surfaces in contact. The rolling friction is lower than the gliding friction.

This perception led to the invention of the wheel.

Then with the beginning of the industrial era new ways were sought even to reduce the gliding friction of the motion of a wheel on its shaft.

The rolling bearing was borne with the idea to enclose round pieces between the hub and the shaft.

From the definition of friction it is clear that the resistance is greater when the surface area in contact greater. If the area could be lesser the frictional resistance can be lesser. The effort to reduce the area of contact is the rolling bearing.

The object of ball and roller bearings is the substitution of pure rolling motion for plain sliding friction.

Fig. 5.6 Taken from The SKF Basic bearing knowledge

Low Starting Friction

The outstanding feature of both ball and roller bearings is their low starting friction, which is practically the same as the running resistance. This renders them particularly well adapted for machinery which has to be frequently started from rest, when they undoubtedly save power. Again, owing to the low starting torque, a motor of less power can be used for a line shaft running in ball bearings.

What is a Rolling Bearing?

The rolling bearings operate on the basis of rolling friction.

Rolling bearings consist of the following main components:

- Rolling elements.
- Inner ring.

- Outer ring.
- Cage.

Rolling bearings are distinguished by the form of the rolling elements into ball and roller bearings.

Rolling Elements

Wide range of balls, cylindrical and needle rollers of carbon chromium steel, stainless steel and unalloyed quality steel (carbon steel) are used as rolling elements. Balls are even made of brass, bronze and aluminum.

Rolling elements are supplied in various classes of accuracy.

Rolling Elements: Designs

Balls

Balls are available from 0.4 to 320 mm diameter.

Material suffixes:

- Hardened stainless steel: RS, RS l, RS2.
- Hardened unalloyed quality steel: B 826, VX 306.
- Unhardened stainless steel: RJ, RJ 2.
- Bronze: LB.
- Brass: LM.
- Unhardened bearing steel (carbon chromium steel).

Cylindrical Rollers From

Rollers are available from 3 mm diameter and 5 mm length to 70 mm diameter and 100 mm length.

Material

Bearing steel (carbon chromium steel) with a hardness from 58 to 65 HRC.

Needle Rollers From

Availability is from 1.5 mm diameter and 5.8 mm length to 5 mm diameter and 49.8 mm length.

Material

Bearing steel (carbon chromium steel) with a hardness from 58 to 65 HRC

The low sectional height makes needle roller bearings particularly suitable for applications where radial space is limited.

These needle roller bearings are available for shaft diameters from 3 to 415 mm.

These are used in trucks (engines and transmissions), in domestic appliances (washing machines and kitchen machines), in agriculture (tractors and harvesters), in most machine tools (lathes, milling and other machines as well as hobby machines). The range of application includes all areas of mechanical engineering.

Balls made of hardened carbon chromium steel are supplied in 7 classes according to DIN 5401. Class I (highest precision) has a tolerance of 0.5 μm.

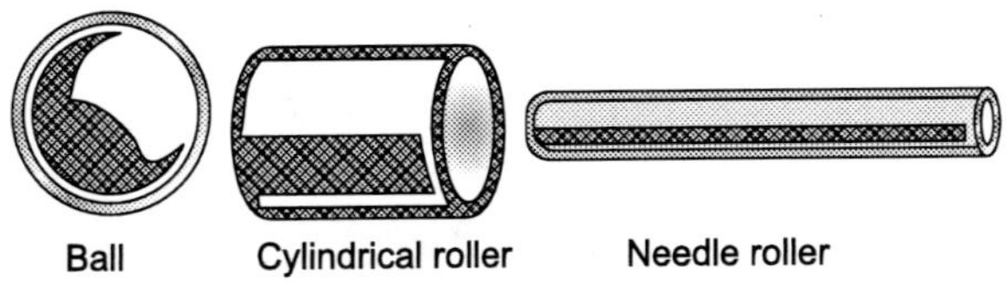

Fig. 5.7 Bearing elements

Well arranged designation systems make the identification easy: Prefix RB designates balls, RC designates cylindrical rollers and RN designates needle rollers.

Besides the prefix and the dimension indicator, the suffixes provide information about material, accuracy, tolerances and for cylindrical and needle rollers also the form of the rollers.

The rolling bearings operate on the basis of rolling friction.

Rolling bearings consist of the following main components:

1. Rolling elements.
2. Inner ring.
3. Outer ring.
4. Cage.

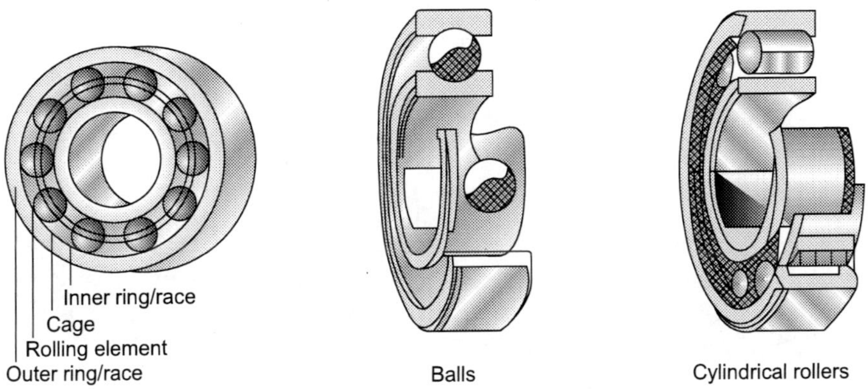

Fig. 5.8

5.4 PARTS OF BEARINGS

Races

The inner and outer races are provided with grooves or race-ways which guide the rolling elements. They are made of high grade chromium steel or chrome-nickel steel. They are hardened, ground and polished.

For any given bore diameter, there are usually two or three sizes of outside diameter width, and the load carrying capacity. The width of these bearings is smaller than the bore diameter. The width (or length) to diameter ratio is much smaller than that of plain bearings. Although principally they are to carry journal loads, the deep groove type of ball races are capable of withstanding the axial thrust.

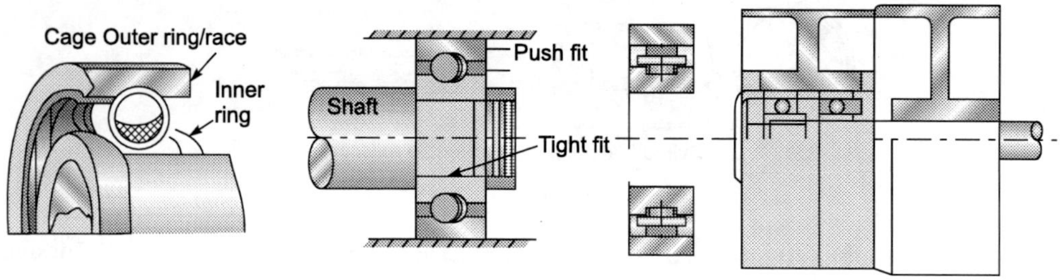

Fig. 5.9 Ball and roller bearings

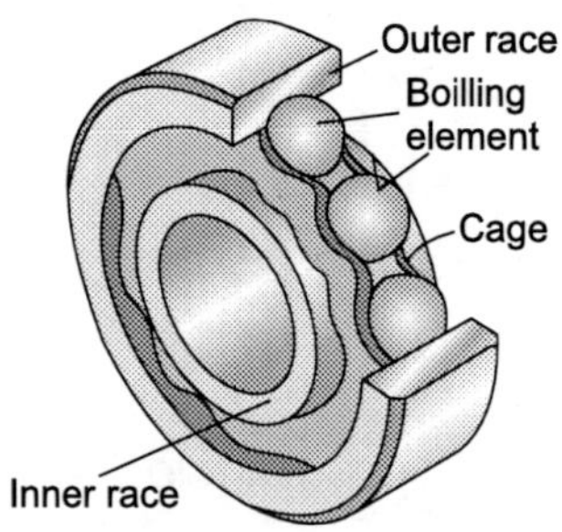

Fig. 5.10 Rolling elements, race cages

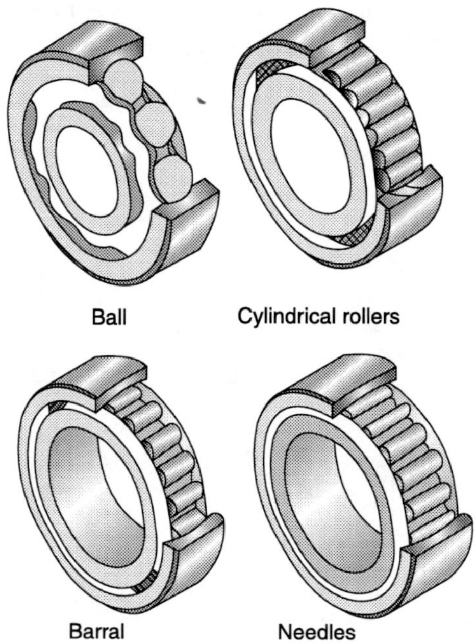

Fig. 5.11 Types of rolling elements

5.5 BEARING COMPONENTS AND MATERIAL

The rings and rolling elements of bearings are generally made from hardened high grade carbon chromium steel having a high degree of cleanliness.

For special applications hardened or even stainless steels may be used.

Each individual bearing receives optimum heat treatment, resulting in hardness normally between 59 and 65 HRC.

Pressed steel or brass cages are normally used for small and medium sized bearings.

Injection-moulded plastic cages are, however, used more and more as standard for small and medium sized bearings.

Large bearings generally have machined cages which are usually made of brass, but even steel. Spheroidal graphite cast iron and light alloys are used for machined cages.

Deep groove ball bearings are also manufactured with a seal or shield on one or both sides. Both shields and seals are secured in a groove in the outer ring and prevent foreign matter getting into the bearing.

Shields are made of steel sheet only and form a narrow gap with the inner ring.

Seals are made of steel sheet with rubber moulded to it and are in contact with the inner ring.

Seals protect bearings better than shields do, but the heat resulting from the friction between the seal lip and the inner ring will reduce the limiting speed by one third.

Bearings with two seals/shields are filled with the correct quantity of grease when they are assembled and do not require lubrication.

5.6 BEARING TYPES

Rolling bearings are distinguished by the form of the rolling elements into ball and roller bearings.

Roller bearings are further subdivided according to the form of the roller:

Radial Bearings

Deep groove ball bearings

Single row

- With shield(s) or seal(s)
- With snap ring groove in outer ring (and snap ring)

Double row

- Self-aligning ball bearings
- With cylindrical or tapered bore
- With seals
- With extended inner ring

Angular Contact Ball Bearings

Single row

- For paired mounting
- Precision bearings

Double row

- With shields or seals
- Four-point contact ball bearings

Cylindrical Roller Bearings

Single row

- NU type
- N type
- NUP type

Double row

- NNU type
- NN type

Four-row

- With cylindrical or tapered bore

Full complement cylindrical roller bearings

- Single row
- Double row
- With or without seals
- Multi-row

Crossed Cylindrical Roller Bearings

Needle roller bearings

- Drawn cup needle roller bearings with open and closed ends
- Needle roller bearings with flanges with or without inner ring with seal(s)
- Needle roller bearings without flanges with or without inner ring
- Needle roller and cage assemblies
- Combined needle roller bearings
- Alignment needle roller bearings

Spherical roller bearings

- With cylindrical or
- Tapered bore

Taper roller bearings

- Single row paired
- Double row
- Four-row

Crossed Taper Roller Bearings

Thrust bearings

Single direction

- With flat housing washer
- With spheric housing washer and
- Seating ring

Double direction

- With flat housing washers
- With spheric housing washers and seating rings (30)

Angular contact thrust ball bearings

- Single direction
- Double direction

Cylindrical Roller Thrust Bearings

Needle roller thrust bearings

Spherical roller thrust bearings

Taper roller thrust bearings

Single direction

Double direction

Y-Bearings

- With eccentric locking ring
- With inner ring extended at one side
- With inner ring extended at both sides with grub screw locking
- With adapter sleeve
- With normal inner ring

Cam Rollers

Narrow design

- With crowned runner surface

Wide design

- With crowned or cylindrical runner surface

Support rollers

- Without axial guidance
- With crowned or cylindrical runner surface
- With or without inner ring
- With or without seals

With axial guidance

- With crowned or cylindrical runner surface.
- Separable design
- Nonseparable design
- With cage
- Full complement

Cam followers

- With crowned or cylindrical runner surface with cage
- Full complement

Ball bearings are subdivided according to the most characteristic design features.

Roller bearings can generally carry higher loads than ball bearings which on the other hand have higher limiting speeds than roller bearings.

Rolling bearings are further subdivided into radial and thrust (axial) bearings. Bearings that are designed primarily to take up loads across (radial to) the shaft are called radial bearings. Bearings designed to take up loads that act mainly along the shaft are called thrust (axial) bearings.

Additionally to the radial bearings, the following thrust bearings exist:

- Thrust ball bearings,
- Angular contact,
- Thrust ball bearings,
- Cylindrical roller thrust bearings,
- Taper roller thrust bearings,
- Spherical roller thrust bearings and
- Needle roller thrust bearings.

Only thrust bearings are described with the word "thrust", i.e., if the description does not contain the word thrust, it is a radial bearing.

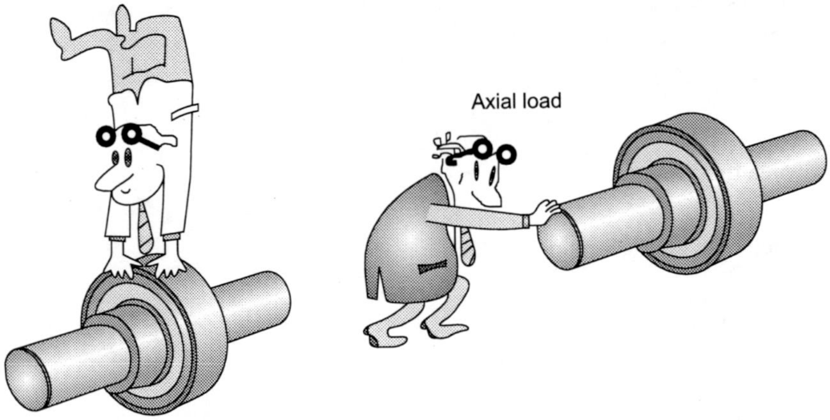

Fig. 5.12

Deep Groove Ball Bearings (Fig. 5.13)

Deep groove ball bearings are one of the most complete in the world. There are thousands of sizes and variants. Deep groove ball bearings are the most frequently used bearing types and can be found in almost all kinds of products in general mechanical engineering, in electrical machinery, air, sea and land vehicles, etc.

Miniature deep groove ball bearings with bore diameter of only a few millimetres are found for instance in instruments and computers while large sizes (over 420 mm outer diameter) are found for instance in rolling mills.

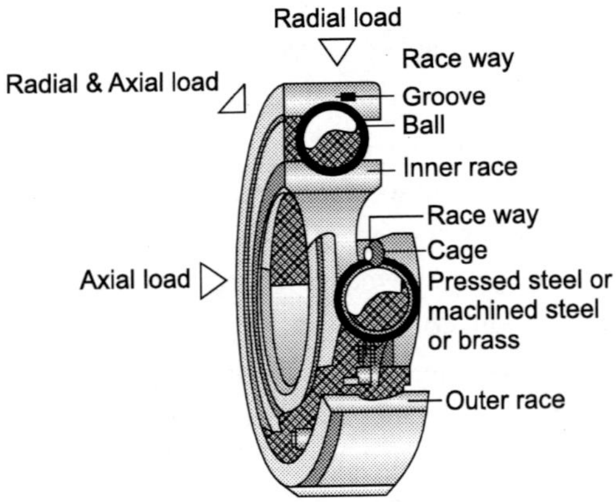

Fig. 5.13 Deep groove ball bearing

Deep Groove Ball Bearings: Designs (Fig. 5.13)

Deep Groove Ball Bearing with one or two shields for shaft diameters from 3 to 120 mm. High sealing effect, low friction, pre lubricated for high velocities, low noise, high temperature range – 30 to + 120°C.

Deep Groove Ball Bearing: with one or two seals for shaft diameters from 6 to 80 mm. High sealing effect also from splashes of water and oil. High stability, vibration resistant, high temperature range – 30 to + 120°C, low noise, pre-lubricated, maintenance free operation Deep groove ball bearings with shields or seals must never be washed.

Deep Groove Ball Bearing: with snap ring groove in outer ring for shaft diameters from 10 to 100 mm. A slit ring can be snapped into the groove to simplify the axial location of the bearing in the housing.

Double Row Deep Groove Ball Bearing: with filling slot for shaft diameters from 10 to 75 mm.

Application Alternatives:

1. If a deep groove ball bearing with normal clearance is not available, a bearing with C3 clearance can be used.
2. If a deep groove ball bearing with pressed steel cage is not available, a bearing with plastic or machined cage can be used.

Self-Aligning Ball Bearings

Self-aligning ball bearings is extraordinarily comprehensive and is not surpassed by any bearing supplier.

These permit minor angular misalignment of the shaft relative to the housing. They are therefore particularly suitable for applications where misalignment can arise from errors in mounting or shaft deflection.

Main fields of application: Agricultural machines, ventilators.

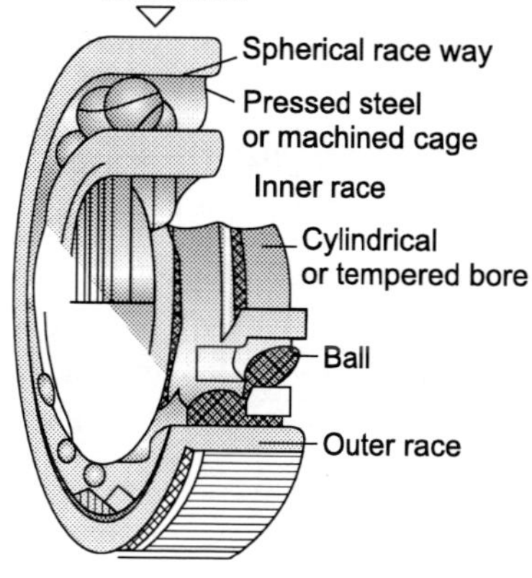

Fig. 5.14A Self-aligning ball bearing

Self-Aligning Ball Bearings–Designs

Self-aligning ball bearings with two seals are supplied filled with correct quantity of lithium base grease. Relubrication in service is not required and the bearings should not be washed or heated before mounting. Available for shaft diameters from 10-60 mm.

These are with tapered bore and adapter sleeve are available with or without seals. Because of the adapter sleeve, it is possible to use drawn shafts. These are available without seals for shaft diameters from 20-100 mm, and with seals for shaft diameters from 20-50 mm **Fig. (5.14 A).**

Self-aligning ball bearings with extended inner ring are used for simple applications with drawn shafts. They are very simple to mount and dismount. These are available for shaft diameters from 20-50 mm **Fig. (5.14 B).**

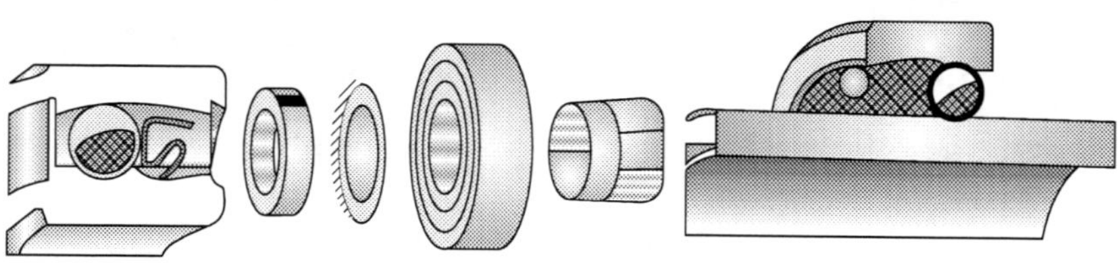

A B

Fig. 5.14 B Self-aligning ball bearing

Angular Contact Ball Bearings

Angular contact ball bearings are very comprehensive and unusually complete. In these ball bearings, the line of the action of the load (load line), at the contacts between balls and race ways, forms an angle with the bearing axis. These bearings are therefore particularly suitable for carrying combined (axial and radial) loads. They can operate at high speeds.

Single Row Angular Contact Ball Bearings are often mounted in pairs. The bearing arrangement can carry higher loads and operate under different load conditions.

Application Examples: Machine tool spindles, in gear boxes and passenger car wheels.

Angular Contact Ball Bearings–Designs

- Angular Contact Ball Bearing, G-Design (Single Row)
- Especially available for mounting in pairs, for shaft diameters from 17-170 mm.

Angular Contact Ball Bearing, Double Row

- With one-piece inner ring and filling slot for shaft diameters from 10-110 mm.
- With split inner ring (D-Design) for shaft diameters from 25 to 70 mm.

Four-Point Contact Ball Bearing

- Single row with split inner ring for shaft diameters from 17-150 mm particularly suitable for high axial loads.

Cylindrical Roller Bearings

Cylindrical roller bearings have high load carrying capacity. They can accommodate heavy radial loads and permit high speed operation.

These are used in most branches of mechanical engineering, particularly in transmissions—both stationary and mobile. From a few millimetre shaft diameter, they are equally common as large size bearings (from 420 mm outer diameter) in rolling mills or in bearing units for rail vehicles.

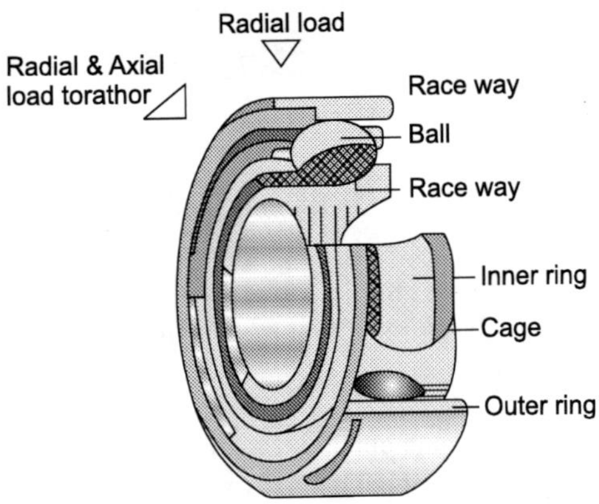

Fig. 5.15 Angular contact ball bearing

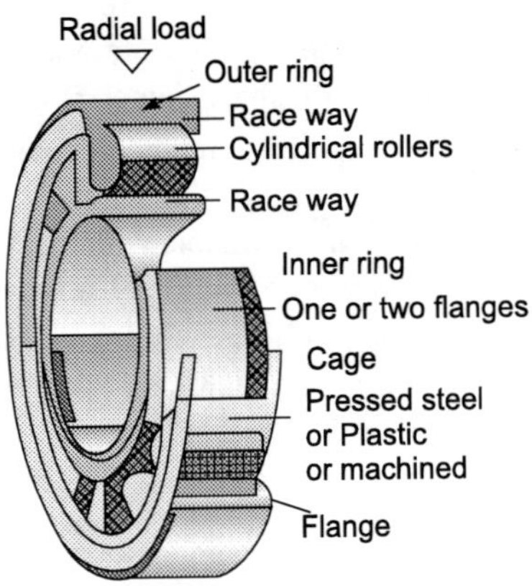

Fig. 5.16 Cylindrical roller bearing

Cylindrical Roller Bearings–Designs

- NU bearings permit axial displacement of the housing relative to the shaft within certain limits in both directions and are therefore used as non-locating bearings (axial tolerance compensation).
- NJ bearings can be used for axial location of the shaft in one direction.
- RNU bearings require hardened and ground shaft as inner rings.
- NUP bearings have one integral and one loose flange on the inner ring and can therefore be used to locate the shaft in both directions.

Double Row Cylindrical Roller

Bearings are mainly used in machine tools (spindles).

- SU bearings which are relatively new can absorb high axial loads due to the spheric roller front and the tapered flanges.
- Full complement cylindrical roller bearings have extremely high load carrying capacity and permit space-saving designs. The absence of a cage results in reduced speed capacity compared to conventional (cage) type cylindrical roller bearings.

Spherical Roller Bearings (Fig. 5.17)

Spherical roller bearings are self-aligning and insensitive to shaft bending and minor angular misalignment between the shaft and the housing. Additionally to radial loads, the bearings can accommodate axial loads in both directions.

Major application areas: Many sectors of general and heavy engineering, stationary transmissions and bearing units for railway axle boxes.

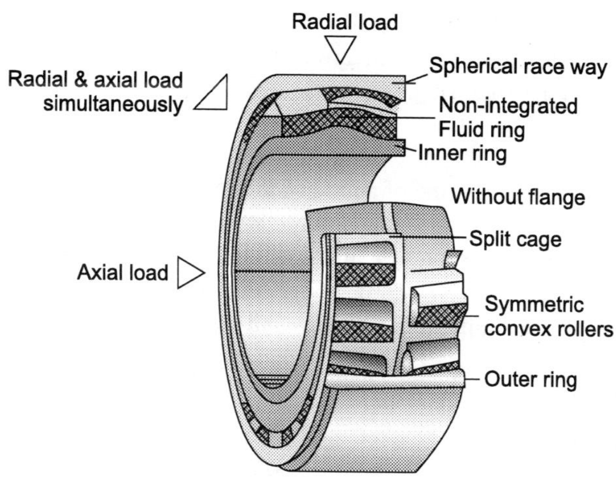

Fig. 5.17 Spherical roller bearing

Spherical Roller Bearings–Designs

- CC-design with cylindrical and tapered bore.

Available for shaft diameters from 20-380 mm.

- CA-design with retaining flanges on the inner ring.

Available with cylindrical and tapered bore for shaft diameters from 200-1180 mm.

Design Characteristics

- Spherical roller bearings with adapter sleeve have tapered bore. They are very easy to mount. Available in CC-design for shaft diameters from 30-300 mm and in CA-design for shaft diameters from 240-450 mm.
- Spherical roller bearings with withdrawal sleeve also have tapered bore. They are particularly easy to dismount. Simply screw a nut on the threaded section of the sleeve and tighten against the bearing until the bearing comes free.

Available in CC-design for shaft diameters from 35-300 mm and in CA-design for shaft diameters from 240-380 mm.

- CAB-design with drilled rollers and pin-type cage.

Available for shaft diameters from 320 to 1120 mm.

Particulars

1. Special spherical roller bearing that can absorb shock loads occurring in vibratory compactors. The cage design is the main difference from a standard bearing. Cage and guide ring is one integral piece. Such bearings are identified by the suffix JA/W33AI5.

   Available for shaft diameters from 40-200 mm.
2. Bearings for bore diameters above 100 mm are delivered as W 33 execution. They have an oil groove and lubricating holes in the outer ring.

Taper Roller Bearings

As for angular contact ball bearings, the line of action of the resultant load forms an angle with the bearing axis. Taper roller bearings are therefore particularly suitable for carrying combined (radial and axial) loads. Bearings with higher contact angle can support higher axial loads.

Main application areas are wheel bearings for passenger cars and trucks, propeller shafts and differentials, mobile and stationary transmissions, railroad axle boxes and as large size bearings in rolling mills.

Design Characteristics

Taper Roller Bearings–Designs

Single row taper roller bearings are available (both in metric and inch series) for shaft diameters from 15-280 mm.

Multi-row taper roller bearings with two, four or six rows are mainly used in heavy engineering applications such as rolling mills. The bearings are only made to order.

Particulars

Single row taper roller bearings can carry axial loads in one direction only. Since the radial load on a Taper roller bearing gives rise to an induced axial load which must be compensated, these bearings therefore are generally mounted against a second bearing - often another taper roller bearing.

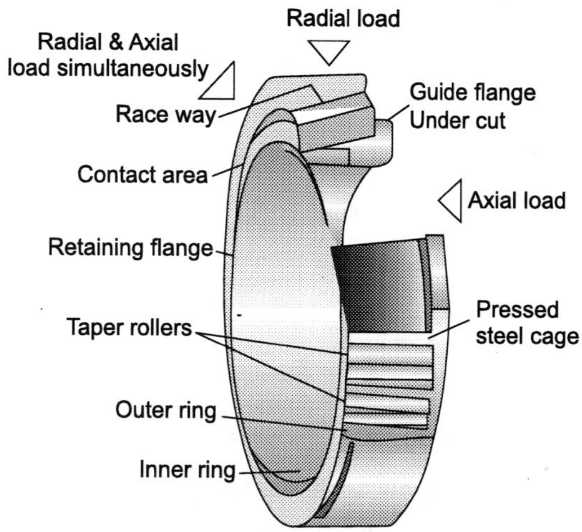

Fig. 5.18 Taper roller bearing

Needle Roller Bearings

The low sectional height makes needle roller bearings particularly suitable for applications where radial space is limited.

These bearings are available for shaft diameters from 3 to 415 mm.

These are used in trucks (engines and transmissions), in domestic appliances (washing machines and kitchen machines), in agriculture (tractors and harvesters), in most machine tools (lathes, milling machines as well as hobby machines). The range of application includes all areas of mechanical engineering.

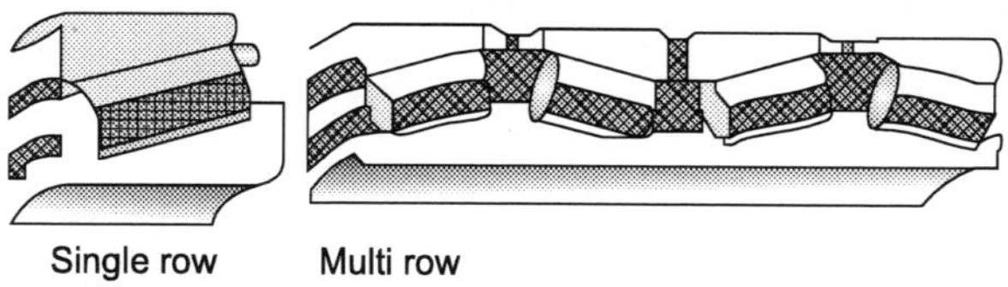

Fig. 5.19 Taper roller bearing

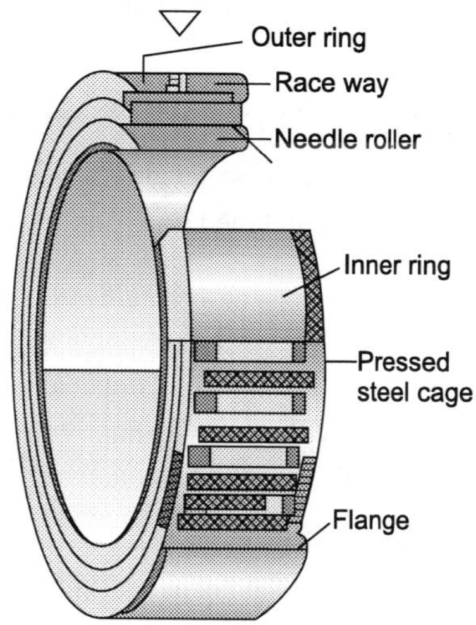

Fig. 5.20 Needle roller bearing

Needle Roller Bearings–Designs

Needle roller bearings—single row or double row, **Fig. (5.21 A)** with or without inner ring, with or without flanges in the outer ring, with or without rubbing contact seals. Used for radial loads only with limited radial mounting space.

Drawn Cup Needle Roller Bearings (Fig. 5.21 B)

Open or closed end.

- Single row or double row, with or without rubbing contact seals with drawn cup, very thin outer ring.

Used for radial loads only for bearing application with very limited radial mounting space.

Needle Roller and Cage Assemblies (Fig. 5.21 C)

- Single row or double row.

Used for radial loads only with minimal radial mounting space.

Needle Roller Thrust Bearings (Fig. 5.21 D)

- With or without bearing washers of steel or steel plate.

Used for axial (thrust) loads only with very limited axial mounting space.

Needle Roller/Ball Bearings (Fig. 5.21 E)

- Combination of a needle roller bearing and an angular contact ball bearing.

Used for combined (radial and axial) loads for fixed bearing applications with very limited radial dial mounting space.

Needle Roller/Thrust Ball Bearings (Fig. 5.21 F)

- Combination of a needle roller bearing and a thrust ball bearing with or without seal.

Used for high combined (radial and axial) loads for fixed bearing applications with limited radial mounting space.

Support Rollers (Fig. 5.21 G)

- With or without axial guidance of the outer ring,
- With or without inner ring, with spherical or cylindrical outer diameter. Ready to mount in all types of cam drives, tracks, rocker arms, etc.

Cam Rollers (Fig. 5.21 H)

- With spherical or cylindrical outer diameter.

Ready to mount in all types of cam drives, tracks, rocker arms, etc.

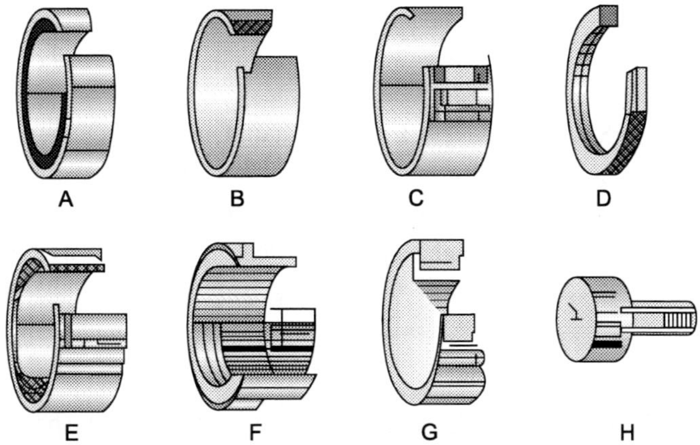

Fig. 5.21 Needle roller bearings

Linear Ball Bearings

These permit unlimited oscillating linear motion and allow particularly simple and economical arrangements to be designed for accurate and precise linear guidance.

Eleven different sizes and three different designs are available.

The complete linear guides for a wide variety of applications can be simply assembled. The accessories include housing of various designs, shaft blocks, shaft supports and precision steel shafts.

Design Characteristics

Linear Ball Bearings–Designs

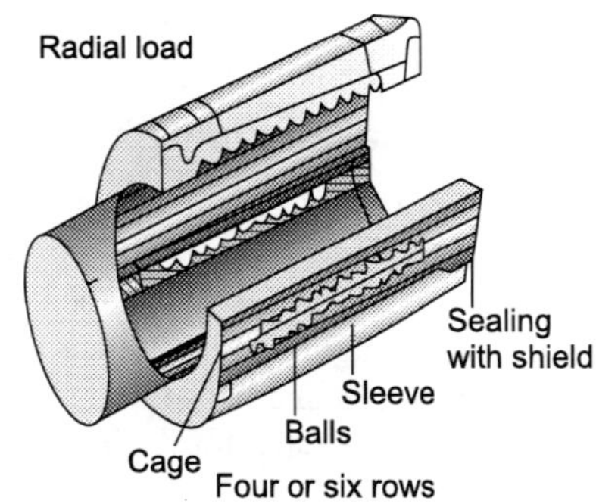

Fig. 5.22 Linear ball bearing

- Linear ball bearings with two shields for shaft diameters from 5 to 80 mm
- High load carrying capacity, rigidity, accuracy, speed and acceleration, low friction and noise. Temperature range from –20° C to + 100° C
- Linear ball bearings with one or two seals for shaft diameters from 5 to 80 mm
- High sealing and wiping action. Maintenance free operation since one grease fill is normally adequate for the life of the bearing. Temperature range is from – 20°C to + 100°C.
- Linear ball bearings with open sleeve for shaft diameters from 12 to 80 mm are recommended for very long arrangements with heavy loads. Open linear ball bearings allow the shaft to be supported either at intervals or along its entire length. In this way, an arrangement having great accuracy and high stiffness can be obtained.

- Linear ball bearings with closed sleeve (type LBAR) for shaft diameters from 5 to 80 mm have very high guiding accuracy due to the considerably closer tolerances on the inscribed diameter of the ball sets than the other types.
- Linear ball bearings with slotted sleeve (LBAS) for shaft diameters from 5 to 80 mm.

When using slotted, adjustable housing, it is possible to adjust the operational clearance as required.

- Compact linear ball bearings for shaft diameters from 12 to 50 mm, with 5 to 9 rows of balls.

Main advantages are space saving, low weight, adjustable working clearance. The sleeve is made of drawn thin-walled hardened steel. The special flexible interlocking feature allows individual adjustment of the working clearance.

Internal Clearance

The bearing internal clearance is defined as the total distance through which one bearing ring can be moved relative to the other under zero measuring loads. Movement in the radial direction is called radial internal clearance; axial movement, axial internal clearance. The internal clearance is standardized.

Rolling bearings are with different internal clearance. The internal clearance will vary from bearing type to bearing type. The most common clearance is normal (with no suffix).

Internal clearance smaller than normal is designated by suffixes C 1 and C 2 while internal clearances larger than normal are designated C 3, C 4 and C 5.

Suffix	**Internal clearance**
C 1	Smaller than C 2
C 2	Smaller than normal
	Normal
C 3	Greater than normal
C 4	Greater than C3
C 5	Greater than C 4

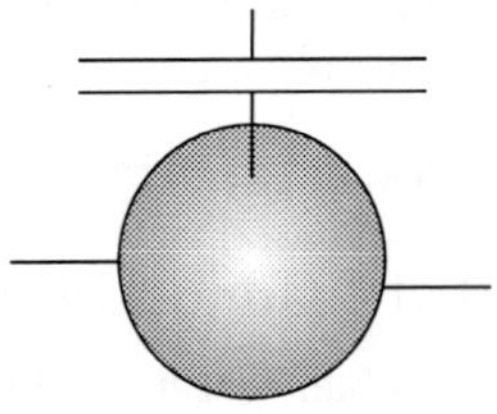

Fig. 5.23 *Internal clearance*

Why Internal Clearance?

1. In order to fit a bearing tightly on a shaft, it is necessary that the shaft is slightly larger than the bore of the inner ring. When the bearing is mounted, the inner ring will be expanded and consequently the space available for the rolling elements will be reduced. (Sometimes the housing/seating are made slightly smaller than the outside diameter of the outer ring resulting in compression of the outer ring and hence further reduction of available space for the rolling elements). The reduction in space available for rolling elements when the bearing is mounted is accommodated by the internal clearance.
2. Steel will expand with increased temperature. In operation the temperature of the various parts of a bearing will not remain uniform and resulting in varying degrees of expansion which is compensated by the internal clearance.

The internal clearance required is dependent upon the mounting fit and the operating conditions. In case of a bearing with greater internal clearance shall always be selected.

Bearing Tolerances

The dimensional and running tolerances for bearings are standardized and divided into different classes of accuracy by ISO (the International Organization for Standardization). In addition to normal tolerances, the ISO standards cover closer tolerances e.g. classes 6 and 5 (corresponding tolerances are designated P6 and P5). For specific applications, such as machine tool spindles, bearings of even greater accuracy (tolerance classes P4, SP and UP).

Standard Normal	High precision
P 6	
P 5	**SP** (special precision)
P 4	**UP** (ultra precision)

Bearing Designations

The bearing designation will describe bearing type, size and deviations from normal execution.

The first digit (or letter) designates the bearing type, the second and third digits indicate the dimension series to which the bearing belongs. The second digit represents the width (or height) series and the third digit the diameter series. The three first digits together designate the bearing series. The two last digits indicate the bore diameter. For bearings with bore diameter between 20 and 490 mm the bearing bore diameter in mm is obtained by multiplying the two last digits by five. For bearings with bore diameter greater than 490 mm the bore diameter is directly indicated after an oblique (Example: 511/500). Special coding applies for bearings with bore diameter less than 20 mm.

Supplementary designations are used to identify bearings which deviate in some ways from the normal design. They may precede the bearing designation (prefixes to identify bearing components) or they may succeed the bearing designation (suffixes to identify internal clearance, tolerances, cage variants, etc.).

The following figure show the composition of bearing designations.

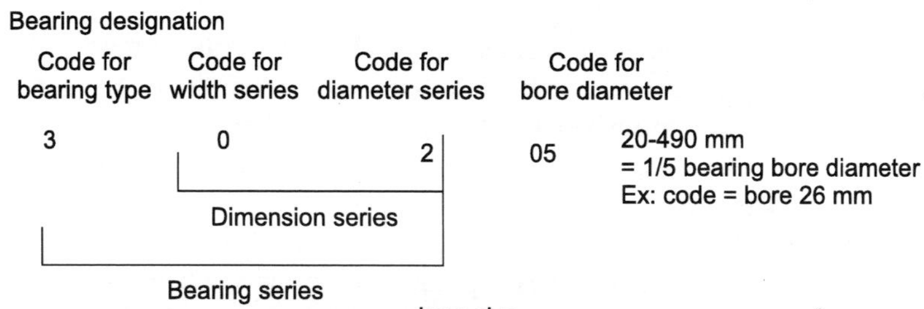

Between 20 to 490 bore code = $\frac{\text{bore size}}{6}$ or Multiply size code by 5.

When bore size is below 20 mm the size digit is not logic

Bore diameter	Size code
10 mm	00
12 mm	01
15 mm	02
17 mm	03

For bearings with bores greater than 490 mm the bore diameter is used, e.g., 630 mm bore would be 630

Fig. 5.24

Load Carrying Capacity and Life

The size of a bearing to be used for an application is initially selected on the basis of its load carrying capacity in relation to the loads to be carried and the requirements regarding life and reliability. Numerical values termed basic load ratings are used in the calculations to express load carrying capacity. Values for the basic dynamic load rating C and the basic static load rating Co are quoted in the bearing tables.

Basic Load Ratings

The basic dynamic load rating C is used for calculations involving dynamically stressed bearings, i.e. when selecting a bearing which is to rotate under load. It expresses the bearing load which will give an ISO basic rating life (defined below) of 10,00,000 revolutions.

The basic dynamic load ratings of bearings have been determined in accordance with the methods prescribed by ISO 281/1-1977 and modified to follow the latest ISO proposals. The values are based on the material and manufacturing techniques used for standard production. They apply to roads which are constant both in magnitude and direction, for radial bearings radial loads, and for thrust bearings axial loads which act centrically.

The basic static load rating Co is used in calculations when the bearings are to rotate at very slow speeds, are to be subjected to very slow oscillating movements, or are to be stationary under load during certain periods. It must also be taken into account when heavy shock loads of short duration act on a rotating (dynamically stressed) bearing.

The basic static load rating is defined in accordance with ISO 76-1987 as the static load which corresponds to a calculated contact stress at the centre of the most heavily loaded rolling element/race way contact.

This stress produces a total permanent deformation of rolling element and race way which is approximately 0,0001 of the rolling element diameter. The loads are purely radial for radial bearings and centrically acting axial loads for thrust bearings.

Life

The life of a rolling bearing is defined as the number of revolutions (or the number of operating hours at a given constant speed) which the bearing is capable of enduring before the first sign of fatigue (flaking, spalling) occurs on one of its rings or rolling elements.

It is, however, evident from both laboratory tests and practical experience that seemingly identical bearings operating under identical conditions have different lives. A clearer definition of the term "life" is therefore essential for the calculation of bearing size. All information presented is based on the life that 90 % of a sufficiently large group of apparently identical bearings can be expected to attain or exceed. This is called the basic rating life and agrees with the ISO definition. The median life is approximately five times the calculated basic rating life.

There are several other bearing "lives".

One of these is the "service life", which is the actual life achieved by a specific bearing before it fails. Failure is not generally by fatigue in the first instance but by wear, corrosion, seal failure etc. Another is "specification life". This is the life specified by an authority and based on hypothetical load and speed data supplied by the same authority. It is generally a requisite L 10 (basic rating life), and it is assumed that the authority has related the specification to experience gained with similar machinery, so that adequate service life will be obtained.

Selecting Bearing Size Using the Life Equations

Bearing life can be calculated with various degrees of sophistication depending on the accuracy with which the operating conditions can be defined.

Basic Rating Life Equation

The simplest method of life calculation is to use the ISO equation for basic rating life which is

$$L_{10} = (C/P)^p \quad \text{or} \quad C/P = L_{10}\ 1/p$$

where L_{10} = basic rating life, millions of revolutions

C = basic dynamic load rating, N

P = equivalent dynamic bearing load, N

p = exponent of the life equation $p = 3$ for ball bearings

p = 10/3 for roller bearings

The related values of load ratio C/ P and life L 10 are set out in the life calculation chart.

For bearings operating at constant speed it may be more convenient to deal with a basic rating life expressed in operating hours using the equation

$$L10h = 1\,00.0\,000/60\ n\ (C/P)^p \quad \text{or} \quad L_{10n} = 1000\,000\ /\ 60\ n\ L_{10}$$

where L_{10n} = basic rating, operating hours

n = rotational speed, r/min

Values of L10h as a function of load ratio

C/P and speed n can be found.

5.7 MAINTENANCE PRODUCTS — MOUNTING AND DISMOUNTING

For a bearing to function properly, it is important that the correct mounting and dismounting method and suitable tools are used. There are three types of mounting and dismounting tools: mechanical, hydraulic and heaters.

Handling Bearings

The working surfaces of the bearings are either honed or very soft; if these surfaces are damaged, the bearing is ruined, therefore:

- Hendee bearings carefully to prevent damage.
- Keep bearings wrapped until fitted, to keep out dirt.
- Protect bearings against corrosion during storage, e. g., steel bearings must be oiled.

Installing Rings

Before any bearing is fitted:

1. Clean the journal or housing thoroughly and the seating of the locating devices.
2. Inspect the surfaces for damage; do not fit bearings to damaged surfaces.

Plain Bearings

These are shells or bushes in which shafts run.

They are of different materials from the shaft, and can be either solid or porous metal.

Note

Porous metal bushes require special handling and fitting, consult manufacturer's manual.

Fit a bush using an arbor press.

Note

A shouldered mandrel must be used, when a bush is fitted, to prevent deformation, use the correct size of mandrel.

1. Clean the housing, the mandrel and the base of the arbor press thoroughly.
2. Coat the housing and the mandrel with light oil. Care must be taken to keep the oil away from the bush and control gear of the motor.
3. Index the base of the arbor press to give the most suitable hole size.
4. Fit the bush on to the mandrel.
5. Align the bush with the housing and lower the ram on to the mandrel so that the ram holds the bush in position, above the housing.
6. Check that the bush in square to the hole.
7. Gently press the bush home with the lever.
8. After fitting, check the internal dimensions.

Note

Some bushes are intended to be reamed after fitting — ordinary hand held reamers will not normally achieve desired alignment – check manufacturer's manual.

Some bushes are secured by locking compounds and only a light push fit is required.

Ball and Roller Bearings

These bearings have balls or rollers, which prevent sliding friction by rolling between the races.

As bearings are used between stationary and revolving machine parts, such bearings have a stationary and revolving race.

Observe whether the outer race is stationary with a revolving shaft, or vice versa.

Normally, the revolving race is a press fit, so it is fitted to the machine part, e.g., a rotor first and the stationary race is light push fit, so, it is assembled with the machine part. Bearings are made of hard, material. Therefore, when force is applied to fit them coat the journal or housing with clean oil. Take care to keep the oil clear of commutator, brushes and control gear of motor.

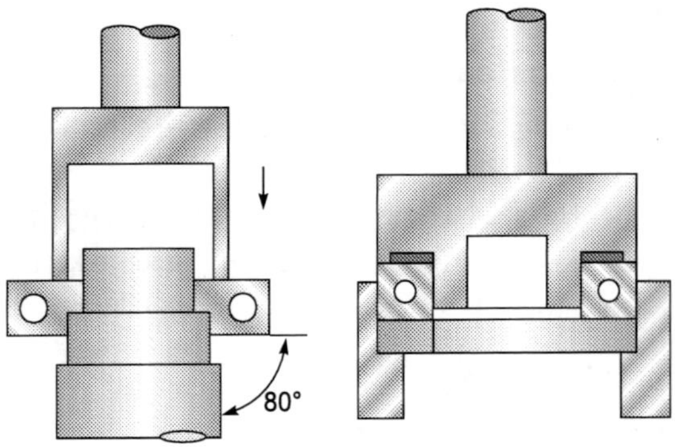

Fig. 5.25A Mountain of bearings

Follow the same procedure as for fitting plain bearings. Force must only be applied to the race in contact with the housing or shaft.

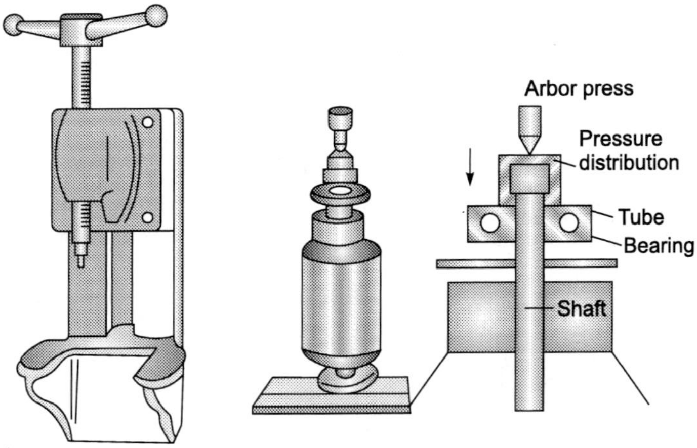

Fig. 5.25 B Using arbor press

Fitting a Bearing Using an Arbor Press

Note

Whenever possible, use an arbor press because it keeps the bearing square to the shaft or housing.

Decide which is the best way to set the job up on the press.

Pressing the Bearing on to the Shaft

Use a table to· distribute the force.

Evenly· keeping the bearing square to the shaft. The tube must only touch the inner race and its ends must be square.

Pressing Shaft into a Bearing

The bearing must be supported by the inner race only.

Pressing a Bearing into Housing

A tube or block must be used so that the force is applied to the outer race only.

Tapping Bearings into Place

Note

Bearings should be only tapped into place when they.cannot be pressed into position.

Decide which is the most appropriate method.

- Using the tube and the skirting block.

This is the best method, since the bearings can more easily be kept square to its seating.

- Using a drift.

Tap evenly around the race being fitted, take care to keep bearing square to seating. The method is useful when seating is in awkward situation. Care to prevent foreign matter from entering the bearing.

- Align the bearing with the seating and start to fit it by hand. A soft mallet may be used if necessary. Never under any circumstance strike any part of a bearing with hard faced hammer
- Tap the bearing home gently. Stopping frequently to check that it is square.

Thrust Races

Thrust bearings are ball or roller bearings specially designed to take end load only.

The revolving race should be a push fit into the housing or on to the shaft, and must rest square against the shoulder. The cage and stationary race should be clear of both the shaft and the housing.

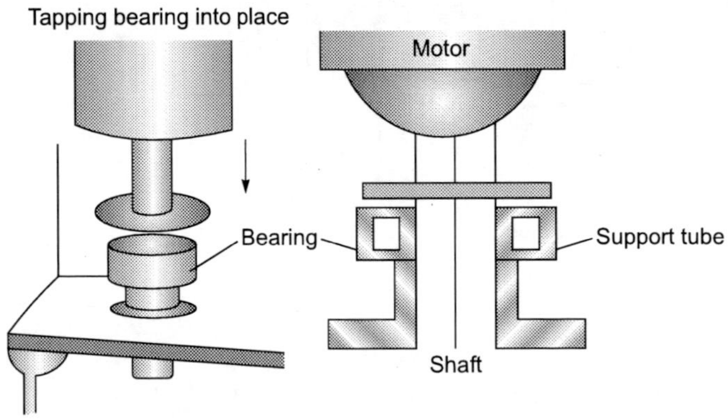

Fig. 5.25 C

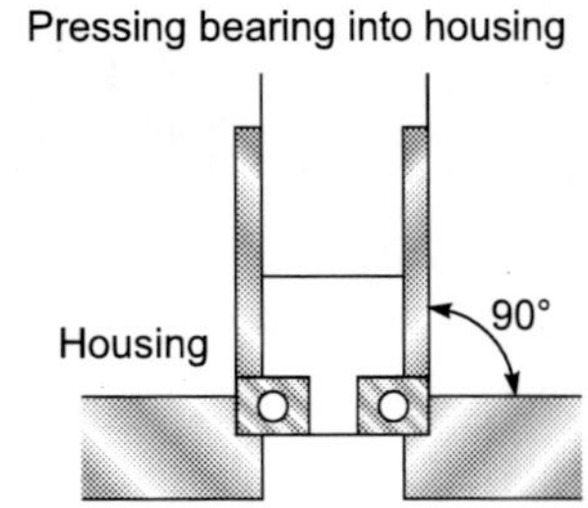

Fig. 5.25 D

The revolving race is, therefore, fitted alone, and the stationary race and case assembled to it.

Fitting the Revolving Race (Fig. 5.26)

Use a dial test indicator to check that the shoulder is square to the shaft.

Fit the race by hand. Press or tap the race home against the shoulder to be sure that it is seating properly.

Note

When fitting huge size bearings, it is an advantage to carry out the assembly with the shaft axis vertical. This enables the race tracks and cage and balls to be maintained concentric without difficulty. If assembled with the shaft horizontal, the cage and the balls may drop and become eccentric to the tracks so that they would be nipped on application of the working load.

As soon as possible, after fitting the bearings should be charged with appropriate lubricant and housing end covers fixed in position.

Installing Tapered Bore Bearings

Most bearings have a parallel bore and fit on to a parallel shaft, but some bearings have a tapered bore and fit on to a tapered sleeve.

When a tapered bore bearing is driven up a taper, the inner ring is stretched to reduce the internal clearance of the bearing to the required dimensions.

These bearings can be fitted to a tapered shaft or on to a tapered sleeve which on to a parallel shaft.

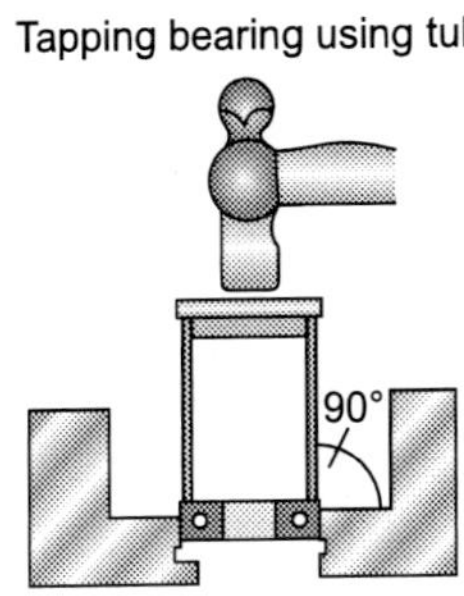

Fig. 5.26 A

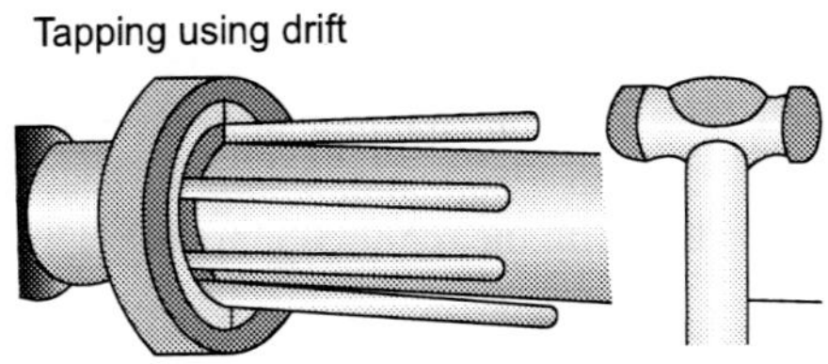

Fig. 5.26 B

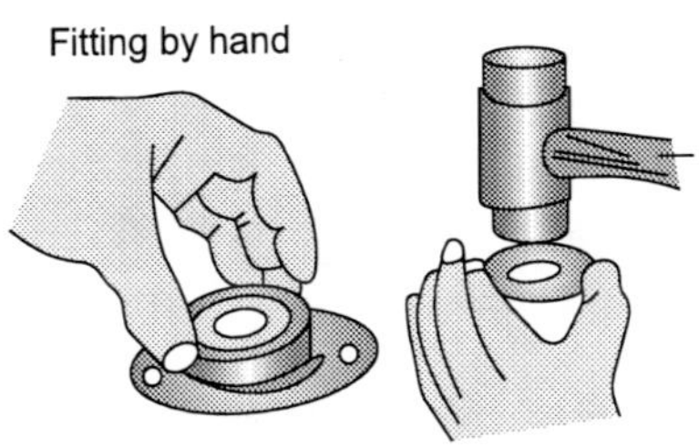

Fig. 5.26 C

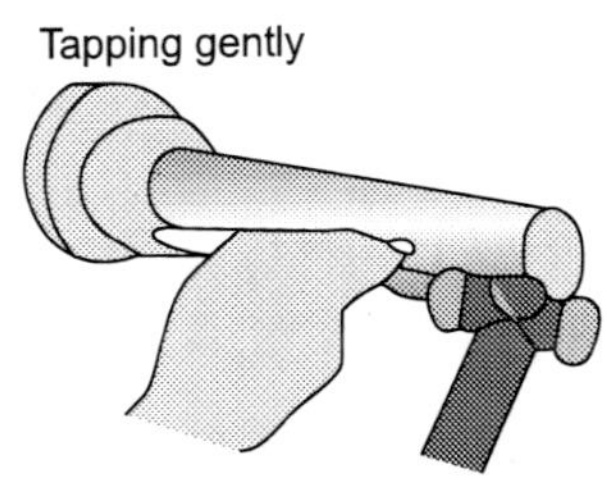

Fig. 5.26 D

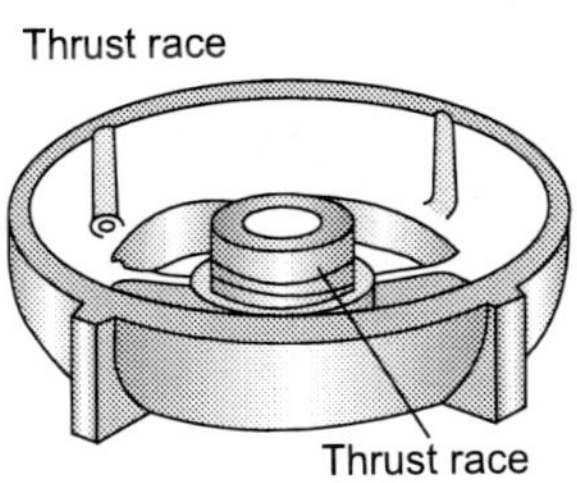

Fig. 5.27 A

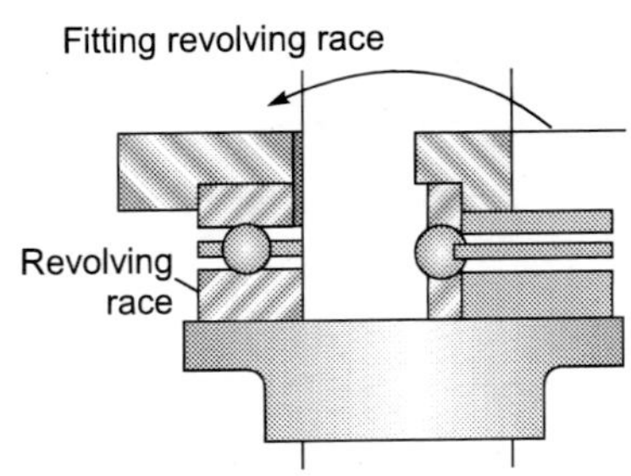

Fig. 5.27 B

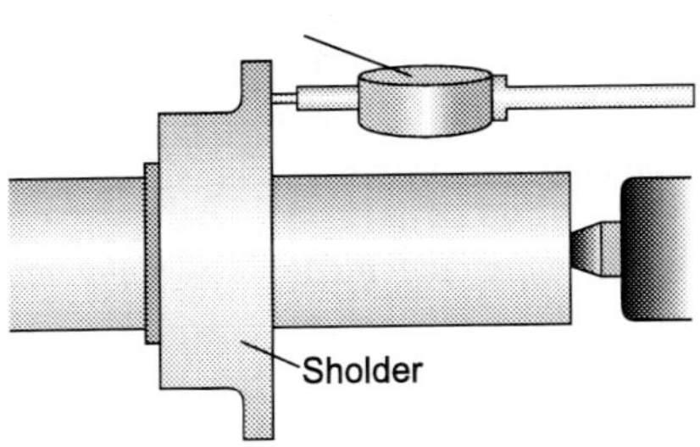

Fig. 5.27 C Dial test indicator

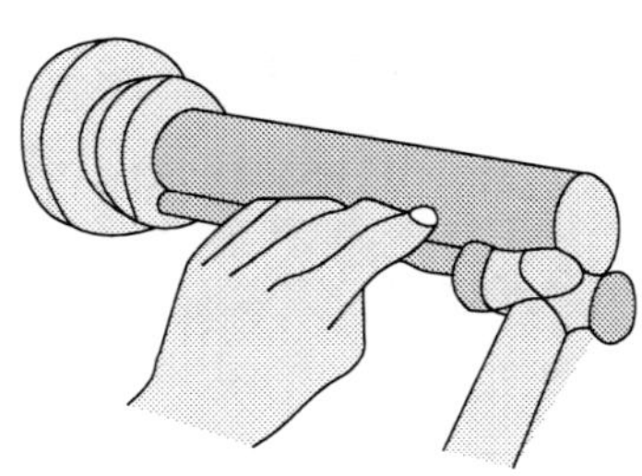

Fig. 5.27 D Tapping against shoulder

Tapered sleeves can be

- Threaded at the narrow end to aid mounting or
- Threaded at the wide end to aid withdrawal.

When a tapered bore bearing must be accurately positioned along a shaft and it is located only in a tapered sleeve, the sleeve must be positioned on the shaft so that the bearing is in the place when the internal clearance has been reduced by the required amount.

Methods of Obtaining Correct Clearance (Fig. 5.29)

The manufacturers indicate the reduction of required amount of clearances and also the axial drive up distance required to obtain the fit.

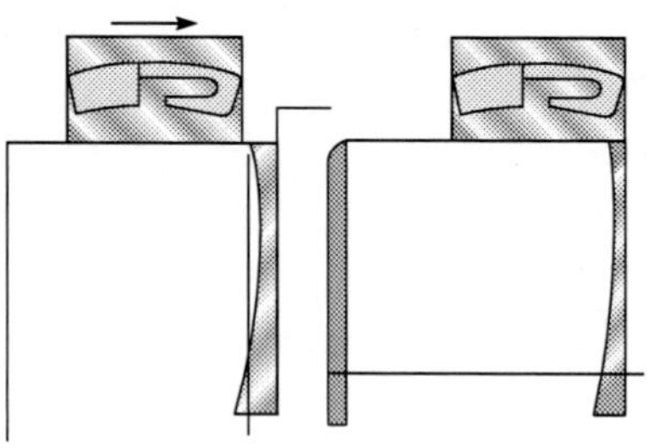

Fig. 5.28 Tapered bore bearing

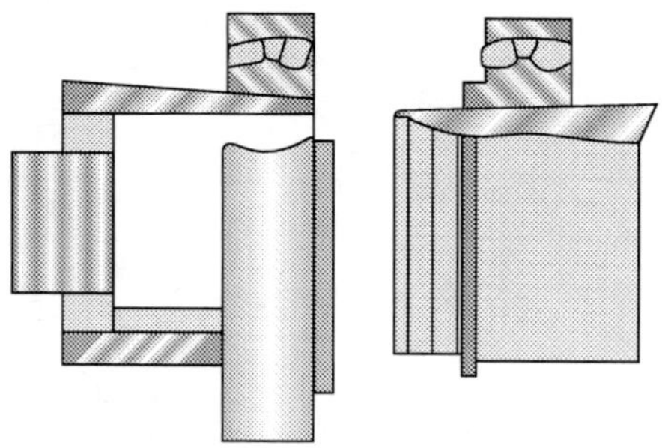

Fig. 5.28 A Taper shaft to taper sleeve

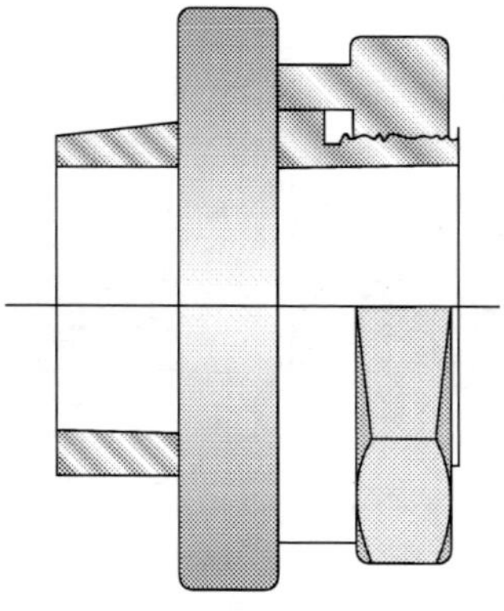

Fig. 5.28 B Taper sleeve threaded

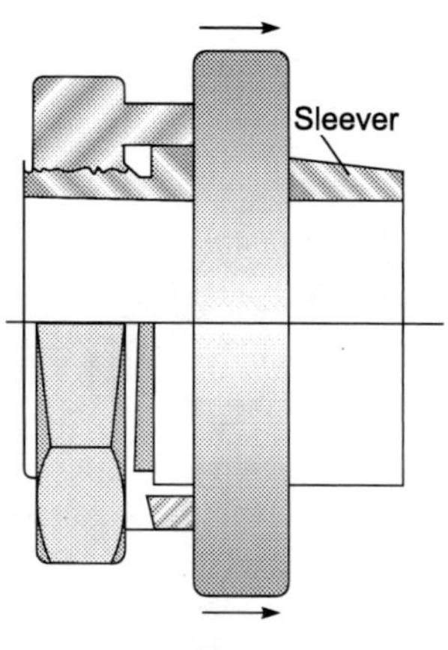

Fig. 5.28 C

Refer to the appropriate table in order to fit the bearing the bearing correctly.

Removal of Bearings

When a new bearing is to be fitted, the seating must not be damaged, therefore:

Oil the surfaces to prevent scratching or scouring during removal

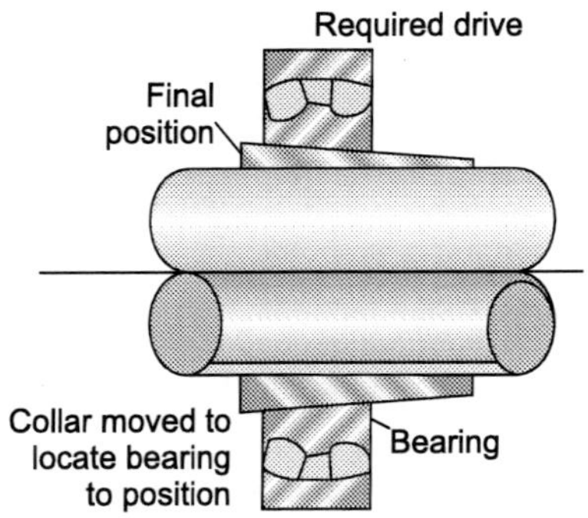

Fig. 5.28 D Location of taper bearing

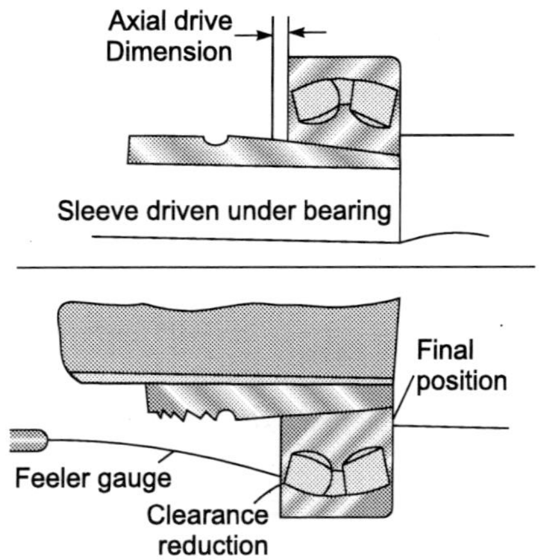

Fig. 5.29 Method of obtaining correct clearance

Care must be taken to keep oil clear of the commuter bushes and switch gear in the motor.

- Keep the bearing square to the seating during removal.
- When the bearing is to be removed, it must not be reused.
- Exert force only on the race being extracted.
- Whenever possible, remove free race and the cage first.
- Never clamp puller to the working surface of a bearing.

Bearing Removal Techniques

Using the arbor press.

Decide the best way to set the job on the press.

- Pressing a shaft out of a bearing. **Fig. (5.31).**
- Pressing a bearing out of the housing **Fig. (5.31 A).**

A tube must be used so that the force is applied evenly to the face.

Using Bearing Pullers (Fig. 5.30)

Note

When using bearing pullers take care to keep the bearing square to the shaft.

Select the most suitable puller.

- Impact pullers suitable for small bearings; the weight must be held when it is tapped against the stop.
- Screw pullers are suitable for moot purposes; take care to keep the puller square when turning the screw.
- Hydraulic pullers are used when large forces are required.

Bearing Dismounting

Dismounting of bearing should be done with proper care using proper tools. If proper tools are not used and right techniques are not adopted, the bearing is likely to be damaged and may lead to premature failure.

While using a puller, the pulling legs of the puller should be placed with the inner race **Fig. (5.30 A).** In certain cases, we use a puller plate **Fig. (5.30 B)** to facilitate the placing of the pulling legs in position so that force is applied on the inner race. Special puller plates **Fig. (5.30 C)** are used along with a two-legged puller so that the pull is applied on the inner race alone.

- Check that the puller is square and bearing only on the correct surfaces.
- Pull the bearing slowly. Stop frequently to check that the pullers are square and bearing only on the correct surfaces.

For detachable inner ring type bearing, the puller legs can be placed with the outer ring as shown in **Fig. (5.30 D)** for dismounting the bearing when the outer ring is having an interference fit in the housing.

A self-aligning ball-bearing can be swivelled as shown in **Fig. (5.30 E)** for fixing the bearing puller to facilitate the dismounting process.

Fig. 5.30A Place pulling edge with inner race

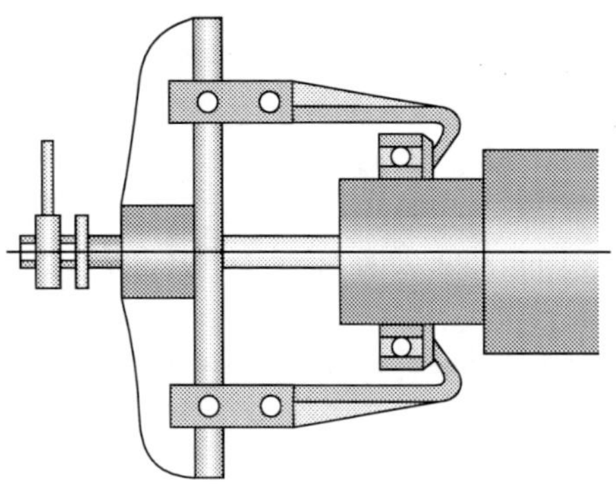

Fig. 5.30 B

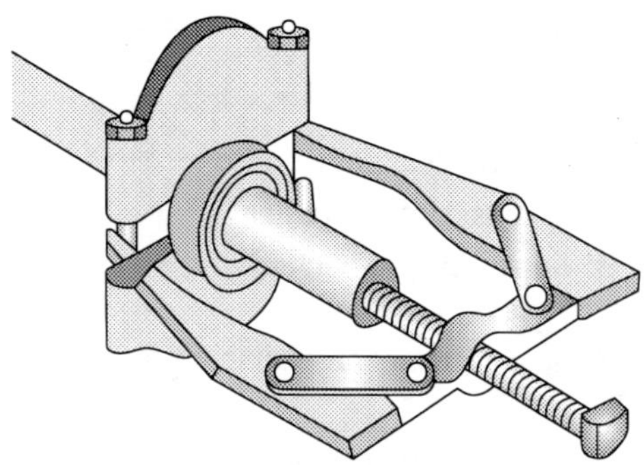

Fig. 5.30 C

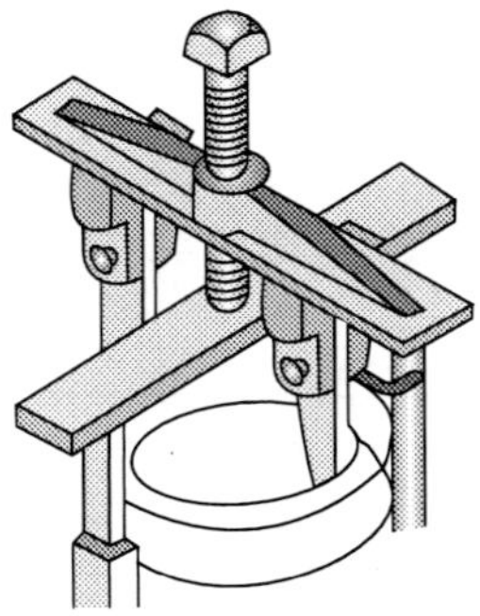

Fig. 5.30 D For detachable inner ring type bearing

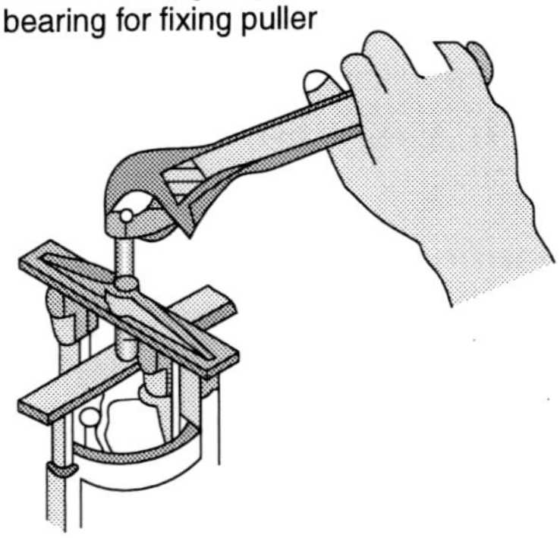

Fig. 5.30 E

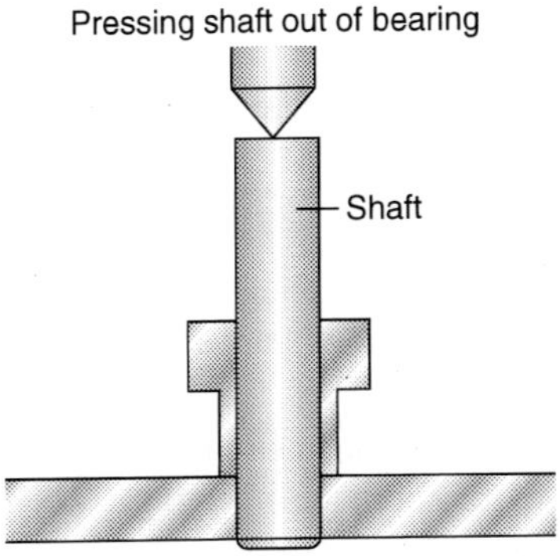

Fig. 5.31

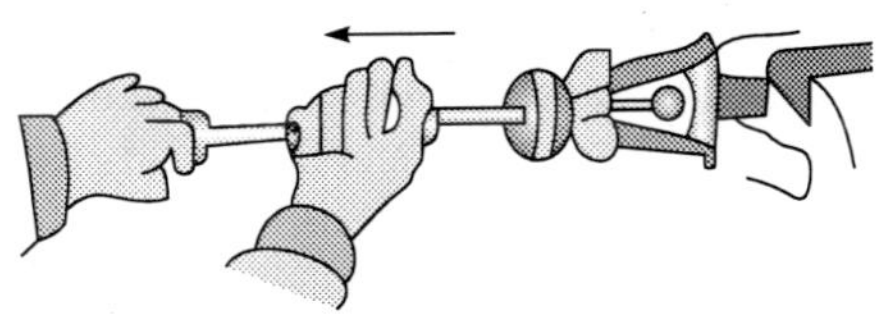

Fig. 5.32

Care and Maintenance

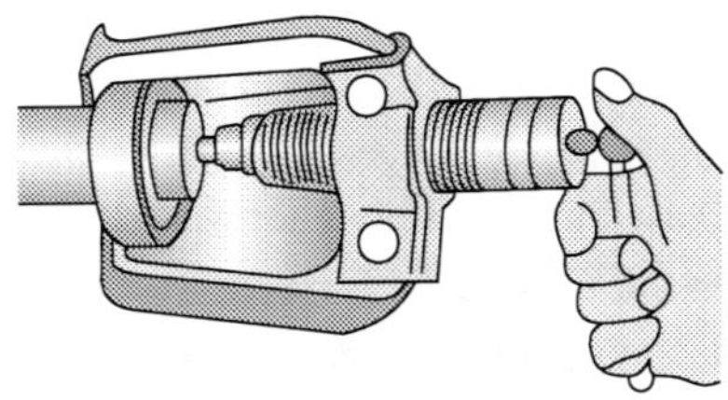

Fig. 5.33 Hydraulic puller

- A good bearing should not be dismantled unless otherwise it is absolutely necessary.
- Bearings should be handled in a dirt/dust free environment. Bearing housing on the shaft should be free from burns or scratches.
- Proper mounting and dismounting tools and correct techniques should be adopted. Provide proper support for the bearing and shaft during disassembly.
- Direct blows should not be given to the bearing.
- Bearing should, not be heated with a naked flame. Before heating ensure that any grease or lubricant does not start a fire.
- Use only the recommended grade and quantity of lubricant for lubrication of the bearing.

Storage of Rolling Bearing

Leave the bearings in their original packing: Fig. (5.34)

When stored the bearings should be kept in their original packing to preserve them from deterioration. Remove the bearing from the package, only at the assembly site immediately prior to mounting.

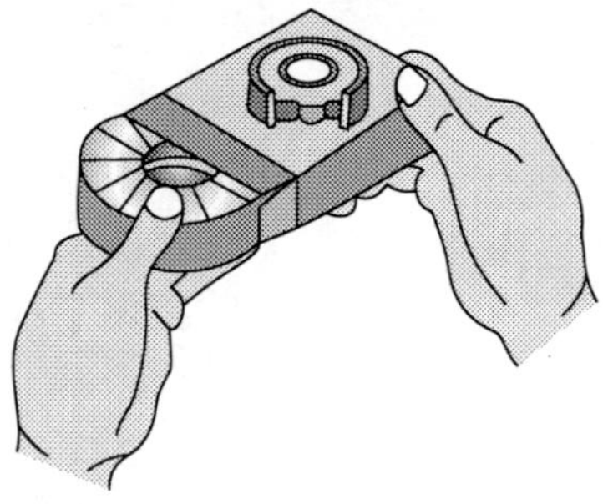

Fig. 5.34 Leave bearing in original packing

Store Large Bearings Flat

Large bearings, because of their relative thin rings, should not be stored standing upright; but laid flat. The support should extend over their whole circumference.

Store Bearings in Dry Rooms

This protection is only effective for longer periods if the bearings are kept in their original packing in a dry and frost-resistant room. Corrosive chemicals such as acids etc., must not be stored along with the bearings.

Rolling Bearing Mounting and Dismounting

The "right" bearing—compare inscription on packing with data on drawing:

Before mounting the fitter should satisfy that the bearing number stamped on the package agrees with the designation given on the drawing and in the parts list.

1. Work planning—know what you are going to do:

Study the shop drawing to acquaint yourself with the design details of the application and the assembly sequence. Phase the individual operation and get reliable information on bearing temperature, mounting and dismounting forces and the amount of grease to be packed into the bearings.

Whenever the installation and removal of rolling bearing necessitates special measures the fitter should be provided with instructions on mountings details, such as the means of transport, the mounting and dismounting equipment, measuring devices, heating facilities and type and quantity of lubricant.

2. Handling of rolling bearings before mounting. Do not wash out anti-corrosive oil

All bearings are preserved, in their original packing, with anti-corrosive oil. The oil need not be washed out when mounting the bearing.

Wash-Out Only Dirty Bearings

Before mounting wash used and dirty bearings carefully with kerosene. The bearings should be oiled and greased immediately after wards.

Do not tamper with rings

Do not perform any modification on the bearing, drilling of lubricating holes, machining of grooves, flats and the like will disturb the stress distribution, resulting in premature bearing failure.

3. Cleanliness in mounting — project disassembled bearings from dust and moisture — watch for cleanliness of shaft and housing:

Absolute cleanliness is essential. Dirt and humidity are dangerous since even the smallest particles penetrating into the bearings will damage the rolling surfaces.

Turned parts must be free from burrs and sharp edges.

After cleaning, the housing bore should receive a protective coating.

4. Surrounding parts check matching parts for dimensional accuracy prior to bearing mounting:

All surrounding parts should be carefully checked for dimensional and geometrical accuracy before starting mounting.

Fits

- Observe ring fits specified on drawing.
- Check shaft and housing tolerances.
- Check out-of-round of shaft and housing seats.

Inspection of Bearing Seats Cylindrical Seats

- Shafts are generally checked with external micrometers.
- Another useful instrument is the dial indicator, snap gauge.
- Bores are measured by means of internal micrometers.

Heating

If tight fits are specified for the inner rings the bearings are heated for mounting. Sufficient expansion is obtained when heated between 80° & 100° C.

- Never cool bearings below - 50° C
- No hearing beyond 80°C for mounting of shielded and sealed.

Bearings

- Accurate control is essential in heating the bearings.

If the temperature exceeds 120° C there is a risk of alteration of bearing grain structure resulting in a drop of hardness and dimensional stability.

- For uniform heating, rolling bearings are neurally immersed.

In an oil bath, this is thermostatically controlled to a temperature of 80°-100° C. The bearing should not be in direct contact with the heat source. The bearings may be suspended in the oil bath. After heating, any oil adhering to the bearing should be well drained off and the fitting surfaces should be carefully wiped clean.

Mounting of Rolling Bearings

Bearings vary greatly in type, design, size and there is no universally applicable mounting method.

With non-separate bearings the mounting and dismounting forces must be applied directly to the ring which is being fitted.

If the bearing is assembled on the shaft pressure must be applied to the face of the inner ring. If the bearing is assembled in the housing the pressure is applied to the outer ring.

Do not subject bearing rings to hammer blows:

Small bearings with cylindrical bore:

- Can be mounted cold.
- Mechanical / hydraulic press is used.

If no other means are available the bearing may be driven on the shaft by gentle taps with a hammer or mallet. An appropriate mounting sleeve of a soft metal should always be used to distribute the mounting force evenly over the entire bearing circumference and to avoid damage.

If the bearing has to be pressed onto the shaft and pushed into the housing at the same time a disk or assembly plate should be used. It should bear against both bearing rings to avoid tilting or jamming. This applies particularly to spherical roller bearings and self-aligning ball bearings. Special precaution should be observed with some self-aligning ball bearings, where the balls protrude beyond the rings. In such cases, the disk must be relieved.

Medium and Large Bearings with Cylindrical Bore

- Heat larger bearings for mounting:

The forces required to press-fit a bearing increase sharply with bearing size.

Fitting of heated rings or bearings requires skill. The outer surfaces of the rings or bearings heated to 80° to 100° C should be wiped dry with a clean cloth when still in heated condition they should be rapidly pushed on the shaft and positioned squarely against the shoulder. It is advisable to use asbestos mittens or non-fraying cloths but never cotton waste.

Larger bearings are generally handled with a crane. In this case the bearing is suspended in mounting grippers or in a rope sling

Heavy bearing outer rings can be mounted with lever.

Bearings with Tapered Bore

Bearings with tapered bore are either fitted directly on the tapered shaft journal or, if the shaft is cylindrical they are mounted on an adapter sleeve or a withdrawal sleeve.

Apply Just a thin oil film to seats of bearing, shaft and sleeve

Forcing the bearing onto the tapered seat expands the inner ring and reduces radial clearance. Hence, the reduction in radial clearance can be used as a measure of the seating condition of the inner ring.

- Do not preload bearing by excessive drive-up.
- Check radial clearance with feeler gauges.

To avoid transport damage of medium and large-sized spherical roller bearings a steel shim is inserted between both sets of rollers and the outer ring and held in position by a clamping band.

- Before measuring radial clearance remove beaming bands and shim.
- Mount shim bearings with shaft nut and hook spanner.
- Use mounting aids for the installation of large bearings.

To avoid tilting the bearing or sleeve into a cocked position the mounting nut should just be tightened enough to make nut and washer bear flush against their mating part. Then the screws of hardened steel evenly spaced around the circumference of the nut—their number depending on the forces required—are diagonally tightened until the required reduction in radial clearance is obtained.

When mounting larger bearings it may be advisable to use a hydraulic press.

Angular contact ball bearings and tapered roller bearings

- These are always mounted in pairs. The clearance of two bearings mounted in opposition is adjusted at mounting, its value depending on operating conditions.
- High loads and high speeds cause a temperature rise at the bearing location. This leads to thermal material expansion and clearance variation.

- If close shaft guidance is required clearance is required to be adjusted by stages. Each adjustment should be followed by a trial run and temp. check.
- Bearings exposed to vibration under low-speed conditions are mounted with zero clearance or even preloaded to avoid the risk of the balls or rollers brimming the race ways.
- Angular contact ball bearings and tapered roller bearings are adjusted against one another by nuts on the shaft, by calibrated shims, by threaded members in the housing or by preloaded springs capable of compensating for preloading in case of thermal expansion.
- A simpler method for correct bearing adjustment is the use of a torque spanner.

Thrust Bearings

With thrust bearings the shaft washers are generally push fits, tight fits being the exception. The housing washers are always loose fits. The shaft washers of double-acting thrust bearings should always be positively locked axially.

Dismounting of Rolling Bearings

- If the bearings are intended for reuse, dismounting should be performed most carefully, it is imperative that the extracting tool should always be applied to the ring to be extracted since otherwise the rolling elements will Brinell the race way.
- With nonseparable bearings either the shaft, with the tight-fitted bearing on it, is withdrawn from the housing or if the outer ring is a tight fit, the housing with the bearing *is* dismounted from the shaft.
- The force required for dismounting *is* generally higher than the force expended for mounting, because, as time passes, the ring becomes embedded on seat.

Dismounting of Small Bearings

- Small bearings are usually dismounted with the aid of mechanical extractors or hydraulic presses. These are applied either directly to the tight-fitted ring or to the adjoining pieces as the labyrinth ring.

Do not subject the bearing rings to hammer blows.

Hydraulic Mounting and Dismounting

With the hydraulic method, a fluid *is* injected between the mating surfaces.

The fluid film overcomes the friction between the mating parts which can then be easily displaced in relation to one another without the risk of surface damage. Fretting corrosion can be dissolved by means of kerosene, paraffin oil or rust-dissolving additives to the oil.

Induction Heating

- For large batch mounting of cylindrical roller bearings the use of induction devices for the heating of the bearing inner ring has proved to be very economical.
- When the induction heaters are used for mounting work care must be exercised not to overheat the rings. The heat-up times are indicated in the operating instructions.
- If no oil ducts are provided in the shaft for hydraulic mounting and induction heating equipment is not available. Flame heat may be used to assist in ring removal.
- Never use welding torch, because of danger of overheating and tempering is possible.
- Use ring burner.

EXERCISE

1. What is the main difference between a shaft and an axle? Why the shafts are made round in cross section?
2. What is a bearing? List the names of bearings used in shafting. How bearings are lubricated? Why?
3. Where are the bearings used?
4. What are the advantages of bearings?
5. What are the kinds of bearings?
6. Of what metal bush bearings are made and where these are used?
7. What is meant by anti-frictional bearing?
8. Describe in brief the kinds of antifriction bearing.
9. What precautions are needed for selection and fitting of bearings?
10. Where are the needle-bearings used?

Lubrication

Chapter 6

Friction is the resistance to motion which takes place when one body is moving upon another, and is generally defined as "that force which acts between two bodies at their surface of contact, so as to resist their sliding on each other." The force of friction, F, bears according to the conditions under which sliding occurs certain relation to the pressure between the two bodies; this pressure is called the normal pressure N. The relation between force of friction and normal pressure given by the coefficient of friction generally denoted by the Greek letter μ. Thus,

$$F = \mu \times N \text{ and } \mu = F/N$$

Example A body weighing 28 kg rests on a horizontal surface. The force required to keep it in motion along the surface is 7 kg find the coefficient of friction.

$$\mu = F/N = 7/28 = 0.025$$

If a body is placed on an inclined plane, the friction between the body and the plane will prevent it from sliding down the inclined surface, provided the angle of the plane with the horizontal is not too great. There will be a certain angle, however, at which the body will just barely be able to remain stationary, the friction resistance being very nearly overcome by the tendency of the body to slide down. This angle is termed the angle of repose, and the tangent of this angle equals the coefficient of friction. The angle of repose is frequently denoted by the Greek letter θ. Thus, $\mu\ L = \tan\theta$.

A greater force is required to start a body from a state of rest than to merely keep it in motion, because the friction of rest is greater than the friction of motion.

6.1 LAWS OF FRICTION

The laws of friction for unlubricated or dry surfaces are summarized in the following statements.

1. For low pressures the friction is directly proportional to the normal pressure, between the two surfaces. As the pressure increases to a high value the friction do not rise as rapidly; but when the pressure becomes abnormally high, the friction increases at a rapid rate until seizing takes place.
2. The friction both in its total amount and its coefficient is independent of the areas in contact, so long as the total pressure remains the same. This is true for moderate pressures only. For high pressures, this law is modified in the same way as in the first case.
3. At very low velocities the friction is independent of the velocity of rubbing. As the velocities increase, the friction decreases.

Lubricated Surfaces: For well lubricated surfaces, the laws of friction are considerably different from those governing dry or poorly lubricated surfaces.

1. The frictional resistance is almost independent of the pressure per square area if the surfaces are flooded with oil.
2. The friction varies directly as the speed, at low pressures; but for high pressures the friction is very great at low velocities, approaching a minimum at about two feet per second linear velocity, and afterwards increasing approximately as the square root of the speed.
3. For well lubricated surfaces the frictional resistance depends, to a very great extent, on the temperature, partly because of the change in the viscosity of the oil and partly because, for a journal bearing, the diameter of the bearing increases with the rise of temperature more rapidly than the diameter of the shaft, thus relieving the bearing of side pressure.
4. If the bearing surfaces are flooded with oil, the friction is almost independent of the nature of the material of the surfaces in contact. As the lubrication becomes "less ample, the coefficient of friction becomes more dependent upon the material of the surfaces.

Influence of Friction on the Efficiency of Small Machine Elements

Friction between machine parts lowers the efficiency of a machine. In the following are given average values of the efficiency, in per cent, of the most common machine elements when carefully made. Ordinary bearings, 95 to 98; roller bearings, 98; ball bearings, 99; spur gears with cut teeth, including bearings, 99; bevel gears with cut teeth, including bearings, 98; belting, from 96 to 98; high-class silent power transmission chain, 97 to 99; roller chains, 95 to 97.

Coefficients of Friction

Chapter on bearings carries coefficient of friction for some common material used frequently in engineering.

Many factors affect friction, and even slight deviations from normal or test conditions can produce wide variations. Accordingly, when using friction coefficients in design calculations, due allowance or factors of safety should be considered and in critical cases, specific tests conducted to provide actual coefficients for material, geometry, and/or lubricant combinations.

Rolling Friction

When a body rolls on a surface, the force resisting the motion is termed rolling friction or rolling resistance. Let W = total weight of rolling body or load on wheel, in kg; r = radius of wheel, in mm; f = coefficient of rolling resistance, in mm. Then: Resistance to rolling, in kg = $(W \times f)/r$.

The coefficient of rolling resistance varies with the conditions. For wood on wood it may be assumed as 0.06×25.4 mm; for iron on iron, 0.02×25.4 mm; iron on granite = 0.085×25.4 mm; iron on asphalt, 0.15×25.4 mm; and iron on wood, 0.22×25.4 mm.

The coefficient of rolling resistance, f, is in mm and is not the same as the ordinary, or sliding coefficient of friction which is a dimensionless ratio between frictional resistance and normal load. As various investigators are not in close agreement on the true values for these coefficients, the foregoing values should only be used for approximate calculation of rolling resistance.

6.2 LUBRICANTS

The concept of supporting a load being moved on a friction reducing film is called lubrication. The material of which the film is made up is defined as a lubricant. Lubricants may be in the form of fluids, such as conventional oils; semisolids, such as grease; and true solids, such as graphite and Teflon. Water, alcohol and even air have been used as lubricants. Factors affecting the selection of the type of lubricant to be used are load, speed, temperature (both ambient and generated) and environmental contamination such as in the food processing industry. Liquid lubricants usually are matched to the application by proper selection, of viscosity and load-carrying ability.

6.3 WEAR

There is no apparent consistent relationship between friction and wear; friction may be high and wear low, or wear may be high and friction low. The principal types of wear are: adhesive, abrasive, and pitting.

Adhesive Wear

As two surfaces slide over each other, wear occurs because of the shearing, deformation, and plucking away of material at points of adhesion; these points of adhesion occur at roughness peaks. The amount of wear is generally proportional to the load and to the distance over which the surfaces have slid, and inversely proportional to the hardness of the surface on which wear occurs.

Abrasive Wear

This occurs when a hard rough surface slides against a softer one, ploughing a series of grooves and removing material; it also occurs when abrasive particles are introduced between sliding surfaces, or when a part is moved through an abrasive medium. Erosion abrasion takes place when abrasive particles impact on surfaces. These particles may be suspended in liquids, carried by air, or flow of their own weight such as sand particles down a chute.

Pitting

This is due to the surface fatigue failure of a material as a result of repeated surface or subsurface stresses that exceed the endurance limit of the material. Pitting is a common mode of failure for gear teeth. Pitting can be non-progressive or destructive. Non-progressive pitting occurs during the initial operation of a machine because of surface fatigue at roughness peaks. As the peaks wear, the surface becomes smoother and hence the load becomes more uniformly distributed. Pitting stops when the stress becomes less than the endurance limit of the material. Destructive pitting, however, is progressive and continues until the surface disintegrates.

6.4 DESIGNING FOR WEAR RESISTANCE

In adhesive wear situations the best results are usually obtained if parts have hard surfaces. Hard surfaces may be obtained by making parts from hard materials or by the application of appropriate surface treatments. Some of the hardest materials are diamond, brittle carbides, and mercantilistic steels. For less severe applications pearlitic and austentitic steels can be used.

Some common surface treatments of metal surfaces for the prevention of wear are chromium and nickel plating, flame plating with tungsten carbide, phosphate treatment, carburizing, nitriding, cyaniding, carbon nitriding, siliconizing, and chromizing. Aluminum and magnesium can be given hard anodic treatments. For moderate cost applications in which normal wear can be tolerated, low or medium carbon steels can be run against cast iron, brass, bronze, aluminum bronze, or nylon. Adequate lubricants should be used. In general, it is best if unlike combinations of materials slide against each other. This is often accomplished by employing different surface treatments; for example, having a nitrided steel part run against one that is carburized.

Abrasive wear can be limited in several ways. Surfaces can be given a higher hardness than the abrasive particles. If a circulating lubricant is used, the abrasive particles can be removed by filtering. A combination of a hard and a soft surface can be used so that the abrasive particles can imbed themselves into the softer material. Typical softer materials used for this purpose are babbitt metals, copperlead alloys, and aluminum alloys.

Pitting is controlled by keeping contact stresses within allowable limits. In the case of gears a hard pinion is often run with a softer gear. As the softer gear wears, there will be better conformity between gear teeth profiles, hence less localized high stresses.

6.5 LUBRICANTS AND LUBRICATION

A lubricant is used for one or more of the following purposes:

1. To reduce friction,
2. To prevent wear,
3. To prevent adhesion,
4. To aid in distributing load,
5. To cool the moving elements,
6. To prevent corrosion.

The range of materials used as lubricants has been greatly broadened over the years so that in addition to oils and greases many plastics and solids and even gases are now being applied for this role. The only limitations of many of these materials are their ability to replenish themselves, to dissipate frictional heat, their reaction to high environmental temperatures, and their stability in combined environments. Because of the wide selection of lubricating materials available, great care is advisable in choosing the material and the method of application. The following types of lubricants are available:

1. Petroleum fluids,
2. Synthetic fluids,
3. Greases,
4. Solid films,
5. Working fluids,
6. Gases,
7. Plastics,
8. Animal fat,
9. Metallic and mineral films,
10. Vegetable oils, and
11. Lubricating oils.

The most versatile and best-known lubricant is mineral oil.

When applied in well-designed applications which provide for the limitations of both mechanical and hydraulic elements, oil is recognized as the most reliable lubricant. Concurrently, it is offered in a wide selection of stocks, carefully developed to meet the requirements of specific applications.

Lubricating oils are seldom marketed without additives blended for a narrow range of applications. Since these "additive-packages" are developed for particular applications it is advisable to consult the sales-engineering representatives of a reputable petroleum company on the proper selection for the conditions under consideration. The following are the most common types of additives:

1. Wear preventive,
2. Oxidation inhibitor,
3. Rust inhibitor,
4. Detergent-dispersant,
5. Viscosity index improve,
6. Defoaming agent, and
7. Pour point depressant.

A more recent development in the field of additives is a series of organic compounds which leave no ash when heated to a temperature high enough to evaporate or burn off the base oil. Initially produced for internal combustion engine applications, they have found ready acceptance in those other applications where metallic or mineral trace elements would promote catalytic, corrosive, deposition, or degradation effects on mechanism materials.

Additives usually are not stable over the entire temperature and shear-rate ranges considered acceptable for the base stock oil application. Because of this, additive type oils must be carefully monitored to insure that they are not continued in service after their principal capabilities have been diminished or depleted. Of primary importance in this regard, is the action of the detergent-dispersant additives which function so well to reduce and control degradation products which would otherwise deposit on the operating parts and oil cavity walls. Since the materials cause the oil to carry a higher than normal amount of the breakdown products in a fine suspension, they may cause an accelerated deposition rate or foaming when they have been depleted or degenerated by thermal or contamination action. In this latter case, the ingestion of water by condensation or leaking can cause markedly harmful effects.

Viscosity index improves serve to modify oils so that their change in viscosity is reduced over the operating temperature range. These materials may be used to improve both heavy and light oil; however, the original stock will tend to revert to its natural state when the additive has been depleted or degraded due to exposure to high temperatures or to the high shear rates normally encountered in the load carrying zones of bearings and gears. In heavy duty installations it is generally advisable to select heavier or more highly refined oil (and one that is generally more costly) rather than to rely on a less stable viscosity index improve product.

These oils are generally used in applications where the shear rate is well below 1000000 reciprocal seconds as determined below:

Shear rate (sec.$^{-1}$) = DN / 60 t where D is the journal diameter in mm N is the journal speed in RPM, and the film thickness in mm.

6.6 TYPES OF OILS

Aside from being aware of the many additives which can be obtained to satisfy particular application requirements and improve the performance of fluids, the designer must also be acquainted with the wide variety of oils, natural, and synthetic, which are available. Each has its own special features which make it suitable for specific applications and which limit its utility in others. Though a complete description of each oil and its application feasibility cannot be given in the following paragraphs, reference to major petroleum and chemical company sales engineers will provide full descriptions and sound recommendations. In many applications, however, it must be accepted that the interrelation of many variables, including shear rate, load, and temperature variations, prohibit precise recommendations or predictions of fluid durability and performance. Thus, prototype and rig testing are often required to insure the final selection of the most satisfactory fluid.

6.7 VISCOSITY

As noted above, fluids used as lubricants are generally categorized by their viscosity at 38 and 95° C. Absolute viscosity is defined as a fluid's resistance to shear or motion - its internal friction in other words. This property is described in several ways, but basically it is the force required to move a plane surface of unit area with unit speed parallel to a second plane and at unit distance from it. In the metric system the unit of viscosity is called the "poise" and in the English system is called the "reyn". One reyn is equal to 68,950 poise. One poise is the viscosity of a fluid, such that one dyne force is required to move a surface of one square centimetre with a speed of one centimetre per second, the distance between surfaces being one centimetre. Kinematic viscosity is related directly to the flow time of a fluid through the viscosimeter capillary. By multiplying the kinematic viscosity by the density of the fluid at the test temperature, one can determine the absolute viscosity. Since, in the metric system, the mass density is equal to the specific gravity, the conversion from kinematic to absolute viscosity is generally made in this system and then converted to English units where required. The densities of typical lubricating fluids with comparable viscosity at 38°C and 95 degrees C are aveilable in standars.

6.8 FINDING SPECIFIC GRAVITY OF OILS AT DIFFERENT TEMPERATURES

The standard practice in the oil industry is to obtain a measure of specific gravity at 15.5° C on an arbitrary scale, in degrees API, as specified by the American Petroleum Institute. As an example, API gravity, PAPI, may be expressed as 27.5° at 60° F.

The relation between gravity in API degrees and specific gravity (grams of mass per cubic centimetres) at 60° F., P60, is: 141.5 /131.5 + PAPI

The specific gravity, P T, at some other temperature, T, is found from the equation:

P T = P60 – 0.00035 (T – 60)

Normal values of specific gravity for sleeve bearing lubricants range from 0.75 to 0.95 at 60° F. If the API rating is not known, an assumed value of 0.85 may he used.

6.9 APPLICATION OF LUBRICATING OILS

In the selection and application of lubricating oils careful attention must be given to the temperature in the critical operating area and its effect on oil properties. Analysis of each application should be made with detailed attention given to cooling, friction losses, shear rates, and contaminants.

Many oil selections are found to result in excessive operating temperatures because of a viscosity that is initially too high, which raises the friction losses. As a general rule, the lightest weight oil which can carry the maximum load should be used. Where it is felt that the load carrying capacity is borderline, lubricity improves may be employed rather than an arbitrarily higher viscosity fluid. It is well to remember that in many mechanisms the thicker fluid may increase frictional losses sufficiently to lower the operating viscosity into the range provided by initially lighter fluid. In such situations also, improved cooling, such as may be accomplished by increasing the oil flow, can improve the fluid properties in load zone.

Similar improvements can be accomplished in many gear trains and other mechanisms by reducing churning and aeration through improved scavenging, direction oil of jets, and elimination of obstacles to the flow of the fluid. Many devices, such as journal bearings, are extremely sensitive to the effects of cooling flow and can be improved by greater flow rates with a lighter fluid. In other cases the load carrying capacity of petroleum oil is affected by pressure, shear rate, and bearing surface finish as well as initial viscosity and therefore these must be considered in the selection of the fluid. Detailed explanation of the factors is not within the scope of this text; however the petroleum companies can supply practical guides for most applications.

Other factors in the selection of oil

1. Compatibility with system materials.
2. Water absorption properties.
3. Break-in requirements.
4. Detergent requirements.
5. Corrosion protection.
6. Low temperature properties.
7. Foaming tendencies.
8. Boundary lubrication properties.
9. Oxidation resistance (high temperature properties).
10. Viscosity/temperature Stability (Viscosity Temperature Index)

Generally, the factors listed above are those which are usually modified by additives described earlier. Since additives are used in limited amounts in most petroleum products, blended oils are not as durable as the base stock and must therefore be used in carefully worked-out systems. Maintenance procedures must be established to monitor the oil so that it may be replaced when the effect of the additive is expected to degrade. In large systems supervised by a lubricating engine sampling and associated laboratory analysis can be relied on, while in customer maintained systems as in automobiles and reciprocating engines, the design engineer must specify a safe replacement period which takes into account any variation type of service or utilization.

Some large systems, such as turbine-power units, have complete oil systems which are designed to filter, cool, monitor, meter, and replenish the oil automatically. In such facilities, much larger oil quantities are used and they are maintained by regularly assigned lubricating personnel. Here reliance is placed on conservatively chosen fluids with the expectation that they will endure many months or even years of service.

6.10 CENTRALIZED LUBRICATION SYSTEMS

Various forms of centralized lubrication systems are used to simplify and render more efficient task of lubricating machines. In general, a central reservoir provides the supply of oil, which is conveyed to each bearing either through individual lines of tubing or through a single line of tubing that has branches extending to each of the different bearings. Oil is pumped into the lines either manually by a single movement of a lever or handle or automatically by mechanical drive from some revolving shaft or other parts of the machine. In either case, all bearings in the central system are lubricated simultaneously. Centralized force-feed lubrication is adaptable to various classes of machine tools such as lathes, planers, and milling machines and to many other types of machines. It permits the use of a lighter grade of oil, especially where complete coverage of the moving parts is assured.

6.11 GRAVITY LUBRICATION SYSTEMS

Gravity systems of lubrication usually consist of a small number of distributing centres or manifolds from which oil is taken by piping as directly as possible to the various surfaces to be lubricated, each bearing point having its own independent pipe and set of connections. The aim of the gravity system, as of all lubrication systems, is to provide a reliable means of supplying of the bearing surfaces with the proper amount of lubricating oil. The means employed to maintain this steady supply of oil include drip feeds, wick feeds, and the wiping type of oil. Most manifolds are adapted to use either or both drip and wick feeds.

6.12 DRIP-FEED LUBRICATORS

A drip feed consists of a simple cup or manifold mounted in a convenient position for filling and connected by a pipe or duct to each bearing to be oiled. The rate of feed in each pipe is regulated by a needle or conical valve. A loose-fitting cover is usually fitted to the manifold in order to prevent cinders or other foreign matter from becoming mixed with the oil. When a cylinder or other chamber operating under pressure is to be lubricated, the oil-cup takes the form of a lubricator having a tightfitting screw cover and a valve in the oil line. To fill a lubricator of this kind, it is only necessary to close the valve and unscrew the cover.

6.13 OPERATION OF WICK FEEDS

For a wick feed, the siphoning effect of strands of worsted yarn is employed. The worsted wicks give a regular and reliable supply of oil and at the same time act as filters and strainers. A wick composed of the proper number of strands is fitted into each oil-tube. In order to insure using the proper size of wicks, a study should be made of the oil requirements of each installation and the number of strands necessary to meet the demands of bearings at different rate of speed should be determined. When the necessary data have been obtained, a table should be prepared showing the size of wick or the number of strands to be used" for each bearing of the machine.

6.14 OIL CONDUCTING CAPACITY OF WICKS

With the oil level maintained at a point 96 to 94 inch below the top of an oil-tube, each strand of a clean worsted yarn will carry slightly more than one drop of oil a minute. A twenty-four-strand wick will feed approximately thirty drops a minute, which is 30 drops a minute, which is ordinarily sufficient for operating a large bearing at high speed. The wicks should be removed from the oil-tubes when the machinery is idle. If left in place, they will continue to deliver oil to the bearing until the supply in the cup is exhausted, thus wasting a considerable quantity of oil, as well as well as flooding the bearing. When bearings require an extra supply of oil temporarily, it may be supplied by dipping the wicks or by pouring oil down the tube from an oilcan or, in the case of drip feeds, by opening the needle valves. When equipment that has remained idle for some time is to be started up, the wicks should be dipped and the moving parts oiled by hand to insure an ample initial supply of oil. The oil should be kept at about the same level in the cup, as otherwise the rate of flow will be affected. Wicks should be lifted periodically to prevent dirt accumulations at the ends from obstructing the flow of oil.

How Lubricating Wicks are made

Wicks for lubricating purposes are made by cutting worsted yarn into lengths about twice the height of the top of the oil-tube above the bottom of the oil-cup, plus 100 mm. Half the required number of strands are then assembled and doubled over a piece of soft copper wire, laid across the middle of the strands. The free ends are then caught together by a small piece folded sheet lead, and the copper wire twisted together throughout its length.

The lead serves to hold the lower end of the wick in place, and the wire assists in forcing the other end of the wick several inches into the tube. When the wicks are removed, the free end of the copper wire may be hooked over the tube end to indicate to which tube the wick belongs to. Dirt from the oil causes the wick to become gummy and to lose its filtering effect. Wicks that have thus become clogged dirt should be cleaned or replaced by new ones. Cleaning is done by boiling the wicks in soda water and then rinsing them thoroughly to remove all traces of soda. Oil-pipes are sometimes fitted with openings through which the flow can be observed. In some installations, a short glass tube is substituted for such an opening.

6.15 WIPER-TYPE LUBRICATING SYSTEMS

Wiper-type lubricators are used for out-of-the way oscillating parts. A wiper consists of an oil-cup with a central blade or plate extending above the cup, and is attached to a moving part. A strip of fibrous material fed with oil from a source of supply is placed on a stationary part in such position that the cup in its motion scrapes along the fibrous material and wipe off the oil, which then passes to the bearing surfaces.

Oil manifolds, cups, and pipes should be cleaned occasionally with steam conducted through a hose or with boiling soda water. When soda water is used, pipes should be disconnected, so that no soda water can reach the bearings

6.16 OIL MIST SYSTEMS

A very effective system for both lubricating and cooling many elements which require a limited quantity of fluid is found in a device which generates a mist of oil, separates out the denser and larger (wet) oil particles, then distributes the mist through a piping or conduit system. The mist is delivered into the bearing, gear, or lubricated element cavity through a condensing or spray nozzle, which also serves to meter the flow. In applications which do not encounter low temperatures or which permit the use of visual devices to monitor the accumulation of solid oil, oil mist devices offer advantages in providing cooling, clean lubricant pressurized cavities which prevent entrance of contaminants, efficient application of limited lubricant quantities, and near-automatic performance. These devices are supplied with fluid reservoirs holding from a few ounces up to several litres of oil and with accommodations for either accepting shop air or working from a self-contained compressor powered by electricity. With proper control of the fluid temperature, these units can atomize and dispense most motor and many gear oils.

6.17 LUBRICATING GREASES

In many applications, fluid lubricants cannot be used because of the difficulty of retention, relubrication, or the danger of churning. To satisfy these and other requirements such as simplification, greases are applied. These formulations are usually petroleum oils thickened by dispersions of soap, but may consist of synthetic oils with soap or inorganic thickeners, or oil with siliceous dispersions. In all cases, the thickener, which must be carefully prepared and mixed with the fluid, is used to immobilize the oil, serving as a storehouse from which the oil bleeds at a slow rate. Though the thickener very often has lubricating properties itself, the oil bleeding from the bulk of the grease is the determining lubricating function. Thus, it has been shown that when the oil has been depleted to the level of 50 per cent of the total weight of the grease, the lubricating ability the material is no longer reliable. In some applications requiring an initially softer and wetter material, however, this level may be as high as 60 per cent.

6.18 GREASE CONSISTENCY CLASSIFICATION

To classify greases as to mobility oil content, they are divided into grades by the NLGI (National Lubricating Grease Institute) these grades, ranging from 0, the softest, up through 6, the stiffest are determined by testing in a penetrometer, with the depth of penetration of a specific cone and weight being the controlling criterion. To insure proper averaging of specimen resistance to the cone, most specifications include a requirement that the specimen be worked in a sievelike device before being packed into the penetrometer (they soften with working). Most often this is done by the original equipment manufacturer with the aid of the petroleum company representatives, but in many cases it is advisable to include the bearing engineer as well. In this general area it is well to remember that shock loads, axial or thrust movement within or on the grease cavity can cause the grease to contact the moving parts and initiate softening due to the shearing or working thus induced. To limit this action, grease-lubricated bearing assemblies often utilize dams or dividers to keep the bulk of the grease contained and unchanged by this working. Successful application of grease depends however, on a relatively small amount of mobile lubricant (the oil bled out of the bulk) to replenish that small amount of lubricant in the element to too large, then a critical delay period (which will be regulated by the grease bleed rate and the temperature at which it is held) will ensue before lubricant in the element can be resupplied. Since most lubricants undergo some attrition due to thermal degradation, evaporation, shearing, or decomposition in bearing area to which applied, this delay can be fatal.

To prevent this from leading to failure, grease is normally applied so that the 'material in the cavity contacts the bearing in the lower quadrants, insuring that the excess originally packed into it impinges on the material in the reservoir. With the proper selection of a grease which does not slump excessively and a reservoir construction to prevent churning, the initial action of the bearing when started into operation will be to purge itself of excess grease, and to establish a flow path for bleed oil to enter the bearing. For this purpose, most grease selected will be of a grade 2 or 3 consistency, falling into the "channelling" variety or designation.

6.19 TYPES OF GREASE

Greases are made with a variety of soaps and are chosen for many particular characteristics. Most popular today, however, are the lithium, or soda-soap grease and the modified-clay thickened materials. For high temperature applications (250° F. and above) certain finely divided dyes and other synthetic thickeners are applied. For all-around use the lithium soap greases are best for moderate temperature applications (up to 225° F.) While a number of soda soap greases have been found to work well up to 285° F. Since the major suppliers offer a number of different formulations for these temperature ranges it is recommended that the user contacts the engineering representatives of a reputable petroleum company before choosing grease. Greases also vary in volatility and viscosity according to the oil used. Since the former will affect the useful life of the bulk applied to the bearing and the latter will affect the load carrying capacity of grease, they must both be considered in selecting grease.

For application to certain gears and slow-speed journal bearings, a variety of grease is thickened with carbon, graphite, molybdenum disulphide, lead, or zinc oxide. Some of these materials are likewise used to inhibit fretting corrosion or wear in sliding or oscillating mechanism and in screw or thread application. One material used as "gear grease" is a residual asphaltic compound which is known as a "Crater Compound." Being extremely stiff and having an extreme temperature-viscosity relationship, its application must also be made with careful consideration of its limitations and only after careful evaluation in the actual application. Its oxidation resistance is limited and its low mobility in winter temperature range make it a material to be used with care. However, it is used extensively in the railroad industry and in other applications where containment and application of lubricants is difficult. In such conditions its ability to adhere to gear and chain contact surfaces far outweighs its limitations and in some extremes it is "painted" onto the elements at regular intervals.

6.20 TEMPERATURE EFFECTS ON GREASE LIFE

Since most grease applications are made where long life is important and relubrication is not too practical, operating temperatures must be carefully considered and controlled. Being a hydrocarbon, and normally susceptible to oxidation, grease is subject to the general rule that, above a critical threshold temperature, each 15° to 18° F rise in temperature reduces the oxidation life of the lubricant by half. For this reason, it is vital that all elements affecting the operating temperature of the application be considered, correlated, and controlled. With sealed-for-life bearings, in particular, grease life must be determined for representative bearings and limits must be established for all subsequent applications.

Most satisfactory control can be established by measuring bearing temperature rise during a controlled test, at a consistent measuring point or location. Once, base line and limiting range are determined, all deviating bearings should be dismantled, inspected, and reassembled with fresh lubricant for retest. In this manner mavericks or faulty assemblies will be ferreted out and the reliability of the application established. Generally, well lubricated grease packed bearing will have a temperature rise above ambient, as measured at the outer race, of from 10 to 50° F. In applications where heat is introduced into the bearing through the shaft or housing, a temperature rise must be added to that of the frame or shaft temperature.

In bearing applications care must be taken to fill the cavity too full. The bearing should have a practical quantity of grease worked into it with the rolling elements thoroughly coated and the cage covered, but the housing (cap and cover) should be no more than 75 per cent filled; with softer greases, this should be no more than 50 per cent. Excessive packing is evidenced by overheating, churning, aerating and eventual purging with final failure due to insufficient lubrication. In grease lubrication, never add a bit more for good luck - hold to the prescribed amount and determine this with care on a number of representative assemblies.

6.21 RELUBRICATING WITH GREASE

In some applications, sealed-grease methods art not applicable and addition of grease at regular intervals is required. Where this is recommended by the manufacturer of the equipment, or where the method has been worked out as part of a development program, the procedure must be carefully followed.

1. Use the proper lubricant - the same as recommended by the manufacturer or as originally applied (grease performance can be drastically impaired if contaminated with another lubricant).
2. Clean the lubrication fitting thoroughly with materials which will not affect the mechanism or penetrate into the grease cavity.
3. Remove the cap (and if applicable, the drain or purge plug).

4. Clean and inspect the drain or scavenge cavity.
5. Weigh the grease gun or calibrate it to determine delivery rate.
6. Apply the directed quantity or fill until grease is detected coming out the drain or purge hole.
7. Operate the mechanism with the drain open so that excess grease is purged.
8. Continue to operate the mechanism while determining the temperature rise and insure that it is within limits. Where there is access to a laboratory, samples of the purged material may be analysed to determine the deterioration of the lubricant and to search for foreign material which may be evidence of contamination or of bearing failure.

Normally, with modern types of grease and bearings, lubrication need only be considered at overhaul periods or over intervals of three to ten years.

6.22 SOLID FILM LUBRICANTS

Solids such as graphite, molybdenum disulfide, polytetra fluoro ethylene, lead, Babbitt, silver, or metallic oxides are used to provide dry film lubrication in high-load, slow-speed or oscillating load conditions. Though most are employed in conjunction with fluid or grease lubricants, they are often applied as the primary or sole lubricant where their inherent limitations are acceptable.

- Foremost importance is their inability to carry away heat.
- Second, they cannot replenish themselves, though they generally do lay down an oriented film on the contacting interface.
- Third, they are relatively immobile and must be bonded to the substrate by a carrier, by plating, fusing, or by chemical or thermal deposition.

Though these materials do not provide the low coefficient of friction associated with fluid lubrication, they do provide coefficients in the range of 0.4 down to 0.02, depending on the method of application and the material against which they rub. Poly tetra fluoro ethylene, in normal atmospheres and after establishing a film on both interfaces has been found to exhibit a coefficient of friction down to 0.02. However, this material is subject to cold flow and must be supported by filler or on a matrix to continue its function. Since it can now be cemented in thin sheets and is often supplied with a fine glass fibre filler, it is practical in a number of installations where the speed and load do not combine to melt the bond or cause the material to sublime.

Bonded films of molybdenum disulfide, using various resins and ceramic combinations as binders, are deposited over phosphate treated steel, aluminum, or other metals with good success. Since its action produces a gradual wear of the lubricant, its life is limited by the thickness which can be applied (not over a thousandth or two in the conventional application). In most applications this is adequate if the material is used to promote break-in, prevent galling or pick-up, and to reduce fretting or abrasion in contacts otherwise impossible to separate.

In all applications of solid film lubricants, the performance of the film is limited by the care and preparation of the surface to which they are applied. If they can't adhere properly, they cannot perform, coming off in flakes and often jamming under flexible components. The best advice is to seek the assistance of the supplier's field engineer and set up a close control of the surface preparation and solid film application procedure. It should be noted that the functions of a good solid film lubricant cannot overcome the need for better surface finishing. Contacting surfaces should be smooth and flat to insure long life and minimum friction forces. Generally, surfaces should be finished to no more than 24 micro-inches, AA with waviness no greater than 0.00002 inch.

6.23 ANTI-FRICTION BEARING LUBRICATION

The limiting factors in bearing lubrication are the load and the linear velocity of the cinders of the balls or rollers. Since these are difficult to evaluate, a speed factor which consists of the inner race bore diameter X RPM is used as a criterion. This factor will be referred to as Si where the bore diameter is in inches and Sm where it is in mm.

In order to be suitable for use in anti-friction bearings, grease must have the following properties:

- Freedom from chemically or mechanically active ingredients, such as uncombined metals or oxides, and similar mineral or solid contaminants.
- The slightest possible tendency is consistency, such as thickening, separation of oil, evaporation or hardening.
- A melting point considerably higher than the operating temperatures.

The choice of lubricating oils is easier. They are more uniform in their characteristics and if resistant to oxidation, gumming and evaporation, can be selected primarily with regard to a suitable viscosity.

6.24 GREASE LUBRICATION

Anti-friction bearings are normally grease lubricated, both because grease is much easier than oil to retain in the housing over a long period and because it acts to some extent as a seal against the entry of dirt and other contaminants into the bearings. For almost all applications, a soda-base grease or a mixed-base grease with up to 5 per cent calcium soap to give a smoother consistency, blended with an oil of around 250 to 300 SSU (Saybolt Universal Seconds) at 38°C is suitable. In cases where speeds are high, say Si is 5000 or over, a grease made with an oil of about 150 SSU at 38°C may be more suitable especially if temperatures are also high. In many cases where bearings are exposed to large quantities of water, it has been found that a standard soda-base ball-bearing grease, although classed as water soluble gives better results than water-insoluble types. Greases are available that will give satisfactory lubrication over a temperature range of – 5 to + 120°C.

Grease should not be allowed to remain in a bearing for longer than 48 months or if the service is very light and temperatures low, 60 months, irrespective of the number of hours of operation during that period as separation of the oil from the soap and oxidation continue whether the bearing is in operation or not.

Before renewing the grease in a hand-packed bearing, the bearing assembly should be removed and washed in clean kerosene, degreasing fluid or other solvent. As soon as the bearing is quite clean it should be washed at once in clean light mineral oil, preferably rust-inhibited. The bearing should not be spun before or while it is being oiled. Caustic solutions may be used if the old grease is hard and difficult to remove, but the best method is to soak the bearing for a few hours in light mineral oil, preferably warmed to about 55°C, and then wash in cleaning fluid as described above. The use of chlorinated solvents is best avoided.

When replacing the grease, it should be forced with the fingers between the balls or rollers, dismantling the bearing, if convenient. The available space inside the bearing should be filled completely and the bearing then spun by hand. Any grease thrown out should be wiped off. The space on each side of the bearing in the housing should be not more than half-filled. Too much grease will result in considerable churning, high bearing temperatures and the possibility of early failure. Unlike any other kind of bearing, anti-friction bearings more often give trouble due to over rather than to under-lubrication.

6.25 OIL LUBRICATION

Oil lubrication is usually adopted when speeds and temperatures are high or when it is desired to adopt a central oil supply for the machine as a whole. Oil for anti-friction bearing lubrication should be well refined with high film strength and good resistance to oxidation and good corrosion protection. Anti-oxidation additives do not harm but are not really necessary at temperatures below about 90° C. Anti-corrosion additives are always desirable. The accompanying table gives recommended viscosity of oil for ball bearing lubrication other than by an air-distributed oil mist. Within a given temperature and speed range, oil towards the lighter end of the grade should be used, if convenient, as speeds increase. Roller bearings usually require an oil one grade heavier than do ball bearings for a given speed and temperature range. Cooled oil is sometimes circulated through an anti-friction bearing to carry off excess heat resulting from high speeds and heavy loads.

Maximum. Temp. Range F	**Optimum Temperature. Range Degree F**	**Speed factor Si Under 1000 Over 1000 Viscosity**	
– 40 to + 100	– 40 to 10	80 to 90 SSU	70 to 80 SSU
– 10 to 100	–10 to 30	100 to 115 SSU	80 to 100 SSU
30 to 150	30 to 150	SAE 20	SAE 10
30 to 200	50 to 200	SAE 40	SAE 30
50 to 300	200 to 300	SEA 70	SAE 60

Why rolling bearings have to be lubricated?

In order to achieve its maximum life in service, a rolling bearing must be lubricated.

The function of the lubricating medium is:

1. To form a film between bearing components which move relative to one another, so that metal contact is prevented. The film must be sufficiently thick if satisfactory lubrication is to be obtained — even under heavy load, high temperature variations and vibrations.
2. To reduce friction and eliminate wear.
3. To protect against corrosion.
4. To seal (in the case of a grease) against impurities, such as dirt, dust, damp or water.

Roiling bearings are normally lubricated by grease or oil and only in very special cases by a solid lubricant.

The selection of a lubricant depends on operating conditions, i.e., in particular on temperature range, speed of rotation and environment.

Grease Lubrication

Grease lubrication is usually chosen for applications in which the bearing operates under normal speed and temperature conditions. Grease has a number of advantages over oil. For example, it enables simpler (i.e. cheaper) installation, offers better adhesion and provides protection against damp and impurities.

Oil Lubrication

Preference is given to oil lubrication, when the speed of rotation or operating temperature makes it impossible to use grease, when heat has to be removed from the bearing assembly by the lubricant and of course when the bearings are mounted in an oil-lubricated construction, for example, a gearbox. 80 to 90 % of all bearing assemblies can be lubricated with grease.

Methods of Applying Lubricant

The following methods are used for efficient lubrication which have been explained in detail are now represented figuratively.

- Gravity feed method
- Force feed method
- Splash method

Gravity Feed Method

There are numerous ways of employing this principle, varying from the simple oil hole to the more elaborate wick and glass-sided drip feed lubricators in which the flow of the oil may be controlled and observed through the glass.

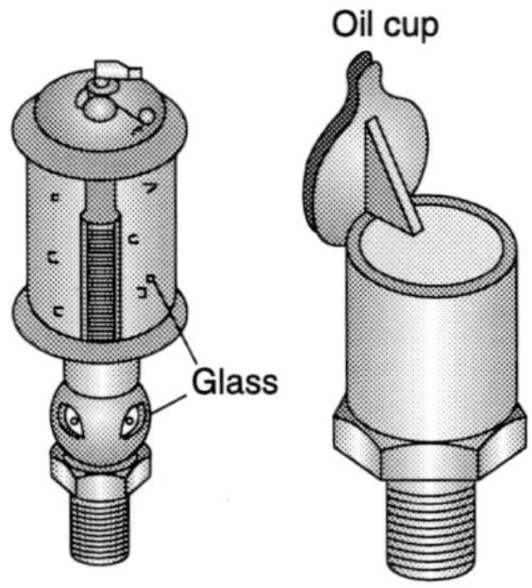

Fig. 6.1 Sight feed lubricator

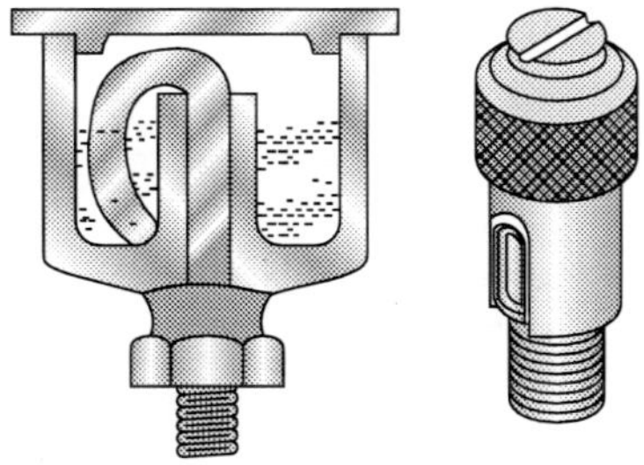

Fig. 6.2 Wick feed lubricator and winkly oiler

Force (Pressure) Feed Method

There are various systems of lubrication employing a pressure feed to the lubricant, and the most important of such systems may be classified roughly into the following.

Continuous feed of oil under pressure to each bearing concerned.

In this method an oil pump driven by the machine delivers oil to the bearings and back to a sump from which it is drawn by the pump.

Pressure Feed by Hand Pump

In which a charge of oil is delivered to each bearing at intervals (once or twice a day) by the machine operator.

Oil or grease gun method: The oil hole leading to each bearing is fitted with a nipple and by pressing the nose of the gun against this and the lubricant is forced into the bearing.

Splash method

In this method, the shaft or something attached to it, actually dips into the oil and a stream of lubricant is continually splashed around the parts requiring lubrication. This method is employed for the gears and bearings inside all gear drives, the lower parts of the gears actually dipping in the oil.

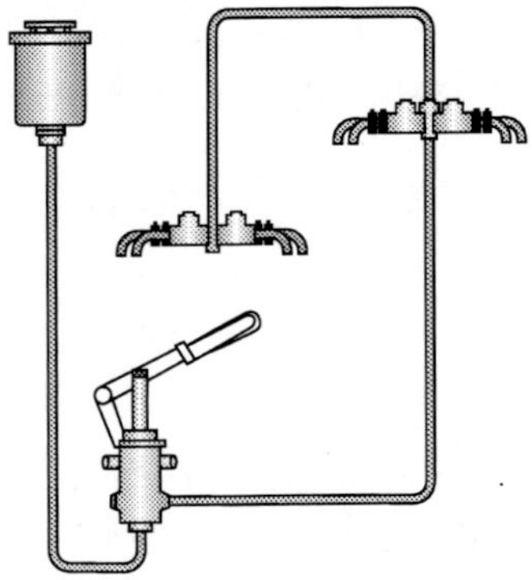

Fig. 6.3 Hand feed lubricator

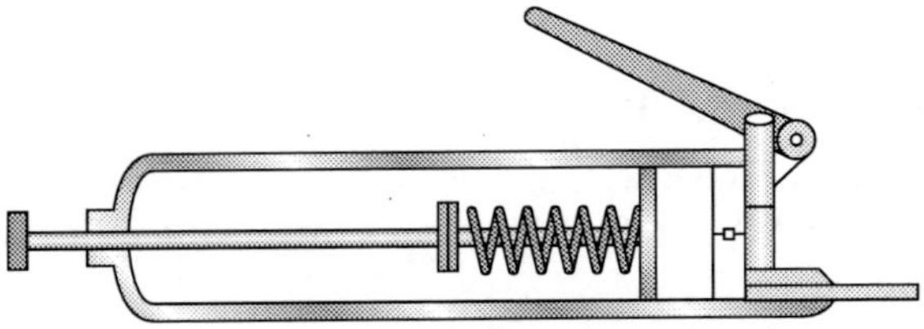

Fig. 6.4 Pressure grease gun

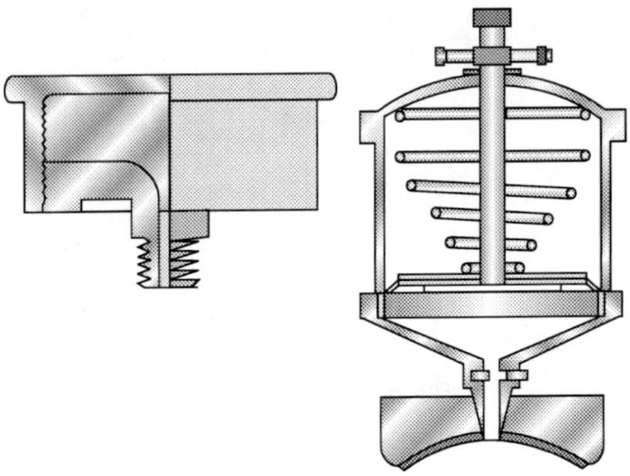

Fig. 6.5 Stuffer screw, drown greaser

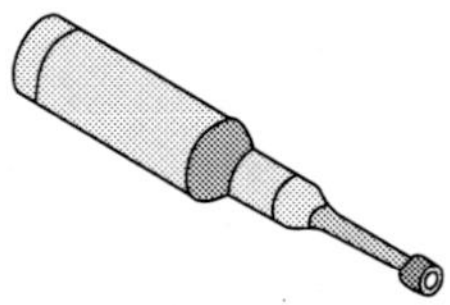

Fig. 6.6 Grease gun

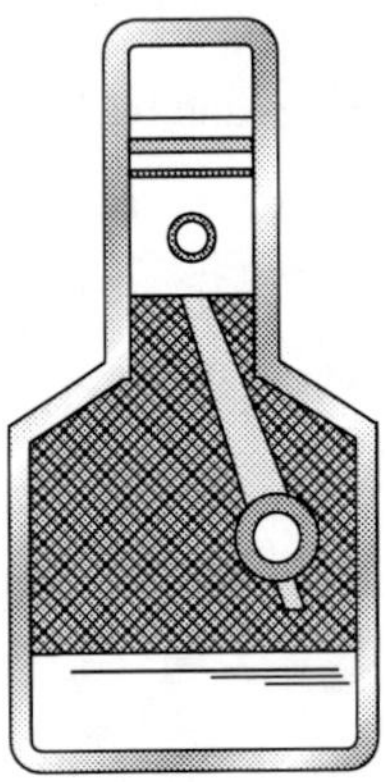

Fig. 6.7 Splash lubricator

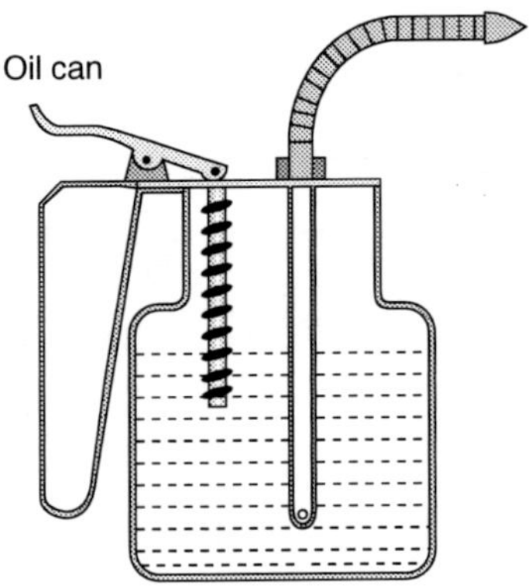

Fig. 6.8

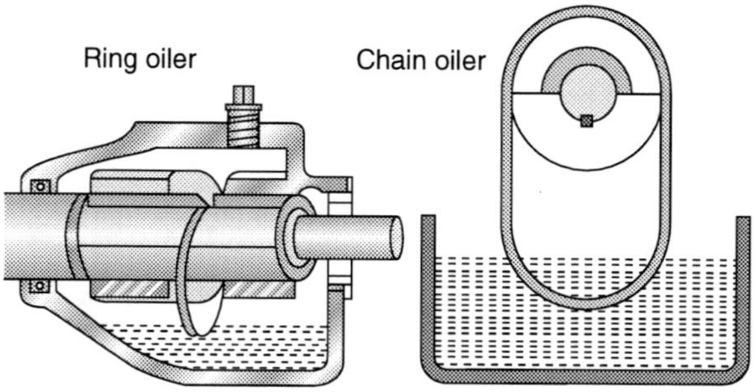

Fig. 6.9

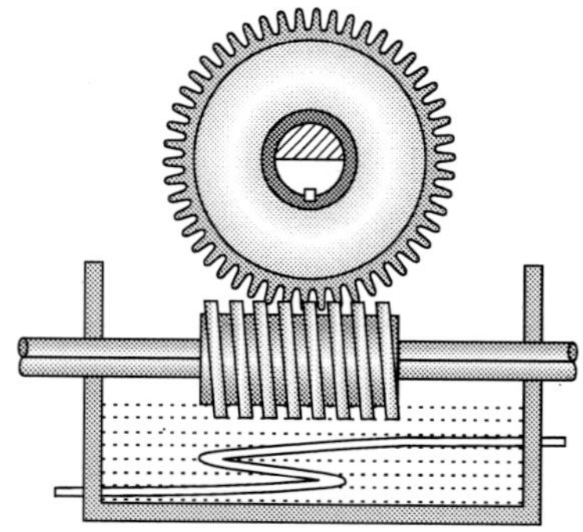

Fig. 6.10 Worm gear bath oiler

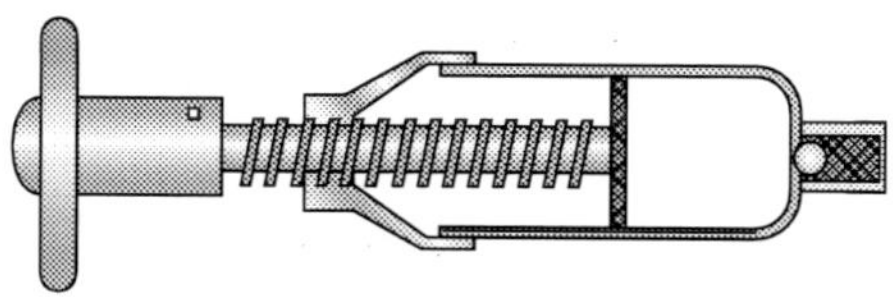

Fig. 6.11 'T' Handle pressure gun

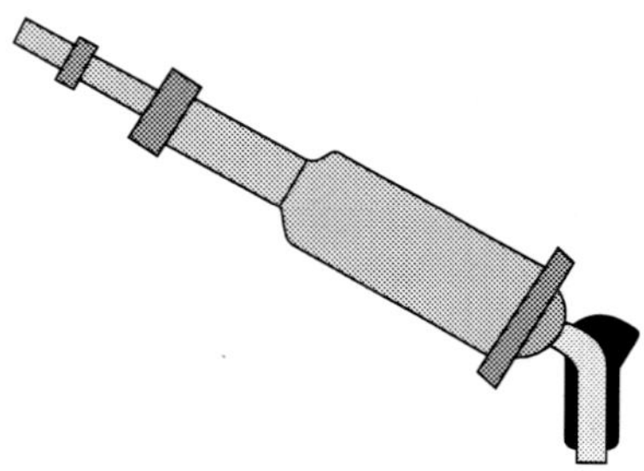

Fig. 6.12 Automatic hydraulic pressure gun

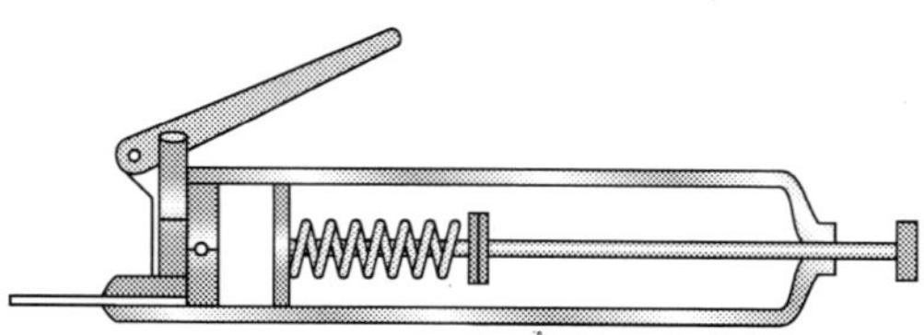

Fig. 6.13 Pressure grease gun

Fig. 6.14

Types of grease guns

The following types of grease guns are used for lubricating machines.

- 'T' handle pressure gun.
- Automatic and hydraulic type pressure gun.
- Lever-type pressure gun.

Lubrication to Exposed Sideways

Fig. 6.15

The moving parts experience some kind of resistance even when the surface of the parts seems to be very smooth.

The resistance is caused by irregularities which cannot be detected by the naked eyes.

The slide ways are lubricated frequently by an oil can.

After cleaning the open gears, oil them and repeat lubrication regularly.

Lubricate bearings

A shaft moving in a bearing is also subjected to frictional resistance. The shaft rotates in a bush bearing or in ball roller bearing, experiencing friction.

Lubricant

(1) Thin layer of this fluid between two machine parts helps in keeping them cool.

(2) It gives long life to machine parts.

(3) It saves machines from rust.

(4) It helps in the smooth running of machines and its parts.

(5) It protects the machine parts from becoming jam.

(7) It helps in keeping the machine fit and in good working condition.

EXERCISE

1. Why lubrication is necessary?
2. What are the different kinds of lubricants?
3. What are the different ways of lubrication?
4. What are the essential properties of a good lubricant?
5. List the types of lubricators.
6. What is the function of coolant?
7. List the properties needed in a good coolant.
8. What are the different types of coolants?
9. What is the difference between a lubricant and a coolant?
10. Why no coolant is used while machining cast iron?

Bibliography

1. Fundamental of the Metal Trade by Erich Wiezorek, expert chief in Berlin – Kreuzberg & Hugo Leben, lecturer, Berlin – Dahlem – Ernst Klett Stuttgart.
2. All about Machine Tools by Heinrich Gerling, Director of Trade, Schwerts (Rukr); Translation by Karl H. Heller, Frankfurt/ Main - George Westermann Verlag, Braunschweig, West Germany.
3. Europa Lehrmittel, Tabellenbuch, Verlag Europa – Lwhrmirrlel, 56, Wuppertal – Barmen, Rauer werth 18.
4. Machine Shop Practice – Volume 1 & 2 by Karl Hans Trecht - Technical & Vocational editor; Industrial Press Inc. 200 Madison Avenue New York.
5. Intermediate Engineering Drawing by A. C. Parkinson, The English Language Book Society & Sir Isaac Pitman & Sons Ltd., London.
6. Elements of Workshop Technology Volume 1 Manufacturing Processes & 2 Machine Tools by S. K. Hajra Choudhry, A. K. Hajra Choudhry, Nirja Choudhry, Media Promoters & Publishers Pvt. Ltd. Mumbai 400 007.
7. Fitter Shop Theory by Kapil Dev, Asian Publishers, 7/31 Ansari Road, Daryaganj, New Delhi-110002.
8. Machine Shop Theory and Practice by Albert M. Wagener and Harlan R. Arthur, D. Van Nostrand Company, Inc. Princeton, New Jersy, New York, Affiliated East West Press Pvt. Ltd.
9. Machine Drawing by K. R. Gopalakrishna, Subhas Store, Avenue Road, Bangalore-560 002, India.
10. Fitter Trade Theoty 1st Year, 2nd Year, Turner Trade Theory, 1st Year, 2nd Year, Machinist Trade Theory, 1st Year, 2nd Year, National Instructional Media Institute, DGET, Ministry of Labour, Govt of India, New Delhi.
11. Engineering Metrology by Er. R. K. Jain, Khanna Publishers, 2 B, Nath Market, Nai Sarak, Delhi-110 006.
12. First Year Workshop Manual by Prof. Vee Ess Prof. of Mech Engg. M. S. R. I. T.

Index